D1324504

Understanding Dynamic Systems: Approaches to Modeling, Analysis, and Design

C. NELSON DORNY

Department of Systems
University of Pennsylvania

PRENTICE HALL, Englewood Cliffs, New Jersey 07632

Library of Congress Cataloging-in-Publication Data

Dorny, C. Nelson.
Understanding dynamic system : approaches to modeling, analysis, and design / by C. Nelson Dorny.

p. cm.
Includes bibliographical references and index.
ISBN 0-13-221839-9
1. Systems engineering. I. Title.
TA168.D67 1993
620'.001'185--dc20 92-5402
CIP

Acquisitions editor: Doug Humphrey
Production editor: Irwin Zucker
Copy editor: Brian Baker
Prepress buyer: Linda Behrens
Manufacturing buyer: David Dickey
Supplements editor: Alice Dworkin

© 1993 by Prentice-Hall, Inc.
A Simon & Schuster Company
Englewood Cliffs, New Jersey 07632

All rights reserved. No part of this book may be reproduced, in any form or by any means, without permission in writing from the publisher.

Printed in the United States of America

10 9 8 7 6 5 4 3 2 1

ISBN 0-13-221839-9

ISBN 0-13-221839-9

90000>

9 780132 218399

Prentice-Hall International (UK) Limited, *London*
Prentice-Hall of Australia Pty. Limited, *Sydney*
Prentice-Hall Canada Inc., *Toronto*
Prentice-Hall Hispanoamericana, S.A., *Mexico*
Prentice-Hall of India Private Limited, *New Delhi*
Prentice-Hall of Japan, Inc., *Tokyo*
Simon & Schuster Asia Pte. Ltd., *Singapore*
Editora Prentice-Hall do Brasil, Ltda., *Rio de Janeiro*

THE PURPOSE OF MODELING IS INSIGHT.

Contents

LIST OF EXAMPLE PROBLEMS

The Example Problems are intended to demonstrate how to apply lumped modeling techniques to complex practical systems. These problems are not simple exercises. Some of them are quite intimidating. Students should not expect to be able to solve them on their own at their current level of understanding.

Preface

I have attempted to embrace the whole of engineering in a unified context. My purpose is to promote system thinking by breaking down unnecessary barriers between disciplines. I have focused on principles that give fundamental insight. Such principles can be carried over to unfamiliar problems.

I introduce the modeling process via mechanical systems because everyone has intuitive familiarity with them. Although an engineer typically views a mechanical system by means of free-body diagrams, that approach does not provide a *whole-system* view. The modeling approach of Section 2.1 does provide a whole-system view and fosters insightful analogies with nonmechanical systems. It also enables the circuit simulator SPICE to simulate the behaviors of complex mechanical systems *without writing system equations.*

The thought process used in the whole-system approach of Section 2.1 is different from the thought process used with free-body diagrams. When an instructor uses the whole-system approach for the first time, the thought process can run against the grain, owing to the habit of free-body thinking. Do not toss the whole-system approach aside too readily. Take the time to understand its advantages. It will be a worthwhile investment.

The Instructors Manual discusses alternative notations and variations on the thought process for mechanical systems. I urge you to inform me of your experience with the approach and to make suggestions for improvement.

I have had to reconcile conflicting terminologies and notations among disciplines. I invite suggestions for changes in terminology, in approach, and in choice of topics. I also welcome users to point out errors and to suggest better examples, applications, and homework problems for future editions.

I use the text for a first course in engineering—a two-semester course for sophomores in systems engineering and other disciplines. The students take differential equations and second-semester physics concurrently. The class focuses on basic concepts in modeling and analysis. The students perform physical experiments to verify the models and apply the analysis techniques. (The Instructors Manual describes some appropriate experiments.) I cover most of chapters 1–4 in the first semester and the rest of the book (except sections 5.2 and 6.4–6.6) in the second semester.

The text has also been used as an integrative course for seniors, primarily in mechanical engineering. At that level, it is possible to address such complexities as linearization, amplifier design, and transport delay. The whole book can be covered in two semesters. In a single semester, it is possible to cover chapters 2–4 with selected topics from chapters 5 and 6.

I am indebted to Alan M. Schneider of the University of California at San Diego and Wen K. Shieh of the University of Pennsylvania. They both made excellent suggestions and were brave enough to teach from the manuscript. Professor Schneider introduced me to the use of 0^- as the lower limit of the Laplace transform (in the manner demonstrated by Robert Cannon of Stanford University). He also showed me the zero-state assumption in my use of transfer functions.

I also appreciate the suggestions of the other reviewers of the manuscript, Stanley Johnson of Lehigh University and Galip Ulsoy of the University of Michigan, and the suggestions of my teaching assistants, Peng Yung Woo and Balaji Srinivasan. My children, Scott, Brett, Jonathan, Jennifer, and Christopher, helped me view the material through student eyes. Each made valuable contributions to the approach or to the notation. Scott is also the author of the theme cartoon shown below.

The unified approach to lumped modeling is patterned after a book by Shearer, Murphy, and Richardson [1.4]. My own urge to simplify—to capture the world in a few basic ideas—is apparent in the book. Simpler usually means better because simpler things are more clear.

I continue to be amazed at the insights I discover while integrating ideas from diverse disciplines. I am convinced that experienced engineers will benefit from reading this book. For most, it will unveil new ways of thinking and enable them to use more creatively the knowledge they already possess.

C. Nelson Dorny
Philadelphia, Pennsylvania

A student who is immersed in models, equations, and solutions is not likely to recognize that their primary function is to help us develop an intuitive feeling for the behaviors of various systems. I have tried to capture that notion in the theme stated on the opening page: *The purpose of modeling is insight.* The statement is a generalization of a similar proposition by Richard Hamming in [1.3]. The following cartoon, which I present to reinforce this theme, is based on a similar cartoon of unknown authorship that I saw at the Westinghouse Research Laboratory in 1967.

THE PURPOSE OF MODELING IS *NOT* IN SIGHT!!

Symbols

We represent a function of time by *upper case or lower case Roman:* e.g., $F_1(t)$, $v_2(t)$, $P_1(t)$, F_i, and v_j.

We use the *lightface Helvetica* symbols A, T, R, and z to denote quantities that apply to all energy domains.

We use the *double subscript* ij to denote the difference in value of a function between point i and point j: e.g., F_{12}, v_{ij}, A_{21}, and P_{10}.

We denote the initial value, final value, and prior value of a function of time by the *italic* subscripts *i*, *f*, and *p*, respectively: e.g., F_{12i}, v_{iji}, A_{21p}, and P_{10f}.

We use various other subscripts to denote particular instances of the subscripted quantity: e.g., v_S, F_n, y_P, y_C, T_1, R_i, R_o, R_I, R_O, J_c, $\mathbf{Y}_O$, and r_+.

We represent the Laplace transform of a function by *bold upper case Roman:* e.g., $\mathbf{F}_1(s)$, $\mathbf{V}_2(s)$, $\mathbf{P}_1(s)$, $\mathbf{F}_i$, $\mathbf{V}_j$, $\mathbf{P}_1$, and $\mathbf{A}_2$.

We use *boldface* to denote a multidimensional function (a mathematical vector function): e.g., $\mathbf{y}_i$, $\mathbf{x}_1$, and $\mathbf{u}$.

We use *boldface with brackets* to denote a matrix: e.g., **[A], [D], [a],** and **[z].**

We use *script* to denote prototype waveforms; specifically, $\mathcal{A}$, $\mathcal{B}$, and $\mathcal{C}$.

We use *bold italic upper case* to denote the phasor representation of a sinusoidal function: e.g., $\boldsymbol{F}_1$, $\boldsymbol{V}_2$, $\boldsymbol{P}_1$, $\boldsymbol{A}_i$, and $\boldsymbol{T}_j$.

We denote a three-dimensional physical vector quantity by a Roman letter with an *overhead arrow:* e.g., $\vec{F}$, $\vec{T}$, and $\vec{S}$.

If a function is held at a constant value over time, we denote that value by the italic subscript c: e.g., F_c, v_c, P_c.

We sometimes use an overhead caret to denote another instance of a particular variable: e.g., F and $\hat{F}$.

We use the symbol 0^- to denote an instant arbitrarily close to, but less than, $t = 0$; similarly, 0^+ denotes an instant just after $t = 0$.

SYMBOL GLOSSARY

a	An acceleration
	A Fourier coefficient
[a]	A matrix of a parameters
A	An area
	An arbitrary constant
	The abbreviation for ampere
	The amplitude of a sinusoid
A	A general *across* variable or potential
$\mathscr{A}$	A prototype waveform
A	A subnet label
b	A conductance-type parameter
b	A friction constant
	A Fourier coefficient
	A birth rate
[b]	A matrix of b parameters
B	A rotational friction constant
	The number of branches in a network
	An arbitrary constant
$\mathscr{B}$	A prototype waveform
B	A subnet label
	The bulk modulus of a fluid
c	A constant
	A coefficient for fluid resistance
	The speed of light
c_p	A specific heat
C	A capacitance (electrical, fluid, or thermal)
$\mathscr{C}$	A prototype waveform
d	A diameter
	A distance
	A death rate
	An amount of delay
dx	A small value of (or increment in) x
D	A diameter
	An arbitrary constant
e	A wall thickness
E	An arbitrary constant
E	An amount of energy
E	Young's modulus
f	A signal (function of time)
f_{OS}	Fractional overshoot
F	A scalar force
	The abbreviation for Farad
$\vec{F}$	A force vector
g	A ground point (electrical or mechanical)
	The gravitational acceleration
	A signal (function of time)
g_t	A transconductance
[g]	A matrix of g parameters
G	A shear modulus
G	An electrical conductance
	A transfer function
	The prefix giga, meaning $\times 10^9$
G_P	A power gain
h	A height or depth
	The response to a unit impulse
	A heat transfer coefficient
[h]	A matrix of h parameters
H	A transfer function or an input-output transfer function
	A quantity of heat
H	The Laplace transform of h (the unit-impulse response)

i	An electric current
	An integer
I	An inertance
$\mathscr{I}m$	The *Imaginary part* operator
j	The square root of −1
J	A moment of inertia
k	A spring stiffness
	A thermal conductivity
	The prefix kilo, meaning $\times 10^3$
K	A torsional stiffness
	The abbreviation for kelvin
K_G	A generator constant
K_T	A torque constant
ℓ	A length
L	An inductance
L	An inductance-type parameter
Lm	The log-magnitude operator
$\mathscr{L}$	The Laplace transformation
m	A mass
m	The prefix milli, meaning $\times 10^{-3}$
	A mass-type parameter
M	The amplitude of a sinusoidal variable
	A mutual inductance
	The prefix mega, meaning $\times 10^6$
M_H	The output/input sinusoidal amplitude ratio for system function H
M_r	The resonant peak magnitude of an input-output frequency response
$\hat{M}_r$	The resonant peak magnitude relative to a corner point
n	The prefix nano, meaning $\times 10^{-9}$
	A turns ratio
	An integer
n_p	The number of poles of a transfer function
n_z	The number of zeros of a transfer function
N	An integer
	The number of nodes in a network
	The *level* (value) of a variable
	A number of turns
N_p	The number of poles at the origin
N_z	The number of zeros at the origin
N_R	Reynold's number
O	An initial orientation
p	An initial position
	The period of a sinusoid
	An integer
	A variable
P	A pressure
P	An amount of power
$\mathscr{P}$	The phasor transformation
q	An electric charge
	A heat flow rate
Q	A volume flow rate
	The quality of a resonance
r	A radius
	The unit ramp function
	The magnitude of a complex number
	A heat-flow coefficient
	A rate of change of *level*
r_t	A transresistance
R	A resistance (electrical, fluid, or thermal)
R	A general resistance (dissipative impedance)
R_i	A general input resistance
R_o	A general output resistance
R_L	A general load resistance
$\mathscr{R}e$	The *Real part* operator
s	The time-derivative operator
	The complex frequency variable
$\vec{S}$	A surface-element vector
t	A time instant
	A thickness
t_r	A rise time
t_d	A time delay
t_0	A particular amount of delay
T	A scalar torsion
T	A general *through* variable or flow
T	A specific time or time increment
$\vec{T}$	A torque vector
u	A source signal
u_S	The unit step function
U_S	The Laplace transform of u_S
v	A velocity
	A phase velocity
	An electric potential (voltage)
+v	The direction of positive motion

V	The abbreviation for volt
	A volume
w	A weight
	A width
W	The abbreviation for watt
x	A displacement
	A distance
	An independent variable
	The real part of a complex number
y	A dependent variable
	The imaginary part of a complex number
[y]	A matrix of y parameters (admittances)
$y^{(n)}$	The nth derivative of y with respect to time
y_p	Parallel admittance per unit length
z	Electrical impedance
z_0	Characteristic impedance
Z	A general impedance to flow
Z_i	A general input impedance
Z_o	A general output impedance
Z_s	Series impedance per unit length
Z_p	Parallel impedance per unit length
[z]	A matrix of z parameters (impedances)
Z	A complex number
z.s.	Zero state
α	A rotational acceleration
	A damping rate
	A resistivity to fluid flow
	A coefficient or proportionality constant
	A real exponent
	A coefficient of migration
	A coefficient of coupling
β	A flow amplification factor
γ	A shear strain
	A propagation constant
	A specific gravity
δ	The unit impulse function
δx	A small increment in x
	The size of an abrupt jump in x
Δy	The difference between two values of y
$\Delta\omega$	The half-power bandwidth of an input-output transfer function
$\boldsymbol{\Delta}$	The Laplace transform of δ
ϵ	A small number
	An electrical permittivity
ζ	A damping ratio
η	A quadratic damper coefficient
	A quadratic fluid resistor coefficient
θ	A rotational displacement
	A temperature
	An argument (angle) of a complex number
κ	A converter constant for a direct converter
κ_g	A converter constant for a gyrating converter
λ	A wavelength
μ	An absolute viscosity
	A time variable
	A magnetic permeability
	A potential amplification factor
	The prefix micro, meaning $\times 10^{-6}$
ν	A wave velocity
ξ	A radiation coefficient
ρ	A density
	An electrical resistivity
	A reflection coefficient
σ	A dummy variable
	The Stefan-Boltzman constant
τ	A shear stress
	A time constant
	A dummy variable
ϕ	An arbitrary constant value of phase
	The phase of a sinusoid
ϕ_H	The phase shift between output and input sinusoids for system function H
ψ	An angle
ω	The frequency of a sinusoid
ω_0	A particular sinusoidal frequency
ω_n	A natural frequency
ω_d	A damped natural frequency
ω_r	The resonant frequency of an input-output transfer function
ω_ℓ	The lower half-power frequency of a resonant system

- ω_n The upper half-power frequency of a resonant system
- Ω A rotational velocity
  The abbreviation for *ohm*
- $+\Omega$ The direction of positive rotation
-  The electrical ground symbol
- ℧ The abbreviation for *siemen* (inverse of ohm)
- $*$ The convolution operator
- $\angle$ The *angle* (or *phase*) operator
- $\arg(\cdot)$ The *angle* operator
- $\underline{\nabla}$ The hydraulic ground symbol
-  The mechanical ground symbol
- $\triangleq$ The symbol for *defined as*
- $\equiv$ The symbol for *is identical to*

CHAPTER

1

Design Insights

This book is about engineering. Engineers design devices and systems that are useful to people. They also invent new ways to build those devices and new ways to operate existing systems. Inventing devices, systems, and procedures is an art that cannot be captured in a few rules. Rather, we learn it by observing and copying the techniques of others and then develop our own style.

Unlike the homework problems found in introductory mathematics and physics courses, the design problems of practicing engineers are not *given*. The solutions are not perfect or unique. An engineering problem begins as dissatisfaction with a current situation—say, with the boredom of long-distance driving. A new device (such as a cruise-control mechanism) is proposed to improve the situation. As the engineer begins to design the device, questions arise that lead to refined objectives, and the design problem becomes more focused. Defining the problem, then, is part of the job of the engineer.

Students resist addressing poorly defined problems. They are confused when they discover that there are two ways to solve the same problem. They have spent years learning *the right ways* to get *the right answers* to precise, narrow problems. But most real problems are vague and have several equally satisfactory solutions. We shall sometimes state a problem in its vague initial form and then sharpen the definition of the problem as the investigation proceeds. It is helpful to our creativity to have more than one way to solve a problem. Therefore, we examine alternative approaches. As we choose from among alternative solutions, we give intuitive reasons for the choices we make.

All the pieces of the universe are coupled. Some of these couplings are quite weak. As we walk across the room, we change the motions of all the stars! But those changes are insignificant. On the other hand, the motions of a passenger can have a strong effect on the stability of an automobile. In designing physical systems, we manage the couplings of the parts, enhancing some and limiting others. For example, in the design of a microwave oven, we enhance the coupling of the microwave energy with the meat in the oven while limiting the coupling of that energy with the cook outside the oven. In the design of an electric heating system, we enhance the thermal coupling of the radiant heaters with the room air and limit the electrical coupling of those heaters with the electric lighting system. We do not want the lights to dim every time the heat comes on. In the design of a computer, we enhance the electrical coupling between the computer and its printer but limit the electrical coupling between the computer and a nearby television set. This book promotes an intuitive understanding of coupling phenomena. That understanding equips us to manage our environment.

The reason we model and analyze systems is *to gain insight.* We solve equations and find numerical values for the system variables only to get that insight. We analyze the behavior of a system under various conditions to get an intuitive feeling for that system and to understand how parameter changes affect its behavior. With this intuitive understanding, we are equipped to *invent new systems.* These new systems are typically modified versions of familiar systems, analogs of familiar systems, or modified versions of such analogs. We call the inventing process **system design**. Use this book to gain insight. That insight will help you choose materials, shape parts, and interconnect parts to produce specified system behaviors.

As we explore a proposed system, we also must look at the system environment to understand clearly what is needed. Different environments require different system behaviors. There is little value in getting the right answer to the wrong question.

Silverman [1.5] asserts that analogical reasoning is a cornerstone of the innovation process. He states, "a well-developed sense of engineering 'intuition' (i.e., analogy to past situations, objects, events) is one of the few universal characteristics of the successful innovator." This book tries to help the reader develop engineering intuition. Trust your intuition! If you find it faulty in some situation, examine your error in thinking and understand it. Then your intuition and your confidence will grow. Throughout the book, we discuss analogies and develop intuitive concepts. We examine the reasonableness of solutions, and we interpret mathematical solutions physically. The homework problems require the student to define problems and interpret solutions.

The systems examined in this book involve a variety of physical phenomena. Yet, a few behavior patterns appear again and again. For example, blood flows from high pressure to low pressure and electric current flows from high voltage to low voltage. These behavioral similarities enable a few simple models to represent many different systems. The similarities also make the behaviors easier to understand and remember. Analogies among the physical phenomena serve as a primary source of intuition and creativity in the design process.[1]

[1]According to Bruner [1.2], "Perhaps the most basic thing that can be said about human memory, after a century of intensive research, is that unless detail is placed into a structured pattern, it is rapidly forgotten."

This chapter reveals *the big picture*. Each concept introduced here is examined in detail in a later chapter. Review this chapter from time to time—whenever you ask yourself, "What are we trying to do?" If you never ask yourself this question, perhaps you should.

Analyzing and Designing Dynamic Systems

According to Ackoff [1.1], a **system** is an object that has multiple parts and that takes on a new characteristic behavior when the parts are combined. He calls this new characteristic behavior the *emerging behavior* of the system. A simple example of emerging behavior is the oscillation that occurs when we connect a mass and spring. Neither the mass nor the spring can oscillate by itself.

A system is **dynamic** if the latest values of its variables depend on *past values* of the energy sources. We can think of a dynamic system as having memory; the effects of the energy sources integrate—accumulate over time. For example, the mass in a spring-mass combination oscillates without an applied force, owing to energy stored in the spring and mass by earlier forces. The spring-mass is a simple *dynamic system*. The mathematical description of a dynamic system is a *differential* equation. Most systems are dynamic.

In contrast, the spring itself is a **static** object. The amount of spring compression at any instant depends only on the spring force at that instant. A pair of springs connected in parallel exhibits a new characteristic: When we apply a force, the pair of springs divides (shares) that force. Hence, the spring combination is a static *system*. The variables in static systems exhibit no memory of past energy inputs. The mathematical description of a static system is a set of *algebraic* equations.

Engineers invent systems to achieve specific purposes. Designing is the specific, detailed part of inventing. It is planning of the form of the system, with specific dimensions and specific materials. We design a pipeline to collect crude oil at one location, to transport the oil under pressure, and to discharge it at another location. Static design of the pipeline means design to withstand the pressure of the pump and the stresses of thermal expansion and contraction. We design a bridge to carry traffic. Static design of a bridge means design to guarantee support of the expected load.

Static design is not enough. If we stop abruptly the flow of oil in the pipeline, we cause severe pressure waves (known as oil hammer). Those waves can destroy the pipeline and damage the environment. Wind gusts and earth tremors excite vibrations in the bridge. The vibrations can be severe enough to destroy the bridge and harm those who use it. Pressure waves and vibrations are dynamic effects. Though they may be incidental to the intended purpose of the system, we cannot ignore them. We must design the system to control them.

In contrast with the static pipeline and the static bridge, automobiles, sound systems, and fighter aircraft must operate *continuously* in a dynamic mode. Many of the principles examined in this book are pertinent to static design. Our focus, however, is on the more subtle dynamic effects.

We analyze a system—that is, break it into its constituent parts—to understand its behavior better. **System analysis** involves four steps: (1) partitioning the system into simple, clearly defined parts; (2) modeling the parts; (3) combining the models of the parts into a model of the whole system; and (4) finding the behavior of the system model. We view the behavior of the model as a predictor of the emerging behavior of the actual system.

For example, we write equations that describe the springiness of an automobile spring, the inertia of an automobile body, and the energy-absorbing capability of a shock absorber. Then we combine and solve those equations to discover (or confirm) the decaying oscillation that occurs when we push down on an automobile fender and release it. The discovered behavior helps us understand automobile suspensions. It also helps us decide whether a particular suspension is satisfactory. Analyzing the equations of the model suggests how to change a suspension design to reduce the size of the oscillation and hasten its decay.

System **design** or **synthesis** is the opposite of analysis. Rather than break a system into parts, we combine parts into a system. The design process must deal not only with *what* a system is to do and *how* it is to do it, but also with *why* the system is to be created. Without knowing the reasons for the system, we cannot define appropriate objectives. Without appropriate objectives, we generate a solution to the wrong problem. **Getting the right question is as important as getting the right answer.** Although this book focuses on what to do and how to do it, it also discusses underlying system objectives and formulates design problems to meet those objectives.

Often, system analysis uncovers behavior that was not anticipated at the beginning of the investigation. For example, using a fan to exhaust the gases from a home furnace can improve the efficiency of the furnace. But forced exhaust depressurizes a home. Depressurization will, in turn, reduce or reverse the draft of a fireplace and make fires burn poorly. Depressurization also can increase the seepage of radioactive radon gas into the home. The discovery of undesirable side effects requires that we revise the goals of the system to account for the bad features. In general, we alternate designing and goal setting until the result is satisfactory.

To discover unforseen side effects, we must look beyond the immediate problem. If the examination of a system is not sufficiently broad during problem definition and system design, side effects are not discovered until the system is in operation. At that point, changes in design are very expensive.

Most systems consist of a *source* (energy-conversion device) acting on a *load.* For example, a crane raises and lowers heavy materials, a furnace raises and lowers the temperature of a building, a concrete pump transports wet concrete to locations not accessible to trucks, and a sound system drives a loudspeaker. Typically, a source is geared or amplified to match the requirements of the load. Often, the action of the source must be carried some distance to the load by a drive shaft, cable, pipe, or similar device. Many systems include sensors to measure quantities (such as engine speed or pump pressure) that are important to the task or to the safety of the equipment and users of the system. Sensors are energy *transducers,* converting actions in one energy domain to actions in another. A tachometer, which converts rotational velocity to an easily displayed electrical voltage, is an example.

When we speak of **system behavior**, we refer to the behaviors of the observable variables of the system. When we drive a car, we *see* velocity and *feel* acceleration. To get a more accurate measurement of velocity, we check the speedometer. We monitor the condition of the vehicle by watching gauges that display the engine temperature, oil pressure, and engine speed. Other variables within the automobile, invisible to the driver, are tracked by a computer and are used to control the behavior of the engine and transmission. Still other variables, which *could* be measured, interact with each other according to natural laws. (For example, the acceleration of the automobile is proportional to the engine torque.) In every dynamic system, there are observable quantities that vary with time and with changes in the energy sources. We call these observable quantities **signals**. It is these signals that reveal the behavior of the system.

Framework for System Models

People understand intuitively, from everyday experience, that water flows downhill, from higher to lower elevation. If the flow is in a filled pipe, the amount of flow is the same at the pipe inlet as it is at the outlet. Since we can measure the flow at either end of the pipe, we say the flow is a **through variable**. To stress the *through* nature of the flow, we denote it by the symbol T.

Elevation is defined relative to a reference point that we designate as zero elevation. Elevation is an **across variable**, in the sense that it can be measured only across the space between two points. Elevation is an example of a more general quantity called a **potential.**[2] To highlight the *across* nature of potentials in general (and elevation, in particular), we represent them by the symbol A. For example, we measure the elevations A_i and A_j of the two pipe ends relative to a reference point. The difference $A_i - A_j$ is the end-to-end drop in elevation. We call it the elevation difference *across* the pipe and denote it by the symbol A_{ij}.

If we increase the tilt angle of the pipe, the pipe flow T increases. Increasing the tilt also increases the end-to-end drop in elevation A_{ij}. The flow T is nearly proportional to A_{ij}. We call the proportionality factor the **impedance to flow** and denote it by the symbol z. That is,

$$A_{ij} = zT \tag{1.1}$$

where A_{ij} is the *drop* in elevation in the direction from point i to point j and T is the flow in the *same* direction.

The impedance z is a property of the pipe. It shows how the pipe behaves. The elevation difference A_{ij} describes the way we *use* the pipe. The flow T is the specific behavior of the pipe when it is used in the manner described by A_{ij}. A pipe of smaller diameter has greater impedance and reduced flow. An obstructing object in the pipe (such as a partly closed valve) also increases the impedance of the pipe and reduces the flow. Equation (1.1) is a **lumped model** of the pipe. That is, it expresses the behavior of the pipe in terms of variables at the ends of the pipe and ignores the behavior at intermediate points.

The interaction between potential and flow is a recurring theme in nature. We see it, for example, in the flow of heat from high to low temperature, in the flow of electric current from high to low voltage, and in the flow of molecules from high to low concentration. In each instance, (1) the flow is a through variable and the potential is an across

[2]The terms *flow* and *potential* are defined mathematically in equations (2.27) and (2.28).

variable, (2) the potential difference between two points or regions is *proportional* to the flow from one to the other, (3) the proportionality factor can be viewed as an impedance to flow, and (4) we can use equation (1.1) to describe what happens.

In general, we refer to a *through variable*—a variable that must have the same value at all points along some path—as a **flow**. We refer to an *across variable*—a variable whose value can differ from point to point along the path—as a **potential**. Thus, in a mechanical damper (such as a door damper or an automobile shock absorber), the compressive force in the damper is a flow, the velocity difference across the damper is a potential difference, and the absolute velocity of one end of the damper (relative to a stationary reference point) is a potential. The ratio of the velocity difference across the damper to the force in the damper is the damper impedance.

In the initial sections of succeeding chapters we examine flows, potentials, and impedances in various contexts. In each context, the impedances represent parts of the system and the flows and potentials are the system **signals**—the observable quantities that reveal the behavior of the system. The lumped model represented by equation (1.1) appears everywhere in the book, but with different symbols for each physical context. In the water-flow example just given, the impedance represents the pipe and the signals are the pipe flow and the elevations at the ends of the pipe. In the mechanical damper, the impedance represents the damper itself and the signals are the velocities at the ends of the damper and the compressive force in the damper. Each passive (energy-free) part of each system exhibits a flow through an impedance from high to low potential. This analogy to *downhill flow* aids our memory as we write the equations that relate the system variables.

Coordinate Systems for Physical Variables

We must define a coordinate system for each measurable quantity in a physical system in order to relate that quantity to the variable that represents it in the model. Many quantities have only **one degree of freedom.** That is, we can represent them by a single scalar variable. We can think of such variables as one dimensional. Most systems can be represented accurately by one-dimensional variables. For a one-dimensional physical quantity, the elements of a coordinate system are:

(1) The definition of the **zero value** of the quantity (corresponding to the origin of the coordinate system);

(2) The definition of **positivity** of the quantity (corresponding to the positive direction on the variable axis); and

(3) The definition of a **unit amount** of the quantity (corresponding to the number 1 on the variable axis).

We can assign each element of the coordinate system arbitrarily. For most physical quantities, such as the one-dimensional displacement of an object, conventional standards exist for the unit amount (e.g., 1 m). We use the International System (SI) of units throughout the book. Conventional standards often exist for the zero value as well, e.g., no displacement. For many quantities, then, defining the coordinate system means defining positivity. For one-dimensional displacement, defining positivity means picking the *positive direction* along the line of action.

We demonstrate the defining of coordinate systems by selecting the elements of coordinate systems for the pipe-flow variables of the previous section. The standard SI unit of length is the meter. Choose the earth's surface as the zero-elevation reference. The *elevation* is then the distance from the surface, with distance *above* the surface chosen as positive. (This is the conventional choice. It would be just as correct to select distance below the surface as positive.) We can measure elevation with a ruler. Let i and j be the endpoints of the water pipe. The elevation of each point is a one-dimensional quantity. The elevation difference A_{ij} is positive if the subscripts are ordered in the downhill direction, with j lower than i.

Exercise 1.1: What changes occur in the elevations of points if we define distance *below* the earth's surface as positive? What changes occur if we choose the center of the earth as the reference?

Water *flow* is a measure of the average motion of water molecules. The SI unit of flow is 1 m^3/s. We recognize flow through a pipe by an increasing accumulation of water at the outlet. Define zero flow to mean zero rate of accumulation. The flow T in a pipe is oriented along the pipe axis. Assign a positive direction for T arbitrarily. Show that direction by an arrow beside the pipe. Then the flow T is positive if the water accumulates at the pipe end that corresponds to the head of the arrow. If the water accumulates at the other end, then T is negative. To measure T, we measure the water that accumulates in a container at the outlet during a measured period.

Exercise 1.2: We can describe the behavior of fluid in a pipe by equation (1.1), which has the subscripts on the elevation-difference variable oriented in the same direction as the arrow for the flow variable. Suppose the i-to-j direction is downhill, but the positive direction for T points uphill. Determine the sign of A_{ij} and the sign of T. Show that the form of the pipe equation is $\mathsf{A}_{ij} = -z\mathsf{T}$.

Models for Energy Sources

The water pipe and the damper just examined are **passive** devices. That is, the signals associated with each device have zero value, unless we energize the device externally. Any object that can be represented by an impedance is passive.

Other devices act as system *energizers.* We call them **active** devices, or **sources**. A human being is an energizer. A human moves objects at will. The mechanical energy that a human inserts into a system is derived from chemical energy stored in the cells of the human body. An internal combustion engine is an energizer that extracts energy from the heat content of fuel. The *power output* of an engine is *controlled* by humans or by stored programs prepared by humans. Electronic amplifiers, electric motors, and hydraulic turbines are other examples of energizers that are controlled or programmed by humans.

Any active device (source) produces a *rise in potential* across the device and a flow through the device in the direction of that rise in potential. This rise in potential in the direction of flow contrasts with the behavior of passive devices. A hydraulic source (a pump) produces a rise in pressure and a flow of fluid through the source in the direction of rise. An electrical source (such as a battery) produces a rise in voltage and a flow of electric current through the source in the direction of rise. A jack is a mechanical source that produces relative motion between its ends and a compressive force in its interior. Choosing a positive direction for velocity makes the velocity at one end higher than the other. We *treat* the compressive force as a flow through the jack in the direction of rise in velocity.

The source flow and the rise in potential across the source are produced simultaneously. We should not think of one as the cause of the other. The balance between flow and potential rise is determined by the load on which the source acts and by limitations of the source itself. For example, suppose a person uses maximum effort to throw an object. If the object is very heavy, then it accelerates slowly, because the person's strength is limited and the force that the person applies to the object is nearly constant. From the viewpoint of the object, the person acts as a source of force. The object accelerates in proportion to that force. On the other hand, if the object is very light, most of the exerted effort is used to accelerate the person's arm. The object has little effect on the acceleration. To the object, the person then acts as a source of acceleration. The force *on the object* is proportional to the acceleration of his or her hand. Although we are accustomed to thinking of a person as a source of force, these examples show that it is equally correct to think of a person as a source of motion. We can view any source in two ways—as a source of potential difference or as a source of flow. Which view is more appropriate depends on the way we *use* the source.

Each source is characterized by a particular trade-off between the flow T through the source and the rise in potential A_{ij} across the source in the direction of flow. We can describe the trade-off by a plot of A_{ij} vs T. We call such a plot the **source performance characteristic.** For example, denote the inlet and outlet of a pump by the numbers j and i, respectively. Then the flow T passes through the pump from point j to point i, the outlet pressure A_i is greater than the inlet pressure A_j, and the rise in potential (pressure) is A_{ij}, a positive number.

A sample source performance characteristic is shown in Figure 1.1. The specific combination of potential rise and flow that the source produces in a system at a particular instant is called the **system operating point.**

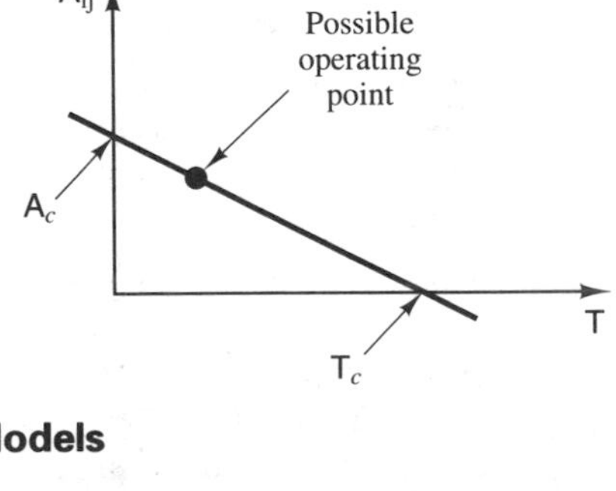

Figure 1.1 Source performance characteristic.

Kinds of Models

A **model** of a system is a simplified representation that mimics some features of the real system. Models are much less expensive to build than are real systems. They are also easier to examine and easier to change. For those reasons, we build and study models of a system before we build the actual system.

There are many kinds of models. Some are mathematical, as illustrated by equation (1.1). The source performance characteristic of Figure 1.1 is a graphical display of that mathematical model. Other models are pictorial, and still others are algorithmic computations which we call simulations. Throughout the book, we examine various types of models in each of these categories.

Equation (1.1) is a mathematical lumped model of a *passive* device (such as the pipe discussed earlier). Figure 1.2(a) shows a pictorial lumped model of the same device. In the pictorial lumped model, the properties of the device are represented by a single quantity, the impedance Z. The two nodes designated i and j represent two points or regions of the device (the ends of the pipe). These nodes are labeled with their potentials, A_i and A_j. The arrow shows the direction of the flow T from node to node through the model.

The pictorial model retains enough physical similarity to the actual device to relate the symbols A_i, A_j, and T to physical quantities. Use of the impedance symbol Z in the block of the pictorial model assumes that the device is described by equation (1.1). Thus, the equation is implicit in the pictorial model, and the meanings of the symbols are implicit in the mathematical model. Hence, it is advantageous to use both models, simultaneously or interchangeably.

The device represented by these two models can be described by a third model, a graph of A_{ij} vs. T similar to the one in Figure 1.1. (See problem 1.2 at the end of the chapter.) This third model implicitly assumes that the across and through relations are as displayed in the pictorial lumped model. None of the three models shows explicitly which points of the physical object correspond to nodes i and j. We can show these physical features of the system in a detailed pictorial model.

We see from the foregoing discussion that models can look quite different, yet be equivalent to each other, in the sense that they are in harmony and convey much the same information. Different types of models focus on different system features. Models need not be equivalent (or even compatible with each other) to be valid and useful. For example, the wave theory of light is not compatible with the particle theory, but both give useful insights about the behavior of light. We use many kinds of models in this book. Do not be disturbed if you cannot easily equate different models of a system. Instead, look for the system insights that each can impart.

It is customary to represent force by an arrow, with magnitude and direction. The arrow (vector) concept is widely used to balance forces in free-body diagrams. But this representation is not well suited to describing the tensile force in a barge tow cable, that is, the force that would be measured by a scale inserted at any point between the barge and the towboat. That force is scalar, with a magnitude but no direction. The cable tension and the associated cable stretching are independent of the orientation of the cable. The force in the coupler between two railroad cars is also scalar. It is either tensile or compressive, depending on whether the cars are pulled or pushed. The line of action of the scalar force in the coupler is constrained by the railroad track and is of no importance to the

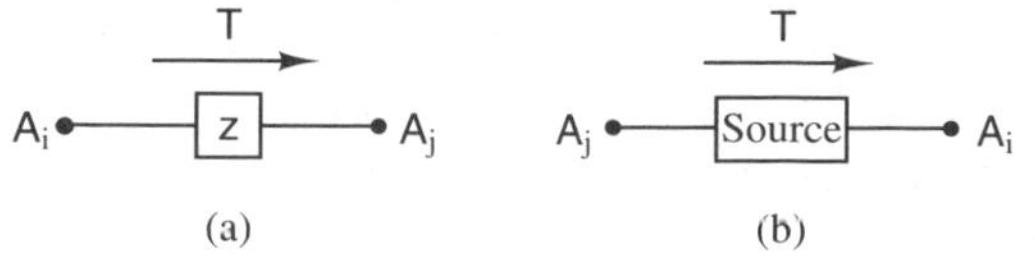

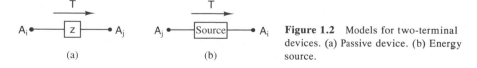

Figure 1.2 Models for two-terminal devices. (a) Passive device. (b) Energy source.

behaviors of the coupler and the two railroad cars it connects. We can use a pair of opposing arrows of equal magnitude to describe the force in these situations. It is more fitting, however, to view the force as a scalar, positive if compressive and negative if tensile. We use this scalar-force model extensively.

A pictorial lumped model of a *source* (perhaps the pump discussed earlier) is shown in Figure 1.2(b). Unlike the passive device of Figure 1.2(a), a source cannot be represented by an impedance alone. The pictorial model conveys information about the meanings of the signals T, A_j, and A_i (the flow and the inlet and outlet pressures), but nothing about how they are related. The relationships between the variables can be described by a performance characteristic, such as the one in Figure 1.1. In problem 1.2 at the end of the chapter, we represent the particular straight-line performance characteristic of Figure 1.1 mathematically.

The two terminals of the pictorial models of Figures 1.2(a) and (b) emphasize that the device signals (a potential difference and a flow) are accessible at two points of the physical device—the two contact points on a light bulb, for example. We call such a pair of access points a **port.** We also call the pair of terminals of the model a port. Throughout chapters 2–5, we represent systems by collections of simple two-terminal pictorial models similar to those in Figure 1.2(a). We call these models *two-terminal elements* or *one-port elements.* Certain devices (amplifiers, transducers, transformers, and transporters) have two ports—an input and an output port. At each port, there is a potential drop across the port and a flow through the port. We examine two-port devices in chapter 6.

Ultimately, we seek *mathematical* models (equations). Equations express relationships precisely. Computer algorithms that solve equations numerically are readily available. We can simulate the behaviors of very complex systems if we model them mathematically. The other types of models help us generate the mathematical models and help us interpret them physically.

We must validate mathematical models experimentally. We check an approximation about which we are uncertain by examining published data or by performing a physical test. We check a whole-system model by comparing the behavior of the model with the behavior of a physical system that is similar to the model. We have to build such a system if one is not already available.

A small amount of modeling effort, if done properly, explains most of the behavior of a system. We need not model a system in great detail to understand its behavior. Any system can be modeled more accurately by increasing the level of detail. As the level of detail increases, however, the effort (and cost) required to add detail and to validate the model grows drastically. The added insight diminishes as the level of detail increases. Thus, the growing effort produces diminishing returns. A model need only be exact enough to accomplish its purpose. To make it more exact is to waste resources.

Simplified Notation for Derivatives

Some observable quantities (signals) are time derivatives or integrals of other observable quantities. For example, velocity v(t) is the derivative of displacement x(t); it is also the integral of acceleration a(t). The equations that describe the behaviors of physical devices often involve derivatives and integrals. We call these equations integro-differential equations. Usually, we manipulate such equations into a form that involves only derivatives. We then call them differential equations. Manipulation of simultaneous differential equations can be confusing.

The derivative of a function y is often denoted y' or $y^{(1)}$; that is, $y' \equiv y^{(1)} \equiv dy/dt$. A similar notation used only for *time* derivatives is $\dot{y}$. This book uses all these notations. To simplify the manipulation of differential equations, we also represent the time-derivative *operation* (d/dt) by the **time-derivative operator** s. If y is a signal (a function of time), we denote its time derivative $\dot{y}$ by the shorthand notation sy, its second derivative $\ddot{y}$ by s^2y, etc.

Differentiation is a *linear* operation. That is, the derivative of a sum is the sum of the separate derivatives, and the derivative of a scalar multiple is the multiple of the derivative. In operator notation, we state these facts as

$$s(y_1 + y_2) = sy_1 + sy_2 \tag{1.2}$$

$$s(ay) = a(sy) \tag{1.3}$$

where y, y_1, and y_2 are signals (functions of time) and a is a scalar constant.

Each term of each differential equation contains a symbol that represents a signal (a function of time). We keep that symbol in the farthest right position in the term. Then, scalar multipliers *multiply signals* from the left, and time-derivative operators *operate on signals* from the left. Except for this restriction on the location of the signal in each term, we manipulate all symbols according to the rules of algebra. When we use the s notation for time derivatives, simultaneous differential equations are no harder to manipulate than simultaneous algebraic equations.

Physical quantities do not change abruptly. Every physical variable corresponds to a smooth curve (or waveform). At each instant, the derivative of that variable is the slope of that smooth curve. Yet some variables change so rapidly relative to others in the same system, that it is convenient to treat them as if they jump abruptly. Therefore, we visualize a system signal y(t) as a smooth curve, perhaps with a few abrupt jumps. If y(t) jumps abruptly at some point in time, we calculate the slope separately to the right and left of that point.

Just as multiplication of a signal by s implies time differentiation, division by s implies time integration. (Any integration inherent in a physical system is definite integration.) The interval of integration is not explicit in the symbol (1/s). Unless we specifically state otherwise, we assume that the integration interval begins at $t = 0$. Then, (1/s)y is a shorthand notation for

$$\int_0^t y(\sigma)\, d\sigma \tag{1.4}$$

where σ is a dummy variable of integration.

The signal y determines the signals sy and (1/s)y. That is, if we know the waveform y(t), we can determine the *slope waveform* $\dot{y}(t)$ and the *integrated waveform* $\int_0^t y(\sigma)\, d\sigma$. If the signal y jumps abruptly at some instant, the slope of (1/s)y changes abruptly at that instant. Problems 1.4 and 1.5 at the end of the chapter examine graphically the processes of differentiation and integration. Note that the operators s and (1/s) are nearly, but not quite, inverses of each other. Although we differentiate the waveform y(t) to produce the slope waveform $\dot{y}(t)$, integrating $\dot{y}(t)$ does not restore y(t). Rather, $\int_0^t \dot{y}(\sigma)\, d\sigma = y(t) - c$, where c is a specific constant.

Exercise 1.3: Let $y(t) = 3t + 2$. Find $\dot{y}(t)$, $\int_0^t \dot{y}(\sigma)\, d\sigma$, and the constant c.

The operator notation s is very important. We use it everywhere in this book. Initially, it is just a notation to simplify the manipulation of differential equations. We expand its use and give it additional physical interpretations in sections 3.4, 5.3, and 5.5.

Test Signals and Characteristic Behaviors

Suppose the driver of a car on a straight highway gives the steering wheel a jerk to one side to avoid an object in the road and then immediately returns the wheel to the original position. If the car does not react (if it hits the object), we say it is sluggish, that is, not very responsive. If the car lurches quickly to the side, we say it is responsive. If, after lurching to the side, the car takes a long time to regain a steady heading, we say that the car is not very stable. If the car returns immediately to a steady heading, we say it is quite stable. If the car is towing a trailer, the jerk on the steering wheel can cause a jackknife oscillation between the car and trailer, and the vehicle can go out of control. We then say that the car (and trailer) is nearly unstable.

Suppose the driver rotates the steering wheel right and left in a sinusoidal pattern (with a rotation angle that looks like a sine wave when plotted against time). Then the car will move from side to side in a sinusoidal pattern. Because it takes time to overcome the inertia of forward motion, the amplitude of that side-to-side motion will get smaller as the wheel is oscillated more rapidly. For a particular rate of wheel oscillation, a very responsive car will have a larger side-to-side amplitude than an unresponsive car will have. If the car were towing a trailer, oscillating the steering wheel at a frequency close to the natural jackknife-oscillation frequency would cause an exaggerated oscillation, indicating a nearly unstable system.

To compare the behaviors of two systems, we energize them with identical source signals and compare their responses. When a test signal is applied to a system, all variables in that system react in concert and show similar features. Two of those features—**responsiveness** and **stability**—have universal importance. As we can see from the preceding example, there are *degrees* of responsiveness and stability. (We assume that *instability* is totally unacceptable.) Two source-signal waveforms have become accepted as standard test waveforms. One, known as a **step function,** is an abruptly applied constant value. A television *on-switch* provides a step function of voltage to the television's electronics. The other waveform is a steady sine wave. The electric current that energizes many of our appliances has a sinusoidal pattern. The unintentional vibration in an electric fan, owing to an imbalance in the fan, is also sinusoidal.

If the source signal (say, the position of a steering wheel) is a step function, the system variable y (say, the direction in which a car is traveling) typically behaves like one of

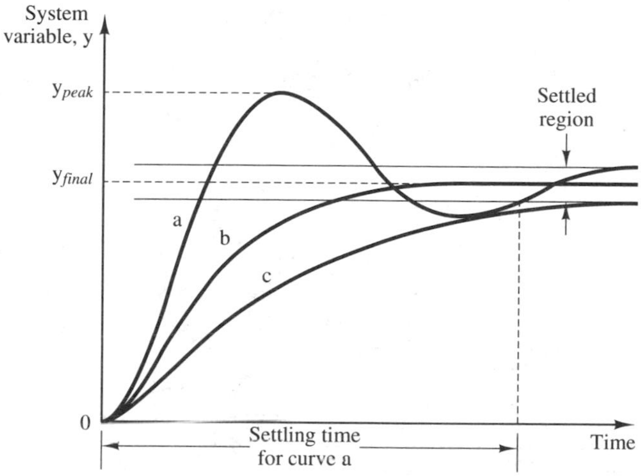

Figure 1.3 Typical responses to a step input.

the waveforms in Figure 1.3. The **fractional overshoot** $(y_{peak} - y_{final})/y_{final}$, indicates the *degree of system stability.* A system with low overshoot is more stable than one with high overshoot. Thus, curve a represents the least stable behavior of the three curves shown. One *might* say that it also represents the most responsive behavior, since the variable takes the shortest time to reach the desired value y_{final}. However, it does not remain at that value until later. The **settling time**—the time after which the system variable does not leave the region designated *sufficiently settled*—is a better *indicator of responsiveness.* The settling time of curve a is longer than that of curve b. Curve c represents the most stable of the three behaviors, but also the least responsive. Curve b exhibits balance between responsiveness and stability.

If the source signal (say, the angle of the steering wheel) varies sinusoidally, the system variable (say, the deviation of the car from the lane) soon varies sinusoidally also. Most physical systems respond well to a *slowly varying* sinusoidal source. (For example, if we oscillate the steering wheel angle slowly, the car moves left and right sinusoidally.) They usually do not respond as well to high-frequency source signals. (The amplitude of side-to-side car motion is smaller for high-frequency oscillations of the steering wheel.) For most physical systems, with fixed input amplitude, the amplitude M of the system variable y changes with frequency like one of the **frequency-response** patterns of Figure 1.4. (These are the same three systems as those shown in Figure 1.3.)

The peak in pattern a means that the system *resonates* for some source frequencies. That is, the amplitude of the sinusoidal system variable is larger for frequencies near the resonant-peak frequency (M_{res}) than it is for low frequencies (M_{low}). The **resonance ratio** M_{res}/M_{low} indicates the *degree of system stability.* A system with a low resonance ratio is more stable than one with a high resonance ratio. Pattern a is thus the least stable of the three patterns shown. (The resonance ratio is 1 for patterns b and c.) We define the **bandwidth** of a system to be the range of source frequencies for which the system shows significant response. (We define *significant response* to be a response which is at least 70% of the low-frequency response M_{low}.) The system of pattern c has the smallest **bandwidth.**

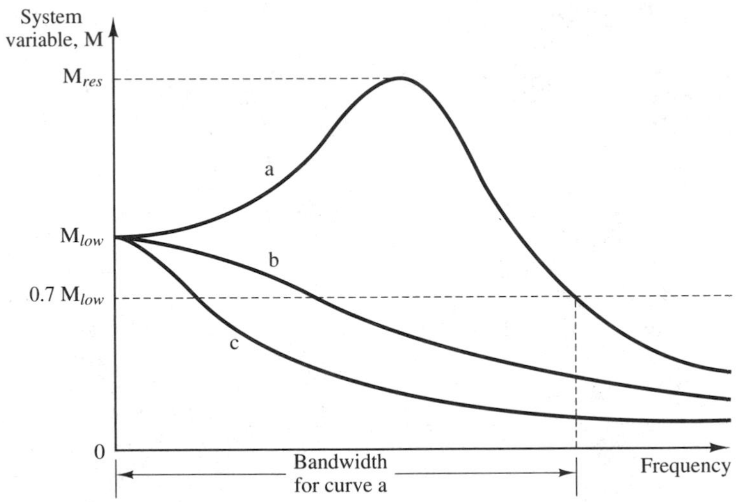

Figure 1.4 Typical frequency-response patterns. (M_{res} is the peak amplitude, M_{low} the low-frequency amplitude.)

Hence, it is the *least responsive* of the three systems—the most limited in ability to follow rapidly varying source signals. Pattern b provides balance between responsiveness and stability.

A frequency-response pattern like those illustrated in Figure 1.4 is sometimes called the *frequency-domain response* of the system. Similarly, a step response like those illustrated in Figure 1.3 is known as the *time-domain response* of the system. We see from the preceding discussion that the time-domain response and the frequency-domain response, although quite different in nature, provide nearly the same information about system stability and responsiveness.

Figure 1.3 shows that each system has a *natural rate of response:* The system variables change to new values no sooner than the settling time. By reducing the abruptness of the source waveform—that, is by turning on the source gradually, over a time interval longer than the settling time—we can cause the system variables to change values more slowly. We cannot make them change more abruptly: If the source signal changes to a new value in less time than the natural settling time for the system, then the system variables cannot keep up with the source. The step function is the ultimate in signal abruptness.

Figure 1.4 shows the same phenomenon as does Figure 1.3, but via sinusoidal source signals. A system with a higher bandwidth can respond to source signals of higher frequency. Thus, it can keep up with signals that are more abrupt. Because of this correlation between a system's bandwidth and its ability to follow abrupt signals, engineers use *source frequency* interchangeably with *source abruptness* to indicate the *stressfulness* of a source signal, that is, the degree of responsiveness required of the system to keep up with the signal.

Since the step input and the sinusoidal input are standard test waveforms for physical systems, we use them throughout the book as source waveforms for mathematical models of systems. To validate a model experimentally, we compare the step response or frequency response (or both) of the physical system with the corresponding response(s) predicted by the model. If the physical response and the response of the model are nearly the same, we say that the model is validated.

The difference we can tolerate between the physical response of a system and the response of a model of the system depends on the situation. The electronic timer that controls a wristwatch should be accurate to at least one part in a million. In contrast, a 20% variation in the flow of water through a shower head would hardly be noticed. Accuracies of 0.1% to 10% are typical for manufactured parts and systems. Heating or wear can change the parameters of devices by as much as 50%. Thus, high accuracy can be expensive and difficult to achieve. It is often cheaper to make a system adjustable.

The Purpose of Each Chapter

This book has two purposes. The first is to show *how* to understand systems, in general. The chapters are organized around this aim. The second purpose is to *understand specific kinds* of systems. To achieve this understanding, the first section of each chapter introduces systems in a new energy domain. We wait to introduce a new domain until after the reader gains confidence in the previous domain.

Chapter 1 gives the big picture. The reader should review it from time to time to maintain perspective. Chapter 2 shows how to find the differential equations that represent a particular system. In chapter 3, we see how to solve those differential equations for a step-function source. We also interpret the solution to understand the system better. Chapter 4 gives ways to simplify system models and understand the system more easily. In chapter 5, we solve the system equations for a sinusoidal source. We use the frequency response to understand the system better and with less effort. Chapter 6 examines two-port coupling devices such as amplifiers and gears. It also shows how to use these devices in systems.

PROBLEMS

1.1 Answer the following questions for one of these devices:
(i) A tow cable; (ii) A fire hose; (iii) An electric drill.
(a) What are the signals in the system; that is, what are the observable variables that reveal the system's behavior?
(b) Which signals are flows and which are potentials?
(c) Is the device dynamic or static?

1.2 **(a)** Express the source performance characteristic of Figure 1.1 as a mathematical relation between A_{ij} and T. (The expression involves the constants A_c and T_c.)
(b) Express the mathematical model given by equation (1.1) as a plot of potential difference vs. flow. Treat z as a constant real number.

1.3 The sun heats a section of highway. The section expands as its temperature increases. The expansion is resisted by the connecting sections, producing compressive force in the heated section. The heating process makes the section act like a mechanical source that can be viewed as a source of compressive force or a source of motion (expansion). Determine the conditions under which it is most appropriate to view the heated section as
(a) A source of compressive force.
(b) A source of motion.

1.4 A velocity waveform is shown in Figure P1.4.

(a) Sketch the corresponding acceleration and displacement waveforms. Carry out the required differentiations and integrations graphically. (What must you assume to find the displacement waveform?) Label the resulting signal diagrams appropriately.

(b) Express the velocity waveform mathematically, as a piecewise-continuous function. That is, state the equation that is appropriate for each time interval.

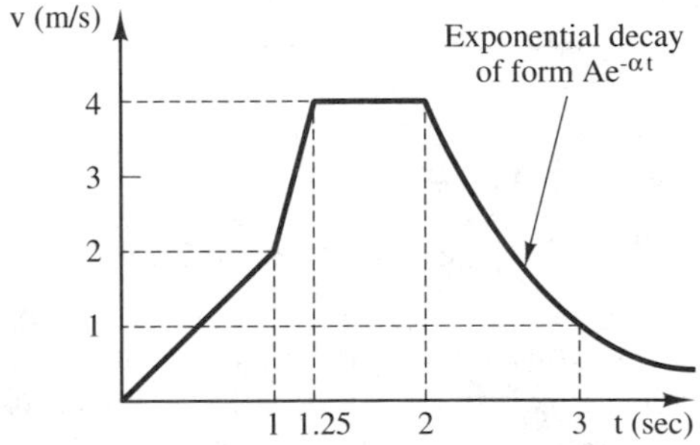

Figure P1.4

1.5 The equation that relates a force and an associated velocity in a particular system is $F = sv + 2v + (1/s)v - 2$, where s is the time-derivative operator. The desired signal v is shown in Figure P1.5.

(a) Sketch the derivative signal sv and the integral signal (1/s)v. Carry out the differentiation and integration graphically. Label the pertinent numerical values on each sketch.

(b) Combine the signals v, sv, and (1/s)v graphically to obtain a sketch of the force signal F that will produce the specified velocity waveform.

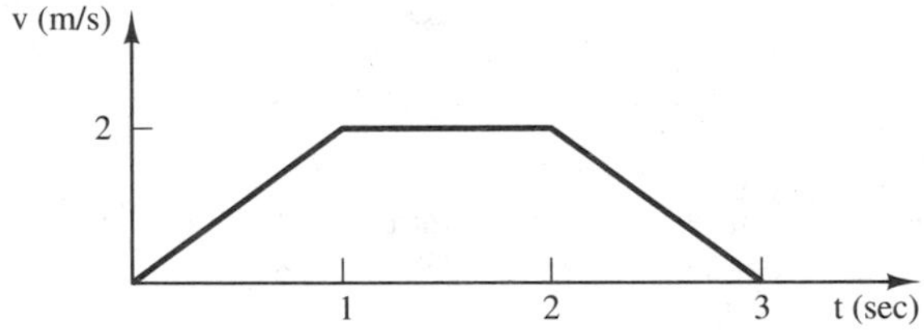

Figure P1.5

REFERENCES

1.1 Ackoff, R. L., *Creating the Corporate Future.* New York: John Wiley & Sons, 1981.

1.2 Bruner, Jerome S., *The Process of Education.* Cambridge MA: Harvard University Press, 1960.

1.3 Hamming, R.W., *Numerical Methods for Scientists and Engineers.* New York: McGraw-Hill Publishing Co., 1962.

1.4 Shearer, J. L., Murphy, A. T., and Richardson, H. H., *Introduction to System Dynamics.* Reading MA: Addison-Wesley Publishing Co., 1971.

1.5 Silverman, B. G., "Toward an Integrated Cognitive Model of the Inventor/Engineer." *R&D Management* 15(2), 1985, pp. 151–8.

CHAPTER

2

Lumped-Network Models of Systems

This chapter shows how to approximate a complicated physical system by a network of simple physical elements (or lumps) and how to represent that lumped network by a set of differential equations. We introduce the *lumping* process in the intuitively familiar context of mechanical motion.

First, we discuss simple models for energy storage (the mass and spring), for energy dissipation (the viscous damper and sliding friction), and for energy supply (force and velocity sources). This discussion uses the intuitively based terminology of flows and potentials introduced in chapter 1. Then we examine the *art* of lumped modeling and the roles of various types of models in the design process. Next, we define precisely the terms *signal, flow,* and *potential,* which serve as unifying concepts throughout the book. We use the line graph, a simplified representation for a lumped model, to clarify the procedures for finding the differential equations that represent the lumped model mathematically. (We examine the *behavior* of the model—the solution of the differential equations that represent the model—in chapter 3.) The lumping process and the procedure for formulating equations are applied to an additional field in the first section of each succeeding chapter.

2.1 LUMPED MODELS FOR TRANSLATIONAL MECHANICAL SYSTEMS

In analyzing and designing the translational motion of mechanical systems, we use three fundamental properties of materials: their mass (or tendency to resist acceleration), stiffness (or tendency to resist compression and stretching), and friction constant (or tendency to resist motion). One of these three properties can dominate the other two in a part of the system. We often choose the material and design the shape and size of an object to enhance one of these three *resisting* phenomena. In other instances, two or all three of the phenomena act inseparably, and the overall behavior is complicated and difficult to predict. In this section, we present a notational framework for modeling mechanical systems. Then we examine the properties of the lumped elements (masses, springs, and dampers) that are the basic building blocks for models of such systems.

Translational Motion—a Potential

All physical objects are in motion, if only because the earth rotates, orbits the sun, and moves with the galaxy. We think of the motion of a point as the velocity of that point with respect to some fixed point. The hypothetical idea that some point in space is stationary forms the basis for a theory of motion. Analyses of the behaviors of systems in motion relative to a fixed point are in close agreement with measured behavior. Associated with the velocity v(t) of a point are two other commonly used motion-related quantities: the displacement x(t) and the acceleration a(t) of the point, relative to the fixed point. The displacement pattern x(t) determines v(t) and a(t); similarly, a(t) determines v(t) and x(t), except for the initial values v(0) and x(0). Velocity is the variable we mentally associate with the term *motion*. Velocity is also the ***central* motion variable,** in the sense that we integrate it to obtain x(t) and differentiate it to obtain a(t). In our models, we usually focus attention on the central motion variables—the velocities of points in the system.

Although the motions of points and the forces on points are three-dimensional vector quantities, most complex mechanical systems can be represented accurately by means of one-dimensional motions and forces. We focus on systems in which each point can move only forward and backward along a single line. We say such a point has one *degree of freedom* of motion.

The point shown in Figure 2.1 translates in one dimension. To establish a coordinate system for the motion of the point, as stated in chapter 1, we select arbitrarily a **positive direction of motion** along the line of action. That direction is denoted +v in the figure. Initially, the point is located a distance p from a fixed reference point. We refer to that reference point as **ground** (denoted g). We define the **displacement** x(t) of the point, for each instant t, as the change in position of the point in the +v direction *from its initial position* p. Zero displacement means no change in position. If the displacement opposes the +v direction, then x is negative. The displacement is expressed in meters.[1] The

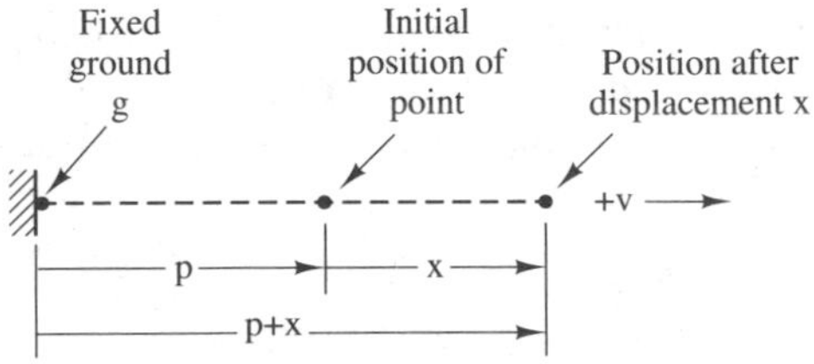

Figure 2.1 Displacement x.

[1]We use the International System (SI) of units throughout this book. Conversion factors that relate British units to SI units are given in Appendix A.

displacement x is independent of the initial position p. Velocities and accelerations of the point are positive if they correspond to displacements in the +v direction.

We typically *display* points at their initial locations. We can represent the motions by arrows that give the positive directions, as shown for points i and j in Figure 2.2. It is not necessary to use the same positive direction for all points. However, in most situations, it is simplest to use a single +v direction for the whole translational system. (For some configurations, such as a rope that changes its angle of motion at a pulley, the positive directions of various points must be adjusted to follow the line of action of the motion.) If we use a single +v direction, the +v direction arrow and the arrows representing v_i and v_j are redundant; the v_i arrow is implied by the +v direction and the label i on the point. Accordingly, we usually use only the single +v direction arrow for the *whole set* of points. Since we focus on the *central* motion variable v, we use the symbol +v to denote the positive direction of motion for all three motion-related quantities, x, v, and a.

The initial position p of a point does not affect its displacement, velocity, or acceleration. It is only a measure of the relative location of the fixed ground. Since the displacement x does not depend on the location of the ground point, we can pick the *location* of the ground arbitrarily. In effect, all fixed points are treated as ground. We need not show the location of the ground in our picture.

We measure the displacement of a point with a calibrated measuring stick fastened to the reference frame (ground) and oriented along the degree of freedom of the point. We align the 0 mark of the stick with the point to be observed (at p) and orient the +1 mark (1 m, 1 cm, etc.) in the positive (+v) direction from the point. In general, the stick must extend in both the positive and negative directions. We measure the displacement of the moving point at any instant by reading the number on the stick beside the point. We view the measuring stick as a *displacement meter,* with two ends or terminals. One end of the meter is attached to the system at the fixed position p. The other end is attached to the moving point. Hence, the meter is connected *across* the measured system.

A particular displacement meter is only suitable for measuring a limited range of displacements. With a typical meter stick, for example, we can directly measure a displacement as small as 1 mm and as large as 1,000 mm. By interpolating between millimeter marks, we can reduce fine measurement to about 0.25 mm. By moving the meter stick, we can extend gross measurement to about 10,000 mm. The ratio of the maximum measurable value to the minimum measurable value of a measuring device is the *dynamic range* of the device. The dynamic range of a meter stick is 1,000–40,000. Although we could increase the dynamic range of a displacement meter by increasing its length, a long measuring stick is inconvenient to manipulate. We would not choose to use a 10-m stick or a 100-m tape to measure displacements as small as 1 mm. If a displacement is too small to measure directly, we magnify the displacement before we measure it. A micrometer is one such magnifier. A magnifying glass is another. Displacements as small as 10^{-5} mm can be measured with an electrical strain gage.

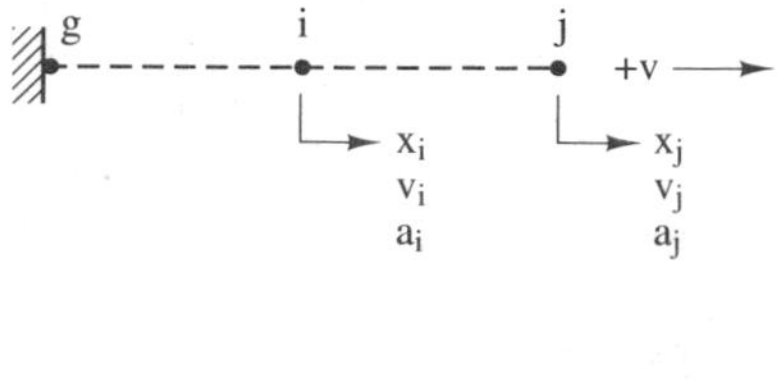

Figure 2.2 Notations for relative translation.

In a dynamic system, we need to know the behavior of the system as a function of time. We can use a clock to determine the time at which each displacement is measured. If the motions are rapid, the measuring and data-recording processes can be automated. Although displacement measurements versus time can be used to compute values of the velocity and acceleration of a point, it is usually more accurate to measure these directly by means of a velocity meter or an accelerometer. These meters also should be viewed as two-terminal devices. (See problems 2.33 and 2.34 at the end of the chapter.)

The *relative* motions of two points are illustrated in Figure 2.2. We denote the displacements of these points by x_i and x_j, respectively. We use a double-subscript notation to denote the **relative displacement** of point i with respect to point j:

$$x_{ij} \triangleq x_i - x_j \tag{2.1}$$

Similarly, the relative velocity v_{ij} and relative acceleration a_{ij} are defined by

$$v_{ij} \triangleq v_i - v_j = \dot{x}_{ij} \tag{2.2}$$

and

$$a_{ij} \triangleq a_i - a_j = \ddot{x}_{ij} \tag{2.3}$$

where the dots mean time derivatives. Note that relative motion can be expressed as a *difference* in the motions of two points *only* if the same positive direction is used for both points.

In general, the motion of one end of an object is different from the motion of the other end. This point-to-point variation is characteristic of a potential, as discussed in chapter 1. Motion variables (x, v, and a) are **across variables**—variables that have different values from point to point across the parts of a system. We measure the *relative motion* (x_{ij}, v_{ij}, or a_{ij}) between two points across the object or across the space that lies between the points. We measure the *absolute motion* of a point across the space between that point and ground. In section 2.3, we demonstrate mathematically that **velocity, displacement, and acceleration are potentials.**

If the relative motion v_{ij} is positive, then $v_i > v_j$. Therefore, the potential-difference variable $\mathbf{v_{ij}}$ **is the drop (or reduction) in velocity from point i to point j.** (If $v_{ij} < 0$, the drop in velocity is negative; in other words, it is a velocity rise.) In general, we view any potential-difference variable as a drop in potential from the point indicated by the first subscript to the point indicated by the second subscript. Thus, the potential differences x_{ij} and a_{ij} represent drops in potential from point i to point j. This potential-drop terminology relates to the intuitive notion of downhill flow introduced in equation (1.1). We use the terminology extensively to determine the correct signs when we write system equations.

Suppose points i and j of Figure 2.2 come closer together, indicating that the object or space between the points is compressing. Since points i and j are ordered in the $+v$ direction, this compression requires that $x_i > x_j$. Thus, the displacement variable x drops (or reduces) in the $+v$ direction for compression. If the displacement variable x in a particular system rises from point to point in the $+v$ direction, then the object or space between the points must be stretching or expanding.

The *relative* displacement x_{ij} should be viewed as an amount of compression if the subscripts denote points that are ordered in the +v direction. That is, a positive value of x_{ij} represents compression of the space between points i and j. The relative displacement x_{ji}, with subscripts ordered in the −v direction, represents an amount of elongation.

In general, **compressive relative motion corresponds to a drop in potential (x, v, or a) in the +v direction.** Since points i and j of Figure 2.2 are ordered in the +v direction, v_{ij} must be positive if the points i and j are moving closer together. A positive value of a_{ij} means that the points i and j are moving closer together or will eventually be moving closer together. We usually focus attention on the *central motion variable* v and express equations in terms of velocities v_i and velocity differences v_{ij}. We speak of v_{ij} as a potential drop or velocity drop if points i and j are ordered in the +v direction. We refer to v_{ij} as a velocity rise if the points are ordered in the −v direction.

Figure 2.3 shows a compressed object and a lumped representation of that object. A **lumped representation** views the object in terms of its behavior at its ends and ignores the behavior at intermediate points. Figure 2.3(a) labels separately the positive directions for the velocities v_i and v_j of the ends of the object. Figure 2.3(b) represents each end of the object by a node. That figure displays the symbols v_i and v_j that represent the node velocities, but uses a +v arrow to show the positive direction for velocities. The associated motion pairs x_i, x_j and a_i, a_j are implicit in both notations. In some instances, we denote explicitly the displacement variables or acceleration variables in order to focus attention on the displacing or accelerating aspects of the motion. More often, we merely number the nodes, *think* in terms of velocities, and recognize the coexistence of displacements and accelerations.

Scalar Force—a Flow

The conventional mechanical model is a free-body diagram that shows force vectors (from the environment) acting at points on the boundary of an object. We show such a model in Figure 2.3(a). The object is acted on by a pair of equal and opposite force vectors. If we cut the object vertically and insert a force scale, that scale senses compression. If we reverse the directions of the force vectors, the scale senses tension. The quantity measured by the scale has an obvious *line of action,* but no orientation; that is, it is not directed from one end of the object toward the other. Thus, the quantity is a *through variable,* the same through the whole length of the object. We call the measured quantity the **scalar force** *in* the object.

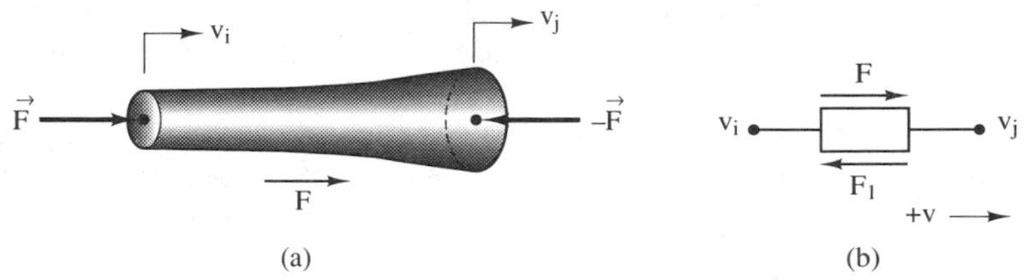

Figure 2.3 Equivalent notations. (a) A compressed object. (b) A lumped representation.

The scalar force in an object is an indicator of what happens to the object as it is subjected to external force vectors. (Another indicator of what happens to the object is the compressive motion across the object.) Scalar force has two opposing senses, compressive and tensile. The distinction between compressive force and tensile force is important. The behavior of a material can change drastically if the sense of the force is reversed. For example, concrete tolerates high compressive force but little tensile force, whereas a steel cable supports high tensile force but no compressive force. The magnitude of the internal scalar force is $|\vec{F}|$—the same as the magnitudes of the two external force vectors. The SI unit for force is the newton, abbreviated N.

It is not possible to measure scalar force at a point in a system without breaking the system at that point to insert a scale. We view the scale as a force meter, with two ends or terminals. The meter is connected in series with the object. In general, a force meter must be able to measure both compressive and tensile scalar forces. (A bathroom scale measures only compressive force.) Example problem 2.1 examines the design of a scalar-force meter.

Since scalar force is a *through* variable, we *treat* it as a flow. For a passive (energy-free) object, inward-pointing force vectors produce compressive force in the object and compressive relative motion across the object. According to the previous discussion of motion, the velocity drops across a compressing object in the +v direction. To produce an analogy to downhill flow, as discussed in chapter 1, we **treat compressive force as a flow in the +v direction,** the direction of velocity drop.[2] Then tensile force is a flow in the −v direction.

We represent the scalar force in an object by the symbol F. We assign a flow direction to F and indicate that direction by an arrow beside the object. We treat a positive value of F as a flow in the direction of the arrow. If F is oriented in the +v direction, it represents compressive force. Then, a negative value of F represents negative compressive force—or tensile force. If F is oriented in the −v direction, it represents tensile force, and a negative value means that the force is compressive.

In Figure 2.3(b), the scalar force is represented as a flow F from i to j. Alternatively, it is represented as a flow F_1 ($\equiv -F$) in the opposite direction. For the +v direction specified in the figure, F is the compressive force and F_1 is the tensile force.

Some users may wish to orient all flow variables in the +v direction. Then all flows represent compressive forces. If we define all scalar-force variables as compressive, then the symbol F beside an element is sufficient to define the sign of the flow F in the element. The flow arrow is redundant.

It might appear that the physical nature of the scalar force depends on the arbitrary +v direction. However, reversing the +v direction does not just reverse the compressive/tensile interpretation of the flow variable; it also reverses the signs of all variables in the system—both motions and scalar forces. Therefore, reversal of the +v direction changes a compressive force to a *negative* tensile force; the physical interpretation of each flow is

[2]We could, instead, draw an analogy between *velocity* and flow. The term *flow* suggests motion, at least in the familiar hydraulic context. However, hydraulic flow is a through variable and velocity is an across variable. As a result, the velocity-flow analogy is more limited in usefulness than the force-flow analogy. The advantage of the force-flow analogy becomes apparent in section 2.3.

unchanged. Of course, the actual nature of the physical scalar force (compressive or tensile) is determined by the interaction of the system parts, not by the way we define positivity of flow. Defining positivity of flows in objects by the arrow directions and positivity of motions of points by the +v direction only determines the signs of the mathematical variables used to represent the scalar forces and the velocities.

We shall use the expression **force on** an object to mean a force vector that acts on a free-body diagram. We shall use the expression **force in** an object to mean a scalar force or flow. Figure 2.4 shows a way to relate the two concepts: we can visualize a scalar force in terms of an equivalent vector force on a free-body diagram. The symbol F in Figure 2.4(a) is a flow in the +v direction. Thus, it represents the compressive force in the object. Visualize the +v end of the object as the reference point (ground), and apply a force vector $\vec{F}$ to the −v end of the object, oriented to agree with the flow direction, as shown in Figure 2.4(b). Then the *force vector* $\vec{F}$ also represents the compressive force in the object. We can identify the force vector with the scalar force and use the same symbol for both quantities.

The flow F_1 in the object of Figure 2.4(c) is oriented in the −v direction. Hence, it represents the tensile force in the object. Again, visualize the +v end of the object as the reference point, and apply a force vector $\vec{F}_1$ to the −v end of the object, oriented to agree with the flow direction, as shown in Figure 2.4(d). Then the force vector $\vec{F}_1$ also represents the tensile force in the object. We can identify the force vector with the scalar force and use the same symbol for both.

In sum, we can identify the flow variable assigned to an object with a force vector that acts at the −v end of the object. The vector is oriented in the same direction as the flow variable. Thus, a flow in the +v direction represents compressive force, and a flow in the −v direction represents tensile force. If we orient all flow variables in the +v direction, then all flows represent compressive forces.

Sign Conventions, Power, and Energy

Let us focus attention on the points of interest in a mechanical system by marking and numbering those points. We designate a single positive (+v) direction for these points and assign a motion variable v_i to each point i. Next, we assign a flow variable F_n to each object n between points and use an arrow beside the object to designate the positive flow direction for F_n. If we order the subscripts on a velocity-difference variable in the +v direction, that velocity difference represents compressive motion. If we orient the arrow for a flow variable in the +v direction, that variable represents compressive force. Reversal of

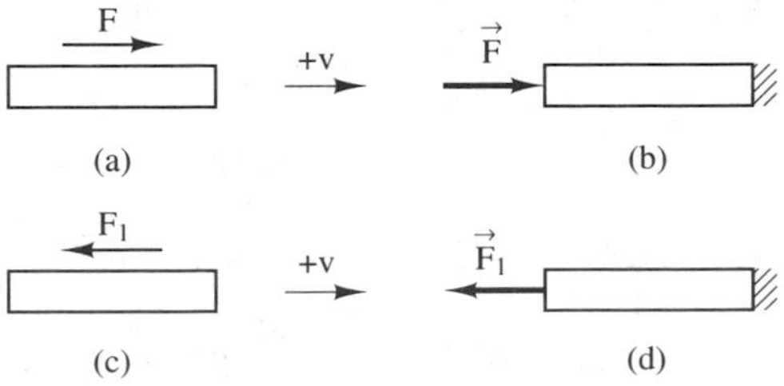

Figure 2.4 Equivalence between force vectors and scalar forces. (a) Compressive force. (b) Compressing force vector. (c) Tensile force. (d) Stretching force vector.

the +v direction reverses the signs and the physical interpretations of all of the variables.

The mechanical energy that an external force vector delivers to an object is the product of the force and the distance the object moves in the direction of the force. The mechanical *power* that the force delivers to the object is the product of the force and the velocity with which the object moves in the direction of the force. The flow F in Figure 2.3 represents compressive force. Since the applied force vectors point inward, it is apparent that F is positive. (The scalar force F is compressive). The force vector $\vec{F}$ applied to point i has magnitude F and the direction of the velocity v_i. Therefore, the power delivered to point i is the product Fv_i of the force and the velocity in the direction of the force. Similarly, the force vector $-\vec{F}$ applied to point j has magnitude F, but the direction of $-v_j$. Therefore, the power delivered to point j is $-Fv_j$. The total **mechanical power** delivered to the object by the two external force vectors is

$$P = Fv_{ij} \tag{2.4}$$

Power is measured in watts (abbreviated W). One watt is 1 N·m/s.

The power formula (2.4) is stated in terms of a velocity drop v_{ij} in the +v direction and a flow F in the +v direction. If we reverse the +v direction, the signs of both F and v_{ij} reverse, so that the power formula itself is unchanged. In general, **the mechanical power delivered to an object is the product of the flow through the object (in some direction) and the velocity drop in the same direction.** That interpretation is consistent with the intuitive expectation that flow will be in the direction of potential drop (chapter 1). Suppose we replace the flow variable F by the flow variable $F_1 \triangleq -F$, oriented to oppose the +v direction (see Figure 2.3(b)). Then, by this intuitive statement of the power calculation, $P = F_1 v_{ji}$, where both the flow F_1 and the velocity drop v_{ji} are oriented in the same direction. Since $v_{ji} = -v_{ij}$, the intuitive calculation produces the same result as equation (2.4).

The two ends of the object of Figure 2.3(a) serve as a **port,** or means of access to the object. The scalar force in the object is produced by actions at the ends. That force is measurable only at the ends. (Otherwise, we must break into the object.) The motion of the object is observed at the ends. The mechanical power delivered to the object by its surroundings is the product of the port variables—the flow F through the port (the pair of ends) and the velocity drop v_{ij} across the port in the direction of flow.

Passive objects do not supply energy. As stated in chapter 1, the variables (or signals) that reveal the behavior of a *passive* object have zero value, unless an external source provides energy to activate the object. Since a passive object cannot deliver energy to its surroundings, the power flow given by equation (2.4) cannot be negative. The flow through the passive object and the velocity drop *in the direction of flow* must have the same sign. The actual flow is in the direction of the actual potential drop for a passive object.

A spring or mass can *store* energy injected by an external source of force or motion. At a later time, the spring or mass can return that stored energy to its surroundings. Averaged over time, however, the energy that a mass or spring can deliver is zero. We say that an energy-storage device is **unenergized** if it contains no stored energy. An unenergized mass or spring is passive. We find in section 4.2 that we can separate the effect of the stored energy from the behavior of the unenergized mass or spring. (We can view the separated energy as an energy source.)

The mechanical energy E delivered to an object during an interval $[t_1, t_2]$ is the time integral of the power; that is,

$$E = \int_{t_1}^{t_2} F v_{ij}\, dt \tag{2.5}$$

where the arguments of the integral—the velocity drop and the flow—must be oriented in the same direction. Energy is measured in N·m. If the signs of F and v_{ij} differ during some time interval, then energy is removed from the object during that time interval. (Energy could be removed from a mass or spring during a time interval if that energy had been stored in the mass or spring during a previous time interval.)

Nothing we have done to this point enables us to predict the values of the motions and scalar forces in the object of Figure 2.3. We must know the details of the system in which the object is imbedded in order to compute the values of the variables at its ends. Once we compute the values of the mathematical variables for a system, we can interpret those values physically. In general, we interpret a positive value of flow as compressive force if the flow is oriented in the +v direction. We interpret a positive drop in potential (displacement, velocity, or acceleration) as a compressive relative motion if the points are ordered in the +v direction.

The Ideal Spring

Suppose we compress the uniform homogeneous object of Figure 2.5 by equal and opposite force vectors. Then there is a compressive force in the object and a compressive relative displacement of the ends of the object. *Compressive force* is represented by a flow F oriented in the +v direction. *Compressive relative displacement* is represented by x_{ij}, with the subscripts ordered in the +v direction. Both F and x_{ij} are positive in Figure 2.5. We define the *compressive stress* in the object as F/A, where A is the cross-sectional area of the object. We define the *compressive strain* (or fractional compression) by x_{ij}/ℓ, where ℓ is the length of the object. (If x_{ij} were negative, the stress and strain would be tensile rather than compressive.) According to Hooke's law, if the strain is not excessive, it is proportional to the stress; that is,

$$\frac{F}{A} = \mathrm{E}\frac{x_{ij}}{\ell} \tag{2.6}$$

where the proportionality factor E is known as Young's modulus.[3] Then

$$F = \mathrm{E}\frac{A}{\ell} x_{ij} \tag{2.7}$$

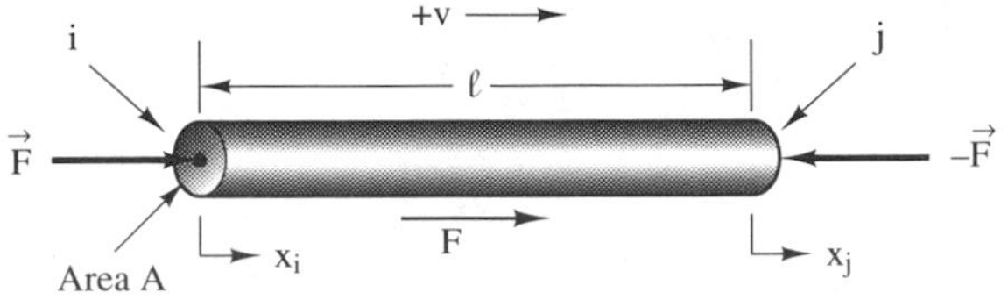

Figure 2.5 A uniform homogeneous object.

[3]Young's modulus E is tabulated for a number of materials in Appendix A. The SI units for E are N/m^2.

The quantity

$$k = \text{E}A/\ell \tag{2.8}$$

is the stiffness of the object.

We are motivated by equation (2.7) to define an **ideal spring** to be a lumped two-terminal element that obeys the formula

$$F = kx_{ij} \tag{2.9}$$

where x_{ij} is the displacement of node i relative to node j and F is the scalar force in the element, oriented in the direction from node i to node j. We call k the **spring constant** or **stiffness** of the spring. The units of stiffness are N/m. The inverse constant 1/k is the **compliance** of the spring. A coil spring approximates the behavior of an ideal spring over a limited range of values of displacement. Accordingly, we use the coil spring symbol of Fig. 2.6 to represent the ideal spring. We note the orientation of the scalar-force variable by an arrow beside the symbol.

Some students are bothered that the double subscript notation in equation (2.9) permits both ends of the spring to move. Perhaps their thinking has been narrowed by simple explanations of the behavior of a spring. A tow rope acts as a spring. One end moves with the vehicle doing the towing, the other end with the vehicle towed. Relative motions between the vehicles cause variations in the tension of the rope.

We view x_{ij} as a potential drop in the direction of flow. Intuitively, equation (2.9) says that the potential drop across the spring in a particular direction is proportional to the flow through the spring in the same direction. This intuitive statement helps us remember how to orient the variables as we write spring equations in complicated mechanical systems. Equation (2.9) does not depend on the +v direction. On the other hand, the physical interpretations of computed values of the flow F and the potential drop x_{ij} as compressive or tensile do depend on the +v direction.

By definition, an ideal spring is frictionless and has no mass. We separate the compliance property of a physical object from its friction and mass properties and *represent it by* an isolated lumped spring. We use the stiffness symbol k beside the ideal-spring symbol (see Figure 2.6) not only to represent the numerical stiffness of the spring, but also to identify the particular spring. We refer to it as spring k.

To determine the behavior of a whole mechanical system, we combine the equations for the various parts of the system. In general, those equations involve a mixture of displacements, velocities, and accelerations. Analysis is simplified, both conceptually and mathematically, if we use only one type of motion variable. We usually use the central motion variables—the velocities of the points in the system. An alternative form of the defining equation (2.7) for the ideal spring is

$$v_{ij} = (1/k)\dot{F} \tag{2.10}$$

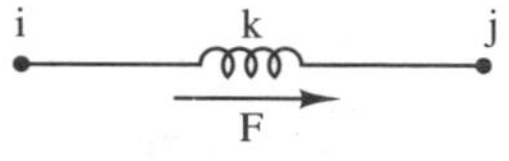

Figure 2.6 An ideal spring.

Intuitively speaking, if the flow F is increasing, then the velocity drops across the spring in the direction chosen for the flow variable. That is, the velocity is less positive on the *outflow* side of the spring.

The manipulation of a number of simultaneous differential equations can be confusing, even if they involve only flows and velocities. Therefore, we introduce one more simplification. As suggested in chapter 1, let the symbol s represent the time-derivative operator. Multiplication by s is a shorthand notation for time differentiation. With this notation, equation (2.10) becomes

$$v_{ij} = (s/k)F \tag{2.11}$$

Equation (2.11) should be interpreted to have the same meaning as equation (2.10). The time-derivative operator s acts on the signal F. The quantity s/k can be thought of as the **impedance** of a translational spring, in the sense discussed in chapter 1. That is, for a given velocity drop across the spring, an increase in impedance produces a decrease in flow F. Some books refer to impedance as *operational impedance* to reflect the fact that s is a differential operator.

The impedance s/k, by itself, describes the ideal spring fully. Intuitively, the velocity drop across the spring in a specific direction is *proportional* to the flow in the same direction. The proportionality factor is the impedance s/k. The parameter k characterizes the particular spring. The s in the numerator indicates that the flow is differentiated; that is, the velocity drop is proportional to the *time derivative* of the flow. We sometimes label the spring symbol of Fig. 2.6 with its impedance rather than its stiffness. The units of translational impedance are m/N·s.

The values of variables in a physical system cannot become infinite. If we do not let the velocity difference across an ideal lumped spring be infinite, then equation (2.10) does not permit the rate of change $\dot{F}$ of scalar force in the spring to be infinite. Accordingly, the scalar force F in the spring cannot change instantaneously. (Otherwise, its time derivative would be infinite.) We use lumped springs to build models of translational systems. No matter how rapidly the other variables in the model change their values, the scalar forces (or flows) in the springs never exhibit abrupt jumps in value, but are continuous with time. The continuity of flow provides the basis for useful analysis and design rules in sections 3.3 and 4.2. We can, of course, do things with models that we cannot do with physical systems. If we choose to *require* the scalar force in a spring to jump instantly, we must permit the velocity difference across the spring to be infinite.

The potential energy that we deliver to an ideal spring by compressing or stretching it from a relaxed state ($x_{ij}(0) = 0$), according to equations (2.5) and (2.9), is

$$\begin{aligned} E(t) &= \int_0^t F v_{ij}\,dt = \int_0^{x_{ij}(t)} k x_{ij}\,dx_{ij} \\ &= \frac{1}{2} k x_{ij}^2 = \frac{1}{2} F^2/k \end{aligned} \tag{2.12}$$

The stored potential energy can be retrieved from the spring by reversing the compression or extension process. The second form of equation (2.12) expresses the stored energy in terms of the flow variable F. Because the scalar force in a physical spring cannot change abruptly, energy cannot be transferred to or from a physical spring instantly.

The stiffness formula given by equation (2.8) provides guidance for the design of springs and stiffeners. By appropriate choices of material (E) and shape (A and ℓ), the engineer can design an object with a desired stiffness. For example, a block of rubber can provide the compliance necessary for a footpad on a washing machine, and a block of styrofoam packing material can provide the compliance needed for the protective shipping of electronic equipment. The freedom of choice in values of k is restricted by practical constraints on thickness, cross section, and price and by the elastic limits of available materials. By making a creative change in the geometry of the object, however, we are usually able to obtain a very significant decrease in stiffness for a given material. Examples of such creative geometries are shown in Figure 2.7, together with the corresponding relationships among the stiffness of the spring, the parameter of the material, and the geometric parameters. The coil spring of Figure 2.7(a) is used in door closers and many automobile suspension systems (see example problem 2.2). The leaf spring common to vehicle suspension systems is essentially a symmetrical pair of cantilever springs similar to that shown in Figure 2.7(b) (see problem 2.3 at the end of the chapter).

The formulas shown in Figure 2.7 apply only for a limited range of values of compression or extension x_{ij}. The spring constants for various geometries are derived in reference [2.1] at the end of the chapter. The stresses in the coil spring are shear stresses, perpendicular to the centerline of the wire. The term *shear* refers to displacements of two neighboring parts of an object parallel to their plane of contact. Hence, the *shear forces* on the neighboring parts act parallel to the plane of contact. The units of the shear modulus G are N/m^2. Values of G are tabulated for various materials in Appendix A.

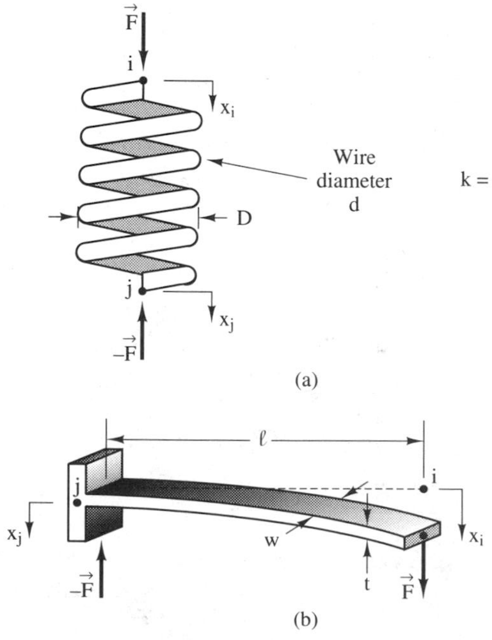

$$k = G\frac{d^4}{8\,nD^3}$$

$$k = E\frac{wt^3}{\ell^3}$$

Figure 2.7 Translational spring geometries. (a) Coil spring; n is the number of turns and G is the shear modulus of the material. (b) Cantilever spring; E is Young's modulus.

No physical spring will have exactly the geometry assumed in the derivation of a particular stiffness formula. We can always measure the stiffness of a particular spring, however. Using an object of known weight w, we stretch or compress the spring and measure the displacement x. Then, according to equation (2.9), k = w/x—limited, of course, by the accuracies of the weight and the displacement measurement. This measurement process can be carried out for a set of weights that produce displacements over a range suitable to the spring. We treat the average value found for k as the stiffness of the spring. For moderate displacements, the measured stiffness will deviate only slightly from the average value. It can deviate drastically for extreme displacements.

Example Problem 2.1: Designing a spring scale.

The displacement of an ideal spring is proportional to the scalar force in the spring. Use this relationship to design a meter that measures scalar force.

Solution: Fasten one end of a translational spring to a frame, and mark the point on the frame beside the free end of the spring with the symbol 0 (see Figure 2.8(a)). Place the fastened end of the spring and frame on the ground. Place a 0.102-kg mass on the other end of the spring, as shown in Figure 2.8(b). Since the weight of this mass is mg = (0.102 kg) × (9.8 m/s^2) = 1 N, the mass compresses the spring with a 1-N force. Mark the point on the stick beside the compressed end of the spring as +1. This mark defines *compressive* force as positive and defines the unit of force as the newton. If the spring is nearly linear, the frame can be marked with other positive and negative numbers consistent with the 0 and +1 marks. To calibrate the meter more accurately, we apply various accurately known positive and negative forces to the spring and label the compressed or extended positions accordingly.

To measure the force in a branch of a system, break the branch at the point where the force is to be measured, and insert the calibrated meter. (It may be necessary to attach an appropriately designed linkage to the ends of the meter in order that the meter fit in the broken branch without changing the system configuration drastically; see Figure 2.8(c).) If the scale shows that the force is positive, the branch of the system is being compressed; if the scale reads negative, the branch is being stretched.

A spring that is appropriate for measuring very large forces will not show measurable displacements for very small forces. Therefore, a spring scale is designed to measure forces in a particular range. The **dynamic range** of a measuring instrument is the ratio of the largest value that it can measure to the smallest value that it can measure. A dynamic range of 1,000 is typical of a simple measuring instrument such as a spring scale.

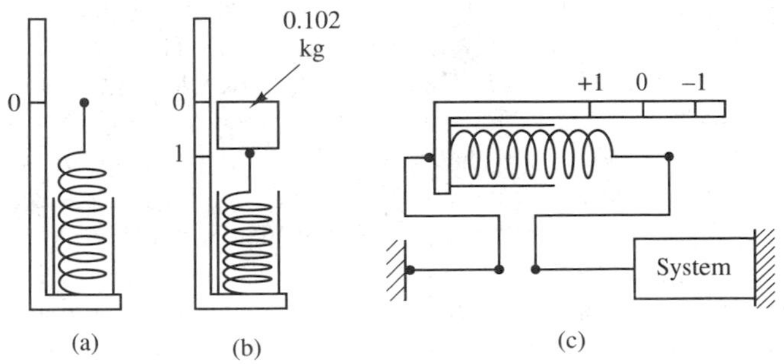

Figure 2.8 Scalar-force meter. (a) Spring scale construction. (b) Scale calibration. (c) Measuring force with the scale.

Example Problem 2.2: Automobile suspension design.
Design the coil spring of Figure 2.7(a) so that it will carry the 3,000-N weight supported by one front wheel of an automobile, under the condition that the automobile is stationary. The shear modulus for steel is $G = 7.59 \times 10^{10}$ N/m^2.

Solution: The relationship between the compressive force F in the spring, the compressive displacement x of the spring, and the stiffness k of the spring is $F = kx$, where $k = Gd^4/8nD^3$. The problem is underspecified. We can choose any combination of the parameters x, d, D, and n that will support the 3,000-N force. In making the arbitrary choices among these parameters, it is necessary to understand the configuration of the wheel, axle, and body of the automobile in the region where the spring will be installed. (Look under the front ends of a few automobiles to gain this understanding.)

Suppose we choose, on the basis of the suspension geometry, D = 10 cm, a spring length of 25 cm (when compressed by the automobile), and a compressed length per turn of 2.5 cm (which implies that n = 10). Then the spring constant is

$$k = \frac{(7.59 \times 10^{10}\ \text{N/m}^2)(d^4)}{(8)(10)(10\ \text{cm})^3} = 94.9\ \text{N/cm}^5 \tag{2.13}$$

and the relation between the displacement x and the wire diameter d is $F = 3{,}000\ \text{N} = (94.9d^4x)\ \text{N/cm}^5$. A typical wire diameter for a vehicle suspension spring is 1.25 cm. For that diameter, the amount by which the spring is compressed is x = 13 cm, a reasonable amount compared with the compressed length of 25 cm. The uncompressed spring length is 38 cm.

The foregoing design focuses on stiffness alone. Practical design has to address a number of other issues as well. For example, the designer must assure that the shear stress (force per unit area) is within the elastic limits of the wire.

The Ideal Damper

Everyone is familiar with the warming effect of rubbing one's hands together. The tendency of materials to resist relative *velocity* between their particles is referred to as friction, and we call the resistive force the friction force. The energy used to cause motion against friction forces is converted to heat. Unlike the energy stored in a moving mass or a compressed spring, this heat energy is seldom recovered.

The friction forces and energy dissipation associated with elastic deformation of a solid (such as compression of a spring) are usually negligible since the particles of material do not move much relative to their neighboring particles. On the other hand, if we break a piece of wire by repeated drastic bending (plastic deformation), a noticeable amount of heat is generated and dissipated. Fluids (liquids and gases) tend to *shear* under stress; that is, layers of the fluid slip continuously relative to each other. The shearing forces and heat dissipation associated with shearing motions can be substantial.

If the friction force that resists the relative motion of two objects is proportional to the relative velocity of the objects, we call that friction *viscous friction*. For low velocities of objects in fluids, the friction between adjacent fluid layers is viscous. Fast motions in fluids produce friction forces that are nearly a quadratic function of relative velocity, as shown in Figure 2.9. The friction force associated with the mutual sliding of two solids is nearly constant. This solid-to-solid friction is called *coulomb friction* (see example problem 2.10).

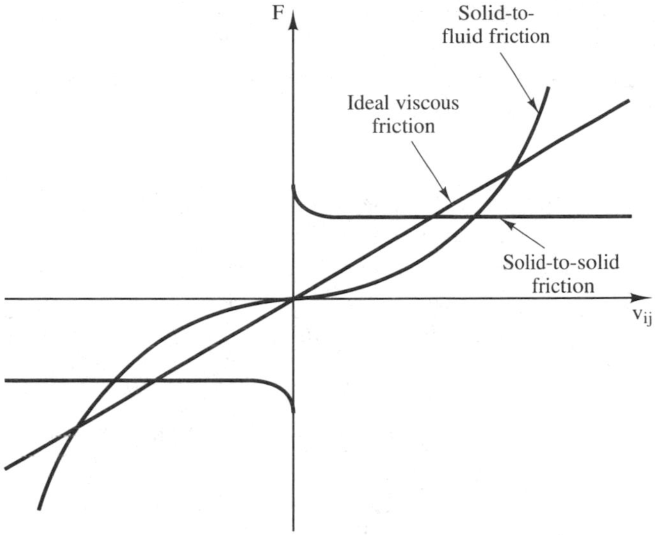

Figure 2.9 Typical friction characteristics; v_{ij} is the relative velocity of two objects.

We prefer to work with *viscous* friction for several reasons. First, a proportional relationship between scalar force F and relative velocity v_{ij} is easier to understand, mathematically and intuitively. Second, engineers are able to design friction devices that exhibit nearly viscous behavior. For example, the friction in lubricated sliding of smooth solids is viscous. Finally, viscous-friction devices are useful. Examples include shock absorbers in automobile suspension systems and speed limiters in door closers.

We define an **ideal damper** to be a two-terminal lumped element that exhibits the ideal viscous friction characteristic. That is, the relative velocity of its terminals is proportional to the *scalar* force in the element. Specifically,

$$F = bv_{ij} \tag{2.14}$$

where F is the scalar force in the damper, oriented in a specific direction, and v_{ij} is the velocity difference across the damper, with subscripts ordered in the direction chosen for F. The proportionality factor b is called the **friction constant** or **damper constant.** It has the units N·s/m.

We represent the ideal damper by the cylinder and piston symbol (or *dashpot*) of Figure 2.10, labeled with the friction constant. (Imagine that the friction results from seepage of fluid past the piston.) We show the orientation of the scalar-force variable beside the damper. Intuitively, the velocity drop v_{ij} across the damper in a particular direction is proportional to the flow through the damper in the same direction. If the chosen direction

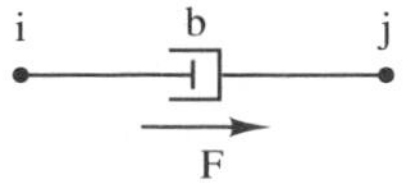

Figure 2.10 Ideal viscous damper.

is the +v direction, the flow represents compressive force and the velocity drop represents compressive motion.

The damper equation (2.14) can be arranged in impedance format, analogous to equations (1.1) and (2.11):

$$v_{ij} = (1/b)F \tag{2.15}$$

We sometimes label the damper symbol with its impedance rather than its friction constant. The damper impedance 1/b is also called the **mechanical resistance** of the damper.[4] The units for mechanical resistance are m/N·s.

Energy consumed by friction is dissipated (lost to heat). The amount of energy dissipated by an ideal damper during the time interval $[t_1, t_2]$, according to equations (2.5) and (2.14), is

$$E = \int_{t_1}^{t_2} Fv_{ij}\, dt = \int_{t_1}^{t_2} bv_{ij}^2\, dt \geq 0 \tag{2.16}$$

Note that E is always positive, regardless of the sign of the relative motion v_{ij}. The power absorbed by the ideal damper is

$$P = bv_{ij}^2 = (1/b)F^2 \tag{2.17}$$

The units of power dissipation are watts (abbreviated W) or N·m/s.

The geometry of one type of viscous damping device is shown in Figure 2.11. It is typical of the *lubricated* sliding of one solid object against another. Point j represents all rigidly connected points on the base plate, and point i represents all rigidly connected points on the top plate. The layers of intervening oil slip continuously relative to each other like cards in a stack. This slipping of layers is called *shearing.*

Shear strain γ is defined as the relative velocity of two parallel fluid layers divided by the thickness of the intervening fluid layers (in m/s/m = s^{-1}). *Shear stress* τ is the

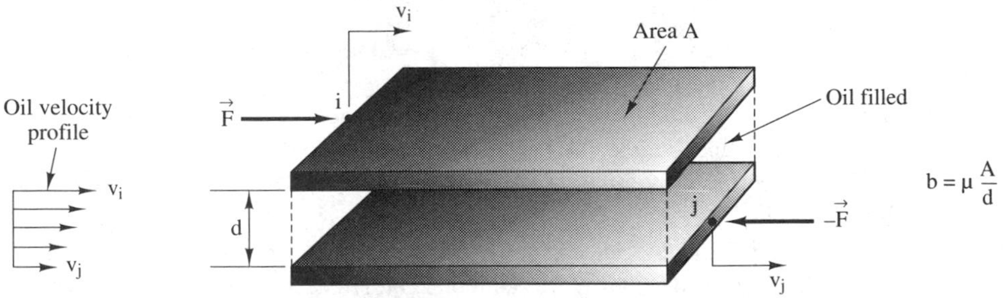

Figure 2.11 A viscous damper; the oil thickness d is exaggerated.

[4]Throughout this book, we use the terms *resistance* and *impedance* to mean resistance or impedance to *flow.* In hydraulic, thermal, and electrical systems, this use agrees with conventional language. In mechanical systems, however, the concept of *resistance to motion* is in common use. The languages developed in these different energy domains are in conflict! We resolve the conflict in favor of the language that appears in most energy domains. Thus, mechanical resistance is a limit to *force:* High damper resistance implies low damper force or high damper velocity.

shearing force (the force that causes the strain) per unit area of shearing surface (parallel to the force), in N/m^2. *Fluid viscosity* μ is the ratio of shear stress to shear strain in the fluid. Thus, viscosity is a measure of the fluid resistance to shearing. The SI units for viscosity are $N{\cdot}s/m^2$.[5]

The horizontal velocities of the oil layers are assumed to vary linearly over the thickness d, from v_i at the top surface to v_j at the bottom surface, as shown in the figure. If the velocities v_i and v_j are constant, then there are no acceleration forces, and the full compressive force F, felt at the points i and j, is applied to shearing of the oil layers. The shear stress on the thin oil film is $\tau = F/A$, where A is the area of each shearing surface. The shear strain in the oil is $\gamma = v_{ij}/d$, measured in m/s per meter of oil thickness. Since the viscosity μ is the ratio of shear stress to shear strain,

$$F = \mu \frac{A}{d} v_{ij} \tag{2.18}$$

Thus, the friction constant for this geometry is

$$b = \mu A/d \tag{2.19}$$

This formula is useful not only as a guide to the design of damping devices, but also as a tool for predicting and limiting the forces caused by (and heat generated by) friction in such applications as the sliding of pistons in cylinders and the sliding of shafts in bushings.

A canoe that is paddled through the water behaves like a viscous damper. The geometry of the canoe and water differ enough from that of Figure 2.11 that we cannot use equation (2.19) to estimate the friction constant b for the canoe. Fortunately, we can measure the friction constant of a damper directly. Suppose that the water is stationary relative to the ground. We use a rope to pull the canoe through the water at a constant velocity. We measure the velocity v of the canoe and the tension F in the rope. According to equation (2.14), the friction constant is $b = F/v$, limited by the accuracy of the measurements. We repeat the process for a range of velocities suitable to the canoe. The measured friction constant will increase somewhat with increasing velocity.

Exercise 2.1: We can use a spring scale to measure the rope tension F. How could we measure the velocity of the canoe?

The Ideal Mass

Each part of a mechanical object has mass and, therefore, resists changes of velocity. The particles of matter that comprise an object are not connected in a completely rigid manner. Under stress, various points in the object exhibit different velocities and accelerations versus time. The motion of each point is influenced by a friction force, a compliance (or springlike) force, and an inertial force. We define an ideal translational mass to isolate the phenomenon of inertia. That is, we separate the inertial force from the two other force-generating phenomena.

[5]The viscosities of various fluids are tabulated in Appendix A.

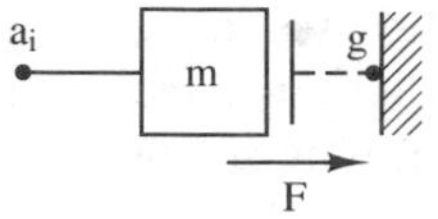

Figure 2.12 An ideal mass.

The rectangular block symbol in Figure 2.12 represents the **ideal mass.** Point i of the figure (labeled with its acceleration a_i) represents all points of the physical system that are treated as a rigidly connected unit, regardless of the geometry of that set of points. We identify the motion of point i of the model with the motions of all of those rigidly connected points. All force vectors that act on those rigidly connected points in the physical system must act on point i of the model.

The dashed line that connects the block to the reference point (labeled point g in the figure) emphasizes that a_i is measured relative to the stationary ground. Accordingly, the lumped mass should be treated as if it has two terminals: terminal i, which is associated with the mass itself, and terminal g, which is associated with the fixed reference. The ground represents all stationary points in the system.

Think of the block symbol as a mass that is rigidly attached to point i. The scalar-force variable F in the figure is oriented from node i to node g. Suppose that direction is the +v direction. Then F represents the compressive force measured by a scale inserted between point i and the rigidly connected mass, and compressive relative acceleration of the two points is represented by a_{ig}, with subscripts oriented in the +v direction. The equation that relates the compressive force and the compressive relative acceleration of the point is

$$F = ma_{ig} = ma_i \tag{2.20}$$

where the proportionality factor m is the mass of the set of rigidly connected points in the physical system. This is Newton's second law. The SI unit for the mass m is the kilogram.

On the other hand, suppose that the +v direction opposes the orientation of the scalar-force variable F. Then $-F$ represents the compressive force between node i and the block mass, a_{gi} represents the compressive acceleration of the two points, and the relation between the scalar force and the relative acceleration is $-F = ma_{gi}$, the same as equation (2.20). Intuitively, this equation says that the flow through the lumped mass in a particular direction is proportional to the potential drop across the mass in the same direction. We use the +v direction only to find the compressive or tensile interpretations of the corresponding physical quantities.

Exercise 2.2: The ground point can be placed at any arbitrary location in the model. Show that redrawing Figure 2.12 with the ground point on the left does not change equation (2.20), but does reverse the physical interpretation of the behavior (compressive or tensile).

It is visually obvious that springs and dampers each have two ends—two terminals that move relative to each other. The two-terminal nature of an ideal mass is less intuitive, perhaps because of the universal use of free-body diagrams in physics. Since the ground terminal is not physically connected to the object, one might be tempted to deny the two-terminal nature of a mass. However, according to Newton's third law of motion, every ac-

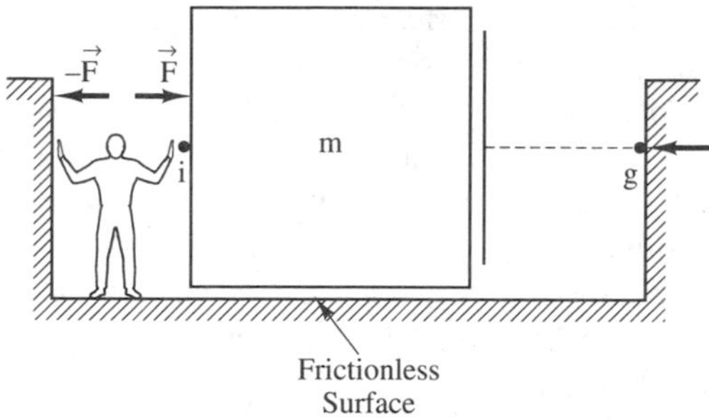

Figure 2.13 Mass as a two-terminal element.

tion has an equal and opposite reaction. As shown in Figure 2.13, one cannot apply a force vector to the mass without applying an equal and opposite force vector to the ground. Since the ground is rigid, the force vector applied to the ground transfers to the point labeled g. Thus, opposing force vectors are applied simultaneously to both terminals of the mass. Put simply, a single force vector cannot be applied in isolation.

It is a matter of personal taste whether one uses a dotted line or a solid line to denote the reference to ground in Figure 2.12. We label the block with its mass m. We *could* draw the block so that it surrounds point i. This change would emphasize point i as the point of action of the mass. But surrounding the point with the block would hamper labeling of the point of action i and would eliminate the connecting line in which we visualize measuring the scalar force. In section 2.2 and later sections, we find that the symbol shown in Figure 2.12, which emphasizes the flow path, facilitates systematic—indeed, automatic—generation and solution of the system equations.

The defining equation for the ideal mass can be rewritten in terms of the central motion variable v as

$$F = m\dot{v}_{ig} \tag{2.21}$$

Intuitively, the flow F through the element is proportional to (and oriented in the same direction as) the drop $\dot{v}_{ig}$ in the rate of change of velocity. This equation can also be expressed in impedance notation, analogous to equations (1.1), (2.11), and (2.15), as

$$v_{ig} = \left(\frac{1}{sm}\right)F \tag{2.22}$$

Equation (2.22) should be interpreted to have the same meaning as equation (2.21). The time-derivative operator s acts on the signal v_i. Therefore, we can interpret 1/s as an integral operator that acts on the signal F. The quantity 1/sm is the **impedance** of the mass. A large value of m means a small impedance. For a given velocity v_i, a large value of m implies a large value of F/s—a large integrated flow. In other words, a large scalar force must be integrated (applied) over time in order for a large mass to achieve the specified velocity. We sometimes label an ideal mass with its impedance rather than its mass.

The values of variables in a physical system cannot become infinite. If we do not let the scalar force F in a lumped-mass model be infinite, then equation (2.21) does not permit the acceleration $\dot{v}_{ig}$ to be infinite. Therefore, v_i cannot change instantaneously. No matter how rapidly various variables in the model of a translational system vary, the velocities of the lumped masses in the model never exhibit abrupt jumps in value, but are continuous with time. This continuity of velocity provides the basis for some useful analysis and design rules which we present in sections 3.3 and 4.2. We can, of course, *require* the velocity of a mass *model* to jump instantaneously; but then we must permit the scalar force in that mass model to be infinite.

A force on a mass delivers or removes translational kinetic energy. Let F denote the flow through the ideal mass of Figure 2.12 during the time interval [0, t]. According to equation (2.5), the energy delivered to the mass of point i during that time interval is

$$E(t) = \int_0^t F v_{ig}\, dt \tag{2.23}$$

From equation (2.21), we obtain

$$E(t) = \int_0^t m\dot{v}_i v_i\, dt = \int_0^{v_i(t)} m v_i\, dv_i$$
$$= \frac{1}{2} m v_i^2(t) \tag{2.24}$$

The stored kinetic energy can be recovered by slowing the mass. Because the velocity of a mass cannot change abruptly, energy cannot be added to or removed from a mass instantaneously.

The displacement of a mass can be measured relative to a moving reference point rather than ground if that reference point has constant velocity. That is, the reference point must be *nonaccelerating* in order for Newton's law to apply. Let x_j denote the displacement of the reference point and x_i the displacement of the mass. Then $a_{ij} = \ddot{x}_{ij} = \ddot{x}_i - \ddot{x}_j$. It is the absolute acceleration a_i (relative to ground) that satisfies Newton's law. If the acceleration $\ddot{x}_j$ of the reference point is not equal to 0, then measurement of $\ddot{x}_{ij}$ alone does not determine the absolute acceleration (relative to ground) of point i. Throughout this text, we assume that the reference point is stationary. Example problem 2.8 explores further the effect of an accelerating reference point.

The mass of a body depends only on the properties of the material and the geometry (shape and size) of the body. Suppose a homogeneous object has uniform cross-sectional area A and length ℓ, as in Figure 2.5. Then the mass of the object is

$$m = \rho A \ell \tag{2.25}$$

where ρ is the density of the material.[6] The engineer can design an object of essentially any desired mass by proper choice of material (ρ) and volume of the object.

In the design of an automobile piston and piston rod, the intention is to transfer the force of an expanding gas to a rotating crankshaft. Because the piston must be accelerated and decelerated repetitively, we keep the masses of the piston and piston rod small so that little force is required to accelerate them. The mass is minimized by the use of relatively

[6]The units for density are kg/m^3. The densities of various materials are given in Appendix A.

thin piston walls and slender piston rods. (A special steel material is used to withstand the high heat and large forces).

The actual mass of an object of complicated shape is difficult to estimate accurately. We can *measure* the mass of an object by measuring its weight w. Then, according to equation (2.20), m = w/g, where g is the acceleration due to gravity. The weight of the object can be measured with a spring scale or by balancing the weight with the weight of a known mass.

Ideal Translational Sources

A *source* is an energizer, a device that can inject energy into a system. No device can *create* energy, of course; all energy is stored. If it is stored in a stressed spring or a moving mass, it can be retrieved directly into a translational system. If it is stored in another form (say, as electrical, hydraulic, or chemical energy), it must be converted to mechanical form as it is retrieved. (See sections 6.2 and 6.3 for discussions of energy conversion processes.)

Chapter 1 notes that there is a rise in potential across a source in the direction of flow. Suppose a stationary person throws an object. Then motion (displacement, velocity, or acceleration) is the potential and scalar force is the flow. Choose the direction in which the object is thrown as the +v direction. When the person's hand is ahead of his or her body (toward the +v direction), the body and arm *push* the object. Then the scalar force along the line of action is compressive. We treat that compressive force as a flow in the +v direction. The person's hand, which contains the thrown object, accelerates faster than his or her stationary foot. Thus, the acceleration variable rises in the +v direction—from foot to hand. When the person's hand is behind his or her body, the person *pulls* the object, and the scalar force is tensile—a flow in the −v direction. Then the acceleration variable drops in the +v direction—from hand to foot—and rises in the −v direction. In both situations, the flow is in the direction of rise in the motion variable.

The relation between the rise a_{ij} in acceleration across the source and the flow F through the source depends on the mass of the object thrown. The acceleration flow tradeoff can be expressed as a source performance characteristic, as illustrated in Figure 1.1. If the object thrown is much heavier than the person's arm, then the acceleration of the object must be low. As a result, the operating point is near the low-acceleration end of the operating characteristic, and the person acts as a source of (nearly) constant force. That force is designated T_c in Figure 1.1. If the object is much lighter than the person's arm, he or she cannot accelerate the object enough to generate much force on it. Therefore, the operating point is near the low-force end of the operating characteristic, and the person acts as a source of (nearly) constant acceleration. That constant acceleration is denoted A_c in Figure 1.1. We develop lumped models that are equivalent to source performance characteristics in Section 4.2. In this section, we examine the simpler situations in which the source behaves either as a source of force or as a source of motion.

We say that an energy source is an **ideal independent compressive-force source** if it maintains a specified compressive-force pattern F(t) regardless of the translational system to which it is connected. (In the preceding example, the person acts like an independent force source only if the object is much heavier than his or her arm.) Figure 2.14 shows a lumped model of the independent force source. It is a *physical* model. That is, it

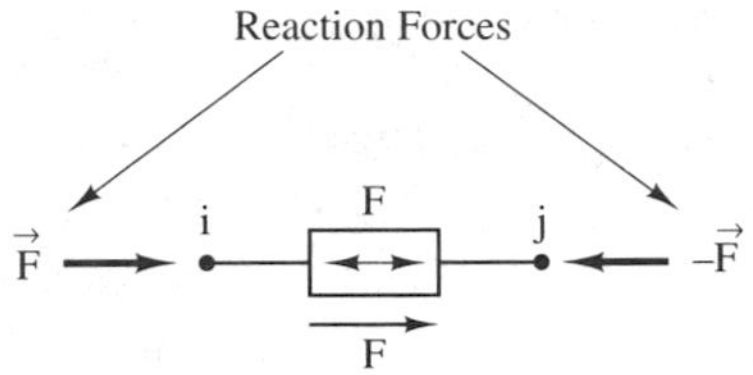

Figure 2.14 Independent compressive-force source; F is the flow in the +v direction.

is an idealization of a physical phenomenon. The two-headed arrow in the source symbol shows that the source *pushes* outward against its environment—that the scalar force it produces is compressive. Of course, if F is negative, the source pulls on its environment and the force is tensile.

The two-headed arrow symbol asserts a *whole-system viewpoint.* It encompasses the effect of the source not only on the load, but also on the portion of the system against which the source is braced. The notation acknowledges that the motion of the load cannot be determined without accounting for the motion of the subsystem that is bracing the source.[7]

We emphasize that the symbol F attached to the source represents *compressive* force. Hence, we treat F as the flow through the source in the +v direction. Think of the compressive-force source as a flow source, and denote the flow by an arrow beside the source, oriented in the +v direction. Then if we reverse the +v direction, we must reverse the arrow that represents the compressive force F.

If point i were fastened to ground, and if F(t) were positive, the compressive-force source of Figure 2.14 would tend to cause positive motion of point j, a rise in potential in the direction of flow. Neither v_{ij} nor the absolute motions of the individual source terminals are determined by the force source alone, however. Rather, they are determined by the compressive-force function F in combination with the reaction properties of the attached system.

Gravity produces a constant tensile force (in the vertical direction) between a mass and the earth, as long as the mass does not move far from the earth's surface. Gravity can be treated as an ideal compressive-force source with $F(t) = -mg$, where m is the mass of the object on which the gravity acts. Gravity can be used in the *design* of force sources. Figure 2.15 illustrates two such designs and ideal-source models that represent them. The lumped mass in each model accounts for the fact that the gravitational force must accelerate the mass as well as the attached system.

The effect of gravity on water (or other liquids) provides additional opportunities for the design of force sources. Figure 2.16(a) shows such a design and its model. The force provided by the piston is derived from the water pressure on the piston. If the height h of the water (sometimes referred to as the *head*) does not change significantly during the motion of the piston, the source provides nearly constant force. A similar system derives force from the pressure in a large pressurized air tank. Figure 2.16(b) shows such a pneumatic force source and its model.

[7]The whole-system viewpoint contrasts with the *piece-of-system viewpoint* inherent in the free-body diagram of mechanics. The free-body diagram views each object in isolation. The interactions among the objects in the system are accounted for by appropriate equations among the force vectors that act on the free bodies.

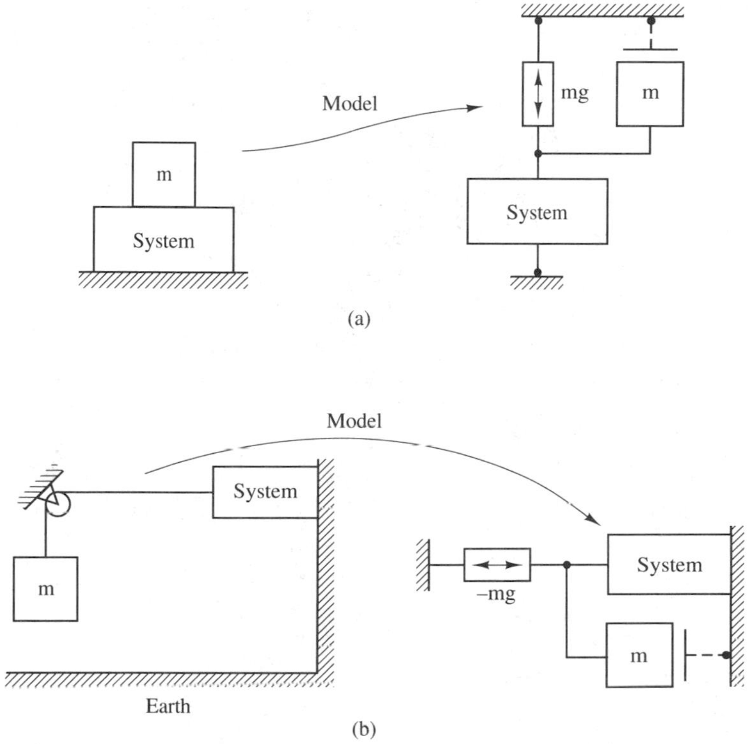

Figure 2.15 Gravity force sources and their models. (a) A source of compressive force. (b) A source of tensile force.

Sections 6.2 and 6.3 examine various energy sources and transducers that can be used to change the energy from one form into another. These devices provide a great variety of mechanical force sources. Electric motors are among the most versatile force sources, hydraulic motors among the most powerful.

We use the term **ideal independent velocity source** for an energy source that produces a **specified relative velocity of separation** v(t) between its two terminals. We represent the independent velocity source by the lumped element shown in Figure 2.17, where v refers to the specified separation-velocity function v(t). The two-headed arrow in the box shows the *physical* behavior of the lumped element. It means that the terminals are separating. Again, the source symbol asserts a *whole-system viewpoint;* it acknowledges that the motions and forces in the system depend on the subsystems connected to *both* ends of the source. The velocity source is an *across* source.

The mathematical equation for the velocity source depends on the +v direction. If we order the subscripts i, j of the relative-velocity variable v_{ij} in the +v direction, then v_{ij} corresponds to *compressive* motion. For the +v direction shown in Figure 2.17, the relative separation velocity (the negative of compressive velocity) is $-v_{ij} = v(t)$. An equivalent statement is $v_j = v_i + v(t)$. If v(t) is positive, then $v_j > v_i$, and the motion is indeed a separation at rate v. The + and − signs beside the source symbol are redundant; they merely emphasize that the velocity is higher on the +v side of the source.

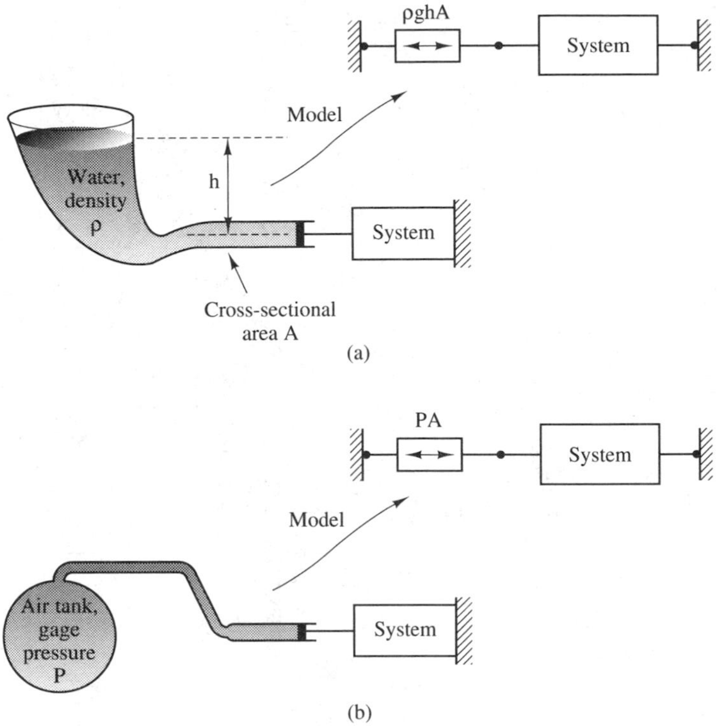

Figure 2.16 Hydraulic and pneumatic force sources and their models; ρ = density of water. (a) Force derived from water pressure. (b) Force derived from air pressure.

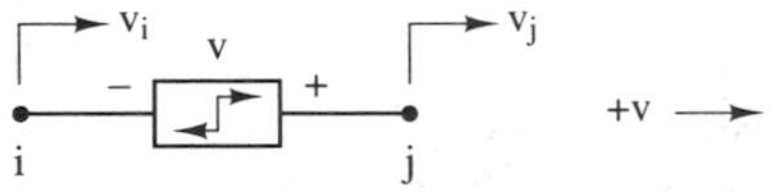

$-v_{ij} = v$ or $v_j = v_i + v$

Figure 2.17 Independent velocity source; the velocity of separation v is the velocity *rise* across the source in the +v direction.

Reversing the +v direction changes the subscript order in the equation that represents the source; it does not change the fact that the terminals *separate* at rate v(t). Of course, if v(t) is negative, the separation velocity is negative.

The *ideal* velocity source maintains the specified relative motion, regardless of the load. The source generates whatever scalar force is necessary to achieve the separation velocity. The scalar force exerted by the velocity source cannot be determined from the specified velocity function v(t) alone. Rather, it is determined by the velocity source in concert with the resistive behavior of the attached system. Since the source demands separation of its terminals, the scalar force within the source is usually compressive, a flow in the +v direction. (It *will* be compressive if the attached system is passive.)

The term *velocity source* is appropriate for a motion source because any motion can be expressed, to within constants of integration, in terms of velocity. For example, the displacement $x(t) = 2$ m can be written $v(t) = 0$ m/s, with $x(0) = 2$ m understood; and the acceleration $a(t) = 3\ \text{m/s}^2$ can be written $v(t) = 3t$ m/s. However, in some instances, velocity does not provide the most convenient form of expression of the motion. For example, an instantaneous (or nearly instantaneous) displacement of position could be expressed as a brief impulse in velocity, but is more conveniently dealt with as a step (or jump) in displacement. Accordingly, we sometimes represent a velocity source not by a velocity statement, but by $-x_{ij} = x$ or $-a_{ij} = a$.

Constant-velocity sources can be constructed by taking advantage of inertia. For instance, a water wheel is rotated by the gravitational force on the water in the buckets attached to the wheel. By making the inertia of the wheel large, we enhance the ability of the wheel to maintain constant velocity under occasional loads. Figure 2.18(a) shows such a velocity source and its model. The clutch mechanism is used to connect and disconnect the velocity source. Inertial principles are also used in pile drivers and other motion sources.

A camshaft rotated at constant velocity acts as an independent *displacement* source (see Figure 2.18(b)). The cam is shaped to produce the desired displacement waveform.

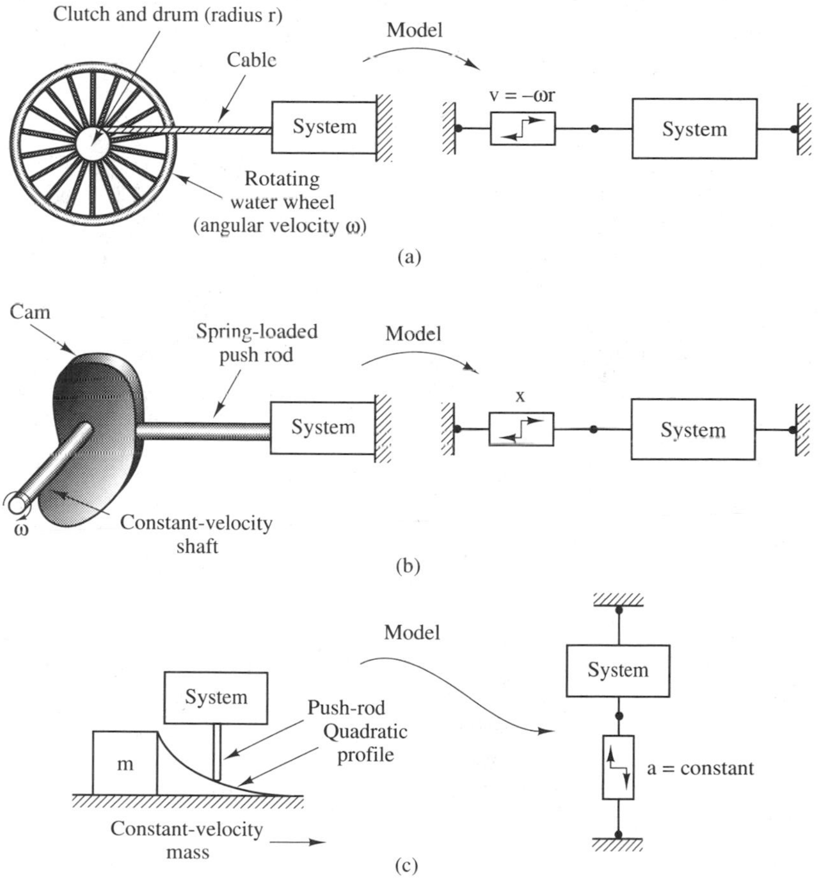

Figure 2.18 Velocity sources and their models. (a) A constant-velocity source. (b) A controlled-displacement source. (c) A constant-acceleration source.

The shaped wedge of Figure 2.18(c), driven by a constant-velocity source, provides constant vertical acceleration. A snowplow has a similar design. In chapter 4, we attach masses, springs, and dampers to ideal velocity sources to represent nonideal sources accurately.

The energy sources, tranducers, and amplifiers examined in sections 6.2 and 6.3 enable us to design velocity sources of almost any size and power. As discussed in chapter 1, there is always a trade-off between force and velocity in any physical source. However, we can design sources that provide nearly independent control of velocity or nearly independent control of force.

Students are accustomed to force sources, but not velocity sources. To understand better the distinction between the two ideal sources, consider the command, "Open the door!" Do you think to yourself, "Apply a force to the door"? Or is your instinct more nearly, "Make the door follow a displacement pattern?" This distinction is the essence of the difference between the force source and the velocity source.

Suppose a car with a manual transmission travels on a hilly road. If the driver holds the pedal steady, the driver-engine combination acts like a constant-force source: The car velocity varies with the slope of the road. If the car is operated under cruise control, on the other hand, the engine-controller combination acts like a constant-velocity source: The amount of force exerted by the engine varies with the slope of the road in order to keep the velocity constant. By appropriate motions of the accelerator and clutch, a driver is able to generate a wide range of controlled-velocity and controlled-force characteristics.

Suppose we position two people on opposite sides of a door. One is told to push the door closed with a specified force. The other is told to open the door slowly. In effect, one acts as a force source, the other as a velocity source. The competition between the two people is represented by the pair of ideal source models in Figure 2.19. A source *usually supplies* energy to a system. But if both of the sources in the figure supplied energy, there would be no place for that energy to go. Instead, one source supplies energy and the other absorbs energy. The one that acts as a supplier produces a potential rise in the direction of flow. The other acts like a passive device, with a potential drop in the direction of flow.

Exercise 2.3: Suppose that both v and F in Figure 2.19 are positive. Show that the flow of compressive force is in the direction of velocity rise in the velocity source, but in the direction of velocity drop in the force source. Use equation (2.4) to show that the velocity source supplies power and the force source consumes power.

In later chapters, we find situations in which we need to represent nonlinear (nonproportional) relations between variables and also linear relations between apparently isolated variables. Inventing a new lumped element for each such instance would make the system models unnecessarily complicated. At the end of section 2.2, we introduce another type of ideal element—the **ideal dependent source**—to portray arbitrary dependencies between system variables.

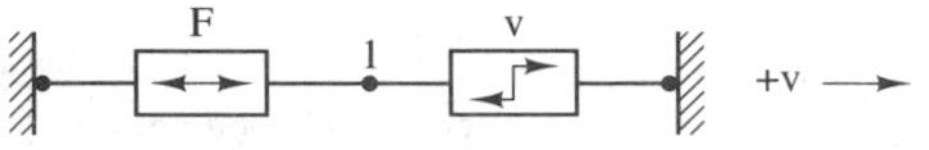

Figure 2.19 Competing sources.

The Lumping Process

We begin this section by using the lumped elements defined in previous sections to construct a lumped model for a particular system, a sled and rider. From this modeling experience we extract a procedure for constructing lumped models for other systems.

Figure 2.20 shows a pictorial drawing of the sled and rider. We are particularly interested in the scalar forces in the neck and lower back of the rider, body areas that tend to experience strain. We consider only motions and forces along the line of motion of the sled. Then the motions and forces are one dimensional. (We can treat vertical motions and stresses separately.)

We redraw the pictorial model as an interconnection of isolated masses, springs, dampers, and sources (Figure 2.21). This lumped model retains much of the geometry of the pictorial model, but separates the various inertial, compliant, and friction phenomena. We keep the number of ideal masses, springs, and dampers small in order to limit the number of variables and differential equations. The lumping process requires judgment and experience. In this case, it may also require experimenting with a sled and rider.

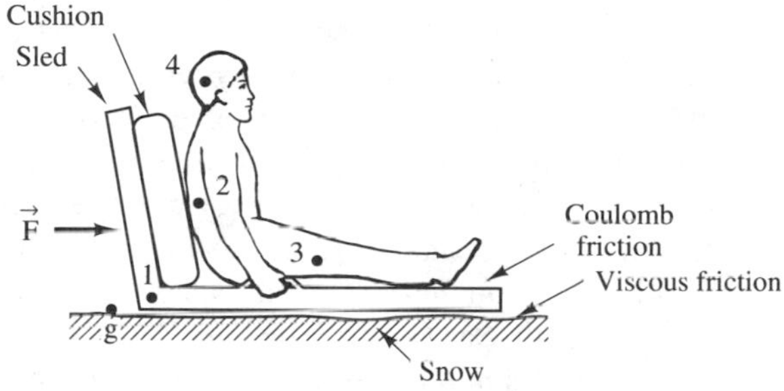

Figure 2.20 Pictorial model of a sled and rider.

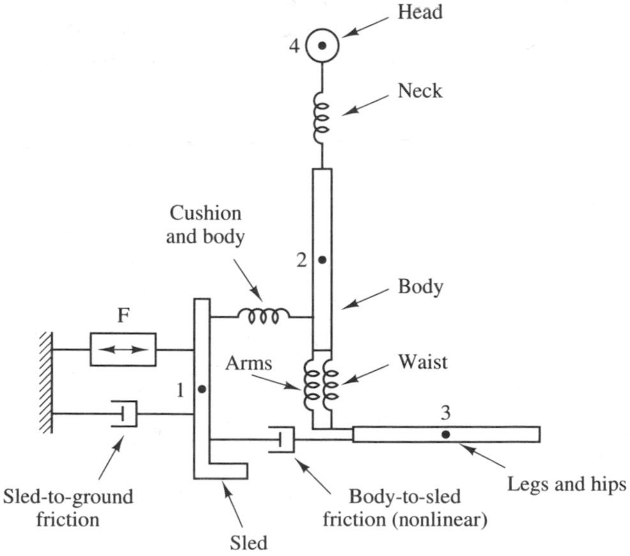

Figure 2.21 Lumped model of sled and rider.

Next, we replace each lumped element of Figure 2.21 by an ideal two-terminal lumped element. In particular, we replace each mass that surrounds its point of action (center of gravity) by a two-terminal mass. The first lumped model (Figure 2.21) retains closer identification with the pictorial diagram of Figure 2.20 than does the second (Figure 2.22). However, the mathematical equations that describe the elements in Figure 2.21 are not obvious, because the graphical notation of that figure has not been uniquely associated with mathematical models. The lumped network of Figure 2.22, on the other hand, is composed of two-terminal elements whose mathematical representations are clearly defined in this section. Hence, the partial loss of physical identification with the pictorial diagram is compensated for by increased ease of mathematical formulation.

At this point, we emphasize the significance of the two-terminal lumped model for mass. One might be tempted to represent each mass of Figure 2.21 by a box that surrounds the numbered node. That is the conventional notation for mass. It is suitable for free-body diagrams and, in fact, is used in most books on lumped modeling. It retains physical similarity to the original system. But it provides no flow path for the inertial force. If we use that conventional notation, we must remember to introduce an inertial force for each equation that includes a mass.

The two-terminal model has three advantages. First, since it treats the inertial force in a way that is consistent with the treatment of other forces in the system, it simplifies the rules for writing equations, so that fewer rules need to be remembered. Second, the two-terminal mass affords a stronger analogy to the corresponding elements in other energy domains (e.g., hydraulic or electrical capacitance). Therefore, it gives easier access to simulation software written for other kinds of systems—software that provides automatic solution of the system equations. Finally, the two-terminal model makes explicit the flow path for inertial force. This explicit flow path clarifies the rules for series and parallel combinations, which we use to simplify models in section 4.2. The conceptual advantage of the explicit flow path for inertial force is discussed in the next section.

We call Figure 2.22 a **lumped network.** We use the term to mean an interconnection of two-terminal lumped elements—elements that correspond to unambiguous mathe-

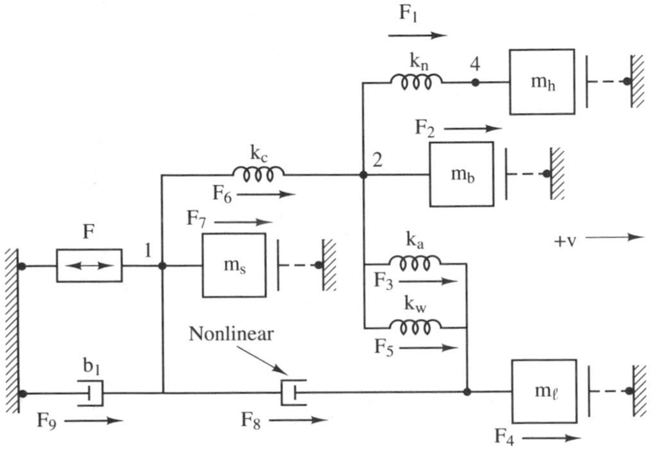

Figure 2.22 Lumped-network model with ideal elements. The vertical displacements shown are for convenience in displaying multiple objects; all motions are horizontal.

matical equations.[8] The lumping process focuses on the pictorial model (Figure 2.20) and the lumped-network model (Figure 2.22). Figure 2.21 is merely a convenient intermediate step in the modeling process for the given example. In general, a lumped network composed of ideal elements can be generated by the following process:

A LUMPING PROCEDURE FOR TRANSLATIONAL SYSTEMS

1. Draw a pictorial diagram.
2. Identify those parts of the system which are rigid relative to their surroundings and which exhibit distinguishable motions or have significant mass. Represent each part by a numbered node. Place the nodes in approximate correspondence with the layout in the pictorial diagram. Keep the number of distinguished parts small; we must eventually determine the motion of each numbered part. One of the parts must be the reference (ground), relative to which all translational motions are measured (see Figures 2.20 and 2.21).
3. Insert lumped (two-terminal) masses between appropriate nodes and the ground node. The ground node need not be isolated at a single location (see Figure 2.22).
4. Represent those portions of the system which are relatively compliant (compared with the rigid masses) by lumped springs that connect two nodes (see Figures 2.21 and 2.22). Keep the number of springs small; we must eventually determine the scalar force in each spring. The number of springs is determined, to a great extent, by the number of nodes selected in step 2.
5. Account for significant friction forces by inserting lumped dampers between appropriate nodes. Friction forces are often nonlinear. (Sled-to-snow friction is nearly viscous; body-to-sled friction is nearly coulomb.) Avoid the use of nonlinear dampers if possible.
6. Insert independent energy sources between nodes as appropriate.
7. Rearrange the lumped network, if necessary, to make it one dimensional (see Figure 2.22). Vertical displacement in the lumped network has no physical meaning; it is just a mechanism for displaying complicated interconnections. All points connected by solid lines have identical motions and are mathematically indistinguishable. (Thus, the connecting lines need not have square corners.)
8. Assign a positive direction (+v) for the network. This direction applies to the motions of all nodes. Establish and orient (by an arrow) a scalar-force variable for each lumped element. If a scalar-force variable is oriented in the +v direction, it represents compressive force; otherwise, it represents tensile force.

When using the foregoing procedure, do not put off identifying the nodes of the model—step 2—until later. To do so invites confusion in steps 3–6.

[8] In chapter 6, we introduce *four-terminal* lumped elements. Each such element is equivalent to a *pair* of mathematical equations. We also apply the term *lumped network* to interconnections that include four-terminal lumped elements.

Restrictions on the manner of system operation or limitations on the information required from the model may suggest further simplifications. For example, if we assume that the rider maintains sufficient contact force with the sled to avoid slipping on the sled, the leg-hip mass can be combined with the sled mass, and the nonlinear body-to-sled friction can be eliminated. If the sled is used on a slope, gravity introduces an additional translational force on each mass in the model. At high speeds, additional viscous friction between the air and the body and sled must be included. On the other hand, if only the sled motion versus time is desired, or if the sled is used on level ground with gentle applied forces, the whole system can be represented by a single mass and a sled-to-snow damper.

The approximation process is an art, learned by observing the models that have been developed and tested by experienced engineers and by personally approximating and testing numerous systems—first simple ones and then some that are more complex. In the process of making these approximations, beginners and experienced engineers alike benefit from drawing analogies with all phenomena in their previous experience. Any approximation about which the engineer is uncertain must eventually be checked by physical testing or by comparison with published experimental data. Ultimately, the full model must be checked by comparing the predicted performance of the system (as determined from the model) with the actual performance of a system that is similar to the model.

In the initial section of each succeeding chapter we develop lumped models for an additional energy domain—rotational, electrical, hydraulic, and thermal. We use the lumping procedure developed for translational systems as a pattern. In Chapter 6 we construct lumped models of electromechanical systems, hydromechanical systems, and other systems in which quantities from different energy domains interact.

Example Problem 2.3: Lumped modeling of a translational system.
A fisherman in a rowboat has just hooked a 4-kg fish 30 m from the boat. The fishing rod is 2 m long and is held in a nearly vertical position by the fisherman. The reel clutch is tight enough that the line does not slip. Make a lumped model of the boat, rod tip, and fish. Choose appropriate values for the model parameters and the source-signal parameters.

The following measured data will assist in determining the parameters. The boat and fisherman have a total mass of 160 kg. A constant 220-N pull on the boat drags the boat and fisherman through the water at 0.8 m/s. A 90-N pull on the tip of the rod deflects the rod tip 30 cm. A 90-N tension in a 100-m length of fishing line stretches the line 2 m. A 4-kg fish typically generates a 130-N tension on the line for 10 s at a time. (Appendix A relates these SI units to other familiar units.)

Solution: The fish exerts a force F(t) against the water. This force is balanced by an acceleration force on the fish, by the viscous drag of the water, and by the tension in the line. Assume that all components of force and motion by the fish are directed parallel to the water's surface and away from the boat. The rod tip bends and the line stretches, both in proportion to the tension in the line. The fisherman, in resisting the pull on the line, transmits the tensile force to the boat. That force accelerates the boat and moves it against the viscous drag of the water.

The elements of the system are shown in partially pictorial form in Figure 2.23(a) and as a lumped model in Fig. 2.23(b). We include the section of line between the reel and the rod tip with the 30 m of line between the tip and the fish. They stretch together. The right-angle turn in the line at the rod tip does not change the one-dimensional nature of the motion of the

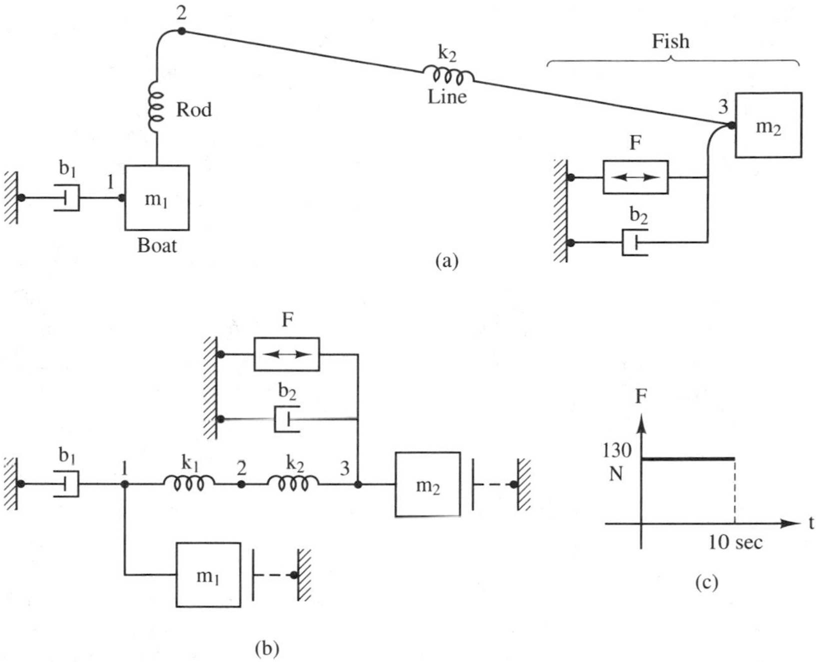

Figure 2.23 Lumped model of fish-boat interaction. (a) A partially pictorial model of the system. (b) A lumped-network model of the system. (c) The source signal.

points on the line. A typical pattern for the compressive-force source F is shown in Figure 2.23(c).

We determine the following lumped-element values from the measured data:

$$
\begin{aligned}
m_1 &= 160 \text{ kg}, \qquad m_2 = 4 \text{ kg}, \\
b_1 &= \frac{220 \text{ N}}{0.8 \text{ m/s}} = 275 \text{ N·s/m} \\
k_1 &= \frac{90 \text{ N}}{30 \text{ cm}} = 300 \text{ N/m} \\
k_2 &= \frac{90 \text{ N}}{(2 \text{ m})[(32 \text{ m})/(100 \text{ m})]} = 141 \text{ N/m}
\end{aligned}
\tag{2.26}
$$

The signal parameters (force and duration) are shown on the source waveform.

No data are given concerning the drag of the water on the moving fish. Therefore, we must *estimate* the friction constant of the damper b_2. If we estimate (from experience in the water) that a 6-N force can pull a large fish through the water at 2 m/s, then we obtain

$$
b_2 \approx \frac{6 \text{ N}}{2 \text{ m/s}} = 3 \text{ N·s/m}
$$

Exercise 2.4: How would you measure the various quantities specified in Example Problem 2.3?

2.2 LINE GRAPHS AND SYSTEM EQUATIONS

In this section, we develop systematic procedures for finding the mathematical equations that describe lumped networks. We also develop another visual model—the *line graph*—that can aid in writing network equations. The line graph focuses attention on the interconnections among the network elements. In section 2.1, we derived simple equations for the lumped elements. The equations that describe the interconnection pattern complete the mathematical model of the network.

We carry out the development of line graphs in the general terminology of potentials and flows. We illustrate the concepts associated with line graphs in the mechanical context of section 2.1. Once the procedures for writing equations are clear, the designer is usually able to write the system equations directly from the lumped network. The processes for generating a lumped network and its corresponding mathematical model are applied to an additional type of system (rotational, electrical, etc.) in the first section of each succeeding chapter.

Line Graphs

A *graph* is a set of points (called **nodes**), together with a set of lines (called **branches**) that connect certain of the nodes. We use the term **line-graph model** (or simply, *line graph*) to mean a graph in which the nodes correspond to the nodes of a lumped network and the branches correspond to the lumped elements of the network. The line graph focuses attention on the interconnections among the lumped elements. The label for a particular branch of the line graph indicates the type of lumped element it represents. Therefore, the label implies a specific relation between the flow through the branch and the potential difference across the branch.

Figure 2.24 shows the simplest meaningful line graph, a pair of nodes connected by a single branch. We label the branch with the symbol for the parameter of the lumped element that the branch represents. In Figure 2.24(a), the label b indicates that the element is an ideal viscous damper. The label b also represents the friction constant for that damper. If we label the branch with a specific value of b (together with its units), as shown in Figure 2.24(b), the units distinguish the damper from other types of lumped elements. In some instances, we shall label the branches of a line graph with the element *impedances* (e.g., s/k, 1/b, 1/sm), rather than the element parameters (k, b, m). The ideal lumped-element symbol can also be drawn on the branch as an unmistakable, though redundant, element designator, as in Figure 2.24(c) (see also Figure P2.26).

Most translational mechanical networks contain a ground node. We label the ground node in the line graph with the symbol g or the velocity $v_g = 0$, as shown in Figure 2.25.

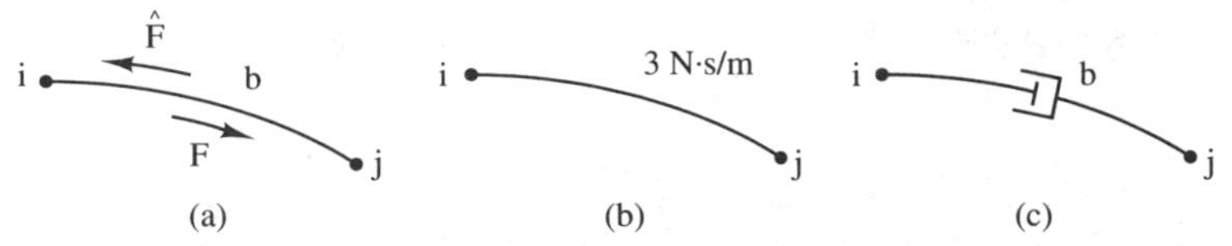

Figure 2.24 Line graph of a viscous damper. (a) Parameter symbol as designator. (b) Parameter units as designator. (c) Lumped-element symbol as designator.

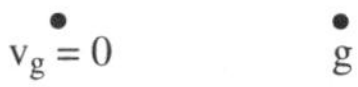

Figure 2.25 Ground node labels.

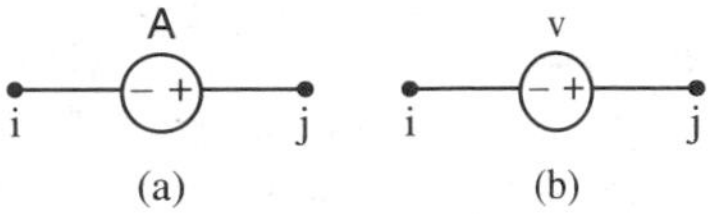

Figure 2.26 Line graph of a potential-difference source; for translation, place the + toward the +v direction. (a) General potential-difference source. (b) Velocity source.

Figure 2.26(a) shows the line-graph symbol we use to represent an independent potential-difference source. It is labeled with the specified source signal. The line-graph representation of the translational velocity source of Figure 2.17 is illustrated in Figure 2.26(b). (Since an independent velocity source is defined as a velocity rise in the +v direction, the terminal on the +v side of the source in the lumped model corresponds to the + terminal of the line graph.

The line-graph symbol we use to represent an independent flow source is shown in Figure 2.27(a). The translational version, the compressive-force source of Figure 2.14, is shown in Figure 2.27(b).

The line graph represents mathematical relationships. There is an across variable (or potential) associated with each line-graph node. In translational mechanical networks, the node potentials are velocities in the +v direction. In line-graph terminology, $v_i > v_j$ means that point i has a higher potential than point j.

PROCEDURE FOR DRAWING LINE GRAPHS

1. Draw and number a set of line-graph nodes, one for each node in the lumped network. If there are multiple reference (or ground) nodes in the lumped network, represent them all with a single reference node in the line graph. Associate one across variable (potential) with each node.
2. Draw and label a line-graph branch for each passive element of the network.
3. Draw and label each source branch. For a translational network, place the flow arrowhead or + symbol on the +v side.
4. Define an oriented flow variable for each branch of the line graph. Show its direction by an arrow beside the branch. If flow variables are already designated on the lumped network, transfer them to the corresponding branches of the line graph.

The Sign Convention for Branch Equations

The equation for any passive translational lumped element has the form $v_{ij} = zF$, where F is the scalar force directed from node i to node j, v_{ij} is the velocity drop in the same direction, and z is the impedance of the element (see equations (2.11), (2.15), and (2.22)). The power delivered to the element by its surroundings is the product of the same variables, Fv_{ij}.

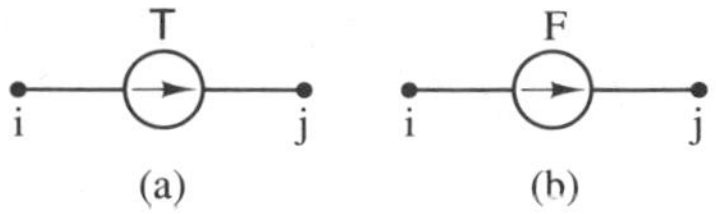

Figure 2.27 Line graph of a flow source; for translation, orient the arrow in the +v direction. (a) General flow source. (b) Compressive-force source.

In chapters 3–5, we examine rotational, electrical, and hydraulic systems. For each type of system, we assign positivity to the potentials and flows in such a way that each lumped-element equation has the form $A_{ij} = zT$, where T is the flow in the direction from node i to node j—the direction of the potential drop A_{ij}. Then the power transferred to that lumped element is the product of those variables, TA_{ij}. We choose the sign convention to provide a single intuitively appealing pattern for all passive objects: flow in the direction of potential drop. In a source, the flow is negative in the direction of potential drop, and the power delivered to the element is negative (see exercise 2.3 and Figure 2.19).

Let us demonstrate how to use the sign convention. Let F be a flow in the direction from node i to node j of Figure 2.24(a). According to the sign convention, the damper equation is $v_{ij} = (1/b)F$. This damper equation is identical to equation (2.15). On the other hand, suppose we let $\hat{F}$ denote a flow from node j to node i. Then, according to the sign convention, the equation for the element is $v_{ji} = (1/b)\hat{F}$. If F represents compressive force, then $\hat{F}$ represents tensile force. Since $\hat{F} = -F$, the damper equation in terms of $\hat{F}$ is *equivalent* to equation (2.14).

Examples of Line-Graph Models

We use the translational network of Figure 2.28(a) to illustrate how to generate a line-graph model. The three fixed nodes of the network are equivalent. We represent them by a single ground node in the line graph. Node 1 is the only movable point in the lumped model. Three branches connect node 1 to the ground node. Two of them are passive dampers, and the other is an active force source. We affix the friction constants b_1 and b_2 to the corresponding passive branches of the line graph. We transfer the flow variables F_1 and F_2 to the line graph with proper orientations relative to the nodes. We use a flow-source symbol to represent the force source in the line graph. We orient the source arrow to correspond to the +v direction in the corresponding branch of the lumped model. Finally, we affix the variable F of the lumped-model source to the line-graph source. (A separate flow-direction arrow is not needed for the source branch of the line graph.)

The line graph of Figure 2.28(b) reveals that the two dampers are connected *in parallel* and that the scalar force F is applied to the combination. The parallel nature of the connection is not as obvious in the lumped network. The lumped network shows that damper 1 is in compression and damper 2 is in tension.

The flows F and F_1 represent compressive forces because they each agree with the +v direction. The flow F_2 represents tensile force. The compressive force F is specified by the source. Balancing compressive forces at point 1 of the lumped network model requires that $F + (-F_2) = F_1$, or $F = F_1 + F_2$. We combine this result with the element

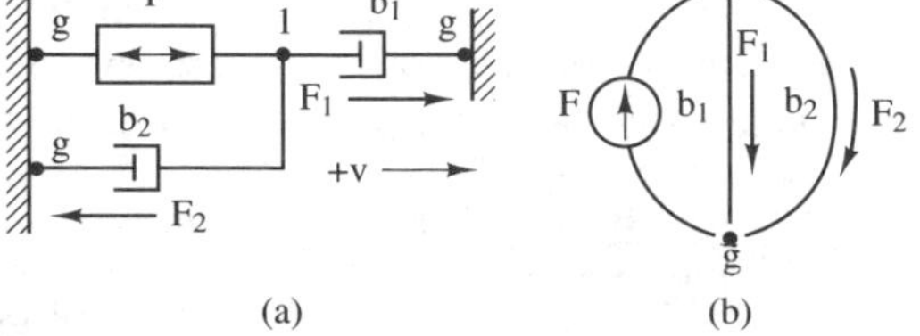

Figure 2.28 A parallel translational network. (a) Lumped model. (b) Line-graph representation.

equations for the two dampers, as determined from equation (2.14): $F_1 = b_1 v_{1g}$ and $F_2 = b_2 v_{1g}$. The ground velocity is $v_g = 0$. The symbol v_1 represents the velocity of node 1 relative to ground. Then $v_{1g} = v_1 - v_g = v_1$. It follows that $v_1 = F/(b_1 + b_2)$, $F_1 = b_1 v_1$, and $F_2 = b_2 v_1$.

The damper b_2 is in fact in tension, since the tensile force F_2 is positive. The two dampers do indeed act *in parallel*. That is, the parameters b_1 and b_2 have identical effects on the motion v_1. As far as the scalar force and velocity *seen* by the source are concerned, we can model the parallel combination of dampers as a single *equivalent* damper with friction constant $b = b_1 + b_2$. If one of the dampers (say, b_1) is much stiffer (larger b_1) than the other, the stiffer damper determines the behavior of the system ($b \approx b_1$).

Our choices of positive directions for the flow variables F_1 and F_2 were arbitrary. Suppose we orient the flow $\hat{F}_2$ in the b_2 branch to agree with the $+v$ direction. Then $\hat{F}_2$ becomes the compressive force in b_2, and we make the substitution $F_2 = -\hat{F}_2$ throughout the preceding equations.

We can state the balance of compressive forces at node 1 of the network by inspection of the flows at node 1 of the line graph. F flows into the node; F_1 and F_2 flow out of the node. We merely express the *conservation of flow* at the node: $F = F_1 + F_2$.

A second network is presented in Figure 2.29(a). A velocity source drives a pair of dampers connected *in series*. The two ground nodes appear as one in the line graph. Therefore, the three branches of the line graph form a single loop. The + symbol of the velocity source in the line graph is placed to correspond to the $+v$ side of the source in the lumped model. The compressive forces in the source and in the two dampers are F, F_1, and F_2, respectively. (We could, with equal validity, let any of the flows oppose the $+v$ direction. Those flows would represent tensile forces in the corresponding elements.) By balancing forces at nodes 1 and 2 of the network (or stating the equivalent *conservation of flow* at nodes 1 and 2 of the line graph), we find that $F = F_1 = F_2$. The element equations that represent the two branches, according to the sign convention, are $F_1 = b_1 v_{12}$ and $F_2 = b_2 v_{2g}$. The mathematical description of the source is $v_{1g} = v$, a velocity drop from node 1 to node g (or a rise from node g to node 1). We solve these five equations for the unknowns: $v_1 = v$, $v_2 = vb_1/(b_1 + b_2)$, and $F = F_1 = F_2 = vb_1b_2/(b_1 + b_2)$.

As we traverse the graph of Figure 2.29(b) from one node to a second node, we find the same drop in velocity, regardless of which path we follow. For example, $v_{1g} = v_{12} + v_{2g}$. We shall find this intuitively obvious continuity of motion useful in analyzing complicated networks of interconnected elements.

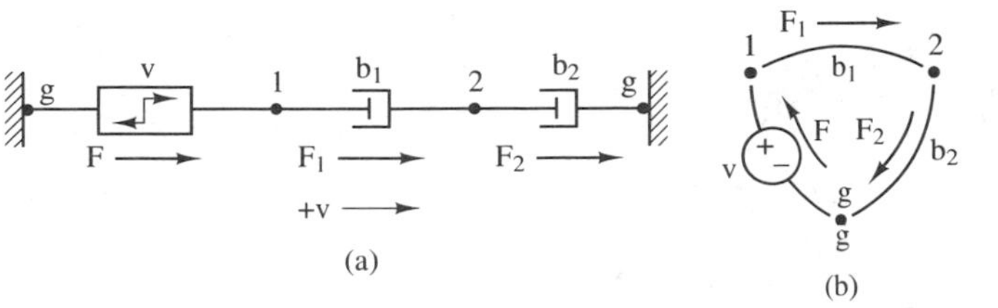

Figure 2.29 A series translational network. (a) Lumped model. (b) Line-graph representation.

The series connection of dampers, if viewed via the pair of terminals $(1, g)$, can be modeled as a single *equivalent* damper with friction constant $b = b_1b_2/(b_1 + b_2)$. If one of the dampers (say, b_2) is much stiffer (larger b_2) than the other, the weaker damper determines the stiffness of the combination ($b \approx b_1$).

Conservation of Flow and Continuity of Potential

The observations made in the two examples of the previous section apply to all translational lumped networks:

1. At each point where terminals of elements are connected, the scalar forces balance (d'Alembert's principle). In terms of the line graph, this fact can be expressed by stating that **the sum of the flows directed into a node is zero.** (A flow directed out of a node is the negative of a flow directed into that node.)
2. **The sum of the drops in velocity encountered in traversing the branches of any closed path** through a network **is zero.** (The concept of a closed path is clear for a line graph. For the statement to make sense in regard to a network, we must assume that all ground nodes are connected.) Similar statements can be made for drops in displacement and acceleration. The motions at the beginning and at the end of a closed path are identical because the two points are the same point. For example, traversing the graph of Figure 2.29(b) in the clockwise direction produces $v_{g1} + v_{12} + v_{2g} = (v_g - v_1) + (v_1 - v_2) + (v_2 - v_g) = 0$.

These conservation laws apply even if the flow through a lumped element is not *proportional* to the potential difference across that element. (For example, the dampers can have a coulomb friction characteristic, rather than a viscous friction characteristic.)

Analogous observations can be made about a broad range of natural phenomena—electrical, mechanical, hydraulic, etc. The effects of interconnecting lumped elements can be expressed as a pair of conservation laws. We state the conservation laws in the following graph-theoretic framework and illustrate them for translational systems.

Suppose a line graph has N numbered nodes with branches between some of the node pairs, as in Figure 2.30. Let B be the number of branches. Assign the variable A_i to the ith node; then $A_{ij} \triangleq A_i - A_j$ is a node difference (an across variable). Assign the through variable T_n to the nth branch, with a specified direction indicated by an arrow beside the branch.

We call the through variables T_n **flows** if the set $\{T_1, \ldots, T_B\}$ obeys the relation

$$\sum_{\text{node}} \pm T_n = 0 \tag{2.27}$$

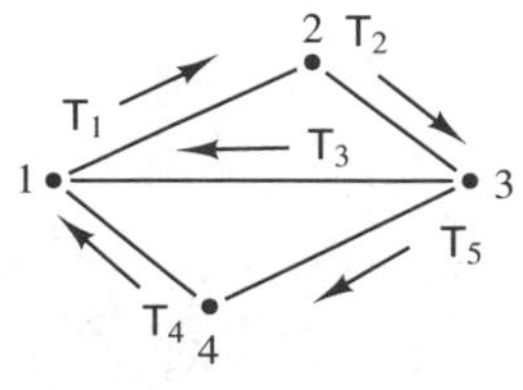

Figure 2.30 A line graph.

for each of the N nodes in the graph, where the sum is taken over all branches connected to the node. The + sign is used for the term T_n in equation (2.27) if T_n is directed *into* the node; the − sign is used if T_n is directed *out of* the node. For example, at node 3 of Figure 2.30, the relation is $T_2 - T_3 - T_5 = 0$. Equation (2.27) states that **flow is conserved** at each node. We sometimes refer to equation (2.27) as the **node equation** or the **node law.**

We call the node variables A_i **potentials** (and the node differences A_{ij} **potential differences**) if the set $\{A_1, \ldots, A_N\}$ obeys the relation

$$\sum_{\text{loop}} A_{ij} = \sum_{\text{loop}} (A_i - A_j) = 0 \qquad (2.28)$$

for every loop (closed path) in the line graph, where the sum is taken over all branches in the loop. Each branch is traversed in the same direction in the loop. For the counterclockwise path around the outside edges of the graph in Figure 2.30, the relation is $A_{43} + A_{32} + A_{21} + A_{14} = 0$. Some books call equation (2.28) the *compatibility condition.* We call it the **loop equation** or the **path law.**

The path law implies that the potential difference (or drop in potential) A_{ij} between nodes i and j can be found by adding the potential differences (or drops) across the branches along any path from i to j; that is, the potential drop A_{ij} from node i to node j is independent of path.

Exercise 2.5: Show that the potential drop A_{24} of Figure 2.30 can be expressed in any of four ways: $A_{21} + A_{14}$; $A_{23} + A_{34}$; $A_{21} + A_{13} + A_{34}$; or $A_{23} + A_{31} + A_{14}$.

In the translational context of section 2.1, we recognize the node law in the balance of oriented scalar-force variables at a node; we recognize the path law in the continuity of motion (x, v, and a) along any path through the network. Hence, the node velocities in a translational lumped network are potentials, whereas the oriented scalar-force variables in the lumped elements are flows.

Differentiating or integrating an equation term by term does not destroy the equality. Accordingly, the derivatives $\dot{T}_n$ and integrals $\int T_n$ are alternative sets of flow variables, and we could express the node law in terms of these derivatives or integrals. Similarly, the derivatives $\dot{A}_i$ and integrals $\int A_i$ are alternative sets of potentials.

In a more abstract sense, we can think of A as a potential function, analogous to gravitational potential or height, which has the value A_i at the ith node of the graph. This analogy is illustrated graphically in Figure 2.31. For the particular potential function

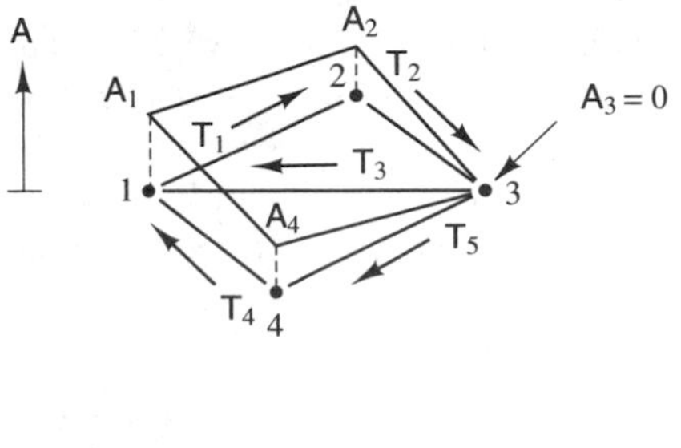

Figure 2.31 A potential function for the line graph of Figure 2.30.

shown in the figure, the flows T_1 and T_2 are *downhill* and the flows T_3, T_4, and T_5 are *uphill*. The path law can be viewed as a statement that the potential function A is continuous along any path.

According to the sign convention for element equations, we express the behavior of each lumped element in terms of an arbitrarily oriented flow variable for the element and the potential drop across the element in the direction of the flow. The element equations and node equations associated with a lumped network determine the values of the potential and flow variables of that network. If the actual flow in a branch is downhill, from higher to lower potential, we say the branch behaves passively. Uphill flow requires energy-producing behavior. If the flows associated with the potential function of Figure 2.31 are all positive, then branches 1 and 2 absorb energy and branches 3, 4, and 5 deliver energy.

Imagine a curve enclosing a subset of the nodes of a graph. It can be shown that the total flow crossing the curve into the enclosed region is zero. (Just add the node relations given by equation (2.27) for the enclosed nodes.) For a curve that encloses nodes 2 and 3 of Figure 2.30, this extended statement of conservation of flow is $T_1 - T_3 - T_5 = 0$. Suppose we select a curve that encloses $N - 1$ (all but one) of the nodes. Then the flows crossing into the enclosed region are identical to the flows leaving the remaining node. That is, the sum of the node equations for the $N - 1$ nodes is the same as the node equation for the remaining node. It follows that only $N - 1$ of the node equations (2.27) are linearly independent; the Nth node equation is redundant.

The System Equations

There are B branch flow variables and N node potential variables associated with a lumped network. Suppose we treat one of the nodes as a reference node and seek the values of all other node potentials relative to the reference node. Then the remaining $B + (N - 1)$ variables—the B branch flows T_n and the $N - 1$ nonreference node potentials A_i—constitute a fundamental set of system variables whose behaviors we must determine.

The $N - 1$ independent node equations (which involve only the flow variables) and the B branch equations (which relate branch flows to potential differences between nodes) form a linearly independent set of equations sufficient to determine the values of the $B + (N - 1)$ variables. We used precisely these two sets of equations to determine the velocities (node potentials) and scalar forces (branch flows) in the two translational examples of the previous section.

We can sometimes formulate a set of equations more efficiently by using one or more loop equations (see equation (2.28)) and by directly incorporating some of the node equations and lumped-element equations. In the network of Figure 2.29(b), for example, let F denote the flow in the *whole loop*. (The node law (2.27), applied at nodes 1 and 2, requires that $F_1 = F_2 = F$.) The path law shows directly that $v_{g1} + v_{12} + v_{2g} = 0$ or $-v + F/b_1 + F/b_2 = 0$. This single equation implies that $F = vb_1b_2/(b_1 + b_2)$. If our interest is in the compressive force F, the network can be described directly by a single loop equation in the single variable F, rather than by a set of two node equations and three element equations.

Associated with a given graph are many closed paths (loops). Associated with each loop is an equation of the form of equation (2.28). It can be shown that only B − (N − 1) of these equations can be linearly independent. (See reference [2.5], for example.) We must choose carefully from among such loop equations to produce a linearly independent set. Incorporating lumped-element equations implicitly can further confuse the process of selecting equations. Therefore, we develop two systematic approaches to writing a necessary and sufficient set of equations. Both approaches incorporate the element equations directly. The two methods are known as the *node method* and the *loop method.*

The single-loop example just discussed shows that the creative formulation of equations not only implicitly incorporates some of the element equations, but also eliminates certain of the network variables. It becomes unclear what constitutes a set of system equations. Throughout the text, we use the expression **system equations** (plural) to signify a linearly independent set of equations that determines fully the behaviors of the variables *which appear in that set of equations.* After solving for the behavior patterns of the variables that appear in the system equations, it is a simple matter to determine the behavior patterns of other variables in the network.

To find the behavior patterns of the variables in a set of system equations, we usually eliminate variables systematically to produce a single higher order differential equation in a single variable. During the elimination process, we progress through a sequence of sets of system equations, each set possessing fewer equations (of higher order) and fewer variables. We call the resulting single equation the **input-output system equation** *for the remaining dependent variable.*

An input-output system equation describes completely the relation between the source signal—the independent variable or input—and a particular network dependent variable—the output. There is such an input-output system equation for each dependent variable of the network. The input-output system equations for different network variables have most features in common. Chapter 3 begins our examination of mathematical and computer-based methods for solving input-output system equations.

Example Problem 2.4: Finding the system equations for a physical system.

A physical system is shown in Figure 2.32(a). The person pushes the object with a force of magnitude F.

(a) Use a conventional free-body diagram to find a mathematical model for the system. Use this model to find the input-output system equation for v_1, the velocity of the mass.

(b) Construct a lumped model of the system. Find a set of system equations from the lumped model. From that set of system equations, find the input-output system equation for v_1.

(c) The system equations found in (b) can also be obtained from a line-graph representation of the lumped model. Compare the use of the line graph, direct use of the lumped model, and the use of free-body diagrams for finding the system equations.

Solution: (a) Figure 2.32(b) shows a conventional free-body diagram for the primary object in the system. Each arrow in that figure shows the presumed direction of action of a force vector on the object. We represent the person that energizes the system by a force labeled F. The symbols F_1 and F_2 denote the spring force and the friction force, respectively. The friction force vector is oriented for motion of the object to the left. Motion of the object to the right would reverse the direction of the friction force and make F_2 negative.

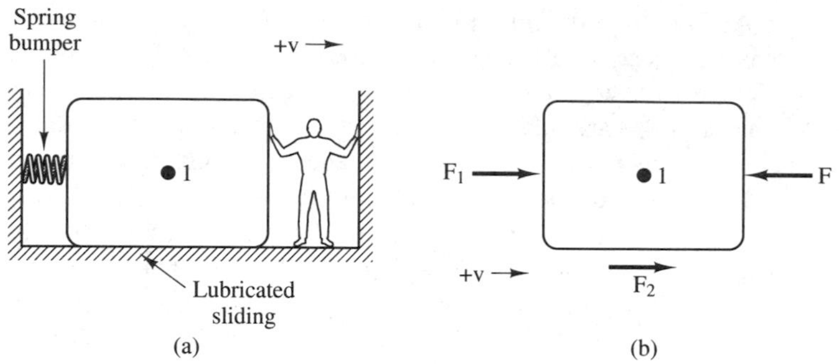

Figure 2.32 A translational system. (a) Pictorial model. (b) Free-body diagram.

Let the symbols k, m, and b represent the spring-bumper stiffness, the object mass, and the viscous-friction constant for sliding of the object, respectively. According to Newton's law, the forces acting on the mass satisfy

$$F_1 + F_2 - F = m\dot{v}_1 \tag{2.29}$$

The spring force is

$$F_1 = -kx_1 \tag{2.30}$$

The viscous friction force is

$$F_2 = -bv_1 \tag{2.31}$$

These three equations, taken together, form a mathematical model for the system. We substitute equations (2.30) and (2.31) into equation (2.29) and differentiate the result with respect to time to obtain the input-output system equation for v_1:

$$m\ddot{v}_1 + b\dot{v}_1 + kv_1 = -\dot{F} \tag{2.32}$$

(b) Each force vector in Figure 2.32(b) is part of an action-reaction pair. It is clear from Figure 2.32(a) that the reactions to the spring, the friction, and the human all act on the ground frame. It is less obvious that the inertial force, $F_3 \triangleq m\dot{v}_1$, also has a corresponding reaction force on the ground frame.

Figure 2.33(a) shows a lumped model for the system. The ideal force source, spring, and damper represent explicitly the mechanisms that produce the force vectors of Figure 2.32(b). The ideal mass manifests the inertial force explicitly and shows its relation to the object, represented by node 1, and to the ground frame. (What are the directions of action of the inertial-force *vectors* on the object and the ground frame?)

The flow variable F shown on the lumped network represents the compressive force F in the human. By the equivalence between force vectors and scalar forces that we introduced in Figure 2.4, it corresponds to the force vector F in the free-body diagram of Figure 2.33(b). The compressive-force variable F_1 and the tensile-force variable F_2 are equivalent to the corresponding force vectors acting on the free body. (The vector F_2 corresponds to tensile force in the lumped model because the ground-node terminal of the damper is placed to the right of node 1; the body—node 1—moves to the left.) The compressive force F_3 corresponds to the inertial-force vector that is implicit in the free-body diagram. In the free-body approach, it is made explicit only in equation (2.29).

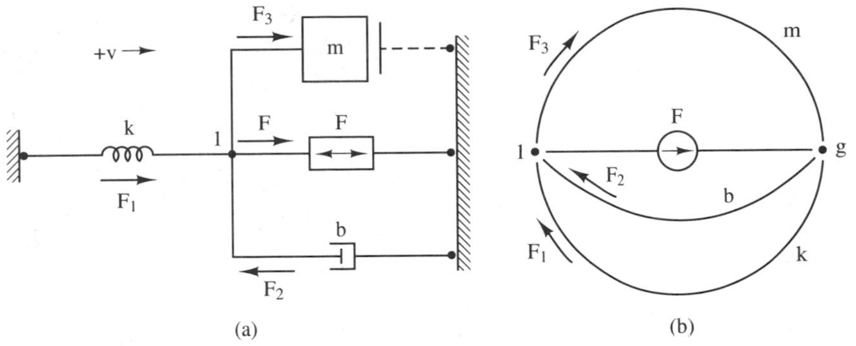

Figure 2.33 A lumped model for the system of Figure 2.32. (a) Lumped-network model. (b) Line-graph representation.

According to the node law given by equation (2.27),

$$F_1 = F - F_2 + F_3 \tag{2.33}$$

The flow through a linear lumped element equals the potential drop across the element in the direction of flow divided by the element's impedance. For the spring, $F_1 = (k/s)v_{g1}$, which is equivalent to equation (2.30). The damper equation is $F_2 = bv_{g1}$, the same as equation (2.31). The equation for the mass is

$$F_3 = msv_1 \tag{2.34}$$

The node law and the three element equations, taken together, constitute the system equations for the lumped network. If we substitute the element equations into equation (2.33) and multiply by s, the result, with s interpreted as the time derivative, is equation (2.32), the input-output system equation for v_1.

(c) The free-body diagram is simple and intuitive. That simplicity is achieved by keeping the spring, damper, and inertial mechanisms implicit. If we show all forces and their relations explicitly, we get the lumped model. Making the model assumptions explicit reduces the likelihood that we will make incorrect assumptions or use incorrect signs in the element equations. The spring, damper, and mass equations can be seen more clearly in the lumped model than in the free-body diagram.

The line graph for the lumped model is shown in Figure 2.33(b). The line graph does not provide any physical insight that is not contained in the lumped model itself. Nor does it aid the writing of element equations or the summing of flows at nodes. It does, however, clarify the closed loops that are formed by the various ground connections. We shall find shortly that these loops can be used to simplify the writing of system equations for some lumped networks. The line graph also provides some economy of notation.

The Line Graph versus the Lumped Model

View the lumped model as a **physical model.** It preserves the physical natures of the observable quantities of the pictorial model. Each node is a lumped version of a physical object. Each node *moves* back and forth along a single physical axis. The relative motions of the nodes are nearly unchanged from the pictorial model. The scalar forces in the branches are lumped versions of physical scalar forces. The compressive or tensile natures of those scalar forces are unchanged from the pictorial model. Each lumped-element symbol denotes a physical trait rather than a mathematical relation.

We associate a velocity *variable* with each node of the lumped model. The +v direction for the lumped model relates the + and − signs of each velocity variable to the back-and-forth motions of the corresponding node. We associate a scalar-force *variable* with each branch. The flow arrow beside the branch relates the + and − signs of that scalar-force variable to the compressive and tensile natures of the physical scalar force in the corresponding branch. (If the arrow is oriented in the +v direction, the variable represents compressive force.)

Physical laws relate the velocity drop across each passive lumped element to the scalar force in that element. A physical law also relates the scalar forces in the branches that connect at each node. By defining velocity and scalar-force variables, we convert these physical laws to equations. We presume a system source is operated in controlled-velocity mode or controlled-force mode. (A human being usually decides, directly or indirectly, which mode and signal waveform to use.) We represent each source by an equation. The full set of equations describes the behavior of the lumped network.

View the line graph as a **mathematical model.** Line-graph nodes do not move. The line graph focuses attention on relations among the system variables rather than on physical behaviors. Line-graph terminology is the terminology of potentials and flows. The difference between lumped models and line-graph models is, perhaps, most distinct for ideal sources. The lumped-model source symbols signify physical behaviors. In contrast, the line-graph source symbols, like the equations that represent the sources, depend on the +v direction. Thus, the line-graph symbols signify mathematical relations. The line graph strips away physical features and retains only mathematical features—the features needed to write the system equations. When we wish to give *physical interpretations* to particular values of potentials and flows, we return to the lumped model.

We can (and do) apply the potential and flow terminology directly to the lumped model. We can also apply the equation-writing techniques (the node and loop methods) of the following sections directly to the lumped model. However, the physical source symbols of the lumped model do not facilitate writing equations. To overcome this limitation, we add the mathematical source notation of the line graph to the lumped model. Specifically, for a velocity source, we mark the + and − ends of the source. For a scalar-force source, we add the compressive-force flow arrow. Then the equation-writing methods can be applied easily to the lumped model.

The Node Method for Writing System Equations

In the node method for writing a set of system equations, we apply the node law to all but one of the N network nodes. At each node, the method incorporates directly the element equations for those branches that are connected to the node. The method focuses attention on the node potentials A_i rather than potential differences. Usually, only node variables appear in the equations. Although the flow variables do not appear explicitly, the flow in any branch can be computed from the element equation for the branch and the values of the potentials at the two end nodes of the branch. The node method requires only $N - 1$ equations.

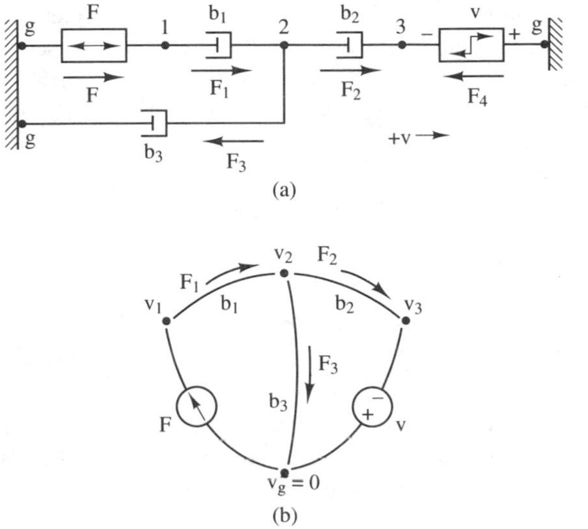

Figure 2.34 A multiloop, multinode translational network.
(a) Lumped-network model.
(b) Line-graph representation.

We apply the node method to the multinode network and corresponding line graph of Figure 2.34. (The source flow F and the source potential rise v are each oriented in the line graph to agree with the +v direction in the lumped model.) Focus attention on three of the four nodes, preferably, nodes with few branches. In this case, we choose nodes 1, 2, and 3. We then write the node equation for each of these three nodes. The result is a set of system equations in terms of the velocities v_1, v_2, v_3, and v_g. We treat the ground node as a reference node ($v_g = 0$).

The source flow F and the source velocity $v_{3g} = v_3 = -v$ are known. According to the lumped-element equation for the b_1 branch, $-F_1 = b_1(v_2 - v_1)$. The node equation for node 1 is $F - F_1 = 0$, or

$$F + b_1(v_2 - v_1) = 0 \tag{2.35}$$

Similarly, the equations for the b_2 and b_3 branches are $-F_2 = b_2 v_{32} = b_2(-v - v_2)$ and $-F_3 = b_3 v_{g2} = b_3(0 - v_2)$. The node equation for node 2 is $F_1 - F_2 - F_3 = 0$, or

$$b_1(v_1 - v_2) + b_2(-v - v_2) + b_3(-v_2) = 0 \tag{2.36}$$

Since v_3 is known, the node equation for node 3 need not be written. (The equation $v_{3g} = v_3 = -v$ serves as a substitute for that node equation.) Equations (2.35) and (2.36) constitute $N - 2$ node equations that determine the remaining $N - 2$ node variables. These two equations can be written directly, by inspection of the line graph. (If the network had not included the velocity source, there would have been $N - 1 = 3$ node equations to determine the $N - 1$ node velocities.) The solution to the set of equations is

$$v_1 = \frac{b_1 + b_2 + b_3}{b_1(b_2 + b_3)}F - \frac{b_2}{b_2 + b_3}v, \qquad v_2 = \frac{F - b_2 v}{b_2 + b_3}, \qquad v_3 = -v \tag{2.37}$$

If we are also interested in the flows, we substitute these node potentials into the lumped-element equations used before. We find F_4 from the node equation for node 3; that is, $F_4 = -F_2$.

We must use the lumped model, rather than the line graph, to interpret the solution physically. F_1, F_2, and F, because they are oriented in the +v direction, represent compressive forces in the corresponding lumped-network elements; F_3 and F_4, which oppose the +v direction, represent tensile forces. We cannot tell which scalar forces are actually compressive or tensile without substituting the numerical values of the element parameters b_1, b_2, and b_3 and the actual source waveforms F(t) and v(t) into equations (2.37). If the actual value of a flow is negative, its physical interpretation is reversed.

THE STEPS IN THE NODE METHOD

1. Select the $N - 1$ network nodes to which the node law will be applied. (In effect, we select a single node to *exclude* from the equations—typically, one with many branches.)
2. Define oriented flows in the B network branches. Write the node equation (2.27) for each of the $N - 1$ nodes, in terms of these branch flows. If a flow source is connected to a node, the flow variable for that source appears explicitly in the equation for that node.
3. For each of the B branches, use the element equation to express the branch flow in the form $\mathsf{T} = \mathsf{A}_{ij}/\mathsf{Z}$, where Z is the branch impedance.
4. Substitute the flow expressions from step 3 into the node equations from step 2 to eliminate the flow variables and leave only node-potential variables. The flow of a flow source is known and can remain in the equations. (If a branch contains a potential-difference source, the flow in that branch cannot be eliminated; it is not controlled by the source. Then that flow is an extra unknown variable in the set of equations. In that case, the known potential difference across that source permits us to use the equation for that source as the extra equation that we need to have a sufficient set. We determine the unknown source flow from a node equation which includes that flow.)

The foregoing steps are simple enough that the engineer usually carries out steps 2, 3, and 4 simultaneously, by inspection of the line graph. A wise choice of nodes in step 1 can sometimes avoid complicated equations. Such wisdom is gained by experience in writing node equations. A detailed example of the node method is given in example problem 2.5.

Exercise 2.6: It is not necessary to use the line graph to write the node equations. They can be written directly from the lumped network of Figure 2.34(a). Repeat the steps used to find equations (2.35)–(2.37), but in terms of the lumped network rather than the line graph. Use the F arrow at the flow source and the + and − symbols at the velocity source to aid the process.

In the absence of potential-difference sources, the node method generates $N - 1$ linearly independent equations in N node-potential variables. Those equations can be solved for $N - 1$ node potentials relative to the remaining node potential. Then any particular potential difference can be computed by taking the difference between two known node potentials. Any branch flow can be found from the element equation for that branch and the known potential difference across that branch. (For mechanical translation, these flows can be interpreted as compressive or tensile forces by comparing their directions with the +v direction on the lumped model.) Although solving the system differential equations might be difficult, the succeeding computation of potential differences and branch flows is straightforward.

Exercise 2.7: Determine the full set of branch flows and corresponding scalar forces for Figure 2.34, in terms of the network parameters and source signals.

Example Problem 2.5: Generating system equations by the node method.
Figure 2.35(a) shows a translational lumped network. The values of the parameters m, k, and b are presumed known. Initially, the spring is uncompressed and the mass is at rest. At $t = 0$, a known velocity function v(t) is applied.
(a) Draw the line-graph representation of the system.
(b) Use the node method to generate a set of system equations.
(c) Eliminate variables from the system equations to obtain the input-output system equation for the variable v_2.
Solution: (a) The line graph is shown in Figure 2.35(b). The flows F_2, F_3, and F_4 represent compressive forces; F_1 is a tensile force.
(b) Because there are three nodes, we must write two node equations. Choose nodes 1 and 2. At node 1, $F_1 + F_2 = 0$. The damper equation is $F_1 = bv_1$. The velocity-source equation does not involve the source flow F_2. Therefore,

$$bv_1 + F_2 = 0 \tag{2.38}$$

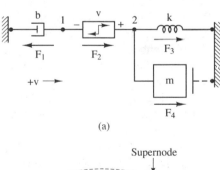

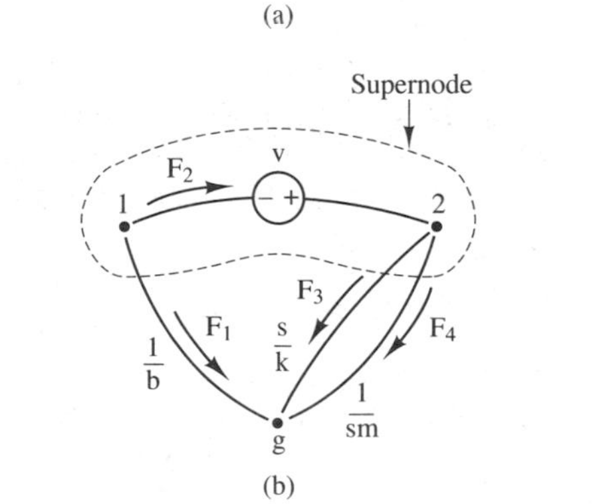

Figure 2.35 A translational system. (a) Lumped network. (b) Line-graph representation.

At node 2, $F_2 - F_3 - F_4 = 0$. The spring and mass equations are $F_3 = (k/s)v_2$ and $F_4 = smv_2$. Therefore,

$$F_2 - (k/s)v_2 - smv_2 = 0 \tag{2.39}$$

The two node equations include three unknown variables (v_1, v_2, and F_2). We must find a third equation before we can eliminate variables and generate the input-output system equation. Since the source waveform is known, the node velocities v_1 and v_2 are related by the velocity-source equation,

$$v_2 = v_1 + v \tag{2.40}$$

This is the third equation.

(c) We solve equation (2.40) for v_1 and substitute into equation (2.38) to obtain $bv_2 - bv + F_2 = 0$. We then solve this equation for F_2, substitute the result into equation (2.39), and rearrange to obtain the input-output system equation

$$s^2v_2 + \frac{bs}{m}v_2 + \frac{k}{m}v_2 = \frac{bs}{m}v \tag{2.41}$$

or

$$\ddot{v}_2 + \frac{b}{m}\dot{v}_2 + \frac{k}{m}v_2 = \frac{b}{m}\dot{v} \tag{2.42}$$

The node method of deriving system equations focuses on the node velocities. The unknown flow F_2 entered the equations only because the velocity source does not directly relate F_2 and v_{12}.

We could have avoided the appearance of F_2 in the equations. Node 1 is a **nonessential node:** The node equation at node 1 is simply $-F_1 = F_2$. We can define, instead, a single flow F_2 that applies to both branches. Then the two branches, together, form a single *essential branch*. How, then, would we form the node equations? The answer is, by the formation of a **supernode,** as illustrated in Figure 2.35(b).[9] The supernode boundary groups the nonessential node with a neighboring **essential node,** one that connects three or more branches. The net flow into the supernode must be zero: $F_2 - F_3 - F_4 = 0$. This single node equation is the only node equation for this example. We substitute the three element equations and the source equation, as before, to produce equation (2.42).

In general, we can use supernodes to eliminate all nonessential nodes. Then the node method generates one node equation for each essential node.

The node method for deriving equations can be applied directly to the lumped model of Figure 2.35(a). The line graph is merely a conceptual aid. For some users, removal of the unique physical features of translational systems makes it easier to write the equations. For other users, the physical features help, rather than hinder, the writing of equations.

The Loop Method for Writing System Equations

In the loop method for writing a set of system equations, we write a set of *independent* loop equations, each based on the loop equation (2.28). The method incorporates the element equations for the branches in each loop in order to eliminate the node variables (po-

[9] The supernode concept is discussed in detail in reference [2.2].

tentials) from the equations. The loop equations usually involve only flow variables. The method could be implemented in terms of the B flow variables defined in the individual branches. However, these variables are usually not linearly independent. We prefer to express the loop equations in terms of an independent set of *circulating-flow* variables. We calculate the branch flows in a separate step. We illustrate the method by an example.

Figure 2.36 shows the line graph for the network of Figure 2.34. Each *window* or *opening* in the graph is referred to as a **mesh.** It can be shown that the number of meshes in a network equals the number of independent loop equations, namely, $B - (N - 1)$, where B is the number of branches and N is the number of nodes (see reference [2.5]). For the network of Figure 2.34, $B - (N - 1) = 2$. Superimposed on the graph of Figure 2.36 are two **circulating flows** (or *mesh flows*), F_a and F_b. The four branch flows can be determined from F_a and F_b: F and F_1 are identical to F_a; F_2 equals F_b; $F_4 = -F_b$; and in the center branch, $F_3 = F_a - F_b$, a statement equivalent to balancing flows at node 2. We can think of the five branch flows as different combinations of the two circulating flows.

In the loop method, one loop equation is written for each mesh. The element equations for the b_2 and b_3 branches of mesh b are $v_{g2} = (F_b - F_a)/b_3$ and $v_{23} = F_b/b_2$. The potential-difference source in mesh b specifies $v_{3g} = -v$. We substitute these element equations into the loop equation for mesh b to eliminate node potentials. The loop equation for mesh b is

$$v_{g2} + v_{23} + v_{3g} = \frac{F_b - F_a}{b_3} + \frac{F_b}{b_2} + (-v) = 0 \tag{2.43}$$

Similarly, the loop equation for mesh a becomes

$$v_{g1} + v_{12} + v_{2g} = -v_1 + \frac{F_a}{b_1} + \frac{F_a - F_b}{b_3} = 0 \tag{2.44}$$

The flow source in mesh a cannot be used to eliminate the node potential v_1. However, since $F_a = F$, the specified flow F of the source determines the circulating flow in mesh a. Consequently, loop equation (2.44) is not needed to determine the circulating flows. Instead, it can be used to determine the velocity v_1 across the flow source. The solution to the set of equations is

$$F_a = F, \qquad F_b = \frac{b_2F + b_2b_3v}{b_2 + b_2}, \qquad v_1 = \frac{b_1 + b_2 + b_3}{b_1(b_2 + b_3)}F - \frac{b_2}{b_2 + b_3}v \tag{2.45}$$

We use the circulating flows to calculate the branch flows. Then the drop in velocity across a branch can be obtained from the branch flow and the element equation for that

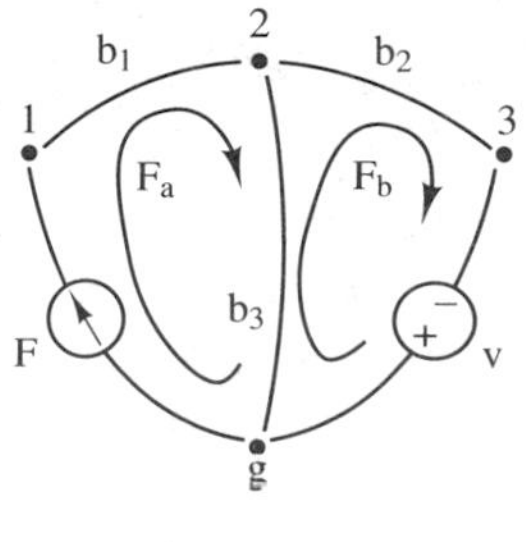

Figure 2.36 Circulating flows for Figure 2.34.

branch. Finally, any node velocity can be computed by summing velocity drops along a path from the node of interest to the reference node.

Exercise 2.8: Compute the branch flows and the remaining node velocity v_2 for the preceding example.

Although meshes (windows) were used in the foregoing example to define the circulating flows and also to specify the loops used for the application of the loop equation, meshes are not the only loops that can be used to define a sufficient set of independent loop equations. In defining the circulating flows or in selecting the loops used for the loop equations, one need only choose a sequence of closed loops, each of which includes at least one new branch and at least one previously used branch. Once all branches are included in the set, a sufficient set of circulating flows (or of loop equations) has been found. The loops used to define the circulating flows need not be the same as the loops used to write the loop equations.

THE STEPS IN THE LOOP METHOD

1. Define and label a sequence of $B - N + 1$ circulating flows T_a, T_b, etc. Typically, one uses meshes to define these flows.
2. Select a sequence of $B - N + 1$ loops, chosen in a manner similar to that in which the circulating-flow paths in step 1 were chosen, and write the loop equation (2.28) for each loop. The loops used for the loop equations need not be the same as those used to define the circulating flows. (Another example is given in example problem 2.6.)
3. For most of the B branches, the lumped-element equation gives the potential difference across the branch in the form $A_{ij} = zT$; express the branch flow T in terms of the circulating flows.
4. Substitute the potential-difference expressions from step 3 into the loop equations from step 2 to eliminate the potential-difference variables and leave only circulating-flow variables. The potential difference of a potential-difference source is known and can remain in the equations. (If a branch contains a flow source, the potential difference across that source cannot be eliminated, because it is not controlled by the source. Then that potential difference is an extra unknown variable in the set of equations. In that case, the known flow of the source specifies one of the branch flows. That branch flow specification serves as the extra equation that we need to have a sufficient set. Then we determine the potential difference across the flow source from a loop equation which contains that potential difference.)

These steps are simple enough that the engineer usually carries out steps 2, 3, and 4 simultaneously, by inspection of the line graph. A wise choice of loops in step 2 can sometimes avoid complicated loop equations. The ability to foresee such complications comes from experience in writing loop equations. A detailed example of the loop method is given in example problem 2.6. It is not necessary to use the line graph to carry out the loop method; it can be carried out directly from the lumped network.

Exercise 2.9: Repeat the steps used to find equations (2.43)–(2.45), but in terms of loop flows drawn directly on the lumped network of Figure 2.34(a).

Example Problem 2.6: Generating system equations by the loop method.
Figure 2.37 shows the same translational lumped network used in Example Problem 2.5. The values of the parameters m, k, and b are presumed known. Initially, the spring is uncompressed and the mass is at rest. At $t = 0$, a known velocity function $v(t)$ is applied.
(a) Define loop variables and use the loop method to generate a set of system equations.
(b) Eliminate variables from the system equations found in (b) to obtain the input-output system equation for F_2, the flow in the source.
(c) Show the equivalence of the loop-generated equations to the node-generated equations of example problem 2.5.
Solution: (a) A pair of loop variables, F_a and F_b, are superimposed on the line graph in Figure 2.37(b). (They need not circulate in the same direction.) We write loop equations for each of the two meshes. For mesh a, the loop equation is $v_{g1} + v_{12} + v_{2g} = 0$, and the element equations are $v_{g1} = F_a/b$, $v_{12} = -v$, and $v_{2g} = (s/k)F_a + (s/k)F_b$. We substitute the element expressions into the loop equation to obtain

$$\frac{F_a}{b} - v + \frac{sF_a}{k} + \frac{sF_b}{k} = 0 \tag{2.46}$$

The loop equation for mesh b is $v_{g2} + v_{2g} = 0$. The element equation for the mass is $v_{g2} = F_b/sm$. We insert the mass and spring expressions into this loop equation and multiply by s (differentiate) to obtain

$$\frac{F_b}{m} + \frac{s^2F_a}{k} + \frac{s^2F_b}{k} = 0 \tag{2.47}$$

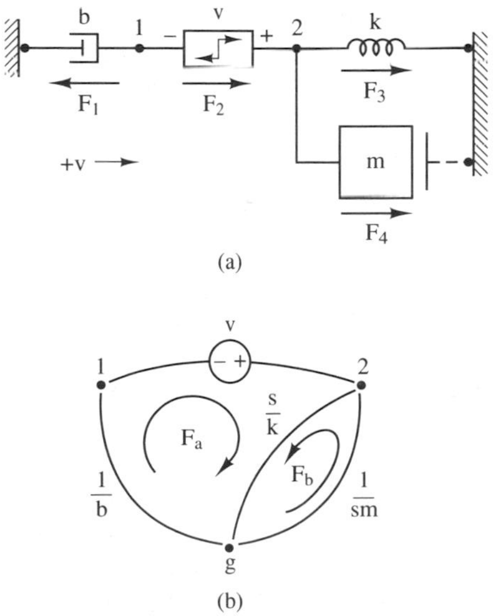

Figure 2.37 A translational system. (a) Lumped network. (b) Line-graph representation.

Equations (2.46) and (2.47) are the loop-generated system equations—two equations in the two unknowns F_a and F_b. We can express any potential or flow in the network in terms of F_a and F_b.

(b) The source flow is $F_2 \equiv F_a$. We multiply equation (2.46) by s and subtract the result from equation (2.47) to obtain

$$\frac{F_b}{m} - \frac{sF_a}{b} + sv = 0 \tag{2.48}$$

We then solve equation (2.48) for F_b and substitute into equation (2.46) to eliminate F_b. The result, rearranged, is the system equation

$$s^2F_a + \frac{b}{m}sF_a + \frac{k}{m}F_a = \frac{kb}{m}v + bs^2v \tag{2.49}$$

or

$$\ddot{F}_2 + \frac{b}{m}\dot{F}_2 + \frac{k}{m}F_2 = \frac{kb}{m}v + b\ddot{v} \tag{2.50}$$

(c) Note that $F_2 = -F_1 = -bv_1 = -b(v_2 - v)$. We substitute this relation between F_2 and v_2 into equation (2.50) and rearrange terms to obtain, once again, equation (2.42). The loop and node methods thus produce identical results. As with the node method, the loop method can be applied directly to the lumped model of Figure 2.37(a).

With the option of using either the node method or the loop method to formulate a set of system equations, which one should we use? We use the one that generates the desired information more directly. If we are interested primarily in the full set of node variables, we would probably use the node method. Interest in the flows would suggest use of the loop method. If we want to compute the full set of node potentials (or potential differences) and branch flows, we would choose the method that involves the fewest simultaneous equations. The node method requires $N - 1$ equations. The loop method requires $B - N + 1$ equations. For either method, the remaining variables are determined one at a time, from the element equations. For each problem you solve, you should spend some time visualizing the expected sequence of manipulations associated with various approaches; it will help you pick an approach that minimizes the manipulations.

Dependent-Source Models

In section 6.2, we examine levers, gears, and motors. There we find a variety of situations in which we need to represent linear relations between two isolated variables. Inventing a new lumped element for every such instance would make the models too complicated. In section 5.1, we encounter a hydraulic lumped element (owing to friction in pipes) for which the element equation is quadratic, not linear. To aid the lumped modeling process, we invent another type of ideal element—the **ideal dependent source**—to portray arbitrary dependencies on other network variables. In particular, dependent sources aid the modeling of transformers, transducers, and amplifiers in chapter 6.

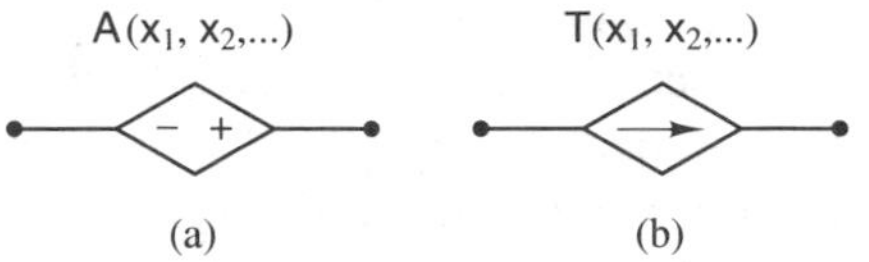

Figure 2.38 Line-graph symbols for ideal dependent sources; the symbols x_i denote potential differences or flows in the network. (a) Potential-difference source. (b) Flow source.

We use diamond-shaped symbols for ideal dependent sources to distinguish them from independent sources. Figure 2.38 illustrates this dependent-source notation for line graphs. The value of a dependent source is a specific function of the values of one or more potential differences or flows in the network. We label the ideal dependent source with its actual value, that is, with the function of the network variable(s) on which it is dependent. These labelings are illustrated in Figure 2.38 (see also Figures 6.19 and 6.20).

2.3 MODELING AND DESIGN

The design process requires that we conceive new system configurations. The ability to think of systems in terms of simple, realistic models is an important aid to creativity. Another such aid is experience with the behaviors of many physical systems. We begin generating that experience by modeling and analyzing dynamic systems in a variety of contexts, namely, mechanical translation and rotation, electrical conduction, fluid flow, heat flow, and various population flows. We also use models and analysis tools to design specific details of specific systems.

A system model mimics the appearance or behavior of a real system. The model is simplified to eliminate nonessential complexity while retaining similarity in the features of interest. For example, the lumped-mass model of Figure 2.12 retains similarity with the inertial behavior of the object it represents, but sacrifices details of appearance, omits the effects of friction between the object and its surroundings, and ignores compliance of the object under compressive forces.

The dynamic systems we study are **time invariant.** That is, the **parameters** of the system models (e.g., the stiffness k, friction constant b, and mass m) do not change with time. The *observable system variables* (velocities, for example) do change with time. We represent a dynamic system by a set of differential equations. For most of the systems examined in this book, those differential equations are linear. Because the systems are time invariant, the differential equations have constant coefficients.

The term *system analysis,* in its broadest sense, means using models of a system to develop an intuitive understanding of the system's behavior. In a narrower sense, system analysis is the process of partitioning a system into parts, discovering the behaviors of the parts and the relationships among the parts, and determining the emerging behavior that results from the interaction of the parts. (The two-terminal elements introduced in section 2.1 are the simplest of parts for translational mechanical systems.) This chapter focuses on modeling—partitioning the system, determining the behaviors of and interconnections among its parts, and writing the equations that characterize the interconnected set of parts. Chapter 3 shows how to determine the emerging system behavior from the system model.

We begin this section by determining the relative advantages and disadvantages of the various types of models that we developed in the previous section. Then we elaborate and refine the concept of a signal—a model for observable system behavior. Next, we discuss the iterative design process of which lumped modeling is a part. Finally, we develop a way to determine how much detail to include in a lumped model.

Types of Models

Throughout the text, we use various types of models to characterize systems. We examine and differentiate these models in this section. First is the pictorial model, typified by Figure 2.20. By a **pictorial model,** we mean a diagram that is geometrically similar to the physical system it represents. In such a model, the physical natures of the measurable variables are apparent. Pictorial models can include an arbitrary degree of detail and, hence, can be arbitrarily accurate.

A second type of model is the lumped-network model, exemplified by Figure 2.22. A **lumped-network model** (or just *lumped network*) is an interconnection of lumped elements that *approximates* the physical system. The network acknowledges only a few system variables, each with only one degree of freedom. Each lumped element in the network corresponds to a specific mathematical relationship between its across variable and its through variable (as in Figure 2.5 and equation (2.7)). We can also state mathematically the way that the lumped elements are interconnected (see section 2.2). Consequently, the lumped network is equivalent to a set of mathematical equations. Yet, like the pictorial model, the lumped-network model is *physical;* that is, the physical nature of each variable is apparent.

A third type of model, introduced in section 2.2, is the line graph. The **line graph** emphasizes the structure of interconnections that determines the interactions among the lumped-network variables. The line graph strips away all geometrical similarity with the physical system; the physical natures of the variables are obscured. In the line graph, we treat *through variables* as flows and *across variables* as potentials. The language of potentials and flows will help us find the equations that relate the variables to each other. The transition from lumped network to line graph requires no approximation: The line graph retains all the mathematical relationships of the lumped network and is in fact *mathematically equivalent* to it.

A **mathematical model** of a system is a set of equations that describes the behaviors of one or more variables of the system. Each element in a lumped network represents a unique mathematical relation between the through and across variables for that element. Each connection of elements establishes a unique mathematical relation between the through variables for those elements. A lumped network determines a unique mathematical model—a set of equations that determines the behaviors of all variables in the network. Since the line graph is mathematically equivalent to the lumped network, it determines the same mathematical model.

The mathematical model retains all the behavioral information of the lumped network and the line graph, but gives up all visual features. The equations in the mathematical model can be rearranged to produce various equivalent descriptions of the system. In section 3.2, we introduce the **operational model,** a graphical visualization of the mathematical operations that relate the variables in the system. There is an operational model for each arrangement of the mathematical equations.

We use the foregoing types of models in various combinations or mixed forms to capture and express an image of the system, much as one uses combinations of words, intonations, facial expressions, and hand motions to communicate ideas. The lumped network serves as a bridge between the physical world and the mathematical world. It retains enough likeness to the physical system to help us develop physical intuition concerning that system, yet it defines its mathematical description fully. The line graph shows clearly the mathematical structure of the model. Clarity of structure makes it easier to write the mathematical equations and strengthens the analogies among the dynamic systems of different energy domains.

The lumping process is the primary creative process in modeling. The line graph and the mathematical model can be interpreted as descriptions of the lumped network, since they correspond to a specific lumping of the original system. Generating the mathematical equations is relatively routine. The line graph is a tool to assist in writing the equations.

In practice, we often use various mixed forms of models rather than a strict sequence of pictorial model, lumped network, line graph, and mathematical model. For example, we use Figure 2.21 as an intermediate step because it aids the lumping process for that particular example. Mixed notations may be used if the designer finds them helpful. If the designer is sufficiently experienced, or if the system to be modeled is simple, the designer can generate the lumped network, the line graph, or even the mathematical model directly, without the aid of intermediate steps.

Types of Signals

A signal is an action that conveys information. In our technical world, we use signals to exchange information with machines. For example, a driver commands a car to accelerate by depressing the accelerator, not unlike using the pressure of one's knees and heels to signal a horse to accelerate. Similarly, we signal a change in direction by a turn of the steering wheel (or motion of the horse's reins). In this text, we use the term **signal** to mean an observable variable, a quantity that reveals the behavior of a system. The displacement, velocity, and acceleration of a point are separate (but related) signals. The compressive force in an object is another signal. We also refer to these time functions as **waveforms, profiles, patterns,** and **trajectories.** We identify the signals of a lumped network with variables of the physical system the network represents. To compare the behavior of the model with that of the system, we calculate the network signals and measure their physical counterparts.

Some signals in a lumped network represent **independent variables**—variables whose values are specified from outside the system. For example, if a driver controls the acceleration of an automobile, that acceleration is an independent variable. We often refer to an independent variable as an **input signal.** The lumped network is represented by a set of differential equations. We solve those equations together with the specified input signals to find the other network signals. We call these other signals **dependent variables, output signals,** or the **network response.**

We usually think of input signals as being *given*. However, source waveforms must be designed. We might first choose output-signal waveforms that constitute good system behavior, then design the network structure, select the lumped-element parameter values, and finally, choose the source waveforms (input signals) to produce the desired output. In such a situation, the term *given* is imprecise. It can refer to the *source*—the *cause* of the output signal—or it can refer to the quantity specified by the designer—the output signal itself. We use terminology that makes clear which meaning is intended in each particular case.

To understand the behavior of a system, we apply a number of different input signals and observe the responses. We compare tests of subsystems with tests of models of those subsystems to see whether the models behave like the physical subsystems they represent; that is, we validate the subsystem models. We examine the behavior of the model of the whole system before constructing the system, so that we can see whether the system will be satisfactory. Finally, we test the actual system to see whether its behavior is satisfactory.

Experience has shown that certain test-signal waveforms are particularly effective at arousing the characteristic behaviors of physical systems and of lumped-network models. These signals are step functions (abrupt changes in value), exemplified by closing a switch; impulse functions, epitomized by a hammer blow; and steady sinusoids, typified by vibrations from an unbalanced motor shaft. Although infinitely abrupt signals cannot be generated in physical systems, infinite abruptness is easier to express mathematically than is very fast change. We can think of an infinitely abrupt signal as an **ideal test signal.** We use an ideal test signal as an input signal to generate the characteristic response of a lumped network. We use a nearly ideal test signal to generate the characteristic response of a physical system.

The System Design Process

Systems should be designed to meet specifically stated performance objectives. Defining appropriate measurable performance objectives is part of the design process. Subsystems (or parts) then can be designed to meet supporting objectives. Ultimately, the user decides whether the performance is satisfactory by observing the system in operation, without regard for the initially stated objectives.

An automobile, for example, is designed to transport passengers and various other loads rapidly under a variety of circumstances. The driver has expectations concerning the acceleration and deceleration capabilities of the vehicle. The capability of the automobile to accelerate depends on the size of its engine, the slope of the road, and the load the automobile is to carry. The last two entities vary for a given automobile, and the first varies from automobile to automobile. Speed limits and stoplights restrict the automobile's motion and exacerbate the driver's felt need for a greater capability to accelerate. The system designer must define the boundaries, inputs, and outputs of the system in order to focus attention on a limited set of design parameters. As part of that design process, the designer specifies the boundaries and performance objectives of subsystems that will affect the system behavior. The designer must always remain aware of the larger system (with its higher objectives).

To illustrate the process of setting objectives, suppose an automobile designer models an automobile as an ideal mass m that is to be accelerated by a constant scalar force F (supplied by the engine) from rest to some maximum velocity (say, $v_{max} = 30$ m/s, or 108 km/hr). This simple model structure (a mass) and simple input-signal structure (constant force) permit the designer to focus attention on the magnitudes of the mass, the acceleration, and the applied force. The designer selects a capability to accelerate that is deemed acceptable to the majority of automobile owners (say, 3 m/s^2, which generates full speed in 10 s). This motion specification is the *given* signal; but it is not the energy source (or input signal). The designer selects a maximum-load mass (including the body of the automobile, the engine and transmission, and the passengers and baggage) that would seem to be acceptable to the majority of automobile owners—say, m = 1,200 kg. Then, according to equation (2.20), the force pattern (the source signal) that will produce the specified acceleration is the constant scalar force F = (1,200 kg) (3 m/s^2) = 3,600 N.

Note the three phases of the design process. First, we model the system by exploring the *situation,* designating the system variables, and proposing specific model and signal structures. Second, we specify performance requirements in terms of the designated system variables. Third, we select specific values for the model parameters and signal-structure parameters to satisfy the performance requirements. In the process, five categories of design results are produced:

1. A lumped network structure, that is, a specific interconnection of ideal lumped elements (a single mass accelerated by a scalar-force source).
2. Specific values for the model parameters (m = 1,200 kg).
3. A source-signal structure (a constant scalar force F_{max}).
4. Specific values for the parameters of the source signals ($F_{max} = 3{,}600$ N).
5. Computed response patterns for the remaining system variables ($\dot{v}_{max} = 3$ m/s^2).

As a result of this high-level design process, the designers of various subsystems have a firm basis on which to carry out their lower level designs. For example, the source-signal parameter F_{max} is used as a performance specification by the designers of the engine and transmission as they determine the engine's torque requirements and the size of the engine and transmission. The body designer uses the 1,200-kg mass as the target mass during the design of a body to carry the engine, passengers, and load comfortably and safely. The design process is iterative. That is, the engine and transmission designed to provide the 3,600-N acceleration force might have a larger mass than assumed during selection of the 1,200-kg maximum load. Consequently, the original designer might have to increase the maximum load, reduce the acceleration requirement, or assume a smaller load of passengers and baggage.

The design process begins with a high-level, simplified model and with *assumptions* about the environment and about the weight of the engine and transmission. Essentially all design processes begin this way. We *break into* the system at some level with a model simple enough that we can conceptualize the parts and their interactions, we make assumptions about the environment and operating conditions, and we begin designing (specifying the model structure, signal structure, objectives, and parameters). As the design proceeds, additional design problems emerge. Results from one portion of the design affect the design of other portions (by specifying inputs or parameters). Once the parameters of a model become *settled,* the model is refined; that is, it is expanded to give a more detailed picture of its operation. Refinement ceases when its cost does not justify the insight it provides about the performance of the system and subsystems.

The purpose of the modeling process is to predict system behavior accurately. In contrast, the purpose of the design process is to invent a system that has a specified behavior. Typically, one designs systems by creating variants of or analogies to previous systems with which one has experience.

Because experience with modeling and with analyzing models is a major source of insight into systems, it is common to find engineers designing system components that mimic closely the behaviors of ideal lumped elements. For example, door closers and automobile suspension systems are typically constructed of discrete springs and dampers. If such a discrete-component system becomes widely used, succeeding generations of the product gradually introduce nonlinearities or meshing of components, which produce incremental improvements in behavior or reductions in cost.

Models for Distributed Systems

We close this discussion of modeling and design by determining the level of detail that we must include in the lumped model of a system to correctly predict the behavior of that system. We find that the required level of detail depends on the abruptness of the source signal. It takes a more detailed model to correctly predict the behavior for a more abrupt source signal.

A homogeneous object that is driven at one end by a velocity source is shown in Figure 2.39(a). The object is compliant and has mass. Its spring and mass properties are intertwined—inseparably distributed along its length. Suppose we model the object by a single spring in series with a single mass, as shown in Figure 2.39(b). We use equations (2.8) and (2.25) to calculate the end-to-end stiffness k and mass m of the object. If the source gradually varies the position x_1 of the end of the spring in the model, the force produced in the spring is small, and the position x_2 of the mass in the model follows the

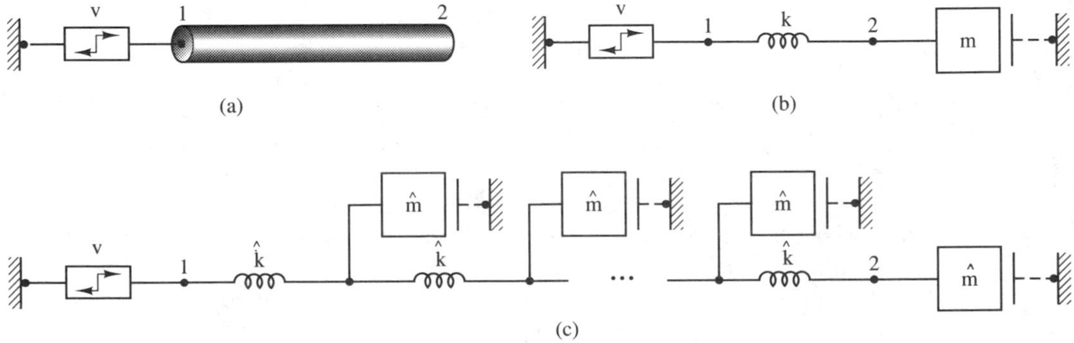

Figure 2.39 A distributed object. (a) A homogeneous object. (b) A simple model of the object. (c) A distributed model of the object.

source motion, as we would expect the physical object to do. Hence, this model is satisfactory for a slowly varying source signal.

The spring-mass model has limited responsiveness, however. If the source is abrupt, the mass cannot follow the source motion. Unlike the physical object, the spring-mass model oscillates at a natural frequency. In section 3.4, we show that the natural frequency of oscillation of the spring-mass model is $\omega_n = \sqrt{k/m}$. We can think of the half-period of the oscillation, $p_n/2 = \pi\sqrt{m/k}$, as the **response time** of the model. Source signals that vary rapidly compared with this response time initiate a wave traveling along the physical object, much like the wave we observe when we jerk the end of a clothesline. The simple spring-mass model cannot simulate a traveling wave.

How should we model the object if the source signal is very abrupt? In that case, we subdivide the object of Figure 2.39(a) into n segments of equal length and represent each segment by a spring-mass model. We then place the models of the segments end to end, as in Figure 2.39(c). We call this chain of submodels a **distributed model** of the object.

According to equation (2.8), the stiffness of each segment of the distributed model is nk. By equation (2.25), the mass of each segment is m/n. The response time (or natural half-period) of each segment model is shorter than the response time of the original spring-mass model by the factor n. Therefore, each segment of the distributed model is n times more responsive than is the original model. That is, the distributed model can respond correctly to signals that are n times more abrupt than the signals to which the original model can respond. The abrupt signals are passed from segment to segment to simulate a traveling wave.

If a sinusoidal velocity source oscillates the end of a homogeneous object at sufficiently high frequency, it initiates a sinusoidal traveling wave. The (spatial) wavelength of the traveling wave is $\lambda = p\nu$, where p is the period of the sinusoid and ν is the velocity of propagation of the wave. (The velocity of propagation is determined by the geometry and material of the object). It has been found by experience that **a distributed object must be subdivided into segments at least 10 times shorter than the wavelength of the propagating wave** in order for lumped models of the segments to represent the dynamic behaviors of those segments accurately. (It takes about 10 uniformly spaced samples per period to expose the shape of a sinusoidal traveling wave.)

What about abrupt signals that are not sinusoidal? Figure 2.40 compares a half-period of a sinusoid with a source signal that has about the same **rise time** or abruptness as the sinusoid. (The rise time need not be defined precisely for this comparison.) One measure of a system's responsiveness is the ability of the system to follow (keep up with)

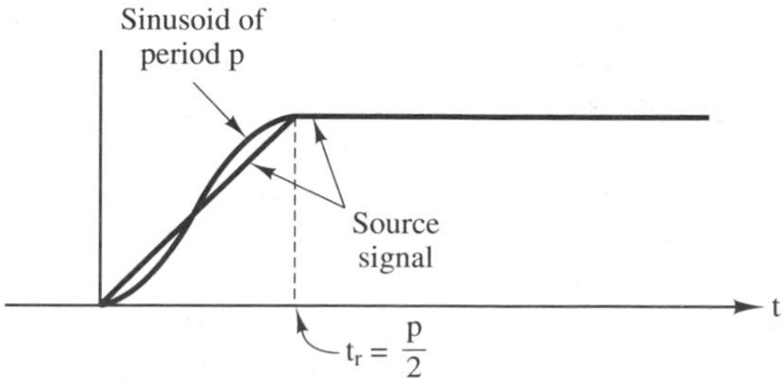

Figure 2.40 Abrupt signal and comparable sinusoid.

a source signal. We view the *stressfulness* of a source signal as the degree of responsiveness required of a system to follow that signal. The two signals in Figure 2.40 are equally stressful ($t_r = p/2$). Engineers often specify the stressfulness of a signal by stating the frequency of an equally stressful sinusoidal signal, even if the signal waveform is not at all similar to a sinusoid.

In order that a distributed model accurately mimic the response of the physical object, we make the physical length of each segment small enough to satisfy the condition

$$\text{segment length} \leq \frac{\lambda}{10} = \frac{p\nu}{10} = \frac{\nu t_r}{5} \tag{2.51}$$

The period p (or rise time t_r) is a measure of the abruptness of the source signal. The velocity of wave propagation ν is determined by the properties of the physical object.

For source signals that are sufficiently abrupt, any object acts in a distributed manner; that is, it acts as if its mass and compliance properties are distributed inseparably along the object. (Friction-induced damping can also be distributed along the object.) The more rapid the source signal, the more segments we must include in the distributed model of that object. If the distributed model satisfies the segment-length criterion given in equation (2.51), then the model behaves nearly the same as the physical object.

The segment-length criterion given in equation (2.51) applies unchanged to all physical phenomena—electrical, mechanical, hydraulic, etc. Example problem 2.7 examines a distributed model for a Slinky®, a wire-coil toy that exhibits wave motions even when the source signals are quite slow. Lumped models for other distributed systems are derived in example problems 3.3, 4.3, 5.5, and 6.2. The behaviors of distributed models are demonstrated in example problems 4.6, 4.7, 5.5, and 6.2.

Figure 4.20 shows the response of a distributed model (of an electrical coaxial cable) that satisfies the segment-length criterion. The slight oscillation in the response is caused by the use of a lumped model. The physical system does not exhibit this oscillation. Using fewer segments will significantly increase the size of the oscillation. Using more segments will decrease the size of the oscillations. However, it takes a great increase in the number of segments to reduce the oscillation significantly. Thus, the segment-length criterion provides an appropriate balance between the accuracy and complexity of the distributed model.

Example Problem 2.7: Lumped modeling of a Slinky—a distributed device.

A toy Slinky is a very soft coil spring (see Figure 2.41(a)). One end of the Slinky is fastened to a support, and the Slinky is allowed to hang vertically. Only vertical forces or motions are used to excite the Slinky. Measurements on a vertically hanging Slinky show that disturbances propagate through the whole length of the Slinky in about 0.5 s. If the top turn of the hanging Slinky is collapsed upward (raising the whole Slinky) and then released, the top turn drops to its original position in about 0.1 s.

(a) Develop a lumped model that can be used to predict the Slinky's vertical translational behavior. How detailed must the model be in order to provide a satisfactory representation of the Slinky?

(b) The wire coil of the Slinky contains 86 turns. The measured total mass of the Slinky is 0.25 kg. The total displacement of the free end of the Slinky from its completely collapsed

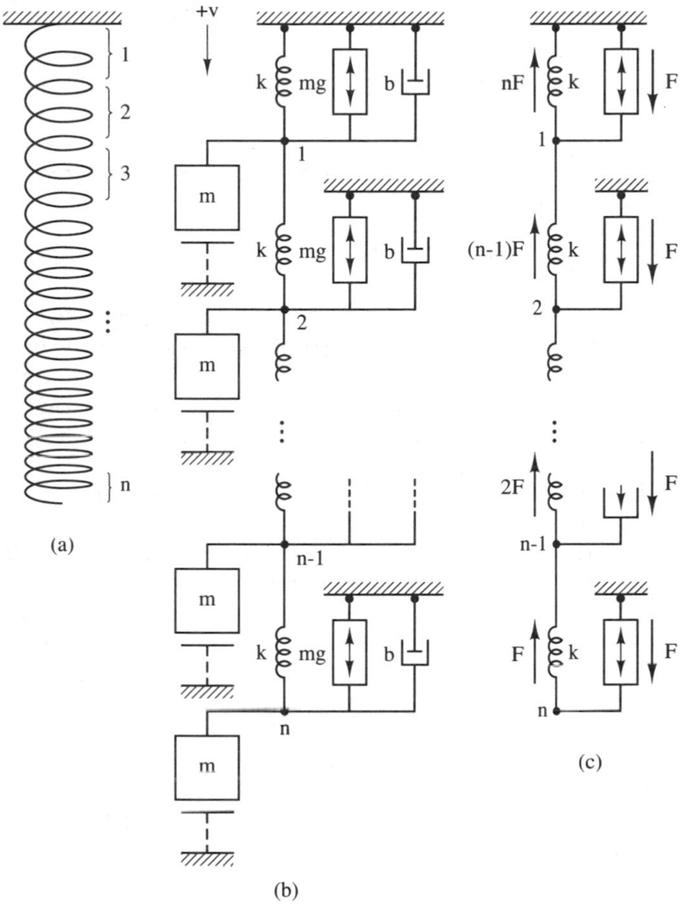

Figure 2.41 Models of a Slinky®. (a) A hanging Slinky. (b) A distributed model of the Slinky. (c) A static model of the hanging Slinky.

position when it is hanging statically is 1.5 m. Use these static measurements to determine the parameters of the model.

Solution: (a) Partition the Slinky into n identical segments. Number the segments as illustrated in Figure 2.41(a). Each segment has mass m and stiffness k. During motion, each segment meets air resistance that is nearly proportional to its velocity. Let b be the friction constant of the segment. Each segment is acted on by a gravitational force of magnitude mg, where g is the gravitational acceleration. In addition, each segment is pulled by the weight of succeeding segments.

A lumped model of the Slinky is shown in Fig. 2.41(b). The physical system has no mass at its end, as does the lumped model. By assigning the mass and gravity force of the ith segment to the ith node, we exaggerate slightly the stretching of each segment. (We could, instead, assign the ith mass and gravity force to the $(i-1)$st node; that assignment would underestimate slightly the stretch of each segment. Both approximations give nearly the same predicted behavior if we use enough segments.) Most of the elements must be referenced to ground. Only the springs are connected between nodes. The iterative series-parallel structure of the model produces traveling-wave behavior. To introduce a specific velocity or force pattern at some point of the Slinky, we attach a suitable source element to the corresponding point of the model.

The model can be made more accurate by increasing the number of segments. However, we must limit the number of elements to keep the number of variables and equations small enough to manage. One way to determine the minimum acceptable number of segments is to examine the behavior of the model for increasing numbers of segments and to stop increasing the number once the behavior ceases to change *significantly.* Or we can pick a segment size that satisfies equation (2.51).

According to the given measurements, the velocity of wave travel along the Slinky is $\nu \approx$ (1 Slinky length)/(0.5 s) = (86 turns)/(0.5 s) = 172 turns/s.[10] The 0.1 s time of drop of the top link corresponds to the rise time t_r of Figure 2.40. According to equation (2.51), if we wish to simulate accurately the travel of a signal that abrupt, we need enough segments to make each segment shorter than $\nu t_r/5$ = (172 turns/s) (0.1 s)/5 = 3.44 turns. For an 86-turn Slinky, we require $n \geq 25$. (Problem 4.17 in chapter 4 uses a network simulation program to examine the response of a 10-segment model to signals with various degrees of abruptness.)

(b) Since the total mass of the Slinky is 0.25 kg, the mass of a single segment is $m = 0.25/n$ kg, and the force source associated with a single segment has the value $F = mg = (9.8\ \text{m/s}^2)(0.25/n\ \text{kg}) = 2.45/n$ N. For $n = 25$, the values are $m = 0.01$ kg and $F = 0.098$ N.

When the Slinky is hanging statically, the masses and dampers of the model have no effect, and the model reduces to the chain of identical springs and forces in Figure 2.41(c). The tensile force in the bottom lumped spring is F, and the drop in displacement across the spring is $x_{n,n-1} = F/k$. Similarly, the force in the next higher spring is 2F, and its displacement is $x_{n-1,n-2} = 2F/k$. The force in the top spring is nF, and its displacement is $x_1 = nF/k$. The total displacement of the bottom of the Slinky model is

$$\begin{aligned} x_n &= x_{n,n-1} + x_{n-1,n-2} + \cdots + x_{21} + x_1 \\ &= [1 + 2 + \cdots + (n-1) + n](F/k) \\ &= [n(n+1)/2](F/k) \end{aligned} \tag{2.52}$$

The measured displacement of the bottom of the actual Slinky while it is hanging statically is $x_n = 1.5$ m. We substitute this measured value of x_n and the value found for F into equation (2.52) to obtain

$$k = \frac{\dfrac{n(n+1)}{2}\,\dfrac{2.45}{n}\text{N}}{1.5\ \text{m}} = 0.82(n+1)\ \text{N/m} \tag{2.53}$$

Thus, increasing the number of segments in the model decreases the mass and gravitational force per segment, but increases the segment stiffness. For $n = 25$, the segment stiffness is $k = 21.3$ N/m.

[10]The wave traverses an equal number of turns per time unit. We measure wave velocity in turns per second rather than meters per second because gravity causes the turns to be stretched unevenly, as shown in Figure 2.41(a). Velocities of traveling waves are examined in equation (6.96) and Table 6.3.

We cannot determine the friction constant b from the static measurements. Its determination requires the measurement of some dynamic quantity, such as the time required for an oscillating Slinky to *quiet down*. Problem 4.18 in chapter 4 uses a network simulation program together with such a measurement to determine b. The damping of a Slinky segment is small enough that, for most purposes, the dampers can be removed from the model. Problems 4.18 and 4.19 validate the Slinky model experimentally.

Exercise 2.10: Show that the end-to-end stiffness of the Slinky model is $k_T = k/n = 0.82(n + 1)/n$ N/m. Yet the total stiffness of the physical Slinky does not depend on n. Explain this discrepancy.

2.4 SUMMARY

The primary subjects of this chapter are a procedure for representing a physical system by a network of lumped elements and two procedures for writing the system equations that describe the lumped network. Section 2.2 introduces the line-graph representation of networks and defines the mathematical concepts of flow and potential in line-graph terminology. Yet, we often *apply* the concepts of flow and potential (or potential difference) directly to the lumped networks. Flows obey the node law (equation (2.27)) at each network node. Potential differences obey the path law (equation (2.28)) around every closed loop in the network.

We orient arbitrarily a flow variable in each network element. The equation that describes a passive element has the form $A_{ij} = zT$, where A_{ij} is the potential drop across the element in the direction from node i to node j and T is the flow through the element from node i to node j. Then, the flow through each passive element is positive in the direction of actual potential drop.

The node method for generating system equations balances the flows at all but one of the network nodes. In this method, the element equations (expressed in flow format) are incorporated directly into the node equations. The loop method for generating system equations equates the potential drop around each network loop to zero. In this method, the element equations (expressed in terms of circulating flows, in potential-difference format) are incorporated directly into the loop equations.

Section 2.1 develops a framework for representing mechanical objects mathematically. We express the system equations for a translational lumped network in terms of the *central* motion variables—the scalar forces in the lumped elements and the velocities at the nodes. All motions in a translational mechanical network are referenced to a stationary point (or ground).

The translational velocity v acts as a potential. It has different values at different nodes of the network. The signs of the node velocity variables v_i are determined by the arbitrarily chosen +v direction for the network. The scalar forces in the lumped elements of a translational network act as flows. If the positive-direction arrow for the scalar-force variable F_n is oriented in the +v direction, then F_n represents compressive force; otherwise it represents tensile force. Reversal of the +v direction reverses the signs of all velocities and flows, but does not change the physical interpretations (compressive or tensile) of the flows and the relative motions.

Section 2.1 also shows how to generate lumped models of translational mechanical systems. A *lumped model* is a network—an interconnection of lumped elements. A two-terminal *lumped element* is a pictorial symbol that represents a specific mathematical relation between a potential difference (or velocity drop) and a flow (or scalar force).

The two-terminal lumped elements can be grouped into two categories: active and passive. Active elements, that is, sources, are able to export power. The power delivered *by* a translational lumped element is the product Fv_{ij} of the flow through the element and the velocity *rise* across the element in the direction of flow. Passive elements cannot export power. Therefore, passive elements never show a rise in potential in the direction of flow.

An *ideal* independent source represents a specified potential-difference waveform or a specified flow waveform. An *ideal passive* element represents a constant proportionality (a linear relation) between a potential difference and a flow. For translational systems, the ideal passive elements are the ideal spring, the ideal damper, and the ideal mass (represented by $v_{ij} = (s/k)F$, $v_{ij} = (1/b)F$, and $v_{ig} = (1/sm)F$, respectively). We view mass as a two-terminal element with one terminal connected to ground. The active elements are the compressive-force source (for which the flow F(t) in the +v direction is specified) and the velocity source (for which the potential rise v(t) in the +v direction is specified).

The linearity assumption is usually excellent for the mass. For the spring, it is a good assumption if the spring displacement is not too large. Linearity can be a poor assumption for the damper. Rather, damper force is often either constant (coulomb friction) or a quadratic function of velocity (see section 5.1).

2.5 MORE EXAMPLE PROBLEMS

Example Problem 2.8: A system with masses in series.

The lumped network in Figure 2.42(a) represents an astronaut on a space station, maneuvering the relative position of a satellite.

(a) Draw the line graph that represents the lumped network.

(b) Write the system equations that describe the behavior of the network.

(c) Solve the system equations in terms of the applied force signal F(t).

(d) Represent the pair of masses as a single equivalent mass; that is, determine the effective mass felt by the astronaut.

Solution: (a) The line graph is shown in Figure 2.42(b). Both masses are referenced to a fixed ground point.

(b) The node equations at nodes 1 and 2 require that $F_1 = F$ and $F_2 = F$. Substitution of the mathematical relations for the masses produces $-m_1\dot{v}_1 = F$ and $m_2\dot{v}_2 = F$ as the pair of system equations.

(c) The pair of differential equations is easily solvable because it involves only one type of ideal lumped element, mass. No interactions among derivatives of different orders appear in the equations. The solution is $\dot{v}_1 = -F/m_1$ and $\dot{v}_2 = F/m_2$; or $v_1(t) = v_1(0) - \int_0^t (F/m_1)\,dt$ and $v_2(t) = v_2(0) + \int_0^t (F/m_2)\,dt$.

(d) From the viewpoint of the astronaut, the pair of masses appears as a single two-terminal element that is characterized by the relation between F and v_{21}. We seek m_{eff} such that $F = m_{eff}\dot{v}_{21}$ (Figure 2.42(c)). The relationship can be obtained from the foregoing pair of solutions:

$$\dot{v}_{21} = \dot{v}_2 - \dot{v}_1 = \frac{F}{m_2} + \frac{F}{m_1} = \frac{F}{m_{eff}}$$

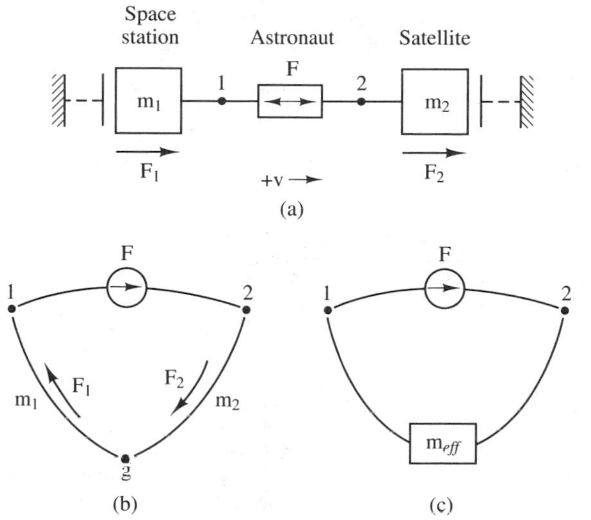

Figure 2.42 Masses connected in series. (a) Lumped model. (b) Line-graph representation. (c) Single-mass equivalent representation.

Therefore,

$$\frac{1}{m_{eff}} = \frac{1}{m_1} + \frac{1}{m_2} \tag{2.54}$$

or

$$m_{eff} = \frac{m_1 m_2}{m_1 + m_2} \tag{2.55}$$

It is apparent from the line graph that the two masses are connected in series. Combining masses in series reduces the effective mass felt by the source, as described by equation (2.55). The effective-mass model masks the motions relative to the fixed ground.

If we view the dynamic interaction from the viewpoint of the astronaut fastened to m_1, rather than from the *nonaccelerating* ground, then v_{21} is the velocity of point 2 (the satellite). One might expect the *relative* motion of the satellite to be determined by its mass according to $F = m_2\dot{v}_{21}$. To the contrary, the correct relation is $F = m_{eff}\dot{v}_{21}$. If $m_1 \gg m_2$, on the other hand, then $|\dot{v}_1| = |-(m_2/m_1)\dot{v}_2| \ll |\dot{v}_2|$, $m_{eff} \approx m_2$, and $F \approx m_2\dot{v}_{21} \approx m_2\dot{v}_2$. That is, if m_1 is large relative to m_2, we can treat m_1 as nonaccelerating.

The concept of an *effective lumped element* can be extended easily to any single type of element. Problem 2.17 at the end of the chapter derives equivalent-spring formulas for springs in series and in parallel. Similar equivalent elements can be derived for dampers. The concept of an equivalent element is developed in detail in section 4.2.

Example Problem 2.9: Frequency dependence of models.
A sinusoidal velocity source is applied to a spring-mass combination in Figure 2.43(a).
(a) Write the equations that describe the behavior of the system.
(b) Determine, in terms of the mass m and stiffness k, the range of values of the source frequency ω for which the spring-mass combination acts like an ideal mass; also, find the range for which it acts like an ideal spring.

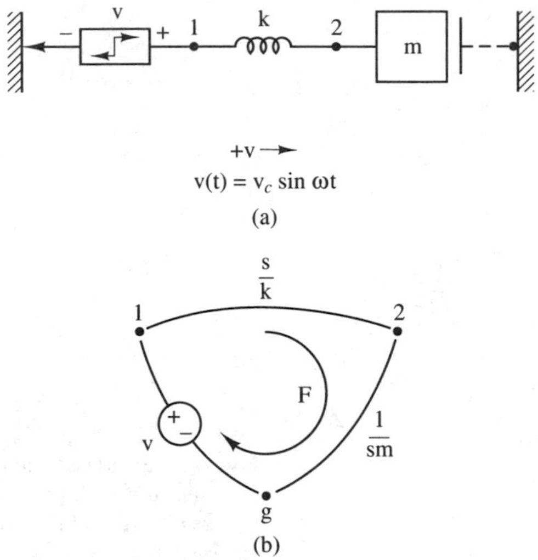

Figure 2.43 Sinusoidal motion of a spring-mass combination. (a) Lumped model. (b) Line-graph representation.

Solution: (a) The line graph is shown in Figure 2.43(b). We seek the conditions under which the spring-mass combination appears *to the source* as essentially a single spring or mass. We relate the source-node velocity v_1 to the compressive force F exerted by the source on the attached network. From the network or the line graph, we find the single loop equation

$$v_{1g} = v_{12} + v_{2g} \quad \text{or} \quad v_1 = \frac{sF}{k} + \frac{F}{sm} \tag{2.56}$$

We then differentiate the system equation (multiply by s) to obtain

$$sv_1 = \frac{s^2F}{k} + \frac{F}{m} \tag{2.57}$$

(b) For the spring-mass combination to act like a mass, it must have an equation of the form $F = smv_1$, or $sv_1 = F/m$. For it to act like a spring, it must have an equation of the form $sF = kv_1$, or $sv_1 = s^2F/k$. The source requires $v_1(t) = v_c \sin \omega t$. The derivative of v_1 is $sv_1(t) = \omega v_c \cos \omega t$. Because the source signal is sinusoidal, the force F must also be sinusoidal, with the same frequency ω. Therefore, $s^2F = -\omega^2F$, and two differentiations of a sinusoid produces the multiplier $-\omega^2$. Then the system equation can be written as

$$sv_1 = \frac{-\omega^2F}{k} + \frac{F}{m}$$

$$\approx \frac{F}{m} \quad \text{if } \omega \ll \sqrt{k/m} \text{ (essentially the mass alone)} \tag{2.58}$$

$$\approx \frac{-\omega^2F}{k} \quad \text{if } \omega \gg \sqrt{k/m} \text{ (essentially the spring alone)} \tag{2.59}$$

We show in section 3.4 that a spring-mass combination resonates (oscillates naturally) at the frequency $\omega = \sqrt{k/m}$. For velocity-source patterns that vary *slowly* compared with the natural frequency, the system acts nearly like the mass alone; the spring is not compressed much. For source patterns that vary *rapidly* compared with the natural frequency, the system acts nearly like the spring alone; the mass remains nearly stationary during compression and stretching of the spring. For intermediate rates of variation, both the mass and the spring contribute to the system's behavior.

Exercise 2.11: Assume that the solution to the system equation (2.57) has the form $F = A\cos\omega t + B\sin\omega t$. Show, by substituting F into equation (2.57), that for $\omega \neq \sqrt{k/m}$,

$$A = \frac{\omega k m v_c}{k - \omega^2 m} \quad \text{and} \quad B = 0\,.$$

(The force amplitude would have to be infinite to sustain the velocity pattern at the resonant frequency.)

Example Problem 2.10: A system with coulomb friction.
A shuffleboard player accelerates the playing disk (of mass m) uniformly from rest to the velocity v_c in 1 second and then breaks contact with the disk. The friction characteristic of the disk sliding on the concrete court is shown in Figure 2.44(a).
(a) Construct a lumped model of the system.
(b) Sketch the source signal that drives the lumped model.
(c) Derive and solve the equations that describe the disk's velocity pattern. Sketch the velocity pattern. Determine the total distance traveled by the disk.
Solution: (a) A lumped model of the system is shown in Figure 2.44(b). The damper is nonlinear. The relation between the velocity difference across the damper and the compressive force in the damper is $F_1 = b_{nl}(v_1) = F_c\text{sign}(v_1)$. That is, the friction force acting on the point 1 is constant in magnitude, but always resists the motion. Another lumped model, identical in behavior, is shown in Figure 2.44(c). In this alternative model, the coulomb friction is modeled by the nonlinear force source $F_{nl} = F_c\,\text{sign}(v_1)$, which is *dependent* on the velocity v_1, rather than by a damper. We represent this force source by the diamond-shaped dependent-source symbol. Neither the dependent force source nor the nonlinear damper are in effect when $v_1 = 0$. Line graphs corresponding to the two lumped models are shown in Figures 2.44(d) and (e). Both line graphs show a switch in the velocity source branch. This switch disconnects the source at the end of the initial 1-second interval.
(b) The velocity function v that drives the network is shown in Figure 2.44(f). It is defined only during the initial 1-second interval.
(c) The lumped network can be analyzed in a piecewise-linear manner. That is, the system equation is linear in each of two successive time invervals. The velocity at the end of the first interval equals the velocity at the beginning of the second interval. The velocity of the disk during the initial 1-second interval (with the switch closed) is $v_1(t) = v(t) = v_c t$, a known function. The differential equation that describes v_1 after $t = 1$ s is determined by the node equation for node 1, with the switch open. It is

$$m\dot{v}_1 + F_c\text{sign}(v_1) = 0 \qquad (2.60)$$

From experience, we know that $v_1 > 0$ during the remainder of the motion. Therefore, $\text{sign}(v_1) = 1$ and we represent the nonlinear friction term by the constant F_c. The differential equation is

$$\dot{v}_1 = -\frac{F_c}{m} \qquad (2.61)$$

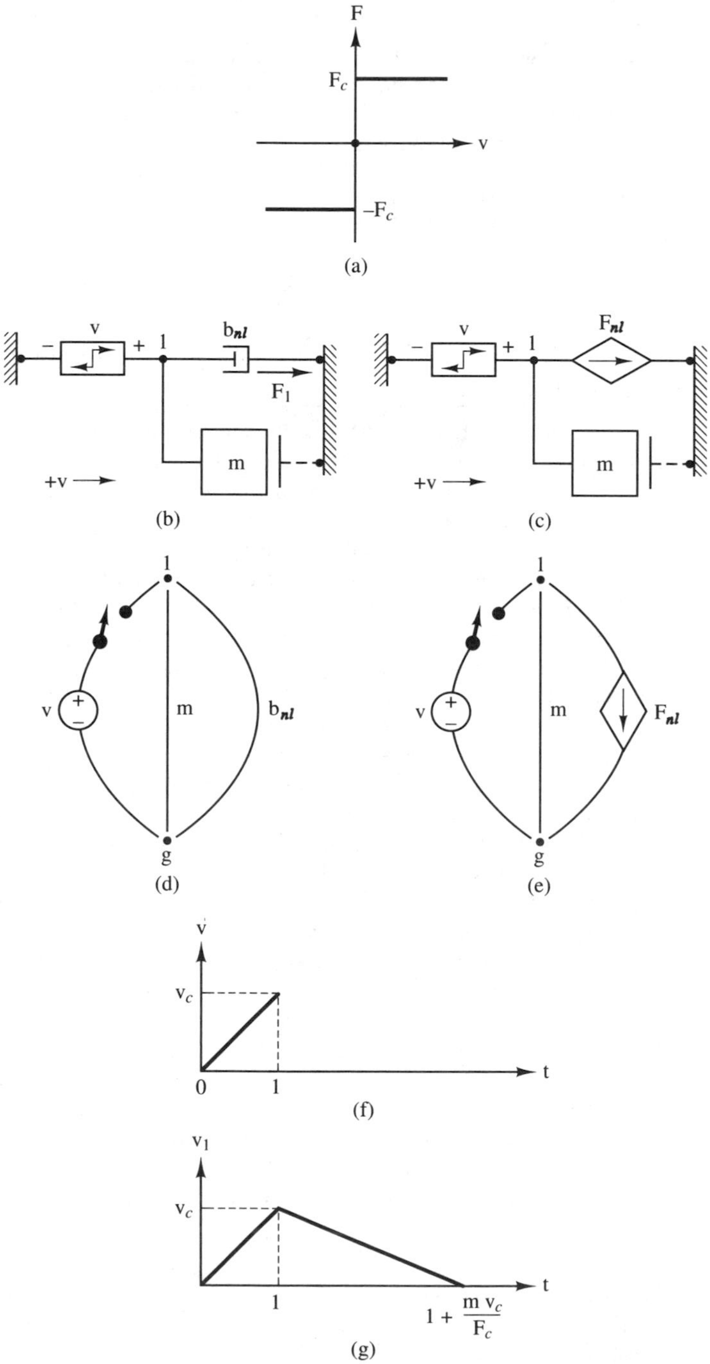

Figure 2.44 Switched source and coulomb friction. (a) Friction characteristic. (b) Lumped model. (c) Alternative lumped model. (d) Line graph for (b). (e) Line-graph for (c). (f) Velocity-source waveform. (g) Disk velocity pattern.

We integrate both sides of equation (2.61) from 1 to t to obtain

$$v_1(t) - v_1(1) = -\int_1^t \frac{F_c}{m}\,dt \tag{2.62}$$

or

$$v_1(t) = v_c - (t - 1)\frac{F_c}{m} \quad \text{for } t \geq 1 \tag{2.63}$$

This linearly decreasing velocity applies only as long as the velocity is nonnegative. Once the disk comes to a stop (at $t = 1 + mv_c/F_c$), the mass has no more kinetic energy, the nonlinear coulomb-friction force function no longer exerts a force on the system, and the disk remains stationary. The complete velocity pattern for the disk is shown in Figure 2.44(g). The total distance traversed by the disk is the integral of the function shown, that is, $(1 + mv_c/F_c)v_c/2$.

2.6 PROBLEMS

2.1 For one or more of the lumped models of Figure P2.1,
- **(a)** Define the +v direction and the through and across variables (signals).
- **(b)** Find a set of equations that describes the lumped elements and balances the scalar forces at the nodes.
- **(c)** Determine the power that the source delivers to the network, in terms of the source signal and the element parameters.

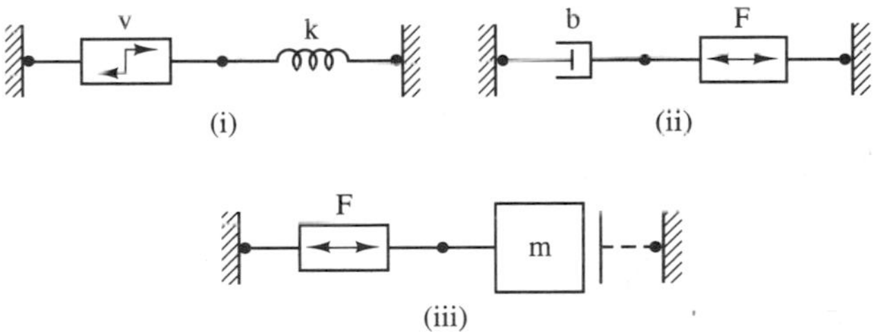

Figure P2.1.

2.2 For one or more of the lumped models of Figure P2.2,
- **(a)** Define the +v direction, the flow variables, and the node potentials.
- **(b)** Find a set of equations that describes the lumped elements and balances the scalar forces at the nodes.

2.3 A simple single-leaf automobile spring is shown in Figure P2.3. The spring is supported rigidly on the axle. Consequently, it can be viewed as a pair of cantilever springs acting in parallel (see Figure 2.7). The spring must carry the 2,200-N weight supported by one rear wheel of an automobile. Young's modulus for steel is E = 193 GPa. Select the length, width, and thickness of the spring to support the automobile appropriately under the condition that the automobile is parked. The solution is not unique. Some *arbitrary* choices must be made in order to design the spring. It would be appropriate to look at the underside of a few automobiles to assist in making these choices.

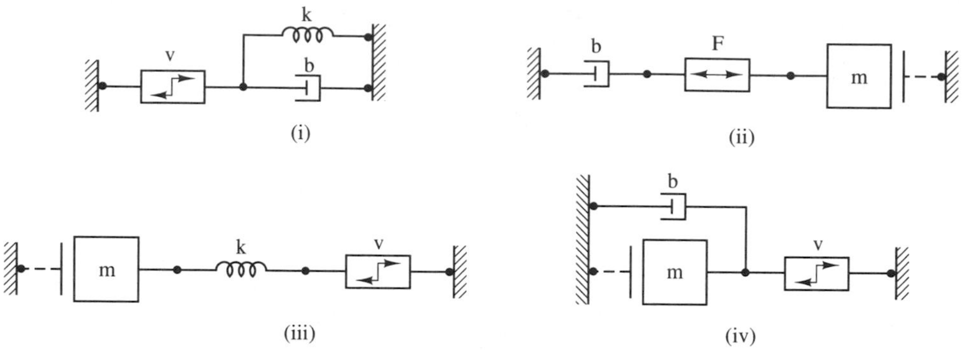

Figure P2.2.

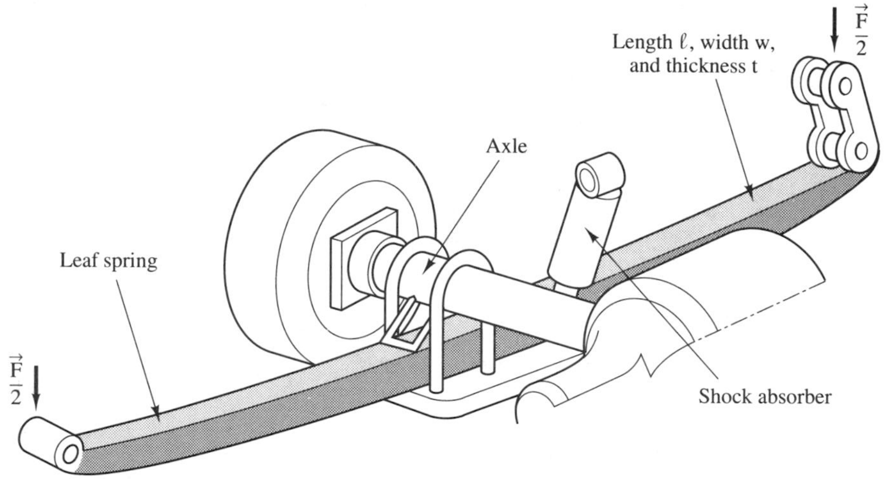

Figure P2.3.

2.4 A particular commercial spring scale is designed to measure small forces. The scale is marked in grams. The distance between the 0-g and 200-g marks on the scale is 45 mm. The scale was tested by suspending a number of precisely calibrated masses. The masses and the corresponding scale readings are as follows:

Calibrated mass (g)	25	50	75	100	125	150	175	200
Scale reading (g)	22	53	69	92	115	141	169	194

(a) Draw the straight-line relation between force (in g) and displacement that corresponds to the marks on the scale. Find the corresponding stiffness k required of the spring in the scale, in g/cm and in N/m, for the scale to produce perfect measurements.

(b) Superimpose the calibration data on the plot from (a). That is, for each scale reading (in g), use the straight line from (a) to find the corresponding scale displacement (in cm).

Then plot the calibrated mass (in g) at that displacement. Use the plotted data to determine a more accurate value of k for the actual spring in the scale.

(c) The largest mass error found in part (b) serves as a bound on the errors in forces measured with the scale. Express that error bound as a percent of full scale, that is, as a percent of the largest force (in g) marked on the scale.

2.5 A steel shaft in a certain system executes a 1-m stroke at a velocity of 10 m/s and then a return stroke at 1 m/s. The cycle is repeated continuously. (Assume that the reversals of motion are instantaneous.) The shaft is supported by two bearings that are lubricated by oil under pressure, as shown in Figure P2.5. The escaping oil is trapped and circulated through a cooler in order to prevent overheating of the bearing. The shaft diameter is 6 cm. The length of each bearing is 4 cm. The shaft-to-bearing clearance is 0.02 mm. The oil viscosity at the nominal operating temperature is $\mu = 0.1$ N·s/m^2.

(a) Determine the friction constant b associated with each bearing.

(b) Determine the total rate of heat generation in the pair of bearings. That is, determine the rate of oil cooling required.

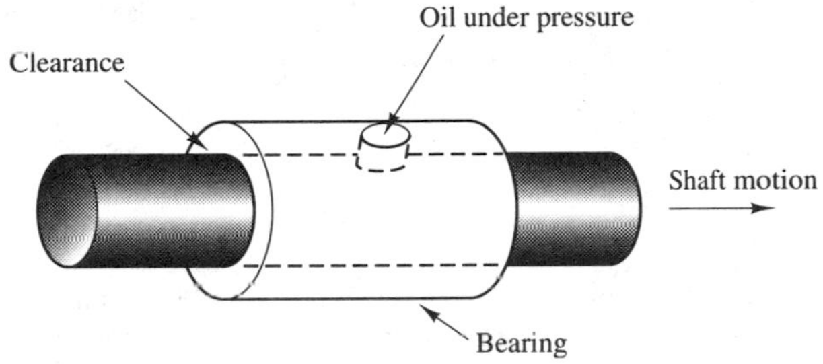

Figure P2.5.

2.6 For one of the lumped networks of Figure P2.2,

(a) Specify the +v direction and the through and across variables.

(b) Draw the line graph that represents the lumped network.

(c) Find a set of system equations from the line graph.

2.7 Derive a simple lumped model that describes the main features of one of the following situations or systems. Include a model for the source that energizes the system. State your assumptions.

(i) Sawing a board with a handsaw
(ii) Mopping the floor
(iii) Swimming
(iv) Pulling a water skier with a boat
(v) Sanding a tabletop
(vi) Painting a wall
(vii) Towing a barge with a tugboat
(viii) A tug of war
(ix) An automobile piston and cylinder
(x) A speedboat riding up a river
(xi) Pulling a sled on snow
(xii) A car climbing a highway slope

2.8 A diagram of a concrete pump is shown in Figure P2.8. Oil enters port 1 at high pressure and leaves port 2 at low pressure, thereby exerting a force on the power piston. That force moves the push rod, forcing the concrete into the delivery tube. Water is drawn into the lubrication chamber during delivery of the concrete, wetting the cylinder before it is refilled with concrete. During the return stroke, the oil pressure ratio at the two ports is reversed, the piston retracts, the flapper at the concrete hopper shifts position to open the hopper and close the delivery tube, concrete is sucked into the cylinder, and the water is forced out of the lubrication chamber. Derive a lumped translational model of the system. Include a model of the force signal that drives the system. Note that the return force is less than the force during the power stroke because the push rod reduces the area on the right side of the power piston. Assume that the steel cylinder and delivery tube are rigid and that all the fluids are incompressible.

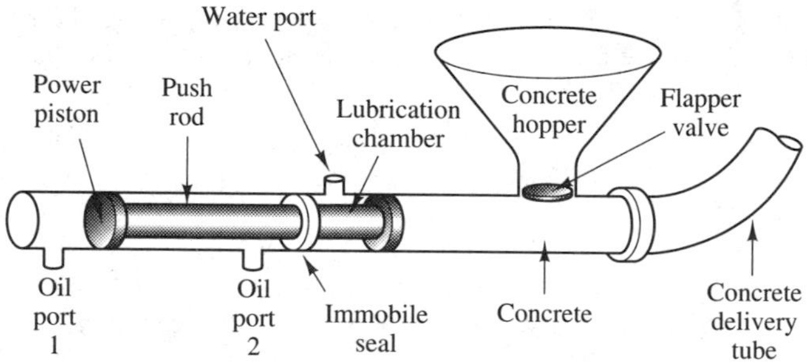

Figure P2.8.

2.9 A diagram of a blowgun is shown in Figure P2.9. The hunter expels a burst of air into the end of the gun. The compressed air exerts a force on the fiber plug that carries the dart.

(a) Make a lumped model of the blowgun and dart.

(b) Estimate values of the parameters of the model and of the source signal.

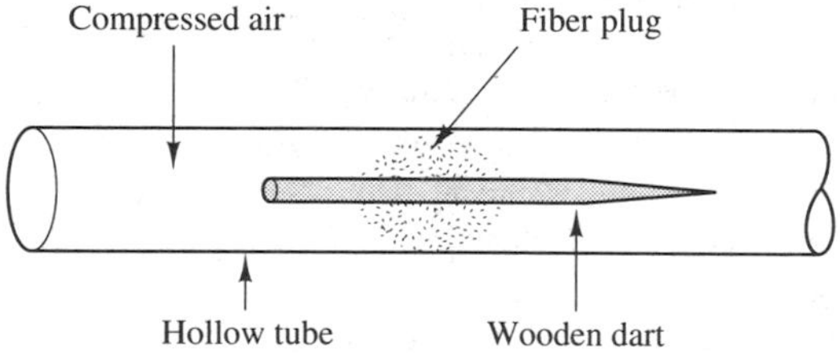

Figure P2.9.

2.10 A person paddles a canoe on a still lake.

(a) Derive a lumped translational model of this dynamic system that contains only two distinguishable velocities (other than ground) and that is applicable during the *power* stroke.

(b) Derive a model that is applicable during the *return* stroke.

(c) Propose appropriate source signals for both parts (a) and (b).

(d) Estimate appropriate values of all parameters in the model.

2.11 A launching system for a model airplane is shown in Figure P2.11. Draw a lumped model that captures the essential motions and forces in the system. State any assumptions you make. Sketch the waveform of the source you use in your model.

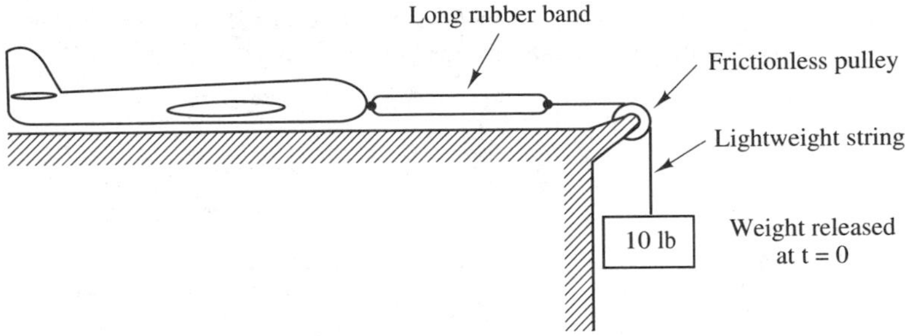

Figure P2.11.

2.12 A barge of mass m_1 floats unhindered down a river. The barge passes a tugboat of mass m_2 that maintains a stationary position on the river. The relative velocity at passing is the river velocity v. As the boats pass, the deckhands connect the boats by means of a long cable. The tugboat maintains its stationary position until the slack in the cable is taken up. Then the tug accelerates its engine and applies a constant tensile force F_c to the cable in an attempt to stop the barge.

(a) Construct a lumped model of the tug-barge interaction. Hint: Treat the river, rather than the earth, as the motion reference (ground).

(b) Draw a line graph that represents the lumped model.

(c) Write a set of system equations that determines the behavior of the barge and the tug.

2.13 A push rod of length ℓ and cross-sectional area A is used to transmit longitudinal forces and motions in a particular system. The rod can be represented by a chain of identical spring-mass pairs, with each pair representing a short segment of the rod. The accuracy of the model increases as the number of segments n is increased. Suppose that the material the rod is made of has density ρ and Young's modulus E.

(a) Sketch the lumped model of the rod. Include the source signal and load reaction at the rod ends. Find the segment stiffness k and segment mass m in terms of the parameters of the material.

(b) State how you would measure k and m.

(c) Draw a line-graph representation of the lumped model found in part (a).

(d) Define a set of potentials and flows for the model. Write the equations that describe the behavior of the ith section of the model in terms of the variables in the neighboring sections.

2.14 A diving board can be viewed as a chain of cantilever springs. (If the board were tapered, the cantilever mass and cantilever stiffness would be different for each segment of the chain.)

(a) Assume a uniform board. Draw and label a lumped model that accounts for the vertical behavior of a single cantilever segment of the chain. Include a damper in the segment to account for the fact that the board does not oscillate indefinitely. State your assumptions.

(b) Draw a line-graph representation of the whole lumped model.

(c) Define a set of potentials and flows for the model. Write the equations that describe the behavior of the ith section of the model in terms of the variables in the neighboring sections.

(d) How would you measure the parameters of the model?

2.15 For one of the networks of Figure P2.15,

(a) Draw the corresponding line graph and define a set of branch flows.

(b) Write a set of system equations.

(c) Solve the system equations for the node velocities and branch flows in terms of the value of the source. Interpret the solution physically.

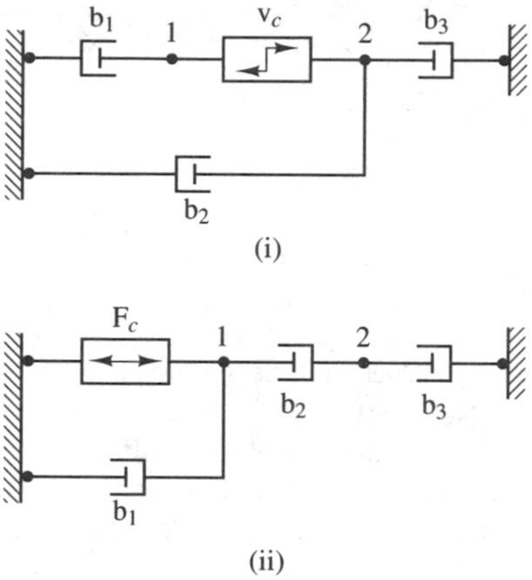

Figure P2.15.

2.16 The force vector $\vec{F}_1$ of Figure P2.16 is a sinusoidal vibration force external to the system. Gravity also acts on the two masses.

(a) Construct a lumped model of the system.

(b) Draw a line-graph representation of the lumped model found in part (a).

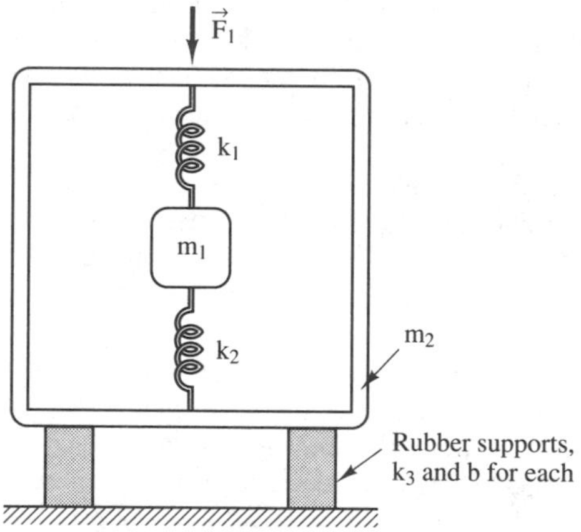

Figure P2.16.

2.17 (a) Let k_a and k_b represent the stiffnesses of two springs that are connected in *parallel* (with end-node motions in common). Show that the parallel combination acts like a single spring with stiffness $k_{par} = k_a + k_b$. That is, the stiffnesses add.

(b) Let k_a and k_b represent the stiffnesses of two springs that are connected in *series* (with scalar forces in common). Show that the series combination acts like a single spring with stiffness $k_{ser} = k_a k_b/(k_a + k_b)$. That is, $1/k_{ser} = 1/k_a + 1/k_b$ (the compliances add).

2.18 A combination of springs is arranged to permit one-dimensional translation, as shown in Figure P2.18. Let F_1, F_2, and F_3 denote the compressive forces in springs 1, 2, and 3, respectively.

(a) Use the equivalent-spring concepts of problem 2.17 to find the effective stiffness k_e of the combination, so that $F = k_e x_{1g}$.

(b) Find F_1, F_2, F_3, x_1, and x_2 in terms of the applied compressive force F and the spring stiffnesses. Use equivalent-spring concepts to avoid solving simultaneous equations. That is, find the values of the variables by using one equation at a time.

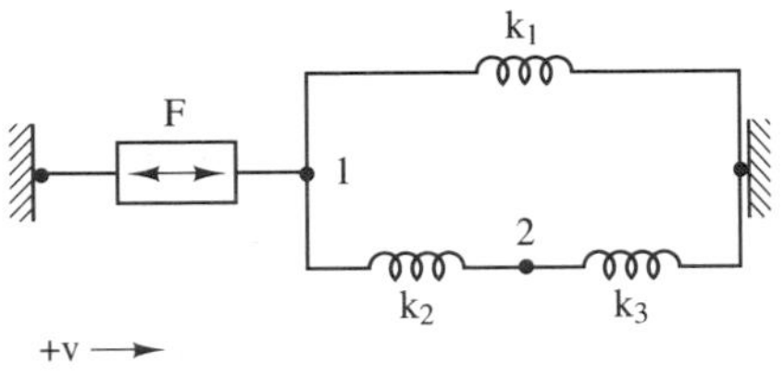

Figure P2.18.

2.19 Suppose the satellite-maneuvering configuration of example problem 2.8 is changed to incorporate a flexible handling device between the astronaut and the satellite, as shown in Figure P2.19.

(a) Write the system equations that describe the network.

(b) Show that the two masses in the system can be replaced by a single effective mass, despite the intervening spring. Construct a modified lumped network that uses the single effective mass instead of the individual masses.

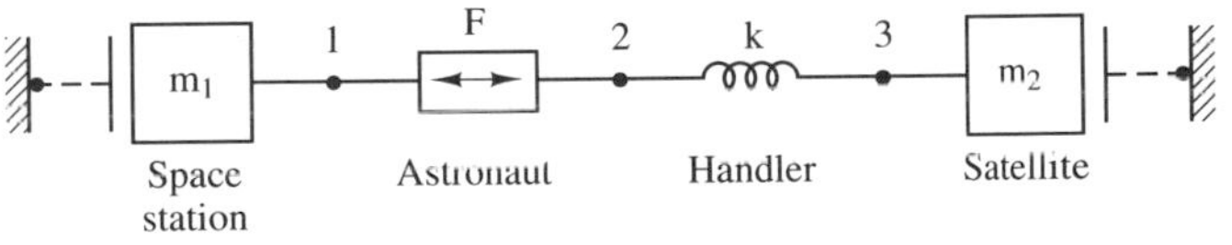

Figure P2.19.

2.20 **(a)** Define flows for the static lumped network of Figure P2.20. Solve for the displacements at the nodes and the scalar forces in the springs. Take advantage of the concept of equivalent springs (see problem 2.17) if you wish.

(b) Suppose the springs are unenergized initially. The sources are applied suddenly at $t = 0$. Determine, as a function of time t, the energy stored in each spring and the power output of each source.

(c) Interpret your results physically. Determine the practical implications of your solution as t becomes large. What are the limitations of the network as a model of a physical system?

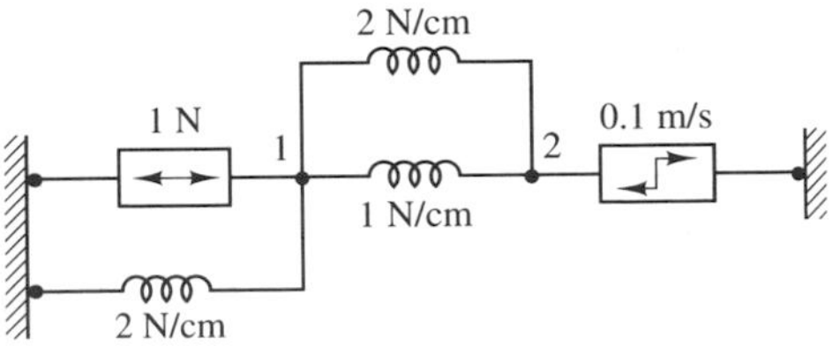

Figure P2.20.

2.21 **(a)** Draw the line graph for the lumped network of Figure P2.21, and solve for the velocities at the nodes and the scalar forces in the dampers.
(b) Determine the power dissipated in each damper and the power output of each source.
(c) Interpret your results physically. Determine the practical implications of your solution as t becomes large. What are the limitations of the network as a model of a physical system?

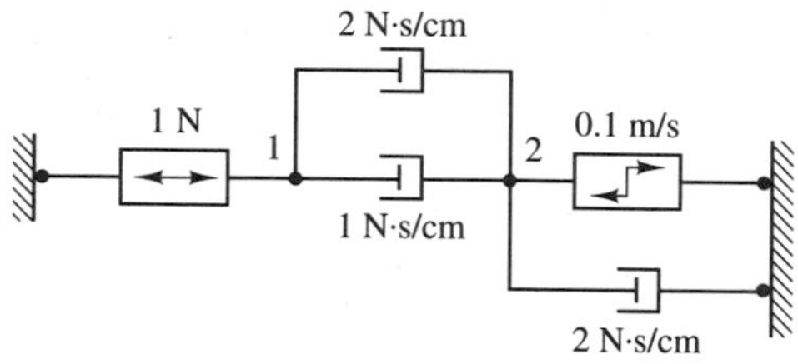

Figure P2.21.

2.22 A lumped network of dampers is shown in Figure P2.22. Assume that the source velocity v and the friction constants are known.
(a) Use the node method to find a set of equations that represents the network.
(b) Solve the equations found in part (a) for the node velocities. Determine the compressive forces in the dampers and in the source.
(c) Interpret the solution physically.

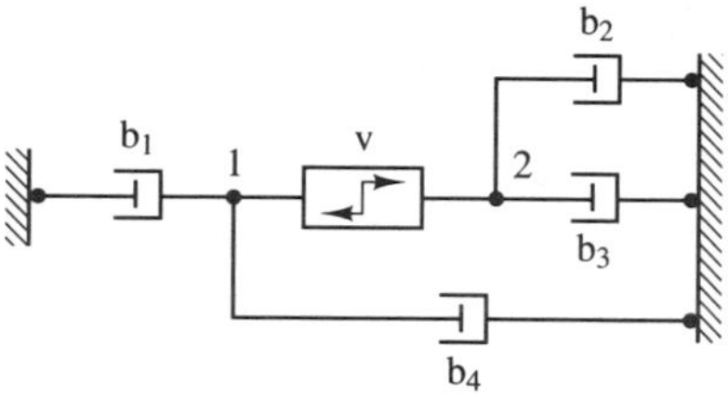

Figure P2.22.

2.23 Assume that the source velocity and the friction constants of Figure P2.22 are known.
(a) Use the loop method to find a set of equations that represents the network.
(b) Solve the equations found in part (a) for the loop flows. Determine the compressive forces in the dampers and the source and the velocity of each node.
(c) Interpret the solution physically.

2.24 A lumped network of springs is shown in Figure P2.24. Assume that the applied force and the spring stiffnesses are known.
(a) Use the node method to find a set of equations that represents the network.
(b) Solve the equations found in part (a) for the node displacements. Determine the compressive force in each spring.
(c) Interpret the solution physically.

2.25 Assume that the applied force and the spring stiffness of Figure P2.24 are known.
(a) Use the loop method to find a set of equations that represents the network.
(b) Solve the equations found in part (a) for the loop flows. Determine the compressive force in each spring and the displacement of each node.
(c) Interpret the solution physically.

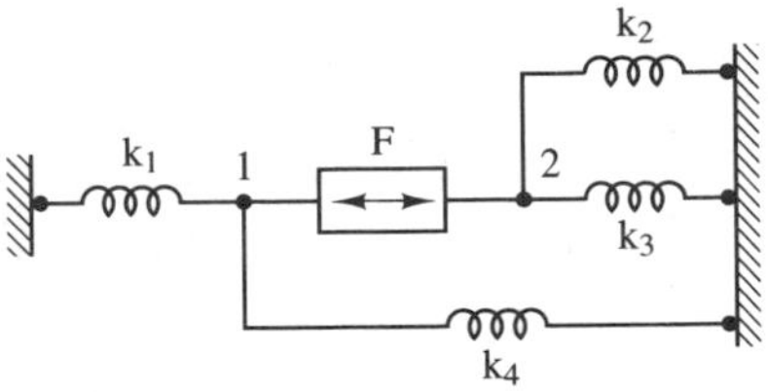

Figure P2.24.

2.26 The line graph of a lumped translational network is shown in Figure P2.26. (The use of lumped-element symbols in the branches is redundant.)

(a) Construct an interconnection of translational lumped elements to correspond to this line graph.

(b) Suppose v is a suddenly applied constant velocity of magnitude v_c. Determine intuitively the nature of each node motion after the motions reach steady state, that is, after the oscillations die away.

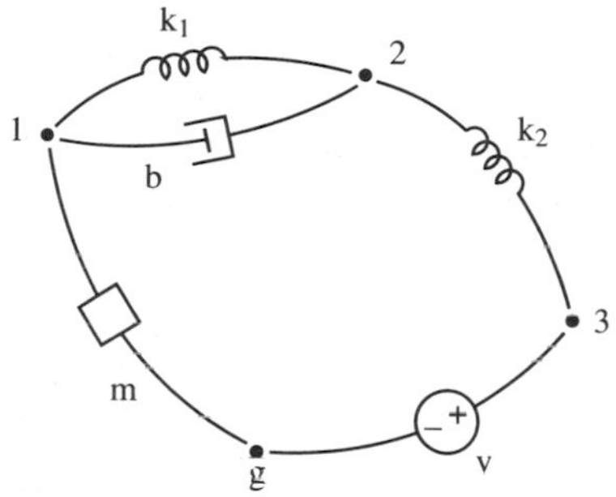

Figure P2.26.

2.27 The translational network of Figure P2.27 is at rest initially, and the spring is uncompressed. At $t = 0$, the two velocity source functions $v_a(t)$ and $v_b(t)$ are applied. Assume that b_1, b_2, and k are known.

(a) Use the loop method to find a set of equations to represent the behavior of the network in terms of circulating loop flows.

(b) State the relations between each of the compressive forces F_1, F_2, F_3, and F_4 and the loop flows used in part (a).

(c) Eliminate variables from the equations in part (a) to form the input-output system equation for F_1.

2.28 The lumped network in Figure P2.28 contains two forces acting in competition with each other.

(a) Find a set of system equations for the network. Which approach is best suited to this problem, the node method or the loop method?

(b) Suppose the mass in the network has the initial velocity $v_2(0) = v_c$. Describe intuitively how the motion of the mass will change with time.

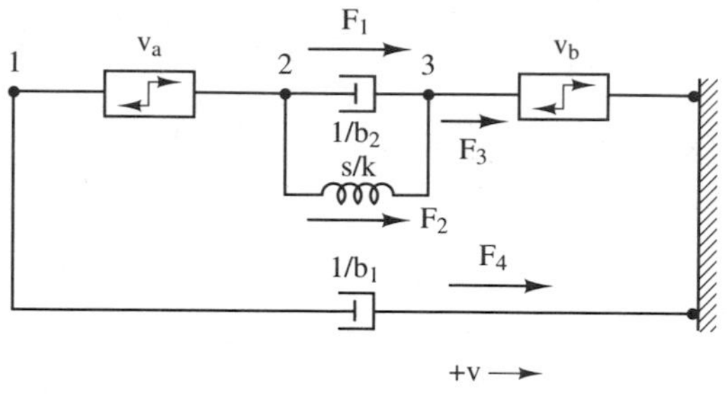

Figure P2.27.

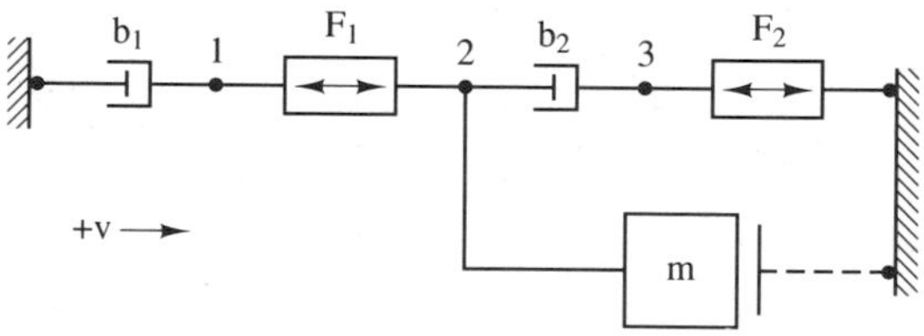

Figure P2.28.

2.29 A lumped translational network actuated by a displacement source x(t) is shown in Figure P2.29. The parameters k, b, m_1, and m_2 are known.

(a) Write a set of node equations that describes the behavior of the network.

(b) Define loop variables, and write a set of loop equations that describes the behavior of the network.

(c) Which method do you prefer for this network? Why?

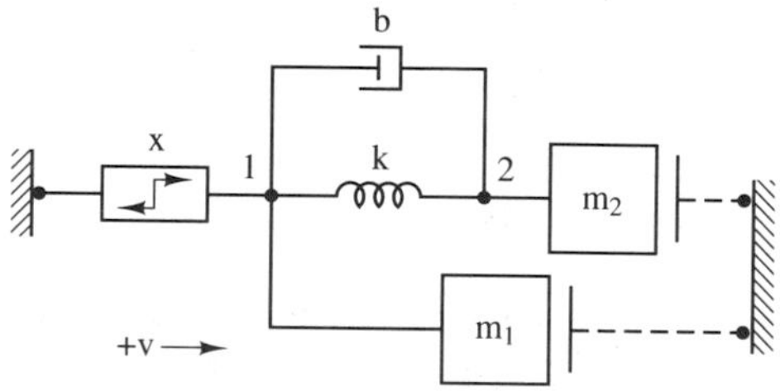

Figure P2.29.

2.30 A lumped translational network is shown in Figure P2.30. The values of the parameters m, k, and b are known. The source function F(t) is also known.

(a) Draw the line-graph representation of the network. Define a flow variable for each branch of the line graph.

(b) Use the node method to generate a set of system equations.

(c) Eliminate all variables except v_1. Express the remaining input-output system equation as a differential equation in the variable v_1.

(d) Define loop variables, and use the loop method to generate a set of system equations. (How can you take advantage of the fact that one loop flow is known?) Eliminate all vari-

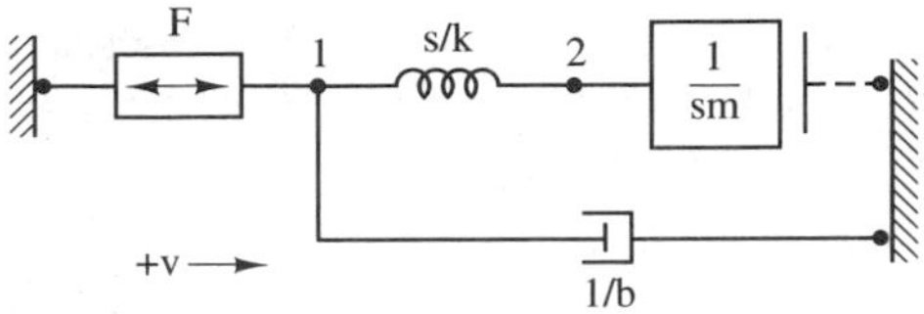

Figure P2.30.

ables except the unknown loop flow. Express the remaining equation as a differential equation. Convert that equation into the input-output system equation for v_1.

2.31 Find a set of system equations that represents the behavior of the lumped model of the fishing situation derived in example problem 2.3.

2.32 A sinusoidal force is applied to a parallel connection of a lumped spring and a lumped damper in Figure P2.32.

(a) Determine the input-output system equation for v_1.

(b) Assume that the solution is sinusoidal, of the form $v_1(t) = A \cos \omega t + B \sin \omega t$. Substitute this form of solution into the node equation, and determine the values of A and B.

(c) Show that for $\omega \ll k/b$, the spring-damper combination acts as an ideal spring, whereas for $\omega \gg k/b$, it acts as an ideal damper. The quantity b/k is known as the time constant of the spring-damper combination.

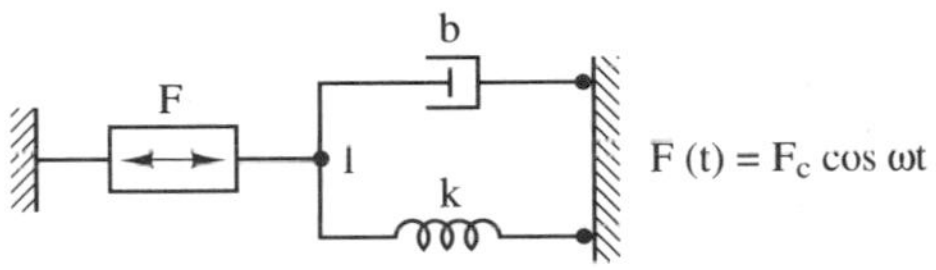

Figure P2.32.

2.33 Figure P2.33 shows the construction of a particular accelerometer. It is used to measure the acceleration of the object to which it is fastened. If the accelerometer is accelerated in the +v direction, the inertial force on the mass m causes a proportional displacement of the attached spring and shaft, relative to the case, in the −v direction. The magnitude of the acceleration can be read from calibrated displacement marks on the shaft. The viscous damping at the two bearings where the shaft penetrates the case damps out the oscillations of the spring-mass after an abrupt acceleration. Design the accelerometer to measure automobile accelerations in the range 0 to 3 m/s^2. That is, select values of the mass m, the stiffness k, and the friction constant b that are appropriate to the given range of acceleration, and explain your choices.

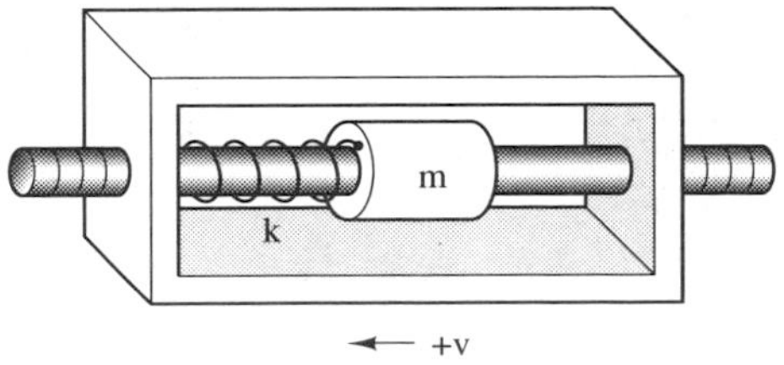

Figure P2.33.

2.34 Propose a means of measuring the velocity of a motorboat relative to the water. Construct a lumped model of the velocity meter, and select appropriate values for the lumped parameters in order that the meter be accurate for velocities in the range 0 to 15 m/s.

2.7 REFERENCES

2.1 Jefferson, T. B., and Brooking, W. J., *Introduction to Mechanical Design.* New York: Ronald Press, 1951.

2.2 Nilsson, James W., *Electric Circuits.* Reading, MA: Addison-Wesley, 1983.

2.3 Ogata, K., *System Dynamics.* Englewood Cliffs, NJ: Prentice-Hall, 1978.

2.4 Perkins, W. R., and Cruz, J. B., Jr., *Engineering of Dynamic Systems.* New York: John Wiley & Sons, 1969.

2.5 Shearer, J. L., Murphy, A. T., and Richardson, H. H., *Introduction to System Dynamics.* Reading, MA: Addison-Wesley, 1971.

2.6 Walther, H., *Ten Applications of Graph Theory.* Boston: D. Reidel Publishing Co., 1984.

CHAPTER

3

Lumped-Network Behavior

The behavior of a lumped model approximates the behavior of the physical system it represents. We gain understanding of the physical system by examining the behavior of its model. Initially, we use a computer to simulate the operation of lumped models. Simulation demonstrates the behavior of a complicated model without requiring much analytical work. The system equations need not be linear, and any source waveform can be used. Through simulation, we are able to examine lumped-model behaviors for practical systems. Simulation does not, however, show how the behavior depends on the parameters of the model.

Most of this chapter is devoted to analyzing lumped networks mathematically. Mathematical analysis relates the network response to the network parameters. It shows how to adjust the network to alter its behavior. We restrict our mathematical analyses to *linear, time-invariant* networks. (Techniques for linearizing nonlinear networks are examined in section 5.1.). Although the analysis techniques of this chapter apply to system equations of any order, most systems behave much like a first-order or second-order network. Therefore, we examine first-order and second-order networks in detail.

In sections 3.3 and 3.4, we find the responses of networks to abrupt source signals (step-function inputs). An abrupt test signal brings out the characteristic behavior of a network. We shall use the phrase *network step response* to refer to the response of a single variable of the network to an abrupt source signal. Step-response waveforms can be expressed in terms of the initial and final values of the network variables. Accordingly, we develop techniques for finding those initial and final values.

Before introducing network solution techniques, we reinforce the lumped-modeling procedures in a new context—mechanical rotation. Transfer of the modeling and equation-formulating techniques of chapter 2 to rotational systems is aided by the terminology of flows and potentials.

3.1 LUMPED MODELS FOR ROTATIONAL MECHANICAL SYSTEMS

In the natural world, we encounter mechanical rotation less often than mechanical translation. Yet we have enough experience with bicycle wheels, clock springs, automobile brakes, and electric motors that we easily conceive of rotational elements that are physically analogous and mathematically equivalent to the translational mass, spring, and damper. In this section, we introduce the notational framework for rotational systems. Then we introduce analogs of the ideal translational elements, derive lumped models of rotational systems, and generate corresponding line graphs and system equations.

Rotational Motion—a Potential

This section addresses the rotation and twisting of physical objects. The points on each slice through the object of Figure 3.1 (perpendicular to the axis of rotation) rotate as a single unit. A slice has one degree of freedom of motion. That is, it can rotate only clockwise or counterclockwise, as seen from the left. The points on a slice have all their motions (rotational displacements, velocities, and accelerations) in common. Let us assign a separate velocity variable to each slice of interest. We describe all motions of a slice in terms of the velocity of the slice.

By analogy to translational motion, we measure all rotational motions relative to a nonrotating reference frame (or ground frame) perpendicular to the axis of rotation (see Figure 3.2). We translate the reference frame parallel to the axis of rotation to measure angles of rotation at various slices of the object. We define the orientation of an object slice to be the angle between a fixed (or **ground**) ray in the reference frame and a particular ray on the slice. We denote the initial orientation of the object ray (and object slice) by O. Let $\theta(t)$ be the **rotational displacement of the ray** (and its object slice) from its initial

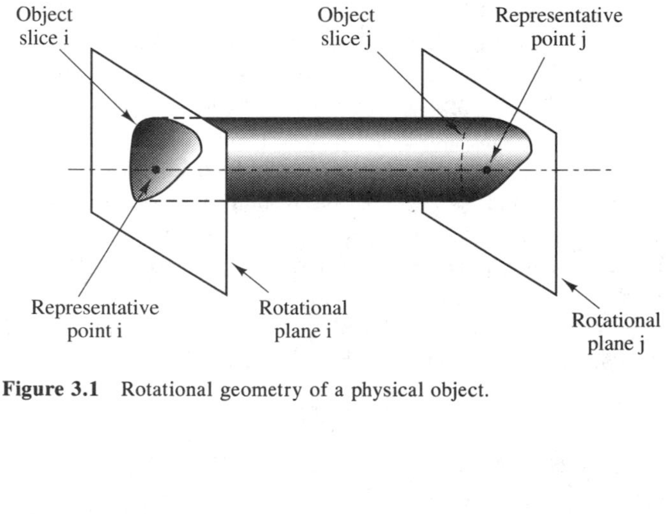

Figure 3.1 Rotational geometry of a physical object.

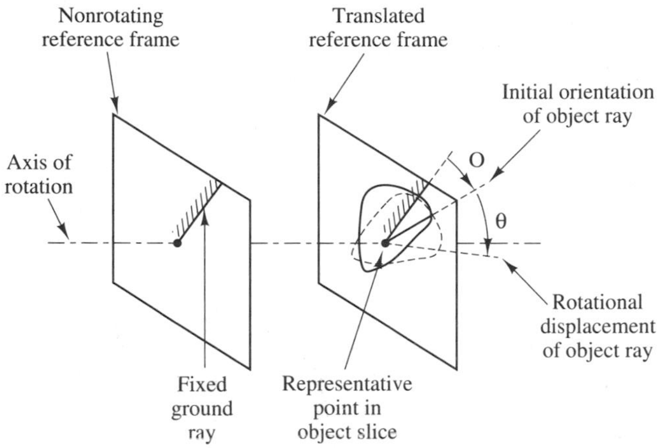

Figure 3.2 Displacement θ relative to a ground ray.

orientation, as a function of time. We measure rotational displacement in radians (a dimensionless unit), in degrees, or in revolutions (multiples of 2π radians).

The initial orientation O of an object slice does not affect the rotational motions of that slice. It is not even necessary that the same ground ray be used as a reference for different object slices. All fixed (nonrotating) rays can be treated as ground rays.

We denote the angular velocity of an object slice by $\Omega = \dot{\theta}$ and the angular acceleration of the slice by $\alpha = \dot{\Omega}$. We use the term *rotational motion* to refer simultaneously to all three quantities of motion—displacement, velocity, and acceleration. Knowledge of any one of these rotational variables as a function of time determines the other two, except for initial values. We focus attention primarily on the *central* motion variables—the angular velocities of the object slices.

To establish positivity of motion of an object slice, as stated in chapter 1, we select a positive direction for rotation of that slice. We display this direction of positive rotation by means of a clockwise or a counterclockwise rotation arrow near the object slice, as shown at the ends i and j of the object in Figure 3.3(a). We usually use the same positive

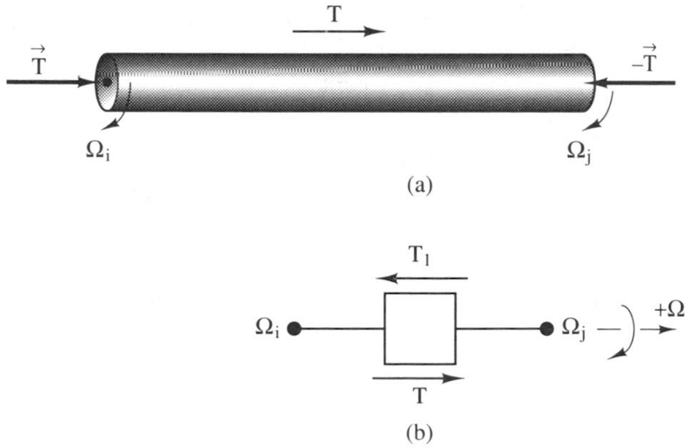

Figure 3.3 Notations for positive torsion and for positive rotational motion. (a) Pictorial model. (b) Lumped model.

rotation direction for all object slices. We denote this single positive rotation direction by a clockwise or counterclockwise rotation arrow, together with the label $+\Omega$. Figure 3.3(b) shows a lumped representation of the object; we use the $+\Omega$ symbol to show the single positive rotation direction. The rotation arrow in the figure is **viewed from the left.** (We draw all rotational figures from this viewpoint.)

The positive rotation symbol in Fig. 3.3(b) also contains a *linear arrow* that is associated with the rotation arrow by the **right-hand rule.** The linear $+\Omega$ arrow serves as a mnemonic aid to interpreting flows physically in lumped models. It also strengthens the analogy between mechanical translation and rotation.

We associate all the points on a specific slice, which rotate together, with a single **representative point** on the axis, as shown in Figures 3.1 and 3.2. We speak of rotation *at that point,* rather than rotation of that object slice. The symbols Ω_i and Ω_j of Figure 3.3(a), for example, denote the rotational velocities at representative points i and j (or the velocities of the corresponding slices i and j). Then we treat each twisting object as if it were one dimensional—with its representative points distributed along the axis of rotation.

We do not show the ground rays in rotational lumped models. Instead, we use a symbol analogous to the translational ground symbol to distinguish the ground frame. The *rotational-ground* symbol should be thought of as a nonrotating frame on the axis of rotation. (An example of this rotational-ground symbol appears in Figure 3.10.)

We represent the **relative rotational displacement** of point i with respect to point j by the double-subscript notation:

$$\theta_{ij} = \theta_i - \theta_j \tag{3.1}$$

The same positive direction must be used for both points to represent relative motion as a difference. We denote the rotational velocity of point i relative to point j by Ω_{ij} and the corresponding relative rotational acceleration by α_{ij}.

It is apparent that θ, Ω, and α are *across variables,* analogous to their translational counterparts in chapter 2; thus, they can be treated as potentials. If the relative motion Ω_{ij} is positive, then $\Omega_i > \Omega_j$. Therefore, **Ω_{ij} is the drop in (rotational) velocity from point i to point j.** (If $\Omega_{ij} < 0$, the drop is negative; in other words, it is a velocity rise.) Rotational displacement and acceleration also act as potentials; θ_{ij}, Ω_{ij}, and α_{ij} are all drops in potential from point i to point j.

We represent the **torque** (or moment of force) acting on a shaft by a vector aligned with the axis of rotation. We use the **right-hand rule** to assign a rotational direction to that vector. The rotational direction assigned to the torque vector $\vec{T}$ at point i of Figure 3.3(a) is the $+\Omega$ direction, clockwise as seen from the left. The torque vector $-\vec{T}$ at point j acts in the $-\Omega$ direction.

To screw a nut on a right-hand screw, the torque vectors that act on the nut and screw must point toward each other. That is, they point *inward.* We call the resulting relative motion *inward-twisting motion.* Unscrewing the nut produces outward-twisting motion. In general, we define inward-twisting motion across an object to be that relative rotation of the object ends which is caused by inward-pointing torque vectors at the ends. Inward-twisting motion is analogous to compressive motion. (Compressive motion requires the action of inward-pointing force vectors.) Outward-twisting motion is analogous to stretching motion.

The inward-pointing torque vectors in Figure 3.3(a) produce inward-twisting relative motion of points i and j. If we order the subscripts on the relative-displacement variable θ_{ij} in the *linear* $+\Omega$ direction, then θ_{ij} is positive. That is, $\theta_i > \theta_j$ corresponds to inward-twisting motion. Therefore, the rotational displacement variable θ drops (or reduces) in the linear $+\Omega$ direction. Similar statements can be made about the velocity Ω and acceleration α: If the relative velocity Ω_{ij} in Figure 3.3 is positive, the pair of velocities Ω_i and Ω_j correspond to inward-twisting motion, and if the relative acceleration α_{ij} is positive, the pair of accelerations α_i and α_j correspond to (or will eventually correspond to) inward-twisting motion. In general, **inward-twisting relative motion corresponds to a drop in potential (θ, Ω, or α) in the linear $+\Omega$ direction.**

Figure 3.3(b) shows a lumped representation of the twisted object. Nodes i and j correspond to the representative axial points of object slices i and j in Figure 3.3(a). The symbols Ω_i and Ω_j represent velocities of rotation (of object slices) about the axis at the representative points. We refer to these variables as node velocities.

Torsion—a Flow

Within the object of Figure 3.3(a), there is a **scalar twisting force** we call **torsion.** Torsion is a *through variable,* analogous to the scalar force in a translational system. The torsion is the same at all points (slices) of the object. To measure that torsion, we must break the system and insert a scale in series with the object. The torsion has magnitude $|\vec{T}|$, the magnitude of the applied torque vector. The SI unit of torsion is the N·m.

The torsion from inward-pointing torque vectors, as in Figure 3.3(a), is opposite in sense to the torsion from outward-pointing torque vectors. The physical difference between these two senses is not as clear as the difference between compressive force and tensile force in translational systems. We use the translational analogy to strengthen the distinction. We define **inward-twisting torsion** to mean torsion owing to inward-pointing torque vectors. An example is the torsion in a right-hand screw as it is screwed into a piece of wood. Inward-twisting torsion is **analogous to compressive force** because compressive force is caused by inward-pointing force vectors. Outward-twisting torsion is torsion owing to outward-pointing torque vectors—the torsion in a screw as it is unscrewed. Outward-twisting torsion is analogous to tensile force.

Since torsion is a through variable, we treat it as a flow. For a passive object, inward-pointing torque vectors produce inward-twisting torsion in the object and inward-twisting relative motion across the object. According to the previous discussion of rotational motion, the rotational velocity drops in the linear $+\Omega$ direction across an inward-twisted object. To produce an analogy to downhill flow, as discussed in chapter 1, **we treat inward-twisting torsion as a flow in the linear $+\Omega$ direction,** the direction of velocity drop. Then outward-twisting torsion is a flow in the linear $-\Omega$ direction.

We represent the torsion in an object by the symbol T. We assign a direction to T and indicate that direction by an arrow beside the object. We treat a positive value of T as a flow in the direction of the arrow. If T is oriented in the linear $+\Omega$ direction, it represents inward-twisting torsion. Then a negative value of T represents negative inward-twisting torsion, or outward-twisting torsion. If T is oriented in the linear $-\Omega$ direction, it represents outward-twisting torsion, and a negative value means that the torsion is inward twisting.

In Figure 3.3(b), the torsion is represented as a flow T from node i to node j. It can also be represented as a flow T_1 ($\equiv -T$) in the opposite direction. For the $+\Omega$ direction specified in the figure, T is inward-twisting torsion and T_1 is outward-twisting torsion.

We shall use the expression **torque on** an object to mean a torque vector that acts on a free-body diagram. We shall use the expression **torsion in** an object to mean a flow. Figure 3.4 relates the two concepts. The symbol T in Figure 3.4(a) is a flow in the linear $+\Omega$ direction. It represents inward-twisting torsion. *Visualize* the end of the object that lies toward the linear $+\Omega$ direction as a nonrotating reference frame, and imagine a torque vector T applied to the other end, oriented in the direction of the flow variable T, as shown in Figure 3.4(b). Then the torque vector T also represents inward-twisting torsion. Hence, it is appropriate to use the same symbol for both the torque and the equivalent torsion.

The flow T_1 in the object of Figure 3.4(c) is oriented in the linear $-\Omega$ direction. Consequently, it represents outward-twisting torsion in the object. Again, let us visualize the linear $+\Omega$ end of the object as the reference frame and imagine a torque vector T_1 applied to the opposite end, oriented in the direction of the flow variable (see Figure 3.4(d)). Then we can identify the torque vector with the flow, and it is appropriate to use the same symbol for both.

In sum, we can identify the flow variable assigned to an object with a torque vector that acts at the linear $-\Omega$ end of the object. That torque vector is oriented in the same direction as the flow variable. Thus, a flow in the $+\Omega$ direction represents inward-twisting torsion, and a flow in the linear $-\Omega$ direction represents outward-twisting torsion. If we orient all flow variables in the linear $+\Omega$ direction, then all flows represent inward-twisting torsion.

Sign Conventions, Power, and Energy

We summarize the sign conventions for rotational mechanical systems as follows. First, we focus attention on the object slices of interest by marking and numbering their representative points on the axis of rotation and assigning a motion variable Ω_i to each point (or object slice) i. We designate a single $+\Omega$ direction as the positive rotation direction for all these points. Then, if we order the subscripts on a velocity-difference variable in the linear $+\Omega$ direction, that velocity difference represents inward-twisting relative mo-

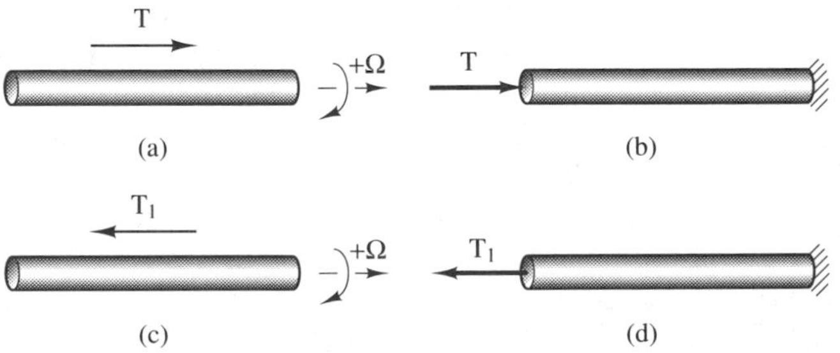

Figure 3.4 Equivalence between torque vectors and torsion variables. (a) Inward-twisting torsion. (b) Inward-pointing torque. (c) Outward-twisting torsion. (d) Outward-pointing torque.

tion. Next, we assign a flow variable T_n to each object n between points, and we use an arrow beside the object to designate the positive flow direction for T_n. If we orient the arrow for a flow variable in the linear $+\Omega$ direction, that variable represents inward-twisting torsion. Reversing the $+\Omega$ direction reverses the signs and the physical interpretations of all variables.

The external environment inserts power into a twisting object by means of opposing torque vectors at its ends. The flow T of Figure 3.3 is positive; that is, T is oriented in the linear $+\Omega$ direction, and the inward-pointing torque vectors produce inward-twisting torsion. The external source exerts a torque on point i that is equal in magnitude to T and is in the direction of Ω_i. Therefore, the power delivered to point i is the product $T\Omega_i$. Similarly, at point j, the external source exerts a torque of magnitude equal to T, but opposing the direction of Ω_j. Therefore, the power delivered to point j is $-T\Omega_j$. The total power delivered to the object by the external source is

$$P = T\Omega_{ij} \tag{3.2}$$

The power delivered is measured in watts (N·m/s).

The power formula (3.2) is stated in terms of a velocity drop Ω_{ij} in the linear $+\Omega$ direction and a flow T in the linear $+\Omega$ direction. If we reverse the $+\Omega$ direction, the signs of both T and Ω_{ij} reverse, and the power formula is unchanged. In general, **the mechanical power delivered to a twisting object is the product of the flow through the object (in some axial direction) and the velocity drop in the same direction.** This interpretation is identical to that of the analogous translational case in equation (2.4).

As in the translational analog, the two ends of a rotational object serve as a **port,** or means of access to the object. The mechanical power delivered to an object by its surroundings is the product of the port variables—the flow T through the port and the velocity drop across the port in the direction of flow.

The energy E delivered to the object during the time interval $[t_1, t_2]$ owing to the torsion in the object and the relative motion across the object is the integral of the delivered power; that is,

$$E = \int_{t_1}^{t_2} T\Omega_{ij}\, dt \tag{3.3}$$

The energy is measured in N·m.

A passive object cannot provide energy. However, a passive rotational spring or inertia can store energy and give it back at a later time. If the signs of T and Ω_{ij} differ during some interval, then energy is removed from the object during that interval.

The Ideal Torsion Spring

Some objects deform (twist) nearly in proportion to the applied torsion. We view these objects as torsion springs. We define an **ideal torsion spring** to be a lumped element that satisfies the equation

$$T = K\theta_{ij} \tag{3.4}$$

where θ_{ij} is the angular displacement of node i relative to node j and T is the torsion directed from node i to node j. We interpret θ_{ij} as the drop in rotational velocity in the direction of the flow T. We call K the **torsion-spring constant** (or **torsional stiffness**); its inverse 1/K is the **torsional compliance.** The units of K are N·m/rad. We use the coil spring symbol of Figure 3.5 to represent the ideal torsion spring.

The equation

$$\Omega_{ij} = \frac{1}{K}\dot{T} \tag{3.5}$$

is an alternative form for the torsion-spring equation, one that focuses attention on the central motion variable Ω. This equation can also be expressed in the impedance form

$$\Omega_{ij} = \left(\frac{s}{K}\right)T \tag{3.6}$$

where the symbol s represents differentiation of the *signal* T with respect to time.

The potential energy E that we deliver to a torsion spring as we twist it from the relaxed state $\theta_{ij}(0) = 0$ is, according to equations (3.3) and (3.4),

$$E(t) = \int_0^t K\theta_{ij}\dot{\theta}_{ij}\,dt = \int_0^{\theta_{ij}(t)} K\theta_{ij}\,d\theta_{ij}$$

$$= \frac{1}{2}K\theta_{ij}^2 = \frac{1}{2}\frac{T^2}{K} \tag{3.7}$$

The energy delivered is always positive, regardless of the sense of twist (inward or outward twisting). The stored potential energy can be recovered by reducing the torsion T and twist θ_{ij}. Because we cannot change the displacement θ_{ij} across a physical spring abruptly, energy cannot be transferred to or from a spring instantly. According to equation (3.7), then, the torsion in a spring cannot be changed abruptly either.

Figure 3.6 shows two common torsion-spring geometries and the corresponding torsion-spring constants. Of course, a physical torsion spring approximates the behavior of an ideal spring only over a limited range of values of the twist angle. One can design a torsion spring with essentially any desired stiffness by an appropriate choice of material, shape, and size. The torsion bar is used in many automobile suspension systems (see problem 3.6 at the end of the chapter). The torsion coil is used for door closers and garage door lifters (see problem 3.12).

A physical spring will not have exactly the stiffness predicted by a design formula. This is because the formulas are approximate and the dimensions and material properties are not uniform. Those dimensions and properties also change with temperature, use, and time. We can always *measure* the stiffness of a particular spring, however. We use accurately known weights and lever arms to apply known torques, and we use accurate angle-measuring devices to measure the spring twists owing to the applied torques. From

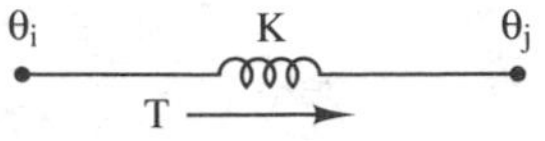

Figure 3.5 Ideal torsion spring.

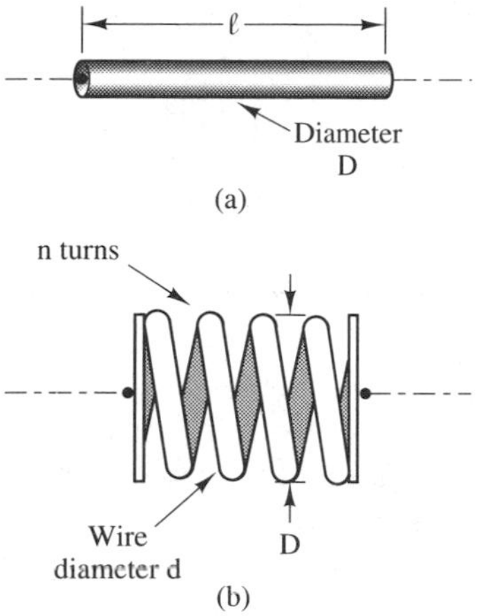

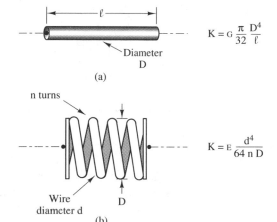

$$K = G\frac{\pi}{32}\frac{D^4}{\ell}$$

$$K = E\frac{d^4}{64\,n\,D}$$

Figure 3.6 Torsion-spring geometries; Torsion-spring constants for these and other rotational geometries are given in reference [3.5]. Values of E and G are tabulated for various materials in Appendix A. (a) Torsion bar (G is the shear modulus). (b) Torsion coil (E is Young's modulus).

measurements of torques and corresponding displacements over a range suitable to the spring, we calculate an average value of stiffness.

The Ideal Rotary Damper

A shaft turning within a bearing encounters rotational friction at the points of contact. Friction also occurs at the surfaces of objects that rotate in fluids. Rotational friction characteristics are similar to the translational characteristics shown in Figure 2.9. Oil lubrication produces a friction characteristic that is nearly proportional to the relative rotational velocity of the parts that rotate against each other. In imitation of lubricated friction, we define the **ideal rotary damper** to be a lumped element that has a torsion-speed characteristic of the form

$$T = B\Omega_{ij} \tag{3.8}$$

where T is the torsion in the element in the direction from node i to node j (see Figure 3.7). According to equation (3.8), the velocity Ω drops across the damper in the direction of the flow T. The **rotational friction constant** (or *damper constant*) B has the units N·m·s/rad. The rotary-damper equation (3.8) can also be expressed in the impedance form

$$\Omega_{ij} = \left(\frac{1}{B}\right)T \tag{3.9}$$

We refer to the impedance 1/B as the *rotational resistance* of the damper. We represent an ideal rotary damper by the *drag-cup* symbol of Figure 3.7. Drag-cup damper design is illustrated in example problem 3.1.

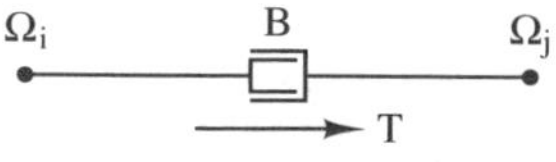

Figure 3.7 Ideal rotary damper.

All friction phenomena are dissipative; that is, they absorb mechanical energy and dissipate it as heat. The power dissipated by relative rotation of the ideal viscous damper, according to equations (3.2) and (3.8), is

$$P = B\Omega_{ij}^2 = \frac{T^2}{B} \geq 0 \tag{3.10}$$

The power absorbed by the damper is always positive, regardless of the direction of relative rotation.

The fluid drive in the automatic transmission of an automobile is essentially a rotary damper. The fluid that connects the engine crankshaft (a source of torque) to the drive shaft (which, in effect, turns the wheels) enables the slippage that produces smooth coupling as we change gears. Suppose, for example, that the engine runs at 360 rad/s (3,400 rpm) and provides 65 N·m of torque to maintain the automobile speed at 30 m/s. According to equation (3.8), if the slippage in the fluid clutch at that speed is 20 rad/s, the friction constant of the clutch must be $B = T/\Omega_{slip} = 3.25$ N·m·s/rad. Then the power absorbed by the damper, according to equation (3.10), is 1,300 N·m/s (1.75 hp, or 1.3 kW). This power appears as heat in the transmission fluid. Accordingly, the need to cool the transmission fluid is apparent. If the car tows a trailer, the clutch slippage increases and the generation of heat becomes more severe.

Example Problem 3.1: Drag-cup damper design.
The cross-sectional geometry of one type of drag-cup rotary damper is shown in Figure 3.8. The viscosity of the oil is μ. Derive the relationship between the friction constant B, the viscosity μ, and the physical dimensions of the device. The relationship can be used to design a damper with a specified friction constant.

Solution: Points 1 and 2 represent the two rigid cups. Suppose the two cups move at constant velocities, Ω_1 and Ω_2, as a result of the applied torque vectors. Then we can ignore the inertias of the two cups. The oil inside cup 2 attaches itself, by friction, to cup 2 and rotates at

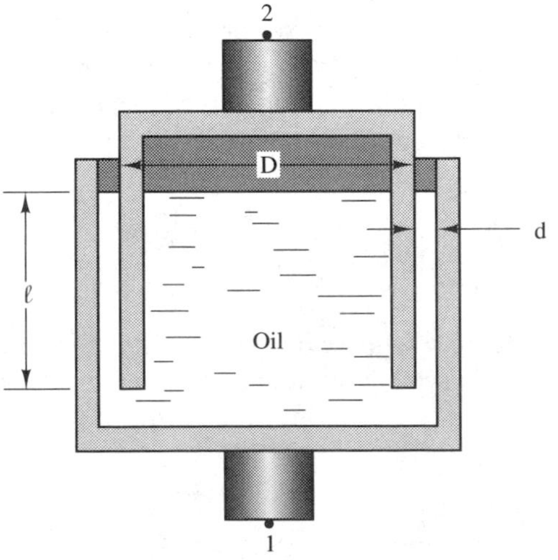

Figure 3.8 Drag-cup rotary damper.

a velocity near Ω_2. Similarly, the oil in the bottom of cup 1 rotates at a velocity near Ω_1. The layers of intervening oil near the center of the device rotate at various intermediate velocities.

The narrow cylinder of oil (of thickness d) between the two cups is sheared in the manner described for the translational viscous damper (see Figure 2.11). The shearing forces oppose relative rotation of the cups. These forces are felt as torsion in the shafts and as torques at points 1 and 2. We may neglect the torsion owing to shearing of oil at the center of the cup because the shearing velocities are low at small radii and the sheared layer of oil is thick. The viscous drag in the cylinder of air between the cups is negligible because the viscosity of air is much less than that of oil.

The cylinder of oil, of length ℓ, can be viewed as equivalent to the layer of oil in Figure 2.11; it has thickness d and total surface area $\pi D\ell$. According to equation (2.18), the shearing force opposing motion at the surfaces of the cups is

$$F = \mu \frac{A}{d} v_{21} = \mu \frac{\pi D\ell}{d}\left(\Omega_{21}\frac{D}{2}\right) \tag{3.11}$$

Consequently, the torsion felt in the shafts is

$$T = F\left(\frac{D}{2}\right) = \mu \frac{\pi}{4}\frac{\ell D^3}{d}\Omega_{21} \tag{3.12}$$

and the damper constant is

$$B = \mu \frac{\pi}{4}\ell\frac{D^3}{d} \tag{3.13}$$

The drag cup can be designed to operate horizontally if cup 1 surrounds cup 2 and the oil is sealed in the cup at shaft 2.

The Ideal Inertia

The particles of any mechanical device resist changes in rotational velocity. The rotational analog of mass is **moment of inertia.** The moment of inertia J of a point mass m revolving at radius r from the axis of rotation is

$$J = mr^2 \tag{3.14}$$

The moment of inertia of a rigid rotating object is obtained by summing (or integrating) the inertias of the individual particles. The moments of inertia for three common geometries are shown in Figure 3.9. The SI unit for moment of inertia is $kg \cdot m^2$. However, it is not unusual to find the literature using the equivalent unit $N \cdot m \cdot s^2$.

The geometries and compositions of many rotational objects are complicated. The rotor of an electric motor, for example, is formed by winding copper wire on a complicated iron and steel frame. Precise calculation of its moment of inertia is impractical. We use moment-of-inertia calculations only to obtain approximate values. Precise values must be measured (see problem 3.22 at the end of the chapter, for example).

We define an **ideal inertia** to be an object in which the particles are connected together rigidly and rotate identically. By Newton's third law, a source that applies a torque to an inertia must also apply an equal and opposite torque to the ground. The rotational

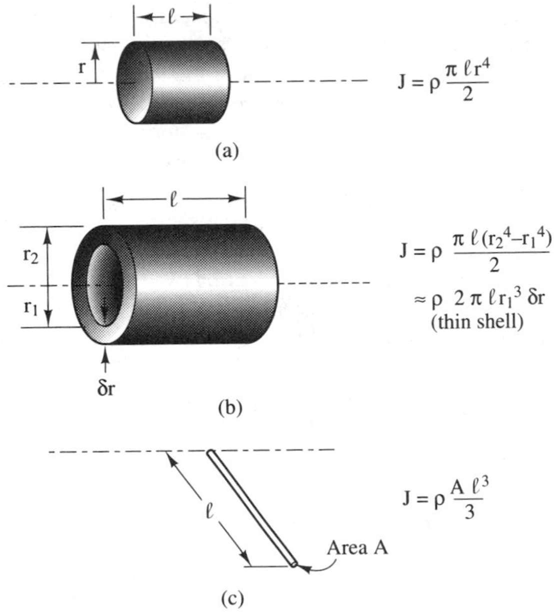

Figure 3.9 Inertias for common geometries (see reference [3.5]); ρ = density. (a) Solid cylinder. (b) Hollow cylinder (shell approximation). (c) Thin rod, about its end.

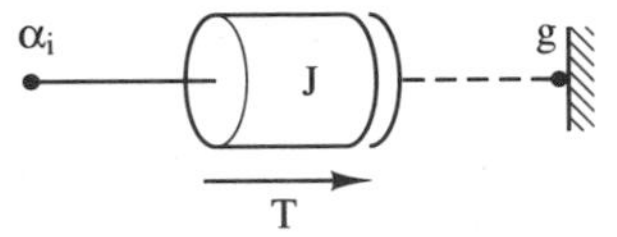

Figure 3.10 Ideal inertia.

velocity of an object is also measured relative to the ground. By analogy to the ideal mass, we treat the ideal inertia as a two-terminal lumped element, as shown in Figure 3.10. The dashed line in the figure emphasizes that the rotational acceleration of point i is measured relative to a fixed reference and that the reaction torque is *felt* at that reference. (See the discussion of ideal mass in section 2.1.) The rigidly attached point i represents all points that are treated as rotating together rigidly, regardless of the geometry of the set of points. The torque vector that accelerates the inertia relative to the fixed ground frame acts on point i. The opposing torque on the ground acts at point g.

Let us orient the torsion variable T from node i to the ground terminal. Then, according to Newton's second law, the torsion T and the relative acceleration $\dot{\Omega}_{ig}$ are related by

$$T = J\dot{\Omega}_{ig} = J\dot{\Omega}_{i} \tag{3.15}$$

where J is the moment of inertia (or just *the inertia*). The subscripts on the motion variable are oriented in the positive direction for the flow T. Hence, Ω_{ig} is a velocity drop in the direction of flow. Equation (3.15) can also be expressed in the impedance form

$$\Omega_{ig} = \left(\frac{1}{sJ}\right)T \tag{3.16}$$

It takes energy to accelerate an object rotationally. That energy is stored in the rotating inertia. The energy delivered to an initially stationary inertia, according to equations (3.3) and (3.15), is

$$E(t) = \int_0^t J\dot{\Omega}_i \Omega_i \, dt = \int_0^{\Omega_i(t)} J\Omega_i \, d\Omega_i$$

$$= \frac{1}{2} J\Omega_i^2 \tag{3.17}$$

The rotational kinetic energy stored in the inertia can be recovered by slowing the rotation. The velocity of an inertia cannot be changed instantly without infinite torsion. Therefore, energy cannot be added to or removed from the inertia instantaneously.

Ideal Rotational Sources

A source injects energy into a system. Hence, we expect the potential to rise across a source in the direction of flow (see chapter 1). We call an energy source an **ideal independent torsion source** if it maintains a specified torsion pattern T(t), regardless of the system to which it is connected. Figure 3.11 shows a symbolic representation of the independent torsion source. The two-headed arrow in the source symbol suggests that the torque vectors applied to the attached system by the torsion source are directed *out of* the source and *into* the attached system. The reaction-torque vectors (exerted on the source) are directed into the source, as shown in the figure. They produce inward-twisting torsion in the source. Therefore, the flow variable T(t) is oriented in the linear $+\Omega$ direction. If point i is fastened to ground, and if T(t) is positive, the torsion source shown in the figure causes positive rotation at point j. If the linear $+\Omega$ direction is from i to j, then T is oriented from i to j, and the velocity rises in that direction.

An **ideal independent rotational-velocity source** is an energy source that produces a specified relative velocity $\Omega(t)$ between its terminals, regardless of the load. Figure 3.12 shows a graphical representation of the independent velocity source. The rotation arrows on the source symbol show the *physical behavior* of the model. The mathematical equation that describes that behavior depends on the $+\Omega$ direction. For the $+\Omega$ direction in

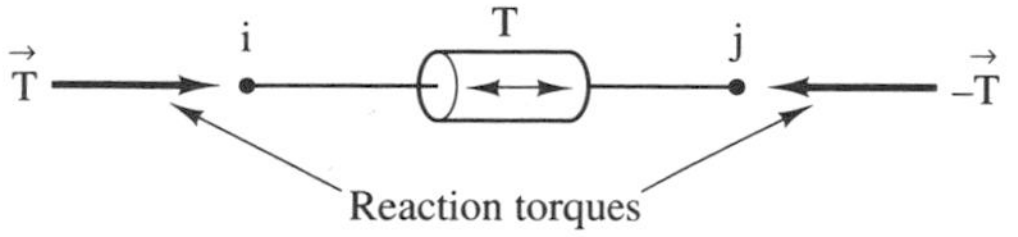

Figure 3.11 Independent torsion source; T is a flow in the linear $+\Omega$ direction.

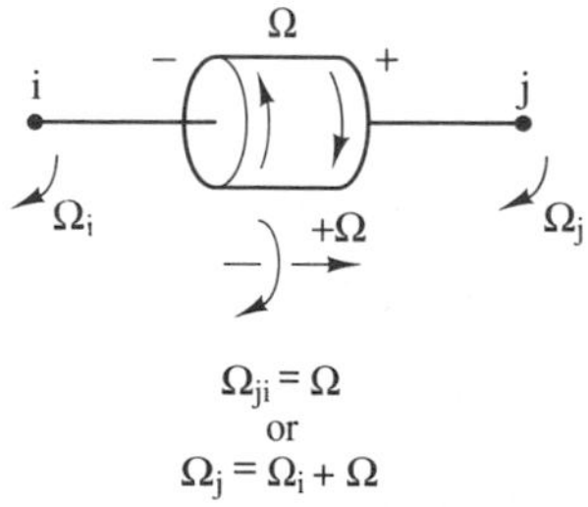

Figure 3.12 Independent rotational-velocity source; Ω is the velocity rise across the source in the linear $+\Omega$ direction.

the figure, $\Omega_j > \Omega_i$, or $\Omega_{ji} > 0$. Thus, the source produces the motion $\Omega_{ji} = \Omega(t)$ or $\Omega_j = \Omega_i + \Omega$. The + and − signs in the figure are a redundant visual reminder that $\Omega_j > \Omega_i$. The source generates whatever torsion is necessary to achieve the specified motion.

Although we refer to the motion source as a velocity source and use the symbol $\Omega(t)$ in Figure 3.12 to specify the motion, a particular pattern of motion might be expressed more easily in terms of displacement or acceleration. Then we label the source with the symbols θ or α to represent the relation $\theta_{ji} = \theta(t)$ or $\alpha_{ji} = \alpha(t)$.

The hanging-weight mechanism in a grandfather clock is an example of an ideal constant-torsion source. A water wheel, because of its large inertia, approximates a constant-velocity source for intermittent loads (see Figure 3.13).

A reciprocating engine is a familiar source of rotational power. In an automobile with a manual transmission, a driver forces the engine to function as a constant rotational-velocity source by holding the speedometer (or tachometer) reading constant. If the driver depresses the clutch pedal enough to cause the clutch to slip, the constant force of sliding friction causes the engine-clutch combination to act like a constant-torsion source. By appropriate motions of the accelerator and clutch, the driver generates a wide range of controlled-velocity and controlled-torsion engine characteristics. Electric motors, hydraulic turbines, gas turbines, and steam turbines are other sources of rotational power (see section 6.3). These sources can approximate ideal-torsion or ideal-velocity sources.

Measurement of Rotation and Torsion

Rotational motion and torsion can be measured by meters that are analogous to the translational meters described in section 2.1 and example problem 2.1. For rotational displacement, we use a protractor ring marked in radians, instead of a stick marked in meters. We

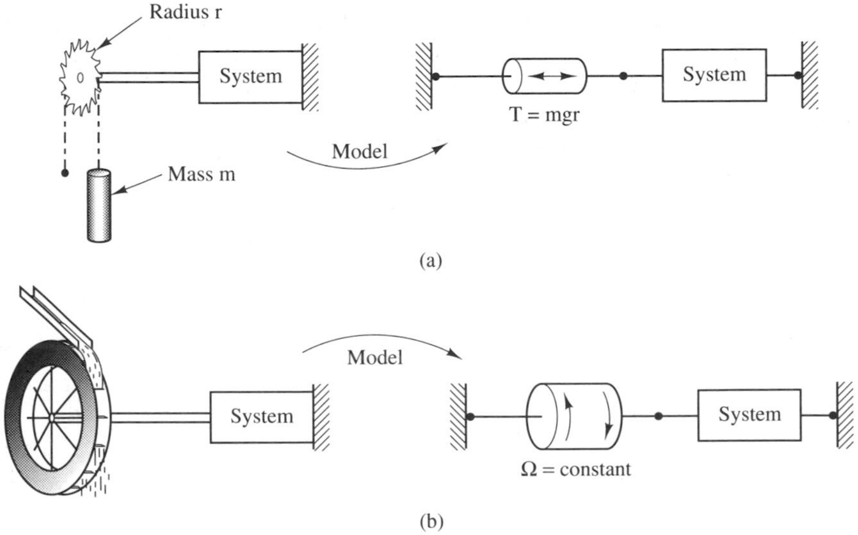

Figure 3.13 Independent rotational sources. (a) Constant torsion. (b) Constant velocity.

center the protractor ring on the axis of rotation of the object slice for which rotation is to be measured. Then we place a mark on the periphery of the object slice. Next, we align the zero angle of the protractor ring with that mark and fasten the protractor to the nonrotating ground. Finally, we measure the displacement of the object slice by reading the angle shown on the protractor beside the mark on the periphery of the slice.

To measure the torsion at a point in the system, we break the system at that point and insert a torsion spring. We then attach a concentric protractor-shaped ring to one end of the spring and a pointer to the other end of the spring. As torsion is produced in the spring, the pointer at the second end sweeps through an angle on the ring. We calibrate the ring in torsion units as follows. First, we label the initial relaxed-spring pointer angle as the 0 angle on the ring. Then, we immobilize one end of the spring and use an accurately known weight and lever arm to apply an *inward-pointing* unit torque (say, 1 N·m) to the other end of the spring. We mark the corresponding rotation angle as +1. If the spring is nearly linear, the ring can be marked with other positive and negative numbers consistent with the 0 and +1 marks. To calibrate the scale more accurately, we can apply various accurately known positive and negative torques to the spring and label the twist angles accordingly.

We can compute rotational velocity and acceleration from a sequence of rotational displacements if we use a clock to measure the times at which those displacements occur. The measurements must be performed rapidly and automatically for this approach to be practical. Computer-monitored optical encoders are often used for this purpose. Rotational velocity and acceleration can also be measured directly by means of a velocity meter or accelerometer. A stroboscope is a light that flashes at an adjustable, accurately known rate. It can be used to measure the velocity of an object that rotates at constant velocity. We simply adjust the flash rate to the fastest value that causes the rotating object to appear stationary without multiple images. Then the velocity (in rpm) equals the flash rate (in flashes per minute). A tachometer-generator produces an electric voltage that is proportional to its speed of rotation. The calibration of a tachometer as a velocity meter is examined in problem 3.5 at the end of the chapter.

Lumped Modeling and Equation Formulating

The processes for lumped modeling, for constructing line graphs, and for formulating system equations are essentially identical for rotational and translational systems. Accordingly, we reiterate the steps in the lumped modeling process for translation, couched in rotational terminology.

THE LUMPING PROCEDURE FOR ROTATIONAL SYSTEMS

1. Draw a pictorial diagram.
2. Identify those parts of the system which are rotationally rigid relative to their surroundings and which exhibit distinguishable rotations or have significant inertia. Represent each part by a numbered node. Place the nodes to correspond approximately to the layout in the pictorial diagram. One of the parts must be the nonrotating ground frame relative to which all motions are referenced.
3. Insert lumped inertias between appropriate nodes and the ground node. The ground node need not be isolated at a single location.

4. Represent those portions of the system which are relatively compliant (compared to the rigid inertias) by lumped torsion springs that connect two nodes.
5. Account for friction by inserting lumped rotary dampers between nodes.
6. Insert independent rotational energy sources between nodes as appropriate.
7. Rearrange the lumped network, if necessary, to make it one dimensional. Vertical displacement in the network has no physical meaning; it is just a mechanism for displaying complicated interconnections. All points connected by a solid line have identical *rotations* and are mathematically indistinguishable.
8. Assign a positive rotation direction ($+\Omega$) for the network. This direction applies to the motions of all the nodes. Establish and orient a flow variable for each lumped element. A flow variable oriented in the linear $+\Omega$ direction represents inward-twisting torsion.

This lumping process is illustrated in example problems 3.2 and 3.3. The process for drawing the line graph that corresponds to a particular lumped model is the same as the process for translational systems (see section 2.2), except that inertia replaces mass and the linear $+\Omega$ direction replaces the $+v$ direction. The nature of each passive branch of the line graph is indicated by one of the rotational lumped-element parameter symbols K, B, or J (or by a number that has the units of one of these symbols). The lumped-element symbol itself can be added to the branch as an unmistakable, redundant element designator. The line-graph source symbols are oriented with the flow arrowhead or the + symbol on the linear $+\Omega$ side.

The node law given by equation (2.27) is d'Alembert's principle for rotational motion. The continuity of rotational motion is given by the path law (equation (2.28)). The loop and node methods for writing system equations are the same as for translational networks. As in the translational case, line graphs clarify the processes for writing the equations. In all equations, we orient the subscripts on the velocity difference for each branch in the same direction as the flow variable for that branch. Example problem 3.2 illustrates the processes for drawing the line graph and writing the system equations.

Example Problem 3.2: Modeling of a log chipper.

The system in Figure 3.14(a) cuts logs into small chips to make paper pulp. The engine accelerates the massive chipper head to high speed. Then the logs are forced against the spinning blades. The high inertia of the head forces the blades through the logs to produce the chips. The fluid clutch provides smooth engagement and disengagement of the engine from the drive shaft.

(a) Derive a lumped model that will represent the rotational behavior of the system during the acceleration of the chipper head and the chipping of a log. Also, model the source waveforms.

(b) Draw the line graph that represents the lumped model, and define appropriate flow variables.

(c) Derive a set of system equations that describes the behavior of the lumped network.

(d) Use the system equations from part (c) to find the input-output system equation for the velocity of the chipper head.

Solution: **(a)** Superimposed on Figure 3.14(a) are four numbered nodes. The one marked g represents the nonrotating ground frame. Node 1 represents the nearly rigid chipper head. As-

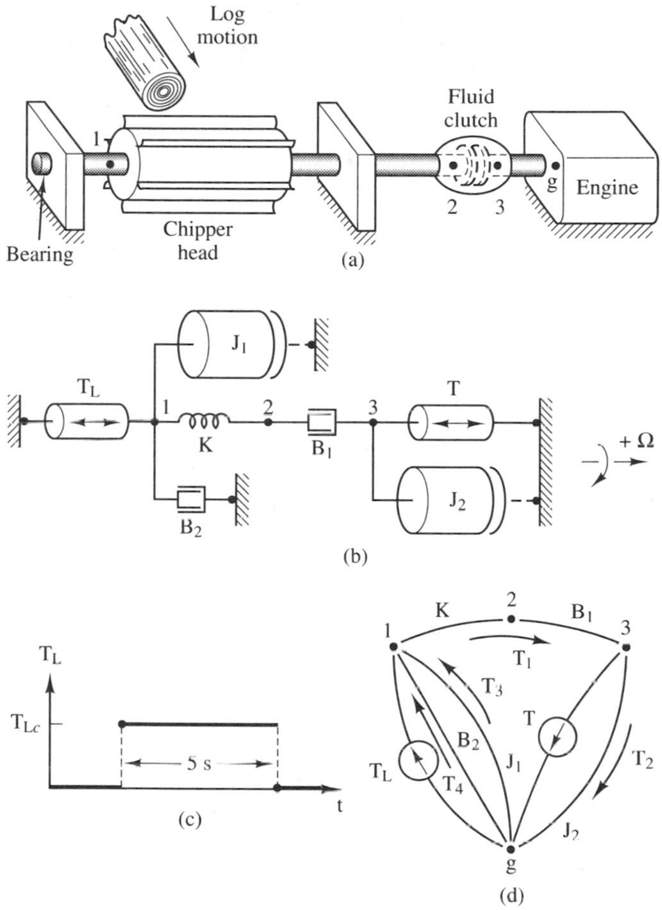

Figure 3.14 Log-chipper model. (a) Pictorial model. (b) Lumped model. (c) Load signal. (d) Line-graph representation.

sume that it rotates without flexing. Node 3 represents one clutch disk, the short shaft that emerges from the engine, and the rotating part of the engine. We assume they rotate identically. Node 2 represents the other disk of the fluid clutch. The long drive shaft between node 2 and the chipper head might not be rigid.

Figure 3.14(b) displays a lumped model of the system. The primary inertia in the system is that of the chipper head, represented by J_1. The inertia J_2 is owing to the rotating part of the engine and the attached drive shaft and clutch disk. The compliant drive shaft is represented by the torsion spring K. The engine provides a specified torque T(t), which we represent by a torsion source. The fluid coupling between nodes 2 and 3 is modeled as a rotary damper with friction constant B_1. We choose to ignore the inertia of the second clutch disk and drive shaft at node 2, because the torsion required to accelerate that inertia is a small part of the torsion in the damper. (That inertia could be included, but the model would be more complicated. To verify that it is negligible, we must calculate the various torsions after we find the motions.) The damper B_2 represents the viscous friction in the bearings at the two ends of the chipper head and the air resistance to the motion of the chipper head. The ends of the lumped elements connected to node 1 rotate with a single velocity pattern. Vertical displacement of the inertias and the damper in the diagram is for notational convenience only.

Suppose that the torque provided by the engine is constant at $T(t) = T_c$, where T_c is large enough to accelerate the chipper head slowly to operating speed. The speed slows during chipping and rises during the intervals between processing of logs. We treat the effect of a log as a torsion source with constant value $T_L(t) = T_{Lc}$ owing to constant blade friction during chipping; $T_L(t) = 0$ during the acceleration between processing of logs. The waveform for T_L is shown in Figure 3.14(c).

(b) The line graph for the lumped model is presented in Figure 3.14(d). We have oriented the flow variables arbitrarily. A single flow variable is defined for the K and B_1 branches, since the branches must have equal flows. These flow variables could be assigned and oriented directly on the lumped model in Figure 3.14(b).

(c) There are four loops, but only three independent nodes. Consequently, we choose the node method to derive a set of system equations, one equation for each of the nodes 1, 2, and 3. At node 1, $T_L + T_4 + T_3 - T_1 = 0$. The element equations are $T_4 = -B_2\Omega_1$, $T_3 = -sJ_1\Omega_1$, and $sT_1 = K\Omega_{12}$. By substituting the element equations into the node equation and rearranging, we obtain

$$(s^2J_1 + sB_1 + K)\Omega_1 = K\Omega_2 + sT_L \tag{3.18}$$

The equations for the B_1 and K branches are $T_1 = B_1\Omega_{23}$ and $sT_1 = K\Omega_{12}$. Equating the flows at node 2 and multiplying the equation by s produces

$$K(\Omega_1 - \Omega_2) = sB_1(\Omega_2 - \Omega_3) \tag{3.19}$$

At node 3, $T_1 - T - T_2 = 0$. The J_2 branch equation is $T_2 = sJ_2\Omega_3$. Substituting the equations for the B_1 and J_2 branches into the node equation yields

$$B_1(\Omega_2 - \Omega_3) - T - sJ_2\Omega_3 = 0 \tag{3.20}$$

(d) The velocity of the chipper head is Ω_1. We eliminate Ω_2 and Ω_3 from the node equations to produce

$$\left[s^3 + \left(\frac{K}{B_1} + \frac{B_2}{J_1}\right)s^2 + \left(\frac{K}{J_1} + \frac{K}{J_2} + \frac{KB_2}{J_1B_1}\right)s + \frac{KB_2}{J_1J_2}\right]\Omega_1 = \left(\frac{s^2}{J_1} + \frac{K}{J_1B_1}s - \frac{K}{J_1J_2}\right)T_L - \frac{K}{J_1J_2}T \tag{3.21}$$

Equation (3.21) corresponds to the differential equation

$$\Omega_1^{(3)} + \left(\frac{K}{B_1} + \frac{B_2}{J_1}\right)\Omega_1^{(2)} + \left(\frac{K}{J_1} + \frac{K}{J_2} + \frac{K}{J_1}\frac{B_2}{B_1}\right)\Omega_1^{(1)} + \frac{KB_2}{J_1J_2}\Omega_1 = \frac{\ddot{T}_L}{J_1} + \frac{K}{J_1B_1}\dot{T}_L + \frac{K}{J_1J_2}(T_L - T) \tag{3.22}$$

The order of the differential equation (three) is the same as the number of energy-storage elements in the system. This equality between equation order and number of energy-storage elements is typical.

Engineers have a variety of techniques for checking their analytical work. One is to **check the units in the final equations to be sure that they balance.** An imbalance of units is a definite indication of error. The location and nature of the imbalance suggest the cause of the error. By working backwards from the point of imbalance, one can usually find the source(s) of error.

Exercise 3.1: Show that equations (3.18), (3.19), and (3.20) can be combined to produce the single third-order differential equation (3.22). Verify that the units of each coefficient are correct.

Example Problem 3.3: Lumped modeling of a distributed speedometer cable.
A speedometer assembly for a motorcycle is shown in Figure 3.15(a) Rotation of the wheel turns the cable within an oiled sheath. The cable turns a cup at the speedometer, which, in turn, drags a second cup by the viscous friction of a film of oil between the cups. A spring restrains rotation of the second cup. An indicator needle is attached to the second cup. We desire that the displacement of this needle be proportional to the angular velocity of the wheel. The needle scale is marked to show the translational velocity of the perimeter of the wheel.

The cable has length 1 m and diameter 2 mm. The inside diameter of the sheath is 0.5 mm larger than the cable diameter. The cable is made of a composite material with density $\rho = 6{,}000\ \text{kg/m}^3$ and shear modulus $G = 7 \times 10^{10}\ \text{N/m}^2$. The viscosity of the oil is $\mu = 0.1\ \text{N·s/m}^2$. The rotating cup is made of aluminum and is 0.3 mm thick, 0.7 cm long, and 2 cm in inside diameter. The spring-restrained cup is made of nylon and is 0.3 mm thick, 0.7 cm long, and 1.98 cm in outside diameter. The thickness of the annular oil film between the cups is 0.1 mm. The indicator needle is an aluminum wire 2.5 cm long and 1 mm in diameter. The coil spring is made of fine steel wire. We choose the stiffness of the coil spring to permit a 0.5-rad rotation of the speedometer needle for a cable velocity of 100 rad/s (motorcycle speed of 97 km/h).

(a) Generate a lumped model of the system that uses a four-segment model of the cable.
(b) Calculate the lumped-model parameters.
(c) Suppose the motorcycle can accelerate from rest to its maximum velocity in 3 s and that this maximum velocity corresponds to a cable velocity of $\Omega_m = 100$ rad/s. In section 6.4 it is shown that a rotational disturbance travels along the cable at the velocity $v = \sqrt{G/\rho} = 3{,}415$ m/s.) Determine how many cable-model segments are needed for the speedometer-cable model to predict correctly the behavior of the speedometer over the normal range of motorcycle speeds. Modify the model and the parameters calculated in parts (a) and (b) appropriately.

Solution: **(a)** We partition the cable into four equal-length segments and assign nodes to the ends of the segments as shown in Figure 3.15(a). We assign to each node the inertia J and viscous friction constant B associated with the length of segment that surrounds the node. Then the values of inertia and damping at nodes 1 and 5 are half the values at the other nodes. We associate a stiffness K with each segment between nodes. Figure 3.15(b) shows the lumped model of the system. The inertia of the rotating cup, J_c, is included with the inertia of the final segment of cable. The inertia of the coil spring is negligible relative to the inertia of the second cup. We shall find that the inertia of the needle is also negligible.

(b) We apply the formulas in Figures 3.6(a) and 3.9(a) to a single segment of cable to find that $K = 0.44$ N·m/rad and $J = 2.36 \times 10^{-9}$ kg·m^2. The geometry of rotation of the cable within the oiled sheath is essentially the same as that of the drag-cup damper in example problem 3.1. According to equation (3.13), the friction constant for each segment of the cable

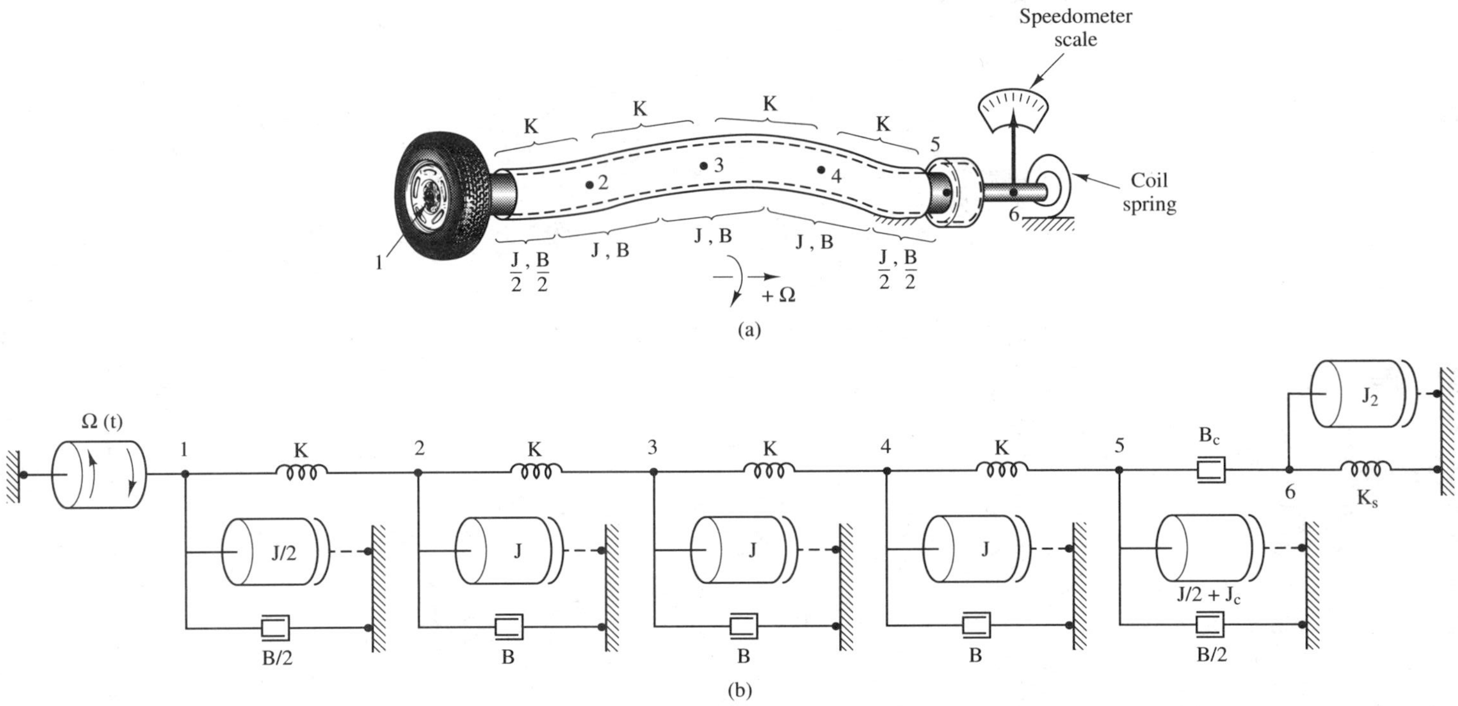

Figure 3.15 Speedometer model. (a) Pictorial model. (b) Lumped model.

is $B = 6.28 \times 10^{-7}$ N·m·s/rad. The friction constant for the drag cup, again from equation (3.13), is $B_c = 4.4 \times 10^{-5}$ N·m·s/rad. According to Figures 3.9(a) and (b), the inertias of the two cups are $J_c = 6.11 \times 10^{-4}$ kg·m^2 and $J_2 = 2.49 \times 10^{-4}$ kg·m^2. The inertia of the needle, by Figure 3.9(c), is $J_N = 1.98 \times 10^{-8}$ kg·m^2, a negligible amount. The stiffness of the coil spring is chosen so that the torsion $T = B_c\Omega_5$ in the drag cup equals the torsion $T = K_s\theta_6$ in the spring for $\Omega_5 = 100$ rad/s and $\theta_6 = 0.5$ rad. Accordingly, $K_s = 8.8 \times 10^{-3}$ N·m/rad.

(c) We select a sinusoidal source signal that has the same maximum acceleration as the highest acceleration expected from the motorcycle. Such a sinusoidal signal is as stressful (as rapidly varying) as any signal the system should encounter. Let p be the period of that sinusoid. Then, according to the segment-length criterion for distributed systems given by equation (2.51), the segment length must be no greater than $p\nu/10$.

Suppose the velocity profile of the accelerating motorcycle is a half period of a sinusoid; that is, $\Omega(t) = \Omega_m[1 - \cos(2\pi t/p)]/2$. According to the statement of the problem, the peak value Ω_m occurs at t = 3 s. Since one half period corresponds to 3 s, the full period must be p = 6 s. Then the maximum acceleration is $\dot{\Omega}_{max} = \pi\Omega_m/p = 52.4$ rad/s^2.

If, instead, the motorcycle accelerates uniformly, the maximum acceleration is $\dot{\Omega}_m = (100 \text{ rad/s})/(3 \text{ s}) = 33.3$ rad/s^2. The two acceleration values are of the same order of magnitude, so both velocity signals vary with about the same rapidity. If we were to reduce the frequency ω of the assumed sinusoidal source signal to produce the more modest 33.3 rad/s^2 maximum acceleration, the corresponding period would be p = 9.4 s.

We choose the more stressful signal with the shorter period, p = 6 s. The velocity of wave propagation is $\nu = 3{,}415$ m/s. Therefore, the segment must be no longer than $p\nu/10 =$ 2049 m. But the whole cable is only 1 m long. Thus, we need only a single segment to represent accurately the behavior of the cable for rotational signals that are within the capability of the motorcycle. Only if we were able to achieve much faster accelerations would we need more segments in the model.

In the single-segment model, it is difficult to determine where to place the cable inertia and damper. If we connect them to node 1 (for which the velocity is specified), they do not affect Ω_5 and θ_6. If we connect them to node 5, they have an exaggerated effect on Ω_5 and θ_6. We choose to connect them to node 5, as shown in Figure 3.16. The symbols J and B in that figure are the parameter values from the four-segment model. Note the reduced cable-segment stiffness and the increased cable-segment inertia and friction constant, as compared with the four-segment model.[1]

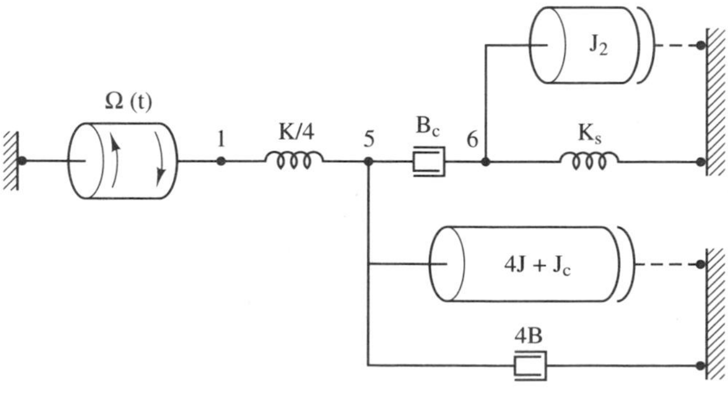

Figure 3.16 Single-segment speedometer-cable model.

[1] The behaviors of these two models are demonstrated in example problem 4.7.

3.2 SIMULATING SYSTEM BEHAVIOR NUMERICALLY

We presented systematic techniques (the loop and node methods) for finding system equations for lumped networks in section 2.2. If the system represented by a network is dynamic, the system equations are differential equations. Because of energy storage in the network, the present values of the system variables depend not only on the present values of the source signals, but also on their past values. In this section, we use computer simulation to analyze dynamic systems numerically. In section 3.3, we examine simple (first-order) dynamic systems mathematically. We treat higher order systems mathematically in section 3.4.

This section introduces operational models. These models structure the system equations to facilitate a numerical solution for the output signals. Readily available computer software can produce a numerical solution of the equations for any form of input signal. Even nonlinear elements, such as the hydraulic resistor of section 5.1, are accommodated easily. (Example problem 5.4 demonstrates the use of computer simulation to solve nonlinear system equations.) Mathematical analysis techniques, on the other hand, are usually restricted to solving *linear* system equations.

The torsion in the network of Figure 3.17(a) is described by the system equation

$$\frac{\dot{T}}{K} + \frac{T}{B} = \Omega \tag{3.23}$$

The velocity Ω_2 is related to the torsion T by the damper equation, $\Omega_2 = T/B$. To understand the behaviors of the network variables T and Ω_2, we must solve the differential equation (3.23).

Suppose the network is *unenergized* prior to $t = 0$. That is, the storage element K contains no energy. Then $T(0^-) = 0$, where $t = 0^-$ denotes the instant *immediately before* the source is turned on. At $t = 0$, we apply to the network the abrupt jump (or step) in velocity shown in Figure 3.17(d).

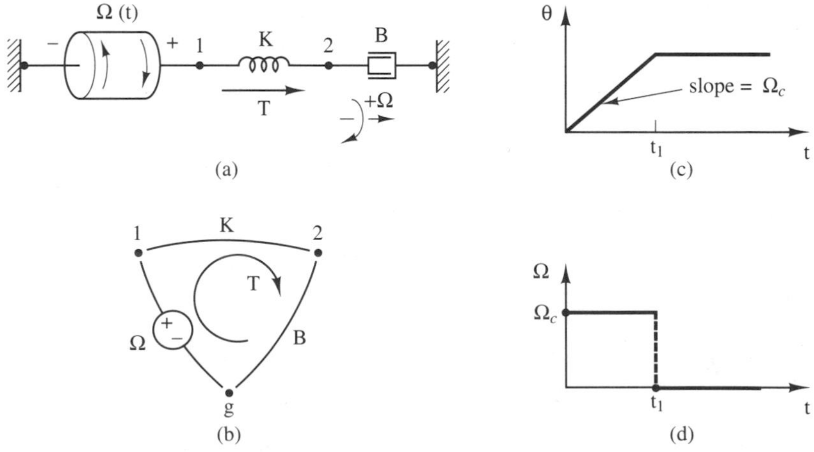

Figure 3.17 A simple rotational network. (a) Lumped model. (b) Line-graph representation. (c) Source displacement pattern. (d) Source velocity pattern.

We assign both θ_1 and θ_2 the value zero at $t = 0^-$. Then at $t = 0^+$, the instant *immediately after* the source is turned on, Ω_1 has jumped to the value Ω_c, but θ_1 is still 0. We should expect the jump in velocity of the source to cause other variables in the network to jump. Those variables associated with energy storage in the spring (T and θ_{12}) cannot jump. But what about the other variables (θ_2, Ω_2, $\dot{\Omega}_2$, $\dot{T}$, etc.)? We can determine the *initial values* of *all* variables (at $t = 0^+$) from the network equations and from the *prior values* of the variables (at $t = 0^-$).

Since the source itself does not undergo an abrupt displacement at the instant $t = 0$, it cannot cause an abrupt displacement of point 2. Hence, $\theta_2(0^+) = 0$. But then the spring equation requires that $T(0^+) = K\theta_{12}(0^+) = 0$. According to the damper equation, $\Omega_2(0^+) = T(0^+)/B = 0$. We use the differential equation (3.23) to compute $\dot{T}(0^+) = K[\Omega(0^+) - T(0^+)/B] = K\Omega_c$. Finally, the damper equation requires that $\dot{\Omega}_2 = \dot{T}/B = K\Omega_c/B$. Thus, we have found the initial values of all network variables and their initial rates of change.

In general, if a network source changes abruptly, every node potential and every branch flow in the network responds with a jump—either itself or in one of its derivatives. In this example, the jumps occur in Ω_1, $\dot{\Omega}_2$, and $\dot{T}$. (A systematic method for determining the sizes of the jumps in the network variables is presented in section 3.3.)

Now let us examine the behavior of the system during a brief increment of time δt after $t = 0$. For $t \geq 0$, the input velocity is $\Omega(t) = \Omega_c$. During the time increment δt, the input moves to $\theta(\delta t) = \Omega_c \delta t$. The velocity of node 1 is specified as $\Omega_1(\delta t) = \Omega(\delta t) = \Omega_c$, a constant. The integral of that constant is $\theta_1(\delta t) = \theta(\delta t) = \Omega_c \delta t$.

If δt is very small, we can approximate the function $T(\delta t)$ by the first two terms of its Taylor series expansion about $t = 0^+$: $T(\delta t) \approx T(0^+) + \dot{T}(0^+)\delta t = K\Omega_c \delta t$. (The last equality follows from the previously computed initial values of $T(0^+)$ and $\dot{T}(0^+)$.) The equation for the spring requires that $\theta_2(\delta t) = \theta_1(\delta t) - T(\delta t)/K = 0$. Finally, the equation for the damper requires that $\Omega_2(\delta t) = T(\delta t)/B = K\Omega_c \delta t/B$. Thus, we have used the initial values of the variables, together with the network equations, to determine the values of the variables at $t = \delta t$.

We repeat these computations again and again to generate numerical values for all network variables at the instants δt, $2\delta t$, $3\delta t$, etc. The numerical errors (owing to the Taylor series truncation) at each step of the process can be kept as small as we wish by using a sufficiently small value of δt. (Of course, the errors cannot be kept smaller than the round-off errors caused by the finite length of a computer word.) This process for computing numerical values of network variables is known as **system simulation.** Simulation is a convenient method for finding network behavior, but it requires specific numerical input signals and specific numerical parameter values.

Exercise 3.2: Determine the values of T, $\dot{T}$, θ_1, θ_2, and Ω_2 at $t = 2\delta t$.

Operational Models

The computations used to find the sequence of numerical values of the variables can be expressed as mathematical operations on signals: scaling, summing, and integrating. We denote these operations graphically by **operational blocks.**

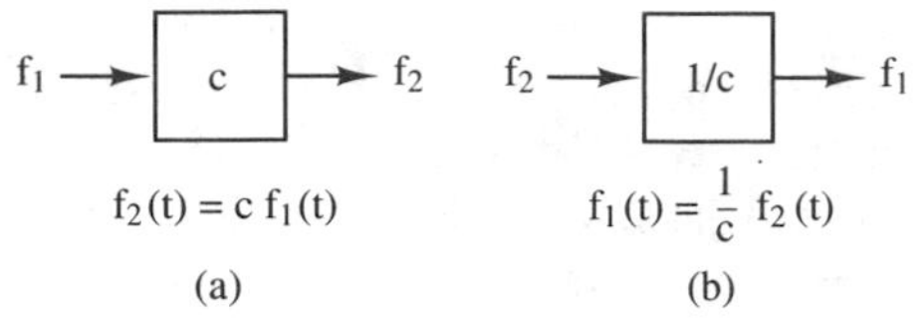

Figure 3.18 Scaling blocks. (a) Multiplication by c. (b) Division by c.

Figure 3.18(a) illustrates the **scaling block.** The output signal $f_2(t)$ is the input signal $f_1(t)$ multiplied by the scale factor c. The fact that the relation $f_2 = cf_1$ can be rewritten $f_1 = f_2/c$ might suggest that the block would also scale the output signal f_2 to produce the input f_1. However, the arrows in the diagram show that this is not the case. The operation of the block applies only in one direction. This directional nature of the operational block contrasts with the nature of the lumped elements of chapter 2 and section 3.1. Each lumped element represents a nondirectional *relationship* between two variables. Each operational block, on the other hand, represents an *operation* that has a cause f_1 and an effect f_2. The scaling block can be thought of as a computer algorithm that multiplies each input sample by the scale factor c to produce the output sample. The algorithm does not operate in reverse. If we wish to carry out the inverse operation, $f_1 = f_2/c$, we must use a different scaling block, one with the scale factor 1/c, as shown in Figure 3.18(b).

Figure 3.19 shows a **summing block.** The output of this block is the indicated sum or difference of the two input signals. In this case, it is obvious that the mathematical operation is directional: One cannot write an algorithm that determines the two inputs from a specified output.

The **integrating block,** or **integrator,** is shown in Figure 3.20. The integrator embodies the essential difference between static and dynamic systems. The integrator is dynamic because its output depends on the past history of the input (via the initial condition). The integrator must be initialized by supplying the initial value of the output variable—in a mathematical sense, specifying the arbitrary constant of integration. We can think of the integrator as a *history block.* The initial condition $f_2(0)$ represents the effect on f_2 of the input history up to (and including) time $t = 0$. Then, the output $f_2(t)$ represents the history up to time t for $t > 0$. The integrator equation is equivalent, except for the initial condition $f_2(0)$, to the differential equation $\dot{f}_2 = f_1$. If f_1 is finite in the neighborhood of $t = 0$, then f_2 cannot jump abruptly at $t = 0$, and we need not distinguish between $f_2(0^-)$ and $f_2(0^+)$.

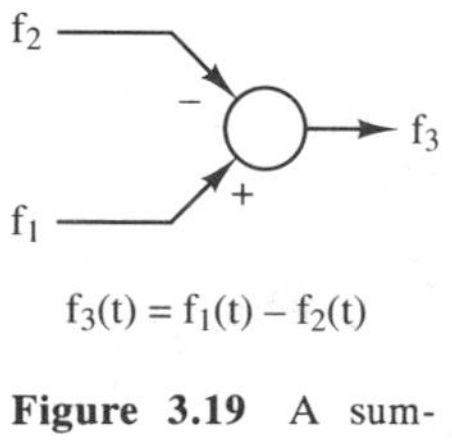

Figure 3.19 A summing block.

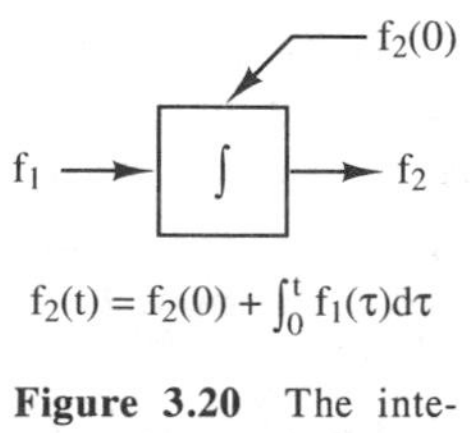

Figure 3.20 The integrating block; the block is sometimes marked with the notation 1/s rather than $\int$.

Any mathematical operation on signals can be represented by an operational block. For example, we can define a differentiating block, a squaring block, or a block that multiplies two signals. We introduce such blocks as the need arises. (Example problem 5.4 uses a square-root block.)

There are two approaches to generating an **operational model** of a network—that is, a model composed of operational blocks. One approach is to represent each network element by an operational model and then interconnect the operational models. The other approach is to write the system equation for the network variable of interest and then use operational blocks to model the mathematical operations of the system equation. The second method produces the simplest operational models and thus is more efficient computationally. The first method, on the other hand, is physically appealing because it retains some similarity to the original network.

We illustrate the computationally efficient approach first. Equation (3.23) is the system equation for the torsion T in the network of Figure 3.17(a). It incorporates all the physical parameters of the network. We rewrite equation (3.23) in the form

$$\dot{T} = -\frac{K}{B}T + K\Omega \tag{3.24}$$

The integral of $\dot{T}$ is

$$T(t) = T(0) + \int_0^t \dot{T}(\tau)\,d\tau \tag{3.25}$$

(We need not distinguish between $T(0^-)$ and $T(0^+)$ because $\dot{T}$ is finite.) The first step in building the operational diagram (or operational model) is to form the integration operation given in equation (3.25), with the highest order derivative $\dot{T}$ as input. Then we use equation (3.24) to construct the signal $\dot{T}$ from the system input signal Ω and the output T of the integrator.

The complete model is shown in Figure 3.21. The model includes a **negative feedback loop.** That is, the integrator output signal T is *fed back* and subtracted from the input signal. Signals circulate around the loop in the direction of the arrows. These signals are integrated and multiplied by $-K/B$ each time they traverse the loop. Typically, operational models of dynamic systems have negative feedback loops that enclose integrators.

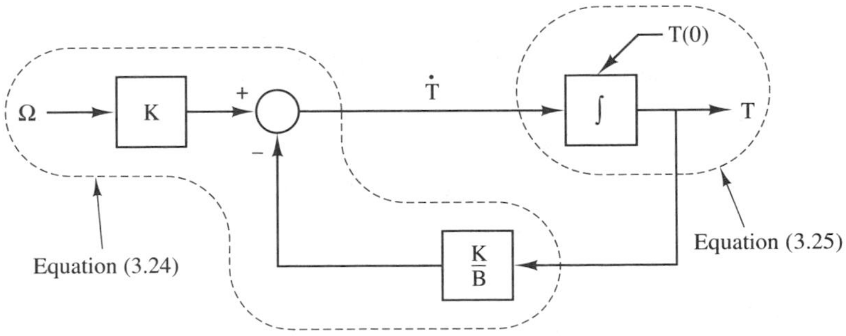

Figure 3.21 An operational model for Figure 3.17.

The operational model makes it clear that the input quantities needed to find a solution are the initial condition T(0) and the source signal $\Omega(t)$ for $t \geq 0$. The initial condition is the initial *state* of the system; it indicates the energy initially stored in the spring. Alternatively, we can view T(0) as representing the *history* of the system owing to the source $\Omega(t)$ for $t < 0$.

The nature of the operational model contrasts sharply with the nature of the lumped-network model (Figure 3.17) for the same system. In a lumped network, there is an *across* variable for each branch and a *through* variable for each node. Each branch represents a mathematical relation between an across variable and a through variable. The graphical layout has a one-dimensional physical interpretation. In the operational model, on the other hand, all variables (across and through) are represented by arrows. The blocks between arrows represent one-way mathematical operations on signals. The graphical layout should be interpreted mathematically rather than physically.

The operational model in Figure 3.21 is not a *complete* description of the network, because it does not generate the values of the motion variables θ_2 or Ω_2. However, each of these variables can be generated by a simple operation on one of the variables in the existing operational diagram; specifically, we attach the operational blocks shown in Figure 3.22 to the operational model in Figure 3.21.

The operational model of Figures 3.21 and 3.22 is not a unique representation of the network: We were somewhat arbitrary in choosing the system equation for T to represent the network. If we use the damper equation $T = B\Omega_2$ to eliminate T from equation (3.23), we produce the system equation for Ω_2:

$$\dot{\Omega}_2 + \frac{K}{B}\Omega_2 = \frac{K}{B}\Omega \tag{3.26}$$

An operational model based on equation (3.26) differs from that of Figure 3.21. However, both operational models have a common structural feature that is fundamental to the network: a single integrator enclosed in a feedback loop, with a loop multiplier $-K/B$.

Exercise 3.3: Draw the operational model associated with equation (3.26).

We summarize the procedure for constructing operational models as follows:

OPERATIONAL-MODEL CONSTRUCTION PROCEDURE

1. Find the system equation for one of the dependent variables.
2. Arrange the equation to express the highest order derivative in terms of lower order derivatives.
3. Use integration operations on the highest order derivative to generate all the lower order derivatives.

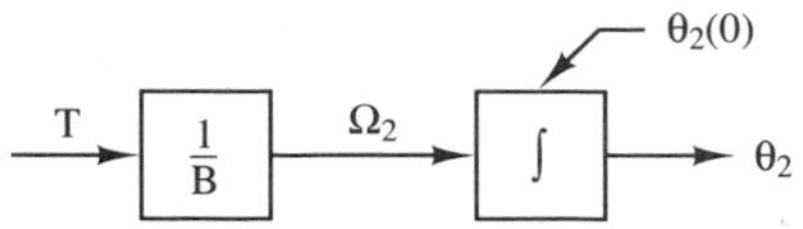

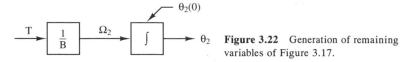

Figure 3.22 Generation of remaining variables of Figure 3.17.

4. Select operational blocks as indicated by the equation in step 2 to generate the highest order derivative from the lower order derivatives. Use the source signal as an input signal.
5. To generate variables of interest that do not yet appear in the model, attach operational blocks that represent lumped-element equations or node flow-balance equations to the diagram obtained from steps (3) and (4).

The operational model of Figure 3.21 does not represent the individual ideal elements of Figure 3.17 explicitly. We can construct an operational model that does represent those elements explicitly by assembling operational models for the individual elements. Figure 3.23 shows such a model. This operational model includes variables for all network nodes and branches, without having to attach the extra operations required by the previous approach. Again, the figure contains a single integrator, and the feedback around that integrator has the scale factor $-K/B$. However, the feedback structure is not in the form that is produced if we use a single system equation to form the operational model.

Example Problem 3.4: An operational model for a multiloop system.
Figures 3.24(a) and (b) show a lumped translational network and its corresponding line graph. Derive the node equations. Use these equations to find the input-output system equation for x_1. Use the system equation to derive an operational model that generates $x_1(t)$. Attach additional blocks as necessary to generate x_2, F_1, F_2, and F_3.
Solution: The node equations, in operator notation, are:

$$\begin{aligned} F - k_1x_1 &= s^2mx_1 + bsx_{12} \\ bsx_{12} &= k_2x_2 \end{aligned} \tag{3.27}$$

We use the second of these equations to eliminate x_{12} from the first, then solve the resulting equation for x_2 in terms of x_1:

$$x_2 = \frac{F - s^2mx_1 - k_1x_1}{k_2} \tag{3.28}$$

We then use equation (3.28) to eliminate x_2 from the second equation of (3.27). The resulting system equation can be expressed in the form

$$\left(s^3 + \frac{k_2}{b}s^2 + \frac{k_1 + k_2}{m}s + \frac{k_1k_2}{mb}\right)x_1 = \left(\frac{s}{m} + \frac{k_2}{mb}\right)F \tag{3.29}$$

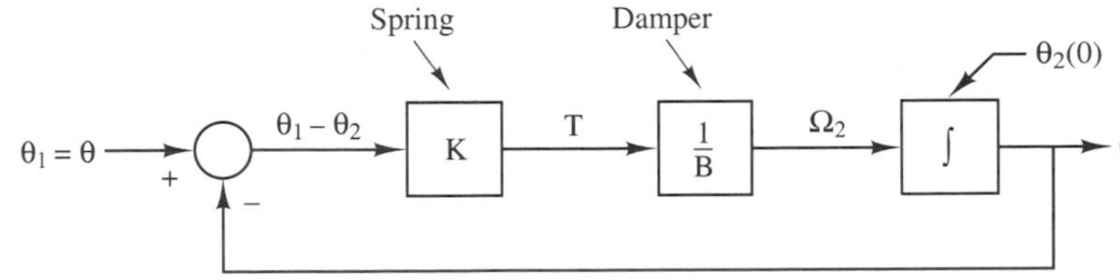

Figure 3.23 Element-based operational model.

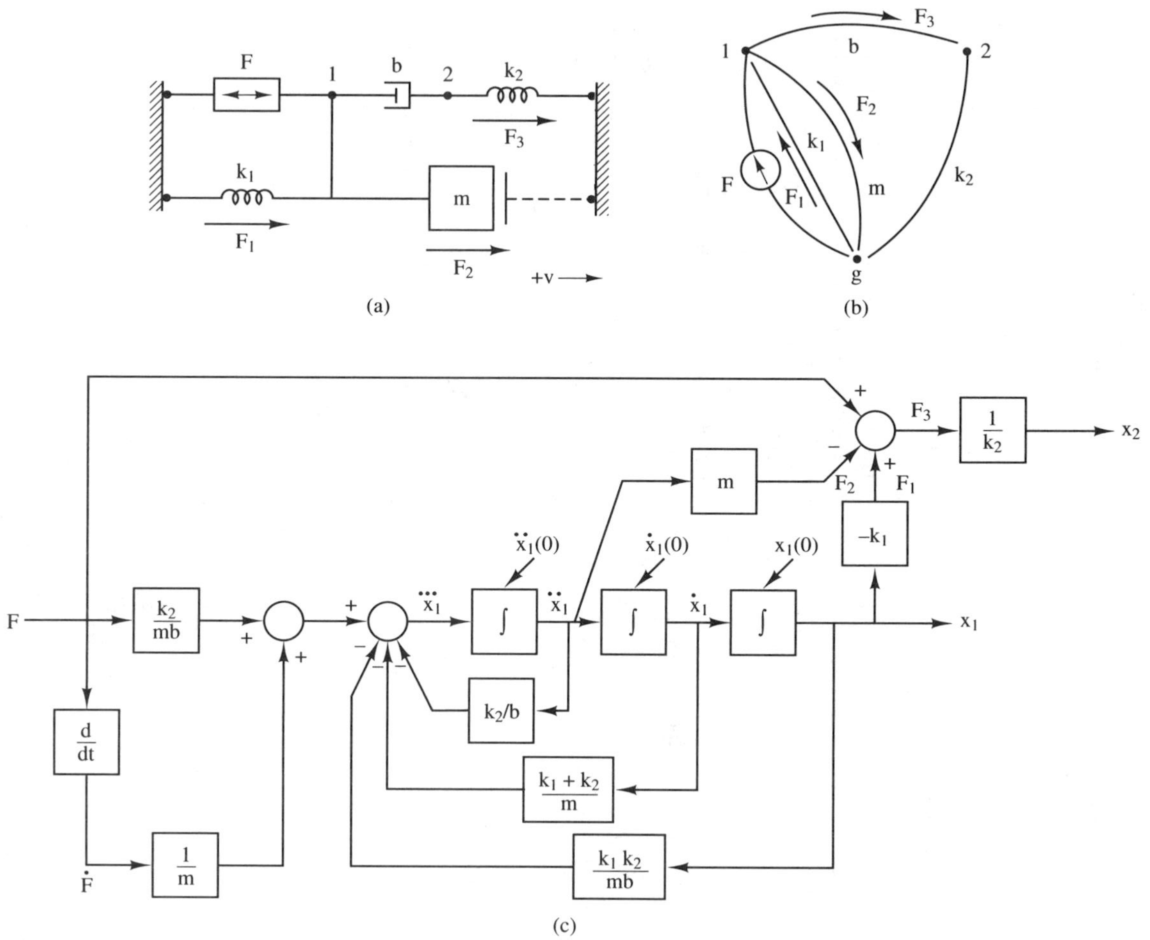

Figure 3.24 Multiloop operational diagram. (a) Lumped network. (b) Line-graph representation. (c) Operational diagram.

By putting this equation in the conventional differential form and solving for the highest order derivative, we obtain

$$\dddot{x}_1 = -\frac{k_2}{b}\ddot{x}_1 - \frac{k_1 + k_2}{m}\dot{x}_1 - \frac{k_1 k_2}{mb}x_1 + \frac{k_2}{mb}F + \frac{1}{m}\dot{F} \tag{3.30}$$

Next, we apply the operational-model construction procedure to equation (3.30) to produce the model shown in Figure 3.24(c). The element equations and the flow balance at node 1 relate the other network variables to x_1: $F_1 = -k_1 x_1$, $F_2 = m\ddot{x}_1$, $F_3 = F + F_1 - F_2$, and $x_2 = F_3/k_2$. We have attached blocks to implement these relations in Figure 3.24(c).

The only input data required to compute the signals in the operational model are the source signal F(t) and the initial values $x_1(0)$, $\dot{x}_1(0)$, and $\ddot{x}_1(0)$. We can never find the behavior without this minimal information, of course. The initial values specify the condition of the network at the instant we apply the source signal. The source signal indicates how we choose to energize the network.

Note that we derive the signal that generates F_2 from the signal $\ddot{x}_1$ that is available in the diagram. We avoid attaching differentiator blocks to the signal x_1 because differentiation magnifies round-off errors.

As a rule, the number of integrators needed to model a network equals the order of the differential equation—in this case, three. The number of integrators equals the number of independent energy-storage elements in the network. Specifying the initial conditions for all of the integrators is equivalent to specifying the initial energy stored in each of the storage elements. That is, we can determine the energy in each storage element from the initial conditions on the integrators. We refer to the required number of integrators as the **order** of the network. Thus, the network of Figure 3.24 is third order.

Exercise 3.4: Show that specifying $x_1(0)$, $\dot{x}_1(0)$, and $\ddot{x}_1(0)$ is equivalent to specifying the initial energy stored in the mass and springs.

The operational model of Figure 3.24(c) differentiates the source signal. Therefore, an abrupt change in the source F would produce a very large, very brief (impulsive) signal at the output of the differentiator block and an abrupt jump in value at the output of the first integrator. The impulsive signal and its integral cannot be computed accurately. Even if we do not use an abruptly changing source, the differentiator magnifies the round-off errors owing to the finite length of a computer word. Hence, we avoid using differentiators in operational diagrams.

The differentiator appears in the operational model because the differential equation has the derivative of the source signal on the right side. We can modify the operational-model construction procedure for system equations that include right-side derivatives so that differentiator blocks do not appear. We first demonstrate the modified procedure for the system equation (3.30) of example problem 3.4. Then we summarize the changes required to avoid using differentiator blocks.

We rearrange equation (3.30) to put the highest order derivatives of both variables (the source F and the output variable x_1) on the left:

$$\dddot{x}_1 - \frac{1}{m}\dot{F} = -\frac{k_2}{b}\ddot{x}_1 - \frac{k_1 + k_2}{m}\dot{x}_1 - \frac{k_1 k_2}{mb}x_1 + \frac{k_2}{mb}F \tag{3.31}$$

Then we use an integrator block to integrate the signal on the left side of equation (3.31), as shown in Figure 3.25. The integrated signal is $\ddot{x}_1 - F/m$. Finally, we divide the source signal F by m and add the result to the output of the first integrator to produce $\ddot{x}_1$. Thereafter, we use integrator blocks to produce the lower order derivatives $\dot{x}_1$ and x_1, as before. Equation (3.31) shows how to form the signal at the input to the first integrator from the source signal and the derivatives of x_1. The resulting operational model is shown in Figure 3.25.

Exercise 3.5: Attach to the operational model of Figure 3.25 the blocks necessary to construct all the other potentials and flows of Figure 3.24(a).

PROCEDURE CHANGES FOR RIGHT-SIDE (SOURCE-SIGNAL) DERIVATIVES:

Replace steps 2 and 3 of the operational-model construction procedure by the following:

2′ Arrange the system equation so that the terms with the highest order derivatives of both variables (the dependent variable and the independent variable) are on the left and the other terms are on the right.

3′ Use integrator operations on the left side of the equation (as arranged in step 2′) to remove the derivative(s) from the *source-signal* term. Then use the original source signal to remove the undifferentiated source-signal term from the integra-

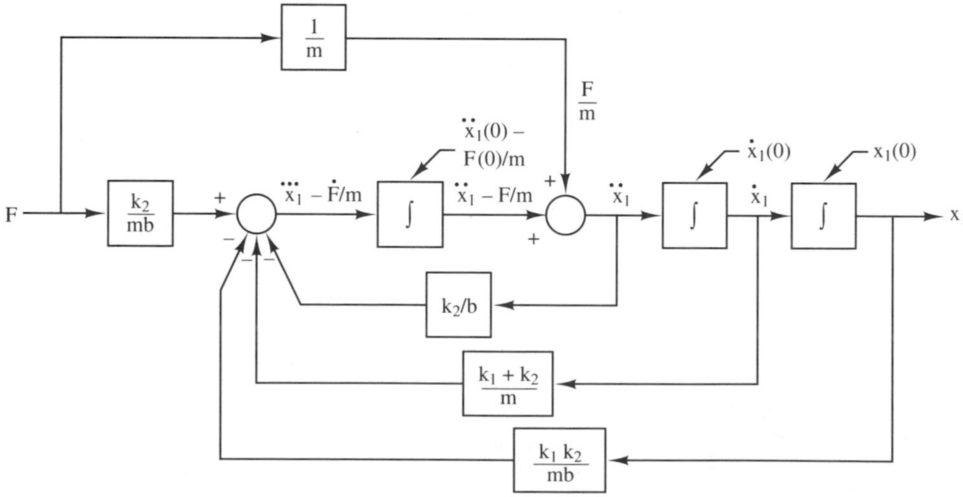

Figure 3.25 An operational model that avoids differentiators.

tor output. Use integrator operations on the remaining dependent-variable signal to generate all lower order derivatives of that signal.

Computer Simulation

A number of commercial computer programs approximate the actions of operational blocks numerically; examples are CSMP, EXTEND, and TUTSIM.[2] These **operational-block simulation programs** compute the values of the variables by time increments, as we demonstrated in finding the values of the variables at $t = \delta t$ and $t = 2\delta t$. However, commercial packages typically provide better integration schemes than the simple two-term Taylor series approximation for $T(\delta t)$ that we used above.

The user of an operational-block simulator must specify the time increment δt and the length T_f of the time interval over which values of variables are to be computed. To keep the error owing to numerical integration small, we must use a small time step δt. However, use of too small a value of δt proliferates computations. A large computation interval T_f (relative to δt) requires that variables be computed for a large number of instants. Thus, we should avoid picking a value of δt that is smaller than necessary or a value of T_f that is larger than necessary.

Experience shows that δt need only be small enough to obtain 10 samples of each significant variation in each signal. For example, if the most rapidly varying signal in the network is a sinusoid with a 1-second period, we only need 10 samples in each quarter period, or $\delta t = 1/40$ sec $= 25$ msec. If the most rapidly varying signal is the exponential $e^{-t/\tau}$, which experiences about 63% of its drop from 1 to 0 in τ seconds, we only need 10 samples in τ seconds, or $\delta t = \tau/10$ seconds. The user must estimate the rapidity of variation of the various signals in the network in order to choose δt. This estimation process requires trial and error or insight and experience. The rapidity of variation of signals in linear networks is discussed in sections 3.3 and 3.4.

We usually simulate network operation for a standard input signal, such as an abruptly applied constant input or a sinusoidal input. A linear network with such an input stabilizes to *steady-state* behavior after a well-defined *transient* interval of time. Computing the values of variables for time instants after the attainment of steady-state behavior contributes no new information about the network behavior. Therefore, it is unnecessary for T_f to be much larger than the *settling time,* i.e., the **length of the transient interval.** The user must estimate the settling time in order to choose an appropriate value for T_f. Again, this estimation requires either trial and error or insight and experience on the part of the user. The settling time is examined in sections 3.3 and 3.4.

Network behavior can be simulated only for a specific numerical input signal, specific initial values of the variables, and specific values for the element parameters. Figure 3.26 shows a plot of the output $T(t)$ for the operational model in Figure 3.21 with the input signal $\Omega(t)$ of Figure 3.17(d), for specific values of K, B, t_1, and Ω_c. The initial torsion is set to the value $T(0^+) = 0$, as determined in the previous section. Note from the plot that $T(t)$ settles *exponentially* at a rate that achieves 63% of the *total amount of settling* in about $\tau = 0.6$ s. The plot was produced by TUTSIM with $\delta t = 0.05$ s ($\approx \tau/10$).

[2] Appendix B gives a brief description of TUTSIM.

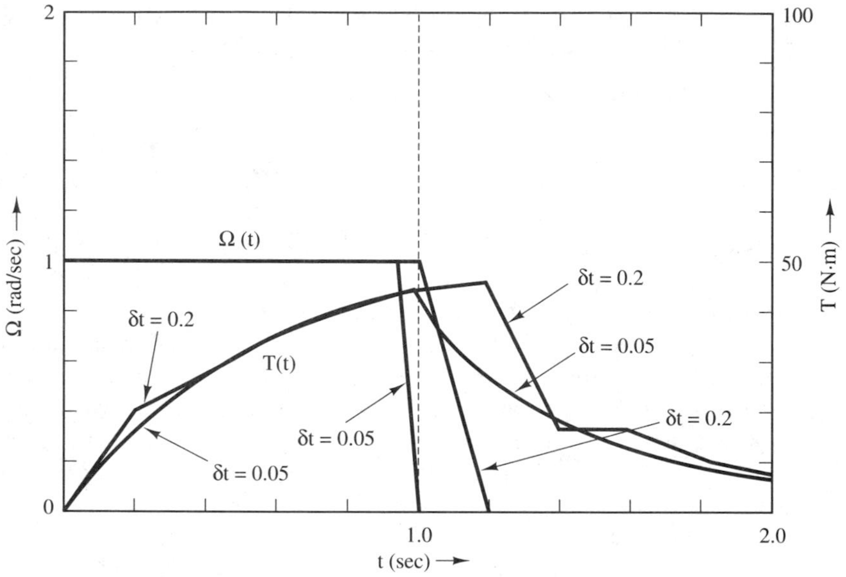

Figure 3.26 Numerical solution for the diagram in Figure 3.21; K = 100 N·m·s/rad, B = 50 N·m·s/rad, Ω_c = 1 rad/s, t_1 = 1 s, and T(0) = 0 N·m.

That choice of time increment is small enough to compute T(t) accurately over the pulse interval, but not small enough to represent accurately the abrupt drop in value of Ω(t) at t = 1 s. Figure 3.26 also shows the simulation for the larger time increment δt = 0.2 s. This time increment, which allows only three samples over the interval of signal *decay,* introduces too much numerical error in the computation of T(t).

Operational-block simulation programs typically provide graphical output such as that in Figure 3.26. Graphical presentation of data gives more insight than does a column of computed values. The total number $T_f/\delta t$ of computed samples can be too large to display on a video screen. For example, if the simulation of the system in Figure 3.21 required that the time increment be no greater than δt = 1 msec for accurate calculation, then a 2-second interval would include 2,000 samples, too many to fit on a typical graphics display. In that case, we maintain the high sampling rate (the small value of δt) to achieve the necessary accuracy, but *display* only every fifth data point, in order that the data fit on the screen.

Exercise 3.6: Use an operational-block simulator to implement the operational diagram of Figure 3.21, and verify the network response shown in Figure 3.26.

Physical Intuition

We analyze a system to develop an intuitive, yet somewhat quantitative, feeling for the system's behavior. It is like examining a car or a house before purchasing it. We cannot understand fully the intricacies and advantages of the system until we have *experienced* its behavior over the complete range of operating conditions. Yet, not wanting to make a major investment without confidence in the purchase, we first expose the character of the item in ways that we expect to concern us. For example, we drive a car over a rough road and around sharp turns. Or we examine the basement moisture and the attic insulation in a house. We also have an expert, someone with experience, help us decide what to look for and how to interpret our observations.

Similarly, in order to test a lumped-network model of a dynamic system, we *stress* the model by determining its response to an abruptly applied input signal, i.e., a *step* input. Such a stimulus exercises two extremes of behavior: the static behavior and the reaction to an abrupt change. Unfortunately, books on modeling and analysis usually focus so much on mathematical techniques, that interpretation of the analysis is forgotten completely. Yet it is only by means of physical interpretations that we develop the breadth of physical insight necessary to create new system designs.

After generating a lumped model and determining its response to a particular signal waveform, we should always interpret the solution physically in terms of the system from which the network is derived. After solving a problem in this text, a student who cannot answer the question, "How does the system behave?" has not finished the problem. We should ask ourselves, "Is this behavior believable?" That is, is it consistent with our physical experience? If so, our confidence in our intuition is strengthened. If not, we should examine the system and the network further in order to educate our intuition better or to find a flaw in our model or simulation.

Figure 3.26 shows the behavior of the lumped network of Figure 3.17. No underlying physical system was discussed. Suppose that the network is a model for a physical rotational spring connected to a physical viscous-friction device. Then, according to the simulation, the velocity source (perhaps an electric motor) abruptly begins to rotate node 1 at a constant velocity, 1 rad/s. (A physical motor could not suddenly jump from rest to a velocity of 1 rad/s, of course. But it could change from rest to 1 rad/s in such a short time, relative to the responsiveness of the spring-damper system, that it *appears* to jump abruptly.) The spring begins to *wind,* opposed by the resistance of the friction device. The torsion in the spring, motor, and damper immediately begins to increase, at about 100 N·m/s for the parameter values in the example. But the rate of increase drops gradually, and the torsion approaches a steady value of about 50 N·m. When the source rotation at node 1 stops abruptly, the manner in which the spring unwinds and relieves the torsion is similar to the winding behavior of the spring. The torsion reduces at a rapid rate initially and eventually approaches zero.

The behavior of node 2, the connection between the spring and the damper, is not shown in the simulation. From the damper equation (3.9), we see that the drop in rotational velocity across the damper must change in proportion to the torsion. When the source *locks* node 1 abruptly, the spring torsion is resisted by continued rotation of the damper against its viscous friction. As the damper continues to turn, it relaxes the spring. The rotational velocity of node 2 decreases to zero in proportion to the torsion. The reader should visualize the physical rotational behavior and compare it with his or her experience.

Model Validation

What values should we use for the parameters of the lumped model? We can use the educated guess of an expert; however, that guess is guided by previous experience—essentially, measurement. We can use design formulas (such as those of Figures 3.6 and 3.9) to compute parameter values. But those formulas require measured material properties (see Appendix A) and dimension estimates that are based on experience. The best way to obtain accurate parameter values for the model of a specific physical system is to make appropriate measurements on the actual physical system. Hence, experimental measurement is essential to accurate modeling. The parameter-measurement process is demonstrated in example problem 3.5.

Once we have obtained a lumped model with specific numerical values for the model parameters, we still must validate the model experimentally. We check the whole-system model by comparing the behavior of the model with the behavior of the physical system that the model represents. Suppose, for example, that we apply the same step input to both the model and the physical system. We measure and compare the same key features of each response. If the physical response and the model response are nearly the same, we say that the model is validated.

But what do we mean by *nearly the same?* It depends on the system. The rate of flow of water through a shower head can vary as much as 1 part in 10 (10%) without our noticing it. Therefore, we need not require the *model* of that hydraulic system to be more accurate than 1 part in 10. But electronic clocks that are accurate to 1 part in 10,000 (0.01%) can gain or lose 4 minutes a month. We might seek, then, to make the model of the clock's timing mechanism more accurate than 1 part in 10,000. We consider a model validated if it behaves the same as the physical system *to the degree that we need.*

Parameter variations of 0.1% to 10% (from the average) are typical for newly manufactured parts and systems. Heating, aging, and use can change a device parameter (and the ensuing behavior of the device) by as much as 50%. Since the parameters of physical systems change so drastically, we usually make systems adjustable. (For example, we put an adjustable valve on a shower so that the user can compensate for changes in water pressure.) As a consequence, we usually do not need lumped models of these systems that are more accurate than 1 part in 100 (1%).

When we compare the behavior of a model with the measured behavior of the physical system that it represents, we must account for the accuracy of measurement. The instruments that we use to measure the behavior of the physical system and to measure the values of the parameters that we use to compute the behavior of the model can be in error by several percent. We can expect agreement between the behavior of the model and the behavior of the physical system only to within the accuracy of the measurements. We must not be misled by the precision (number of decimal places) to which we can read the measuring instruments or by the precision to which we can calculate the predicted behavior. That *precision is no indication of accuracy.*

If a model is found to be less accurate than we desire, we can increase its accuracy by increasing the level of detail of the model. But the added insight diminishes as the level of detail increases. We need not model a system in great detail to understand its behavior. A model need only be exact enough to accomplish its purpose. To make it more exact is to waste resources.

Example Problem 3.5: Experimental validation of a lumped model.
The physical system shown in Figure 3.27(a) is intended to demonstrate the destructive transverse vibrations that earthquakes can induce in a building. The vertical members are of thin flexible steel. They act as transverse cantilever springs. The velocity source represents the vibratory motion of the base that would be caused by an earthquake. We show a lumped model of the system in Figure 3.27(b). We include no damper in the model, because oscillations in this physical model die out very slowly. We performed the following measurements on the physical model.

To determine the mass m, we removed the top bar, found its weight (4 N) with a spring scale, and calculated its mass to be 0.41 kg. We also tested the spring scale by weighing a number of masses of precisely known weight. We found the scale accuracy to be ±5%. Then we reinstalled the top bar.

To determine k, we used the same spring scale to pull the top bar to one side with a scale force of 1 N and measured the resulting deflection. Then we repeated the measurement with the bar pulled to the other side. The average of the two deflections was 3.8 cm. The individual deflections differed from the average by 1 mm. Therefore, we view the measurement as accurate to no more than ±1 mm (or ±2.6%).

To test the system's behavior, we pushed the stationary system abruptly 5 cm to the left, at which point it was stopped abruptly by the barrier. In effect, we executed a step of size 5 cm in the variable $x \equiv \int v$. The bar, labeled as node 2 in the model, oscillated in a manner similar to that shown in Figure 3.28.

We measured the most easily discerned features of the system response to the step input—the displacement at the first peak overshoot of x_2, the time of that first peak overshoot, and the period of oscillation of x_2. We used a meter stick mounted behind the system to measure the displacement and a stopwatch to measure the peak time and the period. The

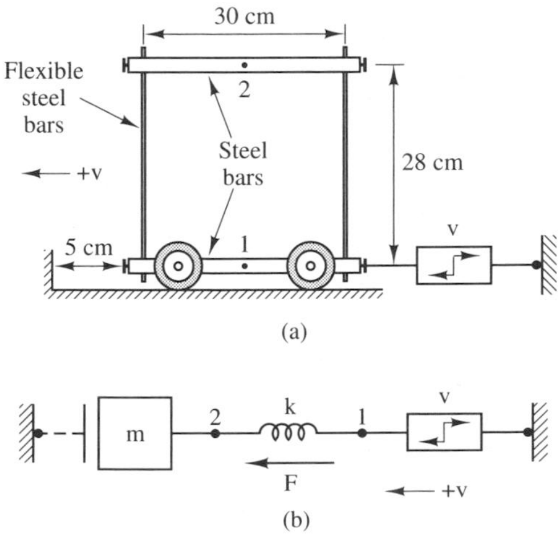

Figure 3.27 A single-story structure. (a) Physical model. (b) Lumped model.

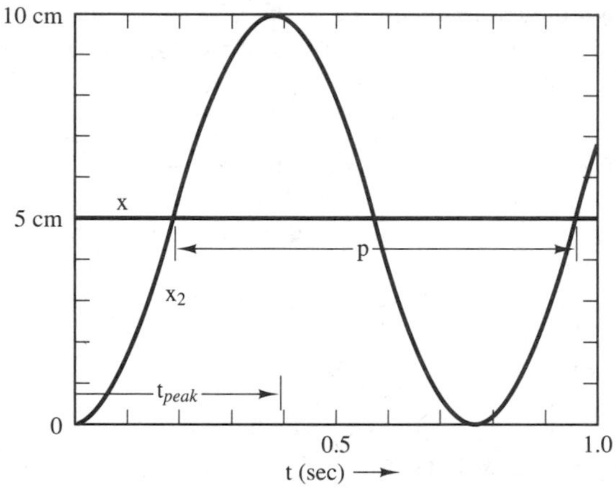

Figure 3.28 Simulated behavior of the system shown in Figure 3.27.

period was measured by timing five cycles of oscillation and dividing by five. The peak time was difficult to measure accurately. The measured values (averages over several trials) were:

Peak value, $x_{2peak} = 9.6 \text{ cm} \pm 0.2 \text{ cm}$ (or $\pm$ 2.1%)

Time of peak, $t_{peak} = 0.35 \text{ s} \pm 0.11 \text{ s}$ (or $\pm$ 31%)

Period of oscillation, $p = 0.76 \text{ s} \pm 0.05 \text{ s}$ (or $\pm$ 6.6%)

The accuracies specified with each measurement were chosen so as to include the range of values that were found on the various trials.

(a) Use the static measurements to determine the parameters b and m of the lumped model.

(b) Simulate the behavior of node 2 of the lumped model for the computed parameter values to predict the key features of the step response—the peak overshoot, the time of peak, and the period of oscillation.

(c) Assess the accuracy of the model.

Solution: **(a)** If we assume that the mass m owes only to the weight of the top bar, then $m = 0.41 \text{ kg} \pm 5\%$. The equation for the spring is $-F = kx_{21}$. When the system was deflected, we measured $F = 1 \text{ N} \pm 5\%$ and $x_{21} = 3.8 \text{ cm} \pm 2.6\%$. Then $k = (1 \text{ N})/(0.038 \text{ m}) = 26.3 \text{ N/m} \pm 7.6\%$.

(b) The input-output system equation for the motion v_2 of the lumped model of Figure 3.27(b) is

$$\left[s^2 + \left(\frac{k}{m}\right)\right]v_2 = \left(\frac{k}{m}\right)v \tag{3.32}$$

Since the source motion is a step function of displacement, it is more appropriately expressed in terms of x(t), rather than v(t). To convert from velocities to displacement, we replace v by sx and v_2 by sx_2 and then divide the equation by s. The result is

$$\left[s^2 + \left(\frac{k}{m}\right)\right]x_2 = \left(\frac{k}{m}\right)x \tag{3.33}$$

We replace s by d/dt and solve for the highest order derivative to obtain

$$\ddot{x}_2 = \left(\frac{k}{m}\right)(x - x_2) \tag{3.34}$$

The corresponding operational diagram, shown in Figure 3.29, was simulated using TUTSIM, with $k/m = (26.3\ N/m)/(0.41\ kg) = 64.1\ rad/s^2$, $x_2(0) = 0$, $\dot{x}_2(0) = 0$, and $x(t) = 5$ cm ($t \geq 0$). The response $x_2(t)$ given in Figure 3.28 shows that the peak overshoot is $x_{2peak} = 10$ cm, the time of the peak is $t_{peak} = 0.39$ s, and the period of oscillation is $p = 0.79$ s.

(c) The simulated time of peak and period are each within the uncertainty range for the corresponding measured value. The simulated peak value, however, differs from the measured peak value by twice the apparent uncertainty in measurement. What might account for this error? The measured values of m and k used in the simulation had uncertainties of 5% and 7.6%, respectively. Therefore, the quantities obtained from the simulated response have roughly the same level of uncertainty. Furthermore, the lumped model of this example does not account for the slow decay in the oscillation of the physical device. A damper with a small friction constant is added to the model to account for this decay in example problem 3.13.

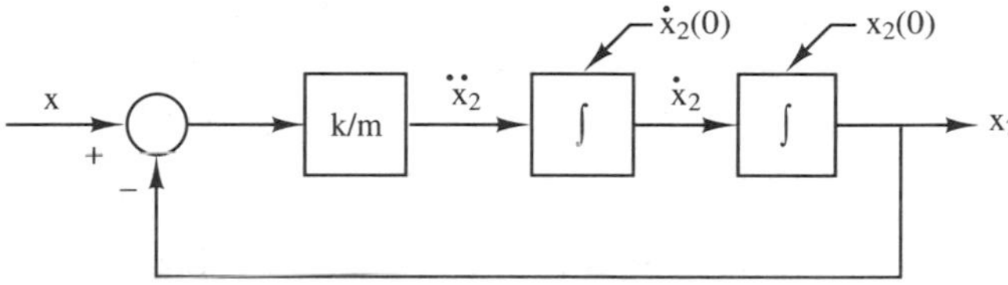

Figure 3.29 Operational diagram for Figure 3.27(b).

Exercise 3.7: How might the dynamic time measurements be performed more accurately?

Exercise 3.8: Show that if ϵ_1 and ϵ_2 are much less than one, then

$$(1 + \epsilon_1)(1 + \epsilon_2) \approx 1 + \epsilon_1 + \epsilon_2$$

$$\frac{1 + \epsilon_1}{1 + \epsilon_2} \approx 1 + \epsilon_1 - \epsilon_2 \tag{3.35}$$

Therefore, the percent error in a product or quotient is the sum of the percent errors of the operands.

3.3 FUNDAMENTALS OF NETWORK BEHAVIOR

The future behavior of a network depends partly on future values of the source and partly on the present energy state of the network, that is, the energy inserted into the storage elements by previous actions of the source. In what follows, we first define and examine the energy state of the network. Then we analyze the behaviors of simple (first-order) networks. We find the way that previously stored energy rearranges itself within the network. We call that natural rearranging process the *zero-input response* of the network. Then we examine the behavior of an unenergized network when the source is suddenly turned on. We call the suddenly applied source signal a *step input* and its effect on the network the *zero-state step response.* Finally, we combine the two responses.

For a step input, it is convenient to express the response of each network variable in terms of its initial and final values. At the end of this section, we derive a systematic technique for finding the initial and final values of all variables in any linear lumped network. Throughout the section (indeed, throughout the book), we assume that the network is time invariant; that is, neither the network structure nor the values of its element parameters change with time. Hence, the system differential equations have constant coefficients.

The Network State

A network has a number of variables. Source variables are *independent variables.* They are not affected by the network. The rest of the variables in a network are *dependent variables.* Their behaviors result from the action of the source on the network. We can think of source signals as *input signals* and dependent-variable signals as *output signals* or *response signals.*

A network is driven, energized, or excited by its energy source. Suppose the source excites the network for a period of time. During this time, energy flows throughout the network. Some of that energy dissipates (turns to heat) in the friction elements (dampers), and some of it energizes the storage elements. Suppose the source is turned off abruptly at an instant we designate as $t = 0$. Because the storage elements are energized, the variables of the network continue to *respond to the source,* despite the fact that it is turned off. The signals keep changing until energy stops transferring from the storage elements to the dissipation elements (friction devices) of the network. This energy decay (after $t = 0$) is the *zero-input response* of the network—the reponse with the source (input signal) set to zero. We can also think of it as the state-induced response—the response owing to the state of the network at the instant the source is turned off. We shall find in section 4.2 that the effect of stored energy on the behavior of a network is the same as the effect of ideal sources inserted in series or parallel with the storage elements. Thus, there is an equivalence between energy sources and stored energy.

For each energy domain, we usually write the system equations in terms of the *central* potential and flow variables. (The central variables for translational networks are velocity and force.) The energy in a storage element is uniquely related either to the central flow through the element or to the central potential difference across the element. We call that *central* flow variable or potential-difference variable the **energy-state variable** for that storage element. The velocity of a mass is the energy-state variable for the mass; the force in a spring is the energy-state variable for the spring.[3] An nth-order network has n independent storage elements and n independent energy-state variables. We call the *value* of an energy-state variable at a particular instant the **energy state** of the corresponding storage element. The energy states of the n storage elements, taken together, constitute the **network state.** The network state is the network "memory" of *past* energy flows.

[3] Since $F = kx_{ij}$ for a spring, we can treat x_{ij} as an alternative energy-state variable for the spring. It is not the central energy-state variable for the spring, however.

We can represent the state of an nth-order network by the values of any n network variables *from which we can determine the energy states* of the individual storage elements. We call such a set of variables a **state-variable set** for the network. (We sometimes refer to an individual variable in such a set as a *state variable.*) The energy-state variables themselves form one set of state variables. If an operational model of the network is constructed without differentiator blocks, the variables at the integrator outputs in that model make up another set of state variables.

We can see the equivalence between the integrator outputs and the energy states in the third-order network of Figure 3.24(a). The energy-state variables are the velocity $\dot{x}_1$ of the mass and the forces F_1 and F_3 in the springs. In the corresponding operational model of Figure 3.25, the integrator outputs are x_1, $\dot{x}_1$, and $\ddot{x}_1 - F/m$. The energy state $\dot{x}_1$ is a direct integrator output. The energy state $F_1 = -k_1x_1$ is proportional to an integrator output. The energy state $F_3 = -F_2 + F + F_1 = -m(\ddot{x}_1 - F/m) - k_1x_1$, a combination of integrator outputs.

The behavior (or response) of a network for $t \geq 0$ depends on the initial network state and on the source signal for $t \geq 0$. We see this dependence in the operational model of Figure 3.25. It has four externally controlled quantities: the source signal and the initial conditions on the three integrators. A change in any one of these quantities will change the future behaviors of all dependent variables in the model. The initial conditions on the integrators represent the initial network state. If the source is zero for $t \geq 0$, then the initial conditions on the integrator still initiate a network response. This response represents a decay of stored energy. In the physical terminology of Figure 3.24(a), the mass gradually slows and the springs gradually decompress until all the stored energy is absorbed by the damper.

We say that a lumped network is **unenergized** (or that it is in the *zero state*) if the values of all the energy-state variables are zero. We call the response of an unenergized network (to the source or input alone) the **zero-state response.** We can think of it as the *source-induced response.* In Figure 3.25, the operational model is initially unenergized if all the integrator initial conditions are zero. The integrator outputs then remain zero, unless an input (source) signal is applied. The response of the unenergized model to a source signal is the zero-state response. On the other hand, we call the response of a network to stored energy alone the **zero-input response.** We can think of it as the *state-induced response;* it is also known as the *natural response* or the *free response.* The response of the operational model of Figure 3.25 to the integrator initial conditions alone is the zero-input response.

For a *linear* network, the behavior of each variable for $t \geq 0$ can be split into a sum of two noninteracting parts. One part—the zero-input response—results only from previously stored energy. The other part—the zero-state response—results only from the continuing source. We show in section 4.3 that we can find these two response components separately and then add them together, a process known as *superposition.* In a nonlinear network, such as the one in example problem 5.4, the effects of the input signal and the network state cannot be treated separately.

It is customary to examine a system's behavior by *turning on* the energy source abruptly. That is, we apply a *step-function input.* We might expect the abrupt change in the source to change the values of all the network dependent variables abruptly, but that is not the case. Values of state variables just will not jump abruptly. The values of a set of state variables for a network correspond to the energy state of the network. An abrupt change in stored energy would require that the source provide infinite flow or infinite potential difference. Similarly, in an operational model, an abrupt jump in an integrator output would require an infinite integrator input. If the operational model does not include differentiators, then the integrator outputs do not jump in response to a step input.

Even though an energy-state variable does not jump abruptly in response to a step input, it does experience a jump in one of its derivatives. When we turn on the force source of Figure 3.24(a), the force F_2 in the mass changes abruptly. The velocity v_1 of the mass does not change abruptly because it is an energy-state variable. However, its derivative $\dot{v}_1$ changes abruptly.

Since some network variables jump in response to a step-function input, we must distinguish between the values of variables before and after the jump. We use the term **prior value** of a variable y to mean the value of y at the instant just prior to the jump. We denote this value by y_p or $y(0^-)$. The **initial value** of y is the value of y just after the jump. We denote it by y_i or $y(0^+)$.

State variables represent the energy state of the network; they do not jump in response to a step input. There is no difference between the initial value and the prior value of a state variable. It is correct, therefore, to speak of the **initial state** of the network when we mean the prior state and to use the notation y(0) to indicate the initial (or prior) value of a **state variable** y.

The behavior of each dependent variable in a network is embodied in a single differential equation—the *input-output system equation* for that variable. It is customary to arrange the terms of a differential equation so that the source-signal terms are on the right side and the response-signal terms are on the left. Each dependent variable of the network has the same differential form on the left side of its system equation. (Transpose $(-K/B)T$ in equation (3.24) to the left side, and compare this equation with equation (3.26), for example.) The right sides of the system equations for different variables of the network contain different combinations of the source signal and its derivatives.

Now look at the third-order system equation (3.29) of example problem 3.4. The right side is a linear combination of the source signal F and its derivative $\dot{F}$. Suppose that F(t) is a step function. If we visualize the step-function waveform as rising from zero to the value F_c over a brief interval of length δt, then the waveform of the derivative $\dot{F}(t)$ is an **impulse**, a large brief pulse of size $F_c/\delta t$ and duration δt. Thus, the waveform of the right side of equation (3.29) includes both a step (owing to F) and an impulse (owing to $\dot{F}$) at $t = 0$. The left side of the equation must then have the same waveform.

Since the left side of the equation includes an impulse (the derivative of a step function), the highest order derivative $\dddot{x}_1$ must include an impulse. But then its integral $\ddot{x}_1$ includes a step; that is, $\ddot{x}_1$ jumps abruptly. The lower order derivatives, $\dot{x}_1$ and x_1, do not jump in response to the abrupt change in the source. (Recall that x_1 and $\dot{x}_1$ are energy-state variables for that network.)

Let us extend these conclusions to a network variable y that has a system equation of order n:

(a) If the input-output system equation for y does not include source-signal (right-side) derivatives, then an abrupt jump in the source signal produces a corresponding jump in $y^{(n)}$, the highest order derivative of y; it does not produce jumps in the lower order derivatives of y, including the zeroth-order derivative, y itself. Hence, the initial values (at $t = 0^+$) of $\{y, y^{(1)}, \ldots, y^{(n-1)}\}$ are the same as the prior values (at $t = 0^-$). These lower order derivatives correspond to the outputs of integrators in an operational model of the network. That is, they form a set of state variables for the network.

(b) If the system equation includes the first derivative of the source signal, then a jump in the source signal produces a jump in $y^{(n-1)}$, the *second-highest* derivative of y; it does not produce jumps in lower order derivatives of y. For example, we see a jump at the output $\ddot{x}_1$ of the first integrator in Figure 3.24(c), owing to the impulse produced by differentiation of the step in F, but no jumps at the outputs of the other integrators.

(c) If the system equation includes higher order derivatives of the source signal, then a jump in the source causes jumps in lower order derivatives of y.

We typically test a network by finding its zero-state step response—the response of the unenergized network to a step function source signal. The input-output system equations for some of the network variables have source-signal (right-side) derivatives (see equations (3.91)–(3.93) and Figure 3.44, for example). Finding the step response is harder for a variable that has source-signal derivatives in its system equation because of the initial jump that occurs in one of its derivatives.

In the next three sections, we examine the decay of the stored energy in linear *first-order* networks. The examination of energy decay demonstrates the exponential nature of solutions to linear differential equations. It also introduces the network *time-constant,* a measure of the rate of exponential decay. The time constant is a *system parameter*—an intuitive indicator of the system's behavior. The two succeeding sections show how to find step responses for linear first-order networks. The step response y(t) depends on the initial condition, that is, the value $y_i = y(0^+)$ just after the source is turned on. We shall find it convenient to express the step response in terms of its final value y_f, the limit of y(t) as $t \to \infty$.

Higher order (nth-order) networks are examined in section 3.4. In these nth-order networks, the input-output system equations for some network variables do include right-side derivatives. If the system equation for y includes right-side derivatives, we must find the initial jumps in value of $\{y, y^{(1)}, \ldots, y^{(n-1)}\}$ in order to find the initial conditions $\{y(0^+), y^{(1)}(0^+), \ldots, y^{(n-1)}(0^+)\}$ that are required to solve for the response y(t) to an input step. Again, we shall find it convenient to express the step response in terms of its final value y_f. Therefore, in section 3.3 we develop methods for finding final values and initial jumps in value for all variables in a lumped network.

An Energy Decay Example

Figure 3.30 shows a lumped model of a door closer. The compressive-force source represents the person who holds the door open against the restoring force of the spring. When the person releases the door, the damper limits the rate at which the spring closes the door. (This model neglects the mass of the door.)

The force source, with strength F_c, holds the spring in a stretched position at $t = 0^-$. Since the door is held stationary, the damper is inactive, with $F_2(0^-) = 0$. The force in the stretched spring is tensile, with $F_1(0^-) = -F_c$. According to the spring equation, the spring displacement is $x_1(0^-) = F_c/k$.

We disconnect the source from the network at $t = 0$. (Disconnecting the source is the same as turning off the force, thereby changing the source value abruptly to zero.) The subsequent network behavior owing to energy stored in the spring is the *characteristic behavior* or *zero-input response* of the network.

Removing the source abruptly causes some of the network variables to change abruptly. Abrupt displacement of node 1 would require infinite velocity v_1 and infinite compressive force F_2 in the damper, an impossibility. Hence, the displacement x_1 is the same just after removal of the source as it was just before release; that is, $x_1(0^+) = x_1(0^-) = F_c/k$. Since the amount by which the spring is stretched does not change abruptly, the spring force does not jump at $t = 0$, and $F_1(0^+) = F_1(0^-) = -F_c$. The variables x_1 and F_1 are energy-state variables, quantities related to the energy stored in the spring. The energy in a storage device cannot be changed abruptly.

After the source is removed, the node equation at node 1 requires $F_2(0^+) = F_1(0^+)$. Thus, the damper force jumps from $F_2(0^-) = 0$ to $F_2(0^+) = -F_c$. This jump in the damper force requires a corresponding jump in the damper velocity, from $v_1(0^-) = 0$ to $v_1(0^+) = F_2(0^+)/b = -F_c/b$.

The node equation for node 1 gives the system equation for the displacement x_1:

$$b\dot{x}_1 + kx_1 = 0 \tag{3.36}$$

(This is just the input-output system equation for x_1, with the source set to zero.) Equation (3.36) is a first-order, linear ordinary differential equation with constant coefficients. We refer to the network, therefore, as a **first-order** linear network. The equation is also homogeneous; that is, its right side is zero.

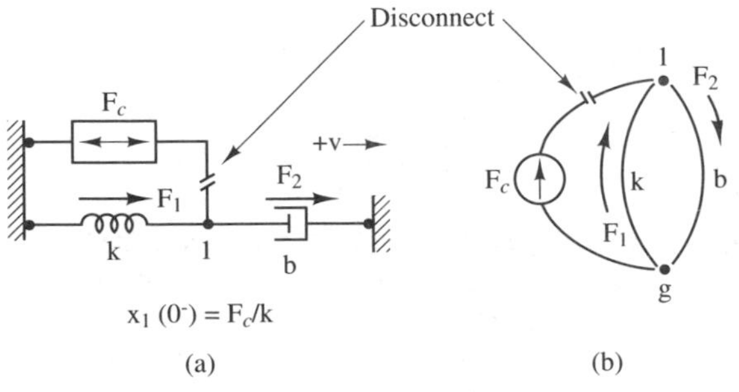

Figure 3.30 A lumped model for a door closer. (a) Lumped model. (b) Line graph.

We can express this equation in the form

$$\dot{x}_1 = -\frac{k}{b} x_1 \tag{3.37}$$

We multiply (3.37) by the differential time increment dt to obtain

$$\left(\frac{dx_1}{dt}\right) dt = -\frac{k}{b} x_1\, dt \tag{3.38}$$

The left side of equation (3.38) is the differential change dx_1 in the motion x_1. We rearrange the equation to the form

$$\left(\frac{dx_1}{x_1}\right) = -\frac{k}{b}\, dt \tag{3.39}$$

Integrating both sides of equation (3.39) over the interval $[0^+, t]$ yields

$$\int_{x_1(0^+)}^{x_1(t)} \frac{dx}{x} = -\frac{k}{b}\int_{0^+}^{t} d\sigma \tag{3.40}$$

or

$$\ln x_1(t) - \ln x_1(0^+) = -\frac{k}{b}\, t \tag{3.41}$$

(We exclude the instant $t = 0$ from the integration interval to avoid concern about jumps in values of network variables at $t = 0$.) Consequently,

$$x_1(t) = x_1(0^+)e^{-(k/b)t}, \qquad t > 0 \tag{3.42}$$

where $x_1(0^+) = F_c/k$.

Exercise 3.9: Use equations (3.42) and (3.37) to show that

$$\dot{x}_1(t) \equiv v_1(t) = v_1(0^+)e^{-(k/b)t}, \qquad t > 0 \tag{3.43}$$

where $v_1(0^+) = -F_c/b$.

Exercise 3.10: Use the damper equation together with equation (3.43) to show that

$$F_2(t) = F_2(0^+)e^{-(k/b)t}, \qquad t > 0 \tag{3.44}$$

where $F_2(0^+) = -F_c$.

The waveforms of x_1 and v_1 are displayed in Figure 3.31. Both signals exhibit the same basic exponential decay for $t > 0$. All dependent variables in a first-order network decay at the same rate. (To verify this fact, sketch $F_2(t)$.) The natural response of each variable is proportional to the initializing force F_c.

Because x_1 is an energy-state variable, we can denote the initial condition by $x_1(0)$ rather than $x_1(0^+)$; that is, the initial and prior values of x_1 are the same. On the other hand, we cannot ignore the distinction between 0^- and 0^+ for the variables v_1 and F_2; both of these variables jump in response to the abrupt change in the source.

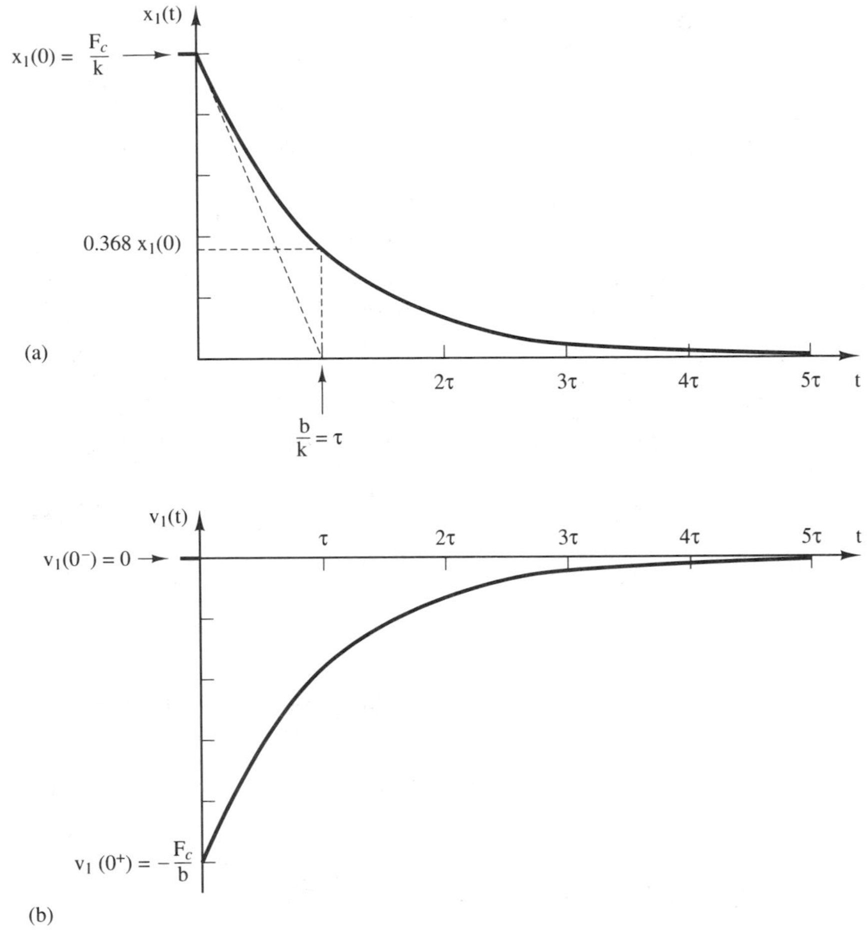

Figure 3.31 Zero-input response of the door-closer model. (a) Door-closer displacement. (b) Door-closer velocity, with abrupt jump.

We model a system and calculate the behavior of the model to develop an intuitive understanding of the system itself. Therefore, we must interpret the solution physically. We visualize the physical behavior of the system in terms of the lumped model, which means that we view the model as physical.

The initial displacement $x_1(0^+)$ of the spring corresponds to an initial compressive force $F_1(0^+) = -F_c$ in the spring. The negative value of the compressive force means that the spring is in tension. This tensile force causes the spring to contract, which movement is resisted by the damper. For $t > 0$, the tensile force $-F_2$ in the damper equals the tensile force $-F_1$ in the spring. The velocity of contraction of the spring decays exponentially in accordance with equation (3.43). The spring (or door) displacement also decays exponentially (see equation (3.42)). Equation (3.44) shows the exponential manner in which the tensile force F_2 in the spring and damper decreases to zero.

The power dissipated in the damper during the natural decay of network signals can be found from any of the expressions

$$P = F_2 v_1 = b v_1^2 = \frac{F_2^2}{b} \tag{3.45}$$

We use the last form, together with equation (3.44) and the spring equation $F(0) = kx_1(0)$, to obtain

$$P(t) = \frac{k^2 x_1^2(0)}{b} e^{-(2k/b)t}, \qquad t > 0 \tag{3.46}$$

(The rate of decay of power dissipation is twice the rate of decay of the signals.) The total energy absorbed by the damper for $t > 0$ is

$$E = \int_{0^+}^{\infty} P\,dt = \frac{k}{2} x_1^2(0) = \frac{F_2^2(0)}{2k} \tag{3.47}$$

This absorbed energy equals the initial energy stored in the spring.

Characteristic Behavior—the Time Constant

Exponential decay is the characteristic behavior of all first-order linear networks, that is, networks in which energy in a single storage element transfers to a single linear energy absorber. In mathematical terms, every first-order, linear, constant-coefficient, homogeneous differential equation has an exponential function of time as its solution. Although we derived equation (3.42) in the context of Figure 3.30, it applies to any quantity that is described by a differential equation of the form of equation (3.37). Thus, the solution to

$$\dot{y} + cy = 0 \qquad (\text{or} \quad \tau\dot{y} + y = 0) \tag{3.48}$$

is

$$y(t) = y(0^+)e^{-ct} \qquad (\text{or} \quad y(t) = y(0^+)e^{-t/\tau}), \qquad t > 0 \tag{3.49}$$

The units of the coefficient c are inverse time. The sign of c determines whether the variable y grows or decays. If the elements that make up the network are passive, then c must be positive and y(t) must decay. In that case, we say that the network is **stable.** A network in which signals *grow* exponentially is said to be unstable.

The reciprocal coefficient $\tau \triangleq 1/c$ is known as the **time constant** of the first-order network. The time constant is a fundamental parameter of the network. In fact, the time constant is the *only* parameter of the first-order *network* (as opposed to parameters of the individual elements of the network). For the specific spring-damper model of the previous section, the network time constant is $\tau = b/k$. The *behavioral property* that τ characterizes, exponential decay, is not possessed by either the spring or the damper alone.

All variables in a first-order network decay with the same time constant. Accordingly, we view that time constant as the natural decay rate *of the network*. It is convenient to express the decay of signals in a first-order network in multiples of the network time constant τ. The *total decay* of the variable y in equation (3.49) is $y(0^+)$. That is, y decays from $y(0^+)$ to 0 as $t \to \infty$. From equation (3.49),

$$y(t + \tau) = y(0^+)e^{-(t+\tau)/\tau} = e^{-1}y(0^+)e^{-t/\tau} = 0.368y(t) \tag{3.50}$$

During any interval of length τ, the value of y reduces by the factor 0.368. Therefore, **63% of the total decay occurs during the first time constant.** During one more time constant (by the time $t = 2\tau$), the variable y reduces by another 63% and reaches the value $y(2\tau) = 0.368\, y(\tau) = 0.135\, y(0^+)$. More than 95% of the total decay occurs within three time constants, 98% within four time constants, and 99% within five time constants. It is customary to define $t_s \triangleq 5\tau$ as the **settling time** of the first-order network. That is, we consider the decay to be essentially complete after five time constants.

The operational model for equation (3.48) is shown in Figure 3.32. Since the source value is zero, the model is energized only by the initial condition on the integrator. It is apparent from the model why we call the characteristic (source-free) response the **zero-input response.** The operational model displays explicitly the key network quantities, τ and $y(0^+)$. The model contains a *feedback signal*—a signal that loops back from the output end of the model toward the input end. Because of the feedback loop, signals *circulate* in the diagram. The output at instant t modifies (is subtracted from) the input at instant t to affect the output at instant $t + \delta t$. The feedback signal for a passive network is always negated, corresponding to a positive coefficient c in equation (3.48). The feedback multiplier is $1/\tau$, the reciprocal of the network time constant.

The zero-input response of each variable y in a first-order linear network is completely characterized by the initial value $y(0^+)$ of that variable and the *network* time constant of decay τ. This statement is demonstrated in the door-closer example of Figures 3.30 and 3.31. Now that we know the nature of the zero-input response of a first-order linear network, we need not solve differential equations to find and plot the signal waveforms. Instead, we write the differential equation for any one of the network variables in the form shown in equation (3.48). From that equation, we read the value of the time constant τ. Then we use equation (3.49) to state the zero-input responses of all the network variables. We find the initial values directly from the network. We demonstrate this procedure in example problem 3.6. We develop systematic procedures for finding the initial values of the network variables in a later section.

A variable in a linear network that has a single storage element and a single energy-dissipation element is represented by a first-order constant-coefficient differential equation. Its characteristic behavior is simple exponential decay. Problem 2.17 in chapter 2 shows that a series or parallel connection of identical elements behaves as a single equivalent element. It should be no surprise, therefore, that a complex interconnection of ele-

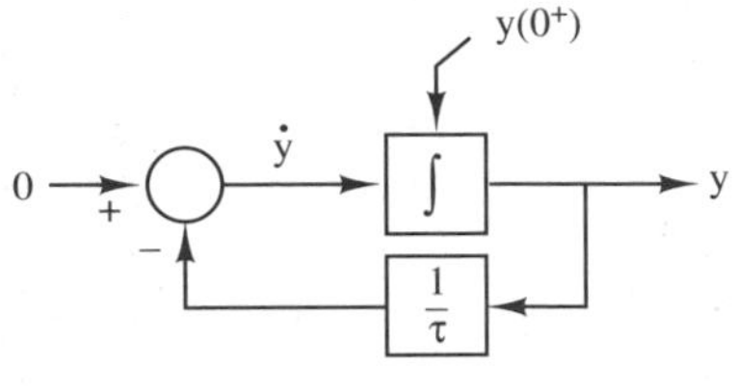

Figure 3.32 Operational model for equation (3.48).

ments can still behave as a first-order network and exhibit simple exponential decay. Example problem 3.7 shows such a case.

Example Problem 3.6: Natural behavior of a first-order network.
In the translational network of Figure 3.33(a), the motion source compresses k with the displacement $x_1(0^-) = x_c$. The source is disconnected at $t = 0$.
(a) Find the motion of the node and the compressive forces in the elements for $t > 0$.
(b) Draw an operational model that generates the motion $x_1(t)$.
Solution: (a) We draw the line graph for the network (for $t > 0$) in Figure 3.33(b). The loop equation for the network is $(s/k)F_2 + (1/b)F_2 = 0$. We rewrite this equation in the form $(b/k)sF_2 + F_2 = 0$ and then compare it with equation (3.48) to find that the network time constant is $\tau = b/k = 2$ s.

Since node 1 cannot move abruptly without changing the energy in the spring, $x_1(0^+) = x_1(0^-) = 6$ cm. Therefore, according to equation (3.49),

$$x_1(t) = 6e^{-t/2} \text{ cm}, \qquad t > 0 \tag{3.51}$$

The initial force in the spring is $F_1(0^+) = kx_1(0^+) = 18$ N. Then equation (3.49) requires that

$$F_1(t) = 18e^{-t/2} \text{ N}, \qquad t > 0 \tag{3.52}$$

Note that $F_2(t) = -F_1(t)$.

We can find $v_1(t)$ by differentiating equation (3.51). Alternatively, we can use the network to relate the initial value of v_1 to the initial value of F_2 and then state $v_1(t)$ directly. The damper equation and flow balance at node 1 require that $v_1 = F_2/b = -F_1/b$. Therefore, $v_1(0^+) = -F_1(0^+)/b = -(18\text{N})/(6 \text{ N·s/cm}) = -3$ cm/s. It follows that

$$v_1(t) = -3e^{-t/2} \text{ cm/s}, \qquad t > 0 \tag{3.53}$$

(b) According to equation (3.48), the differential equation that describes the motion $x_1(t)$ is

$$\dot{x}_1 = -\frac{k}{b}x_1 = -0.5x_1 \tag{3.54}$$

The initial value is $x_1(0^+) = 6$ cm. The corresponding operational diagram is shown in Figure 3.33(c).

Exercise 3.11: The node equation at node 1 requires $-kx_1 = bv_1$. Use this equation to find the initial acceleration $a_1(0^+)$. Then show that

$$a_1(t) = 1.5e^{-t/2} \text{ cm/s}^2, \qquad t > 0 \tag{3.55}$$

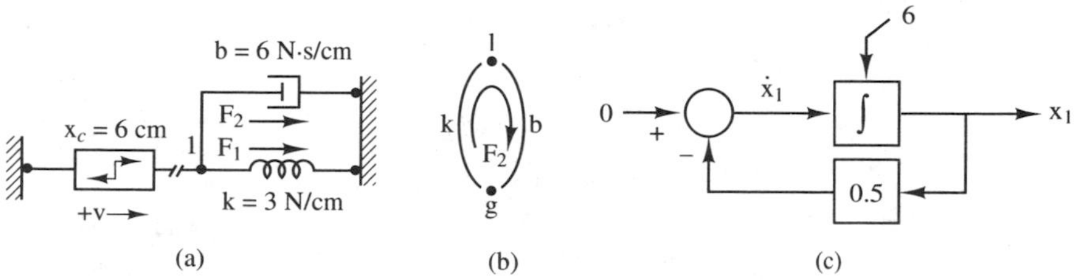

Figure 3.33 First-order characteristic behavior. (a) First-order linear network. (b) Line-graph representation with zero input. (c) Operational model.

The prior values ($t = 0^-$) of F_2, v_1, and a_1 are zero. Thus, these variables jump abruptly at $t = 0$. The prior values of x_1 and F_1 are not zero. These variables are energy-state variables and cannot change abruptly.

Exercise 3.12: Sketch the exponential responses of the variables. Label significant points on the axes, and include the behaviors of the variables at $t = 0$.

Example Problem 3.7: First-order behavior with multiple storage elements.
A rotational network is shown in Figure 3.34. The velocity sources rotate the two inertias at the velocities $\Omega_1 = 20$ rad/s and $\Omega_2 = -5$ rad/s at $t = 0^-$. At $t = 0$, the sources are disconnected. Then the stored kinetic energies are absorbed over time by the connecting damper.

(a) Find the motions $\Omega_1(t)$ and $\Omega_2(t)$ and the damper torsion $T(t)$.

(b) Find the initial energies stored in the inertias (J_1 and J_2) and the energy trapped in the inertias at $t \to \infty$.

Solution: **(a)** The line graph of the network is shown in Figure 3.34(b). We examine the behavior from the viewpoint of nodes 1 and 2. After the sources are disconnected, the two inertias act in series. Let us eliminate the reference (ground) node from the equations. The node equation at node g is $J_2\dot{\Omega}_{2g} = J_1\dot{\Omega}_{g1}$, or $\dot{\Omega}_1 = -(J_2/J_1)\dot{\Omega}_2$. Consequently,

$$\dot{\Omega}_{21} = \dot{\Omega}_2 - \dot{\Omega}_1 = \left(\frac{J_1 + J_2}{J_1}\right)\dot{\Omega}_2 \tag{3.56}$$

We solve equation (3.56) for $\dot{\Omega}_2$ in terms of $\dot{\Omega}_{21}$ and substitute into the inertia equation

$$T = J_2\dot{\Omega}_2 \tag{3.57}$$

to obtain

$$T = \left(\frac{J_1 J_2}{J_1 + J_2}\right)\dot{\Omega}_{21} \triangleq J\dot{\Omega}_{21} \tag{3.58}$$

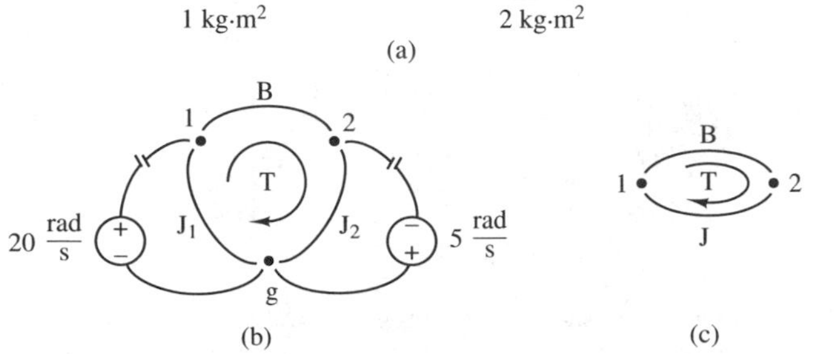

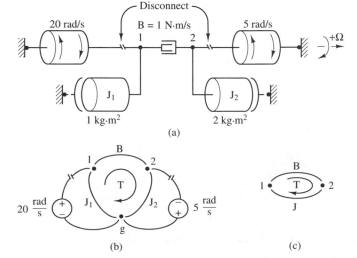

Figure 3.34 First-order energy decay.
(a) Lumped linear network.
(b) Line-graph representation.
(c) Simplified equivalent line graph.

Then we replace the series connection of inertias by the equivalent inertia $J = J_1J_2/(J_1 + J_2) = 0.66\ kg{\cdot}m^2$.

The node velocities Ω_1 and Ω_2 are energy-state variables. They do not change abruptly at $t = 0$. Therefore, the initial velocity drop across the equivalent inertia J is the same as the prior ($t = 0^-$) value, $\Omega_{12}(0^+) = 20 - (-5) = 25$ rad/s. The equivalent network is shown in Figure 3.34(c). The loop equation for this equivalent network is $J\dot{\Omega}_{21} = B\Omega_{12}$, or

$$\dot{\Omega}_{12} = -\frac{B}{J}\Omega_{12} \tag{3.59}$$

Compare equation (3.59) to equation (3.48) to find that the network time constant is $\tau = J/B = 0.667$ sec. According to equation (3.49), then, the velocity difference across the damper is

$$\Omega_{12}(t) = \Omega_{12}(0^+)e^{-(B/J)t} = 25e^{-t/0.667}\ \text{rad/s}, \qquad t > 0 \tag{3.60}$$

It follows that

$$T(t) = B\Omega_{12}(t) = 25e^{-t/0.667}\ \text{N}{\cdot}\text{m}, \qquad t > 0 \tag{3.61}$$

and

$$\dot{\Omega}_1 = -\frac{T}{J_1} = -25e^{-t/0.667}\ \text{rad/s}^2, \qquad t > 0 \tag{3.62}$$

The initial velocity at node 1 is $\Omega_1(0^+) = 20$ rad/s. We obtain the velocity pattern $\Omega_1(t)$ by integrating the acceleration given by equation (3.62) over the interval $[0^+, t]$:

$$\int_{0^+}^{t} \dot{\Omega}_1\, dt = \Omega_1(t) - \Omega_1(0^+) = -25\int_{0^+}^{t} e^{-t/0.667}\, dt$$

$$= 16.68(e^{-t/0.667} - 1)\ \text{rad/s}, \qquad t > 0 \tag{3.63}$$

Thus,

$$\Omega_1(t) = 3.32 + 16.68e^{-t/0.667}\ \text{rad/s}, \qquad t > 0 \tag{3.64}$$

Then, from the equation for the damper,

$$\Omega_2(t) = \Omega_1(t) - \frac{T}{B} = 3.32 - 8.32e^{-t/0.667}\ \text{rad/s}, \qquad t > 0 \tag{3.65}$$

Note that it is the differential equation for Ω_{12} that is in the form of equation (3.48). Therefore, we cannot apply the solution equation (3.49) directly to find Ω_1 and Ω_2. The signal y(t) in equation (3.49) decays to zero, whereas $\Omega_1(t)$ and $\Omega_2(t)$ do not decay to zero.

(b) The initial energies stored in the two inertias are $E_1(0) = J_1\Omega_1^2(0)/2 = 200$ N·m and $E_2(0) = J_2\Omega_2^2(0)/2 = 25$ N·m. (It is not necessary to distinguish 0^+ from 0^-, because Ω_1 and Ω_2 are energy-state variables.) As $t \to 0$, both inertias approach the same velocity, $\Omega_1(\infty) = \Omega_2(\infty) = 3.32$ rad/s, and no more energy is absorbed by the damper. The energy trapped in the inertias is $E_t = (J_1 + J_2)\Omega_2^2(\infty)/2 = 16.68$ N·m. The remainder of the 225 N·m of initially stored energy, 208.3 N·m, is delivered to the damper. (This fact can be checked by integrating $B\Omega_{12}^2$ from 0 to ∞.)

The initial energy stored in the equivalent inertia J is $E(0) = J\Omega_{12}^2(0)/2 = 208.3$ N·m, precisely the amount absorbed by the damper. As $t \to 0$, $\Omega_{12} \to 0$, and all of the energy in the equivalent inertia is transferred to the damper. Only if we acknowledge the ground as the zero-velocity reference do we observe the residual velocity and trapped energy.

Exercise 3.13: Sketch the response functions $\Omega_1(t)$ and $\Omega_2(t)$, and interpret them physically. Do they represent the behavior you would intuitively expect?

Measuring Network Parameters

How might we determine the parameters k and b of a first-order lumped network in order to match the behavior of a physical spring-damper system for which it is a model? The stiffness of the physical spring can be determined from the spring extension caused by a known weight. It is more difficult to measure the velocity needed to determine the friction constant b of the damper. Instead, we displace the physical system as shown in Figure 3.30. Then we use a stopwatch to measure the time for the spring-damper system to relax to 37% of its initial displacement. That measured time is the time constant of the system. The initial displacement and the 37% displacement are not difficult to measure. We then determine the friction constant from the time constant and the stiffness of the spring ($b = k\tau$).

According to equation (3.48), the initial slope of the response y(t) is $\dot{y}(0^+) = y(0^+)/\tau$. That is, the tangent to the curve at $t = 0^+$ drops by the total amount of the decay over a time interval equal to one time constant (see Figure 3.31). Hence, the network time constant can also be determined from measurements of the initial value and initial slope of the curve by the computation $\tau = y(0^+)/\dot{y}(0^+)$. However, accurate measurement of the initial rate $\dot{y}(0^+)$ is difficult.

A Source-Driven Example

We now examine the *forced response* of a first-order linear network—the response to an energy source. In particular, we find the **step response** of an unenergized network, that is, the response to a suddenly applied constant source. Since the network is initially unenergized, the initial values of all the energy-state variables are zero. Therefore, we call the response the **zero-state step response.**

The spring-damper network of Figure 3.35 is unenergized. That is, the source has been *turned off* ($F(t) = 0$) for a long time prior to $t = 0$. The network is totally relaxed: No energy is stored in the spring. At $t = 0$, the source is *turned on*—increased abruptly

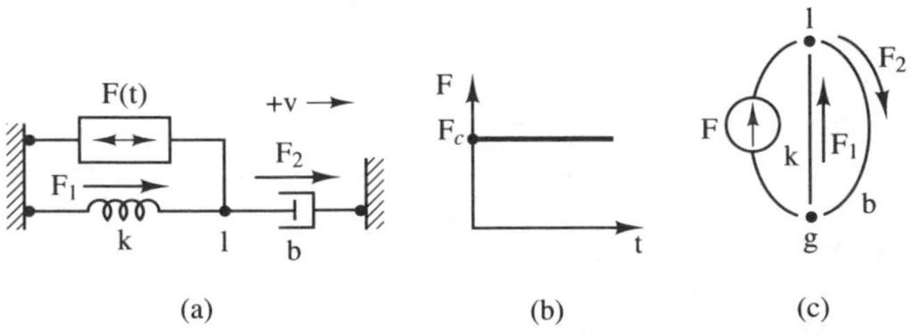

Figure 3.35 Step input to a translational network. (a) Lumped linear network. (b) Step-function input signal. (c) Line-graph representation.

to the constant value $F(t) = F_c$. The differential equation that describes the motion of the network is

$$\dot{x}_1 + \frac{k}{b} x_1 = \frac{F}{b} \tag{3.66}$$

where

$$F(t) = \begin{cases} 0, & t < 0 \\ F_c, & t \geq 0, \end{cases}$$

a step function of size F_c.

First we guess the form of a *particular solution* to equation (3.66), a function $x_{1P}(t)$ which satisfies that equation. Since the right side of the differential equation is the constant F_c/b, we let

$$x_{1P}(t) = D \tag{3.67}$$

where D is some constant.[4] We substitute $x_{1P}(t) = D$ into (3.66) and solve for D. The derivative term vanishes, and $D = F_c/k$.

We now seek a function x_{1C} for which $x_{1P} + x_{1C} = F_c/k + x_{1C}$ is the general solution to equation (3.66). We call x_{1C} the *complementary function* or *characteristic function* for the network. We substitute this general solution form into equation (3.66) to find that $\dot{x}_{1C} + (k/b)x_{1C} = 0$. Therefore, the characteristic function satisfies the homogeneous differential equation—the differential equation with zero right side. We found in the previous section that the solutions to a first-order homogeneous linear differential equation are exponentials of the form $e^{-t/\tau}$. Thus, the characteristic function has the same form as the characteristic response of the network.

The definition of the time constant of decay τ is also based on the homogeneous equation. Hence, we find the time constant for any first-order linear network by comparing the left side of the differential equation with equation (3.48).

We compare equation (3.66) with equation (3.48) to see that $\tau = b/k$ for this example. Therefore, the characteristic function is

$$x_{1C}(t) = Ee^{-(k/b)t} \tag{3.68}$$

Then the *general solution* to the differential equation is

$$x_1(t) = \frac{F_c}{k} + Ee^{-(k/b)t} \tag{3.69}$$

The constant E is commonly referred to as an *arbitrary constant* because equation (3.69) is a solution to the differential equation, regardless of the value of E. There is only one value of E that is consistent with the specified operating conditions, however. We determine separately the value of $x_1(t)$ at the initial instant ($t = 0^+$) and use that value to determine E.

[4] As a particular solution to a linear constant-coefficient differential equation, we can usually use a function that has the same functional form as the *right side* of the differential equation. Or else, we use a sum of that form and its derivatives.

We determine the initial condition $x_1(0^+)$ from the physical nature of the network and knowledge of the prior energy state. At $t = 0^-$, the instant prior to the step, the network is relaxed and $x_1(0^-) = 0$. Since x_1 is an energy-state variable related to the energy in the spring, it does not jump abruptly when the source is applied. Therefore, $x_1(0^+) = x_1(0^-) = 0$. We substitute the initial time $t = 0^+$ and the initial value $x_{1i} \triangleq x_1(0^+) = 0$ into the general solution (3.69) to find $E = -F_c/k$. Therefore, the source-induced response of node 1 to the step input is

$$x_1(t) = (F_c/k)\,[1 - e^{-(k/b)t}], \qquad t > 0 \tag{3.70}$$

The solution is displayed in Figure 3.36. The source causes the spring to stretch rapidly at first, at a rate $\dot{x}_1(0^+) = F_c/b$ limited by the damper. Then x_1 approaches a final value of extension, $x_{1f} = F_c/k$, determined by the spring.

Exercise 3.14: Use the element equations for the network (see Figure 3.35) together with the solution equation (3.70) to show that $v_1(t) = (F_c/b)e^{-(k/b)t}$, $F_1(t) = -F_c(1 - e^{-(k/b)t})$, and $F_2(t) = F_c e^{-(k/b)t}$. Compare the waveforms of these signals with Figure 3.36, and interpret them physically.

One might think that we could use the value of $x_1(t)$ at any instant t to determine E. This is not the case. At $t \to \infty$, the general solution given by equation (3.69) approaches the final value F_c/k, regardless of the value of E. The **final value** of x_1 does not depend on the initial condition; it depends only on the source. The final value cannot be used to determine E. (The final value of x_1 was used to determine the constant D of the particular solution.) The network reaches its final state, for practical purposes, at the end of the transient interval. To determine E, we must use the value of x_1 at some instant in the transient interval. The transient interval is $[0, t_s]$, where $t_s = 5\tau$, the network settling time. We use the initial value (at $t = 0^+$) because the initial energy state of the network is specified in the statement of the problem.

The network in this example need not have been unenergized prior to turning on the source. If the spring were stretched by the prior amount x_{1p} at $t = 0^-$, the initial energy

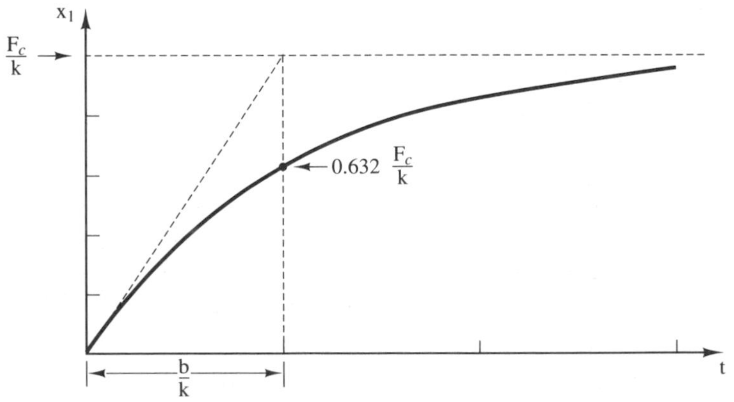

Figure 3.36 Step response for the network of Figure 3.35.

stored in the spring would have been accounted for by a different initial value, $x_1(0^+) = x_{1p}$. Substitution of this initial value into equation (3.69) would have changed the value of the arbitrary constant to $E = x_{1p} - F_c/k$.

The Step Response for First-Order Networks

In general, the step response of a first-order linear network can be expressed in the form

$$y(t) = y_f + (y_i - y_f)e^{-t/\tau}, \qquad t > 0 \tag{3.71}$$

where $y_i \triangleq y(0^+)$ is the initial value of y, τ is the time constant of the network, and y_f is the final value of y, i.e., the value y approaches for $t \gg \tau$. The step response is displayed graphically in this general notation in Figure 3.37. Expressing the step response in this form focuses attention on measurable parameters of the solution, that is, y_i, y_f, and τ.[5] The step response to any variable of the network can be written by inspection, without resorting to the differential equation, once these parameters have been determined. (Compare the waveforms of exercise 3.14 with equation (3.71).)

The step response can be separated into two *time* intervals, namely, a temporary *transient interval,* during which the network variables are changing, and a *steady-state interval,* during which the network is nearly static. The transient response lasts for about 5τ, where τ is the time constant of the network.

If y does not jump abruptly at $t = 0$, then equation (3.71) subdivides y into a zero-input (or state-induced) response $y_i e^{-t/\tau}$ and a zero-state (or source-induced) response $y_f(1 - e^{-t/\tau})$. (If y does jump abruptly, then the initial value y_i depends on the source as well as the prior value. For such a variable, separation of the response into zero-input and zero-state component waveforms is more complicated.) In section 3.4, we call those variables that *do not* jump the *prototype* variables of the network. Every network has some

[5] The network is a lumped model of a physical system. These network response parameters *correspond* to measurable features of the physical system response.

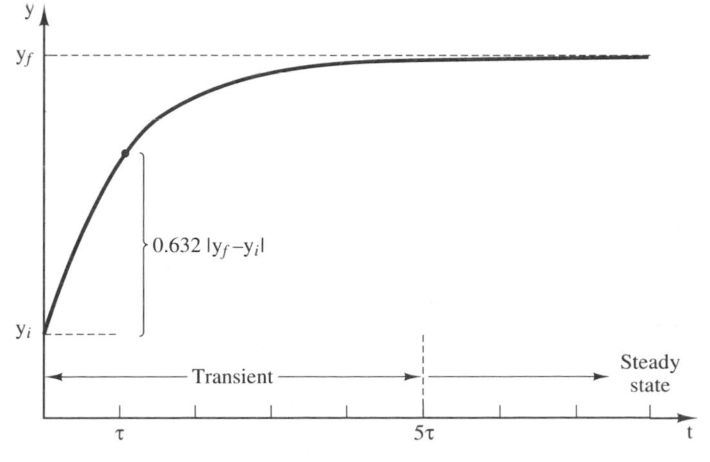

Figure 3.37 Step response form for any first-order linear network.

variables that jump and others that do not (see T and Ω_2 of example problem 3.8, for example).

Example Problem 3.8: First-order response to an abrupt change in the source. Suppose the source for the velocity-driven inertia-damper network of Figure 3.38 has been sustained at the constant value $\Omega = \Omega_c$ long enough prior to $t = 0$ to permit the network to reach steady state. Then at $t = 0$, the source velocity is abruptly increased to $\Omega = \Omega_d$. Find the response of the torsion T to this abrupt change in the source.

Solution: The network is linear and first order. Therefore, the response must have the form of equation (3.71) and Figure 3.37. We must find the initial value T_i, the final value T_f, and the network time constant τ.

The loop equation for the network is $\dot{\Omega} = \dot{T}/B + T/J$, or

$$\dot{T} + \frac{B}{J}T = B\dot{\Omega} \tag{3.72}$$

By inspection—that is, by comparison with equation (3.48)—we see that $\tau = J/B$.

The network is in a steady-state condition (i.e., has unchanging velocities and torsion) during two time intervals—the interval immediately prior to the step and the interval that follows settling of the transient ($t > 5\tau$). During those intervals, the dotted variables of equation (3.72) are zero. Consequently, T must be zero during those intervals. Therefore, the prior value of T is $T_p \triangleq T(0^-) = 0$, and the final value of T is $T_f \triangleq T(\infty) = 0$.

Unlike equation (3.66), the right side of equation (3.72) contains the derivative of the source function. Therefore, that side becomes infinite at the instant of the abrupt change in the input signal $\Omega(t)$. That is, the slope of the function Ω is infinite at $t = 0$. Since the two sides of the equation must be identical functions of time, an abrupt change must also occur in the left-side variable T at that instant. (This abrupt change in T causes the first term of the left side of equation (3.72)—the slope $\dot{T}$ of the function T—to be infinite at $t = 0$, in order that equality be maintained.)

At $t = 0$, the input velocity increases abruptly from Ω_c to Ω_d, a jump of size $\delta\Omega(0) = \Omega_d - \Omega_c$. Since the inertia cannot change velocity abruptly, all of the abrupt change in source velocity must appear across the damper. Hence, the torsion must jump by $\delta T(0) = B\delta\Omega(0)$, and the initial value of the torsion must be $T_i \triangleq T(0^+) = T_p + \delta T(0) = B(\Omega_d - \Omega_c)$.

We use the initial value T_i, the final value T_f, and the time constant τ in equation (3.71), the general solution, to find

$$T(t) = B(\Omega_d - \Omega_c)e^{-(B/J)t} \tag{3.73}$$

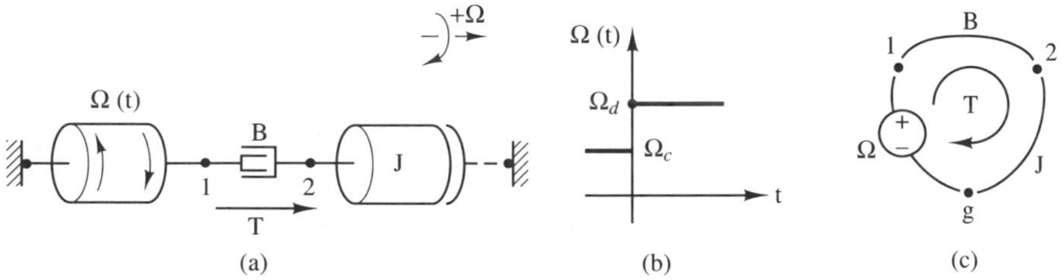

Figure 3.38 Abruptly excited first-order rotational network. (a) Lumped network. (b) Abruptly changing source signal. (c) Line-graph representation.

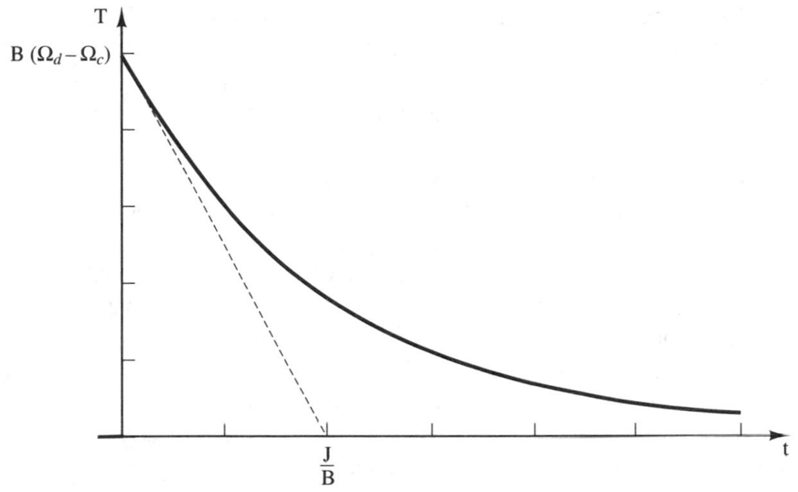

Figure 3.39 Network response for Figure 3.38.

This function is displayed in Figure 3.39.

Exercise 3.15: For the network of example problem 3.8, show that $\Omega_{2p} = \Omega_c$, $\Omega_{2f} = \Omega_d$, and $\Omega_{2i} = \Omega_c$. Use these values to show that $\Omega_2 = \Omega_d + (\Omega_c - \Omega_d)e^{-(B/J)t}$. Sketch this response.

Finding Initial and Final Values of Network Variables

In mathematics books, the boundary values required to solve a differential equation are usually specified by the person who designed the problem and are referred to as *given initial conditions.* In an engineering problem, we are more likely to specify a physical situation or a manner of system operation. We ourselves must determine the initial conditions from the specified situation.

In the step-response analysis of the previous section, we reasoned from physical principles to determine the *prior* steady-state values (at $t = 0^-$), the *jumps* in value (at $t = 0$), the *initial* values (at $t = 0^+$), and the *final* steady-state values (as $t \to \infty$) of the network variables. Then we used the initial and final values in the general solution, equation (3.71), to obtain the step response of the first-order network. Once we know the initial values and final values of the network variables and the time constant of the network, we can sketch the waveforms of all the network variables without resorting to mathematical equations. The step responses of higher order linear lumped networks (described by higher order differential equations) also depend in a well-defined manner on the initial and final values of the network variables, as we show in section 3.4.

In this section, we develop systematic procedures for determining initial and final values of all variables in linear lumped networks. In this development and, indeed, throughout the text, we apply the terms **prior value** (at $t = 0^-$), **initial value** (at $t = 0^+$), **jump value** (the jump in value at $t = 0$, that is, the initial value minus the prior value), and **final value** (the value as $t \rightarrow \infty$) to each network variable. To denote the jump value of a variable, we precede the variable by the symbol δ (e.g., δA_i, δT_n). We denote the other values by the subscripts *p*, *i*, and *f*, respectively (e.g., T_{np}, T_{ni}, and T_{nf}).

A lumped network and its line graph are shown in Figure 3.40. We label the elements with their impedances. We shall learn that the impedance formulas help us find the initial and final values of the network variables.

The network includes two energy-storage elements. Suppose the network is unenergized at $t = 0^-$. Since no energy is stored in the mass and spring, $v_{3p} = 0$ and $F_p = 0$. We apply a step input of size v_c to the unenergized system at $t = 0$; that is, we let $v(t) = v_c$ for $t \geq 0$. We use the network equations and the size of the source jump to determine the initial and final values of the network variables.

We address the steady-state final values first. We write all equations in terms of the *central* flow and potential variables (F and v for translational networks). Then the equations for the energy-storage elements involve time derivatives. For the network of Figure 3.40, the storage-element equations are $F = m\dot{v}_3$ and $\dot{F} = kv_{12}$. By the term **steady state,** we mean that **the *central* variables are not changing.** Therefore, $(\dot{v}_3)_f$ and $(\dot{F})_f$ are zero. Since $(\dot{v}_3)_f = 0$, the element equation for the mass requires that $F_f = 0$. Since $(\dot{F})_f = 0$, the element equation for the spring requires that $v_{12f} = 0$.

The condition $F_f = 0$ means that the flow in the mass is zero in the steady, or final, state. We impose this zero-flow condition on the lumped model or on the line graph by *breaking the branch* that represents the mass. We denote that broken branch by a dotted line in the line graph (see Figure 3.41). In the terminology of water flow in pipes or the flow of electric current in wires, we say that we *open-circuit* the branch. We can think of the broken branch as an *infinite-impedance path,* which prevents flow.

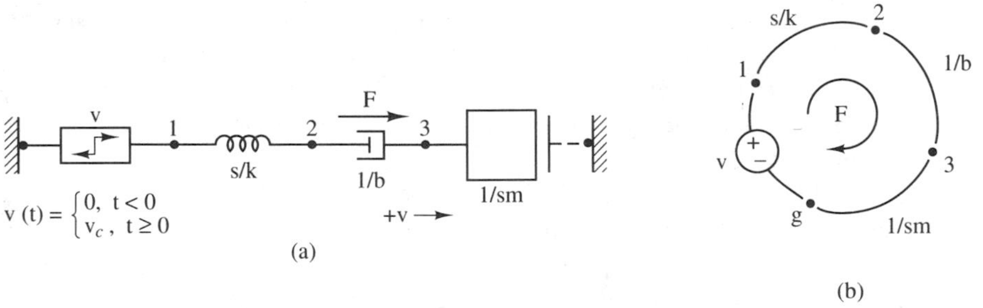

Figure 3.40 A second-order lumped network. (a) Network. (b) Line graph representation.

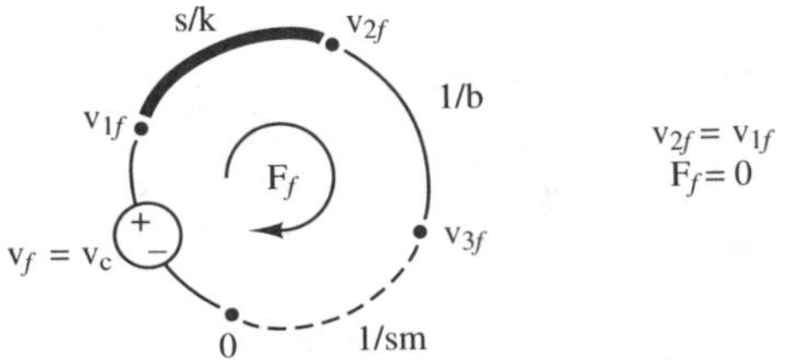

Figure 3.41 Line graph for Figure 3.40 in the steady, or final, state.

Although we have broken the mass branch, we still label the branch with its impedance 1/sm. Note the s in the denominator of the impedance formula; replacing s by 0 implies, correctly, that the impedance behaves as if it were infinite.

The condition $v_{12f} = 0$ is equivalent to $v_{1f} = v_{2f}$. That is, the two ends of the spring have the same velocity in the steady final state. We impose this condition on the lumped model or on the line graph by connecting nodes 1 and 2 with a bold line (see Figure 3.41). The bold line indicates that the nodes have the *same potential.* We borrow from electrical terminology to say that we *short-circuit* the branch. We can think of the bold line as a *zero-impedance path;* that is, the branch cannot support a potential difference without infinite flow.

We label the bold-line spring branch with its impedance s/k. Note the s in the numerator of the impedance formula; replacing s by 0 implies, correctly, that the impedance behaves as if it were zero.

In sum, we modify the network line graph to correspond to the final, steady, state. We express the open-circuit nature of the mass by a dashed branch and the short-circuit nature of the spring by a bold branch. (Similar modifications could be made directly on the lumped model.) All network variables are constant. We attach the subscript *f* to those variables to denote final (steady-state) values. The final velocity of the ground node is zero; the final value of the source velocity is the step size v_c.

We retain the impedance formulas beside the branches. These formulas serve two purposes. First, they provide a simple reminder of the branch equations. Second, the location of the time-derivative operator s helps us open-circuit and short-circuit the branches correctly. We merely let s = 0 in all impedance formulas. Then we short-circuit branches that have zero impedance and open-circuit branches that have infinite impedance.

The replacement of s by 0 does not mean that d/dt = 0, a meaningless statement. We merely take advantage of the fact that the differential operator s appears in such a position in the impedance of each energy-storage device that its replacement by the number zero produces the correct conclusion.

We apply the usual loop and node laws to the final-value line graph to find the final values of the variables. It is evident from Figure 3.41 that the final value of flow is $F_f = 0$, because the open-circuited branch prevents flow. The final velocity of node 1 is $v_{1f} = v_c$. Because of the short circuit between nodes 1 and 2, the identities $v_{2f} = v_{1f} = v_c$ are obvious as well. According to the equation for the damper, $v_{2f} - v_{3f} = F_f/b = 0$, or $v_{3f} = v_{2f} = v_c$.

As this example demonstrates, we can find the steady-state values of all *central* flows and potentials by means of open-circuit and short-circuit rules. We summarize the process in the following general procedure.

PROCEDURE TO CALCULATE FINAL (OR STEADY-STATE) VALUES OF CENTRAL VARIABLES:

1. Replace each ideal source function (A(t) or T(t)) by its steady-state value (A_f or T_f). Place the subscript *f* on *all* network variables to denote final, or steady-state, values. All rates (differentiated variables) are zero. Let s → 0 in the impedance function z(s) for each element of the network.

2. If $z(s) \rightarrow \infty$ for the nth element, the steady-state flow through the element is zero. The open-circuit relation $T_{nf} = 0$ is a simple alternative to the element equation for that element in the steady state. To acknowledge the open-circuit behavior, draw the element with dashed connectors in the lumped model, or represent the branch by a dashed line in the line graph.
3. If $z(s) \rightarrow 0$ for an element between nodes i and j, the steady-state potential difference across the element is zero. The short-circuit relation $A_{ijf} = 0$ (or $A_{if} = A_{jf}$) is a simple alternative to the equation for that element in the steady state. To acknowledge the short-circuit condition, superimpose a bold connector across the element in the lumped model, or represent the branch by a bold line in the line graph.
4. Use the loop and node laws to determine the steady-state values of the network variables of interest. Because the energy-storage elements are represented by simple algebraic equations in the steady state, the node equations and loop equations are algebraic, rather than differential. Solving the equations for the steady-state values is now straightforward. For low-order networks, these values can be found directly by inspection of the network.

If the steady-state equations that result from this procedure are inconsistent, the central variables are not in (and do not have) a steady state. For example, apply a constant-torsion source to a single inertia. The source supplies a specified finite flow, whereas the inertia acts as an open circuit (requiring zero flow). According to the foregoing procedure, these two flows must be equal. In fact, however, the inertia accelerates without ceasing, and the network does not achieve a steady state. (Replace each velocity v_i by the acceleration a_i and each impedance z by sz. Then let $s \rightarrow 0$ to find the steady-state accelerations. Replace v_i by x_i and z by z/s and let $s \rightarrow 0$ to find the steady-state displacements.)

We now address the initial values of the network variables. If the network source is suddenly turned on or off, we see an immediate change in the values of some of these variables. However, an energy-state variable cannot be changed abruptly without an infinitely strong source.

Let us investigate the continuity of energy-state variables via the network of Figure 3.40. The equation for the mass is $F = m\dot{v}_{3g}$. We integrate this element equation over a brief interval around the instant of abrupt change in the source ($t = 0$) to obtain

$$\int_{0-}^{0+} F\,dt = m\int_{0-}^{0+} \dot{v}_{3g}\,dt = m[v_{3g}(0^+) - v_{3g}(0^-)] \tag{3.74}$$

Since F is finite and the interval of integration is infinitesimally small, the integral on the left is zero. We conclude, therefore, that

$$v_{3gi} = v_{3gp} \tag{3.75}$$

We call the mass a **continuous-potential element** because it maintains continuity of the potential difference between its terminals.

We denote an abrupt change in value of a network variable (at $t = 0$) by attaching the symbol δ. For example, $\delta v_3 \triangleq v_{3i} - v_{3p}$. Then equation (3.75) corresponds to

$$\delta v_{3g} \triangleq v_{3gi} - v_{3gp} = 0 \tag{3.76}$$

That is, the initial jump in the potential difference v_{3g} is zero.

In the terminology used for steady state, the mass acts like a *short circuit to abrupt changes* in potential difference. That is, it behaves as a zero-impedance path during the brief instant of the jump. To represent this fact, we replace the mass branch in Figure 3.40(b) by a bold line, as shown in the line graph of Figure 3.42. This modified line-graph element applies only to *abrupt changes* in signals at the specified instant. To make this fact clear, we precede all signals by the symbol δ in the line graph. Of course, $\delta v_g = 0$.

Although we have short-circuited the mass branch in Figure 3.42, we still label the branch with its impedance 1/sm. Again, note the s in the denominator of the impedance formula; replacing s by ∞ implies, correctly, that the impedance behaves as if it were zero.

The equation that characterizes the behavior of the spring in Figure 3.40 is $\dot{F} = kv_{12}$. We integrate this element equation over a brief interval around the instant of abrupt change in the source ($t = 0$), obtaining

$$k\int_{0-}^{0+} v_{12}\,dt = \int_{0-}^{0+} \dot{F}\,dt = F(0^+) - F(0^-) \tag{3.77}$$

Since v_{12} is finite and the interval of integration is infinitesimally small, the integral on the left is zero. Therefore,

$$F_i = F_p \tag{3.78}$$

We call the spring a **continuous-flow element** because it maintains continuity of flow (scalar force).

We again use δ-notation, this time rewriting equation (3.78) as

$$\delta F \triangleq F_i - F_p = 0 \tag{3.79}$$

In the terminology used for the steady state, the spring k acts like an *open circuit (or infinite impedance) to abrupt changes* in flow. We represent the spring by a broken branch in Figure 3.42. We label the open-circuited spring branch with its impedance s/k. Note the s in the numerator of the impedance formula; replacing s by ∞ implies, correctly, that the impedance behaves as if it were infinite to abrupt changes in flow.

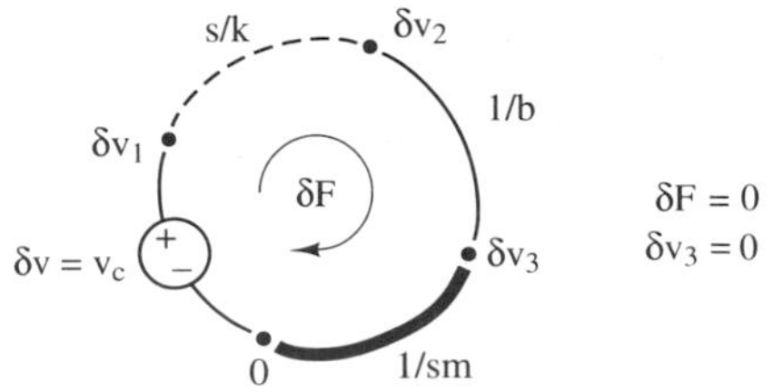

Figure 3.42 Line graph for Figure 3.40 during an abrupt change.

In sum, we modify the network line graph to correspond to the initial abruptly changing condition, as shown in Figure 3.42. (The abrupt-change line graph in the figure would be valid at *any* instant. We are using it at the instant $t = 0$.) We express the short-circuit nature of the mass by a bold branch and the open-circuit nature of the spring by a dashed branch. (Similar modifications could be made directly on the lumped model.) We precede all variables by the symbol δ to denote jumps in value. The jump in value of the ground node is zero; the jump in value of the source velocity is the step size v_c.

The variables associated with the modified line graph are the abrupt jumps in the *central* variables of the network. It is those variables to which the impedance formulas apply. We retain the impedance formulas beside the branches. The location of the time-derivative operator s helps us open-circuit and short-circuit the branches correctly. We merely let $s = \infty$ in all impedance formulas. Then we short-circuit branches that have zero impedance and open-circuit branches that have infinite impedance.

The short-circuited and open-circuited elements shown in Figure 3.42 are represented mathematically by equations (3.76) and (3.79), respectively. The jump at node 1 is the source jump, $\delta v_1 = v_c$. The jump in F is $\delta F = 0$, owing to the open circuit at the spring. Because of the short circuit at the mass, the jump at node 3 is $\delta v_{3g} = \delta v_3 = 0$. According to the equation for the damper, $\delta v_2 - \delta v_3 = \delta F/b$, or $\delta v_2 = 0$.

The network equations relate the jumps in the *rates* (the time derivatives of the central flows and velocities) to the jumps in the flows and velocities themselves. Specifically, $\delta(\dot{F}) = k(\delta v_1 - \delta v_2)$ implies that $\delta(\dot{F}) = kv_c$, $\delta F = m\delta(\dot{v}_3)$ requires that $\delta(\dot{v}_3) = 0$, and $\delta(\dot{F}) = b[\delta(\dot{v}_2) - \delta(\dot{v}_3)]$ produces $\delta(\dot{v}_2) = (k/b)v_c$. Since the *prior values* ($t = 0^-$) of all variables and their derivatives are zero in this example, the *initial values* ($t = 0^+$) of these variables are the same as the jump values, namely, $F_i = 0$, $v_{2i} = 0$, $v_{3i} = 0$, $\dot{F}_i = kv_c$, $\dot{v}_{2i} = (k/b)v_c$, and $\dot{v}_{3i} = 0$.

As the example demonstrates, we can find the jump values of all central flows and potentials in a network from the line graph (or directly from the lumped model). We can find the jump values of derivatives of the variables from the network equations (and the jumps in the variables themselves). These jump values can be added to the prior values to produce the initial values. We summarize the process in the following general procedure.

PROCEDURE FOR FINDING JUMP VALUES OF NETWORK VARIABLES

1. Replace each ideal source value (A or T) by its jump in value (δA or δT) at the instant of interest. Precede all central variables by the symbol δ to denote *jumps* in value. Let $s \to \infty$ in the impedance formula $z(s)$ for each element of the network.
2. If $z(s) \to \infty$ for the nth element, the jump in flow through that element is zero. The open-circuit relation $\delta T_n = 0$ is a simple alternative to the element equation *for this instant.* To indicate the open-circuit condition, draw the element with dashed connectors in the lumped model, or represent the branch by a dashed line in the line graph.
3. If $z(s) \to 0$ for an element between nodes i and j, the jump in potential drop across that element is zero. The short-circuit relation $\delta A_{ij} = 0$ ($\delta A_i = \delta A_j$) is a simple alternative to the element equation *for this instant.* To indicate the short-circuit condition, superimpose a bold line across the element in the lumped model, or represent the element by a bold line in the line graph.

4. Use the loop and node laws to determine the jump values of the *central* variables of interest. Because the energy-storage elements are represented by simple algebraic equations for the specified instant, the node equations and loop equations are algebraic, rather than differential. Solution of the equations for the jump values of the variables is now straightforward. For low-order networks, the jumps in value can be found by inspection of the line graph.
5. The jumps in values of derivatives of the variables can be determined from the jump values of the variables calculated in step 4, together with the element equations and node equations for the network. In these calculations with derivatives, we must use the defining equations for the storage elements rather than the simplified versions. (The defining equations are always valid, even during an abrupt change.) For simplicity, we use the notation $\delta\dot{A}_i$ to mean $\delta(\dot{A}_i)$ and $\delta\dot{T}_n$ to mean $\delta(\dot{T}_n)$.
6. Add the jump values determined in steps 4 and 5 to the prior values of the corresponding variables to determine the initial values of the variables, i.e., the values at the *instant after* the abrupt change in the source occurs.

To remember the rules for open-circuiting and short-circuiting elements, note that we let $s = 0$ to find the conditions for $t \to \infty$. We let $s \to \infty$ to find the conditions for $t \to 0^+$. In section 4.2, we combine impedances of elements to simplify networks. Letting $s = 0$ or $s \to \infty$ in these simplified networks still produces correct steady-state or abruptly changing conditions. The procedures can also be used to find initial and final values for nonlinear lumped networks, such as the hydraulic networks of section 5.1.

Example Problem 3.9: Finding initial and final values.
The torsion source T in the rotational network of Figure 3.43 changes abruptly as shown. Determine the initial ($t = 0^+$) and final ($t \to \infty$) values of all rotational displacements, velocities, and accelerations and of all torsions and rates of change of those torsions.
Solution: Draw the line graph and define branch flows (see Figure 3.43(c)). At $t = 0^-$, the system is in a steady state. Therefore, we let $s = 0$ in the impedance formulas, the inertias act as open circuits, and the springs act as short circuits. The steady-state line graph is shown in Figure 3.43(d). The subscript *p* denotes prior values of the variables. The open-circuit and short-circuit rules apply to the central variables—torsion and velocity. The following values can be found by inspection of the steady-state line graph:

$$\begin{aligned}
&\Omega_{1p} = \Omega_{2p} = 0 && \text{(nodes shorted to ground)}\\
&T_{1p} = T_{5p} = 0 && \text{(open-circuited branches)}\\
&T_{2p} = 0 && (B_1 \text{ short-circuited by } K_1;\ \Omega_{12p} = 0)\\
&T_{3p} = T_{4p} = -T_c && \text{(flows balanced at nodes 1 and 2)}
\end{aligned} \tag{3.80}$$

The prior values of the accelerations $\dot{\Omega}_{1p}$ and $\dot{\Omega}_{2p}$ and all prior rates of change of torsion are zero by definition of the term *steady state.* The steady-state prior flows through the springs are related to the steady-state prior displacements by

$$\begin{aligned}
\theta_{2gp} &= \frac{T_{4p}}{K_2} = -\frac{T_c}{K_2}\\
\theta_{12p} &= \frac{T_{3p}}{K_1} = -\frac{T_c}{K_1}
\end{aligned} \tag{3.81}$$

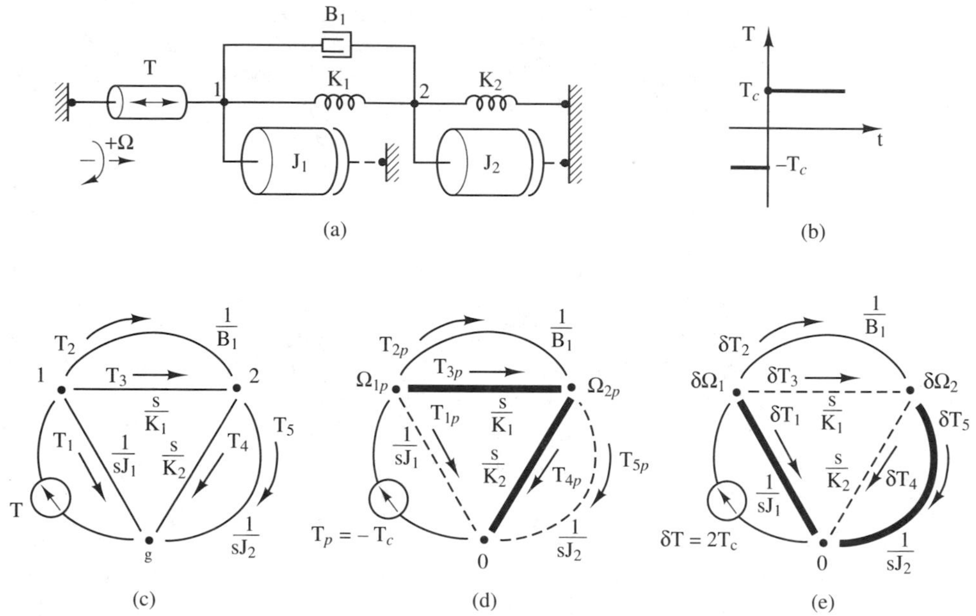

Figure 3.43 Initial-value and final-value simplifications. (a) Lumped network. (b) Abruptly changing source signal. (c) Line-graph representation. (d) Prior steady-state line graph. (e) Line graph for abrupt change.

We solve these algebraic equations to obtain the prior displacements,

$$\theta_{2p} = -\frac{T_c}{K_2}, \qquad \theta_{1p} = -\frac{T_c(K_1 + K_2)}{K_1 K_2} \tag{3.82}$$

The steady-state final condition of the network (as $t \to \infty$) differs from the steady-state prior condition only in a change in the sign of the source. We again use the line graph of Figure 3.43(d), with the source value $-T_c$ replaced by T_c and with the subscript f instead of p; the final values of the variables are the negatives of the values in (3.80) and (3.82).

At $t = 0$, the source increases abruptly by the amount $\delta T = 2T_c$. Therefore, we let $s \to \infty$ in the impedance formulas; the inertias act like short circuits and the springs act like open circuits to the jump in the source. The simplified line graph is shown in Figure 3.43(e). By inspection of the line graph, we find that:

$$\begin{aligned}
\delta\Omega_1 &= \delta\Omega_2 = 0 && \text{(nodes shorted to ground)} \\
\delta T_3 &= \delta T_4 = 0 && \text{(open-circuited branches)} \\
\delta T_2 &= 0 && \text{(because } \delta\Omega_{12} = 0\text{)} \\
\delta T_5 &= 0 && \text{(flow balance at node 2)} \\
\delta T_1 &= \delta T = 2T_c && \text{(flow balance at node 1)}
\end{aligned} \tag{3.83}$$

The jump in acceleration of an inertia J is related to the jump in flow through the inertia by the equation $\delta T = J\delta\dot{\Omega}_{ij}$. Thus, the initial jumps in the node accelerations satisfy $\delta T_1 = 2T_c = J_1\,\delta\dot{\Omega}_{1g}$ and $\delta T_5 = 0 = J_2\,\delta\dot{\Omega}_{2g}$, or

$$\delta\dot{\Omega}_1 = \frac{2T_c}{J_1} \qquad \delta\dot{\Omega}_2 = 0 \tag{3.84}$$

The jump in velocity difference across a torsion spring K is related to the jump in the rate of change of flow through the spring by $\delta\dot{T} = K\delta\Omega_{ij}$. Therefore,

$$\delta\dot{T}_4 = K_2\,\delta\Omega_2 = 0 \qquad \delta\dot{T}_3 = K_1\,\delta\Omega_{12} = 0 \tag{3.85}$$

By the damper equation and by flow balance at the nodes,

$$\delta\dot{T}_2 = B_1(\delta\dot{\Omega}_1 - \delta\dot{\Omega}_2) = \frac{2T_cB_1}{J_1}$$
$$\delta\dot{T}_5 = \delta\dot{T}_2 + \delta\dot{T}_3 - \delta\dot{T}_4 = \frac{2T_cB_1}{J_1} \tag{3.86}$$
$$\delta\dot{T}_1 = \delta\dot{T} - \delta\dot{T}_2 - \delta\dot{T}_3 = -\frac{2T_cB_1}{J_1}$$

Note that $\delta(\dot{T}) \triangleq \dot{T}(0^+) - \dot{T}(0^-) = 0$ in the last equation because T is constant except at $t = 0$. Since the jumps in the flows δT_3 and δT_4 in the springs are zero, the spring relation $\delta T = K\delta\theta$ requires that $\delta\theta_{12} = \delta\theta_2 = 0$; consequently, the jumps in the node displacements are

$$\delta\theta_1 = \delta\theta_2 = 0 \tag{3.87}$$

The initial values of the variables (at $t = 0^+$) consist of the sums of the prior values, equations (3.80) to (3.82), and the jump values, equations (3.83) to (3.87). Hence, the initial values are:

$$\begin{aligned}
&\Omega_{1i} = \Omega_{2i} = 0 && \dot{\Omega}_{1i} = \frac{2T_c}{J_1} \\
&T_{1i} = 2T_c && \dot{\Omega}_{2i} = 0 \\
&T_{2i} = T_{5i} = 0 && \dot{T}_{3i} = \dot{T}_{4i} = 0 \\
&T_{3i} = T_{4i} = -T_c && \dot{T}_{1i} = -\frac{2T_cB_1}{J_1} \\
&\theta_{1i} = -\frac{T_c(K_1 + K_2)}{K_1K_2} && \dot{T}_{2i} = \dot{T}_{5i} = \frac{2T_cB_1}{J_1} \\
&\theta_{2i} = -\frac{T_c}{K_2}
\end{aligned} \tag{3.88}$$

Exercise 3.16: Review the procedures of example problem 3.9, but implement the open-circuit and short-circuit simplifications on the lumped model rather than the line graph.

In section 2.2, we developed systematic loop and node methods for finding the equations that describe a lumped network. By eliminating all but one dependent variable from the network equations, we found the input-output system equation for the remaining dependent variable. The procedures for finding initial and final values of network variables again require that we write the network equations and eliminate variables. However, the open-circuit and short-circuit rules simplify the network and reduce the work of writing equations and eliminating variables. Yet, when we seek the initial jumps in *derivatives* of the network variables, we must use the unsimplified network equations.

In section 3.4, we shall learn how to solve an input-output system equation for the step response of the corresponding dependent variable. The solution process requires that we first find the initial and final values of that dependent variable and certain of its derivatives. Once we have gone to the work of obtaining the input-output system equation for a particular dependent variable, it is a simple matter to find the initial and final values of that variable and all its derivatives without any further writing of equations and eliminating of variables.

It does not matter whether we let $s = 0$ or $s = \infty$ before or after we write the equations and eliminate variables. Therefore, we can find the initial and final values of a particular dependent variable directly from the input-output system equation for that variable, an equation from which all other variables have already been eliminated.

PROCEDURE FOR FINDING FINAL VALUES AND JUMP VALUES FROM THE INPUT-OUTPUT SYSTEM EQUATION

1. Replace the source variable u in the system equation by its constant (step-input) value u_c. Solve the system equation for the output variable y as a ratio of polynomials in s.
2. Let $s \to 0$ in the ratio of polynomials found in step 1. The result is the final value y_f.
3. Let $s \to \infty$ in the ratio of polynomials found in step 1. The result is the initial jump δy.
4. Multiply the equation resulting from step 1 by s. The expression sy on the left represents $\dot{y}$. Let $s \to \infty$. The result is the initial jump $\delta \dot{y}$.
5. Multiply the equation resulting from step 1 by s^p. The expression $s^p y$ on the left represents $y^{(p)}$, the pth derivative of y. Let $s \to \infty$. The result is the initial jump $\delta y^{(p)}$.

Let us apply this procedure to the third-order system equation (3.29) that describes the displacement of node 1 in Figure 3.24(a). Let the source signal F be a step input of size F_c. We replace F by F_c in the system equation and solve for x_1 to find

$$x_1 = \frac{\dfrac{s}{m} + \dfrac{k_2}{mb}}{s^3 + \dfrac{k_2}{b}s^2 + \dfrac{k_1 + k_2}{m}s + \dfrac{k_1k_2}{mb}} F_c \tag{3.89}$$

Let $s \to 0$ in (3.89) to find that $x_{1f} = F_c/k_1$. Let $s \to \infty$ in (3.89) to find that $\delta x_1 = 0$. Multiply (3.89) by s, then let $s \to \infty$ to find that $\delta \dot{x}_1 = 0$. Multiply (3.89) by s^2, then let $s \to \infty$ to find that $\delta \ddot{x}_1 = F_c/m$.

If the values of x_1, $\dot{x}_1$, and $\ddot{x}_1$ just prior to $t = 0$ are all zero, then the initial jumps that we have found are the initial values. We shall find in section 3.4 that we need these initial values to determine the analytical expression for the step response $x_1(t)$. We shall also find it convenient to express the analytical step response in terms of the final value x_{1f}.

Figure 3.24(c) shows one operational model for this system. If we use a step input at $t = 0$, we would choose to initiate the integrators at $t = 0^+$ in order to avoid differentiating the step input and integrating the resulting impulsive signal at the first integrator. Hence, we need the initial values of x_1, $\dot{x}_1$, and $\ddot{x}_1$ at $t = 0^+$ to carry out the simulation.

If we were to use, instead, the operational model shown in Figure 3.25, we could initiate the integrators at $t = 0^-$, because eliminating the source-signal differentiator eliminates the impulsive input to the first integrator and the jump at the output of that integrator. Then we could use the known prior values (at $t = 0^-$) to initiate the integrators and avoid finding the initial jumps in x_1, $\dot{x}_1$, and $\ddot{x}_1$.

Analogous Networks and Physical Insight

The purpose of dynamic system analysis is to develop enough intuitive feeling for the behaviors of physical devices to invent and design new devices with desired behaviors. Changes in the structure of connections and changes in the types of elements produce great variety in physical behavior, even in first-order networks. We develop an intuitive feeling for the behaviors associated with specific connectional structures and types of elements by analyzing, comparing, and physically interpreting the behaviors of a variety of networks.

A K, B network that is analogous to the parallel k, b network of Figure 3.35 is shown in Figure P3.18(iii) at the end of the chapter. The two networks have the same connectional (line-graph) structure. Their element equations have the same mathematical forms. There is no mathematical difference in their behaviors. Their step responses have identical waveforms. The networks differ only in the symbols that represent their variables ($F \to T$, $v \to \Omega$), the symbols that represent their parameters ($k \to K$, $b \to B$), and the physical interpretations of their responses (translation → rotation, compressive force → inward-twisting torsion). If we have analyzed one of these networks, it requires only physical reinterpretation to transfer our understanding of that analysis to the other network. In later chapters, we find that there are electrical networks and hydraulic networks that are also analogous to the k, b network of Figure 3.35—additional networks to which the same analytical insight can be transferred by physical reinterpretation.

What kinds of changes to the first-order k, b example cause significant differences in its behavior? Suppose we change the flow source (F) in Figure 3.35 to a velocity source (v). This change makes mathematical analysis trivial. With the velocity specified across the parallel elements, the flows in the elements k and b can be found separately. Although the velocity source does correspond to a manner of operation of the physical system, the network does not require much attention because the parts do not interact. Together, they do not produce new behavior. The scalar force in the damper has the same step-function waveform as that of the source. (The force in the spring does not approach a steady state, but grows without bound.) The same comments apply to such a source change for the analogous K, B network.

If we change the storage element in Figure 3.35 from a spring to a mass, on the other hand, the change in mathematical form of the element equation causes a great change in network behavior. Replacing the continuous-flow element by a continuous-potential element causes a *reversal in behavior* of all the variables in the first-order network. For example, v_1, which jumped to a positive value and decayed exponentially to zero, now rises exponentially from zero to a positive final value. A *change in type of storage element* always has a great effect on network behavior. That change represents a major physical difference in the underlying system.

Finally, suppose we change the connectional structure of the network. Specifically, we replace the parallel k, b network of Figure 3.35 by a series k, b network. A compressive-force source in a series structure produces the same type of trivial behavior as we found for the velocity-driven parallel structure. Therefore, we switch to a velocity source. In the series structure, there is a common scalar force in the loop rather than the pair of scalar forces of the parallel network. The nature of the response of the intermediate node in the series network depends on the order of placement of the spring and damper. For one order, the node velocity rises exponentially; for the other order, it jumps and then decays exponentially to zero.

It should be apparent that various arrangements of a few network elements can produce a variety of behaviors. For abruptly applied sources, however, those behaviors are always exponential in nature. The physical interpretations of those behaviors are not obvious from the mathematics of the situation. It is important that we interpret the behaviors physically if we are to gain the insight needed to design new systems.

3.4 THE STEP RESPONSE OF A LINEAR NETWORK

The operational models of section 3.2 and the analyses of first-order networks in section 3.3 demonstrate a number of properties of linear networks. We examine these properties for networks of order 2 and higher in this section. We summarize the properties as follows:

1. The order n of the input-output system equation for a network variable is the same as the number of independent energy-storage elements in the network. At any instant, the state of the network is a set of n numbers—the values of n state variables—from which we can compute the energy states of the storage elements.
2. The response of an nth-order linear network is the sum of two independent pieces: the zero-input response, owing to the initial state alone, and the zero-state response, owing to the source alone. For a step input, both response components exhibit the same characteristic wave shapes. Therefore, either component alone gives an intuitive understanding of the network behavior. We use the zero-state step response alone for comparisons.

3. The behavior of an nth-order linear network is characterized by n *system* parameters. (For a first-order network, the system parameter is its time constant τ.) These system parameters correspond to characteristic roots of the homogeneous system equation. The *shape* of the zero-state step response is uniquely related to the system parameters.
4. The behaviors of most nth-order linear networks are dominated by one or two of their characteristic roots. That is, most linear networks behave *nearly* like first-order or second-order networks.
5. Certain *key features* of the zero-state step response(s) of first-order or second-order linear network(s) are uniquely related to the system parameter(s) mentioned in step (3). As a consequence of this relationship, a single measured step response fully discloses the behavior of a physical system that is represented by such a network.

We examined first-order networks in section 3.3. In this section, we determine how to analyze higher order networks, but focus primarily on second-order networks. Second-order networks exhibit much more variety in their response waveforms than do first-order networks. Just as we represent the characteristic response of a first-order network by a single network parameter—the time constant—we represent the characteristic response of a second-order network by two network parameters. For some networks, the two parameters are a pair of time constants. Other networks oscillate, and we must use parameters that describe oscillations—the natural frequency and the damping ratio.

A Second-Order Network

A lumped translational network with two storage elements is shown in Figure 3.44. Suppose the system is unenergized initially. Then $v_{3p} = 0$ and $F_p = 0$. Since the source is zero for $t < 0$, $v_{2p} = 0$ as well. At $t = 0$, the source suddenly jumps from zero to the

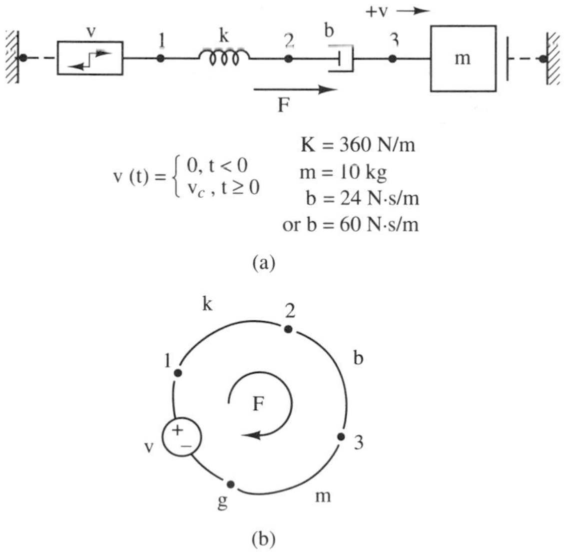

Figure 3.44 A second-order linear network. (a) Lumped network. (b) Line-graph representation.

constant value v_c. We found the jump values, the initial values, and the final values of the variables of this network in the previous section.

The loop equation for the network is

$$\frac{sF}{k} + \frac{F}{b} + \frac{F}{sm} = v \tag{3.90}$$

We multiply by sk and convert to derivative notation to obtain the input-output system equation for F:

$$\ddot{F} + \frac{k}{b}\dot{F} + \frac{k}{m}F = k\dot{v} \tag{3.91}$$

Substituting $F = smv_3$ into equation (3.90) and multiplying by k/m produces the corresponding input-output system equation for v_3:

$$\ddot{v}_3 + \frac{k}{b}\dot{v}_3 + \frac{k}{m}v_3 = \frac{k}{m}v \tag{3.92}$$

Exercise 3.17: Show that the input-output system equation for v_2 is

$$\ddot{v}_2 + \frac{k}{b}\dot{v}_2 + \frac{k}{m}v_2 = \frac{k}{m}v + \frac{k}{b}\dot{v} \tag{3.93}$$

The left sides of equations (3.91), (3.92), and (3.93) all have the same characteristic form. We focus our attention on the velocity v_3 and its input-output system equation (3.92). We shall find $v_3(t)$ for $t \geq 0$.

The **particular solution** y_P to a linear differential equation usually has the same waveform as the independent variable (or source signal) u. For a step (constant) input, then, the particular solution must be constant valued. If the network eventually settles to a steady state, the particular solution must be the final value y_f of the variable.

In this case, the independent variable is the source signal $v(t) = v_c$. We replace each derivative in equation (3.92) by the derivative operator s, let $v = v_c$, and let $s = 0$ to find $v_{3f} = v_c$. Therefore,

$$v_{3P}(t) = v_{3f} = v_c, \qquad t \geq 0 \tag{3.94}$$

To verify equation (3.94), we merely need substitute $v = v_c$ and $v_{3P}(t) = v_c$ into equation (3.92).

It can be shown that, for all linear constant-coefficient differential equations, the *characteristic functions*—the solutions to the homogeneous differential equation—are exponentials. In section 3.3, we found this conclusion to be true for first-order differential equations. (Derivatives of exponentials are also exponentials. The combination of derivatives in the homogeneous equation adds to zero only for exponential functions.)

The homogeneous differential equation is

$$\ddot{v}_3 + \left(\frac{k}{b}\right)\dot{v}_3 + \left(\frac{k}{m}\right)v_3 = 0 \tag{3.95}$$

Every exponential function can be expressed in the form Ae^{pt} for some number p. We substitute $v_3(t) = Ae^{pt}$ into equation (3.95) to obtain

$$p^2Ae^{pt} + \left(\frac{k}{b}\right)pAe^{pt} + \left(\frac{k}{m}\right)Ae^{pt} = 0 \tag{3.96}$$

The common factor Ae^{pt} can be zero only if $A = 0$. We assume that $A \neq 0$; otherwise the hypothetical solution form would be useless. We divide each term of the equation by the common factor to obtain

$$p^2 + \left(\frac{k}{b}\right)p + \left(\frac{k}{m}\right) = 0 \tag{3.97}$$

The polynomial on the left side of equation (3.97) is called the **characteristic polynomial** for the network. Equation (3.97) itself is known as the network **characteristic equation.**

The polynomial in the variable p on the left side of equation (3.97) is the same as the polynomial in s on the left sides of the input-output system equations (3.91)–(3.93). Therefore, we shall use that polynomial in s as the characteristic polynomial. Then the characteristic equation is

$$s^2 + \left(\frac{k}{b}\right)s + \left(\frac{k}{m}\right) = 0 \tag{3.98}$$

We can now write by inspection the characteristic equation corresponding to any linear constant-coefficient differential equation. We need not even convert the equation from operator form to differential form.[6]

The characteristic equation (3.98) has two **characteristic roots,** which we denote s_1 and s_2. (The roots can be complex numbers.) Hence, the network has two exponential solutions: e^{s_1t} and e^{s_2t}. Each of these exponentials satisfies the homogeneous differential equation (3.95). Note that the units of s_1 and s_2 are rad/s.

The general solution to the *homogeneous* differential equation is called the **complementary function.** The complementary function for equation (3.95) is

$$v_{3C}(t) = A_1e^{s_1t} + A_2e^{s_2t} \tag{3.99}$$

where A_1 and A_2 are arbitrary constants with the units m/s. The *general solution* to the input-output system equation (3.92) is the sum of the particular solution and the complementary function; that is,

$$v_3(t) = v_{3P}(t) + v_{3C}(t) = v_c + A_1e^{s_1t} + A_2e^{s_2t}, \qquad t \geq 0 \tag{3.100}$$

We choose the constants A_1 and A_2 so that equation (3.100) satisfies the initial conditions for the network. The abrupt-change line graph for this initially unenergized network is shown in Figure 3.42. In the previous section we used that line graph to show that

[6] This is the third use we have made of the symbol s. (1) It is *defined* as the time-derivative operator d/dt. (2) We *replace* s by ∞ or 0 to obtain the open-circuit and short-circuit conditions that produce the initial or final values of the network variables (see section 3.3). (3) Now we *use* s as the (numeric-valued) variable in the characteristic polynomial.

$v_{3i} \triangleq v_3(0^+) = 0$ and $\dot{v}_{3i} \triangleq \dot{v}_3(0^+) = 0$. (We could also find these initial conditions directly from the system equation (3.92).) We evaluate equation (3.100), the general solution, at $t = 0^+$ to obtain

$$v_3(0^+) = v_c + A_1 + A_2 = 0 \tag{3.101}$$

Differentiating equation (3.100) and evaluating the result at $t = 0^+$ produces

$$\dot{v}_3(0^+) = s_1A_1 + s_2A_2 = 0 \tag{3.102}$$

We solve equations (3.101) and (3.102) for A_1 and A_2 and then substitute their values into equation (3.100) to get the solution $v_3(t)$.

Response for Real Characteristic Roots

Figure 3.44 gives specific numerical values for the network parameters. It shows two different values for the friction constant. For $b = 24$ N·s/m, the characteristic equation (3.98) becomes

$$s^2 + 15s + 36 = 0 \tag{3.103}$$

The characteristic roots of this equation are negative and real: $s_1 = -3$ and $s_2 = -12$. Then the initial-condition equations (3.101) and (3.102) produce

$$A_1 = -\frac{4}{3}v_c, \qquad A_2 = \frac{1}{3}v_c \tag{3.104}$$

The step response at node 3 of the network is

$$v_3(t) = v_c\left(1 - \frac{4}{3}e^{-3t} + \frac{1}{3}e^{-12t}\right), \qquad t \geq 0 \tag{3.105}$$

The response is displayed for a *unit* step ($v_c = 1$ m/s) in Figure 3.45. Associated with each exponential term of equation (3.105) is a time constant. The behavior exhibited in the figure is typical of a stable second-order network with real characteristic roots. The time constant $\tau_2 = 1/12$ s for the third term is much shorter than the time constant

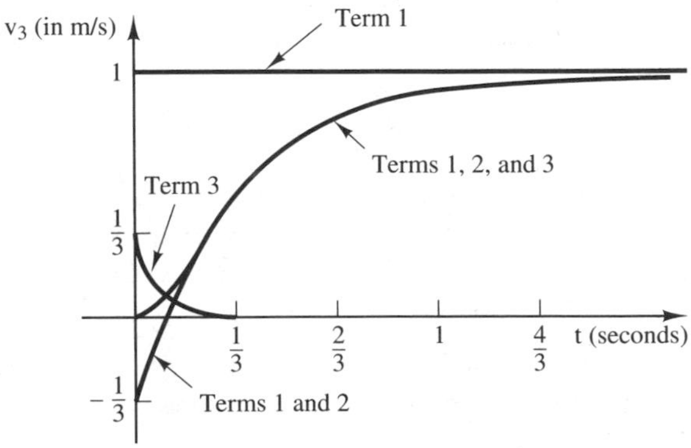

Figure 3.45 Unit-step response for negative, real characteristic roots.

$\tau_1 = 1/3$ s for the second term. The transient effect of the third term is complete by about $t = 5\tau_2 = 5/12$ s.

The transient behavior of the network is dominated by the term with the longer time constant—the one associated with the smaller root of the characteristic equation. The longer of the two time constants also determines the **settling time** t_s, that is, the length of the *transient interval.* The transient is essentially complete by $t_s \triangleq 5\tau_1 = 5/3$ s. The limiting behavior as $t \to \infty$ is described by the first term—the final value.

Exercise 3.18: We can obtain a rough picture of the response waveform shown in Figure 3.45 by using only the initial and final values v_{3i}, $\dot{v}_{3i}$, and v_{3f} and the time constants τ_1 and τ_2. Mark the initial and final values on the graph and imagine how to fill in the rest of the waveform using the values of τ_1 and τ_2.

Each other variable in the network (v_2 and F) has a response that is identical in *form* to equation (3.105), but the initial and final values differ. The response of each other variable can be found in three different ways:

1. By repeating the solution process for the corresponding differential equation—either equation (3.91) or equation (3.93);
2. By writing a response function that has the same *form* as equation (3.105), but basing it on the appropriate initial and final values;
3. By relating the other variable to v_3.

We usually use approach 2. We examine this approach in detail in a later section. As an example of the last approach, we find, from the line graph of Figure 3.44, that $F = m\dot{v}_3$. Therefore, to get F(t), we differentiate the solution equation (3.105) and multiply the result by m (10 kg).

Response for Complex Characteristic Roots

The nature of the transient behavior of v_3 is determined by the roots of the characteristic equation (3.98). If the roots are complex, the behavior differs radically from that shown in Figure 3.45. To demonstrate the difference, we find the response for the friction constant $b = 60$ N·s/m. In this case, the characteristic equation becomes

$$s^2 + 6s + 36 = 0 \tag{3.106}$$

The characteristic roots are $s = -3 \pm j5.2$ rad/s. The general solution, equation (3.100), then becomes

$$\begin{aligned} v_3(t) &= v_c + A_1 e^{(-3+j5.2)t} + A_2 e^{(-3-j5.2)t} \\ &= v_c + e^{-3t}[A_1 e^{j5.2t} + A_2 e^{-j5.2t}] \end{aligned} \tag{3.107}$$

The form of behavior associated with the imaginary exponents is difficult to visualize.

We apply the Euler identity

$$e^{\pm jZ} = \cos Z \pm j \sin Z \tag{3.108}$$

to the two terms in brackets and combine like terms to produce

$$v_3(t) = v_c + e^{-3t}(E_1 \cos 5.2t + E_2 \sin 5.2t) \qquad (3.109)$$

where the new arbitrary constants E_1 and E_2 are related to the previous arbitrary constants by

$$E_1 = A_1 + A_2 \qquad \text{and} \qquad E_2 = j(A_1 - A_2) \qquad (3.110)$$

We now can plot equation (3.109), but it, too, is difficult to visualize.

We convert the sinusoidal equation (3.109) to the single phase-shifted sinusoid

$$v_3(t) = v_c + Ee^{-3t}\cos(5.2t + \phi) \qquad (3.111)$$

by using the identities

$$E = \sqrt{E_1^2 + E_2^2} \qquad \text{and} \qquad \phi = \tan^{-1}(E_2/E_1) \qquad (3.112)$$

The sinusoidal equation (3.111) is easier to plot than is equation (3.109). It is also a function that we can visualize.

We again use the initial conditions to find the values of the constants, here E and ϕ. At $t = 0^+$, equation (3.111) becomes

$$v_3(0^+) = v_c + E \cos \phi = 0 \qquad (3.113)$$

Differentiating equation (3.111) and evaluating it at $t = 0^+$ produces

$$\dot{v}_3(0^+) = -5.2E \sin \phi - 3E \cos \phi = 0 \qquad (3.114)$$

Solving these two equations simultaneously yields

$$E = -1.15 \text{ m/s} \qquad \text{and} \qquad \phi = -0.523 \text{ rad} \qquad (3.115)$$

We substitute these constants into the general solution equation (3.111) to produce

$$v_3(t) = 1 - 1.15e^{-3t}\cos(5.2t - 0.523) \text{ m/s}, \qquad t \geq 0 \qquad (3.116)$$

This response is displayed for a *unit* step input in Figure 3.46. The limiting behavior as $t \to \infty$ is the final value, as in the case of real roots (see Figure 3.45). The steady-state behavior of the system, in general, is not affected by the characteristic roots. The transient waveform is oscillatory. Such behavior is typical of networks with complex characteristic roots.

The negative, real part (= −3) of the complex conjugate root pair determines the time constant $\tau = 1/3$ s of the exponential factor e^{-3t} in the transient term of equation (3.116). That exponential factor (together with its constant multiplier 1.15) acts as a decaying amplitude of the sinusoid. We call it a **decay envelope.** (The equality of τ to τ_2 of the previous case with real roots is accidental.) The transient is essentially complete by the end of the settling time $t_s \triangleq 5\tau = 5/3$ s associated with the decay envelope.

The frequency of oscillation $\omega = 5.2$ rad/s is determined by the imaginary part of the complex conjugate root pair. Since the argument ωt of the sinusoid $\cos \omega t$ progresses through 2π radians in one period, the period p satisfies $\omega p = 2\pi$. Therefore, $p = 2\pi/(5.2 \text{ rad/s}) = 1.21$ s, and the frequency of oscillation is $f = 1/p = 0.828$ Hz.

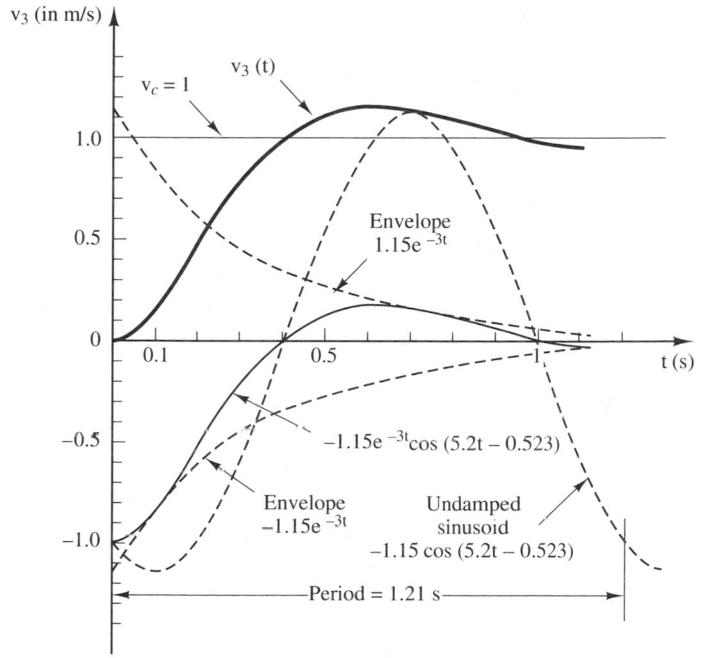

Figure 3.46 Unit-step response for complex conjugate characteristic roots.

Exercise 3.19: The oscillation period p and time constant τ of the decay envelope are useful aids in visualizing waveforms. Sketch the initial and final values v_{3i}, $\dot{v}_{3i}$, and v_{3f} for the preceding example on a graph, and imagine how to fill in the rest of the waveform using only the values of the period p and the time constant τ of the decay envelope.

The sinusoidal portion of the solution equation (3.116) has the form E $\cos(\omega t + \phi)$, where E, ω, and ϕ are the **amplitude, frequency,** and **phase,** respectively, of the sinusoid. Figure 3.47 shows two sinusoids that differ only in phase. These sinusoids are plotted against two abscissa scales—time units (t) and radian units (ωt). Note that ωt and ϕ have the same radian units. We can think of ωt as a quantity of phase that increases with time.

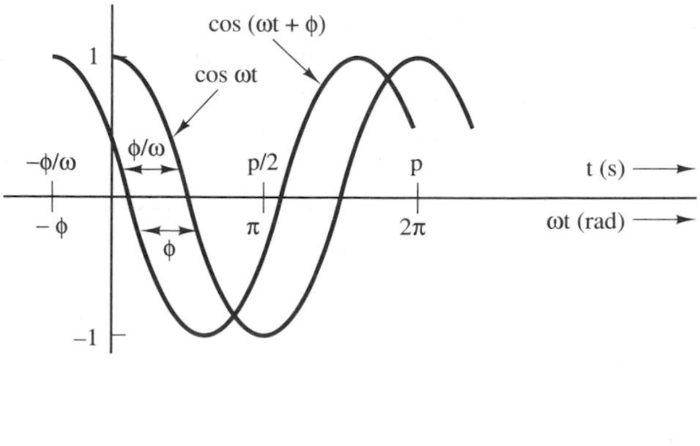

Figure 3.47 Interpretation of positive phase (ϕ = 1.15 rad).

The *unshifted* cosine has a phase $\phi = 0$. The shifted cosine has a positive phase $\phi = 1.15$ rad. A *positive phase* ϕ corresponds to a *leftward shift* of the cosine function, placing it *earlier in time* by ϕ/ω time units and *earlier in phase* by ϕ rad.

The Step Response for an nth-Order Network

The parameters of the elements in lumped models of physical systems are always real numbers. Therefore, the coefficients of the equations that describe the models must be real numbers. If a lumped linear network has n independently acting energy-storage elements, then the input-output system equation for each network variable has order n. Let us express the system equation for the network variable y in operator form (s notation). Then the nth-order polynomial in s which multiplies y on the left side of the equation is the characteristic polynomial, and its zeros are the n characteristic roots, $s_1, \ldots, s_n$.

The roots of a polynomial equation with real coefficients must be real, or else they must occur in complex conjugate pairs. If all the roots of the characteristic polynomial are real, we express the characteristic function (the solution to the homogeneous differential equation) as a sum of n exponentials of the form $A_i e^{s_i t}$, with arbitrary constants $A_1, \ldots, A_n$. If some of the roots occur in complex conjugate pairs, however, it is more convenient to express the corresponding terms of the characteristic function in sinusoidal form, as in equation (3.109) or equation (3.111).

For a step input, the general solution to the system equation for a variable y is the sum of the final value y_f and the nth-order complementary function. We differentiate the general solution $n - 1$ times and evaluate the solution and its derivatives at $t = 0^+$ to obtain n equations that relate the constants $\{A_i\}$ to n initial conditions. The initial conditions and the final value can be found by the open-circuit and short-circuit procedures of section 3.3. We solve the n initial-condition equations for the values of the n constants $\{A_i\}$ to produce the step response y(t).

Some network variables jump in response to abrupt changes in the source, whereas other variables do not. According to the discussion of the network state in section 3.3, if the system equation for the variable y includes no source-signal derivatives, then a jump in the source does not cause a jump in y or its first $n - 1$ derivatives. In that case, the initial values $y_i, y_i^{(1)}, \ldots, y_i^{(n-1)}$ at $t = 0^+$ are the same as the prior values $y_p, y_p^{(1)}, \ldots, y_p^{(n-1)}$ at $t = 0^-$. Then it is appropriate to treat the set $\{y(0), y^{(1)}(0), \ldots, y^{(n-1)}(0)\}$ as the *initial state.* We can find this initial state from the initial energy states of the elements.

If the system equation for y includes the first derivative of the source, then $y^{(n-1)}$ jumps abruptly at $t = 0$. We can find $y_i^{(n-1)}$ by means of the open-circuit and short-circuit rules of section 3.3. Or, we can find a different variable of the network for which the input-output system equation does not include source-signal derivatives, find the response of that variable, and then relate the variable to y.

All potential and flow variables of a particular linear lumped network have the same characteristic behavior and the same characteristic equation. We should think of that equation as the characteristic equation *of the network.* Its roots are characteristic values *of the network.* The characteristic behavior of each variable is a sum of the n characteristic exponential waveforms associated with the n characteristic values of the network. (If any roots are complex, we combine pairs of complex exponentials to produce sinusoidal waveforms.) The behaviors of the various variables in the network differ only in the amount of each characteristic waveform they contain.

Each root of the characteristic equation either is real or is a member of a complex conjugate pair. The real part of each root can be positive, negative, or zero. If the real part of a root is positive, the corresponding characteristic waveform grows exponentially, and the network is unstable. For stable systems, none of the characteristic roots have positive real parts, and each of the characteristic waveforms decays exponentially with time. Each negative real root gives rise to an exponentially decaying component of the response waveform, with a specific time constant of decay. Each *pair* of complex roots gives rise to an exponentially decaying *sinusoidal* component of the response waveform, with a specific period and a specific time constant for its decay envelope.

All variables in a linear network have the same characteristic roots. (The roots belong to the network as a whole.) As a result, the step-response waveforms of all the variables are similar in appearance. The locations of the characteristic roots in the **complex s-plane** are uniquely associated with the *shapes* of the waveforms. Therefore, we focus our attention on the locations of the roots.

Notation for Complex Conjugate Root Pairs

Figure 3.48 relates complex root-pair locations in the s-plane to the *natural-response* waveforms that arise from those root pairs. Each waveform is displayed at the location of the upper root of the corresponding complex pair. Roots that are further into the left half-plane correspond to quicker decay. Roots with larger imaginary components correspond to oscillations at higher frequencies. The decay rate and oscillation frequency interact. As a result, root pairs oriented at smaller angles from the negative real axis correspond to less significant oscillations. The oscillations are negligible if that angle is less than 45°. The oscillations disappear entirely if both roots lie at the same location on the negative real axis.

If a network is first order, it has one real root, usually negative. The corresponding natural response is a simple exponential decay. We could display these first-order decay waveforms along the negative real axis of the s-plane of Figure 3.48. Again, the decay rate would be faster for roots further into the left half-plane.

If the two roots of a second-order network are real, rather than complex, they correspond to a pair of decaying exponentials. In the real-root example of equation (3.105) and Figure 3.45, we discovered that the exponential term with the slowest decay (largest time constant) determines the natural-response waveform.

If a network is of higher order than two, then each complex root pair gives rise to a component waveform of the corresponding shape shown in Figure 3.48. Each real root gives rise to an exponentially decaying component waveform. The natural response of the network is some combination of the component waveforms associated with the individual roots and root pairs.

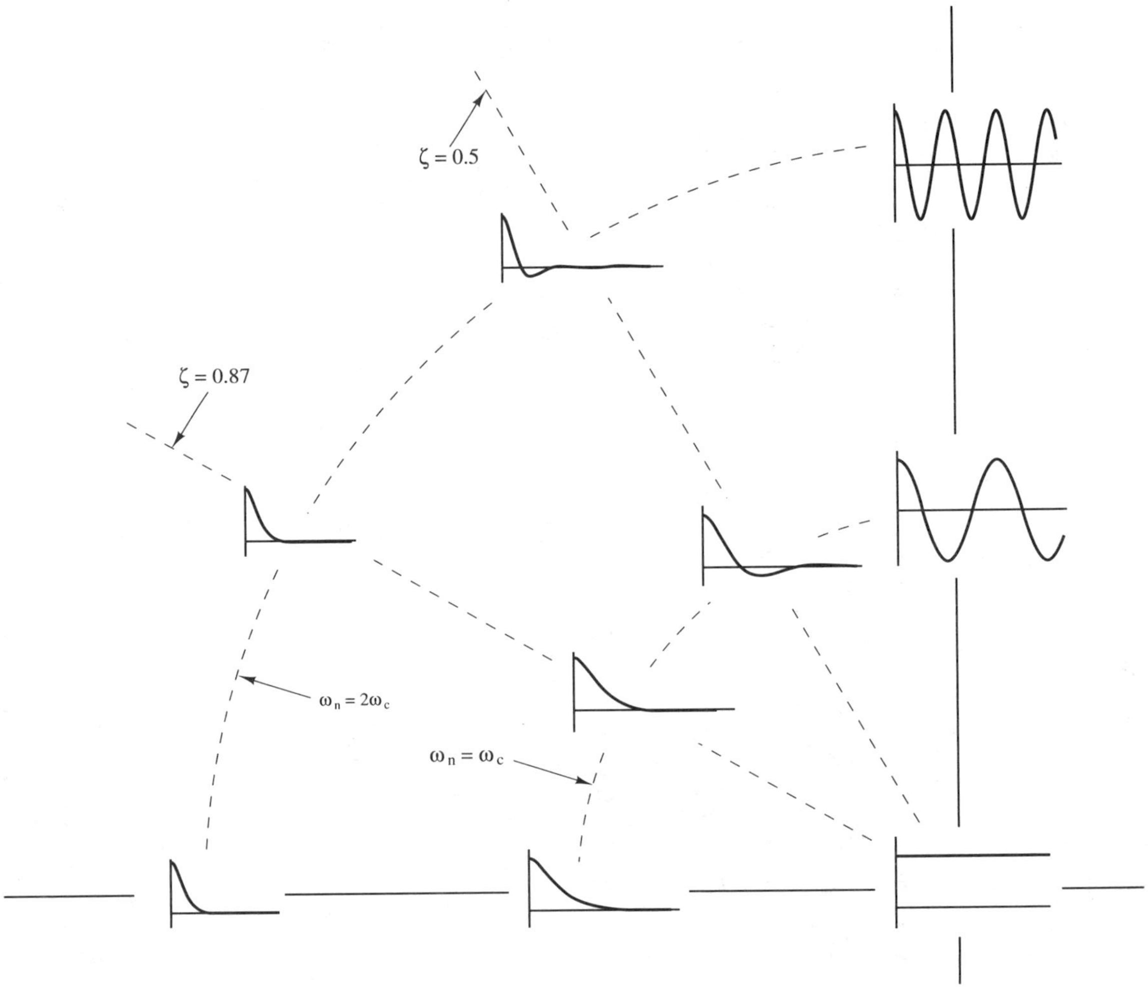

Figure 3.48 Correlation of complex root-pair locations with natural responses. Each waveform is displayed at the location of the upper root of the pair. (Detailed plots of the waveforms are given in Figure 3.53. The notations ζ and ω_n relate to that figure.)

Usually, a network behaves satisfactorily only if all of its characteristic roots are in the left half of the complex s-plane—that is, only if all components of its natural response decay with time. If all of the roots are in the left half-plane, we say that the system is **stable.** (A system is *marginally stable* if it has roots on the imaginary axis.) Roots that are closer to the right half-plane correspond to response terms that decay more slowly. Therefore, after a brief interval, only the response terms owing to the roots closest to the imaginary axis are important. We say that roots close to the imaginary axis **dominate the behavior** of the network. For this reason, most physical systems act as if they are nearly first order or second order.

Since we usually express the response associated with a complex conjugate root pair as a sinusoid, we introduce a notation for complex conjugate roots that we can relate explicitly to sinusoidal waveforms.

The locations of a pair of complex conjugate roots are shown in the left half of the s-plane in Figure 3.49. Because of the nature of mathematical operations with complex numbers, it is convenient to represent the locations in two different ways: *rectangular* ($-\alpha + j\omega_d$) and *modified polar* (magnitude ω_n at angle ψ from the negative real axis). We often mix the two notations.

We call ω_n the **natural frequency.** If the roots lie on the imaginary axis, then ω_n is the oscillation frequency of the natural response, in rad/s. If the roots do not lie on the imaginary axis, then the natural response is *damped;* that is, it decays in amplitude. Since the rate of decay is proportional to α, we call α the **damping rate.** The units of α are rad/s. The *time constant of the decay envelope,* as discussed in connection with Figure 3.46, is $\tau = 1/\alpha$. We call ω_d the **damped natural frequency,** because it is the frequency of the decaying oscillation. Some authors call ω_n the *undamped* natural frequency.

The damping rate α is positive if the network is stable—that is, if the response decays with time. The dimensionless parameter $\zeta = \cos\psi$ is called the **damping ratio.** This name arises from the identity $\zeta = \alpha/\omega_n$, the ratio of the damping rate to the natural frequency of the network. Thus, ζ is a dimensionless measure of the rate of damping in the network.

Figure 3.48 shows that ζ is directly correlated with the *shape* of the response waveform. Therefore, we usually refer to the root locations in terms of ζ. For a given waveform shape (value of ζ), the figure shows that all of the remaining quantities, α, ω_n, and ω_d, are proportional to the waveform *rapidity*—the quickness with which the response plot is enacted along the time axis. We usually relate the rapidity to the polar coordinate ω_n.

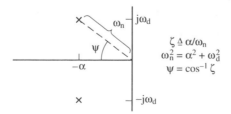

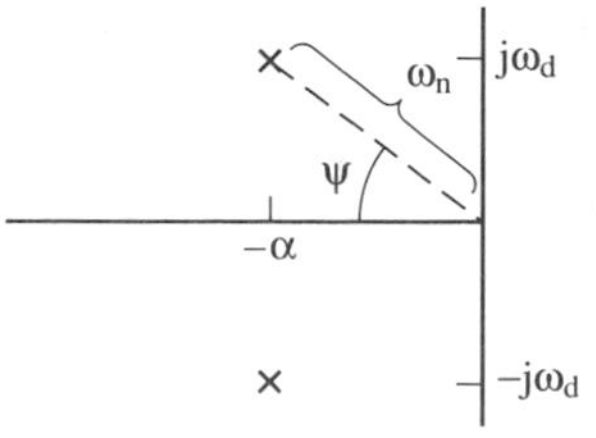

$$\zeta \triangleq \alpha/\omega_n$$
$$\omega_n^2 = \alpha^2 + \omega_d^2$$
$$\psi = \cos^{-1}\zeta$$

Figure 3.49 Notation for conjugate roots in the complex s-plane.

For a second-order network with complex roots, we can express the characteristic equation in the following forms:

$$\begin{aligned}(s - s_1)(s - s_2) &= (s + \alpha - j\omega_d)(s + \alpha + j\omega_d)\\ &= (s + \alpha)^2 + \omega_d^{\,2}\\ &= s^2 + 2\alpha s + \alpha^2 + \omega_d^{\,2}\\ &= s^2 + 2\alpha s + \omega_n^{\,2}\\ &= s^2 + 2\zeta\omega_n s + \omega_n^{\,2} = 0\end{aligned} \tag{3.117}$$

Generally, if the coefficient of the s^2 term of a second-order characteristic equation is normalized to 1, then the constant term depends only on the parameters of the two energy-storage elements of the system—k and m in equation (3.92). ω_n characterizes the frequency (in rad/s) at which energy would oscillate between the two elements if there were no energy-dissipation in the system; thus the name *natural frequency.* The coefficient of the s term, on the other hand, depends on the energy-dissipating element—b in equation (3.92). That coefficient determines α and, hence, sets the rate of extraction (and dissipation) of energy from the storage elements. Intuitively, it is the dissipation of energy that causes the gradual reduction (or damping) of the energy oscillations and, as a consequence, causes decay in the values of all variables in the network.

Exercise 3.20: For the characteristic equation (3.106), show that $\alpha = 3$ rad/s, $\omega_d = 5.2$ rad/s, $\omega_n = 6$ rad/s, $\psi = 1.05$ rad (or 60°), and $\zeta = 0.5$.

If both roots of the characteristic polynomial of a stable second-order system are real, then each root corresponds to a decaying exponential. We usually represent each negative real root in terms of a time constant. The characteristic equation is then expressed as

$$\begin{aligned}(s - s_1)(s - s_2) &= \left(s + \frac{1}{\tau_1}\right)\left(s + \frac{1}{\tau_2}\right)\\ &= s^2 - \left(\frac{1}{\tau_1} + \frac{1}{\tau_2}\right)s + \frac{1}{\tau_1\tau_2} = 0\end{aligned} \tag{3.118}$$

Intuitively, this case with negative, real roots occurs if the energy-dissipating element extracts energy from the network at such a high rate that energy oscillations between the two storage elements cannot occur.

The *pair* of negative, real roots given in equation (3.118) can be expressed in the complex-root notation of equation (3.117). Then, we can put all types of roots for a second-order network into a single notational framework. If a pair of negative, real roots is expressed in the complex-root notation, the natural frequency is $\omega_n = 1/\sqrt{\tau_1\tau_2}$, the damping rate is $\alpha = (1/\tau_1 + 1/\tau_2)/2$, and the damping ratio, $\zeta = \alpha/\omega_n$, is positive, real, and greater than 1. The parameter ω_d is imaginary and does not have an easily identifiable physical interpretation in this nonoscillatory case. A damping ratio ζ greater than 1 signifies real roots and nonoscillatory behavior.

Exercise 3.21: For the characteristic roots $s_1 = -3$ and $s_2 = -12$ discussed earlier, show that $\tau_1 = 1/3$ s and $\tau_2 = 1/12$ s; alternatively, $\alpha = 7.5$ rad/s, $\omega_d = j4.5$ rad/s, $\omega_n = 6$ rad/s, and $\zeta = 1.25$.

For nth-order networks, it is appropriate to use the root notation of Figure 3.49 for each complex conjugate root-pair. Each such root pair gives rise to a decaying sinusoidal component of the response waveform. The frequency of oscillation and the rate of decay of that component can be read directly from the rectangular coordinates, ω_d and $-\alpha$, of the root pair.

A Unified Framework for Second-Order Step Responses

In this section, we examine the various forms of step response for second-order networks. These forms correlate qualitatively and quantitatively with the locations of the roots. We shall find a precise equivalence between the locations of the roots and visible features of the network step response. These visible features are *measurable performance indicators* for the physical system represented by the lumped network. Because of the correlation between the characteristic roots and the performance indicators, we can judge the behavior of a model (and the underlying system) from the characteristic roots alone, without actually computing response waveforms. We shall also be able to use measured features of the step response of the *physical* system to judge the accuracy of the model and its parameters. Or we can use those measured features to compute good values for the parameters of the model.

Each dependent variable in a stable second-order linear network is described by a second-order linear differential equation—an input-output system equation. According to equation (3.117), we can express any such system equation in the form

$$\ddot{y} + 2\zeta\omega_n\dot{y} + \omega_n^2 y = g \tag{3.119}$$

where y is the dependent variable and g, the right side, is a linear combination of the source signal u(t) and its derivatives. We can express the coefficients of the differential equation in this complex-root notation even if the roots are real. The input-output system equations for the various variables of the network differ only on their right sides. All dependent variables in the network have the same characteristic equation (3.117) and the same characteristic roots (see equations (3.91)–(3.93), for example).

Suppose we know the prior state of the network—the values of the energy-state variables. Then we can calculate the prior value y_p. Let us apply a step input to the network. That is, let us hold the source signal u(t) at some constant value u_c for $t \geq 0$. Knowledge of the prior value y_p and the size u_c of the step input is sufficient (together with the system equation) for us to calculate the initial conditions—the initial value y_i and the initial rate of change $\dot{y}_i$. Now, let us use the input-output system equation (3.119), together with the initial conditions, to determine the step response y(t).

Since the source signal is constant, we can use a constant function as a *particular solution* to equation (3.119). Specifically, we use the steady-state final value

$$y_P(t) = y_f \tag{3.120}$$

This final value can be determined from the system equation by setting $s = 0$ in the operator form of the equation. (Of course, different variables y in the network have different right sides g and, thus, different particular solutions.)

If the characteristic roots are complex, the complementary function for equation (3.119) can be expressed as the decaying sinusoid

$$y_C(t) = Ee^{-\alpha t}\cos(\omega_d t + \phi), \qquad t \geq 0 \tag{3.121}$$

where E and ϕ are arbitrary constants and where $\alpha = \zeta\omega_n$ and $\omega_d = \omega_n\sqrt{1-\zeta^2}$, as shown in Figure 3.49. The time constant of the decay envelope is $\tau = 1/\alpha$. The general solution to equation (3.119) is the sum of equations (3.120) and (3.121). We evaluate the general solution and its time derivative at $t = 0^+$ to obtain two equations that relate the arbitrary constants E and ϕ to y_i and $\dot{y}_i$. That process produces the following *step response for the complex-root case:*

$$y(t) = y_f + Ee^{-\alpha t}\cos(\omega_d t + \phi), \qquad t \geq 0 \tag{3.122}$$

where

$$\phi = -\tan^{-1}\left(\frac{\dot{y}_i + \alpha(y_i - y_f)}{\omega_d(y_i - y_f)}\right) \qquad \text{and} \qquad E = \frac{y_i - y_f}{\cos\phi}$$

Exercise 3.22: Evaluate equation (3.122) and its derivative at $t = 0^+$. Solve the resulting equations to verify the preceding formulas for E and ϕ in terms of y_i and $\dot{y}_i$. Then expand the cosine term in equation (3.122) to show that:

$$y(t) = y_f + (y_i - y_f)e^{-\alpha t}\cos\omega_d t + \left(\frac{\zeta(y_i - y_f) + \dot{y}_i/\omega_n}{\sqrt{1-\zeta^2}}\right)e^{-\alpha t}\sin\omega_d t \tag{3.123}$$

where $\alpha = \zeta\omega_n$ and $\omega_d = \omega_n\sqrt{1-\zeta^2}$.

Exercise 3.23: Suppose the initial conditions are zero ($y_i = 0$ and $\dot{y}_i = 0$). Show that the complex-root step response given by equation (3.123) reduces to

$$y(t) = y_f\left[1 - \frac{e^{-\zeta\omega_n t}}{\sqrt{1-\zeta^2}}\sin(\omega_n\sqrt{1-\zeta^2}\,t + \cos^{-1}\zeta)\right], \qquad t \geq 0 \tag{3.124}$$

If $y_i = y_f$, then the argument of $\tan^{-1}$ associated with equation (3.122) is infinite, and E appears indeterminate. In that case, however, equation (3.123) reduces to

$$y(t) = y_f + (\dot{y}_i/\omega_d)e^{-\alpha t}\sin(\omega_d t) \tag{3.125}$$

If the characteristic roots are real, the complementary function is more easily expressed in the time-constant form $y_C(t) = A_1e^{-t/\tau_1} + A_2e^{-t/\tau_2}$. The general solution is the sum of the particular solution given by equation (3.120) and the complementary function. Again, we evaluate the general solution and its derivative at $t = 0^+$ to get two equations to solve for A_1 and A_2. The process produces the following *step response for the real-root case:*

$$y(t) = y_f + A_1e^{-t/\tau_1} + A_2e^{-t/\tau_2}, \qquad t \geq 0 \tag{3.126}$$

where

$$A_1 = \frac{\tau_1}{\tau_1 - \tau_2}[(y_i - y_f) + \tau_2\dot{y}_i] \qquad \text{and} \qquad A_2 = \frac{-\tau_2}{\tau_1 - \tau_2}[(y_i - y_f) + \tau_1\dot{y}_i]$$

Exercise 3.24: Evaluate equation (3.126) and its derivative at $t = 0^+$. Solve the resulting equations to verify the formulas for A_1 and A_2 in terms of y_i and $\dot{y}_i$.

Exercise 3.25: Suppose the initial conditions are zero ($y_i = 0$ and $\dot{y}_i = 0$). Show that the real-root step response given by equation (3.126) reduces to

$$y(t) = y_f\left[1 - \frac{\tau_1}{\tau_1 - \tau_2}e^{-t/\tau_1} + \frac{\tau_2}{\tau_1 - \tau_2}e^{-t/\tau_2}\right], \qquad t \geq 0 \tag{3.127}$$

In equation (3.127), the transient term with the larger time constant is negative and larger in magnitude than the transient term with the smaller time constant. Now suppose that $\tau_1 > \tau_2$. Then the first two terms of equation (3.127) are similar to a first-order network response. The third term provides a small correction to that first-order approximation. If $\tau_1 \gg \tau_2$, we can ignore the term with the smaller time constant τ_2 altogether. The dominance of the *slower* term is illustrated in Figure 3.45 and in example problem 3.10.

Suppose the characteristic roots are real and equal. This is a limiting instance of both the real-root case and the case with complex conjugate roots. Both forms of solution, equations (3.123) and (3.126), are indeterminate because $\tau_1 = \tau_2 = 1/\omega_n$, $\zeta = 1$, and $\omega_d = 0$. In this case, we say the network is **critically damped**. It can be shown that the complementary function for a critically damped network has the form $y_C(t) = D_1e^{-t/\tau} + D_2te^{-t/\tau}$. Once again, we form the general solution y(t) by adding the particular solution given by equation (3.120) to the complementary function, evaluate y(t) and $\dot{y}(t)$ at $t = 0^+$, and use the resulting equations to solve for D_1 and D_2 in terms of y_i and $\dot{y}_i$. The process produces the following *step response for the critically damped case:*

$$y(t) = y_f + e^{-t/\tau}(D_1 + D_2t), \qquad t \geq 0 \tag{3.128}$$

where $D_1 = y_i - y_f$, $D_2 = \dot{y}_i + (y_i - y_f)/\tau$, and $\tau = 1/\alpha$.

Exercise 3.26: Carry out the preceding process to verify the formulas for D_1 and D_2 in terms of y_i and $\dot{y}_i$.

Exercise 3.27: Suppose the initial conditions are zero ($y_i = 0$ and $\dot{y}_i = 0$). Show that the critically damped step response given by equation (3.128) reduces to

$$y(t) = y_f\left[1 - \frac{\tau + t}{\tau}e^{-t/\tau}\right], \qquad t \geq 0 \tag{3.129}$$

The energy-storage elements determine the constant term of the characteristic equation (3.117) (see equation (3.92), for example). Hence, they determine the natural frequency ω_n. The energy dissipation in the network is associated with the s term of the characteristic equation, as demonstrated by the parameter b in equation (3.92). The effectiveness of the energy-dissipating elements is indicated by the damping ratio ζ.[7] The energy-dissipating elements determine whether the characteristic roots are complex or real—whether the step response oscillates or just decays.

Let us examine how a second-order network changes behavior as we change the damping (the value of ζ). We adjust the parameter of the energy-dissipating element to vary the damping ratio ζ from zero to infinity. The natural frequency ω_n remains constant. According to equation (3.117), the roots of the characteristic equation follow the paths in the complex plane shown in Figure 3.50. (Find the roots of the equation for some of the values of ζ shown in the figure.)

Observe, in Figure 3.48, the change in natural response as the roots move along the paths shown in Figure 3.50. The response changes from extreme oscillation (ζ near 0) to fast, smooth settling (ζ near 1) to very slow settling ($\zeta \to \infty$). The fastest settling occurs for **critical damping** ($\zeta = 1$). We say that a network with an oscillatory response ($\zeta < 1$) is **underdamped**. On the other hand, if $\zeta > 1$, the response is unnecessarily slow, and we say that the network is **overdamped**.

The settling time of the network is determined by the time constant of the root(s) closest to the right half-plane. If the roots are complex ($\zeta < 1$), the time constant of the decay envelope is $\tau = 1/\alpha$. Figure 3.50 shows that, as the damping (ζ) is increased from zero, α increases and approaches ω_n. The time constant reduces and reaches the value $1/\omega_n$ at critical damping. **Critical damping** produces the shortest possible settling time for the network. For larger values of damping ($\zeta > 1$), one root moves to the right ($s_1 > -\omega_n$) and produces a time constant τ_1 that is larger than the time constant at critical damping. The other root moves to the left ($s_2 < -\omega_n$) and produces a small time constant τ_2. As ζ becomes large, the larger time constant approaches infinity.

The extreme in low damping is the **undamped network**, in which $\zeta = 0$. According to Figure 3.50, the characteristic roots for an undamped network are purely imaginary and $\alpha = 0$. The complementary function for the undamped network is a pure sinusoid. According to equation (3.123), all variables in the network oscillate at the natural frequency without termination. For an undamped network, it is technically incorrect to refer to the particular solution as the final value, since the network is sinusoidal, rather than constant, in the steady state.

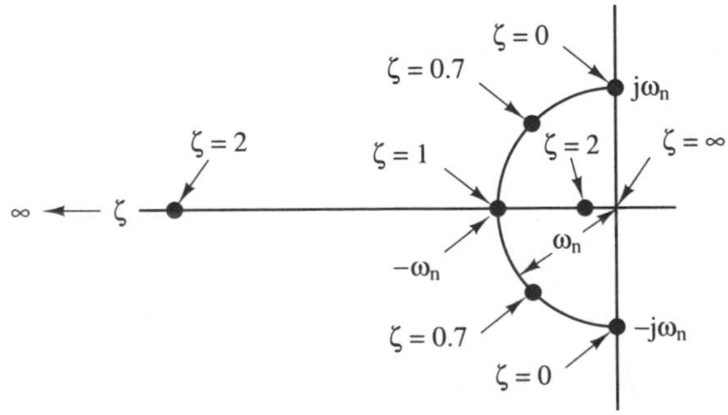

Figure 3.50 Locus of characteristic roots in the s-plane vs. damping ratio.

[7] An increase in ζ corresponds to a *decrease* in the friction constant b in equation (3.92) and in Figure 3.44.

A system in which $\zeta \gg 1$ constitutes the opposite extreme. Such a system is severely **overdamped**. The roots of the characteristic equation (3.117) can be expressed in the form

$$\begin{aligned} s &= -\zeta\omega_n \pm \sqrt{(\zeta\omega_n)^2 - \omega_n^2} \\ &= -\zeta\omega_n(1 \mp \sqrt{1 - 1/\zeta^2}) \end{aligned} \tag{3.130}$$

The square root can be expanded in the binomial series:

$$\sqrt{1 - 1/\zeta^2} = 1 - 1/(2\zeta^2) - 1/(8\zeta^4) - \cdots \tag{3.131}$$

For $\zeta \gg 1$, all terms in the series beyond the first two can be ignored. Then equation (3.130) becomes $s \approx -\zeta\omega_n[1 \mp (1 - 1/2\zeta^2)]$, or

$$\tau_1 = -\frac{1}{s_1} \approx \frac{2\zeta}{\omega_n} \quad \text{and} \quad \tau_2 = -\frac{1}{s_2} \approx \frac{1}{2\zeta\omega_n} \tag{3.132}$$

Since ζ is large, τ_2 is relatively small compared with τ_1. The response associated with τ_2 decays essentially to completion before the term associated with τ_1 begins to decay noticeably. After a brief interval of length $5\tau_2$, the second-order network behaves almost like a first-order network with time constant τ_1. According to equation (3.132), $\tau_1/\tau_2 \approx 4\zeta^2$. Hence, $\zeta > 2$ is sufficiently large for the time constants to be quite different from each other and for the term with the larger time constant $(2\zeta/\omega_n)$ to dominate the response.

A large difference in time constants usually permits portions of the network that are associated only with the faster time constant to be ignored. Network simplification based on differences in time constants is a very useful technique for reducing the complexity of system models. In some instances, it permits the designer to ignore portions of the system altogether (see example problem 3.10).

Example Problem 3.10: Overdamped zero-state step response.

A constant-velocity source is applied suddenly to the unenergized second order network of Figure 3.51. The values of the network and source parameters are shown in the figure. Find and sketch the waveforms $\Omega_2(t)$ and $\dot{\Omega}_2(t)$.

Solution: The responses of all the network variables depend on the locations of the network characteristic roots. The response of each particular network variable y also depends on its final value y_f and its initial values y_i and $\dot{y}_i$. We first focus on finding the values of these quantities. We derive them all from the system equation for Ω_2.

The node equations for the network are

$$\begin{aligned} T_1 &= (K/s)\Omega_{12} \\ T_1 &= B\Omega_2 + sJ\Omega_2 \end{aligned} \tag{3.133}$$

where $\Omega_1 = \Omega$. Therefore, the input-output system equation for Ω_2 is

$$s^2\Omega_2 + \frac{B}{J}s\Omega_2 + \frac{K}{J}\Omega_2 = \frac{K}{J}\Omega \tag{3.134}$$

We find the characteristic equation from the left side of the system equation. For the specified parameter values, it is $s^2 + 12s + 9 = 0$. This equation has roots $s_1 = -0.8$ rad/s and

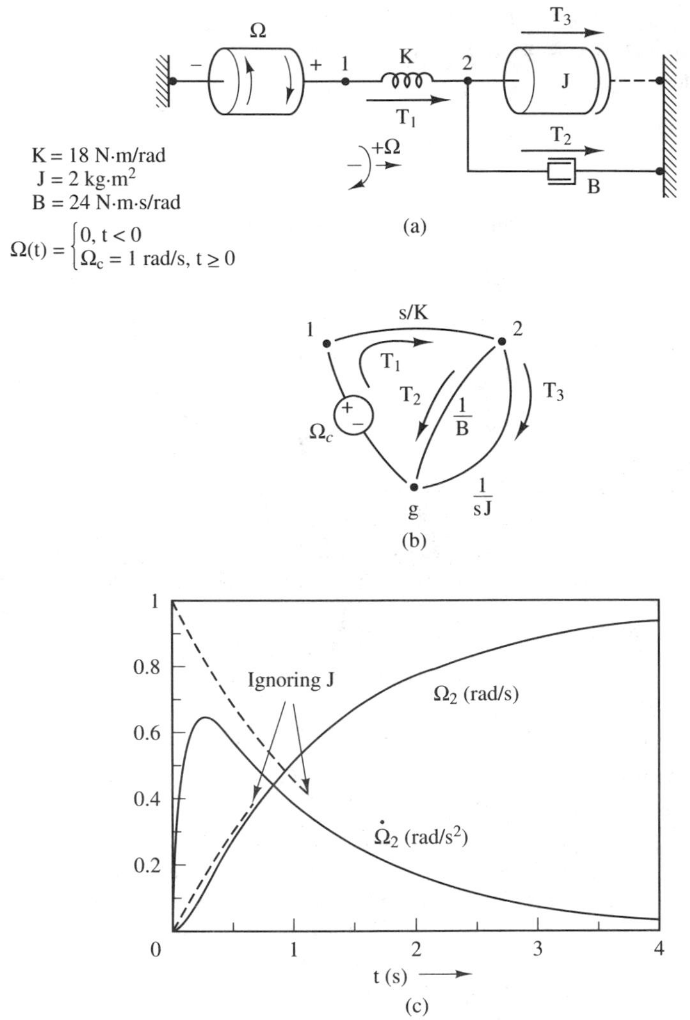

Figure 3.51 Overdamped step response. (a) Lumped model. (b) Line-graph representation. (c) Response waveforms.

$s_2 = -11.2$ rad/s. (These roots correspond to $\zeta = 2$.) The corresponding time constants are $\tau_1 = 1.25$ s and $\tau_2 = 0.089$ s.

Since the network is unenergized, the prior energy states are $T_1(0^-) = 0$ and $\Omega_2(0^-) = 0$. The source Ω is zero for $t < 0$. Consequently, the prior values of all variables in the network are zero.

Now let us find the necessary initial and final values. Throughout this discussion we use the constant-valued source function $\Omega(t) = \Omega_c = 1$ rad/s in the system equation (3.134). We also note that the initial value of each variable equals its initial jump in value because the prior values of all variables are zero.

To find the final value of Ω_2, we let $s = 0$ in equation (3.134) and solve for $\Omega_{2f} = \Omega_c$ rad/s $= 1$ rad/s. To find the initial value (or initial jump in value) of Ω_2, we let $s \to \infty$ in the system equation and solve for $\Omega_{2i} = 0$.

Next, we multiply equation (3.134) by s and solve for $s\Omega_2$. The resulting expression represents $\dot{\Omega}_2$. We let $s = 0$ on the right side of that expression to find the final value $\dot{\Omega}_{2f} = 0$ (as we would expect in steady state). Then we let $s \to \infty$ on the right side of the same expression to find the initial value $\dot{\Omega}_{2i} = 0$.

Finally, we multiply equation (3.134) by s^2 and solve for $s^2\Omega_2$. The resulting expression represents $\ddot{\Omega}_2$. We let $s \to \infty$ on the right side of that expression to find the initial value $\ddot{\Omega}_{2i} = K\Omega_c/J = 9$ m/s^3.

The response $\Omega_2(t)$ for real characteristic roots and zero-valued initial conditions is given by equation (3.127). We apply the foregoing time constants and final value to that equation to obtain

$$\Omega_2(t) = 1 - 1.077e^{-t/1.25} + 0.077e^{-t/0.089} \text{ rad/s} \tag{3.135}$$

This function is shown in Figure 3.51(c). Although the network is second order, the waveform of $\Omega_2(t)$ looks nearly like the exponential response of a first-order network. The third term of equation (3.135), which corresponds to the small time constant, contributes little to Ω_2 because its coefficient is small.

We obtain the expression for $\dot{\Omega}_2(t)$ from equation (3.126) with $y = \dot{\Omega}_2$. We insert the time constants and the values $\dot{\Omega}_{2i} = 0$, $\dot{\Omega}_{2f} = 0$, and $\ddot{\Omega}_{2i} = 9$ m/s^3 into equation (3.126), and find

$$\dot{\Omega}_2(t) = 0.861\, e^{-t/1.25} - 0.861\, e^{-t/0.089} \text{ rad/s}^2 \tag{3.136}$$

Alternatively, we can get equation (3.136) by differentiating equation (3.135). The waveform described by equation (3.136) is plotted in Figure 3.51(c). Although the two terms of that equation correspond to very different time constants, the coefficients on the terms are equal in magnitude. Hence, the component of equation (3.136) associated with the small time constant τ_2 has a major effect on the *initial* behavior of the network. After a very brief interval, however, the effect of the term with the smaller time constant becomes negligible.

According to equations (3.119) and (3.134), $2\zeta\omega_n = B/J$ and $\omega_n^2 = K/J$. From equation (3.132), we find that $\tau_1 \approx B/K$ and $\tau_2 \approx J/B$. Thus, the relatively small value of τ_2 corresponds to a relatively small value of J. Since J is relatively small, the inertia has little effect on the network behavior, except to limit the initial (high) acceleration. Consequently, we might choose to eliminate J from the network altogether, thereby reducing the network to first order. Although ignoring J introduces serious error in the initial value of the acceleration $\dot{\Omega}_2$, it causes little error in any network variables after $t = 5\tau_2 = 0.45$ s (see Figure 3.51(c)).

Exercise 3.27: Use the second equation of (3.133) to generate $T_1(t)$.

Normalized Response Curves

The second-order step-response formulas (3.123), (3.126), and (3.128) each apply for different values of damping ratio ζ. All three formulas, however, depend on two classes of data:

1. A pair of system parameters (from among ζ, ω_n, α, τ, ω_d, τ_1, and τ_2), each of which can be expressed in terms of ζ and ω_n. (These parameters are properties of

the network: they are determined *only* by the network elements and their manner of interconnection.)

2. The signal parameters y_i, $\dot{y}_i$, and y_f. (These quantities are determined by the initial energy state of the network, the size u_c of the source step, and the form of the right side of the system equation.)[8]

We now rearrange the response function y(t) to separate these two classes of data.

Each of the formulas (3.123), (3.126), and (3.128) can be expressed in the form

$$y(t) = y_f\mathcal{A}(t) + y_i\mathcal{B}(t) + \frac{\dot{y}_i}{\omega_n}\mathcal{C}(t) \tag{3.137}$$

where the *normalized waveforms* $\mathcal{A}(t)$, $\mathcal{B}(t)$, and $\mathcal{C}(t)$ are dimensionless and depend only on the network parameters ζ and ω_n. (It would be more accurate to denote them by the less wieldy notation $\mathcal{A}(t; \zeta, \omega_n)$, etc.) We can think of $\mathcal{A}(t)$ as a descriptor of the portion of y(t) owing only to y_f. Similarly, $\mathcal{B}(t)$ and $\mathcal{C}(t)$ describe the portions of the response owing only to y_i and $\dot{y}_i$, respectively.

The step response of any variable of the second-order network can be expressed in the form of equation (3.137). Thus, $\mathcal{A}$, $\mathcal{B}$, and $\mathcal{C}$ characterize the behaviors of *all* variables of that network; different network variables are distinguished only by different initial and final values.

Let us find $\mathcal{A}(t)$. According to equation (3.137), $\mathcal{A}(t) = y(t)/y_f$, under the condition that $y_i = 0$ and $\dot{y}_i = 0$. The formula for y(t) is either (3.123), (3.126), or (3.128), depending on ζ. To compute $\mathcal{A}(t)$, we (1) determine ζ and use it to select the appropriate formula—either (3.123), (3.126), or (3.128); (2) set $y_i = 0$ and $\dot{y}_i = 0$ (as we already did in equations (3.124), (3.127), and (3.129); and (3) divide y by y_f.

To plot $\mathcal{A}(t)$, we express all system parameters in the function in terms of ζ and ω_n. Then we find that the time variable t always appears multiplied by ω_n. Therefore, we plot $\mathcal{A}(t)$ versus the normalized abscissa $\omega_n t$. The normalized waveform $\mathcal{A}(t)$ is plotted for various values of ζ in Figure 3.52. (These curves are obtained from (3.124), (3.127), or (3.129), depending on the value of ζ.) Note that $\Omega_2(t)$ of Figure 3.51(c) is an $\mathcal{A}(t)$ curve for $\zeta = 2$ and $\omega_n = 3$ rad/s. (Compare it with the corresponding curve of Figure 3.52.)

Exercise 3.28: Compare Figure 3.49 and equations (3.117) and (3.118) to show that $1/\tau \equiv \alpha = \omega_n\zeta$, $\omega_d = \omega_n\sqrt{1 - \zeta^2}$, $1/\tau_1 = \omega_n(\zeta + \sqrt{\zeta^2 - 1})$, and $1/\tau_2 = \omega_n(\zeta - \sqrt{\zeta^2 - 1})$.

The final value y_f appears only as a divisor of the ordinate in Figure 3.52. Therefore, although an increase in the magnitude u_c of the step input causes a proportional increase in y_f and in the magnitude of the response, it does not affect the shape of the response.

The **shape** of the response is determined by the damping ratio ζ alone. That is, ζ is the parameter on the $\mathcal{A}$ curves. A small value of ζ (say, $\zeta < 0.3$) implies a low level of

[8] Recall that $y_i = y_p + \delta y$ and $\dot{y}_i = \dot{y}_p + \delta\dot{y}$. If the source signal u(t) is a step, then the absence of $\ddot{u}$ in the system equation implies that $\delta y = 0$, and the absence of $\dot{u}$ implies that $\delta\dot{y} = 0$.

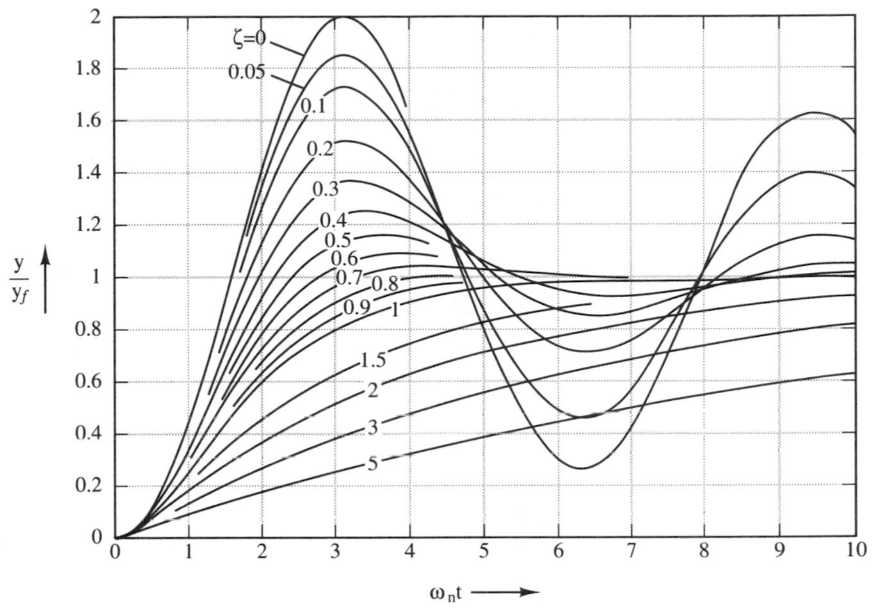

Figure 3.52 Normalized step response $\mathcal{A}(t)$; $y_i = 0$ and $\dot{y}_i = 0$.

stability—in other words, a high fractional overshoot. A high value of ζ (say, $\zeta > 2$) implies good stability, but sluggish response. Thus, ζ, by itself, is a good **indicator of the level of stability of the network.** A value of ζ near unity is a good compromise.

The system parameter ω_n appears as a multiplier of the abscissa. An increase in ω_n causes a proportionate reduction in the time scale, and the system settles more quickly. We can view ω_n as an **indicator of the responsiveness of the network.** A large value of ω_n implies a responsive network.

Most systems are designed so that $0.3 < \zeta < 2$. The curves of Figure 3.52 show that these systems reach the neighborhood of their final value when $t \approx 5/\omega_n$. The manner in which the response approaches the final value depends on ζ. However, we can view $5/\omega_n$ as a *rough measure of the response time,* regardless of the value of ζ.

To derive the normalized function $\mathcal{B}(t)$, we follow a similar procedure. From equation (3.137), $\mathcal{B}(t) = y(t)/y_i$ under the condition that $y_f = 0$ and $\dot{y}_i = 0$. To plot $\mathcal{B}(t)$ for a specific value of ζ, we select the appropriate formula from among equations (3.123), (3.126), and (3.128), set $y_f = 0$ and $\dot{y}_i = 0$, divide by y_i, express all system parameters in terms of ζ and ω_n, and plot the outcome against the abscissa $\omega_n t$. The resulting normalized curves are shown in Figure 3.53.

In a similar fashion, we compute $\mathcal{C}(t) = \omega_n y(t)/\dot{y}_i$ under the condition that $y_f = 0$ and $y_i = 0$. These normalized curves are displayed in Figure 3.54. Note that the $\dot{\Omega}_2(t)$ curve of Figure 3.51(c) is a $\mathcal{C}(t)$ curve for $\zeta = 2$ and $\omega_n = 3$ rad/s; compare it with the corresponding curve in Figure 3.54.

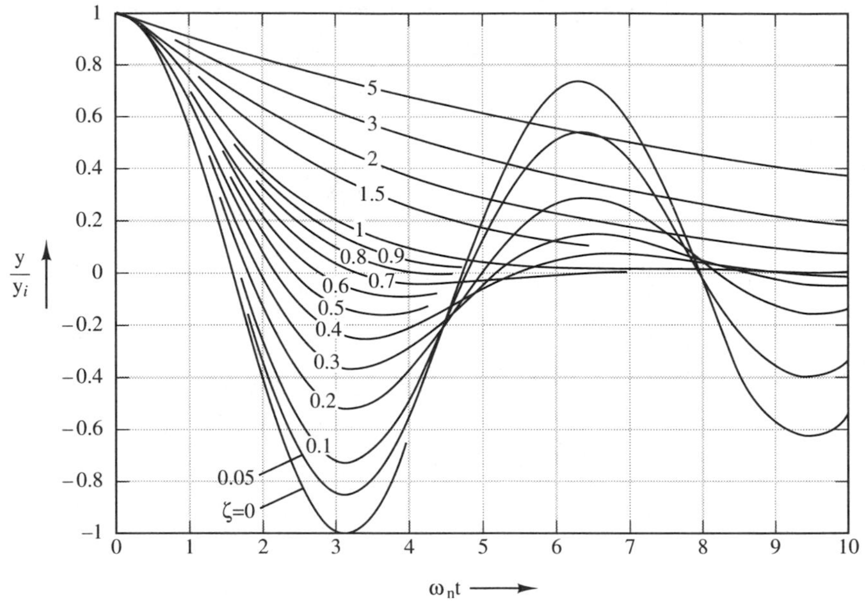

Figure 3.53 Normalized source-free response $\mathcal{B}(t)$; $y_f = 0$ and $\dot{y}_i = 0$.

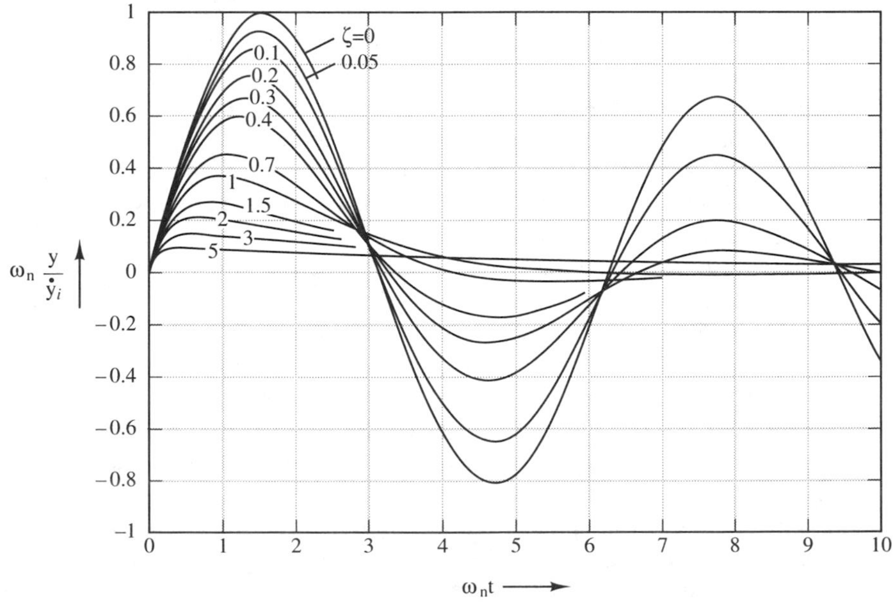

Figure 3.54 Normalized source-free response $\mathcal{C}(t)$; $y_f = 0$ and $y_i = 0$.

Exercise 3.29: Use equation (3.128) together with the identities $\alpha = \omega_n$ and $\tau = 1/\omega_n$ to verify the $\mathcal{A}$, $\mathcal{B}$, and $\mathcal{C}$ curves of Figures 3.52–3.54 for $\zeta = 1$.

Example Problem 3.11: Maximizing network responsiveness.

An unenergized, underdamped, second-order network and its line graph are shown in Figure 3.55.

(a) Determine the response of the variable v_2 to the abrupt application of a 1-N force.

(b) Determine a new value for b_2 that will eliminate the oscillation without unnecessary slowing of the exponential decay. Find the corresponding step response $v_2(t)$.

Solution: **(a)** The response $v_2(t)$ depends on the characteristic roots of the network (ζ and ω_n) and on the initial and final values v_{2i}, $\dot{v}_{2i}$, and v_{2f}. We focus first on finding all of these quantities.

By using the node method and eliminating v_1, we find that the differential equation for v_2 is

$$\ddot{v}_2 + \left(\frac{b_2}{m} + \frac{k}{b_1}\right)\dot{v}_2 + \frac{(b_1 + b_2)k}{b_1 m} v_2 = \frac{k}{b_1 m} F \tag{3.138}$$

For the particular parameter values,

$$\ddot{v}_2 + 1.1\dot{v}_2 + 1.1v_2 = 0.5F \tag{3.139}$$

The characteristic equation is $s^2 + 1.1\,s + 1.1 = 0$. Hence, $\omega_n = 1.05$ rad/s and $\zeta = 0.524$. The $\zeta = 0.524$ curves in Figures 3.52–3.54 show that the response is slightly oscillatory.

For $t < 0$, the network is unenergized and the source is zero. Therefore, the prior values of all the variables are zero. At the instant the 1-N source is applied, the spring acts as an open circuit and the mass as a short circuit. (Let $s = \infty$ in the impedances of the network.) Consequently, $\delta v_2 = 0$, $\delta F_4 = 0$, and $\delta \dot{v}_2 = 0$. It follows that $v_{2i} = 0$ and $\dot{v}_{2i} = 0$.

As $t \to \infty$, the network approaches a steady state, the spring acts as a short circuit, and the mass acts as an open circuit. (Let $s = 0$ in the impedances of the network.) From the line graph, then, $v_{2f} = F_c/(b_1 + b_2) = (1\text{ N})/(2.2\text{ N·s/m}) = 0.455$ m/s.

Since both initial conditions are zero, the response must be a curve from Figure 3.52. We interpolate between the $\zeta = 0.5$ and $\zeta = 0.6$ curves to find the correct shape and transfer that curve to Figure 3.55(c). To set the scales on the two axes, we focus attention on the peak overshoot. The peak value on the $\zeta = 0.524$ curve can be estimated from Figure 3.52 to be $y/y_f \approx 1.16$. Therefore, we label the peak in Figure 3.55(c) with the value $v_2 = (1.16)v_{2f} = (1.16)(0.455\text{ m/s}) = 0.528$ m/s. Other ordinate points can be labeled in proportion to the peak value. We estimate from Figure 3.52 that the peak occurs at $\omega_n t \approx 3.7$ on the $\zeta = 0.524$ curve. Therefore, we label the time of the peak with the value $t = (3.7)/\omega_n = (3.7)/(1.05\text{ rad/s}) = 3.52$ s. Other abscissa points can be labeled in proportion to the peak time.

We can find the functional form of $v_2(t)$ from equation (3.123). For the computed values of ζ, ω_n, and v_{2f}, the step response is

$$\begin{aligned} v_2(t) &= 0.455 + e^{-t/1.82}(-0.455\cos 0.894t + 0.280\sin 0.894t)\text{ m/s} \\ &= 0.455 + 0.534e^{-t/1.82}\cos(0.894t - 0.552)\text{ m/s} \end{aligned} \tag{3.140}$$

This function can be plotted (by hand or computer) to verify that it yields the graphically produced waveform of Figure 3.55(c).

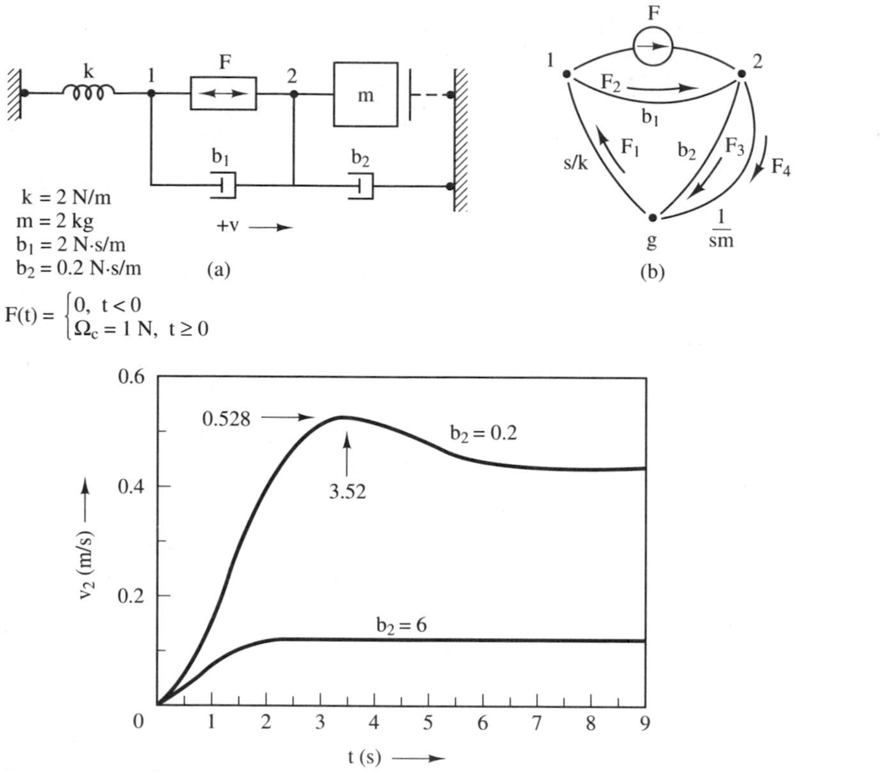

Figure 3.55 Underdamped second-order response. (a) Lumped network. (b) Line-graph representation. (c) Response waveforms.

(b) Critical damping ($\zeta = 1$) produces the shortest settling time, as illustrated in Figure 3.52. We use equation (3.117) to express the second and third coefficients of equation (3.138) in terms of ζ and ω_n:

$$2\zeta\omega_n = \frac{b_2}{m} + \frac{k}{b_1} \tag{3.141}$$

$$\omega_n^2 = \frac{b_1 + b_2}{b_1}\frac{k}{m} \tag{3.142}$$

Setting $\zeta = 1$, substituting the specified values of m, k, and b_1, and solving the two equations gives $b_2 = 6$ N·s/m and $\omega_n = 2$ rad/s.

The change in b_2 does not change the initial values v_{2i} and $\dot{v}_{2i}$. However, $v_{2f} = F/(b_1 + b_2) = 0.125$ m/s. The new step response, according to equation (3.128), is

$$v_2(t) = 0.125 + e^{-t/0.5}(-0.125 + 0.25t) \tag{3.143}$$

This response is displayed in Figure 3.55(c). Observe that the new response rises to its final value in half the time of the old response. As we should expect, this halving of the rise time corresponds to doubling of the responsiveness parameter ω_n. The new response has the smooth shape of the $\zeta = 1$ curve in Figure 3.52. Note that the larger friction coefficient b_2 has also reduced the final velocity of the mass.

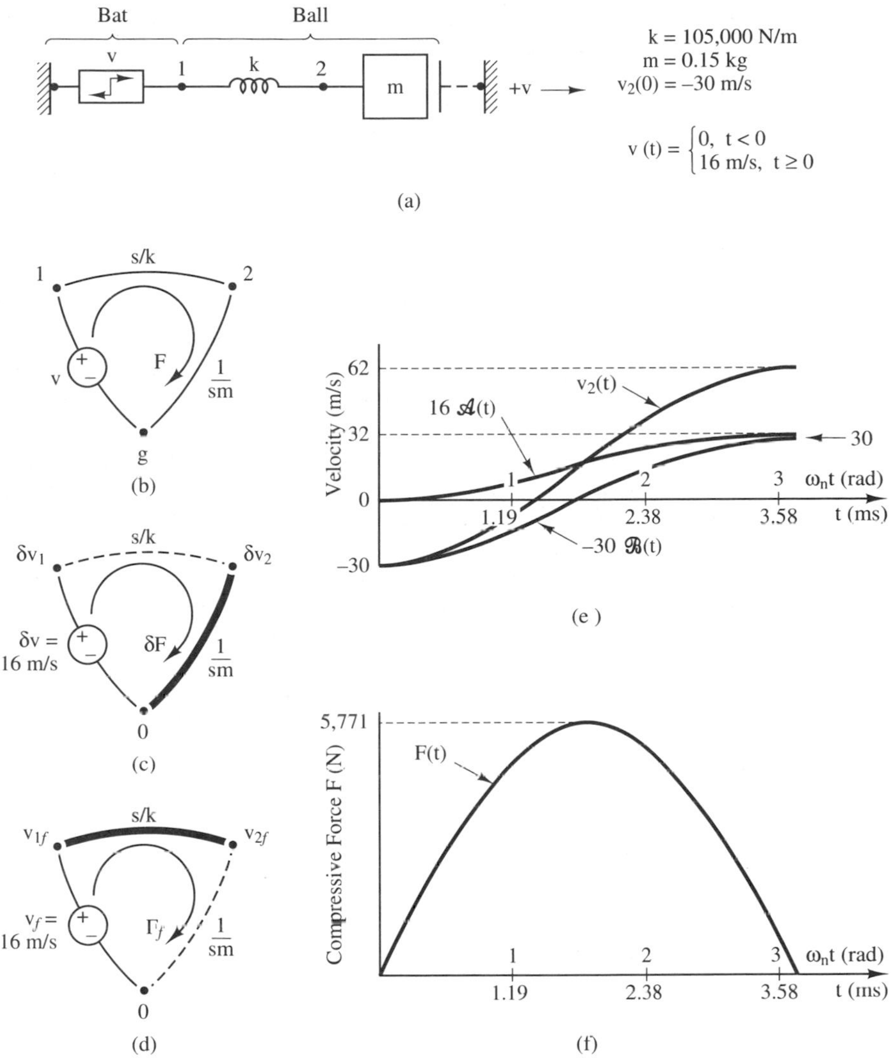

Figure 3.56 Graphical construction of response. (a) Lumped model. (b) Line-graph representation. (c) Abrupt-change line graph. (d) Steady-state line graph. (e) Construction of velocity response. (f) Force response.

Example Problem 3.12: Graphical construction of a step response.
A lumped model of a baseball undergoing impact with a bat is shown in Figure 3.56(a). The initial velocity of the ball is 30 m/s. The bat strikes the ball at 16 m/s. Assume that the bat (together with the arms and body of the batter) is so much heavier and stiffer than the ball, that it can be treated as a constant-velocity source. Treat the ball as a spring (the ball's stiffness) connected to the ball's mass. The line graph for this model is shown in Figure 3.56(b).
(a) Find the energy state of the network at the instant $t = 0$ that the bat makes contact with the ball, in terms of initial values of the central energy-state variables. Find the prior values

(at $t = 0^-$) of the pairs $F, \dot{F}$ and $v_2, \dot{v}_2$. Which of these pairs is a network state vector, unaffected by a jump in the source?

(b) Find the initial values, initial rates of change, and final values of F and v_2. (For the final values, assume that the ball and bat do not break contact.)

(c) Find the system parameters—i.e., the characteristic roots of the network—that apply during impact of the bat on the ball.

(d) Use the normalized response curves of Figures 3.52–3.54 to sketch the velocity $v_2(t)$ and compressive force F(t) that exist during the impact. (These curves apply only during the interval in which $F(t) \geq 0$; once F(t) becomes negative, the ball loses contact with the bat.) Interpret the resulting curves.

Solution: **(a)** The energy stored in the mass of the ball is related to the velocity of the ball, $v_2(0^-) = -30$ m/s. The energy stored in the spring (the compliance of the ball) is related to the compressive force in the spring. Since the spring is not compressed prior to contact with the bat, $F(0^-) = 0$ N. Neither of these energy-state variables changes at the instant of contact.

We assume that the network is in a steady state at $t = 0^-$; that is, $\dot{v}_2(0^-) = 0$ and $\dot{F}(0^-) = 0$. We must determine which of these rate variables is unchanged at the instant of contact, in order to determine which of the pairs of variables acts as a state vector. At the instant of contact between the bat and ball, the abrupt changes in the network are described by the line graph of Figure 3.56(c). The source signal is $\delta v = \delta v_1 = 16$ m/s. The mass acts as a short circuit and the spring as an open circuit to abrupt changes. Accordingly, $\delta F = 0$ and $\delta v_2 = 0$. It follows that $v_2(0^+) = -30$ m/s and $F(0^+) = 0$; that is, the energy state (v_2, F) does not change at contact. According to the mass equation, $F(0^+) = m\dot{v}_2(0^+)$; hence, $\dot{v}_2(0^+) = 0$ m/s^2. The rate variable $\dot{v}_2$ does not change at contact. Because $v_1(0^+) = v_1(0^-) + \delta v_1 = 16$ m/s, the spring equation implies that $\dot{F}(0^+) = kv_{12}(0^+) = 4.83 \times 10^6$ N/s.

The pair $v_2, \dot{v}_2$ is a state vector for the network. Neither of its components is affected by the jump in the source. The pair $F, \dot{F}$, on the other hand, is not a state vector for the network. The abrupt source causes a jump in $\dot{F}$ at $t = 0$. Generation of the input-output system equations for F and for v_2 shows that the derivative of the source appears on the right side of the equation for F but not for v_2.

(b) We have already found the initial values and initial rates of change of F and v_2 (at $t = 0^+$). They are:

$$F_i = 0, \qquad \dot{F}_i = 4.83 \times 10^6 \text{ N/s}, \qquad v_{2i} = -30 \text{ m/s}, \qquad \dot{v}_{2i} = 0 \tag{3.144}$$

Observe that the initial rate $\dot{F}(0^+)$ has two components. One, owing to contact with the bat (the jump in the source), is $\dot{F}(0^+)_{source} = kv_1(0^+) = 1.68 \times 10^6$ N/s; it contributes to the zero-state response. The other, owing to the initial velocity of the ball (the initial state), is $\dot{F}(0^+)_{state} = -kv_2(0^+) = 3.15 \times 10^6$ N/s; it contributes to the zero-input response.

The final values of the network variables are determined from the steady-state line graph of Figure 3.56(d). The values of the network variables associated with this line graph depend only on the source and not on the initial state. The final value of the source (the velocity of the bat) is $v_f = 16$ m/s. From the graph, we determine that

$$F_f = 0 \quad \text{and} \quad v_{2f} = v_{1f} = 16 \text{ m/s} \tag{3.145}$$

Of course, $\dot{F}_f = 0$ and $\dot{v}_{2f} = 0$. These final values presume that the ball and bat will remain in contact forever. Although the ball eventually breaks contact, during the interval of contact the network behaves as if these values will be the final values.

(c) The system parameters are determined by the coefficients of the characteristic equation—the coefficients of the left side of the differential equation. These parameters are not affected by the values of the source variables. We simplify the determination of the characteristic equation by letting the source be zero. (We short-circuit the velocity source in Figure 3.56(b).) The loop equation for the resulting network is $(s/k)F + F/sm = 0$. The characteristic equation is $s^2 = -k/m = -700{,}000$ rad/s^2, and $s = \pm j\ 837$ rad/s. Therefore, according to equation (3.117), $\omega_n = 837$ rad/s and $\zeta = 0$.

(d) According to equation (3.137), the velocity of the ball can be expressed as $v_2(t) = v_{2f}\mathcal{A}(t) + v_{2i}\mathcal{B}(t)$, where $v_{2f} = 16$ m/s and $v_{2i} = -30$ m/s, as determined by equations (3.144) and (3.145). Similarly, the compressive force in the ball can be expressed as $F(t) = (\dot{F}_i/\omega_n)\mathcal{C}(t)$, where $(\dot{F}_i/\omega_n) = 5{,}771$ N. These two responses are weighted sums of the normalized $\zeta = 0$ curves of Figures 3.52–3.54; they are displayed in Figures 3.56(e) and (f). We divide the horizontal scale of each normalized curve by $\omega_n = 837$ rad/s to convert it to time units. We terminate the responses at the instant the compressive force becomes negative; thereafter, the ball loses contact with the bat. (The lumped model must be modified before the ensuing response can be determined.)

In sum, we find the network parameters ζ and ω_n from the network characteristic equation. We use the prior state, the size u_c of the step input, and the short-circuit and open-circuit procedures of section 3.3 to find the initial values y_i and $\dot{y}_i$ and the final value y_f for each network variable y. Then we can get an accurate graph of the step response of each network variable by adding the normalized curves, one from each of Figures 3.52–3.54, appropriately weighted.

If the system equation for y has no derivatives of the source signal u(t), then the step input does not cause jumps in either y or $\dot{y}$ at $t = 0$. In this case, $y_i = y_p$, $\dot{y}_i = \dot{y}_p$, and the pair $(y_i, \dot{y}_i)$ represents the initial state. Then the *zero-state step response* for y is $y_f\mathcal{A}(t)$, and the *zero-input response* for y is $y_i\mathcal{B}(t) + (\dot{y}_i/\omega_n)\mathcal{C}(t)$.

If the system equation for y does include source-signal derivatives, then equation (3.137) does not fully separate the response contributions owing to the step input and the initial state. Rather, $y_i = y_p + \delta y$ and $\dot{y}_i = \dot{y}_p + \delta\dot{y}$, where δy and $\delta\dot{y}$ denote initial jumps owing to the step input. (For any particular variable y, only one of δy and $\delta\dot{y}$ will be nonzero.) Then the zero-state step response is $y_f\mathcal{A}(t) + \delta y\mathcal{B}(t) + (\delta\dot{y}/\omega_n)\mathcal{C}(t)$ and the zero-input response is $y_p\mathcal{B}(t) + (\dot{y}_p/\omega_n)\mathcal{C}(t)$.

Prototype Variables

We use network models to help us understand the behavior of a system. For example, we might want to determine whether a small car is less safe than a large car. To do so, we build a lumped model of the system (the crashing cars) and then analyze the model to see how it behaves. For example, we might investigate whether the smaller car in the crash decelerates more rapidly than the larger car. In this example, then, we would focus our attention on the quickness of response of the two portions of the network.

Figures 3.52–3.54 show that the response components $\mathcal{B}(t)$ and $\mathcal{C}(t)$ give little insight into the *nature* of response that is not already provided by $\mathcal{A}(t)$. To simplify analysis, we usually look for some variable of the network that behaves like the $\mathcal{A}(t)$ curve alone. Every linear network has such variables. We call them *prototype variables* of the network. The response of a prototype variable typifies the behavior of the network as a whole.

Stored energy in a second-order network produces $\mathcal{B}(t)$ and $\mathcal{C}(t)$ response components. To obtain a network response that consists of the $\mathcal{A}(t)$ curve alone, we must begin with an unenergized network. Hence, we seek those variables of the network for which the *zero-state* step response includes no $\mathcal{B}(t)$ or $\mathcal{C}(t)$ components.

An abrupt jump in a network source—that is, an input step—causes abrupt jumps in some of the network variables or their first derivatives. An abrupt jump in a variable initiates a $\mathcal{B}(t)$ component in its response, even if the network is in the zero state. An abrupt jump in the first derivative of a variable initiates a $\mathcal{C}(t)$ component in its response. Therefore, we must find network variables that do not jump themselves or have jumps in their first derivatives in response to an abrupt source.

Suppose the network is nth order. If the system equation for the network variable y has no source-signal (right-side) derivatives, then neither y nor any of its first $n - 1$ derivatives jumps in response to a source step. (According to the earlier discussion of state variables, there *is* a jump in $y^{(n)}$.) Hence, we seek those variables that have no source-side derivatives in their system equations.

We define a **prototype variable** of a network to be a variable for which the right side of the input-output system equation is strictly proportional to the source signal u. The right side does not include $\int u$. (If it did, y(t) would not approach a constant final value for a step input.) Nor does the right side include $\dot{u}$ or higher order derivatives of u. (If the right side included $\dot{u}$, then $y^{(n-1)}$ would jump in response to a step input; if it included a higher order derivative of u, then a lower order derivative of y would jump in response to the step input.) We shall use prototype variables as standard variables by which to judge the behaviors of networks.

If y is a prototype variable of a second-order linear network, then the *zero-state step response* of y is $y(t) = y_f\mathcal{A}(t)$, a waveform from Figure 3.52 alone. For different lumped-element values, the network responds with a different curve from that figure. We need only compare the particular curve y(t) with the other possible curves from the figure to judge the behavior of the network.

Every network has prototype variables. For an nth-order network, a prototype variable and its first $n - 1$ derivatives form a state-variable set. That is, knowledge of the initial values of a prototype variable and its first $n - 1$ derivatives is equivalent to knowledge of the amounts of energy stored in the n independent storage elements of the network. For the network of Figure 3.44, the input-output system equations (3.92) and (3.91) confirm that v_3 and $\int F$ are prototype variables. (Show that knowledge of the initial values of either v_3 or $\int F$ and its derivative is sufficient to determine the initial energy states of the spring and mass.)

The zero-state step response of a prototype variable characterizes the network. For a second-order network, focusing attention on a prototype variable reduces evaluation of the network behavior to appraisal of a particular curve from Figure 3.52. That is, we compare this curve with the other curves from that figure which could occur for different element parameters or which could be produced by a different second-order network.

Time-Domain Performance Indicators

Suppose y is a prototype variable of a second-order linear network. Then Figure 3.52 gives the zero-state step response of y. We use that response to characterize the network as a whole. Note the correlation between the step-response curves of the figure and the locations of the characteristic roots indicated by the parameters ζ and ω_n. (This correlation is enhanced in Figure 3.48.) The damping ratio ζ determines the shape of the response waveform—the *degree of stability* of the network. The natural frequency ω_n determines the time scale—the network *responsiveness.* Once we know the characteristic roots of a second-order network, we can visualize the manner in which the network behaves without ever computing its response. We now develop ways to *measure* network stability and responsiveness.

Suppose a network is linear and second order. Let y be a *prototype variable* of the network. We define the **fractional overshoot** in the zero-state step response y(t) to be

$$f_{OS} \triangleq \frac{y_{peak} - y_f}{y_f} \tag{3.146}$$

Then f_{OS} is an indicator of the *degree of stability* of the network (see Figure 3.57). A network with low fractional overshoot is more stable than a network with high fractional overshoot. The value of f_{OS} is between 0 and 1; it is sometimes expressed as a percentage (0–100%).

It is apparent from the curves of Figure 3.52 that the fractional overshoot for y depends only on the damping ratio ζ of the network. Values of ζ near 0 correspond to values of f_{OS} near 1 and oscillatory behavior. Values of ζ near 1 correspond to values of f_{OS} near 0 and stable behavior. (Values of ζ greater than 1 correspond to an overdamped system and *sluggish response.*) For each curve plotted in Figure 3.52, we can read the value of ζ and estimate the corresponding value of f_{OS}. If we measure f_{OS} for the physical system that is represented by the network, we can find the curve of Figure 3.52 that has the same fractional overshoot and then read the corresponding damping ratio ζ.

The relation between f_{OS} and ζ can be expressed mathematically. The zero-state step response of a prototype variable y for a second-order linear network is equation (3.124).

Exercise 3.30: Equate the derivative of equation (3.124) to zero to show that the time t_{peak} at which the step response reaches its first (highest) peak is

$$t_{peak} = \frac{\pi}{\omega_n\sqrt{1-\zeta^2}}, \qquad 0 < \zeta < 1 \tag{3.147}$$

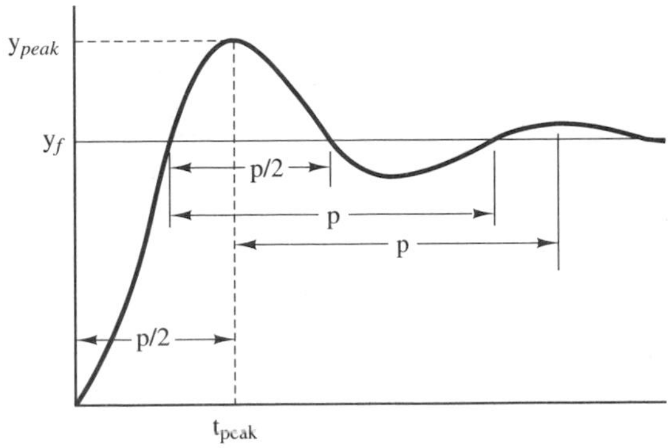

Figure 3.57 Step-response performance indicators for a prototype variable.

Exercise 3.31: Substitute t_{peak} into equation (3.124) to find the peak value y_{peak}. Use this peak value to show that

$$f_{OS} = \exp\left(-\frac{\zeta\pi}{\sqrt{1-\zeta^2}}\right) \tag{3.148}$$

This relation between f_{OS} and ζ is plotted in Figure 3.58.

We can determine ζ from the measured overshoot only for underdamped systems ($\zeta < 1$). If the system is overdamped, we can read ζ from that response curve of Figure 3.52 which has a *shape* similar to the measured response curve. For $\zeta > 2$, the response shape is nearly exponential. Then the behavior is dominated by the response term that has the longest time constant, and the network is nearly first order (see equation (3.126)). We do not need to determine ζ and ω_n if $\zeta > 2$. Instead, we just find the dominant time constant τ from measurements of the physical response. (According to equation (3.132), $\tau \approx 2\zeta/\omega_n$.)

The abscissa of Figure 3.52 is $\omega_n t$. The rapidity of the response y(t), then, is inversely proportional to ω_n. We can treat the measured times of occurrence of various features of the response y(t) as indicators of the *responsiveness* of the network. The peak time t_{peak}, the period of oscillation p, the delay time t_d (the time to reach the value $y_f/2$), and the rise time t_r (the time to rise from $0.1y_f$ to $0.9y_f$) are easily measured features. The settling time of the decay envelope, another indicator of responsiveness, is not as easily measured for an oscillatory signal.

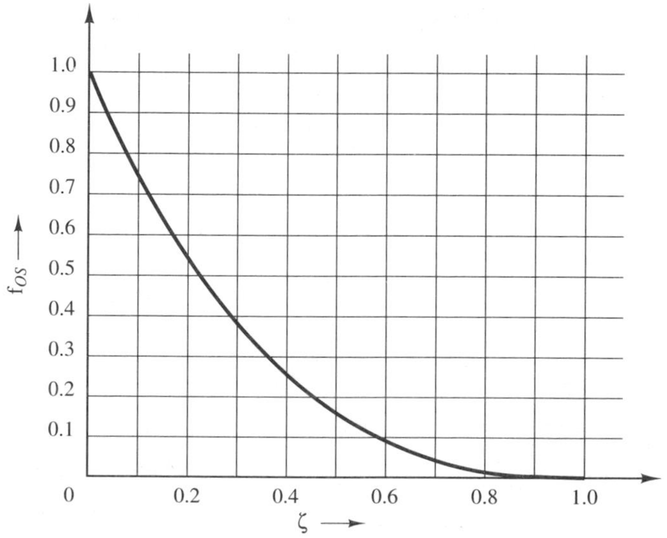

Figure 3.58 Fractional overshoot vs. damping ratio. (Prototype variable of a second-order-network.)

According to equation (3.123), the oscillation period of a system is

$$p = \frac{2\pi}{\omega_d} = \frac{2\pi}{\omega_n\sqrt{1-\zeta^2}}, \qquad 0 < \zeta < 1 \tag{3.149}$$

Note that $p = 2t_{peak}$. Although t_{peak} is a notable feature of the step response, it is difficult to measure accurately because the response curve is nearly flat at the peak. It is easier to measure p. It is shown in reference [3.4] that

$$t_d \approx \frac{1 + 0.7\zeta}{\omega_n} \quad \text{and} \quad t_r \approx \frac{0.8 + 2.5\zeta}{\omega_n} \quad \text{for } 0 < \zeta < 1 \tag{3.150}$$

We treat the fractional overshoot, peak time, oscillation period, rise time, and delay time of the zero-state step response as **time-domain performance indicators** for the network—that is, numerical indicators of the quality of behavior of the network. Corresponding step-response features can be measured directly from the physical system that is represented by the lumped network. We see from the foregoing discussion that the key features of the zero-state step response of a *prototype variable* of a second-order linear lumped model are uniquely related to the system parameters ζ and ω_n of that model. Therefore, we can use measured features of the step response of a *physical* system to select the system parameters for a second-order lumped model for that system.

We can validate a second-order linear lumped model (with specific element values) in two ways: We can compare measured step-response features with the corresponding features derived from the lumped model; or we can compare system parameters (ζ, ω_n) derived from measured step-response features with the system parameters of the lumped model. Example problem 3.13 uses a variation of the second approach.

The input-output system equation for a *prototype* variable y of a second-order network has the form $\ddot{y} + 2\zeta\omega_n\dot{y} + \omega_n^2 y = Du$, where u is the source signal and D is a constant. Let u be a step function of size u_c. When the step response y(t) approaches a steady state, $\ddot{y} \to 0$, $\dot{y} \to 0$, and the system equation approaches $\omega_n^2 y_f = Du_c$, where y_f is the final value of y(t). Therefore, the coefficient of the source signal must be

$$D = \omega_n^2 \frac{y_f}{u_c} \tag{3.151}$$

We can determine the natural frequency ω_n from measured features of the step response (f_{OS} and t_{peak}). We can also measure the size u_c of the source step and the final value y_f of the step response. We can use these measured quantities in equation (3.151) to compute the coefficient D of the right side of the system equation.

It is apparent that we can use zero-state step-response measurements from the physical system to determine the whole input-output system equation for a prototype variable y, without first measuring the parameters of the lumped elements. The system equation is

$$(s^2 + 2\zeta\omega_n s + \omega_n^2)y = \omega_n^2 \frac{y_f}{u_c} u \tag{3.152}$$

where y_f is the final value of y owing to a step input of size u_c. If the model represents the physical system accurately, and if y is a prototype variable of the model, then we can consider the key features (f_{OS}, t_{peak}, and y_f) of the zero-state step response y(t) to be equivalent to the parameters (ζ, ω_n, and D) of the system equation for y.

We can tell from the measured step response y(t) whether or not y is a prototype variable: A jump δy at $t = 0$ indicates the presence of $\ddot{u}$ in the system equation; a jump $\delta\dot{y}$ in the slope at $t = 0$ indicates the presence of $\dot{u}$ in the system equation.

Example Problem 3.13: Finding the system equation from measured *step-response data.*

Example problem 3.5 used steady-state measurements from a physical system to determine lumped model parameters and then used measurements of the dynamic response to validate the model. The model's behavior matched the behavior of the physical system in most respects. However, the physical system showed a decay in oscillation amplitude that did not appear in the simulated behavior.

A new lumped model for the system is shown in Figure 3.59. It differs from the previous model by the addition of a damper. The system equation for the displacement variable x_2 does not have source-signal derivatives. Hence, x_2 is a prototype variable. The measured step-response data from example problem 3.5 are as follows:

Input step size, $x_{1c} = 5$ cm ($\pm$ 2%)

Output final displacement, $x_{2f} = 5$ cm ($\pm$ 2%)

Peak output, $x_{2peak} = 9.6$ cm ($\pm$ 2.1%)

Time of peak, $t_{peak} = 0.35$ s ($\pm$ 31%)

Period of oscillation, $p = 0.76$ s ($\pm$ 6.6%)

(a) Determine directly a second-order input-output system equation for x_2 that is consistent with these measured data.

(b) Find the system equation for the variable x_2 of the lumped model of Figure 3.59. The lumped-element parameters computed in example problem 3.5 are $m = 0.41$ kg ($\pm$ 5%) and $k = 26.3$ N/m ($\pm$ 7.6%). Compare the lumped-model system equation, together with these element parameters, with the results of part (a). What is the value of the damper friction constant b? Is the lumped model valid?

Solution: **(a)** The fractional overshoot is $f_{OS} = (x_{2peak} - x_{2f})/x_{2f} = 0.92$. According to equation (3.148) or Figure 3.58, the corresponding damping ratio is $\zeta = 0.027$. Equation (3.149) relates the natural frequency to the damping ratio and oscillation period. According to that equation, $\omega_n = 2\pi/(p\sqrt{1 - \zeta^2}) = 8.27$ rad/s. The coefficient of the source signal on the right side of the system equation is given by equation (3.151): $\omega_n^2 x_{2f}/x_{1c} = (8.27 \text{ rad/s})^2 (5 \text{ cm})/(5 \text{ cm}) = 68.4 \text{ rad/s}^2$. Therefore, according to equation (3.152), the system equation for x_2 is

$$(s^2 + 0.45\, s + 68.4)x_2 = 68.4x_1 \qquad (3.153)$$

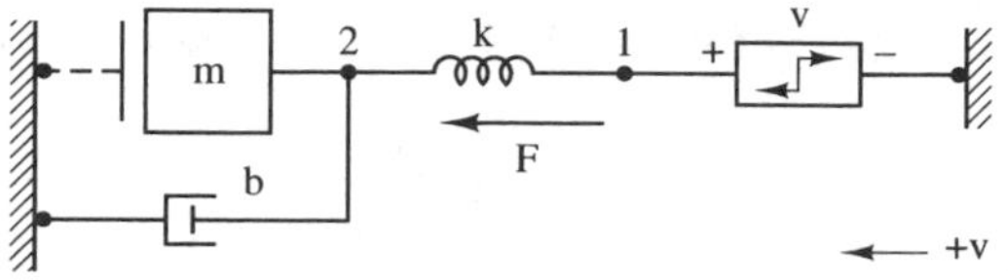

Figure 3.59 Refined model for Figure 3.27.

(We did not use the measurement of peak time to determine ω_n because that measurement was quite uncertain.)

(b) The node equation at node 2 of the lumped model of Figure 3.59 is

$$\frac{v - v_2}{s/k} = \frac{v_2}{1/b} + \frac{v_2}{1/sm} \tag{3.154}$$

We make the substitutions $v_1 = v = sx$ and $v_2 = sx_2$ and rearrange the equation to obtain

$$\left(s^2 + \frac{b}{m}s + \frac{k}{m}\right)x_2 = \frac{k}{m}x \tag{3.155}$$

The measured values of m and k produce the ratio $k/m = 64.1$ ($\pm$ 12.6%). This value agrees with the corresponding coefficient in equation (3.153), which we obtained from the measured step-response data, to within the accuracy of measurement. To this extent, we can consider the model validated.

We derived the coefficient $b/m = 0.45$ rad/s in equation (3.153) from the measured step-response data. We then use that value, together with the measured mass, to compute $b = (0.45 \text{ rad/s})(0.41 \text{ kg}) = 0.185$ N·s/m. (We have no directly measured value of b to which we can compare this result; accordingly, we might wish to measure additional features of the step response and compare them with corresponding values predicted by the model before we consider the model validated.)

Dominant Characteristic Roots

The step response y(t) of a stable nth-order network equals the constant final value y_f plus the sum of n complex exponential terms that decay with time. Equation (3.100) shows this sum of terms for the second-order case. Each complex exponential term is associated with one root of the characteristic equation. The rate of decay of each such term is proportional to the distance of the corresponding characteristic root from the right half of the complex s-plane, as shown in Figure 3.48. Thus, a root closer to the right half-plane indicates a term that decays more slowly.

If one or two of the characteristic roots are three to five times closer to the right half-plane than the other roots are, we say that **the closer roots dominate the behavior** of the network. That is, the terms of the response associated with the *further roots* contribute little to the response because they decay so quickly. The greater the ratio of the distances of two roots from the right half-plane, the stronger is the dominance of the closer root over the further root.

The preceding explanation of dominance deals only with the rates of decay of the various terms. There is another, less obvious, reason for the dominance: If the variable y is a prototype variable of the network—a variable for which the system equation has no source-signal derivatives—then the *magnitudes* of the response terms associated with the various characteristic roots are also lower for roots that lie further into the left half-plane.[9] Hence, the dominance is stronger than that indicated by the decay rates alone.

[9] Source-signal derivatives in the system equation for y influence the *relative* magnitudes of the various response terms that make up y(t). They do not affect the rates of decay of those terms.

As a consequence of the existence of dominant roots, most lumped networks (and physical systems) behave *nearly* as first-order or second-order systems. Hence, the first-order analysis of section 3.3 or the second-order analysis of section 3.4 can be applied *approximately* to most systems. Figures 3.45 and 3.51 show the responses of second-order networks that are dominated by single characteristic roots. Each network approximates first-order behavior, except for a very brief initial interval.

Suppose observation of the characteristic roots of a network shows that one of the roots is *dominated.* Then the response term associated with that root contributes little to the response. If we could eliminate those parts of the system equation that are associated with the dominated root, the equation would be of lower order and its solution would be simpler. As we eliminate the dominated root, however, we must preserve the key feature of the step response, namely, its final value. We demonstrate the approximation process for a specific example and then state a formal approximation procedure.

Let us simplify the system equation for the second-order network of Figure 3.44 (with the damper value $b = 24$ N·s/m). The size of the step input is v_c. The input-output system equation for the prototype variable v_3 is equation (3.92). For the specific element values, the equation is

$$\ddot{v}_3 + 15\dot{v}_3 + 36v_3 = 36v \tag{3.156}$$

We find the final value from equation (3.156) by applying the steady-state conditions $\ddot{v}_{3f} = 0$, $\dot{v}_{3f} = 0$, and $v = v_c$. The result is $36v_{3f} = 36v_c$, or $v_{3f} = v_c$. Note that the final value is determined by the constant term, 36, of the characteristic polynomial on the left side of the system equation.

The system equation (3.156) can be expressed in the factored form

$$(s + 3)(s + 12)v_3 = 36v \tag{3.157}$$

Although the root at $s = -12$ is dominated (somewhat) by the root at $s = -3$, we must not merely cancel the dominated factor $(s + 12)$ from equation (3.157). Such a cancellation would produce a characteristic polynomial that has the constant term 3, leading to the incorrect final value $v_{3f} = 12v_c$.

Before canceling the dominated root, we normalize the constant term of each factor of the characteristic polynomial to unity. That is, we divide the system equation by the product of the characteristic roots. The result is

$$\left(1 + \frac{s}{3}\right)\left(1 + \frac{s}{12}\right)v_3 = v \tag{3.158}$$

Then we cancel the factor owing to the dominated root, $(1 + s/12)$. The simplified, approximate, system equation is $(1 + s/3)v_3 = v$, or

$$\dot{v}_3 + 3v_3 = 3v \tag{3.159}$$

Exercise 3.32: The step response of the original system equation (3.156) is given in equation (3.105) and Figure 3.45. Compare that response with the corresponding step response $v_3(t) = v_c(1 - e^{-3t})$ of the approximate system equation (3.159). Would you say that the root at $s = -12$ is strongly dominated by the root at $s = -3$?

The following procedure eliminates dominated roots from an input-output system equation in order to approximate it by a differential equation of lower order.

APPROXIMATION PROCEDURE FOR SIMPLIFYING A SYSTEM EQUATION

1. Express the system equation in s-notation. Factor the characteristic polynomial on the left side of the equation. Combine conjugate root factors into quadratic terms.
2. Normalize the constant term of each factor of the characteristic polynomial to unity.
3. Determine which characteristic roots, if any, are dominated. Cancel the corresponding factors from the system equation. Arrange the remaining approximate system equation in any desired form.

Once we find a simplified approximation to an input-output system equation, it seems natural to ask whether there is a simplified lumped network that corresponds to that simplified approximation. One way to relate the mathematical changes that we made in the system equation to corresponding changes in the lumped network is to repeat the simplification steps in terms of element parameter symbols instead of numerical values.

If we carry out that process for the example of Figure 3.44 and use the binomial approximation, as in equation (3.131), to eliminate the square root, we find that the neglected root is roughly $-k/b + b/m = -12.6$ rad/s, and the retained root is about $-b/m = 2.4$ rad/s. This fact suggests that, perhaps, we have reduced the network order by eliminating the spring from the network—that we have retained only the mass and damper. Intuitively, if b is small, then the impedance $1/b$ is large, the damper permits little flow (scalar force), the spring cannot deflect much, and $x_2 \approx x_1 = x$.

If we delete the spring from the network (replace it with a short circuit), the resulting system equation is $\dot{v}_3 + (b/m)v_3 = (b/m)v$, which has the characteristic root $-b/m = -2.4$ rad/s. Instead, the approximate system equation (3.159) has the characteristic root $s = -3$ rad/s. Thus, neglecting the dominated root is not quite equivalent to eliminating a specific energy-storage element from the network.

State-Variable Representation of Network Equations

The state of an nth-order network can be represented by any set of quantities that is equivalent to the set of energy states of the n storage elements (see section 3.3)—that is, any set of n quantities from which we can calculate the energy states. Let us choose a set of quantities to represent the network state. We call these quantities the **state variables** for the network. The energy-state variables themselves are an appropriate set of state variables. If the variable x is a prototype variable for the network—that is, if its system equation has no source-variable derivatives—then $(x, \dot{x}, \ldots, x^{(n-1)})$ can serve as a set of state variables. The integrator-output variables for any operational model (for the network) that does not include differentiator blocks are also suitable as state variables. There are as many different sets of state variables as there are ways to structure the integrators in the operational model.

We select a set of state variables and denote them by the symbols $y_1, y_2, \ldots, y_n$. Together, the state variables form a **state vector y,** defined by

$$\mathbf{y} \triangleq \begin{bmatrix} y_1 \\ y_2 \\ \cdot \\ \cdot \\ \cdot \\ y_n \end{bmatrix} \tag{3.160}$$

Let us denote the source signal by the symbol u. Suppose a particular network variable is of interest. We denote that variable by y_O, where the subscript O implies that y_O is the *output* variable. In an operational model of the network, y_O would be obtained as a linear combination of the integrator outputs (state variables) and the source signal.

In general, the system equation for the variable y_O can be expressed in terms of the source signal and the state vector. We express it in matrix notation. We can view the matrix expression as the mathematical equivalent of a particular operational model of the network. We use the network of example problem 3.4 to demonstrate how to write the matrix expression. Then we show the equivalence of that expression to an operational model of the network.

The system equation (3.30) for the displacement variable x_1 of example problem 3.4 includes the source-signal derivative $\dot{F}$ on the right side. (A step input, then, produces an abrupt jump in the second highest derivative $\ddot{x}_1$ of the displacement.) The construction procedure based on the rearranged equation (3.31) avoids the use of differentiator blocks and produces the operational model of Figure 3.25. We select as state variables the outputs of the integrators of that figure; that is,

$$\begin{aligned} y_1 &\triangleq x_1 \\ y_2 &\triangleq \dot{x}_1 \\ y_3 &\triangleq \ddot{x}_1 - F/m \end{aligned} \tag{3.161}$$

The derivatives of these three state variables are the inputs to the corresponding integrators. The operational model of Figure 3.25 shows how these integrator-input variables relate to the other variables in the network. We use equations (3.31) and (3.161) to express the relations in the state-variable notation of equation (3.160):

$$\begin{aligned} \dot{y}_1 &= \dot{x}_1 = y_2 \\ \dot{y}_2 &= \ddot{x}_1 = y_3 + \frac{F}{m} \\ \dot{y}_3 &= \dddot{x}_1 - \frac{\dot{F}}{m} = -\left(\frac{k_2}{b}\right)\ddot{x}_1 - \left(\frac{k_1 + k_2}{m}\right)\dot{x}_1 - \left(\frac{k_1k_2}{mb}\right)x_1 + \left(\frac{k_2}{mb}\right)F \\ &= -\left(\frac{k_1k_2}{mb}\right)y_1 - \left(\frac{k_1 + k_2}{m}\right)y_2 - \left(\frac{k_2}{b}\right)y_3 \end{aligned} \tag{3.162}$$

In matrix notation, these equations are

$$\dot{\mathbf{y}} = [\mathbf{A}]\mathbf{y} + [\mathbf{B}]F \tag{3.163}$$

where

$$[\mathbf{A}] = \begin{bmatrix} 0 & 1 & 0 \\ 0 & 0 & 1 \\ -\dfrac{k_1k_2}{mb} & -\dfrac{k_1 + k_2}{m} & -\dfrac{k_2}{b} \end{bmatrix}$$

and

$$[\mathbf{B}] = \begin{bmatrix} 0 \\ \dfrac{1}{m} \\ 0 \end{bmatrix}$$

Suppose we select the force F_3 in the spring as the output variable y_O. According to the node equation at node 1 of the network (Figure 3.24(b)),

$$\begin{aligned} F_3 &= F - F_1 - F_2 \\ &= F - k_1x_1 - m\ddot{x}_1 \\ &= F - k_1y_1 - m\left(\frac{F}{m} + y_3\right) \\ &= -k_1y_1 - my_3 \end{aligned} \tag{3.164}$$

In matrix notation, equation (3.164) is

$$F_3 = [\mathbf{C}]\mathbf{y} \tag{3.165}$$

where

$$[\mathbf{C}] = (-k_1 \quad 0 \quad -m)$$

We call equation (3.163) the **network state equation;** equation (3.165) is the **output equation** for the variable F_3.

Exercise 3.33: Show that, for the output variable x_2, the matrix of the output equation is

$$[\mathbf{C}] = \left(-\frac{k_1}{k_2} \quad 0 \quad -\frac{m}{k_2}\right) \tag{3.166}$$

In general, the state equation for any linear network is a matrix expression of the form

$$\dot{\mathbf{y}} = [\mathbf{A}]\mathbf{y} + [\mathbf{B}]u \tag{3.167}$$

where $\mathbf{y}$ is some state vector—a vector of state variables—and u is the source signal.[10] Equation (3.167) represents the behavior of an nth-order network as a *first-order* differential equation in n variables.

Suppose we write the output expressions corresponding to equation (3.165) for many dependent variables of the network. (We cannot say *all* dependent variables, because every derivative or integral of a dependent variable is another dependent variable.) If we place the dependent (or output) variables in a column, they form an **output vector** $\mathbf{y}_O$. We can stack the matrix expressions for the individual output signals to form a single matrix output equation. The output vector $\mathbf{y}_O$ can be expressed in terms of the state vector $\mathbf{y}$ by an **output equation** of the form

$$\mathbf{y}_O = [\mathbf{C}]\mathbf{y} + [\mathbf{D}]u \tag{3.168}$$

Exercise 3.34: Suppose the output variables of interest in Figure 3.24 are x_1, x_2, F_1, F_2, and F_3. Show that the output equation for that set of variables is

$$\mathbf{y}_O \triangleq \begin{bmatrix} x_1 \\ x_2 \\ F_1 \\ F_2 \\ F_3 \end{bmatrix} = \begin{bmatrix} 1 & 0 & 0 \\ -\dfrac{k_1}{k_2} & 0 & -\dfrac{m}{k_2} \\ -k_1 & 0 & 0 \\ 0 & 0 & -m \\ -k_1 & 0 & -m \end{bmatrix} \begin{bmatrix} y_1 \\ y_2 \\ y_3 \end{bmatrix} + \begin{bmatrix} 0 \\ 0 \\ 0 \\ 1 \\ 0 \end{bmatrix} F \tag{3.169}$$

Notice that equations (3.167) and (3.168) incorporate all aspects of the linear network into a single pair of equations of standard form. If the network equations are expressed in this state-variable form, techniques of matrix algebra can be used to systematize (and automate) analysis of the network. (For example, the eigenvalues of the system matrix **[A]** are the characteristic values of the network; see reference [3.2] or reference [3.3]). Equations (3.167) and (3.168) are represented visually by a *vector operational diagram* in Figure 3.60. This diagram displays the *conceptual* simplicity of the

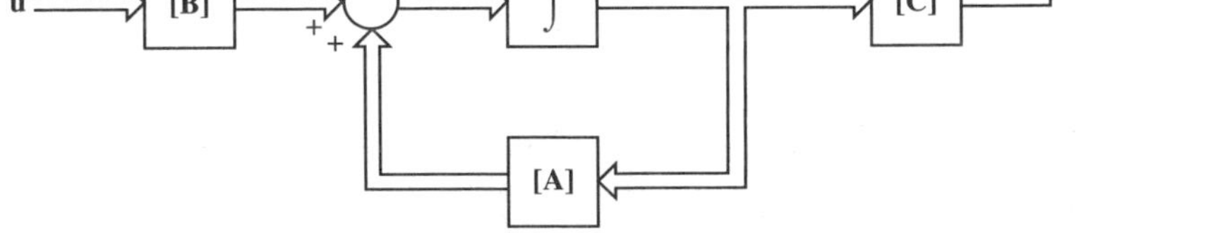

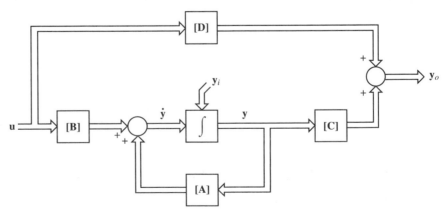

Figure 3.60 Vector operational diagram for state-variable equations.

[10] In section 4.3, we examine systems with multiple sources. To extend equation (3.167) to multiple sources, we merely replace the source signal u by a source-signal vector **u**.

state-variable equations. The nth-order nature of the equations is implicit in the diagram. The complexity of the interconnections among the integrators is hidden in the matrix coefficients.

3.5 SUMMARY

This chapter focuses on finding and interpreting the behaviors of lumped models. The operational model allows computer simulation of the behavior of complicated networks with nonlinear elements and with sources of arbitrary waveform. However, we can simulate a network's behavior only for specific values of the network parameters, for specific initial conditions, and for specific source signals.

The state of an nth-order network can be represented by the values of n state variables. These state variables, taken together, determine the amount of energy there is in each network storage device. A set of energy-state variables, one for each storage device, constitutes one set of state variables. A prototype variable and its first $n - 1$ derivatives is another. The integrator outputs in a well-posed operational model of the network is also a set of state variables.

Most of the chapter is devoted to analyzing the behaviors of *linear* networks. In general, the future behavior of a linear network is the sum of two components, the zero-input response (owing only to currently stored energy) and the zero-state response (owing only to future values of the energy source). The primary features of linear network behavior—responsiveness and degree of stability—are closely related to the characteristic values of the network. We get these characteristic values from the characteristic polynomial—the coefficients of the homogeneous system equation.

Linear network behavior is often dominated by the one or two characteristic roots that are closest to the right half of the complex s-plane. With one or two roots dominant, the behavior of the network is nearly first or second order. First-order behavior is characterized by a time constant. Second-order behavior is characterized by a pair of time constants (overdamped) or by a damping ratio and a natural frequency (underdamped).

For a prototype variable of an underdamped second-order linear network, two features of the zero-state step response characterize the network behavior completely. The fractional overshoot f_{OS}, an indicator of the degree of stability of the network, determines the damping ratio of the network. The peak time t_{peak}, a measure of the network's responsiveness, determines the natural frequency of the network. If the network is a model for an existing physical system, we can determine the damping ratio and the natural frequency of the network from measured values of these features of the step response. Together with the measured final value, the features determine completely the system equation for the prototype variable.

The chapter also introduces lumped models for rotational mechanical systems. All motions in a rotational system are referenced to a nonrotating reference frame (or ground). Rotational velocities of object slices act as potentials. The signs of the rotational motions are determined by the arbitrarily chosen $+\Omega$ direction. The scalar torsions in the parts of the network act as flows. The right-hand rule strengthens the analogy between rotational and translational networks by associating a linear direction with each torsion and each rotational motion. If the positive direction chosen for the flow T_n is the linear $+\Omega$ direction, then T_n represents inward-twisting torsion—the torsion that one would produce by driving a right-handed screw. Reversal of the $+\Omega$ direction reverses the signs of all velocities and torsions, but does not change the physical interpretations of the torsions (i.e., whether they are inward or outward twisting).

We represent rotational systems by interconnections of lumped passive elements and lumped active sources. The passive elements are the torsion spring, the rotary damper, and the inertia (described by $\Omega_{ij} = (s/K)T$, $\Omega_{ij} = (1/B)T$, and $\Omega_{ig} = (1/sJ)T$, respectively). We treat inertia as a two-terminal element with one terminal connected to ground. The active elements are the torsion source, for which the flow T(t) in the linear $+\Omega$ direction is specified, and the velocity source, for which the velocity rise $\Omega(t)$ in the linear $+\Omega$ direction is specified. The power delivered to a rotational lumped element is the product $T\Omega_{ij}$ of the flow through the element and the velocity drop across the element in the direction of flow.

3.6 PROBLEMS

3.1 For one of the lumped models of Figure P3.1:

(a) Specify and label the $+\Omega$ direction and the through and across variables.

(b) Find a set of system equations in terms of the specified variables.

(c) Determine the power that the source delivers to the network, in terms of the source signal and the element parameters.

(d) Find the energy that the source delivers in 1 second if the source signal has a constant value denoted by subscript *c*.

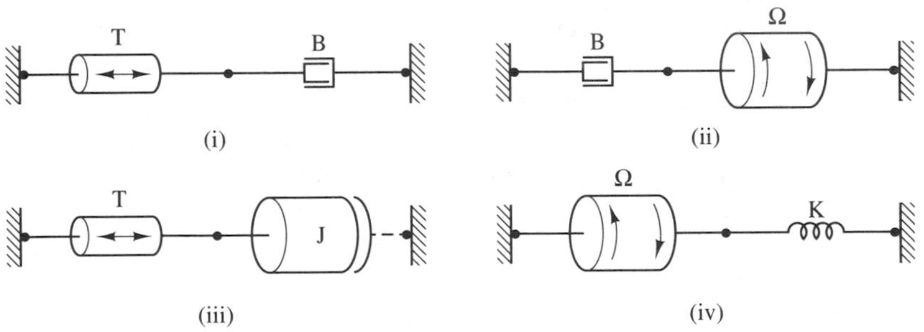

Figure P3.1

3.2 For one of the lumped models of Figure P3.2:

(a) Specify and label the $+\Omega$ direction, the flow variables, and the node-potential variables.

(b) Find a set of system equations directly from the lumped network, in terms of the specified variables.

(c) Eliminate variables from the set of system equations to find the input-output system equation for one of the dependent variables.

3.3 If the source signal in a linear lumped network is sinusoidal, then all network variables eventually become sinusoidal with the same frequency ω. For one of the lumped networks of Figures P3.1, P3.2(ii), or P3.2(iv):

(a) Specify and label the $+\Omega$ direction and the through and across variables.

(b) Find the input-output system equation for one dependent variable of the network.

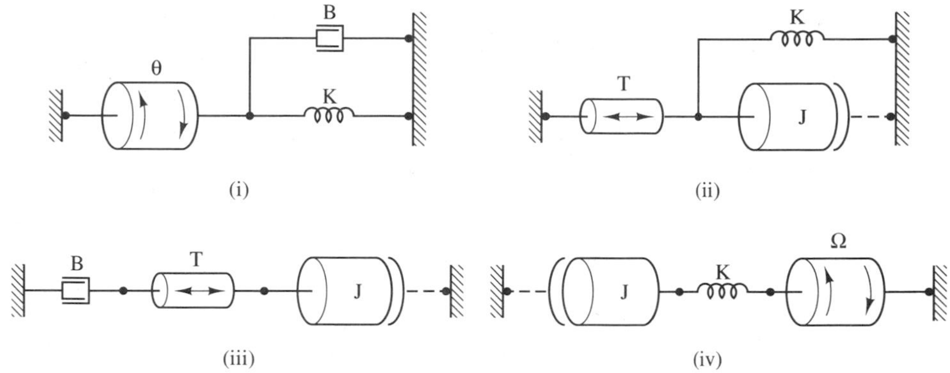

Figure P3.2

(c) Let the source signal be sinusoidal with frequency ω. Use the subscript c to denote the amplitudes of sinusoids. Substitute a sinusoidal form of solution $x_c \sin(\omega t + \phi)$ into the system equation from part (b) to find the amplitude x_c and phase ϕ of the sinusoidal dependent variable, in terms of the element parameters and the source-signal amplitude and frequency.

(d) Interpret the answer to part (c) physically. Is it believable?

3.4 The handle of the torque wrench shown in Figure P3.4 flexes as torque is applied. As it flcxcs, the pointer attached ncar thc grip shows the angle of bend on a scale. Deflection-angle marks with 1° spacings are scribed on the scale. As the wrench applies torque to a bolt, the needle deflects to indicate the amount of torque being applied. The wrench was tested by fastening the socket rigidly in a vice, balancing an accurate 5-kg mass at measured distances along a steel tube placed ovcr thc handle, and reading the deflection angles on the scale. Deflection angles could be read no more accurately than +0.1°. The lever arm lengths were accuratc to ±2 mm. The following measurements of lever-arm length vs. deflection angle were taken:

Length (cm)	90.2	73.8	62.1	31.2
Angle (°)	9.2	7.2	5.8	3.2

(a) Plot the data, and determine an appropriate estimate of the stiffness K of the wrench, in °/kg·cm and in rad/N·m.

(b) Determine what torque value to scribe beside the 10° mark.

(c) Estimate the torque accuracy of the wrench as a percent of full scale, that is, as a percent of the torque value at 10° deflection.

3.5 The tachometer-generator of Figure P3.5 produces an electrical voltage that is proportional to the rotational velocity of the instrument's shaft. The device can be used as a rotational velocity meter by attaching the shaft to the rotating object, measuring the voltage produced, and then multiplying that voltage by a generator constant K_G, measured in rad/volt·s. The tachometer was calibrated as follows. A hand-held adjustable-speed electric drill was used to turn the shaft at constant speed. A voltmeter was used to measure the voltage produced by the generator. The stated accuracy of the voltmeter was ±3% of the full 20-volt scale. The flash rate of a stroboscope was adjusted to make the rotating shaft appear stationary. Then the flash rate equaled the

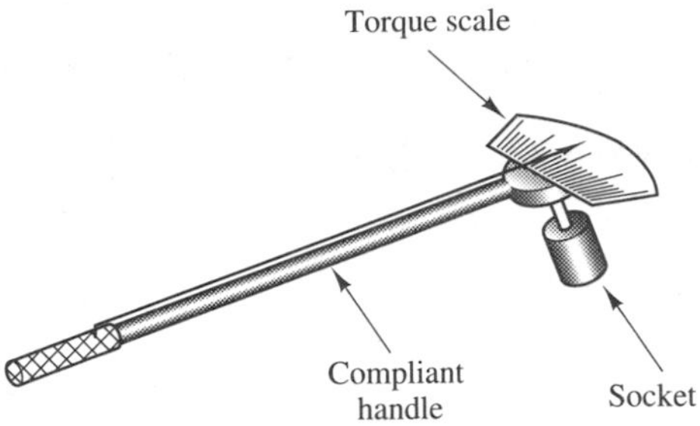

Figure P3.4

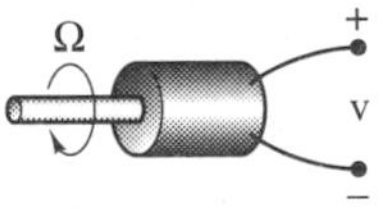

Figure P3.5

rotation rate. The stated accuracy of the stroboscope was ±280 flashes per minute. The following data for flash rate vs. generated voltage were taken:

Rate (fpm)	2,825	2,300	1,825	1,300	925
Voltage	18.6	15.3	11.5	8.6	6.4

(a) Plot the data, and determine an appropriate estimate of the generator constant K_G, in rad/volt·s.

(b) Estimate the accuracy of the velocity meter, as a percentage of the velocity that corresponds to 20 volts.

3.6 A torsion-bar automobile suspension system is shown in Figure P3.6. The spring must support 230 kg, half the mass of the rear end of the automobile. The shear modulus for steel is $G = 7.6 \times 10^{10}\ \text{N/m}^2$. Select an appropriate length and diameter of the torsion bar so that it will support the automobile under the condition that it is stationary. The design process requires some judgment; the solution is not unique.

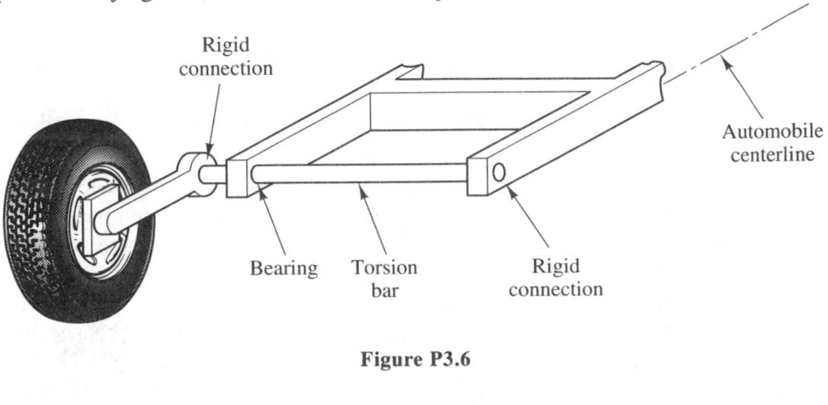

Figure P3.6

3.7 A heavy disk is fastened to a wall by means of a shaft, as shown in Figure P3.7. The shaft and disk are made of steel ($\rho = 7.88 \times 10^3$ kg/m^3, $G = 7.6 \times 10^{10}$ N/m^2). The device provides a stable mounting surface for an electric motor that generates intermittent torques at the disk surface.

(a) Draw a lumped rotational model for the system.

(b) Calculate the values of the parameters of the ideal elements used in the lumped model.

(c) Find the differential equation that describes the rotational motion of the disk.

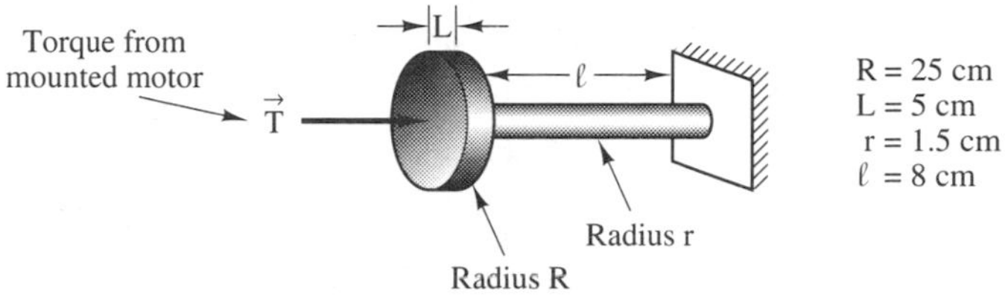

Figure P3.7

3.8 A rotational system for accelerating a flywheel is shown in Figure P3.8.

(a) Construct a lumped model of the system.

(b) Generate a set of system equations, and determine the differential equation that describes the rotational velocity of the flywheel.

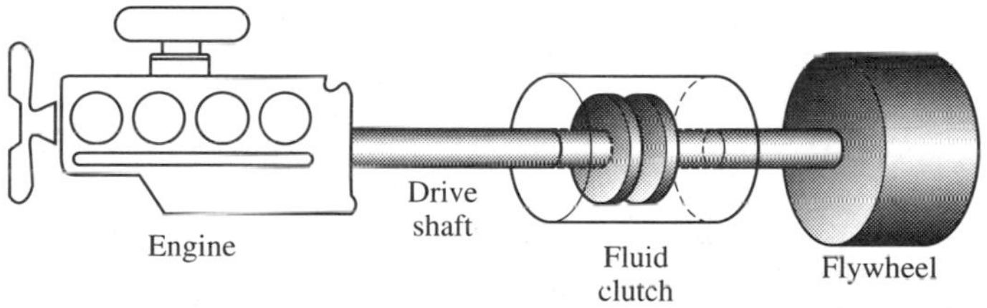

Figure P3.8

3.9 Derive a simple rotational lumped model that describes the main features of one of the following situations or systems. Include a model for the source signal that energizes the system. State your assumptions.

(i) Drilling a hole with a hand-held electric drill.
(ii) Sawing a board by pushing it past a stationary rotary saw blade.
(iii) A rotating lawn sprinkler.
(iv) An electric fan.
(v) A toy yo-yo.
(vi) A lawn mower (rotary or reel).
(vii) A roller skate or skateboard.
(viii) A record player.
(ix) A windlass for raising a weight.
(x) The timing gear of a watch.
(xi) A merry-go-round.

3.10 Construct a rotational lumped model that will exhibit the dynamic characteristics of a bicycle. Estimate the values of the parameters in the model. Also, estimate the shape and magnitude of the source waveform. For simplicity, assume that the bicycle gear is set for one-to-one rotation of the pedals and rear wheel. (*Hint*: Do not confuse the forward translational motion of the bicycle and rider with the rotational motion of the wheels. The rear wheel of the bicycle converts the translational inertial force of the rider to an equivalent rotational inertial force felt at the axis of the wheel. Treat the axis as fixed, and let the ground move relative to it.)

3.11 A lumped rotational model is shown in Figure P3.11.

(a) Determine the effective spring constant *felt* by the torsion source, in terms of K_1, K_2, and K_3.

(b) Solve for all torsions and motions (deflections) in the system, in terms of the spring constants and the source torsion T.

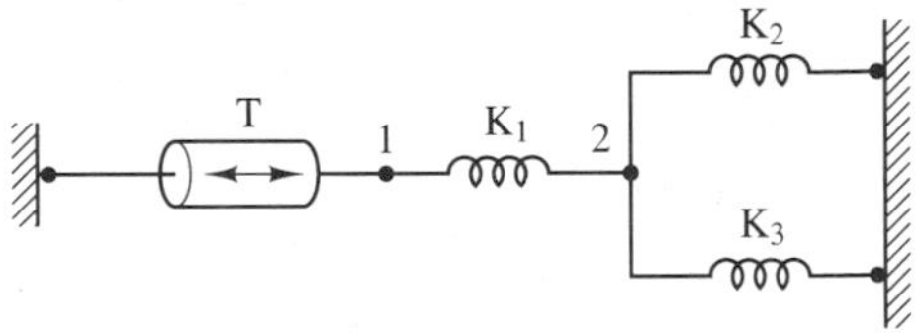

Figure P3.11

3.12 One-half of a symmetrical mechanism for lifting a garage door is shown in Figure P3.12. The shaft and the two coil torsion springs are made of steel ($G = 76 \times 10^9$ N/m^2, $E = 193 \times 10^9$ N/m^2). We wish the torsion spring to support 25% of the weight of the door when the door is at its raised position (in which the rest of the weight is supported by an overhead track) and to support 60% of the weight of the door when the door is 3 feet below its raised position. Friction in the motor drive holds the door stationary in its raised or lowered position when the motor is off.

(a) Draw a rotational lumped model for the system. Model the velocity source v(t) and the weight of the door as rotational sources.

(b) Determine the smallest value of D_s for which the twisting of the shaft will be negligible.

(c) Specify the spring constant K for each of the coil springs. Specify a set of values for the number of turns n, spring diameter D, spring wire thickness d, and cable pulley diameter D_c that meet the stated requirements. Also, specify the number of revolutions the springs should be prestressed when the door is in the fully lowered position (6 feet below the raised position).

3.13 The lumped model of a rotational system is shown in Figure P3.13. Suppose T is a suddenly applied constant torsion. Determine intuitively the natures of the rotational motions and torsions in steady state, after the network variables settle.

3.14 For the lumped model of Figure P3.13:

(a) Use the node method to write a set of system equations.

(b) Use the loop method to write a set of system equations.

(c) Relate the equations found in parts (a) and (b); that is, show that they agree.

3.15 For the second-order translational network of Figure 3.44 and equation (3.92):

(a) Draw an operational diagram that generates the waveform of v_3. Attach additional operational blocks as necessary to generate the waveforms of v_1, v_2, and F.

(b) Use an operational-diagram simulator to compute the behavior of the variable v_3 for the parameter values given in Figure 3.44, with $b = 24$ N·s/m. Compare the resulting waveform with the analytical solution given in equation (3.105) and Figure 3.45.

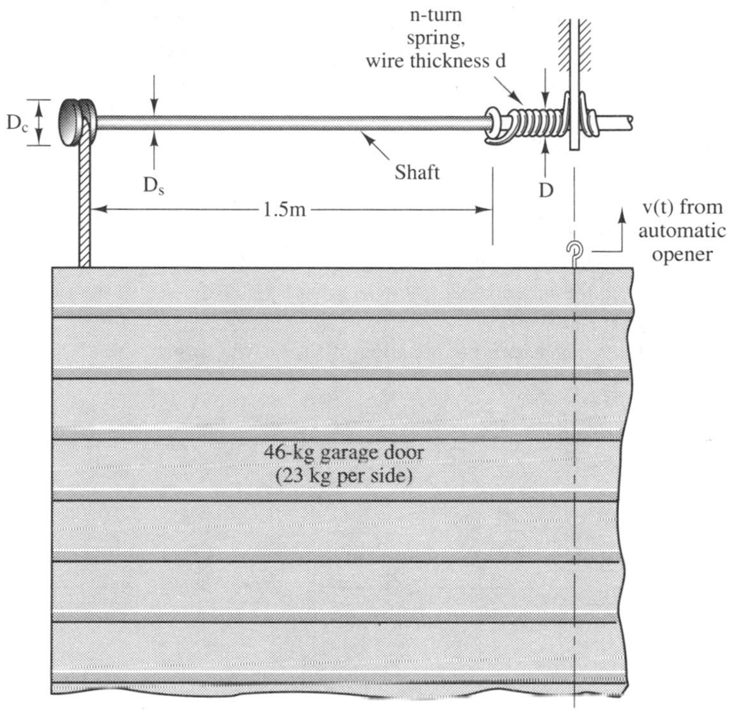

Figure P3.12

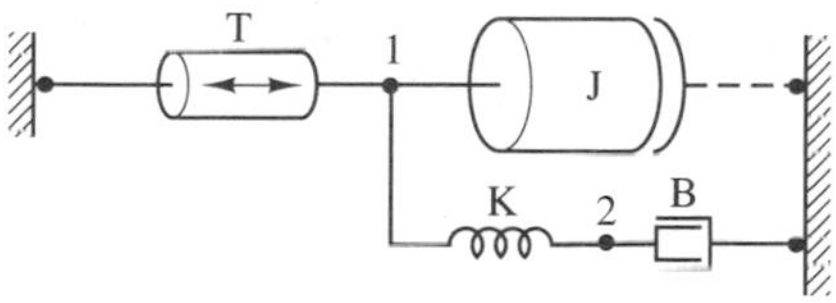

Figure P3.13

(c) Simulate the behavior of the variable v_3 if the damper value is changed to b = 60 N·s/m, and compare the resulting waveform with the oscillatory analytical solution shown in equation (3.116) and Figure 3.46.

3.16 **(a)** Draw a simplified version of the lumped model of the speedometer of Figure 3.15 that uses only two segments. (One node is internal to the speedometer cable.) Determine the new element values. (*Hint:* How are the new values related to the values K, B, and J of the original model.)

(b) Write a set of equations that characterizes the behavior of the simplified lumped model.

(c) Draw an operational diagram that can generate the values of the variables of the lumped model. (*Hint:* Solve all differential equations from part b for their highest order derivatives, draw their operational diagrams, and interconnect the diagrams.)

(d) Simulate the operational diagram numerically for the element values found in part (b) of example problem 3.3. Use $\Omega(t) = 10t$ rad/s (constant motorcycle acceleration from 0 to 97 km/hr in 5 s) as the input signal. Plot the angle of the speedometer needle and the input velocity Ω that it represents as outputs.

3.17 **(a)** Draw a complete operational diagram for one of the networks shown in Figure P3.18 or Figure P3.21. The diagram should produce all network variables as outputs.

(b) Use an operational-block simulation program to simulate operation of the diagram obtained in part (a). Assume: Initially unenergized; a unit step input; $K = 1$ N·m/rad, $B = 3$ N·m·s/rad, $J = 1$ N·m·s²/rad; $k = 1$ N/m, $b = 2$ N·s/m, and $m = 3$ N·s²/m.

3.18 In each first-order linear network of Figure P3.18, all variables are zero at $t = 0^-$, and at $t = 0$ the source jumps to a constant positive value, denoted by the subscript *c*. For one of these networks:

(a) Find the initial and final values of all network variables.

(b) Find the network step response; that is, find and sketch the time variations of *all* variables in the network.

(c) Give a physical interpretation of the answer to part (b).

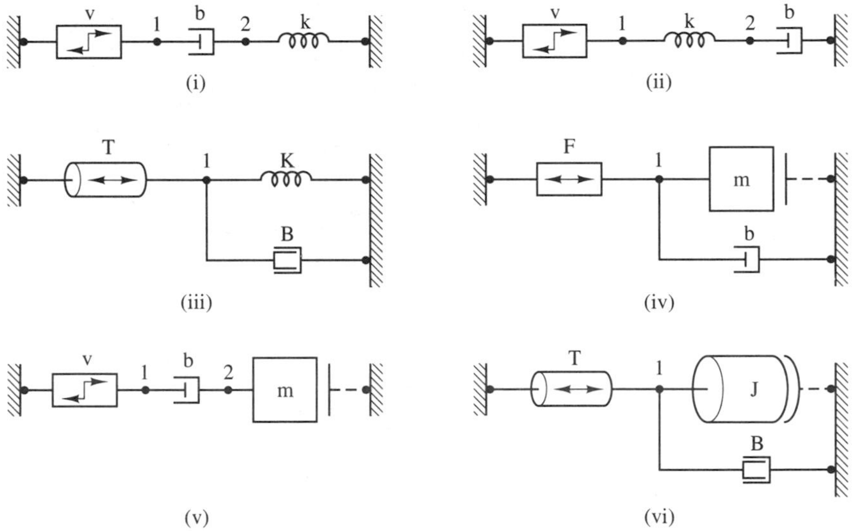

Figure P3.18

3.19 For one of the networks of Figure P3.19:

(a) Find the input-output system equation for the flow variable.

(b) Combine elements to produce a simpler lumped model that has the same flow behavior as the original model.

3.20 The masses in Figure P3.20 are originally stationary relative to the inertial reference (ground). The motion source applies the velocity $v = 1$ m/s to the network until the masses achieve their steady-state velocities. Then $v_1 = v_2$, $v_3 = v_4$, and $v_3 = v_2 + 1$ m/s. At $t = 0$, the source instantly halts; that is, $v = 0$ for $t \geq 0$.

(a) Determine the differential equation that describes the force F exerted by the velocity source for $t \geq 0$.

(b) Determine and sketch $F(t)$ for $t \geq 0$.

(c) Find the system equation for v_4. Use it to determine $v_4(0)$ and the motion $v_4(t)$ relative to ground for $t \geq 0$.

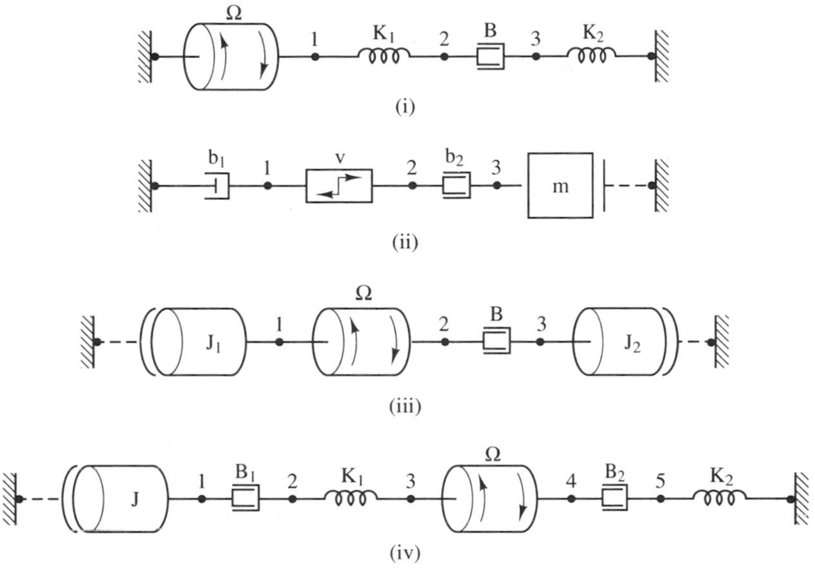

Figure P3.19

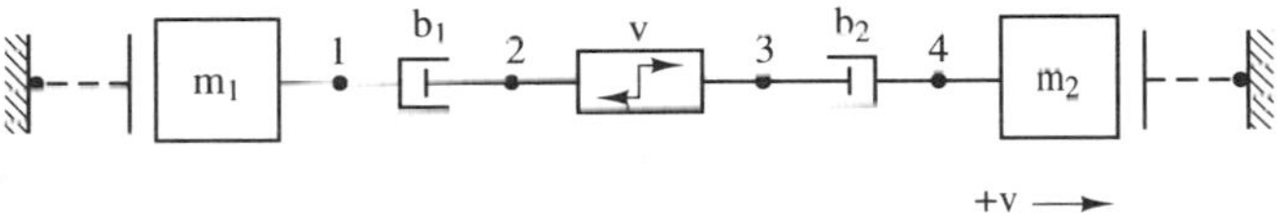

Figure P3.20

3.21 In each second-order linear network of Figure P3.21, all variables are zero at $t = 0^-$, and at $t = 0$ the source jumps to a constant positive value denoted by the subscript *c*. For one of the networks:

(a) Find the differential equation that characterizes the response of the velocity of the inertia or mass.

(b) Find the initial and final values of the velocity and acceleration of the inertia or mass, in terms of the element parameters and the constant source value.

Do *only one* of parts (c), (c′), and (c″).

(c) Let $K = 1$ N·m/rad, $B = 1$ N·m·s/rad, $J = 1$ N·m·s^2/rad, $k = 1$ N/m, $b = 1$ N·s/m, and $m = 1$ N·s^2/m. Find the characteristic values of the network. Find and sketch the waveform of the velocity of the inertia or mass for a *unit-valued* step input.

(c′) Use the alternative values $K = 1$, $B = 3$, $J = 1$, $k = 1$, $b = 3$, and $m = 1$ in part (c).

(c″) Use the alternative values $K = 1$, $B = 2$, $J = 3$, $k = 1$, $b = 2$, and $m = 3$ in part (c).

(d) Give a physical interpretation of the analytical solution of the network.

3.22 The rotor of an electric motor is complicated. Its moment of inertia is difficult to calculate. We can find the inertia of a particular rotor by the following experiment. We suspend the rotor by its end on a steel guitar string. (Denote the rotor inertia by J_1 and the rotational stiffness of the string by K.) We twist the string, let the rotor oscillate about its axis sinusoidally, and measure the period p_1 of the oscillation. Next, we replace the rotor on the guitar string by a

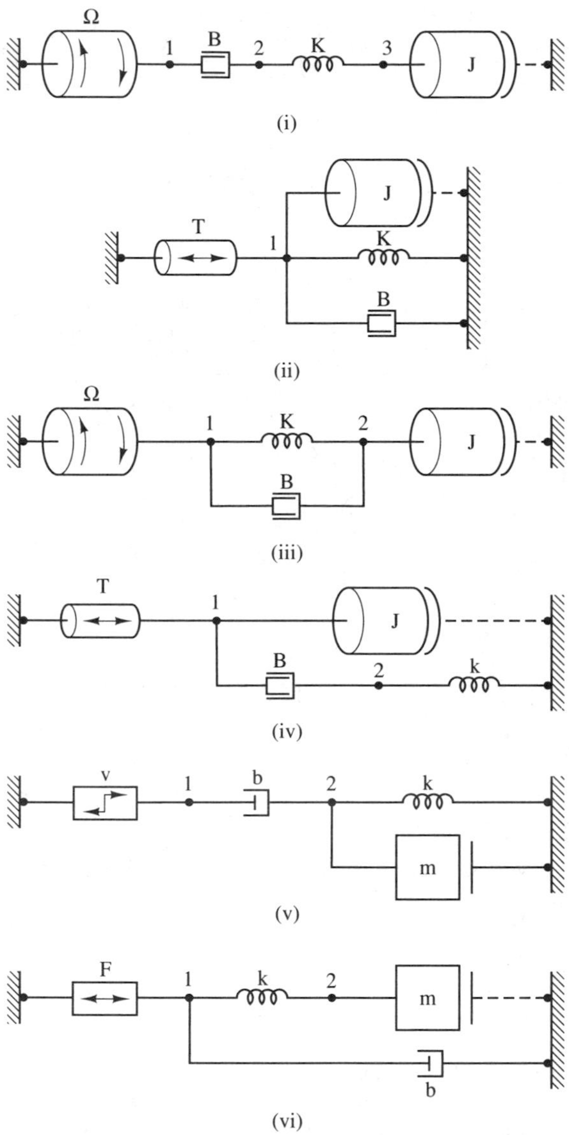

Figure P3.21

cylindrical object of comparable size, but of known (or easily calculated) inertia J_2, and measure the new period p_2 of oscillation. Determine the rotor inertia J_1 as a function of the measured periods p_1 and p_2 and the known inertia J_2.

3.23 Modify example problem 3.12 so that the bat is represented as an energized 3-kg mass, rather than as a velocity source. Assume that the bat velocity is 16 m/s. The response of the ball during the interval of impact is a zero-input response.

(a) Find the lumped model and its line-graph representation.

(b) Find the system equation for the velocity of the ball during the interval of impact.

(c) Find the characteristic roots for the network, and use them to determine the general form of the zero-input response of the ball during the impact.

(d) Find the required initial and final values of the velocity of the ball, and use these values to determine the arbitrary constants in the zero-input response found in part (c).

3.24 The input-output system equations for the variables F, v_3, and v_2 of the second-order network of Figure 3.44 are given in equations (3.91)–(3.93). The network is initially unenergized. Initial and final values of the network variables can be found from the input-output system equations or from Figures 3.41 and 3.42. The characteristic equation of the network is shown in equation (3.103) for b = 24 N·s/m (overdamped) and in equation (3.106) for b = 60 N·s/m (underdamped). The response $v_3(t)$ for a unit-step input is shown in Figure 3.45 for b = 24 N·s/m and in Figure 3.46 for b = 60 N·s/m. Use the general solution form given by equation (3.137), together with the graphical response curves of Figures 3.52–3.54 to sketch the zero-state step response for one of the following cases:

(i) For the variable v_3, with b = 24 N·s/m.
(ii) For the variable v_3, with b = 60 N·s/m.
(iii) For the variable F, with b = 24 N·s/m.
(iv) For the variable F, with b = 60 N·s/m.
(v) For the variable v_2, with b = 24 N·s/m.
(vi) For the variable v_2, with b = 60 N·s/m.

3.25 Consider the second-order network shown in Figure 3.44. The input-output system equations for the variables F, v_3, and v_2 of that network are given in equations (3.91)–(3.93).

(a) Determine at least two prototype variables for the network.

(b) What are the two energy-state variables for the network? Find the relations between each prototype variable from part (a) and these energy-state variables.

3.26 The zero-state unit-step response for the prototype variable $v_3(t)$ of the underdamped network of Figure 3.44 (b = 60 N·s/m) is shown in Figure 3.46.

(a) Find the fractional overshoot f_{OS}, the peak time t_{peak}, and the final value v_{3f} from the response in the figure.

(b) Use the measured features from part (a) to compute the coefficients of the input-output system equation for v_3.

(c) Compare the coefficients found in part (b) with the coefficients computed from equation (3.92) with the parameter values of Figure 3.44.

3.27 For one of the input-output system equations (i) through (iv):

(a) Find the values of the key features of the zero-state unit-step response. Use equations (3.147)–(3.152) as appropriate.

(b) Use the features found in part (a) to sketch the unit-step response.

(i) $(s^2 + 4s + 13)v_1 = 3F$
(ii) $(s^2 + 625s + 2{,}225)x_2 = 2{,}225x_1$
(iii) $(s^2 + s + 5)T = 10\Omega$
(iv) $(s^2 + 25s + 425)F = v$

3.28 For one of the networks shown in Figure P3.21:

(a) Use the energy-state variables for the storage elements to form a state vector **y**, and find the corresponding network state equation.

(b) Find the output equation for the velocity of the inertia or mass.

3.29 For one of the networks shown in Figure P3.21:

(a) Select a prototype variable x for the network. Use that variable and its derivative as a state vector **y**.

(b) Find the corresponding network state equation and the output equation for the velocity of the inertia or mass.

3.30 For one of the networks shown in Figure P3.21:

(a) Find the input-output system equation for the velocity of the inertia or mass.

(b) Let m = 1 kg, k = 200 N/m, J = 1 kg·m^2, and K = 200 N·m·s/rad. Find a value of the friction constant for which one characteristic root is approximately 10 times farther into the left half-plane than the other.

(c) Using the value of the friction constant found in part (b), approximate the system equation of part (a) by a first-order system equation; that is, eliminate the dominated-root factor from the system equation. Sketch the unit-step response of the simplified system equation.

(d) Remove one storage element from the network to obtain a first-order network that behaves similar to the second-order network. Sketch the unit-step response of this first-order network.

3.7 REFERENCES

3.1 Boas, Mary L. *Mathematical Methods in the Physical Sciences.* New York: John Wiley & Sons, 1966.

3.2 DeRusso, P. M., Roy, R. J., and Close, C. M., *State Variables for Engineers.* New York: John Wiley & Sons, 1966.

3.3 Dorny, C. Nelson, *A Vector Space Approach to Models and Optimization.* New York: John Wiley & Sons, 1975.

3.4 Kuo, Benjamin C., *Automatic Control Systems,* 6th ed. Englewood Cliffs, NJ: Prentice Hall, 1991.

3.5 Parrish, A. *Mechanical Engineer's Reference Book.* London: Butterworth, 1973.

3.6 Shearer, J. L., Murphy, A. T., and Richardson, H. H., *Introduction to System Dynamics.* Reading, MA: Addison-Wesley, 1971.

3.7 *TUTSIM User's Manual.* Palo Alto, CA: APPLIED i, 1985.

CHAPTER

4

Equivalence and Superposition in Linear Networks

Our design creativity depends on our ability to relate the system to be designed to similar systems with which we are familiar. The primary hindrance to good system design is difficulty in obtaining a clear mental image of the system to be created. Conceptual simplicity of the system model is a major advantage to the design process.

The fundamental premise of this chapter is that simple systems are easier to design, analyze, and interpret than are complicated systems. If we simplify the lumped network that represents a system, we can better understand the system itself. For example, we make a network less complicated by combining series or parallel elements into a single equivalent element. Suppressing the less important variables makes it easier to solve the system equations and interpret their solutions. Simplifying a network and its equations is an art. This chapter focuses on that art.

A system can be thought of as an energy source acting on a load. Suppose we choose a pair of network terminals—referred to as a port—at which we separate the network into two halves, the source half and the load half. We can represent the whole subnet on the load side of the port by a single equation that involves only the port variables—the potential drop across the port and the flow through the port. This single equation can be interpreted physically as a simple equivalent representation of the load subnet. We are able to find a similar simple equivalent representation of the whole subnet on the source side of the port, using the same pair of port variables. Hence, every network, no matter how complicated, can be viewed as a connection of two simple subnets. Section 4.2 helps us develop facility with equivalent models for linear subnets.

Many networks contain more than one energy source. If the system equations are linear, the response owing to multiple sources is the sum of the separate responses to the individual sources. We use this fact to simplify analysis, by finding the response to one source at a time. We call such an approach superposition. In section 4.3, we find that we can decompose a complicated source waveform into a sum of simpler waveforms. Then we compute the response to the complicated waveform as the sum of the responses to its simpler constituents. Again, we divide the problem into pieces in order to keep it simple. Section 4.3 develops these superposition techniques.

We begin the chapter by examining lumped models, the lumping process, and the writing of system equations for electrical systems. In addition to opening a new domain of application for the modeling techniques of chapter 2, designing and analyzing electric circuits serves to reinforce the equation-solving techniques of chapter 3. We also introduce a powerful software tool, SPICE, which enables us to simulate network behavior without determining the system equations. We apply the tool to both electrical and mechanical networks.

4.1 LUMPED MODELS FOR ELECTRICAL SYSTEMS

Electric Potential and Current

In early experiments with electricity, it was discovered that rubbing a glass rod with a silk cloth caused a transfer of electric charge from the one to the other. The two objects became oppositely charged and attracted each other. Benjamin Franklin assigned the charge on the glass as positive. We continue to use that choice for positivity of charge today. It was learned later that it is the negatively charge particles—electrons—that actually move in such experiments.

The motion of electric charge is the basis for describing all electrical phenomena. We measure quantity of electric charge in coulombs (abbreviated C). Some objects (conductors) allow charge to move easily, whereas others (insulators) do not permit motion of charge. The geometry of a system and the properties of its materials determine the way in which charge flows in that system.

An electrical system contains distributions of positive and negative charges. Because of the attraction between charges of opposite sign and the repulsion between charges of like sign, it takes force (and work) to displace charge in an electrical system. We define the **electric potential** v at a point in the system as the work required to move a unit *positive* test charge to that point from a chosen reference point. (The choice of a positive test charge constitutes the definition of positivity of electric potential.) The required work can be positive or negative. We assign the potential zero to the reference point and refer to that reference point as **ground**. The work required to move the test charge does not depend on the path through which it is moved from the reference to the point of interest.

We can express the electric potential at a point in terms of a test charge of arbitrary amount q. The potential v is the ratio

$$\mathrm{v} = \frac{E}{\mathrm{q}} \tag{4.1}$$

where E is the energy, in joules, required to move the charge to the point from the zero reference. The unit of potential is the volt, abbreviated V; 1 volt = 1 joule/coulomb.

It is customary to call the electric potential at a point the **voltage** at the point. The difference in potential between two points is the voltage *between* the points. The potential is typically different at the two ends of an electrical object. Therefore, voltage is an *across* variable.

We select an arbitrary or convenient point in the system as the zero-potential reference and measure all potentials (voltages) relative to that reference point. We denote the reference by the **electrical ground** symbol labeled g in Figure 4.1(a). (This ground symbol is analogous to the mechanical ground symbols of sections 2.1 and 3.1.) We sometimes denote an electrical ground point by means of the label g; or we use both the symbol and the label redundantly.

Figure 4.1(a) shows an electrical object. Suppose one end of the object is an equipotential (equal-voltage) surface with potential $v_{1g} = v_1 - v_g = v_1$, relative to ground. Suppose the surface at the other end of the object has potential $v_{2g} = v_2$. Then, between the two equal-voltage ends are other equal-voltage surfaces. In fact, we can visualize the object as a stack of parallel equal-voltage surfaces. It takes work to move positive charge from a low potential to a high potential. Conversely, positive charge flows naturally from a high potential to a low potential. The natural flow of charge is perpendicular to the equal-voltage surfaces.

We represent the object of Figure 4.1(a) by a lumped model in Figure 4.1(b), with the two equal-voltage end surfaces represented by points 1 and 2. The lumped model does not account for the internal behavior of the object. The voltage between points 1 and 2 is $v_{12} = v_1 - v_2$. If $v_{12} > 0$, then $v_1 > v_2$. Hence, we can view v_{12} as the *drop in voltage* from point 1 to point 2 of the model. It is also the drop in voltage from surface 1 to surface 2 of the object itself. If v_{12} is negative, then the voltage rises from surface 1 to surface 2.

The rate of flow of positive charge through a specified surface is called the **electric current** through the surface. A flow of positive charge in one direction is indistinguishable from an equal flow of negative charge in the opposite direction. (In most situations, it is only negative charge that flows.) Suppose that the material adjacent to the sides of the object of Figure 4.1(a) is an electrical insulator, such as air, which does not permit charge

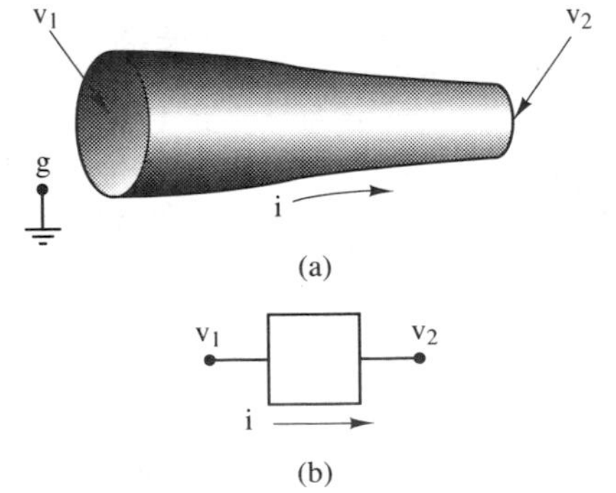

Figure 4.1 Electric system notation. (a) An electrical object. (b) Lumped model of the object.

to flow. Then the object forms a one-dimensional channel for the flow of charge. In most electrical systems in this book, the flow of charge is restricted to such one-dimensional channels.

Current is a *through variable.* That is, the current is the same at each cross section of a flow channel. Let i denote the current in a particular channel of the circuit. Assign one of the two directions along the channel as the **positive direction for i.** Denote that direction by an arrow beside the channel, as shown in Figure 4.1. Then i is the rate at which positive charge flows in the indicated direction. A negative value of i corresponds to either a flow of negative charge in the direction of the arrow or a flow of positive charge in the opposite direction. In differential form,

$$i = \frac{dq}{dt} \tag{4.2}$$

where dq is the incremental charge that flows past a cross section during the time increment dt. The unit of electric current is the ampere, abbreviated A or amp. Thus, 1 ampere = 1 coulomb/s.

In a passive device, the flow of charge is always downhill, from a higher to a lower voltage. (This downhill flow is a result of the way we have defined positivity for voltage and current.) As a quantity of charge flows through a drop in voltage, it delivers energy to the object through which it flows. Suppose that the current i indicated in Figure 4.1 is positive and that the circuit element is passive. Then the voltage v_1 must be larger than v_2. According to equations (4.1) and (4.2), the power delivered to (or absorbed by) the element is

$$P = \frac{dE}{dt} = \frac{dE}{dq}\frac{dq}{dt} = v_{12}i \tag{4.3}$$

Electrical power is measured in watts (abbreviated W); 1 watt = 1 joule/s. A device that can move charge through a *rise* in voltage is called a *source.* According to equation (4.3), the power absorbed by a source is negative.

Ideal Sources, Resistors, and Switches

An electrical source produces a rise in voltage and a flow of current in the direction of the voltage rise. We define the **ideal independent voltage source** to be a lumped element that maintains a specified voltage pattern between its terminals, regardless of the amount of current which that voltage causes to flow through the attached system. The symbol we use to represent an independent voltage source is shown in Figure 4.2(a). The voltage at the positive side of the source is higher than the voltage at the negative side by the amount v(t). An alternative symbol that is often used for a *constant-value* voltage source (or battery) is shown in Figure 4.2(b). The side of the symbol with the larger crossbar has the higher potential.

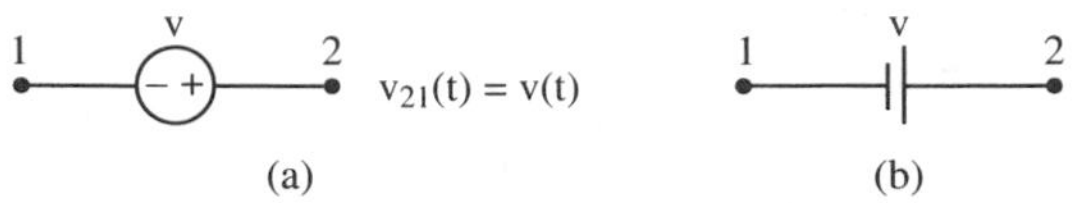

Figure 4.2 Independent voltage source symbols. (a) General-waveform source. (b) Constant-voltage source.

The **ideal independent current source** is a lumped element that delivers a specified current pattern to the attached system, regardless of the voltage drop which that current produces across the system. The symbol for the independent current source is shown in Figure 4.3. The current in the direction of the arrow is i(t).

All materials resist the flow of electric charge. Materials with very low resistance to flow are called *conductors*; those with very high resistance are called *insulators*. Metals are conductors, whereas glass, ceramics, and plastics are insulators. We represent the resistive behavior of a material by the lumped resistor shown in Figure 4.4. We define the **ideal resistor** to be a lumped two-terminal element that behaves according to the formula

$$v_{12} = Ri \tag{4.4}$$

where the proportionality factor R is the **resistance** of the resistor. The subscripts on the voltage difference are oriented in the same direction as the positive direction for the current i in the element. That is, the voltage drops across the resistor in the direction of current flow. Resistance is an *electrical impedance*—a ratio of potential difference to flow, as discussed in chapter 1. Equation (4.4) is known as *Ohm's law*. The unit of resistance is the **ohm** (volt/amp). The symbol Ω is often used in place of the word *ohm*. The reciprocal of the resistance is the **conductance** $G \triangleq 1/R$. The unit of conductance is the *siemen* (amp/volt), sometimes symbolized by ℧.

Figure 4.5 relates the resistance of a homogeneous object to the geometry of the object and the electrical resistivity of the material the object is made of. The resistance formula shown in the figure assumes that the current is distributed uniformly over the cross section of the object. If the length of the object is increased, the current encounters more material and the resistance increases. By contrast, increased cross-sectional area provides additional flow paths and reduced resistance. If the current were supplied to the ends of the object by thin wires, the current at the ends would concentrate near the regions of contact with the wire. This *end effect* would increase the resistance of the object.

Electrical resistances for other geometries can be found by the procedure used in example problem 4.4. By an appropriate choice of material and geometry, one can design an object with a lumped resistance of any desired value. However, in the design of certain

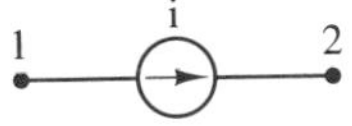

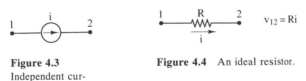

Figure 4.3 Independent current source.

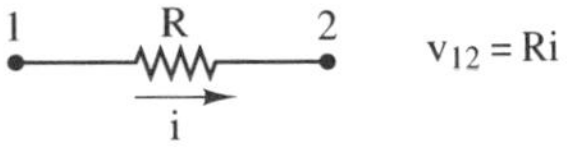

Figure 4.4 An ideal resistor.

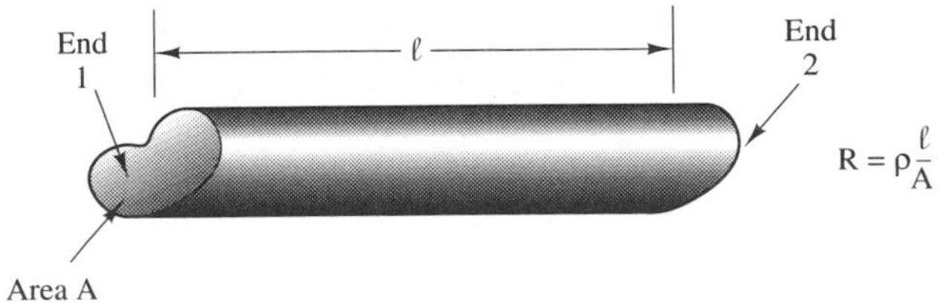

Figure 4.5 Electrical resistance; ρ = resistivity, in ohm·m. (Appendix A provides a list of resistivities for various materials.)

electrical elements, such as sources and inductors, one cannot make the resistance of the element small without nearly eliminating the desired electrical property of the element.

According to equations (4.3) and (4.4), the power absorbed by the lumped resistor is

$$P = v_{12}i = i^2R = \frac{v_{12}^2}{R} \geq 0 \tag{4.5}$$

This power is transformed into heat. From the electrical viewpoint, therefore, the resistor is a dissipative (energy-consuming) element. The generation of heat causes a temperature rise within the resistor and a flow of heat to the environment. Many useful electrical devices are designed specifically to convert electrical energy to heat. Stoves, toasters, irons, and space heaters rely on resistive energy conversion. On the other hand, heat damages the components in radios, televisions, and sound systems, and cooling schemes are needed to protect them.

Figure 4.6 shows an **ideal switch**, a model for the device used to start and stop action or change the nature of action in electric circuits. The ideal switch offers no resistance to the current when it is closed. We view the closed switch as a *short circuit*, which we represent mathematically by $v_{12} = 0$. The open switch has infinite resistance. We call it an *open circuit* and represent it by $i = 0$.

In many countries, the electric utility system generates and distributes sinusoidal voltage-source signals that have fixed amplitude. These utility systems are known as ac (*alternating current*) systems. In contrast, some utility systems use constant (nonsinusoidal) voltage. These are called dc (*direct current*) systems. In both types of utility systems, the electrical outlets act as nearly ideal voltage sources. The term *ac* has become synonymous with *sinusoidal*, and *dc* has become synonymous with *constant*. Thus, we speak of the ac voltage of an electrical outlet and the dc voltage of a battery. Among electrical engineers, the term *dc* is sometimes used loosely to mean *nearly constant*, and the term *ac* is sometimes used to mean *time varying* (as opposed to nearly constant).

Analysis of a Resistive Circuit

Figure 4.7(a) shows a pictorial model of a flashlight. It includes two 1.5-V dry-cell batteries, a resistive light bulb, a slide switch, and a conducting case. The batteries act as sources. Current flows out of the positive terminal of the top battery, through the bulb, the switch conductor, and the case, and into the negative terminal of the bottom battery. We call the conducting path a *circuit*.

Figure 4.7(b) shows a lumped-network model of the flashlight. (An electrical lumped-network model is also commonly known as an **electric circuit**.) The pair of batteries is treated as a constant 3-V source in this model. (Section 4.2 presents a more accurate model of the battery that includes source resistance.) The resistance of the light bulb filament is small (2 ohms). The current through the resistor heats the filament enough that it glows. We presume that the case acts as a perfect conductor. The switch is presumed to act as either an open or a short circuit, depending on its position.

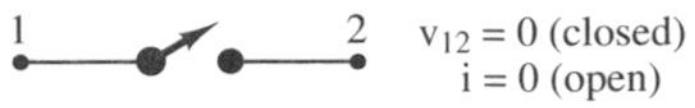

Figure 4.6 An ideal electrical switch.

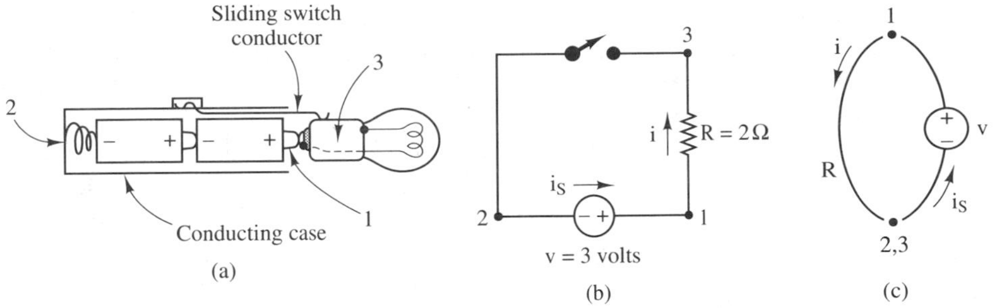

Figure 4.7 Flashlight models. (a) Pictorial model. (b) Lumped model. (c) Line-graph representation.

Most people know that a poor contact somewhere in the flashlight assembly causes the flashlight to give little or no light. A poor contact can be modeled by inserting an additional *contact resistance* at an appropriate point in the circuit, typically at the switch or at the coil spring at the bottom of the flashlight. It does not take much contact resistance to reduce the current through the filament below the value needed to produce light-generating heat.

Figure 4.7(c) shows a line-graph representation of the network. (The figure represents the electrical network only when the switch is in the closed position.) We place beside each branch of the line graph a symbol (e.g., R) that indicates the kind of element the branch represents. The resistor symbol of Figure 4.7(b) could be superimposed on the branch, producing a mixed notation. The line-graph structure is little different from, and therefore has little advantage over, the structure of the conventional circuit diagram. Indeed, if we used the mixed notation, there is no real difference between the two.

We can orient arbitrarily the positive direction for current in each branch of the lumped network. Since we expect the current to flow from negative to positive in the source, we orient the source-current variable i_S in that direction. Because the circuit of Figure 4.7 has only one loop, it is natural to orient both currents in the same direction, as in Figure 4.7(b). (We could, instead, define a single *loop* current, as discussed in section 2.2). **Kirchhoff's current law** for lumped electric circuits states that the sum of the currents entering any node in a circuit is zero. This current law is identical to equation (2.27), the mathematical definition of a set of flows.

According to section 2.2, one node equation, together with the two equations that describe the lumped elements of the circuit, are necessary and sufficient to determine all of the potentials and flows in the circuit. These equations are:

$$\begin{aligned} i &= i_S && \text{(current law)} \\ v_{12} &= Ri && \text{(resistor)} \\ v_{12} &= v = 3 \text{ volts} && \text{(dc source)} \end{aligned} \tag{4.6}$$

The solution to this set of equations is $i = i_S = v/R = (3\text{ V})/2\ \Omega) = 1.5$ A. The power absorbed by the bulb is $P = i^2R = 4.5$ W.

We usually use the element equations to express the node currents directly in terms of node voltages when writing node equations. That procedure constitutes the node method for writing system equations (see section 2.2). The voltage source in this example does not lend itself well to the node method: The current in the source cannot be expressed in terms of node voltages. We illustrate the node method for a multiloop, multinode resistive network in example problem 4.1.

Kirchhoff's voltage law for electric circuits states that the sum of the voltage drops around any closed path in a circuit is zero. This voltage law is identical to the path law given by equation (2.28), the mathematical definition of potentials and potential differences in networks. Since the node voltages obey the path law, the voltage drop from one node to another is the same by any path through the network.

We apply the voltage law together with the element equations to find the following description of the single-loop circuit of Figure 4.7(c):

$$\begin{aligned} v_{21} + v_{12} &= 0 && \text{(voltage law)} \\ v_{21} &= -3\text{ V} && \text{(dc source)} \\ v_{12} &= Ri && \text{(resistor)} \end{aligned} \tag{4.7}$$

Inserting the source equation and the resistor equation into the voltage-law equation yields $3\text{ V} = Ri$, or $i = 1.5$ A.

A single current circulates around the loop. We denote that current by i in equations (4.7). We can use the element equations to express unknown potential drops directly in terms of loop currents when writing loop equations. This procedure constitutes the loop method for writing system equations (see section 2.2). The loop method is illustrated for a multiloop resistive network in example problem 4.1.

Example Problem 4.1: The node and loop methods for an electrical network.
Figure 4.8(a) shows a lumped electrical network and defines currents in each of its branches.
(a) Use the node method to solve for v_2; find the branch currents.
(b) Define loop currents, and use the loop method to solve for the branch currents; find v_2.
Solution: **(a)** The two node equations and the four branch equations are:

$$\begin{aligned} i_S &= i_1 && \text{(node 1)} \\ i_1 &= i_2 + i_3 && \text{(node 2)} \\ v_{1g} &= v_1 = v && \text{(source)} \\ \left.\begin{aligned} v_{12} &= R_1 i_1 \\ v_{2g} &= R_2 i_2 \\ v_{2g} &= R_3 i_3 \end{aligned}\right\} & && \text{(resistors)} \end{aligned} \tag{4.8}$$

We solve each of the resistor equations for the corresponding current, substitute for these currents in the equation for node 2, and replace v_1 by v to produce

$$\frac{v}{R_1} - \frac{v_2}{R_1} = \frac{v_2}{R_2} + \frac{v_2}{R_3} \tag{4.9}$$

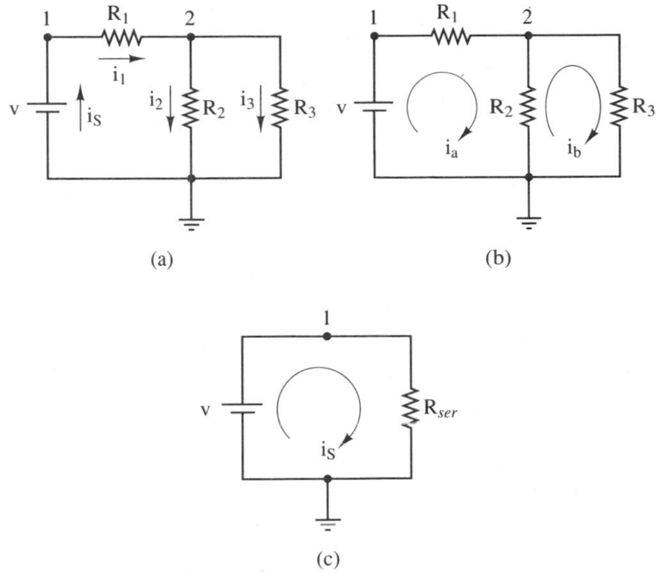

Figure 4.8 An electrical network. (a) Lumped circuit. (b) Circulating loop currents. (c) Simplified equivalent circuit.

The procedure that generated equation (4.9) is the node method for writing the system equation. From equation (4.9) we can determine that the voltage at node 2 is

$$v_2 = \left(\frac{R_2R_3}{R_1R_2 + R_1R_3 + R_2R_3}\right) v \tag{4.10}$$

We solve the first resistor equation for i_1, then substitute the source equation and equation (4.10) into the result to determine that

$$i_1 = \left(\frac{R_2 + R_3}{R_1R_2 + R_1R_3 + R_2R_3}\right) v \tag{4.11}$$

Then $i_S = i_1$. Finally, we use equation (4.10) with the last two resistor equations to show that

$$i_2 = \left(\frac{R_3}{R_1R_2 + R_1R_3 + R_2R_3}\right) v, \qquad i_3 = \left(\frac{R_2}{R_1R_2 + R_1R_3 + R_2R_3}\right) v \tag{4.12}$$

(b) Figure 4.8(b) defines two loop currents for the network. In the loop method for writing system equations, we equate the sums of voltage drops around two independent loops to zero. One set of loop equations for this network is

$$\begin{aligned} v_{g1} + v_{12} + v_{2g} &= -v + R_1i_a + R_2(i_a - i_b) = 0 \\ v_{g2} + v_{2g} &= R_2(i_b - i_a) + R_3i_b = 0 \end{aligned} \tag{4.13}$$

Then,

$$i_b = \left(\frac{R_2}{R_1R_2 + R_1R_3 + R_2R_3}\right) v \quad \text{and} \quad i_a = \left(\frac{R_2 + R_3}{R_1R_2 + R_1R_3 + R_2R_3}\right) v \tag{4.14}$$

Compare the currents in Figures 4.8(a) and (b) to see that $i_1 = i_S = i_a$, $i_2 = i_a - i_b$, and $i_3 = i_b$. It then follows that $v_2 = R_3 i_b$. Compare these results with the results of the node analysis.

Exercise 4.1: Derive equation (4.10) from equation (4.9).

Exercise 4.2: Verify equations (4.12).

Exercise 4.3: The current entering node 2 of Figure 4.8(a) is i_1.
(a) Show that $v_{2g} = i_1 R_{par}$, where

$$\frac{1}{R_{par}} = \frac{1}{R_2} + \frac{1}{R_3} \tag{4.15}$$

That is, if two resistors R_2 and R_3 are connected in parallel, the conductance of the parallel combination equals the sum of the conductances of the individual resistors.
(b) Show that $v_{1g} = i_1 R_{ser}$, where

$$R_{ser} = R_1 + R_{par} \tag{4.16}$$

That is, if two resistors R_1 and R_{par} are connected in series, the resistance of the series combination equals the sum of the individual resistances.

Example Problem 4.2: Analysis by series and parallel combination.
For the lumped electrical network shown in Figure 4.8(a):
(a) Use series and parallel combination, as summarized in equations (4.15) and (4.16), to replace the resistor network by a single equivalent resistor between nodes 1 and g. Then find i_S directly.
(b) Find the branch currents and v_2, one variable at a time, by decomposing the equivalent resistor into its constituent resistors.
Solution: **(a)** According to equation (4.15), $R_{par} = R_2R_3/(R_2 + R_3)$. Then, from equation (4.16), $R_{ser} = R_1 + R_{par} = (R_1R_2 + R_1R_3 + R_2R_3)/(R_2 + R_3)$ (see Figure 4.8(c)). Finally, $i_S = v/R_{ser} = v(R_2 + R_3)/(R_1R_2 + R_1R_3 + R_2R_3)$.
(b) Note that $i_1 = i_S$. Then the equation for the resistor R_1 implies that $v_2 = v_1 - R_1 i_1 = v - R_1 i_S$. From the equations for the resistors R_2 and R_3, we find that $i_2 = v_2/R_2$ and $i_3 = v_2/R_3$. Substitute for i_S in the equation for v_2, then substitute for v_2 in these equations. Compare the results with the results of the node method in example problem 4.1.

Measurement of Electrical Signals

A *voltmeter* is a two-terminal device that measures the voltage drop between its terminals. We calibrate a voltmeter by measuring the voltage across a standard cell. For example, the difference in electric potential across a Weston standard cell is 1.018 volts. The cell terminal with the higher potential is marked with a + symbol.

An oscilloscope is a versatile voltmeter. In the oscilloscope of Figure 4.9, an electron beam strikes a phosphor-covered screen and creates a lighted spot in the center of the screen. The beam passes between a vertically stacked pair of conducting plates. The terminals of the oscilloscope are connected to these plates. We connect the terminals to those points in the electrical system between which the voltage difference is to be measured. Then that voltage difference appears across the plates and deposits charge on the plates. The charged plates deflect the beam (and the lighted spot on the screen) in proportion to the voltage difference.

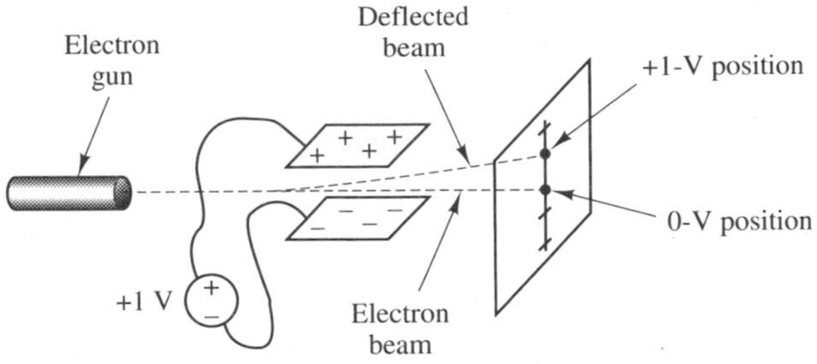

Figure 4.9 Oscilloscope voltage measurement.

We mark a voltage scale on the oscilloscope screen. We calibrate the scale as follows. First, we short-circuit the plates (produce zero voltage) and mark the position of the lighted spot as the zero-volt position on the screen. Then we choose one of the oscilloscope terminals as the reference (ground). We connect the oscilloscope terminals to a standard 1-V cell, with the negative terminal of the cell connected to the ground terminal of the oscilloscope. Next, we mark the position of the lighted spot as the +1-volt position on the screen. Additional positive and negative voltage marks can be placed on the scale consistent with the 0 and +1 marks. This voltage-measuring device is nearly ideal; that is, the oscilloscope does not significantly distort the voltage difference being measured.

The oscilloscope can be used to plot the waveform of a rapidly changing voltage signal if that waveform is repetitive. In front of the electron gun, we place a second pair of conducting plates, oriented perpendicular to the first pair. To those plates, we apply the *sawtooth* voltage waveform v_{sweep} shown in Figure 4.10. This waveform repeatedly sweeps the lighted spot horizontally across the screen at a constant rate. As the voltage to be measured varies the position of the lighted spot in the vertical dimension, the sweep signal sweeps the spot across the screen in the horizontal dimension. The visual effect of the two simultaneous signals is a plot of the measured waveform versus time.

Of course, the images owing to successive horizontal sweeps are superimposed on the screen. This waveform visualization is useful only if the waveform of the voltage to be measured is repetitive at the same rate as the sawtooth. Commercial oscilloscopes generate the sawtooth sweep voltage internally and allow the user to synchronize the horizontal sweep with the periodic variations of the vertical signal.

The oscilloscope can also be used as an *ammeter* to measure current. First, we pass the current i from a standard (known) current source through a resistor of resistance R. Then we connect the oscilloscope terminals across the resistor to measure the voltage drop Ri. The deflection of the lighted spot on the screen is proportional to the known current flowing through the resistor. If the current source is a +1-amp source, we mark +1 at the location of the lighted spot when the source is connected. (We need not know the resistance R.) We mark 0 at the position on the screen when the source is disconnected. Additional positive- and negative-current marks can be placed on the screen consistent with the 0 and +1 marks.

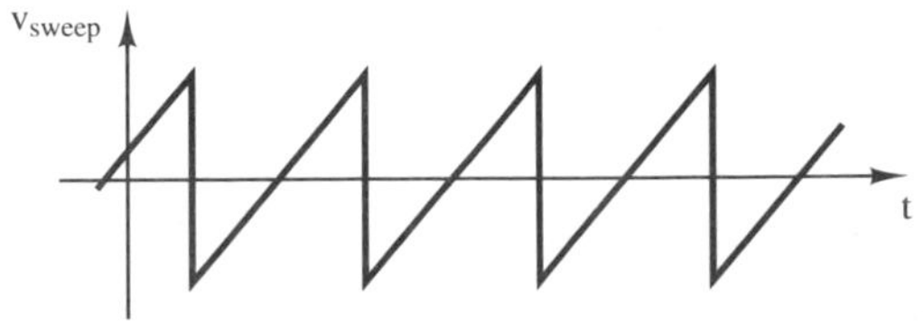

Figure 4.10 Sawtooth horizontal-sweep signal.

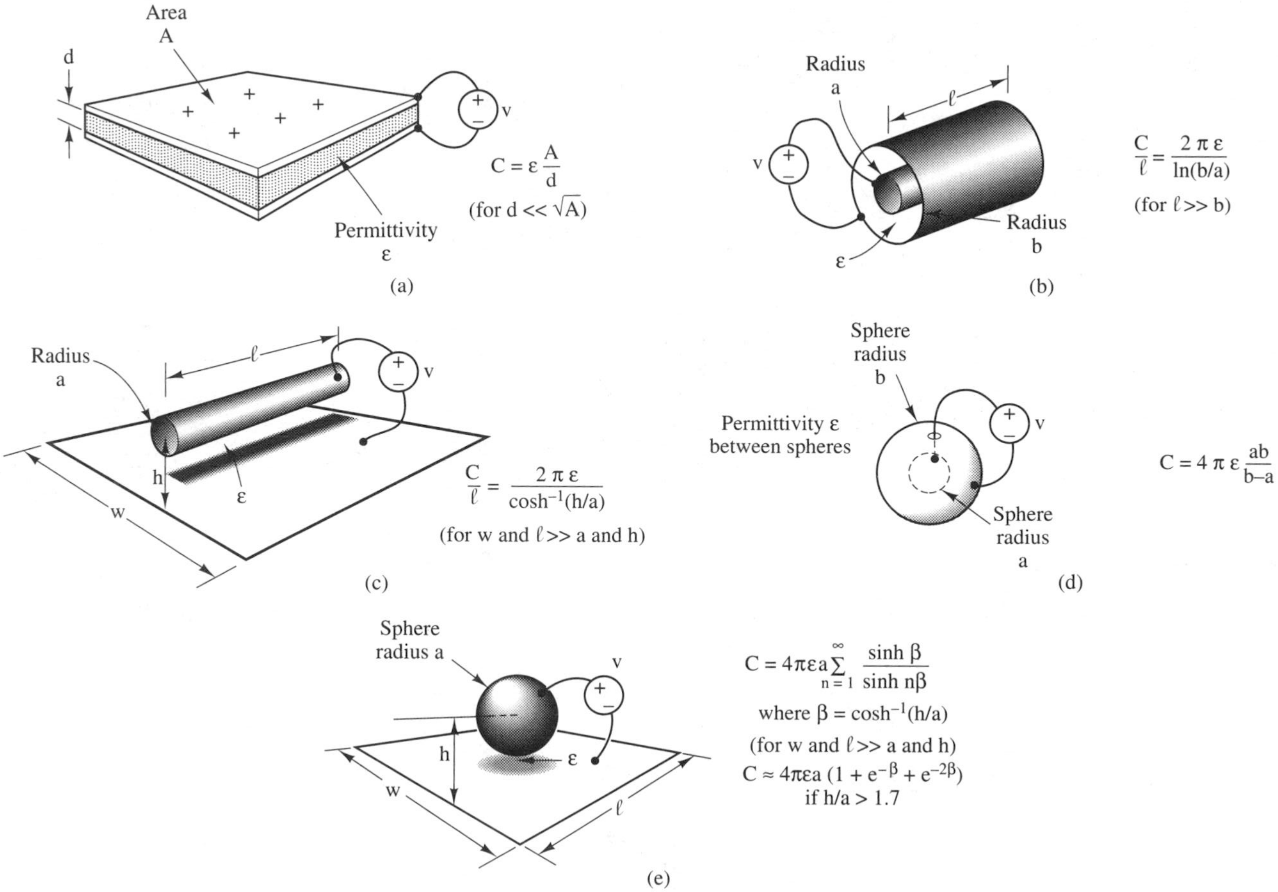

Figure 4.11 Capacitive electrical structures; ϵ is the electric permittivity of the dielectric material through which the electric flux flows. In air, $\epsilon = 8.85 \times 10^{-12}$ F/m. (See references [4.1], [4.6], [4.7] and [4.8]; the permittivities for various materials are listed in Appendix A.) (a) Parallel plates. (b) Coaxial cylinders. (c) Cylinder and plane. (d) Concentric spheres. (e) Sphere and plane.

To measure the current in a branch of a circuit, we must break that branch and insert the resistor between the broken ends so the current flows through it. The inserted resistance must be small, or it will change the measured current significantly.

In practice, rather than recalibrate the oscilloscope as a current meter, we insert a small resistor of *known* resistance R into the circuit branch. (Or we use a known resistor that is already in the branch.) We use the oscilloscope to measure the voltage drop v across the resistor. Then we *calculate* the current v/R through the resistor.

Problem 4.20 at the end of the chapter examines the operation of a d'Arsonval galvanometer. The galvanometer is fundamentally a current sensor. The problem expores how to use the galvanometer to measure both current and voltage.

The Ideal Capacitor

Suppose two pieces of conducting material are separated by an insulating (or *dielectric*) material, as shown in Figure 4.11(a). Let us use a voltage source to establish a potential difference between the conductors. The voltage source deposits a positive charge on the conductor of higher potential and a negative charge on the conductor of lower potential. The amount of charge deposited depends on the voltage applied, the geometry of the conductors and dielectric, and the dielectric material. The relationship can be expressed as

$$q = Cv_{12} \tag{4.17}$$

where q is the amount of charge deposited on each conductor, v_{12} is the voltage drop between plates 1 and 2, and C is the **capacitance** associated with the configuration of the conductors and dielectric. The capacitance indicates the capability of the conductor pair to store charge.

The capacitance associated with particular configurations of conductors and dielectrics can be found from electric field theory. The capacitances for several configurations are shown in Figure 4.11. The capacitance of a device can also be approximated by the lumping process examined later (see example problem 4.4).

In general, conductor pairs that approximate parallel-plate geometry have significant capacitance. If the conductors have a large surface area and are closely spaced, the capacitance is high. The unit of capacitance is the farad, abbreviated F. One farad equals one coulomb per volt, an extremely large unit of capacitance. Practical devices typically have capacitances in the picofarad (pF) to microfarad (μF) range.

We use the derivative of equation (4.17) to define the **ideal capacitor;** it is a lumped electrical element that obeys the relation

$$i = C\dot{v}_{12} \tag{4.18}$$

where C is the capacitance in farads, i is the current in amps, and the subscripts on the voltage difference are ordered in the positive direction for current through the device. Hence, $\dot{v}$ drops in the direction of the actual current flow, and positive i implies that v_1 is increasing relative to v_2. Figure 4.12 shows the symbol that we use to represent the ideal capacitor. The capacitor equation (4.18) can also be expressed in the notation

$$v_{12} = \left(\frac{1}{sC}\right)i \tag{4.19}$$

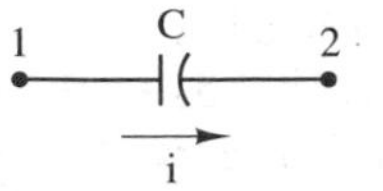

Figure 4.12 An ideal electrical capacitor.

where s is the time-derivative operator introduced in chapter 1. The quantity 1/(sC) is called the *impedance* of the capacitor. Equation (4.19) should be interpreted the same as equation (4.18). The operator s can be interpreted as differentiation of the voltage variable v_{12} or as integration of the current variable i.

The voltage at a point, its derivative with respect to time, and its integral over time are analogs of the mechanical motion variables—velocity, acceleration, and displacement, respectively. All three electrical variables obey the path law. Similarly, electric current is the analog of compressive force: di/dt and $\int i$ obey the node law. Throughout this text, we focus attention on *central* flow and potential variables. In the electrical context, these central variables are i and v. Thus, we express the capacitor equation in terms of i and v.

Charge cannot be transported through the insulating gaps between the conductors of a capacitor (see Figure 4.11). Yet, at the terminals of the device, one observes a flow of charge. We refer to this apparent current, which results from the displacement of charge, as *displacement current.* The displacement current that passes through the terminals of a capacitive structure is indistinguishable from the *conduction current* that flows through a resistive object such as a wire. According to equation (4.18), the displacement current exists only if the voltage across the device is changing with time; the magnitude of the current is proportional to the rate of change of that voltage.

It takes work to charge the plates of a capacitor. The energy required to deposit a quantity of charge is the integral over time of equation (4.3), that is,

$$E(t) = \int_0^t v_{12} i \, dt = \int_0^q \left(\frac{\sigma}{C}\right) d\sigma = \frac{q^2}{2C} = \frac{Cv_{12}^2}{2} \tag{4.20}$$

where we substitute $v_{12} = q/C$ and $i = dq/dt$ into the first integral and then use the dummy charge variable σ to avoid confusion between the limit q(t) and the variable of integration. The energy, measured in joules, is stored in the electrostatic field that exists in the dielectric between the opposing groups of charge. This stored energy is retrievable. For example, if the charged capacitor is disconnected from the voltage source and connected to a resistor, the stored charge flows through the resistor from the high-voltage side of the capacitor to the low-voltage side (see example problem 4.10).

The amount of stored energy cannot be changed instantly. That is, neither the capacitor voltage v_{12} nor the charge q can jump abruptly without infinite capacitor current. Thus, a capacitor acts as a *short circuit* ($\delta v_{12} = 0$) to abrupt changes in the source (see section 3.3). Both q and v_{12} determine the energy in the capacitor. Since v_{12} is one of the *central* variables of the network, we call it the *energy-state variable* for the capacitor. That is, of the two central variables (v_{12} and i) associated with the capacitor, it is the voltage v_{12} that uniquely determines the energy state of the capacitor.

The Ideal Inductor

Associated with every electric current is a magnetic field that surrounds the current-carrying conductor. The strength of this magnetic field is proportional to the magnitude of the current. The magnetic field contains stored energy. Energy cannot be added to or removed from the magnetic field instantly. Hence, the current in a conductor cannot be changed abruptly. By Faraday's law, an increase in the electric current through a conducting device induces a voltage drop across the device in the direction of the current. The proportionality factor L is called the **inductance** of the device. The unit of inductance is the henry, abbreviated H; 1 henry = 1 volt·s/amp.

Figure 4.13 shows several conductor geometries and the associated inductances. Coiled-conductor geometries have strong magnetic couplings, high energy-storage capacity, high inductance, and high voltage drops associated with changes in current. A simple wire loop, on the other hand, has relatively little inductance, unless the loop is very large. We can design a device of any inductance by the appropriate choice of materials and geometry.

We define an **ideal inductor** to be a lumped element in which the current and the voltage drop between the terminals are related by the formula

$$v_{12} = L\frac{di}{dt} \tag{4.21}$$

where L is the *inductance* of the element. We represent the lumped inductor by the symbol shown in Figure 4.14. The symbol has the high-inductance coil geometry.

We can also express the inductor equation (4.21) in the impedance notation as

$$v_{12} = (sL)i \tag{4.22}$$

This equation should be interpreted to mean the same as equation (4.21). That is, the time-derivative operator s acts on the flow variable i. In equations (4.21) and (4.22), the subscripts on the voltage difference are ordered in the positive direction for current in the inductor. That is, the potential v_{12} drops in the direction of the flow i if i is increasing.

The energy stored in an ideal inductor, from equation (4.21) and the integral over time of (4.3), is

$$E(t) = \int_0^t \left(L\frac{di}{dt}\right) i\, dt = \int_0^{i(t)} L\sigma\, d\sigma = \frac{Li^2}{2} \tag{4.23}$$

The energy stored in an inductor cannot be changed abruptly. The inductor current is the *energy-state variable* for the inductor. It cannot jump abruptly without infinite voltage across the inductor. As a consequence, an ideal inductor acts as an *open circuit* ($\delta i = 0$) to abrupt changes in the value of the source (see section 3.3).

A coil that has many turns of small-diameter wire (in order to achieve high inductance) necessarily has significant resistance as well. Practical devices that exhibit significant inductance usually should be modeled by an ideal inductor in series with an ideal resistor. The use of a ferromagnetic material for the core of a coil can increase the inductance of the coil by a factor as large as 10^6 relative to other core materials. However, the

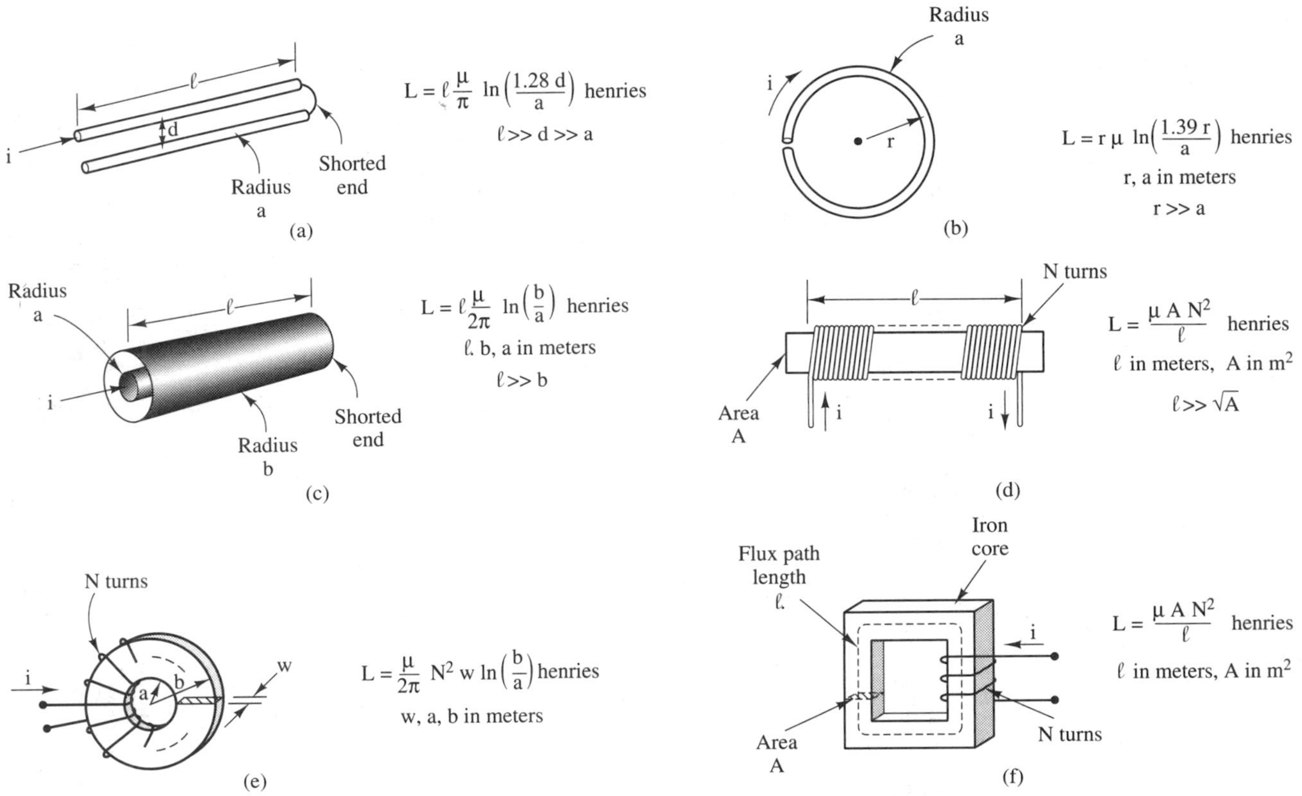

Figure 4.13 Inductive electrical structures; μ is the permeability of the material through which the magnetic flux circulates. For most materials $\mu = 4\pi \times 10^{-7}$ H/m. (a) Parallel wires. (b) Wire loop. (c) Coaxial cable. (d) Solenoid. (e) Toroid. (f) Iron-core inductor. (See references [4.2] and [4.3]; the permeabilities for various materials are listed in Appendix A.)

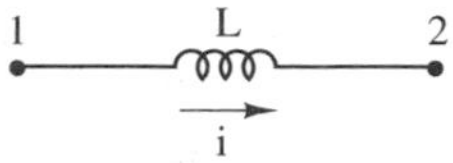

Figure 4.14 An ideal inductor.

permeabilities of ferromagnetic materials vary for high magnetic field strengths. Thus, the unusually high inductance is not constant, but changes with the current.

The Lumping Process

The lumping process for an electrical system is basically the same as that for a mechanical system. However, electrical systems differ in some respects from their mechanical counterparts. For example, a capacitor, which is the electrical counterpart of a mass, need not be connected to the ground terminal. Since electrical systems experience physical flows of charge, lumped-circuit diagrams are physically similar to the corresponding line graphs. Hence, there is no need to articulate a procedure for drawing line graphs for electrical networks. The conventional ideal source symbols for electrical systems are identical to the corresponding line-graph symbols. It is a matter of personal taste whether or not to remove the various lumped electrical element symbols in favor of the simple branch labels of the line graph. We summarize and illustrate the lumping process here. Other illustrations are given in example problems 4.3 and 4.4.

THE LUMPING PROCEDURE FOR ELECTRICAL SYSTEMS:

1. Draw a pictorial diagram of the system.
2. Identify and number the conductors or other *equipotential surfaces*—that is, surfaces over which the potential is the same. Represent each numbered surface by a numbered node. Place the nodes in approximate correspondence with the layout in the pictorial diagram.
3. Represent dissipative conducting paths by lumped resistors between nodes.
4. Represent conductors of great length or of coiled geometry by lumped inductors between nodes. Long conductive paths usually also have significant series resistance; if so, represent them by a series resistor-inductor connection.
5. Insert lumped capacitors between nodes that represent conductors that approximate a parallel-plate geometry.
6. Insert independent voltage or current sources between appropriate nodes.
7. Redraw the network to obtain a convenient layout. All points connected by a solid line have the same potential. Establish and orient a current variable for each branch of the network.

To illustrate the modeling process, let us examine the electrostatic discharge that often occurs as a person walks about a dry house in the winter. The human body is a relatively good conductor; so is the earth on which the house sits. Various metal objects around the house (doorknobs, stoves, water pipes, electrical switch boxes, etc.) are very good conductors. Each metal object and each human body in a particular room can be considered a node in an electrical model of the room. Most of the objects are separated by insulating materials (wood, plaster, and fabrics). Some of the objects (such as the electrical outlets and light switches) are electrically grounded to a cold-water pipe (and hence, to the earth) by means of wires in the walls; these should be represented by a single ground node. Between each pair of conductors separated by insulators, we insert a lumped capacitor. The capacitance is larger between conductors that are large and close together than between conductors that are small or far apart.

A simple electrical model of a room with one person and one other ungrounded object (the doorknob) is shown in Figure 4.15. The human-to-ground capacitance $C_1 = 150$ pF is typical. (See reference [4.4], or insert typical dimensions into the formula of Figure 4.11(c) or (e).) The capacitance-to-ground of a typical metal object such as a doorknob is on the order of $C_2 = 10$ pF. The human-to-object capacitance depends on the relative locations of the person and object, but it is small. Accordingly, we neglect it.

As the person walks across a dry carpet, the process of contacting and separating induces an imbalance of charge within the insulating shoes and carpet and a slight flow of charge across the carpet-to-shoe boundary. As a consequence, a charge difference (and associated voltage) builds up across the capacitor C_1 between the human body and ground. A 10-kV human-to-ground voltage is not unusual. We represent the process whereby charge is built up by a current source.

If the capacitors and current source alone formed an adequate model of the room, the voltage on the person would build without limit. Fortunately, this is not the case. The electrical insulation of the carpet and shoes is not perfect; rather, there is a large, but finite resistance across C_1, on the order of $R_1 = 10^{11}$ ohms. As the electrostatic voltage rises, the discharging current also rises, ultimately limiting the voltage.

If, after the buildup of charge, the person touches a grounded object such as a light-switch grounding screw, the capacitor C_1 discharges quite rapidly by sparking at the point of contact. Specifically, when the separation between conductors becomes small enough to produce a voltage gradient of 30 kV/cm across the air gap, the air in the gap ionizes and a spark jumps across the gap. The brief current can be as large as 100 A. If the person touches an ungrounded object (the doorknob) instead, the discharge is a sharing of energy between capacitors C_1 and C_2, the current is smaller, and the sparking is less severe. The resistors labeled R_c in the figure represent the ionized air path. The switch represents the process of contacting either the light switch or the doorknob. An increase in humidity in the room can reduce R_1 to 10^8 ohms or less and thereby eliminate any significant buildup of voltage.

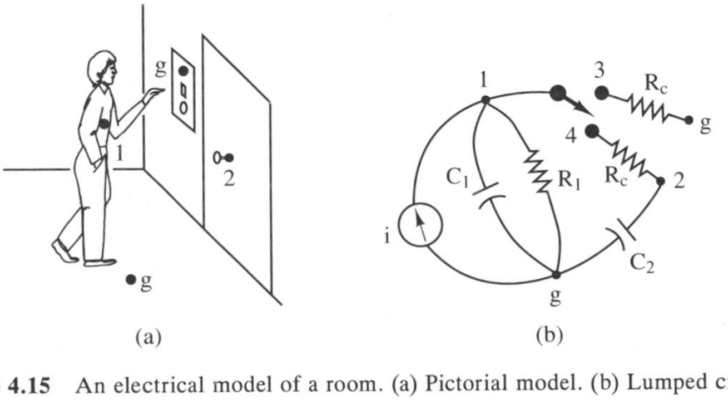

Figure 4.15 An electrical model of a room. (a) Pictorial model. (b) Lumped circuit model.

Example Problem 4.3: Lumped modeling of a coaxial cable—a distributed device. The coaxial cable pictured in Figure 4.16(a) is intended to carry electrical signals from one point to another (say, from a computer to a printer) without changing the signals significantly. (The ideal source v_S and resistor R_S represent the physical source. The resistor R_L represents the electrical load.)

(a) Construct a lumped linear network that approximates the behavior of the coaxial cable.

(b) Determine the parameters of the network as functions of the materials and geometry of the cable and the number of segments used in the model.

(c) Determine the segment length in the cable model for which the model will be accurate for a 2-meter length of cable with signals which vary as rapidly as 25 MHz. (This frequency is typical of computer signals.)

(d) A standard RG-58 communications cable has dimensions $r_1 = 0.432$ mm and $r_2 = 1.47$ mm; the conductors are copper and the dielectric is polyethylene. Compute the values of the lumped parameters for a 2-m length of RG-58 cable.

Solution: **(a)** According to the lumping procedure presented earlier, we must identify conductors or surfaces of distinguishable electric potential. In a coaxial cable, there is a clear distinction between the outer (grounded) conductor and the inner conductor. If the source voltage were constant, a pair of nodes, one for each conductor, would be sufficient. Suppose, however, that the source signal changes abruptly. Concentric cylinders act like a capacitor, which resists rapid change in voltage and stored charge. It takes time for charge to distribute itself along the cable in response to an abrupt change in the source voltage, just as an abrupt disturbance at one end of a clothesline takes time to travel to the other end. At any instant, there is a traveling wave distributed along the cable.

We break the cable into segments short enough that the variation in voltage along a single segment is negligible. Then we can assume that each segment of each conductor has a single potential. (See the dashed partitions and numbered nodes in Figure 4.16(a).) Each node represents the potential of that portion of the inner conductor which lies between the neighboring partitions. According to the node numbering scheme of Figure 4.16(a), there are n nodes, $n - 2$ segments of length $\ell/(n - 1)$, and 2 half-length segments. Since the outer conductor is grounded at both ends, we represent it by a single node. (Compare the segmented cable to the nodes in the lumped network of Figure 4.16(b).)

Each segment of the concentric cylinder acts as a lumped capacitor. These lumped capacitors are labeled C in Figure 4.16(b). Since the end nodes represent half-length segments, they are assigned only half the capacitance of the other nodes. If the insulator between the two conductors is imperfect, then there is a conducting path from the inner conductor to the outer conductor in each segment. We account for this *leakage* of current by inserting the *leakage resistances* (or shunt resistances) R_2 in the lumped model.[1] Because the end segments are half-length segments, the leakage resistances for those segments are doubled.

Current flows through the inner conductor from node to node and returns through the outer conductor. Current in conductors encounters inductance and resistance. In the lumped model, we represent the inductance and resistance of each segment by a series connection of an inductor L and resistor R. The nodes labeled a_i are artificial, in the sense that they do not represent physical points in the system.

[1] The term *shunt* denotes a circuit element that diverts some of the current from the main path by providing a parallel path.

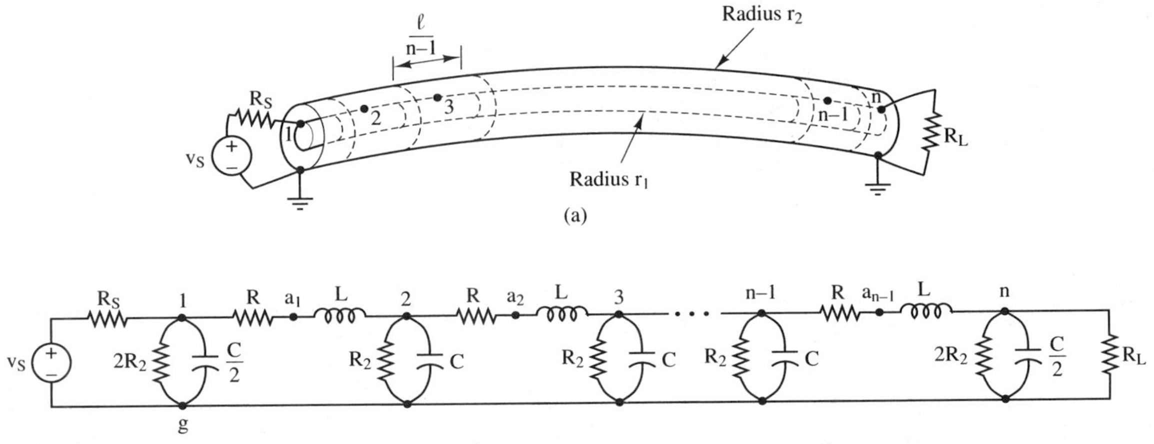

Figure 4.16 A lumped model for a coaxial cable. (a) Pictorial model. (b) Lumped circuit model.

(b) The capacitance for a coaxial conductor geometry is given in Figure 4.11(b). The capacitance for a single segment length is

$$C = \frac{\ell}{n-1}\frac{2\pi\epsilon}{\ln(r_2/r_1)} \tag{4.24}$$

where ϵ is the permittivity of the dielectric in the cable. Figure 4.11 indicates that this formula is accurate only if the coaxial segment is about 10 times longer than the diameter of the outer conductor. The purpose of this condition is to guarantee that *end effects* (electric field distortions at the ends of the segment) are relatively small. Therefore, the restriction applies only to the end segments. For the values given in parts (c) and (d), the restriction is satisfied for the end segments.

According to Figure 4.5, the resistance R associated with the inner conductor in each node-to-node path is

$$R = \frac{\ell}{n-1}\frac{\rho}{\pi r_1^2} \tag{4.25}$$

where ρ is the resistivity of the conductor. The outer cylindrical conductor is also resistive. We assume that it has a much greater cross-sectional area than the inner conductor has. It is also in parallel with a (supposedly) perfectly conducting ground path. Therefore, we ignore the resistance of the return path.

The inductance of a coaxial geometry is given by the formula in Figure 4.13(c). We divide that inductance in proportion to the segment lengths to give

$$L = \frac{\ell}{n-1}\frac{\mu}{2\pi}\ln\left(\frac{r_2}{r_1}\right) \tag{4.26}$$

where μ is the permeability of the dielectric.

The leakage resistance is usually negligible. Example problem 4.4 shows that the leakage resistance is

$$R_2 = \frac{n-1}{\ell}\frac{\rho_2}{2\pi}\ln\left(\frac{r_2}{r_1}\right) \tag{4.27}$$

where ρ_2 is the resistivity of the dielectric.

(c) According to the segment-length criterion for modeling distributed systems, equation (2.51), the segment length must be no longer than $p\nu/10$, where p is the period of the 25-MHz sinusoidal signal and ν is the velocity of propagation of the signal along the cable. For a velocity of propagation equal to the speed of light ($\nu = 3 \times 10^8$ m/s), the segment length $\ell/(n-1)$ must satisfy

$$\frac{\ell}{n-1} \le \frac{p\nu}{10} = \frac{3 \times 10^8 \text{ m/s}}{(25 \times 10^6 \text{ Hz})(10)} = 1.2 \text{ m} \tag{4.28}$$

(We show in section 6.4 that the precise velocity of propagation, assuming that the series and leakage resistances are negligible, is $\nu = 1/\sqrt{LC} = 1/\sqrt{\mu\epsilon} = 2 \times 10^8$ m/s. Hence, the segments should be no longer than 0.8 m.) We pick n = 4, which produces three segments and a segment length of 0.67 m.

(d) From Appendix A, the resistivity of copper is $\rho = 1.7 \times 10^{-8}\ \Omega\cdot\text{m}$; polyethylene has resistivity $\rho_2 = 10^{15}\ \Omega\cdot\text{m}$, permittivity $\epsilon = 2 \times 10^{-11}$ F/m, and permeability $\mu = 1.26 \times 10^{-6}$ H/m. We substitute these material-related constants and the dimensions of the RG-58 cable into the preceding equations to find that

$$\begin{aligned} C &= \frac{205}{(n-1)}\ \text{pF} = 68.4\ \text{pF} \\ R &= \frac{0.058}{(n-1)}\ \Omega = 0.0193\ \Omega \\ L &= \frac{489}{(n-1)}\ \text{nH} = 0.163\ \mu\text{H} \\ R_2 &= 9.78 \times 10^{13}(n-1)\ \Omega = 2.93 \times 10^{13}\ \Omega \end{aligned} \tag{4.29}$$

To assure ourselves of the accuracy of the lumped model in representing the electrical behavior of the cable, we should compare the behavior of the model to the behavior of a physical cable. First, we attach a voltage source to one end of a 2-m length of RG-58 cable and attach a resistive load to the other end. Then we apply a voltage-source signal at the input end and measure the voltage pattern across the load at the output end. (We can use an oscilloscope to measure the rapidly varying voltages.) Finally, we simulate the behavior of the model in response to the same form of source signal. The differences between the measured and simulated behaviors are a measure of the model accuracy.

Example Problem 4.4: Finding capacitance and resistance by lumped modeling.
Use the lumping procedure, together with series and parallel combination, to derive:
(a) The formula given in Figure 4.11(c) for the capacitance of a length of coaxial cable; and
(b) The formula for the leakage resistance of a length of coaxial cable.
Solution: **(a)** A cross section of a coaxial cable is shown in Figure 4.17(a). We focus attention only on its capacitative behavior. Suppose the outer conductor is grounded and the inner conductor is held constant at voltage v. Then the potentials of points in the dielectric are between zero and v, with points near the outer conductor having potentials near zero and those near the inner conductor having potentials near v. Because of the circular symmetry, each circular path centered on the axis of the cable is an *equipotential path* with all of its points at the same potential, as shown in the cross section of the figure. The cable has concentric cylindrical equipotential surfaces.

We choose a number of equipotential surfaces, about equally spaced between the two conductors, as shown in the figure. Then we number the surfaces, as required in step 2 of the lumping process. It follows that each numbered equipotential surface corresponds to a node in the lumped model. Each neighboring pair of equipotential surfaces approximates a parallel-plate geometry. We treat each pair as if it were a capacitor. Therefore, we insert a lumped capacitor between each corresponding pair of nodes, as shown in the lumped model of Figure 4.17(b).

Since no conducting plates exist at these equipotential surfaces, one might ask how charge can be deposited on the plates. Each surface acts as the positive plate for one capacitor and as the negative plate for the neighboring capacitor, so the charges cancel. Therefore, we do not need conductors at the equipotential surfaces.

The formula for the capacitance of a parallel-plate capacitor is given in Figure 4.11(a). We use it to find the capacitance C_p between nodes $p-1$ and p. The two plates of the ca-

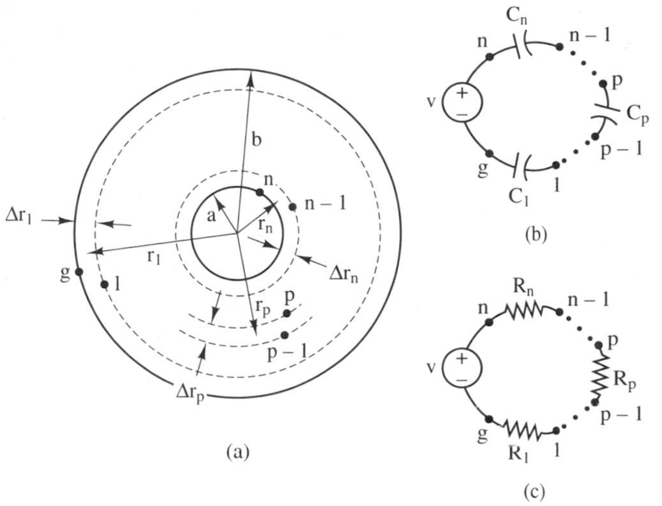

Figure 4.17 Coaxial cable model. (a) Cable cross section. (b) Capacitance model. (c) Resistance model.

pacitor C_p have different radii. We use the average radius r_p to approximate the areas of the plates. The capacitance is

$$C_p = \frac{\epsilon 2\pi r_p \ell}{\Delta r_p} \tag{4.30}$$

where ℓ is the length of the cable and Δr_p is the distance between the cylindrical plates. If we use enough nodes to make Δr_p small compared with r_p, then the error in this lumped approximation will be insignificant.

The lumped model in Figure 4.17(b) shows that the concentric capacitors are connected in series. Let C denote the total capacitance between the inner and outer conductor of the cable. It can be shown that the series connection satisfies the formula

$$\frac{1}{C} = \frac{1}{C_1} + \frac{1}{C_2} + \ldots + \frac{1}{C_n} = \frac{\Delta r_1/r_1 + \Delta r_2/r_2 + \ldots + \Delta r_n/r_n}{2\pi\epsilon\ell} \tag{4.31}$$

(See exercise 4.4.) If the ratio b/a of the outer radius to the inner radius is not unusually high, then equation (4.31) accurately determines the capacitance C of the coaxial cable with only a few lumps (small n).

We find an exact expression for the total capacitance C by letting the number of nodes n approach infinity (and $\Delta r_p \to 0$) in equation (4.31). The sum approaches an integral over the radius r from its smallest value a to its largest value b:

$$\lim_{n\to\infty} \sum_{p=1}^{n} \frac{\Delta r_p}{r_p} \longrightarrow \int_a^b \frac{dr}{r} \longrightarrow \ln\left(\frac{b}{a}\right) \tag{4.32}$$

Therefore, the total capacitance of the coaxial cable is

$$C = \frac{2\pi\epsilon\ell}{\ln(b/a)} \tag{4.33}$$

where a and b are the inner and outer radii of the cable, respectively.

(b) Let us hold the inner conductor at a constant voltage v and focus attention only on leakage current from the inner conductor to the outer conductor through the insulating dielectric. The formula for resistance given in Figure 4.5 assumes that the current is distributed uniformly over the cross section of the object through which it flows and that the cross section is the same for the whole flow path. The symmetrical coaxial geometry satisfies the first requirement, but not the second.

The current flow is perpendicular to the equipotential surfaces. Again, let us use the n equipotential surfaces of Figure 4.17(a) to define the nodes of the lumped model. Since current flows from surface to surface, we insert a lumped resistor between neighboring nodes, as shown in the lumped model of Figure 4.17(c). We make n large enough (Δr_p small enough) that neighboring cylinders (which form the end plates of one resistor) have almost the same cross-sectional area. Then we use the formula of Figure 4.5 to compute the lumped resistance. We obtain

$$R_p = \frac{\rho \Delta r_p}{2\pi r_p \ell} \tag{4.34}$$

where ρ is the electrical resistivity of the dielectric. According to equation (4.16), the effective resistance of resistors connected in series is the sum of the resistances. Therefore, the total effective resistance R between the inner and outer conductors is

$$R = R_1 + \dots + R_n \tag{4.35}$$

If the ratio b/a of the outer radius to the inner radius is not unusually high, then equations (4.34) and (4.35) accurately determine the leakage resistance of the coaxial cable with only a few lumps (small n).

We find an exact expression for the total resistance R by letting the number of nodes n approach infinity (and $\Delta r_p \to 0$) in equations (4.34) and (4.35). The sum approaches an integral over the variable r from its smallest value a to its largest value b. The result is shown in exercise 4.5.

The total leakage resistance R acts in parallel with the total capacitance C of the cable. The models shown in Figs. 4.17(b) and (c) should be connected electrically only at nodes 1 and n. The displacement current through the capacitors and the conduction current through the resistors combine only at those two points.

Exercise 4.4: Suppose two capacitors C_1 and C_2 are connected in series. Then the same current flows through both of them. Use the capacitor impedance equation (4.19) to show that the series connection of capacitors behaves as a single capacitor of capacitance C_{ser}, where

$$\frac{1}{C_{ser}} = \frac{1}{C_1} + \frac{1}{C_2} \tag{4.36}$$

Exercise 4.5: Use (4.32) to show that the exact leakage resistance between the conductors of the coaxial cable is

$$R = \frac{\rho}{2\pi\ell} \ln(b/a) \tag{4.37}$$

where a and b are the inner and outer radii of the cable, respectively.

Finding and Solving the System Equations

The procedures for finding the system equations that represent an electrical network are identical to the procedures introduced for mechanical lumped networks in section 2.2. The equation-solving procedures of chapter 3 apply to all dynamic networks. We summarize these procedures here and demonstrate their use in example problem 4.5.

The node law given by equation (2.27) for electrical networks is known as **Kirchhoff's current law.** In the node method for finding the system equations, we write the node equations for all nodes except one. We use the branch element equations to express each branch current in terms of node voltages. Then the system equations are expressed in terms of node voltages.

In each element equation, we order the voltage-difference subscripts in the direction of the current through the element. This ordering produces the correct sign in each equation. In the literature on electric circuits, one often sees the terminals of a circuit element labeled with + and − signs to indicate an assumed direction of voltage drop across the element. Then the current in the direction from + to − is used in the element equation. These + and − signs play the same role as our ordering of the subscripts on the potential-difference variables.

Equation (2.28), the path law for electrical networks, is known as **Kirchhoff's voltage law.** In the loop method for finding the system equations, we define a set of loop flows, one for each mesh in the network. We then apply the path law to the voltage drops around each loop. Next, we use the branch equations to express the voltage drop across each element in terms of loop currents. Finally, we express the system equations in terms of loop flows.

Example Problem 4.5: Analysis of an electrical network.
Figure 4.18 shows a second-order electrical network.

(a) Use the node method to find the system equations.

(b) Use the loop method to find the system equations.

(c) Find the input-output system equation for the node voltage v_5.

(d) Let $R_S = 50\ \Omega$, $R_1 = 100\ \Omega$, $R_2 = 12\ \Omega$, $R_L = 25\ \Omega$, $L = 10$ mH, and $C = 1\mu F$. Use the input-output system equation from part (c) to find, analytically, the zero-state response of the node voltage $v_5(t)$ to a 10-V step input v_S.

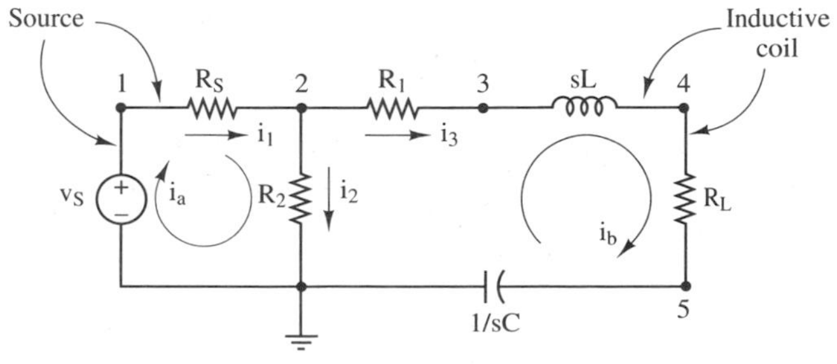

Figure 4.18 A second-order electrical network.

Solution: **(a)** A set of branch currents is identified in the figure. The currents entering each node sum to zero. We use the element equations to express the currents in terms of node potentials. The voltage at node 1 is known, $v_1 = v_S$. Therefore, we do not need the node equation for node 1. The node equations for nodes 2–5 are as follows:

$$
\begin{aligned}
\frac{v_S - v_2}{R_S} + \frac{-v_2}{R_2} + \frac{v_3 - v_2}{R_1} &= 0 \\
\frac{v_2 - v_3}{R_1} + \frac{v_4 - v_3}{sL} &= 0 \\
\frac{v_3 - v_4}{sL} + \frac{v_5 - v_4}{R_L} &= 0 \\
\frac{v_4 - v_5}{R_L} + \frac{-v_5}{1/(sC)} &= 0
\end{aligned}
\tag{4.38}
$$

The node method is not well suited to this five-node network, because it produces many simultaneous equations.

(b) Two loop currents are identified in the figure. Each loop equation sums the voltage drops around one of the loops to zero. We use the element equations to express unknown node voltages in terms of loop currents. The loop equations are:

$$
\begin{aligned}
-v_S + R_S i_a + R_2(i_a - i_b) &= 0 \\
R_2(i_b - i_a) + \left[R_1 + sL + R_L + \frac{1}{sC}\right] i_b &= 0
\end{aligned}
\tag{4.39}
$$

(c) The node voltage v_5 is the voltage drop across the capacitor. Therefore, $v_5 = i_b/sC$. Since the loop equations are fewer in number than the node equations, we use the loop equations to find i_b. We arrange equations (4.39) in the form

$$
\begin{aligned}
(R_S + R_2)i_a - R_2 i_b &= v_S \\
-R_2 i_a + \left(R_1 + R_2 + R_L + sL + \frac{1}{sC}\right) i_b &= 0
\end{aligned}
\tag{4.40}
$$

If we solve the second equation for i_a, substitute the result into the first equation and solve for i_b, and then divide by sC, we obtain

$$
v_5 = \frac{R_2 v_S}{s^2 LC(R_S + R_2) + sC[R_S(R_1 + R_2 + R_L) + R_2(R_1 + R_L)] + (R_S + R_2)} \tag{4.41}
$$

Equation (4.41) represents a second-order linear differential equation, the input-output system equation for v_5. Even though the loop method produces only two equations, the manipulations necessary to find the system equation for v_5 are messy and tedious. In the next section, we examine equivalent-network methods that simplify both the mathematics and our conceptual image of the network.

(d) Since equation (4.41) includes no source-signal derivatives, v_5 is a prototype variable of the network. Therefore, $(v_5, \dot{v}_5)$ represents the network state. These state variables do not jump abruptly. Hence, the zero prior state means that $v_{5i} = 0$ and $\dot{v}_{5i} = 0$.

We substitute the specified parameter values into equation (4.41) to produce

$$(s^2 + 13{,}470s + 10^8)v_5 = 1.94 \times 10^7\ v_S \tag{4.42}$$

From the characteristic polynomial on the left side of equation (4.42), we obtain the characteristic roots

$$s = -6{,}735 \pm j\ 7{,}392 \text{ rad/s}. \tag{4.43}$$

These roots correspond to $\alpha = 6{,}735$ rad/s, $\omega_d = 7{,}392$ rad/s, $\zeta = 0.673$, and $\omega_n = 10{,}000$ rad/s. The damping ratio shows that the network is slightly underdamped. According to equation (3.137), the solution can be expressed as $v_5(t) = v_{5f}\mathcal{A}(t)$, where $\mathcal{A}(t)$ is that normalized step-response curve from Figure (3.52) which corresponds to $\zeta = 0.673$. A glance at that figure is sufficient to obtain an intuitive grasp of the behavior of the network.

We find v_{5f} from the system equation (4.41). According to section 3.3, we merely let $s = 0$ in that equation, let $v_S = 10$ V, and solve for v_5 to obtain $v_{5f} = [R_2/(R_S + R_2)]v_S = 1.94$ V.

To find the response waveform $v_5(t)$, we simply multiply the ordinate of the appropriate curve ($\zeta = 0.673$) of Figure 3.52 by 1.94 volts. We also must divide the normalized abscissa by $\omega_n = 10{,}000$ rad/s to convert it to a time scale.

The analytical expression for the zero-state step response of an underdamped second-order network is given in equation (3.124). For this particular example, the expression is

$$v_5(t) = v_{5f}\left[1 - \frac{e^{-\alpha t}}{\sqrt{1 - \zeta^2}}\sin(\omega_d t + \cos^{-1}\zeta)\right] \tag{4.44}$$

where $\alpha = 6{,}735$, $\omega_d = 7{,}392$, and $\cos^{-1}\zeta = 0.833$ rad. Therefore,

$$v_5(t) = 1.94[1 - 1.35e^{-6{,}735t}\sin(7{,}392t + 0.833)] \text{ volts} \tag{4.45}$$

Although this expression is precise, it gives less intuitive understanding of the behavior than can be obtained from Figure 3.52.

The specific response for the particular values of the element parameters gives insight concerning the behavior of the network. It does not, however, suggest which element parameters affect the decay rate, the fractional overshoot, or the period of oscillation. To discover how these time-domain performance indicators are related to the element parameters, we must repeat the calculations we have just performed in terms of the coefficients of equation (4.41).

Exercise 4.6: Derive equation (4.41) in the manner indicated in part (c) of the solution to example problem 4.5.

Exercise 4.7: Letting $s = 0$ in an impedance produces the limiting behavior associated with steady-state operation. Let $s = 0$ in the impedances of the network of Figure 4.18 and use the simplified network that results to determine the final value v_{5f} for the source voltage $v_S = 10$ V.

Automating Network Analysis with SPICE

The computer program SPICE was developed to simulate numerically the behaviors of electrical networks. Various versions of that program have been marketed. (One of these, PSPICE, is described briefly in Appendix C.) Unlike computer programs that are based on operational models, SPICE-based programs do not require the user to find the system equations. The user need only specify the nature of each branch element (resistor, capacitor, inductor, voltage source, etc.), the way those elements are interconnected (via numbered nodes), the source waveforms, and the initial energy states of the storage elements. Thereafter, generation of the time response of the node voltages and branch currents is automatic.

These programs make an initial estimate of the node voltages and use the element equations to calculate the branch currents that correspond to those node voltages. Next they adjust the node voltages iteratively until the currents balance at each node. Then they change the node voltages to maintain current balance over a sequence of time steps.

SPICE-based programs compute the behaviors of complicated electrical networks with little effort on the part of the user. Of course, the specific network responses produced by these programs cannot provide the user with the same insight obtainable from analytical solutions to the system equations. However, a network simulation program is an easily used tool for experimenting quickly with the behaviors of realistic, but complicated, lumped models. We demonstrate this ease of use in example problem 4.6.

Example Problem 4.6: SPICE-based simulation of coaxial cable behavior.
A four-segment lumped model of a coaxial electrical cable is derived in example problem 4.3. The model is shown in Figure 4.16(b). The values of the lumped-element parameters for a specific cable are given in equations (4.29). The leakage resistance R_2 is high enough to be ignored. The series resistance R is small, but we retain it in the model. Let the source resistance be $R_S = 50\ \Omega$, and let the load resistance be $R_L = 50\ \Omega$. (Section 6.4 shows that $50\ \Omega$ is the *characteristic impedance* for this cable; signals propagate most effectively if the source and load match the characteristic impedance.) The resulting seventh-order network is shown in Figure 4.19.

(a) Use a SPICE-based program to determine the response $v_4(t)$ of the load voltage to an abrupt input $v_S(t)$ that rises sinusoidally from 0 to 10 V in 20 ns. (This rise corresponds to a half-cycle of a 25-MHz sinusoid.)

(b) Interpret the computed response.

Solution: **(a)** We use the network simulation program PSPICE. The data file that corresponds to the network of Figure 4.19 is shown in Table 4.1.[2] The subscript 0 denotes the ground node. All potentials are calculated relative to ground. It is easy to correlate the input data with the lumped network. It is apparent that preparing the data requires no more than specifying the network connection structure and stating numerical values for the element parameters and source signal. This particular data file does not specify which variables are to

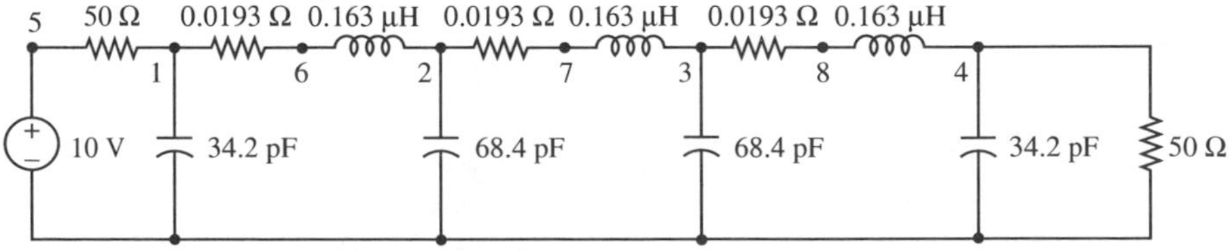

Figure 4.19 The coaxial cable model.

[2] A brief description of PSPICE is given in Appendix C.

TABLE 4.1 PSPICE DATA FILE

```
RG-58 Cable Response
RS     5    1     50ohms
R1     1    6     .0193ohms
L1     6    2     .163uhenries
R2     2    7     .0193ohms
L2     7    3     .163uhenries
R3     3    8     .0193ohms
L3     8    4     .163uhenries
C1     1    0     34.2pfarads
C2     2    0     68.4pfarads
C3     3    0     68.4pfarads
C4     4    0     34.2pfarads
RL     4    0     50ohms
vS     5    0     PWL(0s 0V 2ns .24V 4ns .95V 6ns 2.1V 8ns 3.5V
+ 10ns 5V 12ns 6.5V 14ns 7.9V 16ns 9.05V 18ns 9.76V 20ns 10V
+ 40ns 10V)
.TRAN        1ns   40ns UIC
.PROBE
.END
```

be plotted. Rather, the .PROBE command indicates that all computed quantities are to be stored. Then a *postprocessor* program (entitled PROBE) is used to designate variables for plotting and to arrange the plots. The abrupt source signal $v_S \equiv v_5$ and the computed load-voltage response v_4 are shown in Figure 4.20.

(b) The jump in the source signal appears at the load after a delay of about 10 ns. This delay corresponds to the propagation velocity of 2×10^8 m/s and the cable length of 2 m used in example problem 4.3. The response at an intermediate node (node 3) is also displayed in the figure. It is identical to the response at node 4, except for the reduced delay.

The amplitude of the output signal is only 5 volts, half that of the input. This halving of the source signal is a *voltage-divider* effect. If the cable were removed and the source and its 50-Ω resistance were connected directly to the load, the output signal would again be reduced by half. According to section 6.4, the input of the cable acts as if it were a 50-Ω resistor because the load resistance is the same as the 50-Ω characteristic impedance of the cable. Thus, the signal that *enters* the model of the cable at node 1 is a 5-V signal. We see evidence of the truth of this statement in the fact that the propagating signal has the same 5-V level at nodes 3 and 4 of the cable model.

The shape of the response at the cable output is not quite identical to that at the input: There is a slight decaying oscillation. A physical cable would not introduce this oscillation. It is caused by the lumped approximation. If we double the rate of rise of the step input, this slight oscillation quadruples in amplitude. If we halve the rate of rise, however, the oscillation reduces only slightly. It is observations such as this that led to equation (2.51) as the segment length criterion for lumped models of distributed systems.

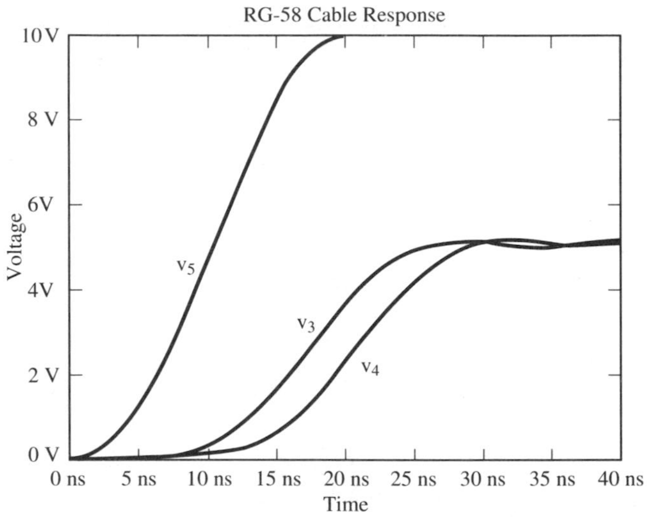

Figure 4.20 PSPICE simulation of coaxial cable behavior.

Exercise 4.8: Carry out the same SPICE-based simulation as in example problem 4.6 for source signals that are twice as abrupt and half as abrupt as the source signal in that problem.

If we jerk a clothesline, a wave travels along the line, reflects from the end, and returns. A similar phenomenon occurs in the coaxial cable if the load resistance R_L does not match the 50-Ω characteristic impedance of the cable. (This phenomenon is explained in Section 6.4.) If we were to change the load resistor to a much different value (say, 5 Ω), the signal would reflect from the load and return, at reduced amplitude, toward the source. If we were also to make a big change in the source resistance R_S, the reflected signal would bounce back and forth along the cable, perhaps giving the impression of oscillation at the load. These reflection phenomena are real; they appear in the physical system if the source and load resistors are mismatched.

Exercise 4.9: Repeat the simulation of example problem 4.6 with the load resistance changed to 5 Ω.

Although SPICE-based programs were developed for electrical networks, and the terminology used to interact with the programs is electrical, the algorithms used in the programs are based on the node law of equation (2.27) and on the forms $v_{ij} = Ri$, $i = C\dot{v}_{ij}$, and $v_{ij} = L\,di/dt$ of the element equations. These mathematical relations are identical in form to those used for lumped models of mechanical networks. SPICE-based programs can analyze mechanical networks (or other analogous networks) with the same ease as electrical networks. SPICE-based simulation is demonstrated for a complicated rotational network in example problem 4.7.

Example Problem 4.7: SPICE-based simulation of a mechanical speedometer.
A four-segment lumped rotational model of a motorcycle speedometer cable is derived in example problem 3.3. However, the problem shows that a single-segment model also can represent the behavior of the speedometer adequately for typical motorcycle acceleration capabilities. The motorcycle wheel accelerates from rest to 100 rad/s in 3 s.
(a) Use a network simulation program to demonstrate the speedometer behavior for both the four-segment and single-segment models in response to *uniform* acceleration of the motorcycle.
(b) Assume that the velocity of the wheel increases sinusoidally, as illustrated in Figure 2.40. Simulate the behavior of the speedometer for an acceleration that *requires* the four-segment model. Interpret the results.
Solution: **(a)** The lumped models for the speedometer are shown in Figures 3.15 and 3.16. As the input velocity profile for the simulation, we choose $\Omega(t) = 33.3\,t$ (a uniform acceleration of 33.3 rad/s) for the first 3 s, followed by a constant velocity $\Omega(t) = 100$ rad/s for $t > 3$ s.

We apply this velocity signal first to the single-segment model. We use the PSPICE network simulation program. The data file is shown in Table 4.2. Ground points are denoted by the subscript 0. PSPICE computes directly only the *central* variables—Ω and T (analogous to v and i)—of the model. We use the postprocessor PROBE to integrate Ω_6 and produce the desired output signal θ_6.

The velocity Ω_1 of the wheel and position θ_6 of the speedometer needle are displayed in Figure 4.21. The angle of the speedometer needle follows the velocity of the wheel correctly, *on the average*. However, there is a bothersome oscillation of the needle owing to the interaction of the inertia J_2 with the spring K_s. Friction in the needle-support structure, which we have ignored, should eliminate this oscillation. The response of the four-segment model to the same source signal is indistinguishable from that of the single-segment model.

TABLE 4.2 PSPICE DATA FILE

```
Single-Segment Speedometer Model
L       1    5      9.09:rad/Nm
*L is analogous to compliance 4/K
C       5    0      6.11E-4:kgm2
*C is analogous to 4J + JC
R       5    0      3.98E5:rad/Nms
*R is analogous to 1/(4B)
RC      5    6      2.27E4:rad/Nms
LS      6    0      3.114:rad/Nm
C2      6    0      2.49E-4:kgm2
v       1    0      PWL(0s 0:rad/s 3s 100:rad/s 5s 100:rad/s)
*v is analogous to omega
.TRAN        0.1s   5s     UIC
.PROBE
.END
```

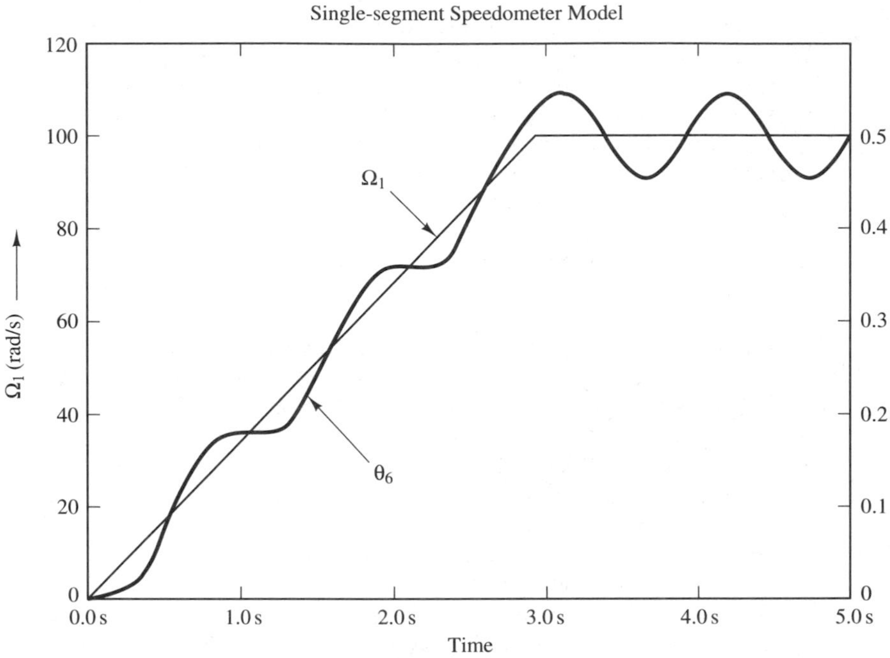

Figure 4.21 Computed response of the single-segment speedometer model.

(b) Example problem 3.3 specifies the length of the speedometer cable as 1 m. Hence, each segment of the four-segment model represents a 0.25-m length of cable. According to equation (2.51), the segment-length criterion for modeling distributed systems, a segment length $p\nu/10$ is barely short enough to adequately represent the behavior of the cable for a sinusoidal source signal of period p. Section 6.4 (Table 6.3) shows that the velocity of propagation of mechanical disturbances along the cable is $\nu = 3{,}415$ m/s. Therefore, the period of the sinusoidal source waveform can be no smaller than $p = (10)(0.25 \text{ m})/(3{,}415 \text{ m/s}) = 0.73$ ms.

We choose an input that rises sinusoidally from 0 to 100 rad/s during the 0.37-ms half-period and then remains constant at 100 rad/s thereafter. We applied such a signal to the four-segment model and used PSPICE to compute the behavior of the model. The simulated response is shown in Figure 4.22. The source signal at node 1 travels along the cable at 3,415 m/s and arrives at node 3 (halfway through the cable) after a delay of about 0.15 ms, as it should. About 0.15 ms later, the propagating velocity signal reflects from the end of the cable and returns toward the source. Evidence of this reflection appears at node 3 at $t \approx 0.45$ ms, about 0.3 ms after the signal first became evident at node 3. Thereafter, we observe reversals in the signal at node 3 every 0.3 ms as the signal bounces back and forth between the two ends of the cable. These reflections would occur in the physical cable; they are not a result of the lumped approximation.

Node 5 hardly moves during the 1.2-ms interval shown in the figure, owing to the high value of J_c. Neither the velocity waveform Ω_5 (at the end of the cable) nor the speedometer needle displacement pattern $\theta_6(t)$ shows any effect of the reflections as time proceeds. Within a second or so after the input velocity stops changing, the speedometer needle rises

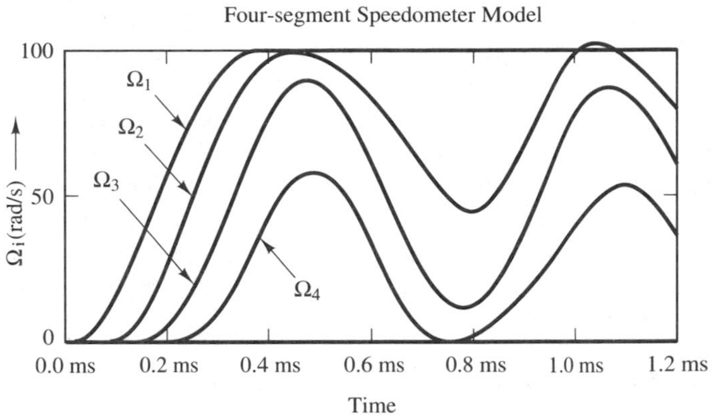

Figure 4.22 Computed response of the four-segment speedometer model.

smoothly to the correct value. Thus, the speedometer is able to track the slowly changing signals from the wheels of the vehicle satisfactorily.

The electrical cable examined in example problem 4.6 does not exhibit any reflections of signals from the load and source. That is because the source and load impedances are properly matched to the characteristics of the cable. By contrast, the effective impedance of the drag-cup, needle, and spring combination at the end of the speedometer cable does not match the *characteristic impedance* of the cable. But this does not matter if the source signal changes so slowly that reflections disappear before they are noticed. That is what happens in the physical speedometer. Matching of cable characteristics is discussed in section 6.4.

4.2 EQUIVALENT LINEAR MODELS

We discovered in section 2.2 that a combination of dampers connected in series or parallel can be replaced by a single damper that is equivalent in behavior to the combination. That relationship is shown to apply as well to springs in problem 2.17 and to electrical capacitors in exercise 4.4. In this section, we extend the equivalent-element concept to *linear subnets.* We show that any interconnection of linear lumped elements and sources that lies between two nodes of a network can be replaced by a simple *equivalent subnet*—a single source in series or parallel with a single impedance. As viewed at the two nodes, this equivalent subnet is indistinguishable from the original connection of elements: The two subnets are represented by the same equations.

Each element in a lumped network represents a unique equation between the through and across variables at its terminals. Each connection of elements at a node corresponds to a unique node equation. Hence, mathematical manipulation of the network equations corresponds to pictorial manipulation of the elements and their interconnections. In this section, we learn to simplify the network representations of systems by interchanging equivalent subnets pictorially.

We begin by enhancing the analogies among the lumped elements of different physical domains. Clear analogies improve our ability to transfer insights from one field to another. *Impedance* is a key unifying concept in these analogies.

Impedance is a conceptually simple lumped representation for a linear relation between the flow through an object and the potential drop across the same object. We shall extend this simple lumped representation to any two linearly related network variables. We shall call this general lumped representation a *transfer function.* An *input-output* transfer function relates a dependent variable of a network directly to an independent source variable. It is the simplest of system representations. It is equivalent to the input-output system differential equation.

Parameter Analogies

Table 4.3 lists the potential and flow variables and compares corresponding ideal-element impedances for the translational, rotational, and electrical networks examined in earlier sections and also for the hydraulic and thermal networks treated in sections 5.1 and 6.1. The format of the table makes explicit the analogies among the parameters of elements of different energy domains.

Impedance is defined only for passive linear lumped objects. In general, a two-terminal passive linear lumped object is described by a linear differential relation between the flow through the object and the potential drop across the object. Let T denote the *central* flow variable and A the *central* potential variable. Then we can express the linear relation for an object in the form

$$A_{ij} = zT \tag{4.46}$$

TABLE 4.3 ELEMENT IMPEDANCES

Energy Domain	Central Flow Variable	Central Potential Variable	Continuous-Potential Element	Dissipative Element	Continuous-Flow Element
Translational	F	v	$\frac{1}{sm}$	$\frac{1}{b}$	$\frac{s}{k}$
Rotational	T	Ω	$\frac{1}{sJ}$	$\frac{1}{B}$	$\frac{s}{K}$
Electrical	i	v	$\frac{1}{sC}$	R	sL
Hydraulic	Q	P	$\frac{1}{sC_f}$	R_f	sI
Thermal	q	θ	$\frac{1}{sC_t}$	R_t	None
Steady (s = 0) Behavior	T	A	Open Circuit (T = 0)	Unchanged (R)	Short Circuit ($A_{ij} = 0$)
Jump ($s \to \infty$) Behavior	δT	δA	Short Circuit ($\delta A_{ij} = 0$)	Unchanged (R)	Open Circuit ($\delta T = 0$)

where the flow variable T is oriented from node i to node j, in the direction of the potential drop A_{ij}. We call z the **operational impedance** (or just the *impedance*) of the object. We use the shorthand expression *impedance* z to mean an object that has impedance z. (We sometimes use the symbol y and the term *admittance* for the reciprocal of impedance.)

We use the word *operational* because the symbol s (or 1/s) in an impedance represents differentiation with respect to time (or integration over time) in the equation associated with that impedance. We manipulate the variable s as if it were a number. Operational notation makes the manipulation of equations easier by eliminating the need to think about differentiations and integrations. At any point in the manipulations, we can replace s by d/dt and 1/s by $\int$ dt to obtain the corresponding integro-differential equations.

Electrical resistance R is a measure of resistance to electric current—that is, to flow of charge. The dissipative impedances 1/b and 1/B are mechanical resistances to flow (scalar force). Each friction constant, b or B, is a conductance—the reciprocal of a resistance. For a fixed potential difference, a high friction constant means high flow and a high resistance means low flow. When we wish to refer to dissipative impedances in general, we will use the general resistance notation R.

The compliances 1/k and 1/K are analogous to electrical inductance L. If the potential difference is fixed, a high compliance or inductance implies a low *rate of change* of flow. Compliance and inductance are measures of the **tendency to hold flow constant.** A high stiffness, k or K, means that flow (scalar force) can change rapidly.

Each of the parameters m, J, and C is a measure of resistance to changes in potential. These parameters thus embody the concept of **inertia**; that is, for a fixed flow through the element, a large value of m, J, or C implies a low rate of change in the potential difference across the element. The parameters m, J, and C are measures of the **tendency to hold potential constant.**

Each of m, J, and C can also be thought of as a **capacitance**—a measure of **ability to accumulate flow.** For a fixed potential difference across the element, a large value of the parameter implies a large integrated flow through the element. Both high inertia and high capacitance correspond to low impedance.

Table 4.3 also summarizes the behaviors of the continuous-potential and continuous-flow elements under steady and abruptly changing conditions. The behaviors are simpler under these special conditions. Consequently, the behaviors of the system equations also are simpler under these special conditions. As noted in the table, setting $s = 0$ in the impedances of a network corresponds to open-circuiting continuous-potential elements and short-circuiting continuous-flow elements. Conversely, letting $s \to \infty$ short-circuits continuous-potential elements and open-circuits continuous-flow elements.

Since the network system equations correspond to linear combinations of impedances, we can find the effect of open- and short-circuiting elements *after* we have determined the system equations. To do this, we express the system equations in operational form (s notation) and then let $s \to 0$ for steady-state conditions or let $s \to \infty$ for abruptly changing conditions.

Transfer Functions

Impedance represents a *linear* potential-flow relation. That is, z(s) is a shorthand notation for the linear equation $A_{ij} = z(s)T$. Thus, the impedance of an element and the mathematical equation that represents the behavior of that element are mathematically equivalent models. We can relate *any* two variables of a linear network by an *operational equation* that is similar to an impedance equation. In this section, we broaden the concept of impedance to the more general concept of a transfer function.

Consider the subnet shown in Figure 4.23. The variables of the subnet (i, v_1, v_2, and v_g) are related *to each other* only by the resistor equation, the capacitor equation, and the node equation for node 2. The rest of the network has no effect on the relations among *these* variables. The relation between each pair of variables is shown in the figure, under the assumption that the capacitor is uncharged. If the capacitor were charged, we would have to specify *separately* its initial energy state (the capacitor voltage) to describe fully the future relations among the variables.

Suppose y_1 and y_2 are two network variables related by a *linear* differential equation. If we write that differential equation in operator (s) notation and solve the operator equation for the variable y_2 in the form

$$y_2 = G(s)y_1 \tag{4.47}$$

then G(s) is a ratio of polynomials in s. The numerator of G represents a differential operator on y_1, and the denominator represents a differential operator on y_2. We call G(s) the *transfer function* from y_1 to y_2; that is, multiplication by G(s) converts *from* y_1 *to* y_2. Depending on the nature (potential or flow) of y_1 and y_2, the transfer function G(s) can have units of impedance, units of admittance (the reciprocal of impedance), or no units at all (i.e., it can be dimensionless). The equations in Figure 4.23 have the form of equation (4.47). The transfer function in the first equation has units of impedance. What are the units of the transfer functions in the other two equations?

We formally define the **transfer function** from y_1 to y_2 by

$$G(s) = \left.\frac{y_2}{y_1}\right|_{z.s.} \tag{4.48}$$

where the notation *z.s.* means *zero initial state*. Then the inverted operator $\hat{G}(s) \triangleq 1/G(s)$ is the transfer function from y_2 to y_1. The relation between y_1 and y_2 is determined fully only with knowledge of the initial energy states of the storage elements that affect that relation. Just as we use *zero-state* behaviors to compare networks (see section 3.4), we also use *zero-state* transfer functions to compare relations between pairs of network variables.

The transfer functions G and $\hat{G}$ are equivalent in the sense that they both represent the same zero-state linear differential equation. They differ only in conceptual interpretation: We view G as an operator on the signal y_1 and $\hat{G}$ as an operator on the signal y_2. Equation (4.47), of course, can be interpreted either in terms of G or in terms of $\hat{G}$.

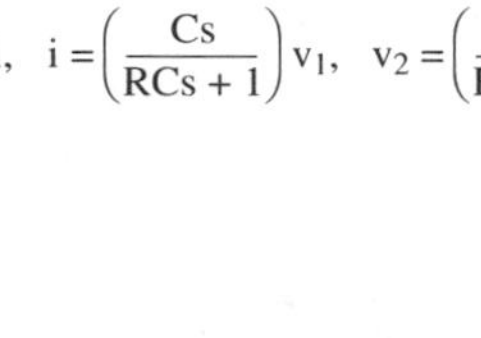

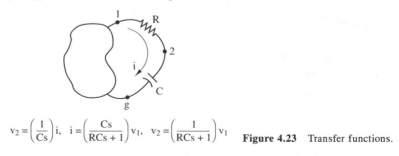

$$v_2 = \left(\frac{1}{Cs}\right)i, \quad i = \left(\frac{Cs}{RCs+1}\right)v_1, \quad v_2 = \left(\frac{1}{RCs+1}\right)v_1$$

Figure 4.23 Transfer functions.

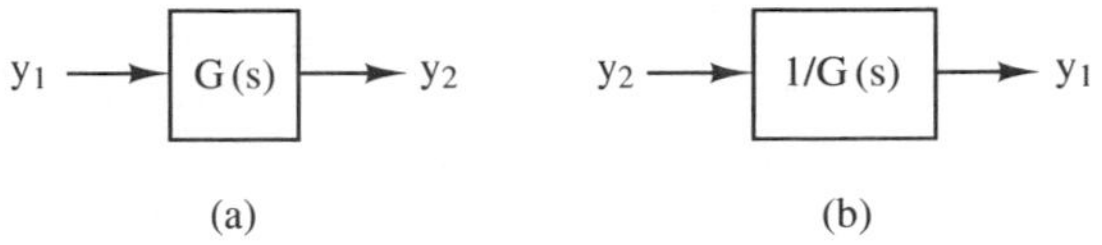

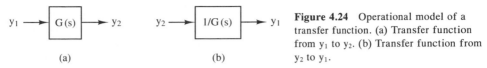

Figure 4.24 Operational model of a transfer function. (a) Transfer function from y_1 to y_2. (b) Transfer function from y_2 to y_1.

Figure 4.24 shows operational models of the transfer functions G(s) and $\hat{G}(s)$. Recall from section 3.2 that an operational model acts on signals only in the direction of the arrows. Engineers sometimes construct and manipulate operational models to help them conceptualize relations among variables, even if they do not intend to simulate the operations numerically.

Section 2.2 uses loop and node rules, and then elimination of variables, to find the input-output system equation that relates a particular dependent (or output) variable to the source (or input) variable. This input-output system equation can be expressed in the operator form

$$y_O = H(s)y_I \tag{4.49}$$

where y_O is the output variable, y_I is the input variable, and H(s) is the **input-output transfer function** for the variable y_O. The input-output transfer function is a special case of equation (4.48) in which the input function is the source function, i.e., the independent variable.[3] The input-output transfer function for the variable y_O is also known as the **system transfer function** for y_O.

Exercise 4.10: Show that the system transfer function for the variable v_5 of equation (4.42) is

$$H(s) = \frac{1.94 \times 10^7}{s^2 + 13{,}470\ s + 10^8}$$

In section 3.3, we found that the coefficients of the homogeneous differential equation (the left side of the input-output system equation), with the differential operator s treated as a complex number, are the coefficients of the characteristic polynomial of the equation. The roots of that polynomial are the characteristic values for the network. Therefore, **the denominator of the system transfer function is the characteristic polynomial of the network.** Its roots, the characteristic values, are called the **poles** of the transfer function. (A pole of H(s) is a value of s for which H(s) becomes infinite.) The roots of the numerator of the system transfer function are called the **zeros** of the transfer function. An s in the numerator of H(s) represents a right-side derivative in the input-output system equation. A system transfer function with no zeros corresponds to an input-output system equation with no right-side (source-signal) derivatives. Therefore, a *prototype variable* of a linear network is a variable for which the system transfer function has no zeros (see section 3.4).

[3] The symbols G and H are commonly used to represent transfer functions. Either symbol can be used to represent an arbitrary transfer function or an input-output transfer function.

For any dependent variable of a particular network, the system transfer function is equivalent to the input-output system equation. Thus, from one of these, we can determine the other by inspection. For a prototype variable y of the network, according to section 3.4, there is also an equivalence between the key features of the zero-state step response (the fractional overshoot f_{OS}, the time of peak t_{peak}, and the final value y_f) and the system differential equation. The equivalence among the transfer function, the system equation, and the time-domain performance indicators is examined further in sections 5.3 and 5.4.

Series and Parallel Combination

The symbol that represents a lumped element is a visual shorthand for an equation that relates the through and across variables for that element. Similarly, a lumped network is equivalent to the set of equations that represents the relations among the network variables. Manipulation and simplification of the equations that represent the network, then, correspond to manipulation and simplification of the lumped network itself.

From example problems 2.8 and 4.4 and problems 2.17–2.21 at the end of chapter 2, we find that any two ideal (linear) elements of the same kind that are connected in series or in parallel can be replaced by a single equivalent ideal element of the same kind. This equivalence between an ideal element and a series or parallel combination of ideal elements extends in a natural way to series and parallel connections of mixed types of *linear* lumped elements. The equivalence relations are a consequence of the nature of the interconnection between the elements.

The creative use of *equivalent elements* can simplify lumped models and reduce the effort of analysis. Since the complexity of a network is a conceptual barrier during the design process, the ability to think in terms of simplified equivalent networks during system conceptualization and design is valuable. In this section, we summarize the equations that characterize series and parallel connections of lumped elements in a way that applies to all types of potential-flow networks. We use the impedance notation of equation (4.46) to describe the lumped elements.

Impedance is a simple way of representing a *linear* potential-flow relation. Nonlinear effects, such as coulomb friction or turbulent fluid resistance, cannot be represented by an impedance. Equivalence relations do exist for series and parallel connections of nonlinear lumped elements—that is, elements for which the potential drop is not proportional to the flow. But the nature of the equivalence relation is different for each form of nonlinearity (see problem 5.3 at the end of chapter 5). Here, we treat linear elements only.

Suppose two linear lumped elements are connected **in series** (end to end), as illustrated in Figure 4.25(a). According to the node law, the flows in the elements are identical. Also, by the path law, the total potential drop divides across the pair of elements in the form $A_{13} = A_{12} + A_{23}$. From equation (4.46), $A_{12} = z_a T$ and $A_{23} = z_b T$. Therefore, $A_{13} = (z_a + z_b)T$. That is, the impedances of series elements add directly to form the impedance of the equivalent element:

$$z_{ser} = z_a + z_b \tag{4.50}$$

We show the equivalent impedance in Figure 4.25(b).

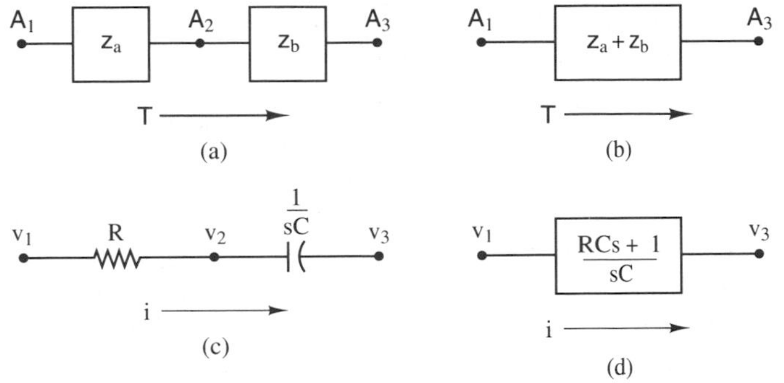

Figure 4.25 Combining series impedances. (a) Impedances connected in series. (b) The equivalent impedance. (c) Electrical elements connected in series. (d) The equivalent electrical impedance.

Let us examine the impedances listed in Table 4.3. If two series elements are of the same type, and if the element parameter is in the numerator of the impedance formula, the parameter of the equivalent element is the sum of the parameters of the individual elements. Thus, $R_{ser} = R_a + R_b$ and $L_{ser} = L_a + L_b$. On the other hand, if the element parameter appears in the denominator of the impedance formula, as it does for electrical capacitors and translational dampers, the inverse of the equivalent parameter equals the sum of the inverses of the individual parameters. For example, $1/b_{ser} = 1/b_a + 1/b_b$, or $b_{ser} = b_a b_b/(b_a + b_b)$.

Exercise 4.11: Use equation (4.50) and the appropriate impedance formulas from Table 4.3 to show that a series connection of electrical capacitors C_a and C_b can be replaced by the single equivalent capacitor $C_{ser} = C_a C_b/(C_a + C_b)$.

Now suppose that the series elements are from the same physical domain, but are not of the same type.[4] Then equation (4.50) still applies. For example, suppose that element a is a resistor and element b is a capacitor, as shown in Figure 4.25(c). Then the impedance of the equivalent element is

$$Z_{ser} = R + \frac{1}{sC} = \frac{RCs + 1}{sC} \tag{4.51}$$

But how should we interpret this impedance? According to equation (4.46), it means

$$v_{13} = \frac{RCs + 1}{sC}\, i \tag{4.52}$$

or

$$Csv_{13} = RCsi + i \tag{4.53}$$

[4] Only elements from the same physical domain can be connected directly. Connecting lumped elements from different domains—say, electrical and rotational—requires transducers (see chapter 6).

Since multiplication by the operator s corresponds to differentiation with respect to time, both equations (4.51) and (4.53) should be interpreted as

$$C(\dot{v}_1 - \dot{v}_3) = RC\frac{di}{dt} + i \tag{4.54}$$

Suppose next that two lumped elements are connected **in parallel** (both ends in common), as illustrated in Figure 4.26(a). According to the path law, the potential drops across the elements are identical. Also, by the node law, the total flow into the pair divides according to $\mathsf{T} = \mathsf{T}_a + \mathsf{T}_b$. From equation (4.46), $\mathsf{T}_a = \mathsf{A}_{12}/\mathsf{Z}_a$ and $\mathsf{T}_b = \mathsf{A}_{12}/\mathsf{Z}_b$. Therefore, $\mathsf{T} = \mathsf{A}_{12}(1/\mathsf{Z}_a + 1/\mathsf{Z}_b)$. That is, the admittances of parallel elements add directly to form the admittance of an equivalent element according to

$$\frac{1}{\mathsf{Z}_{par}} = \frac{1}{\mathsf{Z}_a} + \frac{1}{\mathsf{Z}_b} \tag{4.55}$$

The impedance of the equivalent element is

$$\mathsf{Z}_{par} = \frac{\mathsf{Z}_a\mathsf{Z}_b}{(\mathsf{Z}_a + \mathsf{Z}_b)} \tag{4.56}$$

We show this equivalent element in Figure 4.26(b).

Let us refer again to the impedances of Table 4.3. If two parallel elements are of the same type, and if the element parameter is in the numerator of the impedance formula, the inverse of the equivalent parameter is the sum of the inverses of the individual parameters. Thus, parallel electrical resistors combine according to $R_{par} = R_aR_b/(R_a + R_b)$. On the other hand, if the element parameter is in the denominator of the impedance formula, the

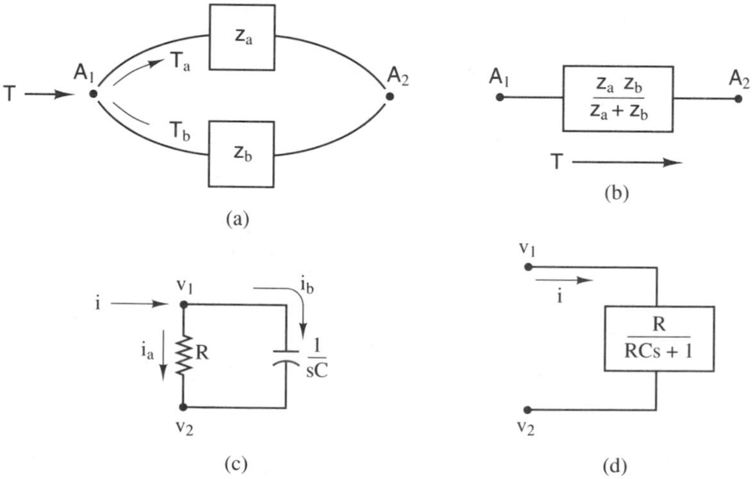

Figure 4.26 Combining parallel impedances. (a) Impedances connected in parallel. (b) The equivalent impedance. (c) Electrical elements connected in parallel. (d) The equivalent electrical impedance.

equivalent parameter is the sum of the individual parameters. Hence, parallel rotational dampers combine according to $B_{par} = B_a + B_b$.

Exercise 4.12: Use equation (4.55) and the appropriate impedance formulas from Table 4.3 to show that parallel electrical capacitors combine according to $C_{par} = C_a + C_b$, parallel translational springs combine according to $k_{par} = k_a + k_b$, and parallel electrical inductors combine according to $L_{par} = L_aL_b/(L_a + L_b)$.

Exercise 4.13: A capacitor and a resistor are connected in parallel in Figure 4.26(c). Show that the equivalent impedance of the parallel combination is $z_{par} = R/(RCs + 1)$. This impedance should be interpreted to mean $RC\dot{v}_{12} + v_{12} = Ri$, where i is the current flowing into the parallel combination and v_{12} is the voltage drop across the parallel combination in the same direction as the current.

Ideal sources connected in series or in parallel can be replaced by a single equivalent source. The process of combining the sources is usually trivial because ideal flow sources can only be connected in parallel and ideal potential-difference sources can only be connected in series; otherwise, their connection would violate the assumption that they are ideal (and independent). Accordingly, the values of the equivalent sources are the sums of the values of the individual sources.

Networks can be simplified drastically by sequentially combining their series or parallel elements using the equivalent-impedance principles embodied in equations (4.50) and (4.55) (or (4.56)). In point of fact, we often use this simplification process to derive the input-output system equation for a network, as we illustrate in example problem 4.8. Delta or wye interconnections of elements sometimes do not lend themselves to series or parallel combination. Problem 4.25 at the end of the chapter derives a delta-wye transformation of impedances that can be used to simplify networks containing such interconnections.

Example Problem 4.8: Simplification of a mechanical impedance network.
A lumped model of the interaction between a fish and a floating boat is derived in example problem 2.3. The model is repeated in Figure 4.27(a).

(a) Predict the order of the input-output system equation for the fish motion v_3.

(b) Simplify the model. Represent the fisherman, boat, and line as a single *external* impedance z_{ext} that is felt by the fish at port 3, 0. Express z_{ext} in terms of m_1, k_1, k_2, and b_1. Also, find the effective impedance z_{fish} of the fish itself. Interpret the simplified model physically.

(c) Use the simplified model, together with the element parameters found in example problem 2.3, to find the system equation for v_3.

Solution: **(a)** Because the lumped model contains four energy-storage elements, we might expect the order of the system to be four. However, the two springs are connected in series and act as one. Thus, the system is third order, and the input-output system equation for each dependent variable, including v_3, is third order.

(b) A line-graph representation of the network is shown in Figure 4.27(b). The branches are labeled in impedance notation. The line-graph notation, by obscuring the mechanical nature of the network, makes the simplification steps in this example more clearly applicable to other energy domains. On the other hand, the network and the succeeding simplified equivalent representations could be displayed just as well in translational notation.

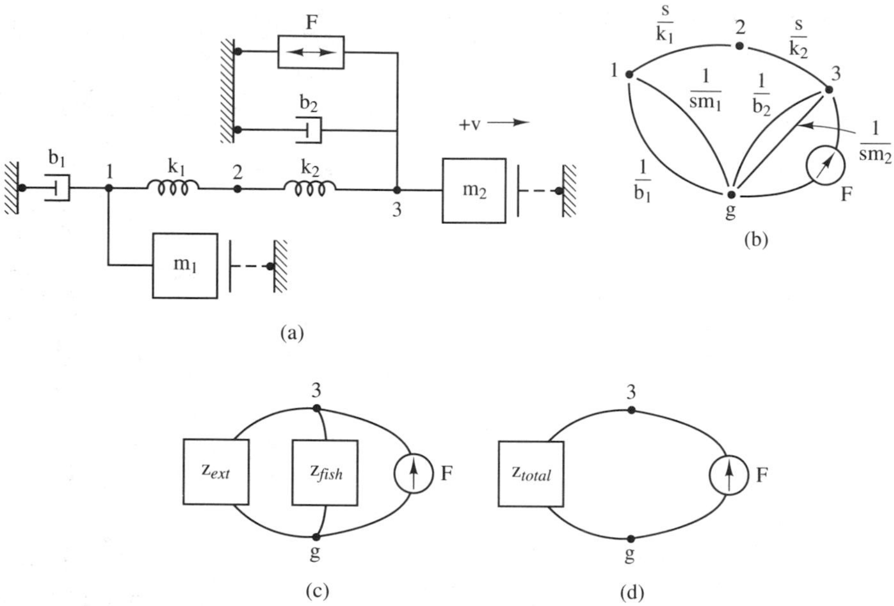

Figure 4.27 A fish-boat network. (a) Lumped model. (b) Line-graph representation. (c) Equivalent simplified model. (d) Impedance energized by the fish.

The form of the simplified model is shown in Figure 4.27(c). We find z_{ext} by a sequence of series and parallel combinations. The parallel damper and mass between nodes 1 and g can be replaced by an equivalent impedance z_1 that satisfies

$$\frac{1}{z_1} = \frac{1}{1/b_1} + \frac{1}{1/(sm_1)} \quad \text{or} \quad z_1 = \frac{1}{sm_1 + b_1} \tag{4.57}$$

The series connection of springs between nodes 1 and 3 is equivalent to a single spring of impedance $z_2 = s/k$, where $k = k_1k_2/(k_1 + k_2)$. Then, we combine z_1 and z_2 in series to produce

$$z_{ext} = \frac{1}{sm_1 + b_1} + \frac{s}{k} = \frac{s^2m_1 + sb_1 + k}{k(sm_1 + b_1)} \tag{4.58}$$

The lumped elements that constitute the *fish* impedance are the fish mass m_2 and the water-friction damper b_2 connected in parallel between nodes 3 and g. We combine the corresponding impedances in parallel to produce

$$\frac{1}{z_{fish}} = \frac{1}{1/(sm_2)} + \frac{1}{1/b_2} \quad \text{or} \quad z_{fish} = \frac{1}{sm_2 + b_2} \tag{4.59}$$

When the fish is free, it exerts its force F only on z_{fish}. The force is large enough to accelerate the fish and sustain its velocity. Once the hook is set, the additional impedance z_{ext} acts in parallel with z_{fish}. Since $m_1 \gg m_2$ and $b_1 \gg b_2$, $z_{ext} \ll z_{fish}$. Most of F is consumed by the smaller impedance z_{ext} (or larger admittance, $1/z_{ext}$) and is unavailable to sustain the fish's motion.

(c) Let the effective impedance of the parallel combination of z_{ext} and z_{fish} be z_{total} (see Figure 4.27(d)). Then

$$F = \frac{v_3}{z_{total}} = \left(\frac{1}{z_{ext}} + \frac{1}{z_{fish}}\right)v_3 \tag{4.60}$$

We substitute the formulas for the impedances z_{ext} and z_{fish} into equation (4.60) to produce

$$\begin{aligned} F &= \frac{k(sm_1 + b_1)}{s^2m_1 + sb_1 + k} v_3 + (sm_2 + b_2)v_3 \\ &= \frac{s^3m_1m_2 + s^2(m_2b_1 + m_1b_2) + s[k(m_1 + m_2) + b_1b_2] + k(b_1 + b_2)}{s^2m_1 + sb_1 + k} v_3 \end{aligned} \tag{4.61}$$

Then,

$$m_1m_2v_3^{(3)} + (m_2b_1 + m_1b_2)v_3^{(2)} + [(m_1 + m_2)k + b_1b_2]v_3^{(1)} + k(b_1 + b_2)v_3 = m_1F^{(2)} + b_1F^{(1)} + kF \tag{4.62}$$

We substitute the parameter values $m_1 = 160$ kg, $m_2 = 4$ kg, $b_1 = 275$ N·s/m, $b_2 = 3$ N·s/m, $k_1 = 300$ N/m, and $k_2 = 141$ N/m to get $k = 95.9$ N/m and

$$640v_3^{(3)} + 1{,}580v_3^{(2)} + 16{,}553v_3^{(1)} + 26{,}660v_3 = 160F^{(2)} + 275F^{(1)} + 95.9F \tag{4.63}$$

Realistic Source Models

The lumped linear models of sections 2.1, 3.1, and 4.1 use sources that are ideal—sources that produce either an independent flow or an independent potential difference. In a physical system, however, a source can approximate ideal behavior only if the system does not require much power from the source and does not demand too rapid a *change* in the flow from the source or in the potential difference across the source. We now develop source models that are more realistic, but more complicated. In the process of this development, we show that every *linear* source can be expressed in two simple equivalent forms.

Figure 4.28(a) shows a voltmeter, a current meter (or ammeter), and an adjustable resistor connected together to measure the voltage-current relation for a flashlight battery. In the actual circuit, the *load* resistance R was changed repeatedly to draw various values of current from the battery. For each value of R, the voltage v_{12} across the battery and the current i through the battery were measured. (The current passing through the voltmeter and the voltage drop across the current meter were negligible.) The resulting data are displayed as + marks in Figure 4.28(b). Note that the current flows through the battery in the direction of potential *rise,* as we would expect for a source.

If the battery had been an ideal voltage source, v_{12} would have remained constant at the 1.5 volts shown on the battery label, regardless of the amount of current drawn from the battery by the resistive load. However, every practical voltage source has only limited ability to maintain a constant voltage. If the resistance of the load is small, a constant-voltage source must provide a large current—and hence, much power—to that load. A physical source responds, in part, by lowering the voltage at its terminals, as observed in

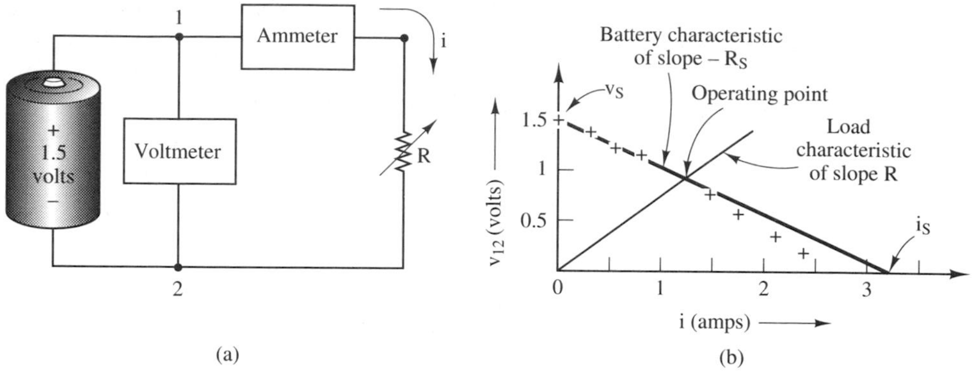

Figure 4.28 Measurement of the v-i characteristic of a flashlight battery. (a) Measurement circuitry. (b) Measured data and straight-line approximation.

Figure 4.28(b), thereby lowering the demand for power. (When the load causes a significant change in the terminal voltage of the source, we say it is *loading* the source.)

The data in the figure can be approximated by a straight line. (The particular straight line shown in the figure is a good approximation for battery currents less than 1.5 A.) Let us denote the voltage and current intercepts of that straight line by v_S and i_S, respectively, and denote the slope by $-R_S$. Then the equation that describes the straight-line *operating characteristic* of the source can be expressed in either the form

$$v_{12} = v_S - R_S i \tag{4.64}$$

or the form

$$i = i_S - \frac{v_{12}}{R_S} \tag{4.65}$$

where $v_S = R_S i_S$. We refer to the **source parameters** v_S, i_S, and R_S as the **open-circuit voltage,** the **short-circuit current,** and the **internal source resistance** of the battery, respectively. For the straight-line approximation in the figure, their values are $v_S = 1.52$ V, $i_S = 3.15$ A, and $R_S = 0.48\ \Omega$.

Although equation (4.65) correctly describes the behavior of the battery in terms of a short-circuit current, the model is not accurate if we actually short-circuit the battery. (In fact, the battery cannot be short-circuited for very long without destroying it.) The particular parameter values associated with the straight line of Figure 4.28(b) are accurate only if the source is used in a system that draws less than 1.5 amps from the battery.

Exercise 4.14: What would be a better set of model parameters for the battery of Figure 4.28 if it is used in an application that draws current in the range $1.5\text{ A} < i < 2.5\text{ A}$?

Figures 4.29(a) and (b) show linear lumped models that correspond to equations (4.64) and (4.65) respectively. These two models should be thought of as equivalent lumped representations of the electrical source of Figure 4.28(a). According to the series

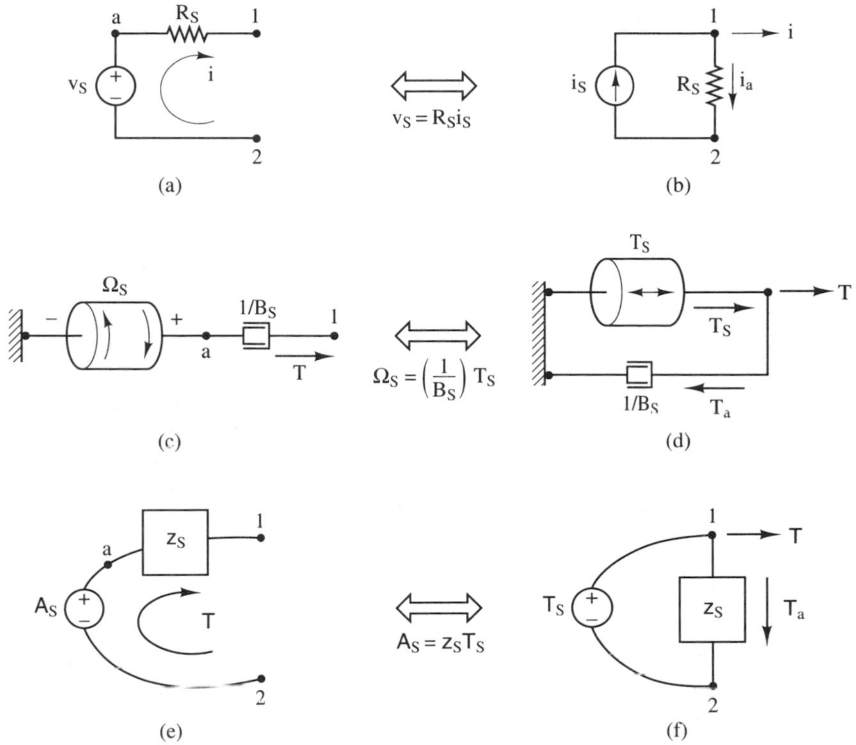

Figure 4.29 Equivalent source-model pairs for linear sources. (a) Voltage-source version. (b) Current-source version. (c) Velocity-source version. (d) Torsion-source version. (e) Potential-difference source version. (f) Flow-source version.

voltage-source model of Figure 4.29(a), the open-circuit voltage v_S of the battery divides. Part of it ($v_{a1} = iR_S$) drops across the internal source resistance. The rest ($v_{12} = iR$) drops across the load resistor R, which is external to the battery. According to the parallel current-source model of Figure 4.29(b), the short-circuit current i_S of the battery divides. Part of it ($i_a = v_{12}/R_S$) flows through the internal source resistance. The rest ($i = i_S - i_a$) flows through the load resistor R.

We emphasize that these two lumped models are precisely equivalent. That is, we can replace one by the other without affecting the behavior of the lumped network that is attached between nodes 1 and 2. The terminal voltage v_{12} and the source current i are the same for both models. We can make such equivalent-model substitutions whenever we find it convenient to switch from series to parallel form or vice versa. Note that node a of the series model is an artificial node: The voltage v_a does not correspond to a physical point in the source. Similarly, the current i_a is not identified with a physical current in the source.

If the internal resistance R_S of the battery were very small, the battery's operating characteristic in Figure 4.28(b) would be nearly horizontal. Then we would call the source a **voltage source,** and the series representation of Figure 4.29(a) would be most appropriate. We might even ignore the small series source resistance and use an *ideal* voltage-source representation. (If we do choose to ignore the small series resistance, we cannot replace the resulting ideal voltage source with an equivalent current-source model. Try it!)

If the internal resistance of the battery were very large, on the other hand, the battery's operating characteristic would be almost vertical. Then we would call the source a **current source,** and the parallel representation would be most appropriate. We might even ignore the large parallel source resistance and use an ideal current-source representation. (If we do use the ideal current-source model, we cannot replace it with a voltage-source equivalent.)

In Figure 4.28(a), a load of resistance R is connected across the battery. The v-i graph of the equation for that load resistor is labeled as the *load characteristic* in Figure 4.28(b). The two straight lines of this figure are the v-i characteristics of the pair of terminals 1 and g as required by the source model and the load model, respectively. The intersection of the *source line* and the *load line* is the system's **operating point,** the solution to the pair of equations. Changing the load resistance R can be viewed graphically as changing the slope of the load characteristic. The change in slope changes the operating point.

The two equivalent source representations describe the behavior of the battery only as seen at its terminals. The models do not indicate how much power is generated (converted from chemical form) or how much of that generated power is dissipated internally (converted to heat within the battery). The models should not be used to evaluate the power dissipated within the physical source; it is not possible to determine the internal workings of the source by external electrical measurements alone.

Exercise 4.15: Show that the equivalent source models of Figures 4.29(a) and (b) give the same value for the power delivered by the battery to the load resistor. Show that they do *not* give the same values for the power produced by the ideal source (v_S or i_S) or the power consumed by the source resistance R_S.

A water wheel is a rotational source. The water in the buckets of the wheel produces a constant torque. An unloaded wheel would accelerate without bound if it were not for bearing friction and air friction. Because that friction increases with speed, the *unloaded* wheel reaches a constant maximum equilibrium speed. If an external load is applied, some of the torque produced by the weight of the water is consumed by that load, and the speed drops. By operating the water wheel with various friction loads and measuring the torque and speed of the *load,* we can obtain an Ω-T operating characteristic that is analogous to the v-i operating characteristic for the battery in Figure 4.28. If that Ω-T operating characteristic is approximately a straight line, we can represent it by two different, but equivalent, lumped models that are analogous to the electrical models. Those models are shown in Figures 4.29(c) and (d). The torsion-source model agrees with intuition; the velocity-source model is just as correct.

A water wheel is a massive device. Its inertia J_S cannot be ignored. Figure 4.30(a) shows that inertia attached to node 1 of the torsion-source model. Since the damper and

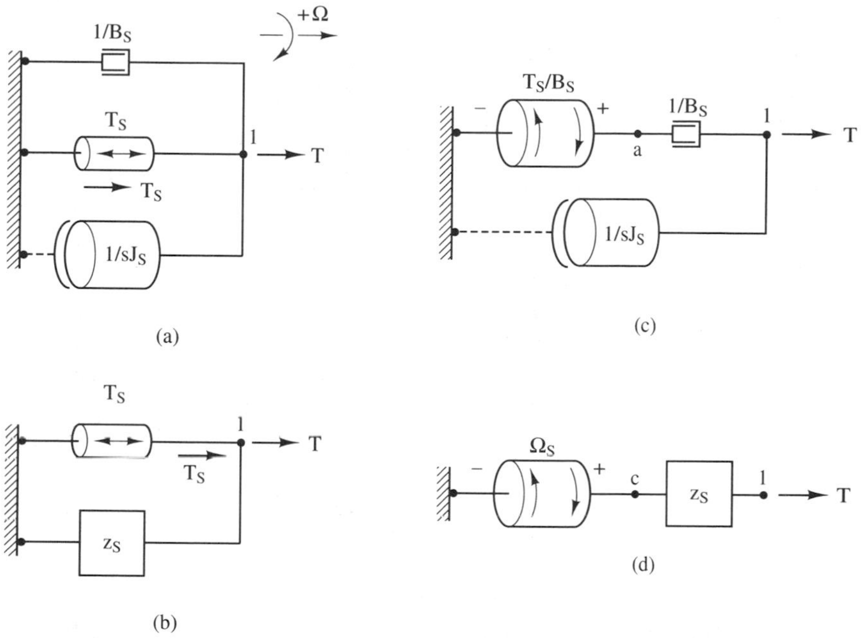

Figure 4.30 Equivalent water-wheel models. (a) Torsion-source model with wheel inertia. (b) Simple equivalent torsion-source model. (c) Velocity-source model with wheel inertia. (d) Simple equivalent velocity-source model.

inertia are connected in parallel with each other, we can replace them by the single equivalent source impedance $z_S = 1/(sJ_S + B_S)$, as shown in Figure 4.30(b). The inertia-free model of Figure 4.29(d) is a special case of the refined model of Figure 4.30(b), for $J_S = 0$.

The series and parallel models of Figures 4.29(c) and (d) are equivalent, with $T_S = B_S\Omega_S$. We attach the wheel inertia J_S to node 1 of the equivalent series model in Figure 4.30(c). Since the refined model of Figure 4.30(a) can be represented by the simple parallel equivalent model of Figure 4.30(b), we should expect to find a corresponding series equivalent representation of the form shown in Figure 4.30(d). In fact, we can, with the modified source signal $\Omega_S \triangleq T_S/(sJ_S + B_S)$. (Show that Figures 4.30(c) and (d) have the same linear relation between Ω_1 and T.) The inertia-free model of Figure 4.29(c) is a special case of Figure 4.30(d), with $J_S = 0$. We examine such equivalence relations in detail later, in the section on Thévenin and Norton equivalent subnets.

It is apparent that equivalent series and parallel source models exist for all types of linear sources. The equations that describe these two equivalent models are, respectively,

$$A_{ij} = A_S - z_S T \tag{4.66}$$

and

$$T = T_S - A_{ij}/z_S \tag{4.67}$$

where A_S is the open-circuit ($T = 0$) potential difference across the source, T_S is the short-circuit ($A_{ij} = 0$) flow in the direction of potential *rise,* z_S is the source impedance, and $A_S = z_S T_S$. The corresponding linear lumped representations are shown in line-graph form in Figures 4.29(e) and (f).

In the steady state, all energy-storage elements behave as open circuits or short circuits, and the equivalent impedance of any subnet is resistive; that is, it does not contain the differential operator s. Consequently, we can plot the *steady-state* potential-difference vs. flow for any physical source, as we did for the battery. The steady-state operating characteristic of a linear source is a straight line.

If a source is nonlinear, the steady-state operating characteristic of that source is a curved line. It usually can be approximated by a straight line over a restricted range of operation—a limited range of values of the potential difference and flow at the terminals of the source. The two equivalent linear source models can still be used, as long as the signals at the terminals do not depart from the range for which the approximation is valid. (See the discussion of linearization in section 5.2.)

The reason for attaching a source to a load is to make the load signals exhibit some desired pattern or behavior. A physical source must drive not only the intended load, but also the passive elements inherent in the source. These elements reduce the capability of the source to cause the desired behavior in the load signal. We can think of the source elements as *degrading elements.* The source impedance describes the collective behavior of these degrading elements. To optimize the desired behavior in the load signal, we must find an appropriate match between the source and load impedances.

Example Problem 4.9: Source-load impedance matching.
Figure 4.31 shows a realistic circuit model for the flashlight of Figure 4.7—a model that includes the resistance of the battery. The open-circuit battery voltage v_S and the battery resistance R_S are fixed. The purpose of the flashlight is to generate light. Assume that the amount of light is proportional to the amount of power consumed by the bulb (resistance R). Choose R so that it maximizes the power transferred from the battery to the bulb.
Solution: According to equation (4.5), the power delivered to the bulb is $P = i^2R$. But the current i depends on the total resistance driven by the source v_S. According to the loop equation, $i = v_S/(R_S + R)$. Therefore, $P = v_S^2R/(R_S + R)^2$. The necessary condition for a maximum is $dP/dR = 0$.

Exercise 4.16: Carry out the differentiation dP/dR and solve the resulting equation to show that P is maximized for $R = R_S$. The load resistance must match the source resistance. This result is known as the *condition for maximum power transfer.*

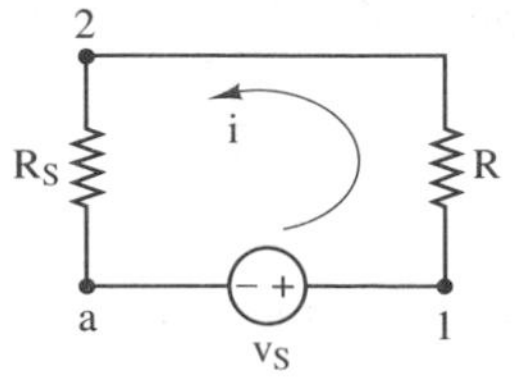

Figure 4.31 A flashlight model.

Choosing a load that matches the source is not the only way to optimize the behavior of a load signal. Another way is to insert an appropriate matching device, such as a gear, a lever, or a transformer, between the source and the load. For example, suppose we wish to use the water wheel of Figure 4.30 to accelerate a heavy load of inertia J. Suppose that the water wheel is initially stationary and that we want maximum load acceleration. If we connect the water wheel directly to the load, the inertias of the source and load combine, and the combination accelerates slowly.

Let us insert a gear between the water wheel and the load to increase the load torque. Then the load inertia appears (to the source) to be reduced. However, the load end of the gear accelerates more slowly than the source end. It can be shown that the load acceleration is maximum for that gear ratio which makes the load inertia *appear* to match the source inertia.

Equivalent-Source Models for Energized Storage Elements

This section examines equivalent representations for energized storage elements—elements that contain stored energy. We discover that stored energy acts like an ideal energy source. Analysis of a network is simpler if we replace stored energy by an equivalent source. The equivalent source can then be treated as a step input to a zero-state network. We examine stored energy in the electrical context first. Then we extend the energy-replacement concept to all types of lumped networks.

A lumped capacitor maintains continuity of potential. That is, the voltage across the capacitor is continuous with time; it cannot change abruptly without infinite current. This fact is signaled by the presence of the derivative on the potential difference in the element equation $i = C\dot{v}_{12}$. At an arbitrary time t_1, the voltage across the capacitor is $v_{12}(t_1)$. We denote this fact graphically by the notation of Figure 4.32(a). Integration of the element equation over the time interval $[t_1, t]$ produces

$$\int_{t_1}^{t} i\,dt = C\int_{t_1}^{t} \dot{v}_{12}\,dt = C[v_{12}(t) - v_{12}(t_1)]$$

or

$$v_{12}(t) = v_{12}(t_1) + (1/C)\int_{t_1}^{t} i\,dt \tag{4.68}$$

The first term on the right side of equation (4.68) is the voltage across the capacitor at the initial time t_1. If $v_{12}(t_1)$ is not zero, the capacitor contains stored electrical energy at that instant. The second term on the right side of the equation describes the voltage behavior of an uncharged capacitor. Thus, equation (4.68) can be viewed as the mathematical description of a constant voltage source of voltage $v_{12}(t_1)$ in series with an uncharged capacitor, as shown in Figure 4.32(b). For $t \geq t_1$, the relations between i and v_{12} are indistinguishable for Figures 4.32(a) and (b). Consequently, we can replace a capacitor charged to voltage v by an equivalent subnet consisting of the uncharged capacitor in series with a constant voltage source of voltage v. We can use this equivalent uncharged-capacitor model in any subsequent analysis of the system.

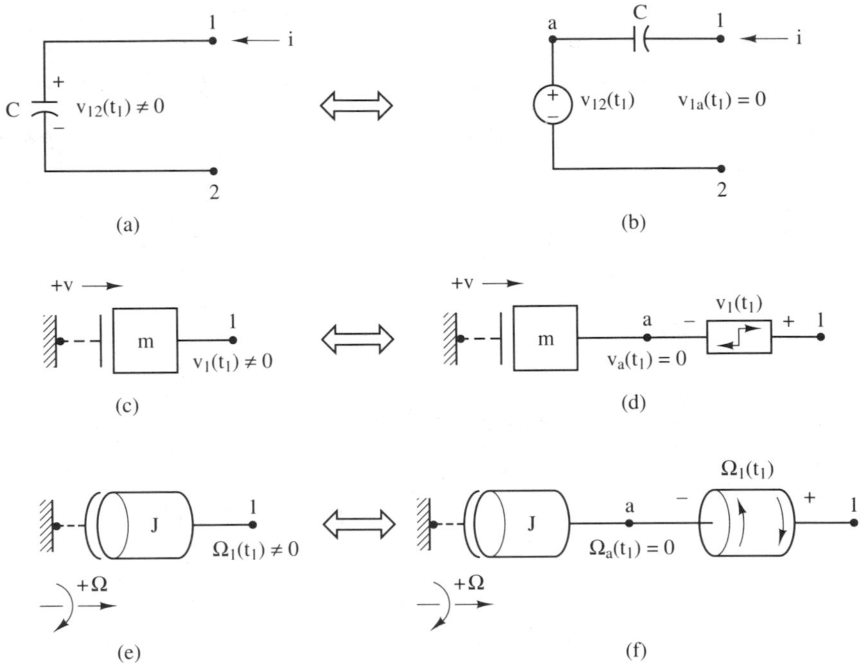

Figure 4.32 Equivalent models for continuous-potential elements. (a) A charged capacitor. (b) An equivalent model with an uncharged capacitor. (c) A moving mass. (d) An equivalent model with an unenergized mass. (e) A rotating inertia. (f) An equivalent model with an unenergized inertia.

The placement of the equivalent voltage source below the uncharged capacitor was arbitrary. The source and uncharged capacitor can be interchanged without changing equation (4.68) or the behavior of the capacitor as seen at terminals 1 and 2. Note that this equivalent model for the charged capacitor is a special case of the model of a linear potential-difference source described by equation (4.66) and Figure 4.29(e).

The circuit to which the capacitor is attached can cause (or permit) current to flow through the capacitor over time. As a consequence, the voltage v_{12} across the capacitor can change over time. The node labeled a in Figure 4.32(b) is an artificial node. It does not correspond to an identifiable point in the physical system. At any instant, the voltage v_{12} across the actual capacitor is the sum of the initial voltage $v_{12}(t_1)$ and the additional accumulated voltage $v_{1a}(t)$, as we illustrate in example problem 4.10.

The energy in any continuous-potential element can be represented as a separate potential-difference source. According to Table 4.3, mass and inertia are the translational and rotational analogs of electrical capacitance. Figures 4.32(c)–(f) show equivalent models for these analogous mechanical elements. In each case, the equivalent model uses a potential-difference source to represent the stored energy. For these mechanical elements, we do not place the equivalent source on the grounded side of the unenergized element. Rather, we reference lumped masses and inertias directly to ground, as discussed in sections 2.1 and 3.1.

Exercise 4.17: Derive the equivalent *stationary-mass* model for the moving mass of Figure 4.32(c) by carrying out an integration equivalent to that shown in equation (4.68).

An energized storage element that maintains continuous flow also can be represented by an equivalent unenergized subnet. Let us examine the representation for an electrical inductor. Integration of the equation $v_{12} = L\, di/dt$ for the inductor of Figure 4.33(a) over the time interval $[t_1, t]$ produces

$$i(t) = i(t_1) + (1/L)\int_{t_1}^{t} v_{12}\, dt \tag{4.69}$$

The first term of equation (4.69) is the current in the inductor at the instant t_1. The second term describes the current flow in an unenergized inductor. According to this equation, the current-carrying inductor can be replaced by an unenergized inductor in parallel with a constant-current source of value $i(t_1)$, as shown in Figure 4.33(b). This equivalent subnet can be used in any subsequent analysis of the network. Similarly, the energy in any

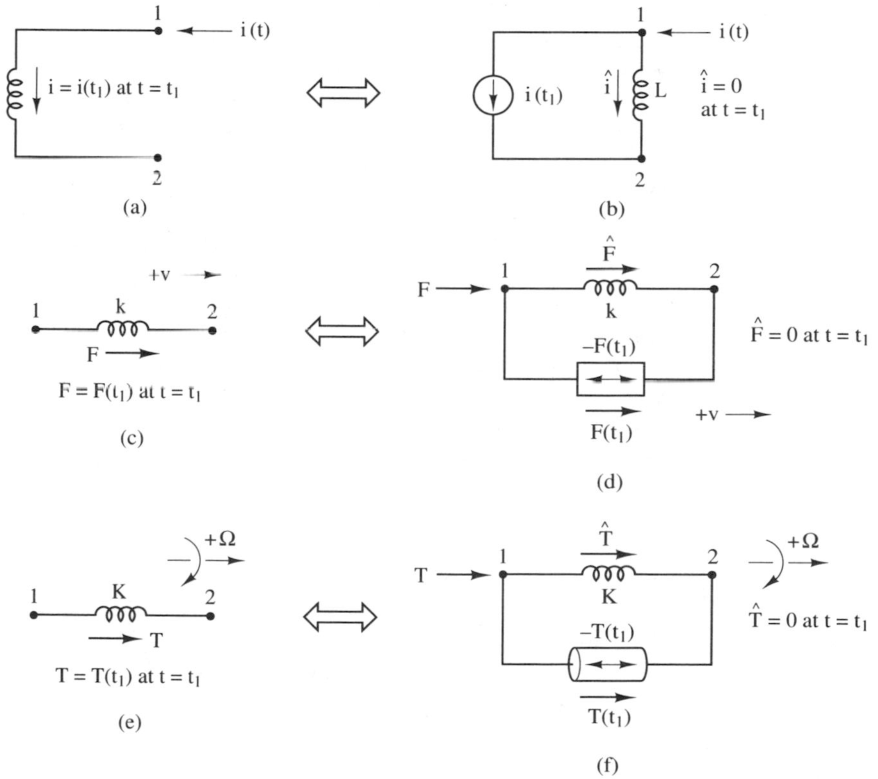

Figure 4.33 Equivalent models for continuous-flow elements; F is compressive force and T is inward-twisting torsion. (a) An energized inductor. (b) An equivalent model with an unenergized inductor. (c) An energized spring. (d) An equivalent model with a relaxed spring. (e) An energized rotary spring. (f) An equivalent model with an untwisted spring.

other continuous-flow element can be replaced by a parallel constant-flow source. Equivalent models for energized springs in the translational and rotational contexts are shown in Figures 4.33(c)–(f).

We found in section 3.4 that the response of a network is the sum of two parts, the zero-state response and the zero-input response. The zero-state response is produced by a source. The zero-input response is produced by stored energy. We now see that energy stored in a lumped element can be interpreted as a constant source. Therefore, if we wish, we can convert all stored energy into equivalent sources and analyze all responses by the zero-state approach.

Example Problem 4.10: Equivalent model for a charged capacitor.
A capacitor previously charged to v volts is connected in series with an uncharged capacitor and a resistor, as shown in Figure 4.34(a). The switch closes at $t = 0$.
(a) Determine the current through the resistor and the voltages across the two capacitors for $t \geq 0$.
(b) How much energy is trapped in the pair of capacitors as $t \to \infty$?
Solution: **(a)** We seek the natural (zero-input) response of the network. First, we replace the initial capacitor energy in C_1 by an equivalent source, as shown in Figure 4.34(b). At $t = 0$, the switch closes, suddenly applying that source to the *unenergized* network.

Let $s \to \infty$ to see that the capacitors initially act as short circuits. Therefore, the initial ($t = 0^+$) capacitor voltages are the same as the prior ($t = 0^-$) capacitor voltages: $v_{1i} = v$ and $v_{2i} = 0$. The initial current through the resistor is $i_i = (v_{1i} - v_{2i})/R = v/R$.

Now let $s = 0$ to see that the capacitors act as open circuits in the final steady state. The final current is $i_f = 0$, and $v_{12f} = i_f R = 0$. The final voltages on the *individual* capacitors are not obvious from Figure 4.34. The voltage v of the equivalent source appears across the series connection of capacitors; that is, $v_{2f} - v_{af} = v$.

According to the rules for combining impedances, the series connection of capacitors is equivalent to a single capacitor of capacitance $C_e = C_1C_2/(C_1 + C_2)$ (see Figure 4.34(c)). Since the capacitors C_1 and C_2 and the equivalent capacitor C_e carry the same current, the stored charge (the integral q of current) is the same for C_1, C_2, and C_e. Hence, as $t \to \infty$, the voltages approached by the various nodes satisfy

$$q_f = C_2 v_{2gf} = C_1 v_{gaf} = C_e v \tag{4.70}$$

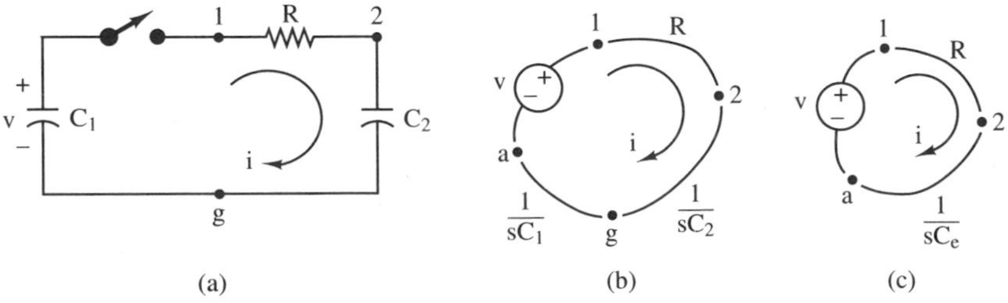

Figure 4.34 Use of an equivalent uncharged model. (a) A circuit with one charged capacitor. (b) An equivalent circuit with uncharged capacitors. (c) An equivalent circuit with a single capacitor.

According to equation (4.70), with $v_g = 0$, the final values of the voltages at nodes 2 and a are $v_{2f} = C_1 v/(C_1 + C_2)$ and $v_{af} = -C_2 v/(C_1 + C_2)$. Thus, the voltage v distributes across the two capacitors in inverse proportion to their capacitances.

The final voltage on the actual (charged) capacitor C_1 of Figure 4.34(a) is the voltage that appears between terminals 1 and g. It is $v_{1f} = v + v_{af} = C_1 v/(C_1 + C_2)$, the same as the final voltage on C_2.

Since the two capacitors act as a single equivalent capacitor, the network is first order. We find the time constant τ of the network from the coefficients of the homogeneous (zero-source) system equation. If we short-circuit the source (set the equivalent source of Figure 4.34(c) to zero), then the loop equation becomes

$$iR + \left(\frac{1}{sC_e}\right)i = 0 \tag{4.71}$$

or

$$s + \frac{1}{RC_e} = 0 \tag{4.72}$$

Therefore, $\tau = RC_e$.

According to the general first-order response equation (3.71), the current in the network is

$$i(t) = i_f + (i_i - i_f)e^{-t/\tau} = \left(\frac{v}{R}\right)e^{-t/RC_e} \tag{4.73}$$

Similarly, the voltage across the actual (charged) capacitor C_1 is

$$\begin{aligned} v_1(t) &= v_{1f} + (v_{1i} - v_{1f})e^{-t/\tau} \\ &= ve^{-t/RC_e} + \left(\frac{C_1 v}{C_1 + C_2}\right)(1 - e^{-t/RC_e}) \end{aligned} \tag{4.74}$$

and the voltage v_2 across the capacitor C_2 is

$$v_2(t) = \left(\frac{C_1 v}{C_1 + C_2}\right)(1 - e^{-t/RC_e}) \tag{4.75}$$

(b) The energy originally stored in the capacitor C_1, owing to the voltage v across the capacitor, was $E = C_1 v^2/2$. There was no energy in C_2. As $t \to \infty$, the voltage v redistributes across the pair of capacitors—or across the equivalent capacitor C_e. The final stored energy is $E = C_e v^2/2$, less than the initial energy because $C_e < C_1$. A second approach to finding the remaining stored energy is to use the final voltages on the two individual capacitors to determine the individual stored energies and then add the results.

Thévenin and Norton Equivalent Subnets

In our earlier examination of realistic sources, we found that any linear source can be represented in two simple equivalent forms. Suppose we partition a linear network into two halves, the source half and the load half. We now show that each half can be represented in the same two simple equivalent forms that we found for sources. We shall find that interchanging these two model forms is the basis of a powerful technique for simplifying networks.

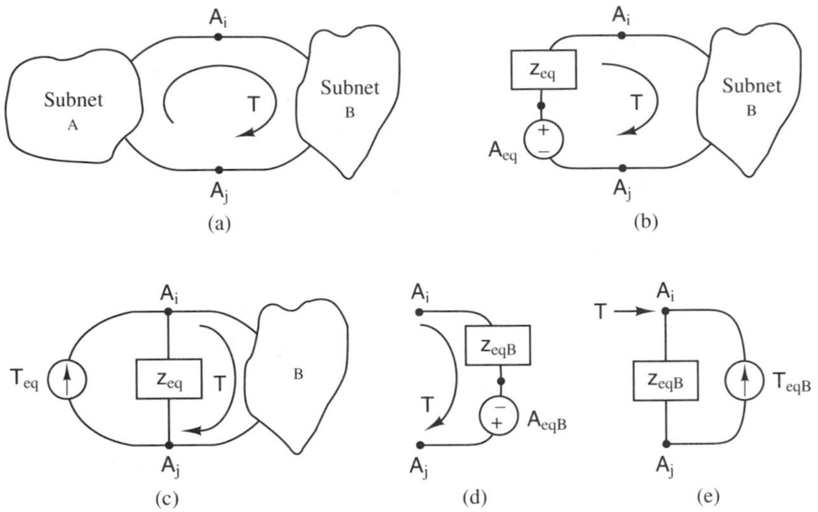

Figure 4.35 Subnets connected at a port. (a) A network partitioned into two subnets. (b) Subnet A replaced by its Thévenin equivalent. (c) Subnet A replaced by its Norton equivalent. (d) Thévenin equivalent of subnet B. (e) Norton equivalent of subnet B.

Let us partition an arbitrary linear lumped network into two subnets at a pair of terminals, as illustrated in Figure 4.35(a). We view the pair of terminals as a **port** through which the two subnets affect each other. At any instant, one of the subnets delivers power to the other. That is, the flow at the port is in the direction of potential rise in one of the subnets and in the direction of potential drop in the other. Thus, we can view one of the subnets as a source subnet and the other as a load subnet.

Suppose we write the linear element equations and node equations for subnet A.[5] Then we eliminate all subnet variables from those equations, except for the port variables A_{ij} and T. We are left with a single linear equation that relates A_{ij} to T. We express that equation in the general linear form

$$A_{ij} = A_{eq} - z_{eq}T \tag{4.76}$$

We call A_{eq} the **equivalent potential** of the subnet and z_{eq} the **equivalent impedance** of the subnet. Both quantities can be functions of the time-derivative operator s.

Equation (4.76) corresponds to the simple series subnet shown on the left side of Figure 4.35(b). Thus, *any* lumped linear subnet that interacts with the rest of the network via a single two-terminal port behaves *at that port* as an ideal potential-difference source A_{eq} in series with an impedance z_{eq}. If the subnet contains no sources, then it behaves as an impedance alone. It is customary to refer to this series representation of a linear subnet as a **Thévenin equivalent** subnet, in recognition of the engineer who is credited with using it in 1883. We also refer to the equivalent potential A_{eq} as the **Thévenin source** for the subnet.

[5] Most lumped elements have two terminals. However, we introduce four-terminal lumped elements in chapter 6. The equivalent network ideas of this section still apply if the linear subnet includes four-terminal elements.

The linear equation (4.76) can be rearranged to the form

$$T = T_{eq} - \frac{A_{ij}}{Z_{eq}} \tag{4.77}$$

where

$$A_{eq} = Z_{eq} T_{eq} \tag{4.78}$$

We call T_{eq} the **equivalent flow** of the subnet. Equation (4.77) corresponds to the simple parallel subnet of Figure 4.35(c). Thus, *any* lumped linear subnet that is connected to the rest of the network via a single two-terminal port behaves *at that port* as an ideal flow source T_{eq} in parallel with the impedance Z_{eq}. This flow-source representation is customarily called the **Norton equivalent** subnet, again, in honor of an early user. We also refer to the equivalent flow T_{eq} as the **Norton source** for the subnet.

Exercise 4.18: Derive equations (4.77) and (4.78) from equation (4.76).

Equations (4.76)–(4.78) assume that the port flow T passes *through* subnet A from node j to node i, as specified in Figure 4.35(a). The equations also assume that the equivalent sources A_{eq} and T_{eq} are oriented (relative to nodes i and j) as shown in Figure 4.35(b) and (c), respectively. Rather than memorize these orientations and the corresponding equivalent-network formulas, we shall find it easier to draw specific equivalent networks of appropriate form and then use conventional network analysis to find the values of the equivalent sources and impedances that produce equivalence to the original subnet.

Exercise 4.19: Thévenin and Norton equivalent representations of subnet B are shown in Figures 4.35(d) and (e), respectively. The orientations of A_{eqB} and T_{eqB} were chosen arbitrarily. Show that the two equivalent representations are described by the equations $A_{ij} = -A_{eqB} + Z_{eqB}T$ and $T = -T_{eqB} + A_{ij}/Z_{eqB}$. Also, show that $A_{eqB} = -Z_{eqB}T_{eqB}$ for these particular equivalent networks.

PROCEDURE FOR FINDING A THÉVENIN OR NORTON EQUIVALENT NETWORK

1. Select the subnet to be replaced. Establish and orient the flow variables for that subnet, including the flow at the port that connects the subnet to the rest of the network.
2. Select a Thévenin or Norton equivalent for the subnet. Choose an orientation for the Thévenin or Norton source.
3. Write the equations that represent the subnet. Eliminate all variables except the potential difference and flow at the subnet port to produce a single subnet equation. Also, write the equation for the Thévenin or Norton equivalent subnet. Compare the equation for the Thévenin or Norton equivalent with the equation for the subnet to determine the values and signs of the equivalent source (A_{eq} or T_{eq}) and the equivalent impedance Z_{eq}.

Let us demonstrate this procedure by an example. Figures 4.36(a) and (b) show a simple translational network and its line graph. Let us find the Thévenin equivalent of the subnet that interacts with the mass. That subnet includes the force source and the damper.

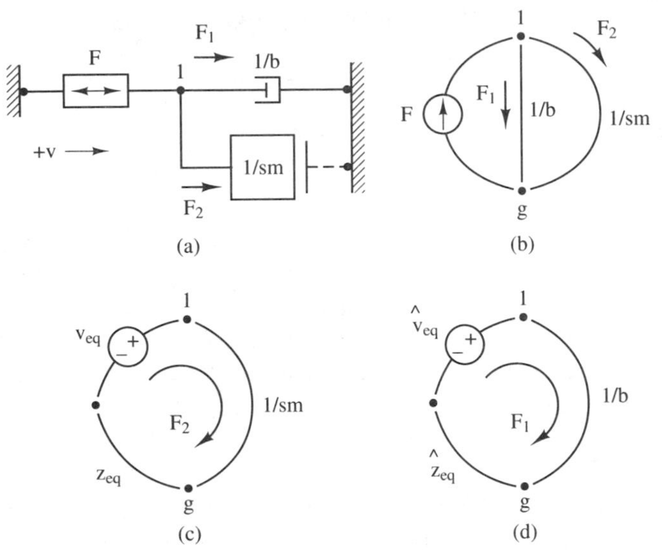

Figure 4.36 Equivalent translational subnets. (a) Lumped network. (b) Line-graph representation. (c) The Thévenin equivalent for the parallel source and damper. (d) The Thévenin equivalent for the parallel source and mass.

As viewed in the translational network, it may seem that the source-damper subnet is already in Thévenin (series) form. The line graph, however, makes clear that the subnet is in Norton (parallel) form. The subnet port parameters are v_{1g} and F_2.

Figure 4.36(c) shows a Thévenin equivalent for the subnet. The equation that represents the Thévenin equivalent is

$$v_{1g} = v_{eq} - F_2 z_{eq} \tag{4.79}$$

The Thévenin source could have been oriented with the + sign at the bottom rather than the top. Then the sign in front of v_{eq} would have been reversed. If the order of the equivalent source and equivalent impedance were interchanged in the model, equation (4.79) would not change.

To write the equations for the subnet, imagine concealing the mass branch of the line graph in Figure 4.36(b). (We must not use element or source parameters from the concealed portion of the network.) The subnet is represented by the single node equation

$$F - F_2 = bv_{1g} \tag{4.80}$$

Compare equations (4.79) and (4.80) to find that $z_{eq} = 1/b$ and $v_{eq} = F/b$.

Exercise 4.20: It can be seen from the forms of v_{eq} and z_{eq} that the Thévenin equivalent of the parallel force source and damper corresponds to a series velocity source and damper. Draw and label the translational network that corresponds to Figure 4.36(c).

Now let us find, instead, the Thévenin equivalent of the subnet of Figure 4.36(b) that interacts with the damper. That subnet includes the force source and the mass. The subnet port parameters are v_{1g} and F_1. Figure 4.36(d) shows a Thévenin equivalent for this subnet. The equation that represents the Thévenin equivalent is

$$v_{1g} = \hat{v}_{eq} - F_1\hat{z}_{eq} \tag{4.81}$$

To write the equations for the subnet, we conceal the damper branch of the line graph. The subnet is then represented by the node equation

$$F - F_1 = smv_{1g} \tag{4.82}$$

Compare equations (4.81) and (4.82) to find that $\hat{z}_{eq} = 1/(sm)$ and $\hat{v}_{eq} = F/(sm)$. The equivalent impedance corresponds to a mass. The waveform of $\hat{v}_{eq}$ is the integral of the force-source waveform.

Exercise 4.21: Draw and label the translational lumped network that corresponds to Figure 4.36(d). Compare it to the network that corresponds to Figure 4.36(c).

The foregoing discussion demonstrates that the equivalent sources A_{eq} and T_{eq} and the equivalent impedance z_{eq} can be functions of s. We then interpret s as d/dt. It is still correct to refer to A_{eq} and T_{eq} as signals. However, the waveforms of the equivalent-source signals are not the same as the waveforms of the original network sources.

In the example presented, we easily associated the mathematical quantities A_{eq}, T_{eq}, and z_{eq} with conventional lumped elements. For more complicated subnets, A_{eq}, T_{eq}, and z_{eq} can be complicated functions of the time-derivative operator s. (We encountered such complicated functions of s in the equivalent-source network of Figure 4.30(d). That network had the Thévenin source $\Omega_S \triangleq T_S/(sJ_S + B_S)$ and the equivalent impedance $z_S = 1/(sJ_S + B_S)$.) Then, the elements of the Thévenin and Norton equivalent networks do not each correspond to a single ideal lumped element. In that event, we merely label the equivalent network (line graph or lumped model) with the derived functions A_{eq}, T_{eq}, and z_{eq}.

Open- and Short-Circuit Methods for Finding Equivalent Networks

The Thévenin equivalent potential A_{eq} for a subnet is the **open-circuit potential difference** of the subnet. That is, if we open-circuit the port (compel $T = 0$), then the potential difference A_{eq} appears across the port terminals. Accordingly, one method for finding A_{eq} for a subnet is to open-circuit the port and then determine the resulting potential difference across the port.

The Norton equivalent flow T_{eq} for a subnet is the **short-circuit flow** of the subnet. That is, if we short-circuit the subnet (compel $A_{ij} = 0$), then $T = T_{eq}$. Accordingly, one method for finding T_{eq} for a subnet is to short-circuit the port and then determine the resulting flow through the port.

If we zero the sources in the subnet (i.e., short-circuit each potential-difference source, open-circuit each flow source, and remove the energy from each storage element), then both equivalent networks reduce to the impedance z_{eq}. Accordingly, one way to find the effective impedance of a linear subnet is to zero all the sources and use series and parallel combination to reduce the remaining subnet to a single impedance z_{eq}.

We will use open-circuiting and short-circuiting techniques, when they are convenient, to help find equivalent networks. Open-circuiting and short-circuiting are particularly effective for converting between Thévenin and Norton forms.

PROCEDURE FOR CONVERTING BETWEEN THÉVENIN AND NORTON FORMS

1. The equivalent impedance Z_{eq} is the same for both the Thévenin and Norton forms. To convert from the Thévenin to the Norton form, change the series impedance to a parallel impedance. To convert from the Norton to the Thévenin form, change the parallel impedance to a series impedance.
2. Suppose the subnet is in Thévenin form. Then A_{eq} and Z_{eq} are known. Short-circuit the port, and find the flow $\mathsf{T}_{eq} = \mathsf{A}_{eq}/\mathsf{Z}_{eq}$ through the short circuit. Orient the Norton source so that the short-circuit flow through the port is in the same direction as it is for the short-circuited Thévenin form. (Or reverse both the orientation and the sign of T_{eq}.)
3. Suppose the subnet is in Norton form. Then T_{eq} and Z_{eq} are known. Open-circuit the port, and find the potential difference $\mathsf{A}_{eq} = \mathsf{Z}_{eq}\mathsf{T}_{eq}$ across the open circuit. Orient the Thévenin source so that the open-circuit potential drop across the port is in the same direction as it is for the open-circuited Norton form. (Or reverse both the orientation and the sign of A_{eq}.)

Let us use this procedure to find the Thévenin equivalent of the Norton form that interacts with the damper in Figure 4.36(b). The Norton source signal is F. The equivalent impedance is 1/(sm). Therefore, the Thévenin equivalent source signal is $v_{eq} = F/(sm)$. To open-circuit the port, we remove the damper. Since the open-circuit potential of the Norton form drops from node 1 to node g, we orient the Thévenin source with the + mark toward node 1, as shown in Figure 4.36(d). Then the potential of the open-circuited Thévenin equivalent drops from node 1 to node g.

Figures 4.32(a) and (b) show a charged capacitor and an equivalent network in which the capacitor is uncharged. This equivalent network is the Thévenin equivalent at instant t_1. We determine, by inspection, that the Thévenin voltage (the open-circuit voltage of the subnet) is the capacitor voltage $v_{12}(t_1)$. If we zero the subnet sources, the subnet should behave at its port like Z_{eq}. The charge on the capacitor acts as a source. We eliminate that stored charge, and the corresponding voltage across the capacitor becomes zero. Then we find, by inspection, that $\mathsf{Z}_{eq} = 1/(sC)$—an uncharged capacitor.

The Norton equivalent of the charged capacitor is not a convenient representation. The Norton source (the short-circuit current) for the charged capacitor is zero, except for a brief infinite impulse. That impulse instantly discharges the capacitor during short-circuiting. Then, during circuit operation, the Norton source instantly recharges the uncharged capacitor.

How, in general, should one compute the values of open-circuit and short-circuit port variables? The straightforward approach is to write loop or node equations for the subnet and eliminate all variables except the remaining unknown port variable. This approach is not much different than the general procedure for finding Thévenin and Norton equivalent networks. However, if we do a few series-parallel source transformations and series or parallel combinations to simplify the subnet before writing the equations, we might be able to reduce the total amount of work. Selecting appropriate simplifications is an art that is acquired by practice. It is usually worthwhile to visualize several reduction sequences first and then carry out one that appears likely to minimize effort.

Sequential Simplification to Thévenin or Norton Form

The electrical network of Figure 4.37 is partitioned into two subnets, denoted A and B, that interact at the port 2,5. Subnet A is a source, represented in Thévenin form. Subnet B includes a source in Norton form. The remaining three resistors connect these two sources.

Two subnets are equivalent if the voltage-current relationships at their corresponding ports are the same. We seek to represent subnet B by a Thévenin equivalent subnet. We obtain that equivalent subnet by successively applying equivalent-source transformations and combining series or parallel elements.

The port of subnet B_1 of Figure 4.37 is to the left of nodes 3 and 4. We use a Norton-to-Thévenin transformation to convert the parallel-source model to series form, as shown in Figure 4.38(a). We make this change in the structure of the subnet to put the 10-Ω source resistance in series with the 3-Ω and 5-Ω resistors of the original circuit. Then those resistors can be combined to simplify the network.

The port of subnet B_2 is to the right of nodes 2 and 5. Figure 4.38(b) shows an equivalent of subnet B_2 that includes the Thévenin equivalent from Figure 4.38(a). Let us interchange the order of the voltage source and the 5-Ω resistor, an interchange that does not affect the behavior seen at the terminals of subnet B_2. We then replace the series combination of three resistors by the equivalent resistor $R = 3\ \Omega + 10\ \Omega + 5\ \Omega = 18\ \Omega$ to put subnet B_2 in Thévenin form. Finally, we convert the Thévenin form of subnet B_2 to Norton form so that the resistor can combine easily with the parallel 20-Ω resistor between nodes 2 and 5 of Figure 4.37. This sequence of transformations of subnet B_2 is illustrated in Figure 4.38(b).

The foregoing transformations simplify subnet B to the form shown in the leftmost circuit of Figure 4.38(c). We replace the parallel resistors in that circuit by a single 9.5-Ω resistor between nodes 2 and 5. Then we transform the Norton form to the required Thévenin equivalent, as originally specified.

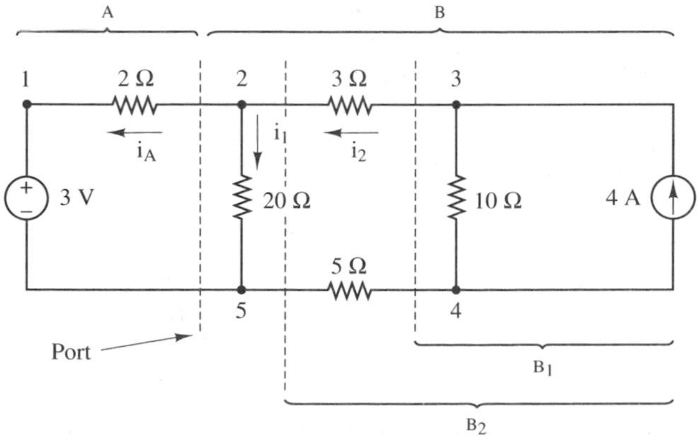

Figure 4.37 A partitioned electrical network.

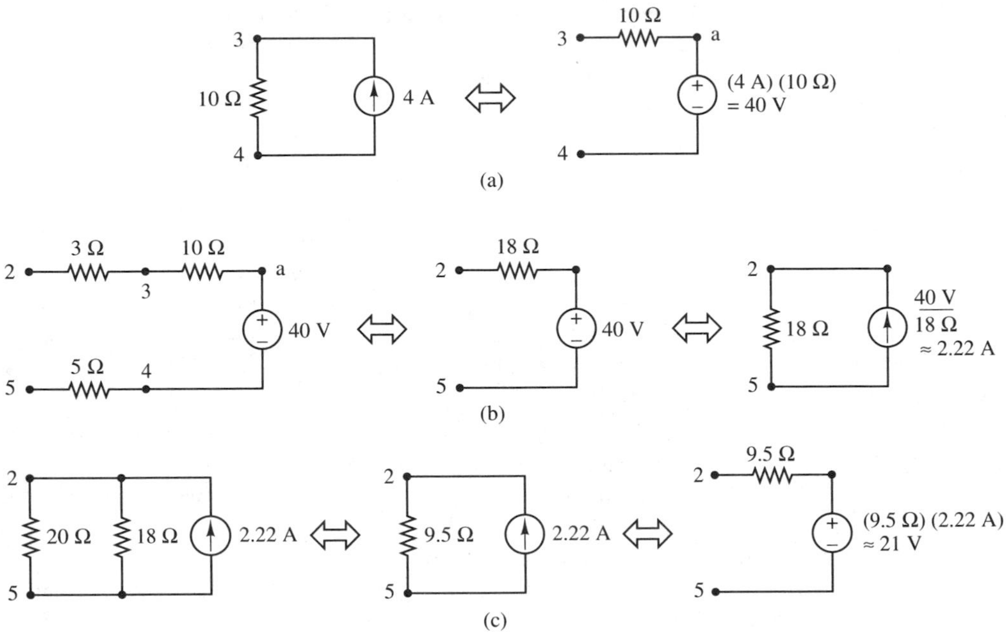

Figure 4.38 Equivalent networks. (a) Equivalent forms for subnet B_1. (b) Equivalent forms for subnet B_2. (c) Equivalent forms for subnet B.

In this final, simplified form, we can see that the source in subnet A receives power from subnet B, because the source in the Thévenin equivalent representation of subnet B is larger than the source in subnet A. We can also see that the resistance of the source in subnet A is not well matched to the resistance of subnet B. Suppose the network is *intended* to transfer power from subnet B to subnet A. Then the significant mismatch in the resistances of the subnets implies that the power transfer is not as high as it could be. (See the discussion of impedance matching in example problem 4.9.) These facts are not obvious by inspection of the original network.

Exercise 4.22: The equivalent network concept equips us to compute the values of all network variables without generating and solving simultaneous equations. Use the Thévenin equivalent of subnet B (Figure 4.38(c)) to show that the port variables in Figure 4.37 have the values $i_A = 1.57$ A and $v_{25} = 6.13$ V. Work backwards through the sequence of subnet models in Figure 4.38 to show that $i_1 = 0.307$ A, $i_2 = 1.88$ A, $v_{45} = -9.38$ V, and $v_{35} = 11.8$ V.

An arbitrarily complex linear subnet composed of sources and impedances can be reduced to either Thévenin or Norton form by a similar sequential simplification process. *The rest of the network need not be linear.* The sequential simplification of a network with energy-storage elements is demonstrated in example problem 4.11. It is appropriate to visualize any linear subnet (mechanical, hydraulic, etc.) in Thévenin or Norton form from the outset. We can obtain either representation by a sequence of steps similar to the one we have just examined.

If we are interested only in the *impedance* of a subnet, the sequential process can be simplified further. (For example, we might wish to determine the impedances of a source and load to see how well they match.) We can see from the preceding sequence of transformations that Thévenin and Norton equivalent sources arise from energy sources in the subnet. If we zero the subnet sources first, we can find z_{eq} by series and parallel combination alone. Zeroing the sources reduces the effort required to find z_{eq}.

For example, suppose we isolate subnet B of Figure 4.37 and set the current source to zero. (We open the 4-A branch.) Then the network observed at the port 2, 5 is a 20-Ω resistor in parallel with a series of three resistors. By inspection, the three series-connected resistors can be replaced by a single 18-Ω resistor. The equivalent resistance of the subnet is the parallel combination, $R_{eq} = (18\Omega)(20\Omega)/(18\Omega + 20\Omega) = 9.5\ \Omega = z_{eq}$.

Example Problem 4.11: Analysis by network simplification.
The input-output system equation for the voltage v_5 of Figure 4.18 is derived by loop and node methods in example problem 4.5. Derive the same input-output system equation by a sequence of Thévenin and Norton transformations and series or parallel combinations.

Solution: A line graph for the network, with its branches labeled in impedance notation, is shown in Figure 4.39(a). (The network and the succeeding simplified equivalents could be displayed in electric circuit notation. The line-graph notation, by hiding the electrical nature of the network, makes the simplification steps more clearly applicable to other energy domains.) We use a sequence of transformations to reduce the left side of the network to a simple loop. We retain node 5 and the ground node in the simplified network in order to keep v_5 in the loop equation.

We replace the Thévenin source model on the left by a Norton equivalent, and then we combine parallel resistors and convert the resulting Norton form to Thévenin form, as shown in Figure 4.39(b). The result is a single-loop circuit. We reverse the order of the impedance sL and R_L (without affecting v_5) to obtain the circuit of Figure 4.39(c). Finally, we combine the series resistors, as shown in Figure 4.39(d). Only nodes 5 and g are nodes of the original line graph.

Figure 4.39(d) defines the loop current i. (This is the current in the capacitor of the original network.) The loop equation is

$$v_e = \left(R_T + sL + \frac{1}{sC}\right) i \tag{4.83}$$

We use the capacitor equation $i = sCv_5$ and the equations for v_e and R_T from Figure 4.39 to substitute for i, v_e, and R_T in equation (4.83). Then we multiply by $(R_2 + R_S)$ to get the system equation

$$[(R_2 + R_S)LCs^2 + (R_2R_S + (R_2 + R_S)(R_1 + R_L))Cs + (R_2 + R_S)]v_5 = R_2v_S \tag{4.84}$$

Compare this result with equation (4.41).

Once we become familiar with the simplification procedure, we can skip some of the steps to make the process easier. For example, we can convert the first Thévenin form to a Norton form and combine the resulting parallel impedances all in one step. Also, we

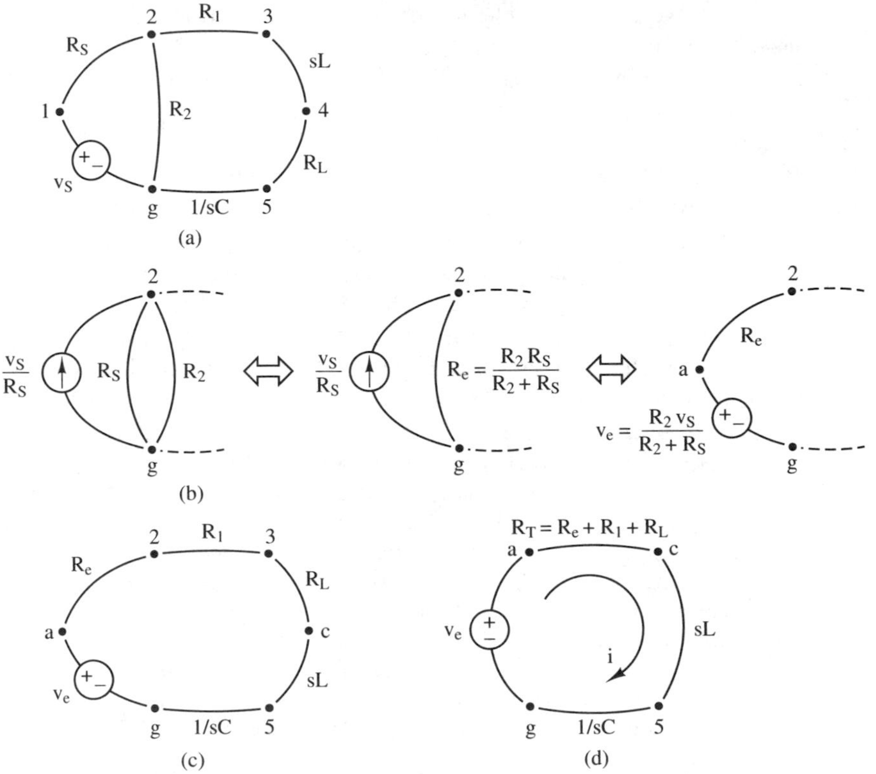

Figure 4.39 Network simplifications. (a) Line graph for Figure 4.5. (b) Simplification of the left side. (c) Simplified equivalent network. (d) Another equivalent network.

can combine the series impedances of Figure 4.39(c) while we interchange sL and R_L. If we perform the simplifications with specific parameter values, rather than parameter symbols, then the reduction in network complexity is more dramatic because the equivalent parameters do not become so complicated as we proceed through the sequence of steps.

Steady-State Operating Characteristics

Let us partition the battery circuit of Figure 4.28(a) into two halves at the port 1,2. (Ignore the voltmeter and ammeter.) Then the source characteristic of Figure 4.28(b) describes the left half of the circuit (the battery) and the load characteristic describes the right half (the resistor). These two straight lines fully define the behavior of the circuit *as seen from the port 1, 2*. Their intersection uniquely determines the circuit's operating point—the values of the two port variables v_{12} and i.

In a similar fashion, we can partition any network into two halves at the port i, j, as in Figure 4.35(a). Then we can represent each half by an equation that relates the port variables A_{ij} and T. Each equation can be viewed as the **operating characteristic** of the corresponding subnet. The two operating characteristics form a set of system equations for the network. The solution (A_{ij}, T) of the equations constitutes the network behavior as seen at the port.

If the operating characteristic (equation) for a subnet is linear, that subnet can be represented in either Thévenin or Norton form. Typically, the equivalent sources and equivalent impedances associated with the two subnets are functions of the time-derivative operator s. Hence, the port signals A_{ij} and T are usually functions of time.

If all the network elements are resistive (dissipative) and the network sources are constant, then the equivalent impedances and equivalent sources for the two subnets are real numbers, rather than functions of s. In that case, the network is not dynamic, and the two operating characteristics correspond to two lines on a graph of A_{ij} vs T. (The battery and load characteristics of Figure 4.28(b) are an example.) The intersection of the two lines is the network *operating point* (as seen at the port that connects the two subnets).

If a subnet includes energy-storage elements, we cannot fully characterize the subnet equation by a graph that relates the port variables. On the other hand, for constant sources in a steady state (s = 0), energy-storage elements behave as open circuits or short circuits. Then the equivalent impedances of the two subnets are resistive and the equivalent sources are real valued. Therefore, we can always represent the **steady-state operating characteristics** of the two subnets by two graphs that relate the port variables. (The network sources must be held constant for the network to achieve a steady state.) In Figure 4.40, we show such steady-state graphs in the notation of Figure 4.35. Since the graphs are straight lines, each graph has both a Thévenin and a Norton representation.

If we change the (constant) value of the source signal for one of the subnets, we cause a change in the intercepts (the Thévenin and Norton source values) for the corresponding straight line, but no change in the slope of the line (the equivalent resistance of the subnet). If we change the value of a dissipative element in a subnet, we change the equivalent resistance of the subnet and the slope of the corresponding straight line. Such changes in slopes and intercepts move the network operating point. We shall find in section 6.3 that designing the steady-state operating characteristic is an important step in the design of power amplifiers.

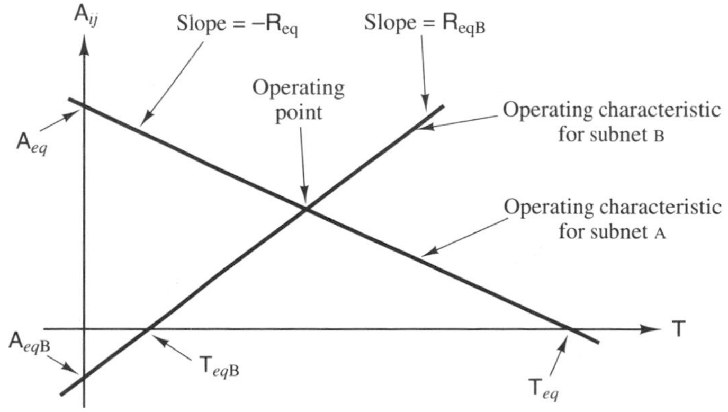

Figure 4.40 Steady-state operating characteristics.

4.3 THE PRINCIPLE OF SUPERPOSITION

The analysis in section 3.4 shows that the response of a network is the sum of the zero-state and zero-input responses. The zero-input response is the response owing to initial stored energy. In the analysis, we found the response to the sources and the response to the initial energy state separately and then added them to obtain the total response. Section 4.2 shows that energy stored in an element can be interpreted as an additional constant source. Thus, the zero-input response is the response to stored-energy *sources*. We add the response owing to stored-energy sources and the response owing to true (continuing) sources to obtain the total response.

This license to treat the individual causes of response separately is known as the **principle of superposition**. By that principle, we can determine the response of a *linear network* to each source or energized storage element separately and then add the results to obtain the response when all of the (continuing) sources and stored energies act simultaneously. We can even break a complicated individual source waveform into a sum of simpler components and determine separately the response to each component waveform. Superposition reduces the complexity of computations and simplifies the manner in which we conceive of the behavior of linear networks. We derive and demonstrate the superposition principle in this section.

Multiple Sources

In section 3.4, we showed how to solve an input-output system equation for a linear network with a single source. (We found solutions in detail only for step inputs, i.e., constant sources.) Some networks have multiple sources. In section 4.2, we learned that the energy in a storage element is equivalent to an additional constant source. We now show that the response of a multiple-source linear network is the sum of the responses to the individual sources, each acting alone. The effects of the sources are independent of each other because the network is linear.

Suppose that y is a variable of a linear network that has two sources. Let u_1 and u_2 denote the two source signals. Since the network is linear, the element equations can be expressed in impedance notation and the system equations involve polynomials in s. Let us eliminate all the dependent variables except y from the system equations. Then we arrange the resulting input-output system equation in the conventional form, with the dependent-variable terms on the left side and the source-signal terms on the right. The coefficients of the left side now form the characteristic polynomial. We group the terms on the right side so as to separate the u_1 terms from the u_2 terms. Then we divide the input-output system equation by the characteristic polynomial to express y as a function of s, u_1, and u_2. The resulting system equation must have the form

$$y = H_1(s)u_1 + H_2(s)u_2 \tag{4.85}$$

where $H_1(s)$ and $H_2(s)$ are input-output transfer functions from u_1 and u_2, respectively, to y.

Both transfer functions have the same denominator—the characteristic polynomial for the network. The roots of that denominator are the characteristic values for the network. The numerator of H_1 contains the right-side coefficients for u_1 and its derivatives. The numerator of H_2 contains the right-side coefficients for u_2 and its derivatives.

Suppose that $u_2(t) = 0$; that is, we turn off the second source. Then, just like the system equations of section 3.4, the system equation has a single input, and we can find the response by the methods of that section. We denote the solution by $y_1(t)$. Then, according to equation (4.85), y_1 is the solution to the equation $y_1 = H_1u_1$. Next, we let $u_1(t) = 0$, and solve for the corresponding solution $y_2(t)$. This solution satisfies $y_2 = H_2u_2$. Summing these two equations shows that $y_1 + y_2$ equals the right side of equation (4.85), or $y = y_1 + y_2$. Hence, we can find the component of y owing to each source separately and then add the components to get the total solution. The two sources affect the network variables independently.

The same logic applies to any number of sources. Because an energized storage element is equivalent to a constant source, the logic also applies to energized storage elements. To find the response of a linear network with any number of sources and any number of energized storage elements, we break the problem into pieces. Then we solve for the response to each source and stored energy separately and add the results.

We used the foregoing principle of superposition in section 3.4 when we decomposed the network response into a zero-input response and a zero-state response. The principle depends on the linearity of the network being analyzed. A nonlinear element cannot be represented in impedance notation. A network with a nonlinear element cannot be expressed in terms of the polynomials in s that make up the right side of equation (4.85). The hydraulic system of example problem 5.4 illustrates the difficulties caused by nonlinear elements.

Simplification of Analysis by Superposition

The network of Figure 4.41(a) has two constant sources and two energized storage elements. First, we will use this network to show the effect of multiple sources. Then we will use it to show how superposition simplifies the solution process. The initial energy states of the storage elements are F_{1c} and v_{3c}. We replace these energy states by equivalent constant sources in the network line graph, Figure 4.41(b).

The line graph has three essential nodes—that is, nodes with more than two branches. The potentials of the remaining nodes, 1 and a, are known relative to the essential nodes g and 3. Therefore, we need only write the node equations for two essential nodes. We choose

$$\begin{aligned} \text{Node 2:}\quad & bv_{12} = \left(\frac{k}{s}\right) v_{23} + F_{1c} \\ \text{Node g:}\quad & bv_{12} = F_c + smv_{ag} \end{aligned} \tag{4.86}$$

From these two equations, together with the conditions $v_1 = v_c$, $v_{3a} = v_{3c}$, and $v_g = 0$, we find that v_3 satisfies

$$\left(s^2 + \frac{k}{b}s + \frac{k}{m}\right) v_3 = \frac{k}{m}v_c - \frac{k}{mb}F_c - \frac{s}{m}F_c + \frac{s}{m}F_{1c} + \frac{sk}{b}v_{3c} + s^2v_{3c} \tag{4.87}$$

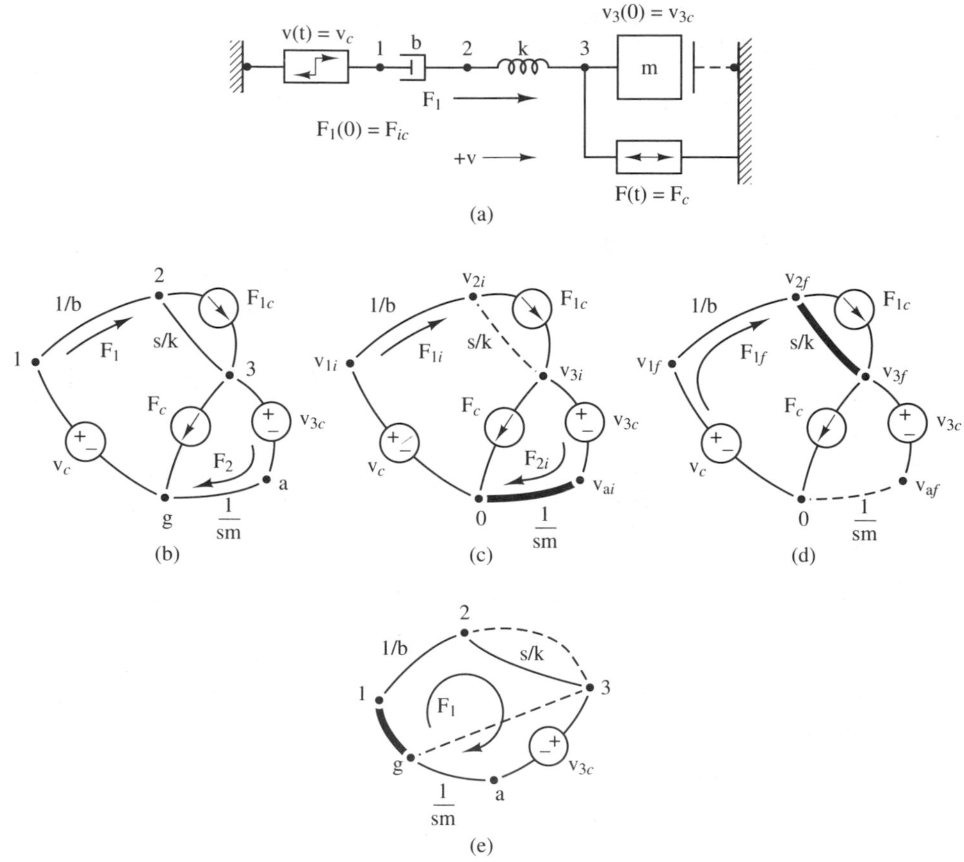

Figure 4.41 A multisource network. (a) Lumped network. (b) Zero-state line-graph representation. (c) Initial-condition line graph. (d) Final-condition line graph. (e) Line graph for energy state v_{3c} alone.

The source variables v_c, F_c, F_{1c}, and v_{3c} all appear on the right side of the system equation. (They should all be treated as constant-valued functions, rather than numbers.) Furthermore, each appears linearly, in separate terms. Because system equations are *linear* combinations of the element equations, two different source variables cannot appear in the same term on the right side.

Since all the sources variables are constant, the differentiated terms—those multiplied by some power of s—disappear from the right side of the system equation for $t > 0$. Note that only truly independent sources, as opposed to sources derived from energy states, continue to affect the response in the steady state. Hence, those sources determine the final values. The sources derived from energy states influence the transient behavior via initial conditions.

We get the characteristic equation by setting the right side of equation (4.87) equal to zero. We solve that equation for ζ and ω_n. Since we use the system equation only to find the characteristic equation, it would be more efficient to set all sources to zero in the line graph before deriving the characteristic equation.

We apply the open-circuit and short-circuit rules of section 3.3 to obtain the initial-value line graph, Figure 4.41(c). This line graph is used to calculate the *initial jumps* in the values of the variables. The initial jumps must be added to the prior values to get the initial values. However, since the energy states have been converted to equivalent constant sources, the prior values of all storage elements are zero. Therefore, rather than use the δ (jump) notation, we label the variables of Figure 4.41(c) directly as initial values. We use this graph to find the initial conditions v_{3i} and $\dot{v}_{3i}$.

We observe the initial value

$$v_{3i} = v_{3c} \tag{4.88}$$

by inspection. The initial rate $\dot{v}_{3i}$ is determined from the mass equation $F_{2i} = m\dot{v}_{ai} = m\dot{v}_{3i}$ (since $v_3 - v_a = v_{3c}$, a constant). The node equation at node 3 requires that $F_{2i} = F_{1c} - F_c$. Therefore, the initial rate is

$$\dot{v}_{3i} = \frac{F_{1c} - F_c}{m} \tag{4.89}$$

Since the source signals are all constant valued (step inputs), the particular solution is equal to the final value. We determine the final value of v_3 from the final-value line graph, Figure 4.41(d). Observe that $v_{3f} = v_{2f} = v_c \quad (1/b)F_{1f}$. The node equation at node g implies that $F_{1f} = F_c$. Therefore,

$$v_{3f} = v_c - \frac{F_c}{b} \tag{4.90}$$

We see in equations (4.88)–(4.90) that the source magnitudes and energy states appear linearly, in separate terms of the initial and final values. This separation of individual sources and states is a consequence of the linear combining used to determine the initial and final values. Both of the true-source values appear in the final value. In general, for constant sources, we should expect each source magnitude to appear in the final value of some network variable. True-source magnitudes *might* also appear in the initial values for some of the network variables.

In contrast to the true-source values, neither of the energy-state values appears in the final-value equation (4.90). In general, we should expect each energy-state value to appear in some initial condition, but energy states usually do not appear in the final values.

The total response of the network to the set of constant sources and energy states, according to equation (3.137), is

$$\begin{aligned} v_3(t) &= \left(v_c - \frac{F_c}{b}\right)\mathscr{A}(t) + v_{3c}\mathscr{B}(t) + \left(\frac{F_{1c} - F_c}{m\omega_n}\right)\mathscr{C}(t) \\ &= v_c\mathscr{A}(t) - F_c\left[\frac{\mathscr{A}(t)}{b} + \frac{\mathscr{C}(t)}{m\omega_n}\right] + v_{3c}\mathscr{B}(t) + F_{1c}\left(\frac{1}{m\omega_n}\right)\mathscr{C}(t) \qquad \text{for } t \geq 0 \end{aligned} \tag{4.91}$$

The response functions $\mathcal{A}$, $\mathcal{B}$, and $\mathcal{C}$ of equation (4.91) depend on the characteristic roots ζ and ω_n. Observe that the total response is a sum of the responses owing to the separate sources and energy states. There is no interaction between the sources. If only one source is nonzero, the component of the solution owing to that source is not changed. We can find each component separately without changing the result. If the true-source functions had not been constant, the particular solution would be different, but the total response would still be a sum of responses owing to the separate sources and states.

The preceding analysis demonstrates that the total response is a sum of terms owing to the separate causes of the response. However, the analysis does not *take advantage* of the principle of superposition. Finding the individual components of response separately is simpler, both conceptually and computationally.

Let us return to the line graph of Figure 4.41(b). To find the response only to the energy stored in the mass, we short-circuit the potential-difference source v_c, open-circuit the flow source F_{1c}, and open-circuit the flow source F_c, as illustrated in Figure 4.41(e). To obtain the characteristic equation, we short-circuit the remaining source v_{3c} of Figure 4.41(e) and, by inspection, write the loop equation:

$$\left(\frac{s}{k} + \frac{1}{b} + \frac{1}{sm}\right)F_1 = 0 \tag{4.92}$$

Then we multiply the equation by ks and equate the left-side polynomial to zero.

To find the final value of v_3, we open-circuit the mass and short-circuit the spring in Figure 4.41(e). By inspection, $v_{3f} = v_{2f}$. Then, because $F_{1f} = 0$, $v_{2f} = v_{1f} = 0$.

To find the initial values, we short-circuit the mass and open-circuit the spring in Figure 4.41(e). We treat the variables in the line graph as initial values rather than jumps, because the prior state is zero. By inspection, $v_{ai} = 0$ and $v_{3i} = v_{3c}$. According to the mass equation, $F_{1i} = m\dot{v}_{ai} = m(\dot{v}_{3i}) = 0$, or $\dot{v}_{3i} = 0$.

Finally, we apply the general solution equation (3.137) to produce

$$v_3(t) = v_{3c}\mathcal{B} \tag{4.93}$$

where the particular $\mathcal{B}$ curve is determined by the roots ζ and ω_n of the characteristic equation.

This simplified procedure generates the part of the response (4.91) that results from the initial energy in the mass. The other three parts of the response can be determined even more easily, because we need not determine the characteristic equation again.

Exercise 4.23: Find the initial values of v_3 and $\dot{v}_3$ and the final value of v_3 for each of the sources v_c, F_c, and F_{1c} of Figure 4.41(b) *acting alone*. Verify that the initial and final values associated with the individual sources and energy states sum to equations (4.88)–(4.90).

Exercise 4.24: Use superposition to find the response of the variable F_1 to the simultaneous sources and energy states of Figure 4.41(a).

The following procedure summarizes how to use superposition to find the response of an energized *linear* network to a number of simultaneous sources.

SUPERPOSITION PROCEDURE

1. Represent the energy in each storage element by an equivalent constant source. (Then the network is in the zero state.) Use equivalent-network concepts to simplify the network if desired.
2. Zero all sources (by open-circuiting flow sources and short-circuiting potential-difference sources) and find the characteristic equation and characteristic roots.
3. Focus attention on a single source and zero all other sources. Let $s \rightarrow \infty$ in the network impedances to find the initial value(s) of the variable(s). For a constant source, let $s = 0$ in the network impedances to find the final value(s) of the variable(s). Then find the response of the network variable(s) of interest to the single nonzero source.
4. Repeat step 3 for each of the sources in the model. Sum the responses owing to the various sources to obtain the response for all sources acting simultaneously.

Source Decomposition

In the previous section, we simplified network analysis by dealing with multiple sources individually. In this section, we show that we can simplify network analysis further by breaking a single *complicated* source into a sum of simpler sources. Then we can deal with each simpler source separately.

Suppose, for example, that the velocity-source signal v(t) of Figure 4.41(a) is the pulse waveform shown in Figure 4.42(a). The latter figure shows that the pulse can be treated as the sum of two step functions:

$$v(t) = v_a(t) + v_b(t) \tag{4.94}$$

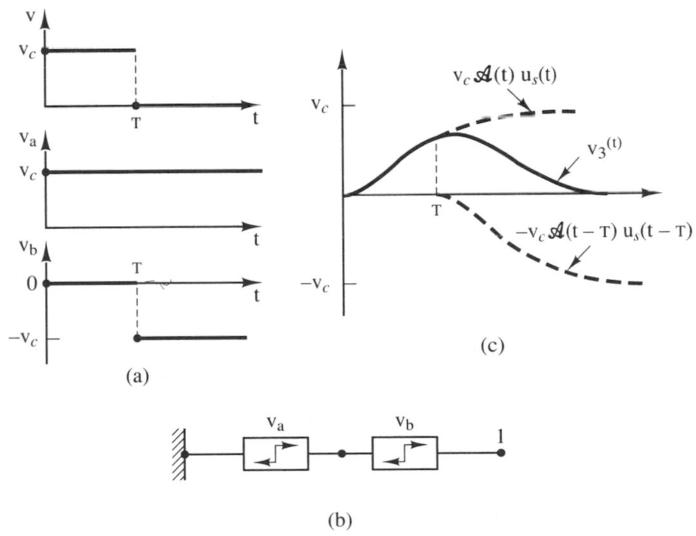

Figure 4.42 Decomposition into simpler sources. (a) Decomposition of a pulse waveform. (b) Decomposition of a source into a sum of sources. (c) Response to the pulse waveform.

The step function v_b is inverted and delayed by τ seconds relative to the step function v_a. Therefore, we can replace the velocity source symbol of Figure 4.41(a) by the series connection of velocity-source symbols shown in Figure 4.42(b). In fact, we can decompose *any* signal into a sum of components in this manner. Only if the network is linear, however, does the principle of superposition permit us to take advantage of this decomposition. For a linear network, we can find the responses to the two different step inputs separately and then add the responses.

We shall denote the **unit step function** by

$$u_s(t) = \begin{cases} 0, & t < 0 \\ 1, & t \geq 0 \end{cases} \tag{4.95}$$

Then we can express the step functions of Figure 4.42(a) and equation (4.94) as $v_a(t) = v_c u_s(t)$ and $v_b = -v_c u_s(t - \tau)$. The factor $u_s(t - \tau)$ provides a way to denote, in a simple equation, that a function is zero until $t = \tau$.

Exercise 4.25: Let the waveform of the force source of Figure 4.41(a) be a pulse of magnitude F_c and duration τ. Show that this pulsed source can be represented by a parallel connection of force sources, each of which has the form of a step input.

Most of the networks that we consider in this book are **time invariant.** That is, the lumped elements do not change with time, and the differential equations have constant coefficients. If the source in an unenergized time-invariant system is delayed by τ seconds, the system merely responds τ seconds later. The response changes from $y(t)$ $(t \geq 0)$ to $y(t - \tau)$ $(t \geq \tau)$. Section 3.4 showed how to find the zero-state step response of a linear network. We can use the time-shift notation to convert that zero-state step response into the response for the delayed-step input.

The response $v_3(t)$ of the network of Figure 4.41(a) to the step function $v_c u_s(t)$ is given by the first term of equation (4.91) for $t \geq 0$. It is zero for $t < 0$. We can express it as $v_c \mathcal{A}(t) u_s(t)$. Therefore, the response of node 3 to the step function $v_a(t) = v_c u_s(t)$ is also $v_c \mathcal{A}(t) u_s(t)$. The response of node 3 to the negated and delayed step function $v_b(t) = -v_c u_s(t - \tau)$ is $-v_c \mathcal{A}(t - \tau) u_s(t - \tau)$, also negated and delayed by τ seconds. By superposition, the response of node 3 to the pulse waveform $v(t)$ is the sum of the two step responses:

$$v_3(t) = v_c[\mathcal{A}(t) u_s(t) - \mathcal{A}(t - \tau) u_s(t - \tau)] \tag{4.96}$$

This response is illustrated in Figure 4.42(c) for the case of critical damping ($\zeta = 1$) and for $\omega_n = 2/\tau$.

Exercise 4.26: Sketch $u_s(t)$, $u_s(t - 1)$, $e^{-t} u_s(t)$, $e^{-(t-1)} u_s(t - 1)$, and $-e^{-(t-1)} u_s(t - 1)$.

We have shown how to decompose a source waveform into a sum of simpler waveforms. The concept aids network analysis by enabling us to handle a larger set of source waveforms. We can easily find the response to any waveform that decomposes neatly into a few step functions.

Response to an Impulse

In this section, we expand further the set of source waveforms we can handle by finding the response to an impulse—a strong, brief burst of energy. We show that the impulse response is the derivative of the step response. We can find either response from the other. The ability to find the response to one source waveform from the response to another stems from the linearity of the network.

Let us apply to the unenergized network of Figure 4.43(a) the brief scalar-force pulse F(t) shown in Figure 4.43(b). The pulse has width τ and area A. (The area of the pulse is sometimes referred to as the *impulse*. It has the units of momentum, N·s = kg·m/s; since the pulse is brief, its area equals the change in momentum of node 1.) The rectangular pulse F can be decomposed into the sum of two step functions:

$$F(t) = \frac{A}{\tau}u_s(t) - \frac{A}{\tau}u_s(t - \tau) \text{ newtons} \tag{4.97}$$

where u_s is the *unit step* function of equation (4.95).

The system equation for the translational motion of the network of Figure 4.43(a) is

$$\dot{v}_1 + \frac{b}{m}v_1 = \frac{F}{m} \tag{4.98}$$

The initial velocity of the unenergized network is $v_{1i} = 0$. The final velocity for the unit step source signal ($F = u_s$ newtons) is $v_{1f} = 1/b$ m/s. Therefore, the *unit*-step response of the unenergized network is

$$v_1(t) = \frac{1}{b}(1 - e^{-bt/m}) \text{ m/s} \tag{4.99}$$

By the principle of superposition, the response of the network to the difference of step functions in equation (4.97) is

$$\begin{aligned} v_1(t) &= \frac{A}{\tau b}(1 - e^{-bt/m}) \text{ m/s}, \qquad 0 \le t < \tau \\ &= \frac{A}{\tau b}(1 - e^{-bt/m}) - \frac{A}{\tau b}(1 - e^{-b(t-\tau)/m}) \text{ m/s}, \qquad t \ge \tau \end{aligned} \tag{4.100}$$

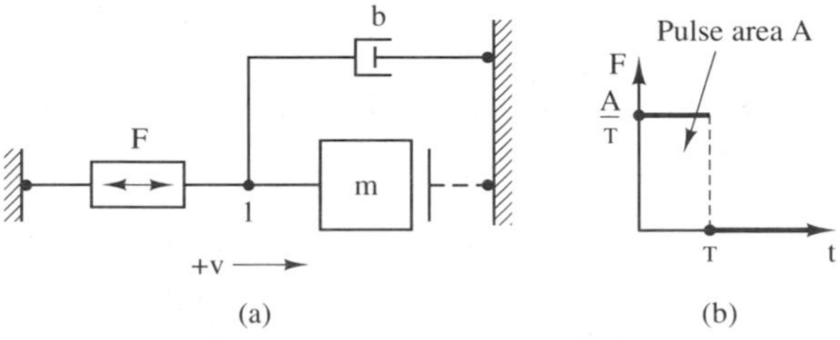

Figure 4.43 A network with an impulsive source. (a) Lumped network. (b) Source waveform.

The portion of v_1 for $t \geq \tau$ can be expressed as

$$v_1(t) = \frac{A}{\tau b}[e^{-b(t-\tau)/m} - e^{-bt/m}]$$

$$= \frac{A}{\tau b}e^{-bt/m}[e^{b\tau/m} - 1] \text{ m/s}, \qquad t \geq \tau \tag{4.101}$$

According to the Maclaurin series expansion of the exponential,

$$e^{b\tau/m} - 1 = \frac{b\tau}{m} + \frac{1}{2!}\left(\frac{b\tau}{m}\right)^2 + \cdots \tag{4.102}$$

If we make the duration τ of the pulse much smaller than the time constant of the system, i.e., $\tau \ll m/b$, then the series expansion given in equation (4.102) can be approximated by its first term alone. Except for the brief interval during which the pulse is applied, then, the response can be expressed as

$$v_1(t) \approx \frac{A}{\tau b}e^{-bt/m}\left[\frac{b\tau}{m}\right] = \frac{A}{m}e^{-bt/m} \text{ m/s}, \qquad t \geq \tau \tag{4.103}$$

The response of the network to the pulse source is illustrated in Figure 4.44.

After the end of the brief interval $0 \leq t < \tau$ during which the pulse is applied, the response depends on the area A of the pulse, but not on its amplitude or its duration τ. As we reduce τ, we change very slightly the instant at which the response begins to decay, but we do not change the decay waveform. If we were to change the *shape* of the source pulse F(t) without changing its area (impulse), we would find that the response $v_1(t)$ is the same as equation (4.103), except in the brief pulse interval. (See problem 4.39 at the end of the chapter.)

In general, we apply the term **impulse** to any source waveform that is **much shorter in duration than the smallest time constant** of the network to which it is applied. We usually ignore the behavior of the network during the interval of application of an impulse because that interval is so brief. Since the *area* of the impulse determines the magnitude of the network response, we treat the impulse area as the magnitude of the impulse.

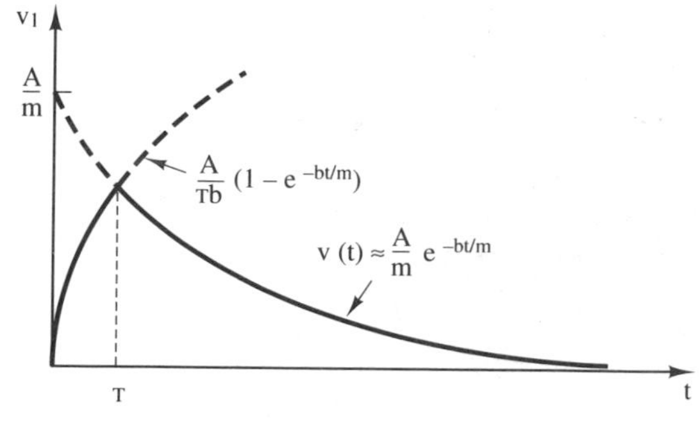

Figure 4.44 Response to an impulsive source.

We use the notation $\delta(t)$ to denote an **ideal unit impulse**—a waveform of infinitesimal duration and unit area. Specifically, the ideal unit impulse satisfies

$$\delta(t) = 0, \qquad t \neq 0 \tag{4.104}$$

and

$$\int_{0-}^{0+} \delta(t)\, dt = 1 \tag{4.105}$$

Mathematically, we shall treat every impulse as if it were ideal. We denote an impulse of area (or magnitude) A by $A\delta(t)$. Then $\delta(t - \tau_1)$ is a unit impulse at the instant τ_1.

Since an impulse is very tall and narrow relative to the scales against which we display the network response, we typically **portray an impulse by a vertical arrow.** The height of the arrow is the area of the impulse, and the location of the arrow represents the *instant of application* of the impulse. Figure 4.45 gives several examples of unit impulses and shows the corresponding arrow notation.

The natural (zero-input) response of the network (Figure 4.43) owing to an initial velocity $v_1(0)$ is

$$v_1(t) = v_1(0)e^{-bt/m} \text{ m/s}, \qquad t \geq 0 \tag{4.106}$$

The impulse response of the network, equation (4.103), is essentially the same as the natural energy decay for the initial velocity $v_1(0) = A/m$. In other words, by applying an impulsive force of area $A = mv_1(0)$, we can *instantaneously* produce the initial condition $v_1(0)$. By instantaneously, we mean that the impulsive source inserts the energy $\frac{1}{2}mv_1^2(0)$ into the mass during the brief interval τ of the impulse, as shown in Figure 4.44. (Of course, to insert the energy over a very short interval, we need a pulse of very large magnitude.)

We define the **unit-impulse response** h of a network to be the *zero-state* response *per unit* of impulse input. The response y to an impulse of size (area) A is $y(t) = Ah(t)$. Hence, h has the units of y/A. In general, the units of h are (output units)/(impulse

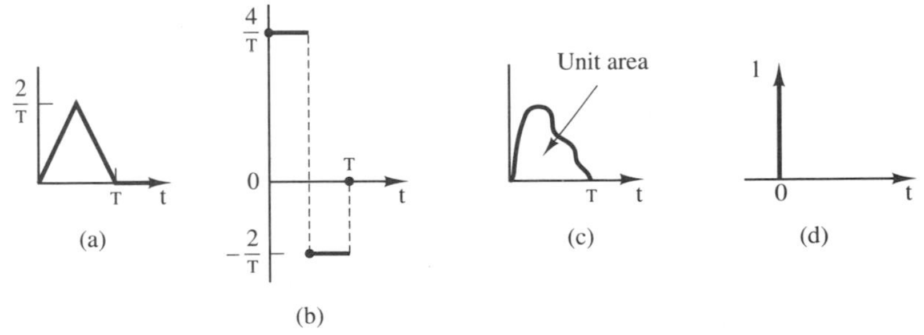

Figure 4.45 Unit impulses. (a) A triangular impulse. (b) An impulse with opposing components. (c) An impulse with irregular shape. (d) Arrow notation for a unit impulse.

units) = (output units)/(input units)·s). For the network described by equation (4.98), the unit-impulse response is

$$h(t) = \frac{v_1(t)}{A\ N\cdot s} = \begin{cases} 0, & t < 0 \\ \dfrac{1}{m} e^{-bt/m}\ m/(N\cdot s^2), & t \geq 0 \end{cases}$$

or

$$h(t) = \frac{1}{m} e^{-bt/m} u_s(t)\ m/(N\cdot s^2) \quad \text{for all } t \tag{4.107}$$

where the unit step notation $u_s(t)$ is used to permit writing of the discontinuous impulse response as a single expression.

Differentiated and Integrated Inputs

We now use superposition to show that differentiation or integration of an input signal produces the same effect—differentiation or integration—in the network response. This knowledge expands the variety of source waveforms for which we can generate analytical responses.

The integral of the unit impulse function is the unit step function

$$u_s(t) = \int_{0^-}^{t} \delta(\sigma)\, d\sigma \tag{4.108}$$

Conversely, the derivative of the unit step function is the unit impulse function $\dot{u}_s(t) = \delta(t)$. These are statements about mathematical functions without physical units.

The derivative with respect to time of an input function u(t) is defined as the limit as $\Delta t \to 0$ of the difference quotient

$$\frac{u(t + \Delta t)}{\Delta t} - \frac{u(t)}{\Delta t} \tag{4.109}$$

According to the principle of superposition, if the response of a linear lumped network to the input u(t) is y(t), then the response of that network to the difference-quotient input (4.109) is

$$\frac{y(t + \Delta t)}{\Delta t} - \frac{y(t)}{\Delta t} \tag{4.110}$$

that is, the difference quotient of the response y. If we take the limits of equations (4.109) and (4.110) as $\Delta t \to 0$, we see that the response to the differentiated input $\dot{u}(t)$ is the differentiated response $\dot{y}(t)$.

We can interpret this statement about $\dot{y}$ and $\dot{u}$ in two ways. In the first interpretation, we say the rate of change of y is related to the rate of change of u in the same way that y is related to u. In the second interpretation, we treat u as the *waveform* of the input (without units) and y as the *waveform* of the output (without units). Then the input waveform $\dot{u}$ produces the output waveform $\dot{y}$. In the second interpretation, assigning of units is separate from determining of waveforms.

If we replace the waveform of the input to any linear lumped network by its derivative, the waveform of the response of that network is also replaced by its derivative. Thus, for the translational system of Figure 4.43, the waveform of the unit-impulse response given by equation (4.107) is the derivative of the waveform of the unit-step response given by equation (4.99). It can be shown that the reverse property is also true. That is, if we replace a source waveform by its integral over time, the response waveform is replaced by its integral over time. (For example, the waveform of the unit-step response is the integral of the waveform of the unit-impulse response.) The integral of the unit step function is called the **unit ramp function** r(t) (see Figure 4.46). If we integrate the waveform of the unit-step response, we obtain the waveform of the **unit-ramp response**—the response to the unit ramp input. In Table 4.4, we summarize these useful relations between input and output waveforms for linear networks.

Generally, we will encounter little conceptual difficulty if we think of all these relations as pertaining to *waveforms,* without units. When we use the relations to analyze a lumped network, we can determine units separately. If we wish to include units in the relations among the unit-step response, unit-impulse response, and unit-ramp response, we must define the unit-step response as the response *per unit* of input step size, the unit-impulse response as the response *per unit* of input impulse area, and the unit-ramp response as the response *per unit* rate of increase of input.

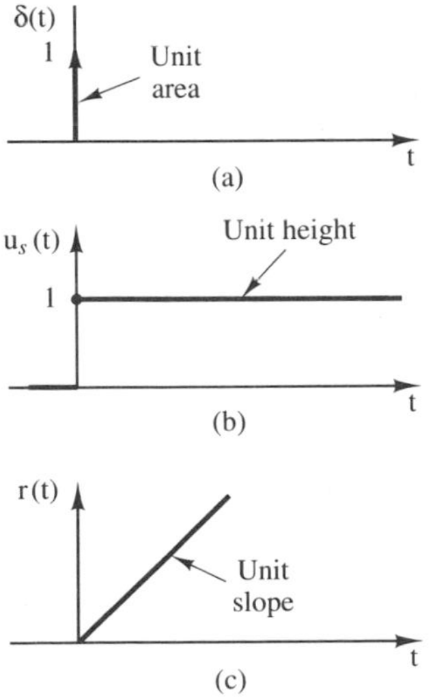

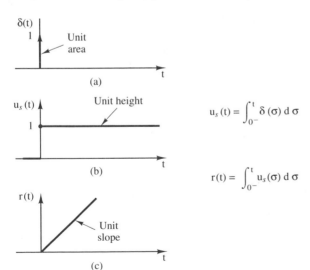

$$u_s(t) = \int_{0^-}^{t} \delta(\sigma)\, d\sigma$$

$$r(t) = \int_{0^-}^{t} u_s(\sigma)\, d\sigma$$

Figure 4.46 Source function relationships. (a) Unit impulse function. (b) Unit step function. (c) Unit ramp function.

TABLE 4.4 INPUT-OUTPUT RELATIONSHIPS

	Input Waveform		Output Waveform
If	$u(t)$	$\rightarrow$	$y(t)$
then	$\dot{u}(t)$	$\rightarrow$	$\dot{y}(t)$
and	$\int u(t)$	$\rightarrow$	$\int y(t)$
	$u_s(t)$	$\rightarrow$	unit-step response
	$\delta(t)$	$\rightarrow$	unit-impulse response $= \frac{d}{dt}$(unit-step response) $\triangleq h(t)$
	$r(t)$	$\rightarrow$	unit-ramp response $= \int$(unit-step response)

Response to Arbitrary Waveforms by Convolution

In this section, we show that the process of decomposing the source waveform, finding individual responses, and then summing the responses can be used for any waveform. As a result, we are able to use the network *step* response discussed in section 3.4 to produce the response to any waveform.

Suppose we approximate an arbitrary source function u(t) by a sequence of narrow pulses, as shown in Figure 4.47(a). We can use superposition to find the zero-state response of a linear network to that sequence of pulses. If the width T of the pulses is small compared with the time constants of the network, each pulse can be treated as an impulse. The magnitude of the impulse at the instant nT is the area of the corresponding pulse, u(nT)T. We represent that impulse by an arrow of height u(nT)T in Figure 4.47(b). The response of the network to that particular impulse is u(nT)T h(t − nT), where h(t) is the unit-

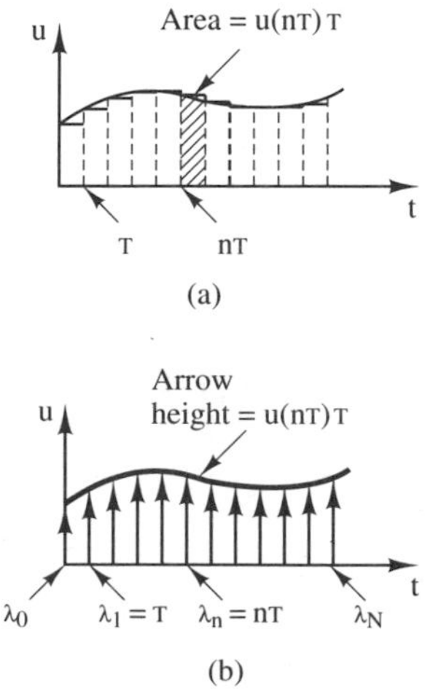

Figure 4.47 Approximation of a function by an impulse stream. (a) Approximation by brief pulses. (b) Approximation by ideal impulses.

impulse response of the network. Then the network response to the sequence of impulses is the sum of the sequence of impulse responses.

Let us compute the value at instant t of the response $v_1(t)$ of the network of Figure 4.43 to an arbitrary input function F(t). For this network, the unit-impulse response h(t) is equation (4.107), and the magnitude of the nth impulse is $F(n\tau)\tau$. Suppose that the interval [0, t] includes N full pulses. That is, $N\tau \leq t < (N + 1)\tau$. Then $v_1(t)$ includes responses from the first N impulses in the sequence. Specifically,

$$v_1(t) = \sum_{n=0}^{N} \left(F(n\tau)\tau \frac{1}{m} e^{-b(t-n\tau)/m} \right) \text{ m/s} \tag{4.111}$$

In general, for t in the interval $N\tau \leq t < (N + 1)\tau$, the response y(t) of a linear network to the source waveform u(t) is

$$y(t) = \sum_{n=0}^{N} u(n\tau)h(t - n\tau)(\tau) \tag{4.112}$$

Define $\lambda_n \triangleq n\tau$ and $\Delta\lambda \triangleq \lambda_n - \lambda_{n-1} = \tau$. Let the pulse width τ (or $\Delta\lambda$) approach zero. Then, since $N \approx t/\Delta\lambda$, $N \to \infty$, and the summation becomes an integral. That is,

$$y(t) = \sum_{n=0}^{N} u(\lambda_n) h(t - \lambda_n) \Delta\lambda \to \int_0^t u(\lambda) h(t - \lambda) \, d\lambda \tag{4.113}$$

We refer to the integral in equation (4.113) as the **convolution** of u and h. We sometimes denote it by the simpler notation $u * h$. That is, if h(t) is the unit-impulse response of the variable y of a *linear* network, then the zero-state response y(t) to the source signal u(t) is

$$y(t) = \int_0^t u(\lambda)h(t - \lambda) \, d\lambda \qquad \text{or} \qquad y = u * h \tag{4.114}$$

The zero-state response y of a *linear* network to a unit impulse input characterizes the behavior of y completely. We can use that unit-impulse response, without any further reference to the network, to determine the response of the same variable y to any other input waveform. Since we can obtain the unit-impulse response by differentiating the zero-state unit-step response, the zero-state step-response analysis of section 3.4 provides the information necessary to obtain the response to any source waveform. However, the convolution operation required to compute that waveform can be quite complicated (see example problem 4.12).

PROCEDURE FOR FINDING THE ZERO-STATE RESPONSE FOR AN ARBITRARY INPUT

1. Find the zero-state unit-step response of the network for the variable y of interest, as discussed in section 3.4.
2. Differentiate the zero-state unit-step response to produce the unit-impulse response h(t).
3. **Convolve** the arbitrarily shaped input waveform u with the unit-impulse response h. The result is the zero-state response y(t) to the input u(t) (see example problem 4.12).

Exercise 4.27: Let $\mu = t - \lambda$ in equation (4.114). Show that the convolution integral is symmetric in its arguments. That is, show that

$$y(t) = \int_0^t u(t - \mu)h(\mu)\, d\mu \tag{4.115}$$

and hence, that $u * h = h * u$.

Suppose that $y = u * h$, where u and h are any two functions that are differentiable and integrable. Then it can be shown, using equations (4.114) and (4.115), that

$$\begin{aligned} \dot{y} &= \dot{u} * h = u * \dot{h} \\ \int y &= \left(\int u\right) * h = u * \int h \\ y &= \dot{u} * \int h \end{aligned} \tag{4.116}$$

Exercise 4.28: Suppose u is the input function for a linear network and h is the unit-impulse response. (Then $\int h$ is the unit-step response.) Interpret each of the statements in equation (4.116).

Example Problem 4.12: Calculation of the network response by convolution.
The circuit of Figure 4.48(a) is driven by a voltage source that has the triangular waveform shown in Figure 4.48(b). The output signal of interest is $v_3(t)$.
(a) Determine the zero-state unit-impulse response of the network.
(b) Use the convolution integral to determine the zero-state response of the network to the triangular input waveform.
Solution: **(a)** To determine the zero-state unit impulse response of the network, we first find the zero-state unit-step response—that is, the response $v_3(t)$ for $v_S(t) = u_s(t)$ volts. The network is first order. From the loop equation with the source set to zero, we determine that the time constant is $\tau = (R_S + R)C$. To find the initial value for the zero state (uncharged capacitor), let $s \to \infty$ (short-circuit the capacitor). Then $v_{3i} = 0$. To find the final value associated with the unit step input, let $s = 0$ (open-circuit the capacitor). Then the current is zero, and $v_{3f} = v_S = 1$ V. Hence, the unit-step response is

$$v_3(t) = \begin{cases} 0, & t < 0 \\ 1 - e^{-t/\tau} \text{ volts}, & t \geq 0 \end{cases} \tag{4.117}$$

The unit-impulse response is the derivative of equation (4.117), i.e.,

$$h(t) = \begin{cases} 0, & t < 0 \\ \dfrac{1}{\tau} e^{-t/\tau} \text{ rad/s}, & t \geq 0 \end{cases} \tag{4.118}$$

We can ignore the units of (4.117) and just use (output units)/((input units)·s) = volts/(volt·s) = rad/s as units for h. Alternatively, we can use for the unit-step response $v_3(t)$ the normalized units: (output units)/(input units) = volts/volts (dimensionless). Then differentiation will produce directly the correct units (rad/s) for h.

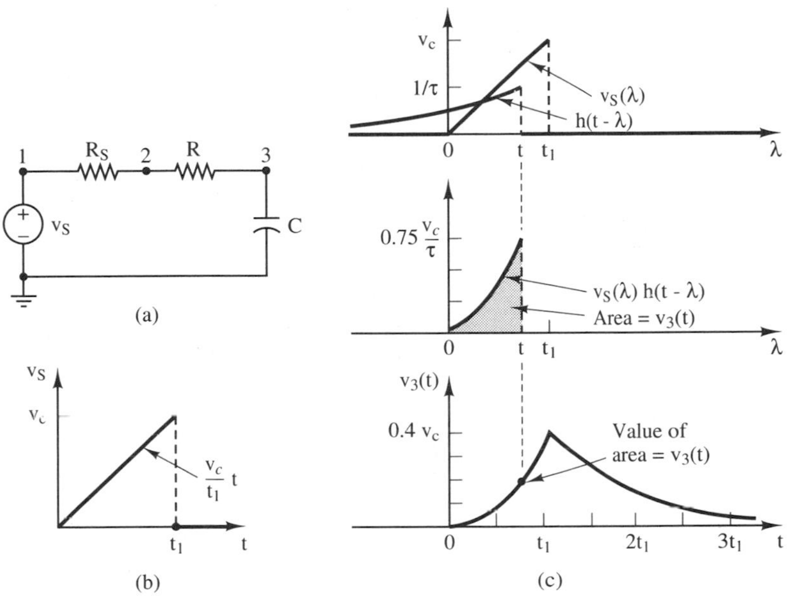

Figure 4.48 Convolution of functions. (a) Lumped circuit. (b) Source waveform. (c) Graphical interpretation of the convolution integral.

(b) According to equation (4.114), the zero-state response to the input waveform $v_S(t)$ is

$$v_3(t) = v_S * h$$
$$= \int_0^t v_S(\lambda)h(t - \lambda)\,d\lambda \qquad (4.119)$$

where $d\lambda$ is expressed in seconds. The functions in the integrand are

$$v_S(\lambda) = \begin{cases} (v_c/t_1)\lambda \text{ volts} & \text{for } \lambda \text{ in } [0, t_1] \\ 0 & \text{elsewhere} \end{cases} \qquad (4.120)$$

and

$$h(t - \lambda) = \begin{cases} (1/\tau)e^{(\lambda - t)/\tau} \text{ rad/s} & \text{for } \lambda \geq 0 \\ 0 & \text{elsewhere} \end{cases} \qquad (4.121)$$

Notice that the integral treats t as a constant; the variable in the integrand is λ. The first line of Figure 4.48(c) shows the functions $v_S(\lambda)$ and $h(t - \lambda)$ vs. λ for the case in which the duration t_1 of the input signal equals the network time constant τ and for a value of t smaller than t_1.

The function $h(t - \lambda)$ is perhaps better viewed as $h(-\lambda + t)$. Convolution requires reversing the function $h(\lambda)$ about the origin (on the λ axis) to produce $h(-\lambda)$, time-shifting the reversed function to the right by t units to produce $h(-\lambda + t) = h(t - \lambda)$, multiplying that function by the input function $v_S(\lambda)$, and then integrating the product.

The second line of Figure 4.48(c) shows the product function $v_S(\lambda)h(t - \lambda)$ vs. λ for a particular value of t. The area under the product curve corresponds to the integral in equation (4.119) for that value of t. The process—reverse, shift, multiply, and integrate—must be carried out for each value of t.

For $t < 0$, the integrand is zero and $v_3(t) = 0$. For t in $[0, t_1]$, the integrand has the form shown on the second line of Figure 4.48(c), and $v_3(t)$ is the area under the curve. For $t > t_1$, the curve $h(t - \lambda)$ extends to the right beyond t_1, but the integrand is nonzero only in the interval $[0, t_1]$. Therefore, we must carry out the integration in three pieces. First,

$$\text{For } t < 0: \qquad v_3(t) = 0 \text{ volts} \tag{4.122}$$

$$\begin{aligned} \text{For } 0 \le t < t_1: \qquad v_3(t) &= \int_0^t \left(\frac{V_c}{t_1}\lambda\right)\frac{1}{\tau}e^{(\lambda - t)/\tau}\,d\lambda \\ &= \frac{V_c}{t_1\tau}e^{-t/\tau}\int_0^t \lambda e^{\lambda/\tau}\,d\lambda \\ &= \frac{V_c}{t_1}e^{-t/\tau}\left[e^{\lambda/\tau}(\lambda - \tau)\right]_0^t \\ &= \frac{V_c}{t_1}t - \frac{V_c}{t_1}\tau(1 - e^{-t/\tau}) \text{ volts} \end{aligned} \tag{4.123}$$

For $t \ge t_1$, we merely change the limits of integration in equation (4.123) to $[0, t_1]$ and obtain

$$\begin{aligned} \text{For } t \ge t_1: \qquad v_3(t) &= \frac{V_c}{t_1}e^{-t/\tau}\left[e^{\lambda/\tau}(\lambda - \tau)\right]_0^{t_1} \\ &= \frac{V_c}{t_1}(t_1 - \tau)e^{-(t - t_1)/\tau} + \frac{V_c\tau}{t_1}e^{-t/\tau} \text{ volts} \end{aligned} \tag{4.124}$$

The response waveform described by equations (4.122)–(4.124) is shown on the third line of Figure 4.48(c) for $t_1 = \tau$.

Exercise 4.29: Sketch $h(\lambda)$ and $h(-\lambda)$. Sketch $h(t - \lambda)$ and the product $v_S(t)h(t - \lambda)$ for $t = t_1/2$ and for $t = 2t_1$.

Exercise 4.30: The source signal in Figure 4.48(b) can be expressed as

$$v_S = \frac{V_c}{t_1}[r(t)u_s(t) - r(t - t_1)u_s(t - t_1)] = \frac{V_c}{t_1}\int_0^t [u_s(t) - u_s(t - t_1)]\,dt$$

where r(t) is the unit-ramp response and $u_s(t)$ is the unit-step response. We can find the response to this source signal by integrating the unit step response found in equation (4.117), then subtracting a delayed version of the same integrated response. Carry out this process to verify equations (4.122)–(4.124).

4.4 ANOTHER EXAMPLE PROBLEM

Example Problem 4.13: Effects of nonlinearity.

The techniques of this chapter—network simplification and superposition—depend on the linearity of the network being analyzed. Suppose we were to treat a nonlinear network as if it were linear. This example problem shows how much error we could expect to incur.

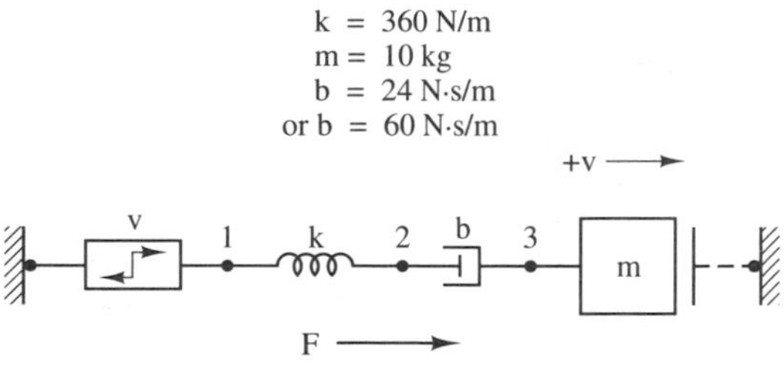

Figure 4.49 A second-order network.

The second-order linear translational network of Figure 3.44 is reproduced in Figure 4.49. For the friction constant b = 24 N·s/m, the network is overdamped and has the characteristic roots s = −3 rad/s and −12 rad/s. For b = 60 N·s/m, the network is underdamped, with s = −3 ± j5.2 rad/s. The zero-state unit-step responses for these two friction constants (copied from Figures 3.45 and 3.46) are shown in Figure 4.50.

In physical systems, it is damping (friction) phenomena that are most likely to deviate from linearity. Figure 4.51 shows the quadratic damper relation $F = \eta v_{23}^2$. A quadratic relation is typical for the motion of solid objects immersed in liquids or gases. The figure also shows a *linear* approximation to the quadratic damper that has the same force for $v_{23} = 1$ m/s. (It can be shown that for the unit step input used to produce the responses of Figure 4.50, v_{23} remains in the region between 0 and 1 m/s.)

(a) Replace the linear damper of Figure 4.49 by the quadratic damper of Figure 4.51. Find the input-output system equation for v_3, and draw an operational diagram that produces v_3.

(b) For b = 24 N·s/m and for b = 60 N·s/m, find the value of the quadratic-damper coefficient η that produces the same damper force as the linear model at 1 m/s.

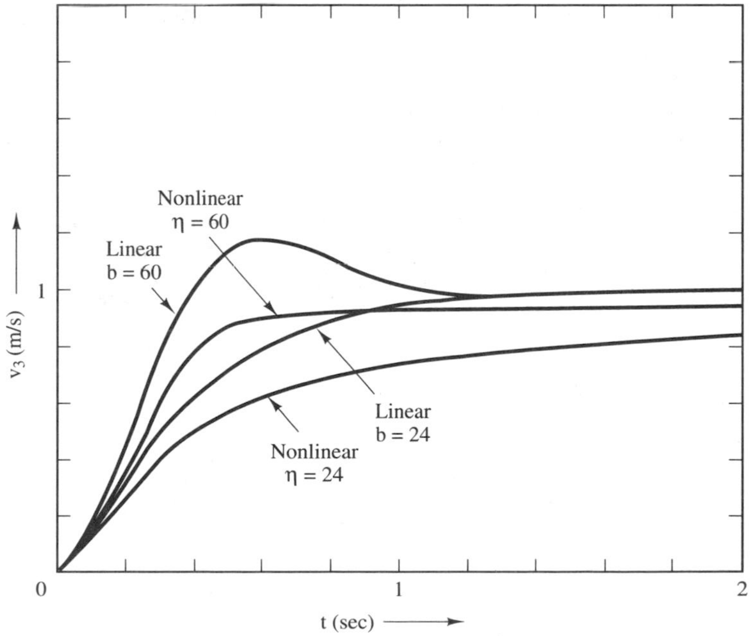

Figure 4.50 Unit-step responses for linear and nonlinear dampers.

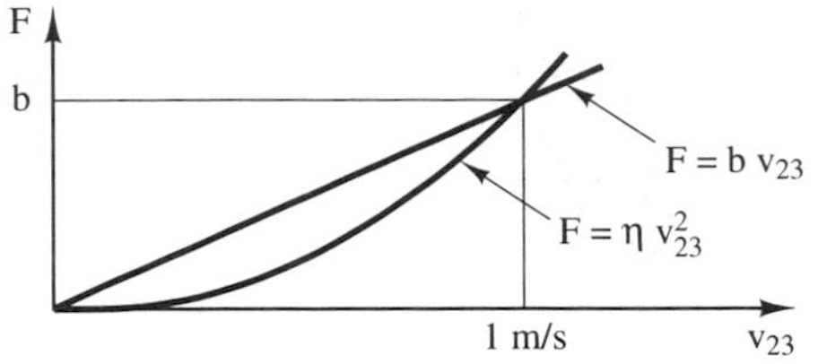

Figure 4.51 Quadratic and linear damper characteristics.

(c) For each value of b, use an operational-block simulator to find the zero-state unit-step response of the nonlinear network. Compare the linear and nonlinear responses.

Solution: **(a)** The loop equation for the nonlinear network is

$$v = (s/k)F + \sqrt{\frac{F}{\eta}} + \left(\frac{1}{sm}\right)F \tag{4.125}$$

Substituting the mass equation $F = smv_3$ into equation (4.125) and rearranging yields

$$\ddot{v}_3 + \sqrt{\frac{k^2}{m\eta}\dot{v}_3} + \frac{k}{m}v_3 = \frac{k}{m}v \tag{4.126}$$

Note the nonlinear form of the second term of equation (4.126). Figure 4.52 shows an operational diagram that produces v_3.

(b) We equate the forces in the linear and nonlinear damper models to find that

$$\eta = \frac{b}{v_{23}} \tag{4.127}$$

We substitute $v_{23} = 1$ m/s to find that $\eta = b$ N·s^2/m^2. Therefore, $\eta = 24$ N·s^2/m^2 or $\eta = 60$ N·s^2/m^2.

(c) The operational diagram of Figure 4.52 was simulated with TUTSIM. The zero-state unit-step responses for the nonlinear dampers are superimposed on the linear-network responses in Figure 4.50. The effect of the nonlinearity is substantial. First, this particular nonlinear friction characteristic tends to make the network more stable. The reason can be understood intuitively: As the damper velocity increases, the friction increases; therefore, high velocities are inhibited. Second, the settling time is increased drastically. The intuitive reason for this increase is that the damping force becomes very small for the low damper velocities that exist as the network approaches a steady state.

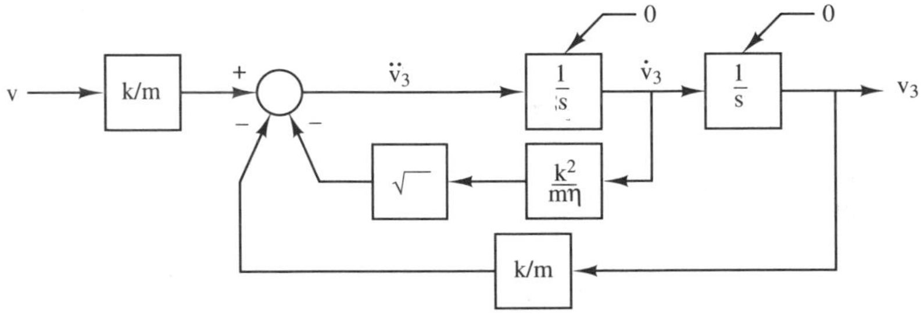

Figure 4.52 Operational diagram for equation (4.126).

This example illustrates the extent to which response analyses can be in error if we falsely assume linearity. Physical systems *can* be designed to have linear (viscous) friction characteristics. Often, the characteristics lie between linear and pure quadratic curves. That is, they are partly viscous and partly quadratic. In some instances, the friction characteristics can change abruptly, as in the solid-to-solid (coulomb) characteristic of Figure 2.9. In all but the abruptly changing case, we can linearize the model (see section 5.2). Such linearization permits the use of linear techniques without incurring the errors demonstrated in this example.

4.5 SUMMARY

Every linear network has a number of simple equivalent representations. Each network source can be handled individually if the network is linear. Every source is equivalent to a *set* of sources with simple waveforms. This chapter examines equivalence of networks, decomposition of waveforms, and superposition and uses them as tools in the art of network simplification. These tools depend on the linearity of the network. Networks that are not completely linear can usually be linearized (see section 5.2).

Stored energy acts like a constant source. The energy in a storage element can be represented as an ideal source in series or parallel with a zero-state element. Accordingly, every network can be represented as a zero-state network.

The transfer function is an s-notation representation of the zero-state relation between two network variables. The impedance of a subnet is a special case of a transfer function. The input-output transfer function for an independent variable is equivalent to the input-output system equation for that variable. For a step input, letting $s = 0$ in an impedance or transfer function converts it to the final steady-state relation between the variables. For a zero-state network, letting $s \to \infty$ in an impedance or transfer function converts it to the initial ($t = 0^+$) relation between the variables.

Every linear subnet has two equivalent representations: the Thévenin source A_{eq} in series with the equivalent impedance Z_{eq} and the Norton source T_{eq} in parallel with the equivalent impedance Z_{eq}. These two equivalent subnets are illustrated in Figures 4.35(b) and (c) and described mathematically in equations (4.76) and (4.77).

A complicated waveform can be treated as a sum of simple waveforms. Hence, every signal has many equivalent representations. The principle of superposition replaces a difficult problem (finding the response of a linear network with multiple sources and complicated waveforms) by a sequence of simpler problems (finding the zero-state responses to single inputs with simple waveforms). The responses to the simpler problems are added to get the total response.

The art of using the equivalence and decomposition tools efficiently is acquired by observation and practice. This art is illustrated in the examples throughout the remainder of the text.

The chapter also introduces electrical systems. Electric current is a flow and voltage is a potential. The derivatives and integrals of current and voltage also act as flows and potentials, respectively. Section 4.1 shows how to generate lumped models of electrical systems. The ideal passive elements are the ideal capacitor, the ideal resistor, and the ideal inductor (described by $i = C\dot{v}_{ij}$, $v_{ij} = Ri$, and $v_{ij} = L di/dt$, respectively). The active elements are the ideal current source (for which the current i(t) in a particular direction is specified) and the ideal voltage source (for which the voltage drop v(t) in a particular direction is specified). We define the directions of the currents in the elements arbitrarily and write the system equations in terms of those arbitrarily defined currents. The sign convention for writing element equations is to order the voltage subscripts in the same direction as the current. Then current flows in the direction of the drop in voltage. The power delivered to an electrical element is the product of the current through the element and the voltage drop across the element in the direction of the current.

SPICE-based electric-network simulation tools can be used to determine responses for all types of linear networks. These tools require no mathematical manipulation. Therefore, we are able to analyze networks of high complexity.

4.6 PROBLEMS

4.1 For one of the lumped electric circuits of Figure P4.1:

(a) Define the currents and node voltages.

(b) Find a set of system equations.

(c) Solve for the values of the variables.

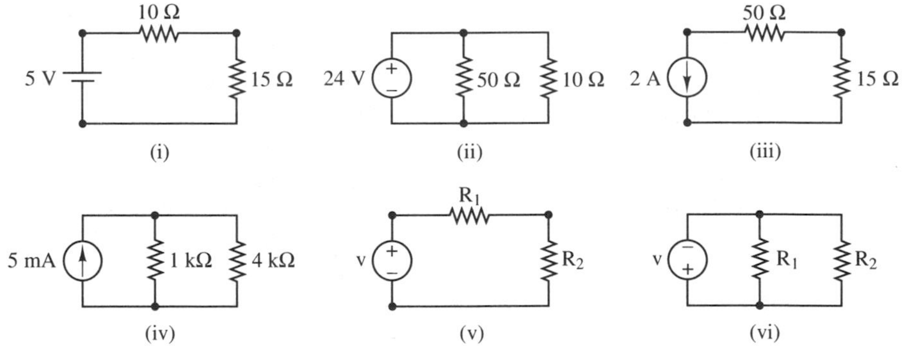

Figure P4.1.

4.2 For one of the lumped electric circuits of Figure P4.2:

(a) Define the currents and node voltages.

(b) Find a set of system equations.

(c) Find the input-output system equation for the voltage across the resistor.

(d) Find the initial and final values of the voltage across the resistor if the network is initially in the zero state and the source signal is constant with a value denoted by the subscript c.

4.3 Figure P4.3 is a model for a flashlight. The resistor $R_1 = 1\ \Omega$ represents the internal resistance of the battery; $R_2 = 1\ \Omega$ represents the bulb. The user connects a second identical bulb, R_3, to the battery in an attempt to double the light output.

(a) Calculate the power that is delivered to the bulb R_2 before R_3 is connected.

(b) Calculate the power delivered to the pair of bulbs R_2 and R_3 after connecting the second bulb. What do you conclude?

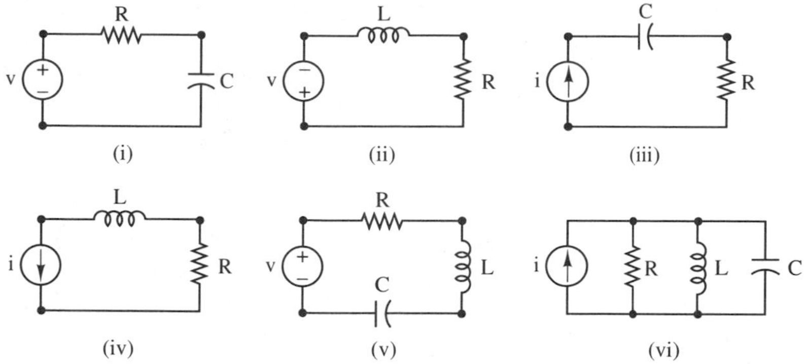

Figure P4.2.

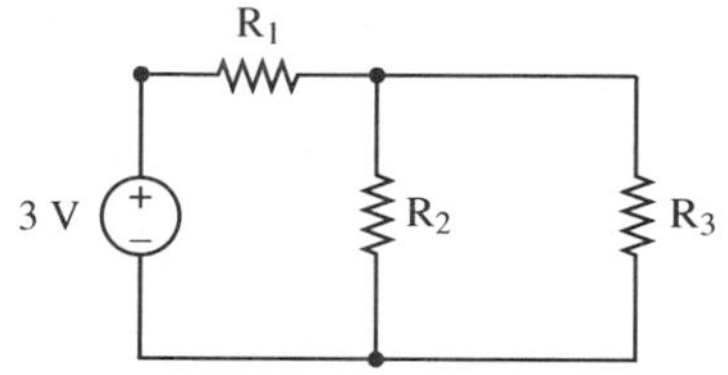

Figure P4.3.

4.4 Design the voltage divider of Figure P4.4 to reduce the 12-V source voltage to 6 V. The unloaded (open-switch) voltage v_{21} at the output of the voltage divider is to be 6 V. When a load R_L is applied, the voltage must not drop below 4 V. The smallest load resistor R_L ever to be connected to the divider is 2,000 Ω.

(a) Specify the minimum values for R_1 and R_2.

(b) For the values of R_1 and R_2 found in part (a), find the power dissipated in the voltage divider (R_1 and R_2) under no-load condition and under the condition of maximum load current ($R_L = 2000\ \Omega$).

(c) Find that fraction of the power delivered by the source which is dissipated in the load R_L when the load current is maximal ($R_L = 2000\ \Omega$). How is this fraction changed if the values of R_1 and R_2 are doubled?

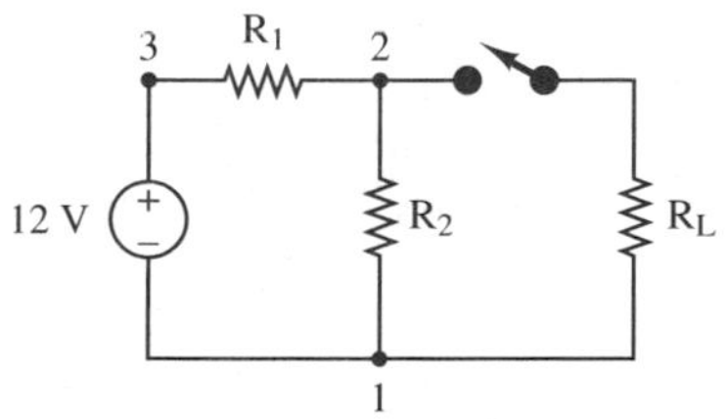

Figure P4.4.

4.5 Electric power utility companies in the United States provide residential customers with an ac voltage source that has the waveform $v_S = v_{max} \sin(2\pi ft)$, where $f = 60$ Hz and $v_{max} = 163$ V. Suppose this voltage is connected to a resistor R (perhaps a heater or a lamp).

(a) Find the waveform of the current in R.

(b) Find the average power delivered to R. (*Hint:* Find the energy delivered in one cycle of the sinusoidal voltage.)

(c) Find the dc voltage v_{rms} that would provide the same power to R as that found in part (b). What is the relation between v_{max} and v_{rms}? (We call v_{rms} the *root-mean-square equivalent* of v_S. It is v_{rms} that the utility company quotes as the voltage level delivered to the customer; ac voltmeters are calibrated in rms values.)

4.6 Find the node voltages and branch currents for one of the networks of Figure P4.6.

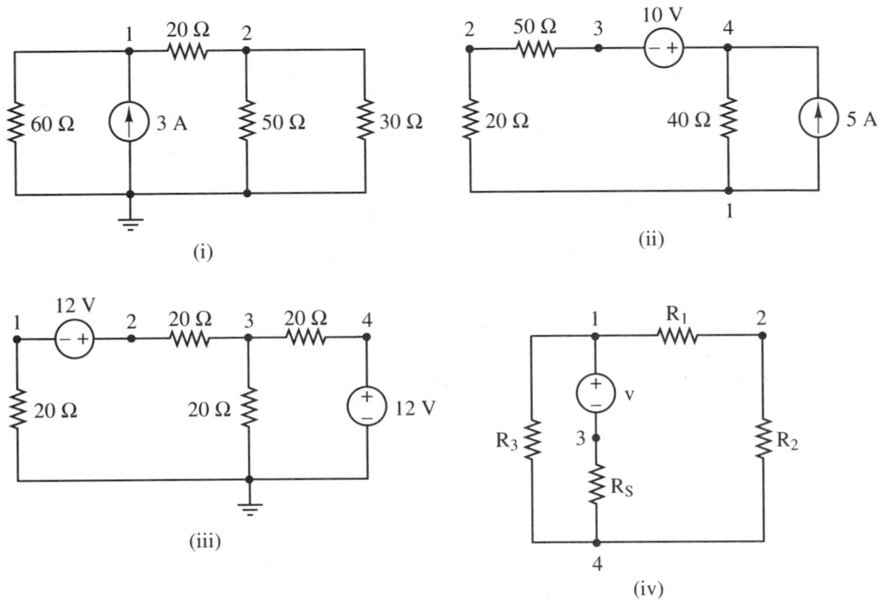

Figure P4.6.

4.7 The Wheatstone-bridge circuit of Figure P4.7 is designed to measure the resistance R accurately. Precise values are known for R_1, R_2, and the variable resistor R_3. The current meter is very sensitive. After R is attached to the bridge, the value of R_3 is adjusted until the bridge is balanced (the current through the meter is zero). Then the fact of balance is used to calculate R. The battery voltage does not affect the measurement.

(a) Calculate the value of R in terms of R_1, R_2, and R_3 when the bridge is balanced.

(b) A commercial bridge typically has switches that change the values of R_1 and R_2. Suppose that R_1 and R_2 are accurate to 0.1% and R_3 can be adjusted continuously from 0 Ω to 1000 Ω with an accuracy ±0.5 Ω. Suppose that $R_1 = 1$ Ω, $R_2 = 1000$ Ω, and

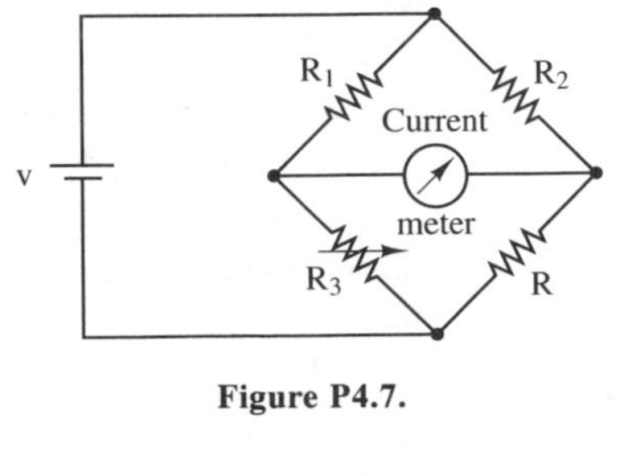

Figure P4.7.

$R_3 = 1000\ \Omega$. What is the percent uncertainty in the measured value of R? What is the uncertainty if $R_3 = 10\ \Omega$?

4.8 Derive a simple lumped electric-circuit model that describes the main features of one of the following situations or systems. Include a model for the source that energizes the system. Estimate all parameter values. State your assumptions.

(i) A battery-driven electromagnet made from 100 turns of copper wire wound on a nail.

(ii) A cloud above the earth during an electrical storm.

(iii) A hand-held hair dryer that has an electric fan motor (with an armature winding) connected in parallel with a resistance heater.

(iv) A 1,500-watt radiant heater.

(v) A 20-m length of two-conductor wire that connects a 115-V circuit-breaker box to an electric outlet in a room; a 1,200-W heater and a 250-W lamp are connected (in parallel) to the outlet.

4.9 An electric circuit composed of discrete physical elements lays on an insulating plastic film in the bottom of a five-sided steel box (no top), as illustrated in Figure P4.9. In addition to the obvious lumped electrical parameters associated with the circuit, there are unintended *stray* capacitances between the copper wires and the box. These stray capacitances can have a significant effect on the circuit behavior if the source signal varies rapidly. The circuit is drawn approximately to scale. The box is 1 cm deep. The wire diameter is 1 mm. The height of the wires above the bottom of the box varies between 1 mm and 2 mm.

(a) Generate a lumped model for the circuit that includes the stray capacitances.

(b) Estimate the values of the stray capacitances.

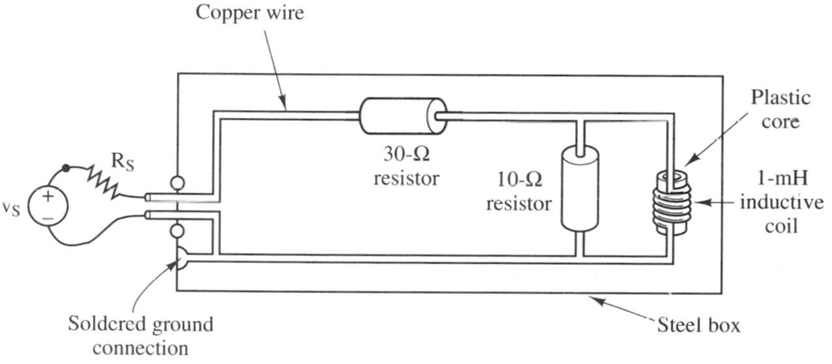

Figure P4.9.

4.10 A 7,675-volt ac power line is strung from pole to pole over the 5-km distance from an electric utility substation to a transformer in a particular neighborhood. (The transformer reduces the voltage from 7,675 volts to the 240-volt level that is used in the neighborhood.) The power line is a copper wire 4 mm in diameter which is carried on poles at a height of 9 m above the ground. Treat the ground as the return conductor. The substation acts as a voltage source $v_S(t) = 7{,}675\sqrt{2}\,\sin(277t)$ volts. The transformer (to which all the electrical equipment in the neighborhood is connected) acts as a slightly inductive load; that is, it acts like a resistor in series with an inductor. Generate a lumped model for the power line. Estimate the values of the elements in the power-line model. (*Hint:* The plane midway between the wires in Figure 4.13(a) acts as an equipotential surface.)

4.11 An integrated circuit is constructed by modifying the surface of a chip of silicon. The modifications produce various degrees of conductivity on various portions of the surface to form a useful electric circuit. Figure P4.11 shows a simple integrated circuit chip. Develop a lumped model for the circuit. Estimate the values of the lumped element parameters. (Note that 1 mil $\triangleq$ 0.001 inch $= 2.54 \times 10^{-5}$ m.)

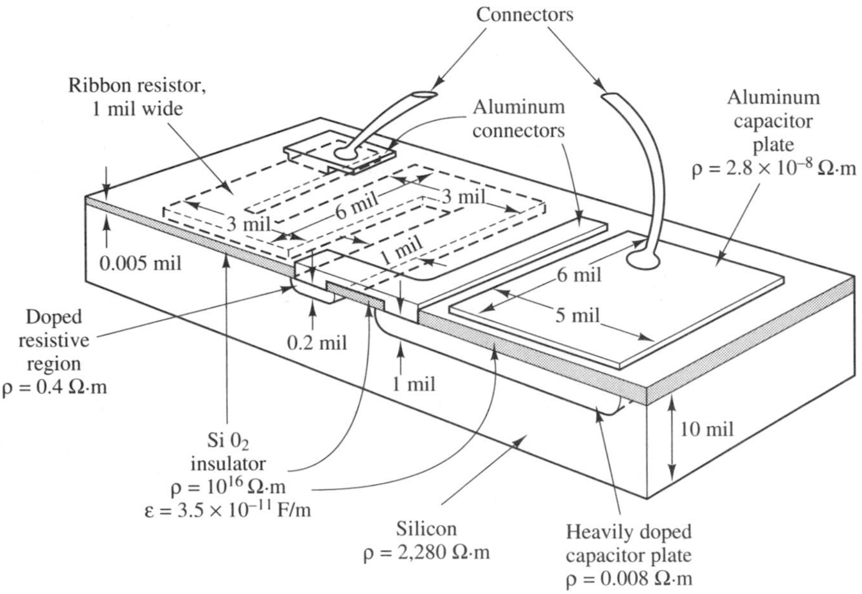

Figure P4.11.

4.12 Find the response $v_1(t)$ for one of the first-order networks of Figure P4.12. Assume that the energy-storage element is initially unenergized and the source is applied abruptly with the value shown.

4.13 For one of the second-order networks of Figure P4.13:

(a) Find the characteristic equation and the characteristic roots. (*Hint:* Begin with symbols for the element parameters; substitute values later.)

(b) Find the initial values v_{1i} and $\dot{v}_{1i}$ and the final value v_{1f} under the assumption that the energy-storage elements are unenergized initially and the source is applied abruptly with the value shown.

(c) Find the mathematical expression for the response $v_1(t)$ from the quantities found in parts (a) and (b).

(d) Sketch the response $v_1(t)$. Use the response curves of Figures 3.52–3.54 as a guide if they are helpful.

(e) *Optional:* Use a SPICE-based program to determine the response $v_1(t)$ directly from the network for a unit-valued source.

4.14 For one of the networks of Figure P4.14:

(a) Use the node method to find a set of system equations; find the input-output system equation for v_1.

(b) Use the loop method to find a set of system equations; find the input-output system equation for v_1.

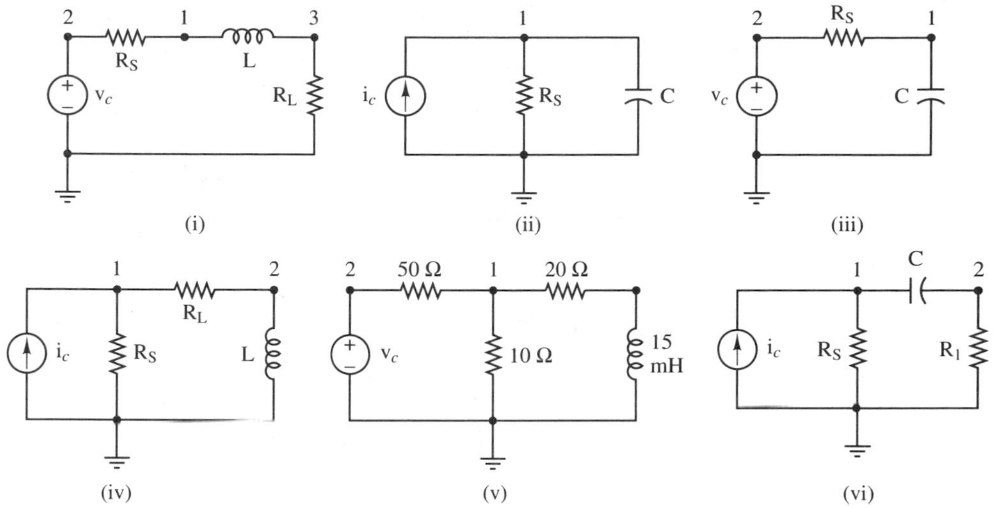

Figure P4.12.

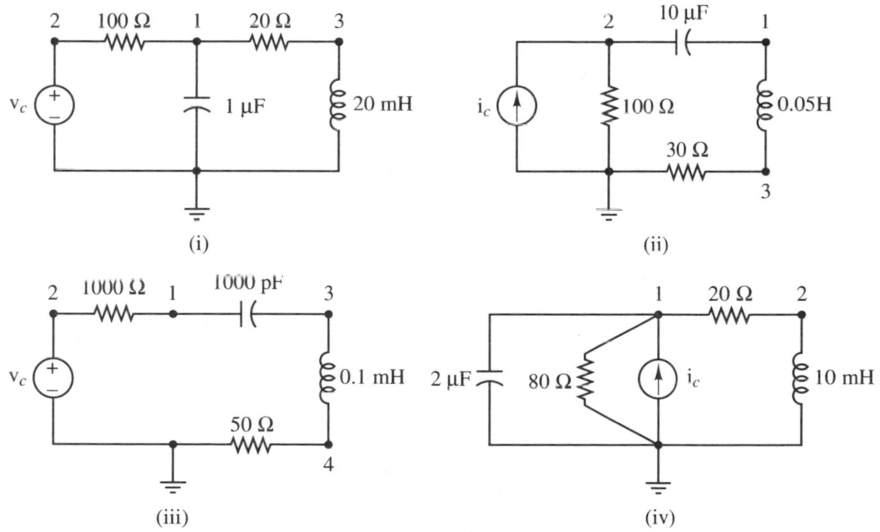

Figure P4.13

4.15 A physical circuit and a corresponding lumped model are shown in Figure P4.15. We used an impedance meter to measure the resistance, capacitance, and inductance of the physical elements. The results were R = 100.1 Ω, R_L = 18.7 Ω, C = 4.83 nF, and L = 9.86 mH.

We temporarily removed the physical circuit from the signal generator and replaced it with an adjustable resistor. We measured the voltage output v_1 of the generator and the current

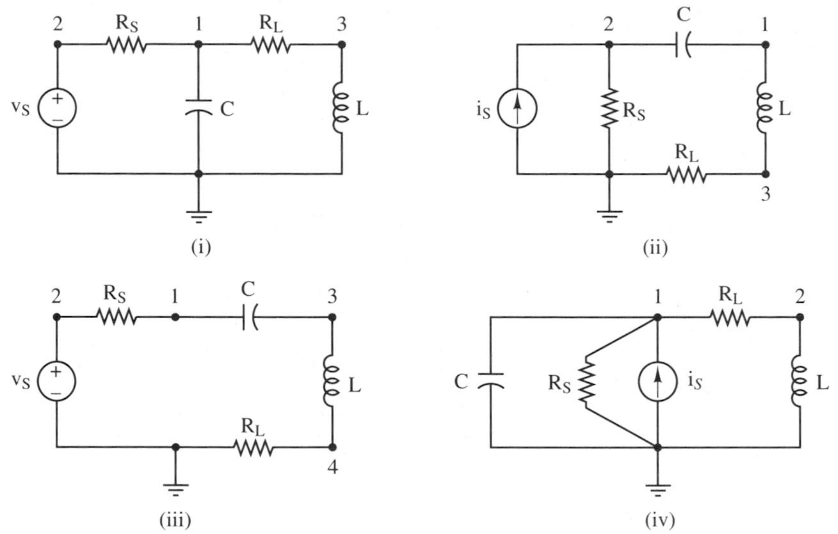

Figure P4.14

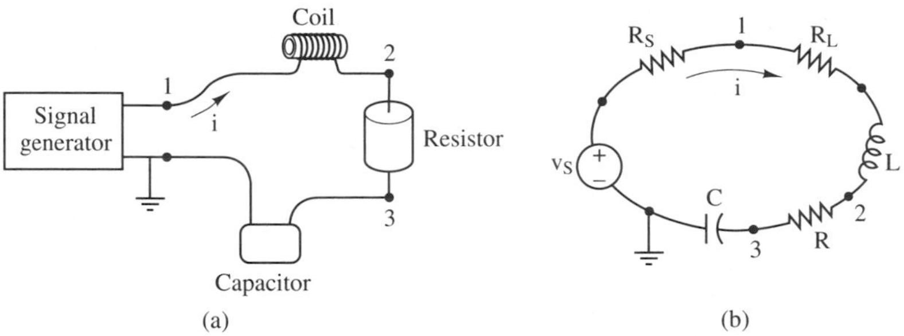

Figure P4.15. (a) Pictorial model. (b) Lumped model.

i emerging from the generator for various settings of the resistor. In that way, we obtained the following data:

v_1 (volts)	6.9	6.4	5.6	4.5	2.95
i (mA)	69	80	93.3	113	148

Then we used the signal generator to apply a voltage step to the unenergized physical circuit, and observed the step response $v_3(t)$ on an oscilloscope. The response looked underdamped and second order. The peak value of $v_3(t)$ was 41.2 V. The final value was 23.5 V. The period of the oscillation was 44 μs.

(a) Use the v-i data to select an appropriate value for the source resistance R_S.

(b) Use the lumped model with the measured parameters to find the input-output system equation for the capacitor voltage v_3. Determine the characteristic roots.

(c) Use the measured values for the key features of the step-response to determine the characteristic roots and the input-output system equation.

(d) Compare the results of parts b and c. What do you conclude about the adequacy of the lumped model?

4.16 A lumped model of a fish-boat interaction is derived in example problem 2.3 (Figure 2.23). The system differential equation and the associated initial conditions for the velocity v_3 of the fish are derived in example problem 4.8 (see Figure 4.27). Inspection of the lumped model and its parameters suggests that the boat (node 1) will remain almost stationary during the fish's attempted flight.

(a) Assume that node 1 is stationary, and derive the new system equation for v_3.

(b) Insert the parameter values into the system equations for the two models. Use an operational-block simulator or network (SPICE-based) simulator to find the responses of both models if F is a 10-s, 130-N pulse. Compare the two responses.

4.17 Figure 2.41(b) shows a lumped model for the translational behavior of a vertically hanging Slinky. Use that model without the dampers. According to example problem 2.7, the mass per segment of the model is $m = 0.25/n$ kg, the gravity force per segment is $F = mg = 2.45/n$ N, and the stiffness per segment is $k = 0.82(n + 1)$ N/m, where n is the number of segments in the model.

Insert a velocity source between the support and the top spring of the model, as indicated in Figure P4.17(a). You are to simulate the behavior of the Slinky using a SPICE-based network simulation program. For each simulation, the Slinky is initially hanging motionless.

(a) Determine the initial energy states (the forces in the springs and the velocities of the masses), as a function of n, in preparation for the simulations.

(b) Find the response of a 10-segment version of the Slinky model to the faster of the velocity-source waveforms of Figure P4.17(b).

(c) Find the response of a 10-segment model to the slower waveform of Figure P4.17(b).

(d) Find the response of a 2-segment model to the fast waveform.

(e) Find the response of a 2-segment model to the slow waveform.

(f) What do you conclude about the relationship between the rapidity of variation of the source and the number of segments required to obtain accurate mimicking of the Slinky's behavior?

4.18 Figure 2.41(b) shows a lumped model for the translational behavior of a vertically hanging Slinky. The mass per segment of the model is $m = 0.25/n$ kg, the gravity force per segment

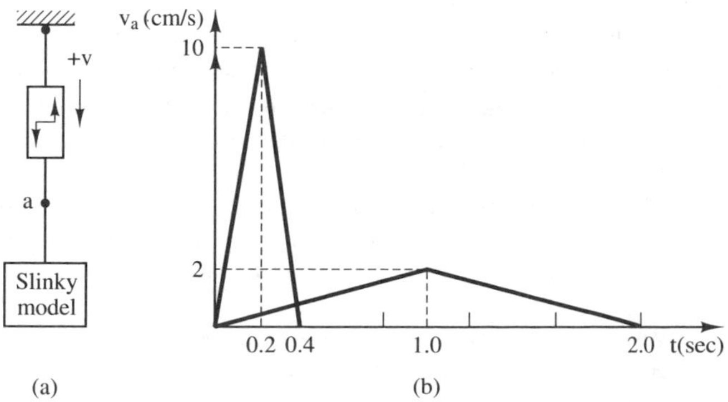

Figure P4.17. (a) Slinky model. (b) Source waveforms.

is $F = mg = 2.45/n$ N, and the stiffness per segment is $k = 0.82(n + 1)$ N/m, where n is the number of segments in the model. The friction constant b per segment is small and unknown.

A Slinky was fastened to a support, then collapsed against the support and released. The Slinky exhibited a slow, large-amplitude oscillation similar to that of a single mass on a spring. The first peak displacement was measured to be 2.95 m. The period of oscillation was 1.65 s. The final value of the displacement (when the Slinky hung motionless) was 1.5 m. The decay-envelope time constant (the time for the peak variations *about the final value* to decrease to 37%) was 30 s.

(a) Assume the oscillating behavior of the Slinky can be adequately represented by a single-segment (second-order) Slinky model without a damper. Find *analytically* the period of oscillation and the peak displacement that should occur when the Slinky is released. Compare those values with the measured values.

(b) Use a SPICE-based network simulation program to simulate the behavior when the Slinky is released. Use a 20-segment Slinky model with the friction constants set to zero. All energy states are initially zero. Compare the period of oscillation and peak displacement of the simulated response with the measured values.

(c) Add to each segment of the 20-segment Slinky model a damper with friction constant $b = 0.0005$ N·s/m. Adjust the value of b until the simulated response decays at about the same rate as the measured response.

4.19 Figure 2.41(b) shows a lumped model for the translational behavior of a vertically hanging Slinky. The mass per segment of the model is $m = 0.25/n$ kg, the gravity force per segment is $F = mg = 2.45/n$ N, and the stiffness per segment is $k = 0.82(n + 1)$ N/m, where n is the number of segments in the model. Ignore the dampers.

A Slinky was hung from a support. The top loop of the Slinky was collapsed against the support (raising the whole stretched Slinky by 3 cm) and then was released. Subsequently, a wave was observed to travel down the Slinky. The wave reflected back and forth from top to bottom for about 10 round trips before it died away. By looking down the center of the Slinky, we were able to see the wave clearly enough to time its velocity. It traveled the length of the hanging Slinky (1.5 m) in 0.55 s.

(a) Use a network simulation program to simulate the traveling wave behavior of a 20-segment Slinky model. (*Hint:* Use the forces in the segment springs during the stationary hang as initial conditions. The force in the top spring must be reduced to account for initial raising of the top loop. To observe the traveling wave, simulate the *displacement* at the top, middle, and bottom of the Slinky.)

(b) Evaluate the simulated response found in part (a). What was the wave velocity? Compare it with the measured velocity. Why does the bottom of the Slinky drop by more than the other segments? How accurate is the model in predicting the Slinky's behavior?

4.20 The needle of the d'Arsonval galvanometer shown in Figure P4.20 rotates in proportion to the current in the coil. The resistance of the coil is $R_c = 50\ \Omega$. The current in the coil is 1 mA when the needle reads full scale.

(a) We can use the galvanometer to measure larger currents by putting a resistor in parallel with the coil. Choose the size of the parallel (or *shunt*) resistor R_p so that the meter will read full scale for 15 mA. What effect does the resulting ammeter have on the circuit being measured?

(b) We can use the galvanometer to measure voltages by putting a resistor in series with the coil. Choose the size of the series resistor R_s so that the meter will read full scale for 15 V. What effect does the resulting voltmeter have on the circuit being measured?

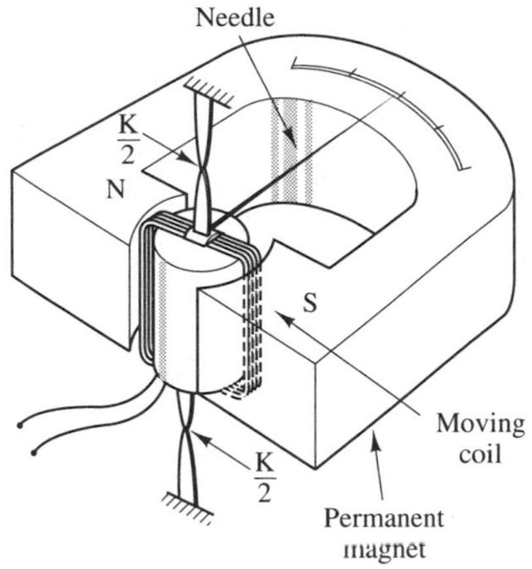

Figure P4.20.

4.21 In the electric circuit of Figure P4.21, the voltage source v_S is constant. R_S is the source resistance, and R_L is the resistance of the inductor. The switch is connected to node 2 until the circuit reaches a steady state. At $t = 0$, the switch is moved from node 2 to node 4. (The switch is a make-before-break switch.)

(a) Determine the current i_1 in the inductor at $t = 0^-$.

(b) Represent the energized inductor for $t \geq 0$ by an unenergized inductor and a constant-current source. Draw the network lumped model for $t \geq 0$.

(c) Determine the current i_2 in R_2 for $t \geq 0$.

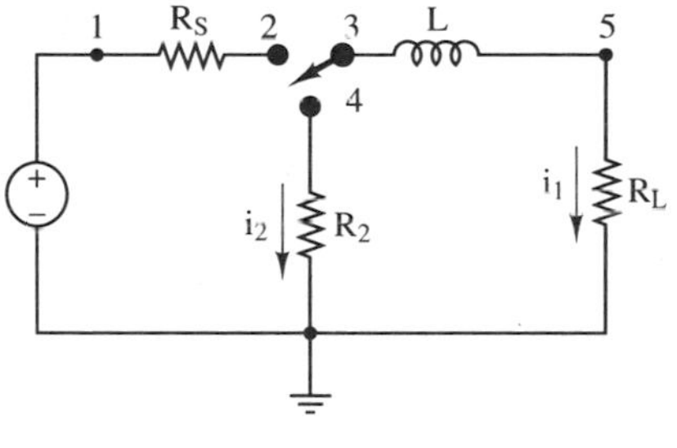

Figure P4.21.

4.22 A certain flashlight uses two 1.5-V batteries. Use the battery measurement data of Figure 4.28(b) to determine the resistance of each battery. Find the resistance of the light bulb necessary for maximum transfer of power to the bulb (maximum light output). Determine the amount of power delivered to the bulb under these optimum conditions. Suppose, on the other hand, that a 2-ohm bulb is used. Find the amount of power delivered to the bulb and the fraction of the bulb power that is lost owing to impedance mismatch.

4.23 An electric motor is used to turn paddles (B_L) that mix a chemical product (see Figure P4.23). The transmission of power to the paddles is hampered by the motor friction (B_m) and bearing friction (B_b). The size of the paddles determines the load friction constant B_L. Select B_L (as a function of the other parameters) so that maximum power is delivered to the paddles during steady mixing.

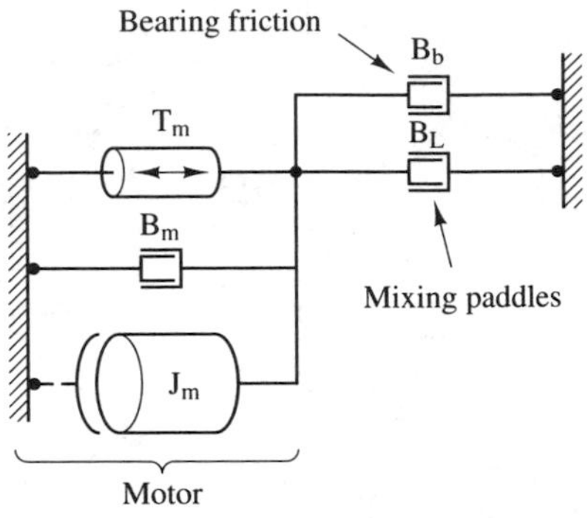

Figure P4.23

4.24 An electric motor is used to accelerate and decelerate a turntable in an amusement park. (Patrons attempt to stay on the turntable as it accelerates.) We can change the inertia J_L of the turntable (with its load of people) by changing the diameter of the turntable. The motor can be represented as a lumped torsion source with parameters T_m, J_m, and B_m. Choose the inertia J_L that will maximize the initial acceleration of the wheel. Express J_L in terms of the source parameters.

4.25 **Delta** and **wye** connections of impedances are shown in Figure P4.25. The delta connection can be replaced by an equivalent wye connection, and vice versa. Use the path law and the node law to verify the following equivalence relations:

$$Z_1 = \frac{Z_b Z_c}{Z_a + Z_b + Z_c} \qquad Z_a = \frac{Z_1Z_2 + Z_2Z_3 + Z_3Z_1}{Z_1}$$

$$Z_2 = \frac{Z_c Z_a}{Z_a + Z_b + Z_c} \qquad Z_b = \frac{Z_1Z_2 + Z_2Z_3 + Z_3Z_1}{Z_2}$$

$$Z_3 = \frac{Z_a Z_b}{Z_a + Z_b + Z_c} \qquad Z_c = \frac{Z_1Z_2 + Z_2Z_3 + Z_3Z_1}{Z_3}$$

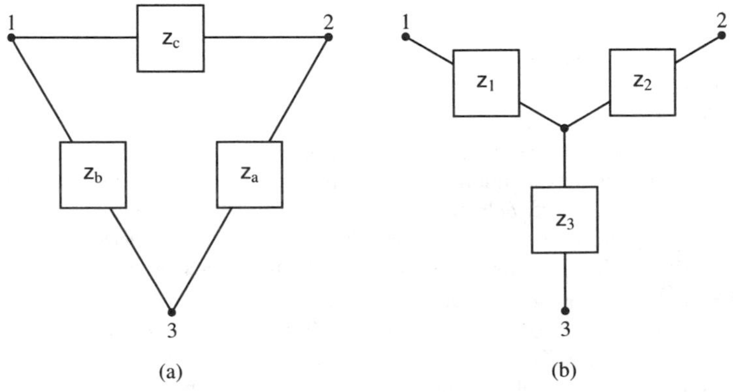

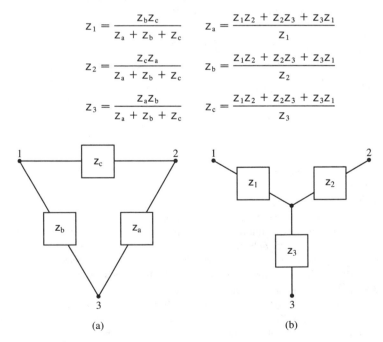

Figure P4.25 (a) Delta connection. (b) Wye connection.

4.26 For one of the electrical networks of Figure P4.26:

(a) Draw and label the Thévenin equivalent of the subnet *seen* by the load. (*Hint:* Use symbols first; substitute values later.)

(b) Find the corresponding Norton equivalent of the subnet.

(c) Use part (a) or part (b) to find the differential equation for the load current if the load is a 15-ohm resistor.

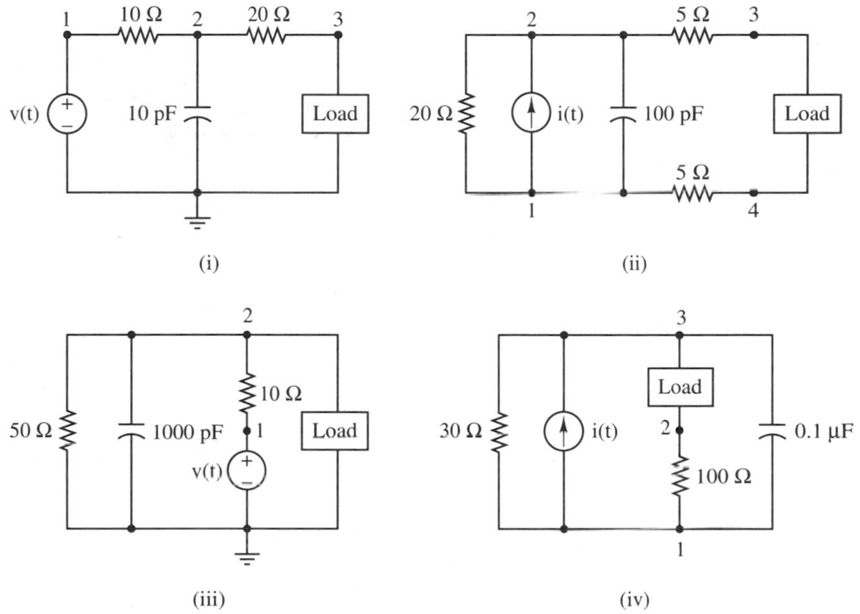

Figure P4.26.

4.27 For one of the translational networks of Figure P4.27:

(a) Draw and label the Thévenin equivalent of the subnet seen by the load. (*Hint:* Use symbols first; substitute values later.)

(b) Find the corresponding Norton equivalent of the subnet.

(c) Use part (a) or part (b) to find the differential equation for the scalar force in the load if the load is a spring of stiffness 30 N/m.

4.28 An electrical network is shown in Figure P4.28.

(a) Find the Norton equivalent for the two-terminal subnet that lies to the right side of the dashed lines.

(b) Use the Norton equivalent found in part (a), together with the subnet to the left of the dashed lines, to express the voltage v_1 as a function of the source current i_s.

(c) Superimpose the v_1-vs.-i characteristic of the Norton equivalent found in part (a) and the corresponding characteristic for the subnet that lies to the left of the dashed lines. Label the operating point and all other pertinent quantities.

4.29 A spring network is shown in Figure P4.29.

(a) Find the Thévenin equivalent for the subnet that lies to the right of node 2. (The ground forms the second terminal of this subnet.)

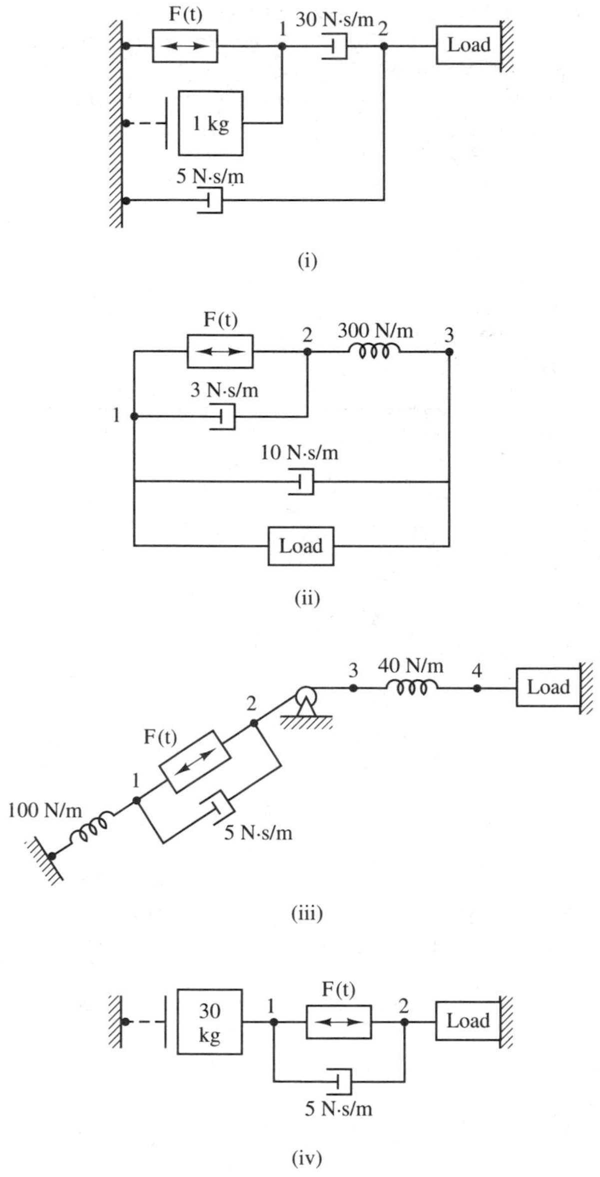

Figure P4.27.

(b) Write the equation that relates x_2 and F_2 as required by the source to the left of node 2. Write the equation that relates x_2 and F_2 as required by the Thévenin equivalent obtained in part (a). Sketch the two straight-line equations on a single graph. Label the operating point and any other significant quantities.

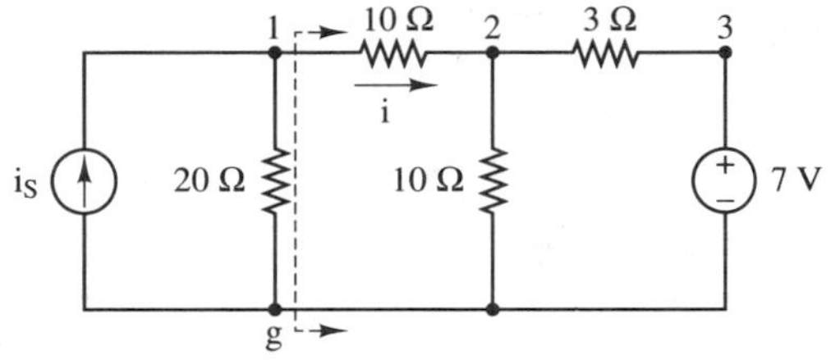

Figure P4.28.

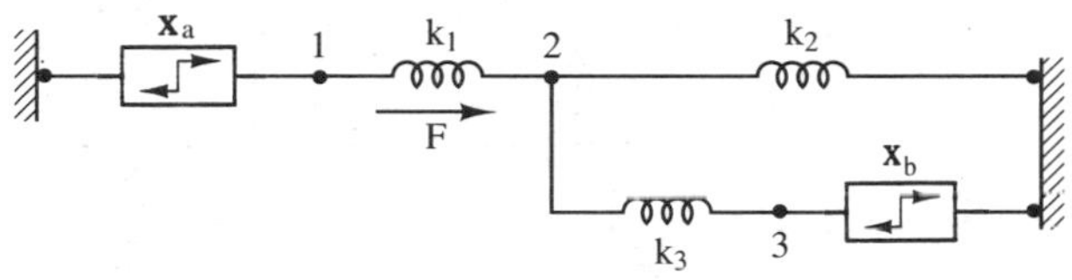

Figure P4.29.

4.30 The electrical network of Figure P4.30 is designed to deliver power to a variable resistive load R_L. The resistor pair R_1, R_2 acts as a source *stiffener*. That is, by proper choice of R_1 and R_2, we cause the effective resistance *seen* by the load to be relatively small and, hence, make the source approximate an ideal voltage source. The stiffener also protects the source by limiting the source current i_S.

(a) Find the Thévenin equivalent of the subnet to the left of the dashed line. Suppose $R_S = 50\ \Omega$. Use the Thévenin equivalent to state the restriction on R_1 and R_2 for which the resistance *seen* by the load will be no more than 2 Ω.

(b) Find the equivalent impedance *seen* by the source v. Suppose $v = 20$ V and $20\ \Omega \leq R_L \leq 200$ kΩ. Use the equivalent impedance to state the restriction on R_1 and R_2 for which the source current i_S will be no more than 200 mA.

(c) Design the stiffener. That is, find values for R_1 and R_2 that satisfy the restrictions found in parts (a) and (b). The answer is not unique.

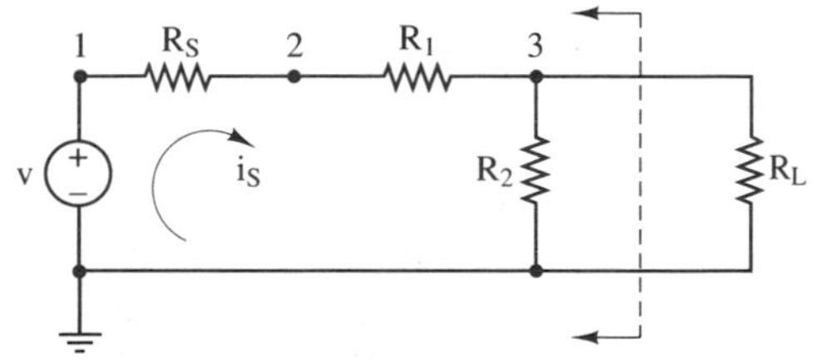

Figure P4.30.

4.31 Figure P4.31(a) shows an electrical subnet.

(a) The port variables i and v_{14} are related by a straight line, as shown in Figure P4.31(b). Find the intercepts v_{eq} and i_{eq} and the slope of the line.

(b) Represent the subnet by a Thévenin equivalent subnet.

(c) Represent the subnet by a Norton equivalent subnet.

(d) A variable-load resistor R_L is connected to the port. R_L varies from 3 Ω to 7 Ω. Superimpose the two extremes of the load line on the subnet port characteristic. Determine the range of values of the port voltage.

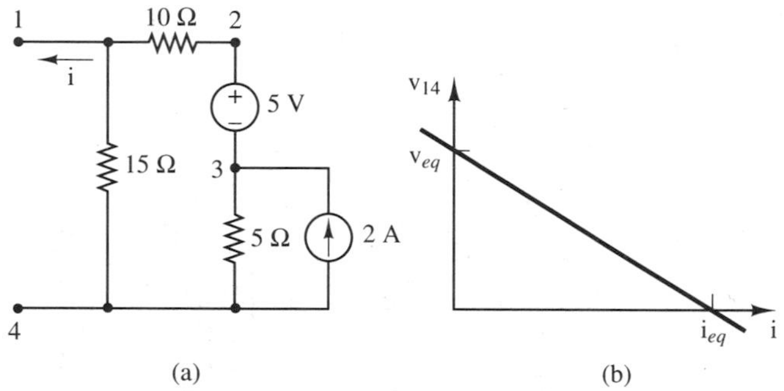

Figure P4.31. (a) An electrical subnet. (b) The v-i characteristic for the subnet.

4.32 Figure P4.32(a) shows a lumped rotational subnet that represents the power train of an automobile with a fluid clutch. (The two terminals of the subnet are node 2 and ground.)

(a) Find the Thévenin equivalent of the subnet.

(b) Find the Norton equivalent of the subnet.

(c) The load connected to node 2 exhibits coulomb friction with the load characteristic shown in Figure P4.32(b). Superimpose the straight-line subnet characteristic and the coulomb-friction load characteristic, and determine the operating point graphically.

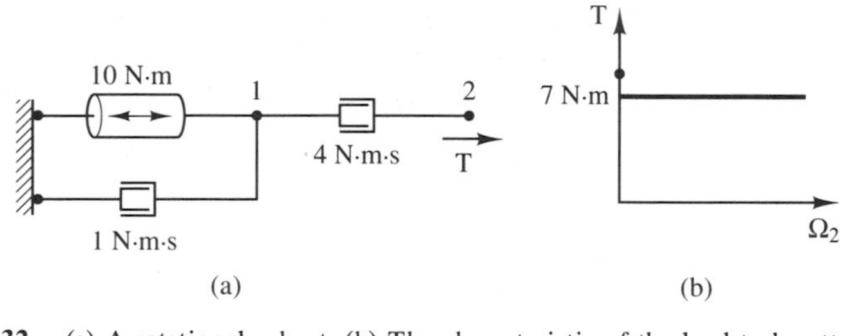

Figure P4.32. (a) A rotational subnet. (b) The characteristic of the load to be attached to the subnet.

4.33 Partition the translational network of Figures 4.36(a) and (b) into two halves at the port 1, g, with the mass in one half and the force source and damper in the other.

(a) Write the two equations that describe the two halves in terms of the port variables v_{1g} and F_2.

(b) Draw the simplified lumped model and line graph that correspond to steady-state operation.

(c) Assume that the source F is constant. Sketch and label the steady-state operating characteristics for the two subnets, as in Figure 4.40.

(d) Find the steady-state operating point, both analytically, by the equations from part (a), and graphically, by the steady-state operating characteristics from part (c).

4.34 Figure P4.34 shows an electrical network with two sources. v(t) is a step of size 10 V; i(t) is a step of size 1 A.

(a) Find the response $v_3(t)$ to the simultaneous sources.

(b) Find the response $v_3(t)$ if the current source is delayed 25 μs before it is turned on.

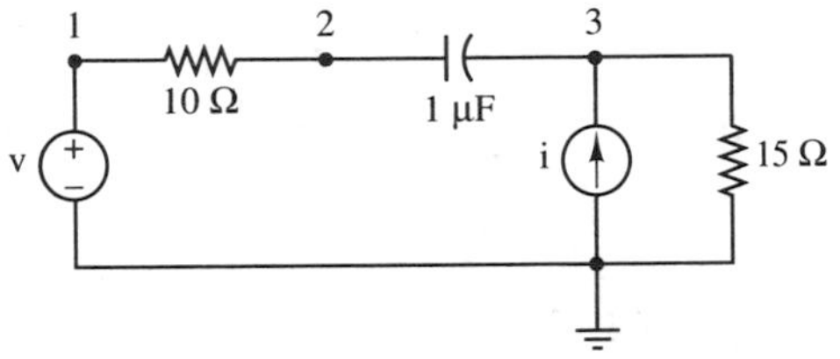

Figure P4.34.

4.35 Figure P4.35 shows a spring and damper acted on at one end by a Thévenin equivalent subnet and at the other end by a Norton equivalent subnet. F(t) is a step of size 10 N; v(t) is a step of size 5 m/s. Find the response $v_{12}(t)$ of the spring-damper to the pair of simultaneous sources.

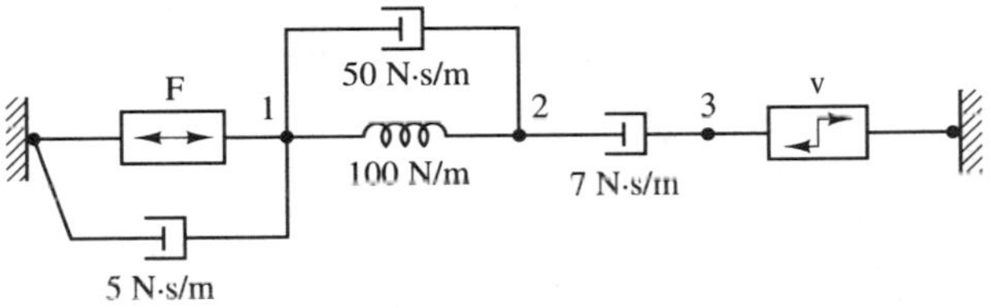

Figure P4.35.

4.36 The lumped network of Figure P4.36 represents a gondolier poling his gondola. The gondola starts from rest. The gondolier applies a constant force F = 350 N for 8 s and then removes the force from the pole (F = 0). Determine the response $v_2(t)$.

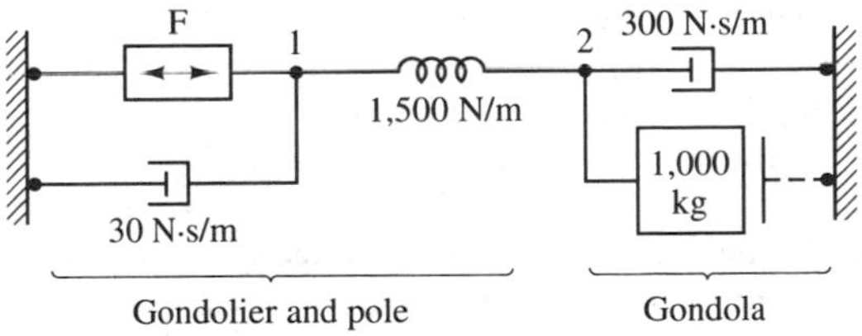

Figure P4.36.

4.37 Find and sketch the zero-state response $v_2(t)$ of the electric network of Figure P4.37(a) to the reversing-pulse waveform of Figure P4.37(b).

4.38 Find and sketch the zero-state response $v_2(t)$ of the electric network of Figure P4.38 to the specified reversing pulse.

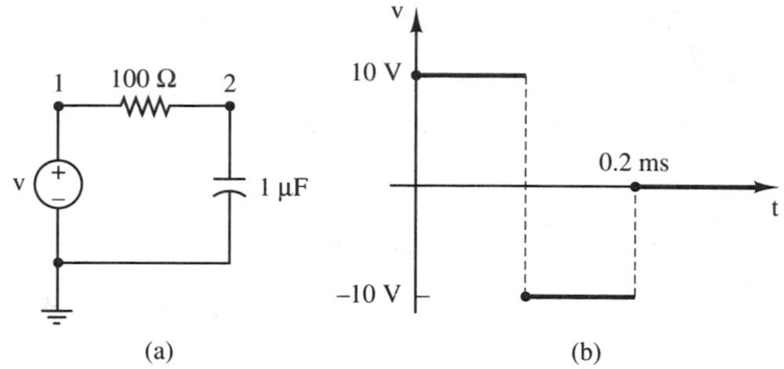

Figure P4.37. (a) Electric circuit. (b) Source waveform.

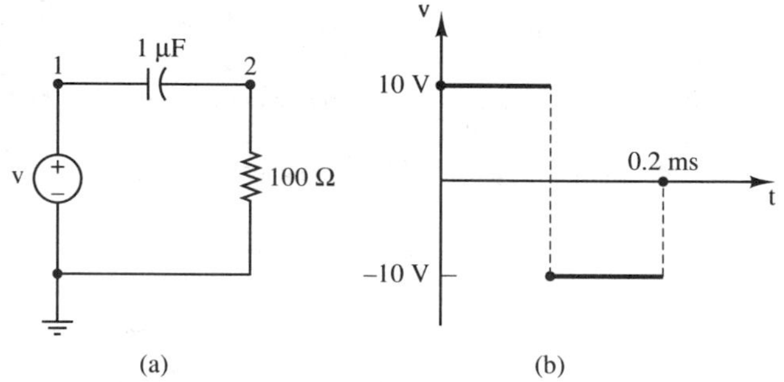

Figure P4.38. (a) Electric circuit. (b) Source waveform.

4.39 Find and sketch the zero-state response $v_1(t)$ of the network of Figure 4.43(a) to the unit impulse of Figure 4.45(b). Assume that $\tau \ll m/b$. Compare the response with Figure 4.44 for $A = 1$ and $t > \tau$.

4.40 The waveform shown in Figure P4.40(a) is called a *square wave.* It is zero for $t < 0$, after which it switches repetitively between $v = 0$ and $v = v_c$. The voltage source in the circuit of Figure P4.40(b) has such a waveform. The capacitor is unenergized at $t = 0^-$. The pulse width τ equals the time constant RC of the network.

(a) Determine the first cycle of the response $v_2(t)$, over the interval $0 \le t < 2\tau$, by decomposing the first cycle of the source into a sum of step functions and using the principle of superposition. Sketch the response.

(b) Determine the first cycle of the response $v_2(t)$, over the interval $0 \le t < 2\tau$, without using the source decomposition employed in part (a). Rather, find the zero-state response to the step input at $t = 0$, and determine the state at $t = \tau$. Then use that state to find the zero-input response of the energized system for $\tau \le t < 2\tau$.

(c) After some time, the response $v_2(t)$ approaches a time-varying steady-state waveform. Determine that waveform. (*Hint:* Assume that the network has been energized to the unknown voltage $v_2(n\tau) = v_{2c}$ at $t = n\tau$. Determine the response of the energized system to the next cycle of the square-wave input, and find the value of $v_2((n + 2)\tau)$ in terms of v_{2c}. Either the method of part (a) or the method of part (b) can be used. Then determine the value of v_{2c} for which $v_2((n + 2)\tau) = v_2(n\tau)$.) Sketch the steady-state response waveform.

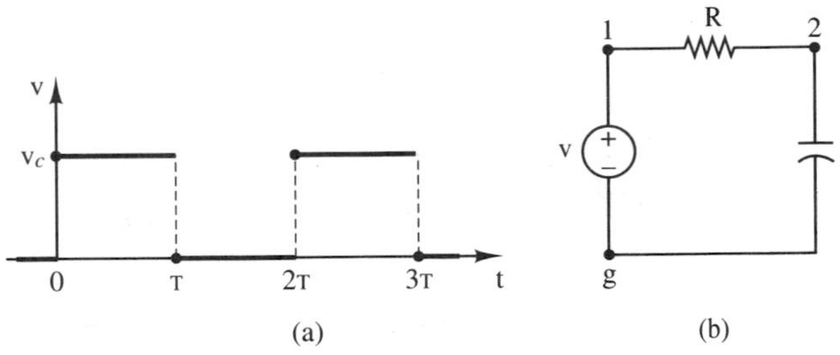

Figure P4.40. (a) Square-wave source signal. (b) Electric circuit.

4.41 For the electrical network shown in Figure P4.37(a):

(a) Find the zero-state response of node 2 to a *unit* impulse source signal v(t).

(b) Find the zero-state response of node 2 to a 5-V, 0.1-ms pulse by convolving that pulse with the impulse response found in part (a).

4.42 For the electrical network is shown in Figure P4.38(a):

(a) Find the zero-state response of node 2 to a *unit* impulse source signal v(t).

(b) Find the zero-state response of node 2 to a 10-V, 0.1-ms pulse by convolving that pulse with the impulse response found in part (a).

4.43 The rotational network of Figure P4.43(a) is driven by the torsion pulse shown in Figure P4.43(b).

(a) Determine the zero-state response of the variable Ω_1 to a unit impulse input.

(b) Use the convolution integral to determine the zero-state response $\Omega_1(t)$ to the rectangular pulse input in terms of the parameters of the source and the network.

(c) Sketch the zero-state response found in part (b) for the case in which $t_1 = \tau$ and $t_2 = 2\tau$, where τ is the time constant of the network.

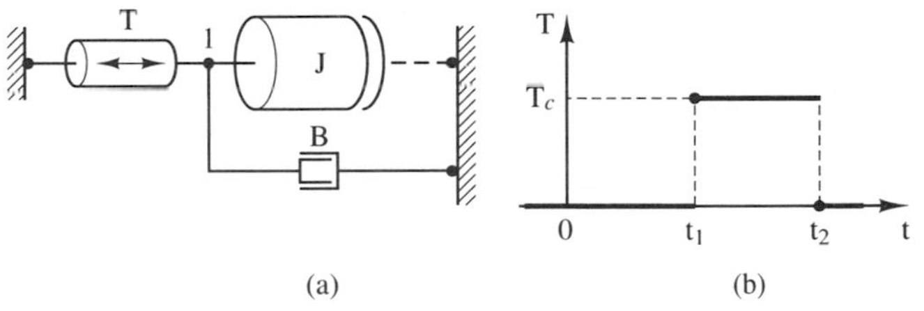

Figure P4.43. (a) Rotational network. (b) Source waveform.

4.44 The electrical network of Figure P4.44(a) is driven by the current waveform shown in Figure P4.44(b).

(a) Determine the zero-state response of v_1 to a unit impulse input.

(b) Use the convolution integral to determine the zero-state response $v_1(t)$ to the rectangular pulse input in terms of the parameters of the source and the network.

(c) Sketch the zero-state response found in part (b) for the case in which $T = \tau$, where τ is the time constant of the system.

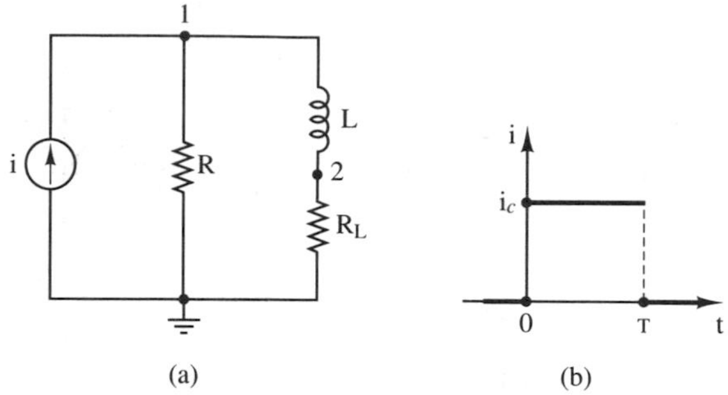

Figure P4.44. (a) Electric circuit. (b) Source waveform.

4.7 REFERENCES

4.1 Gray, Dwight E., ed., *American Institute of Physics Handbook.* New York: McGraw-Hill, 1957.

4.2 Grover, Frederick W., *Inductance Calculations.* New York: D. Van Nostrand Co., 1946.

4.3 Halliday, David, and Resnick, Robert, *Fundamentals of Physics,* 2d ed. New York: John Wiley and Sons, 1986.

4.4 Horvath, T., and Berta, I., *Static Elimination.* Chichester, U.K.: Research Studies Press, John Wiley and Sons, Ltd., 1982.

4.5 Nilsson, James W., *Electric Circuits.* Reading, MA: Addison-Wesley, 1983.

4.6 Smythe, William R., *Static and Dynamic Electricity,* New York: McGraw Hill, 1950.

4.7 Weast, Robert C., ed., *Handbook of Chemistry and Physics,* 68th ed. Boca Raton, FL: CRC Press, 1987.

4.8 Westman, H. P., ed., *Reference Data for Radio Engineers,* 4th ed. New York: International Telephone and Telegraph Corp., 1956.

CHAPTER

5

Frequency-Response Models

In chapter 3, we examined the step responses of linear networks. The step input exposes the characteristic behaviors of the network. There is one characteristic behavior, however, that is not aroused by an abrupt source: Resonance. Networks with this characteristic respond sympathetically to an input that is periodic at or near the resonant frequency.

In this chapter, we examine the response to sustained sinusoidal inputs. We determine the network behavior as a function of the *frequency* of the test signal. Most networks respond well to slowly oscillating inputs, but not to rapidly oscillating inputs. We view such networks as *low-pass filters.* They filter out rapid input-signal variations, but preserve the slow variations. The response of a network as a function of the frequency of a sinusoidal input is termed the *frequency response* of the network.

We find that the frequency response of a lumped *linear* network characterizes the behavior of that network as completely as does either its system equations or its zero-state step response. The system equations, the step response, and the frequency response are three equivalent characterizations of the network. Yet each gives different insights concerning the underlying physical system.

We begin the chapter by examining hydraulic systems. Although hydraulic networks apply and reinforce all the linear-network concepts of the previous chapters, they also introduce unavoidable nonlinearities. We use these nonlinearities to pinpoint the limitations of linear-analysis methods. We then devise a linearization technique that makes it possible to apply these methods to most nonlinear systems.

5.1 LUMPED MODELS FOR HYDRAULIC SYSTEMS

Piping systems are used for gathering, transporting, and distributing fluids (water, natural gas, oil, sewage, etc.). Home plumbing systems and airflow systems for heating, cooling, and ventilating are familiar examples of piping systems. Fluids also provide a versatile medium for transmitting signals and transporting energy, as in hydraulic braking systems and automobile cooling systems. We direct this section specifically to **hydraulic** systems, in which the moving fluid is a liquid, rather than a gas. However, many of the concepts of the section extend, at least qualitatively, to pneumatic systems.

Pressure and Flow Rate

A quantity of fluid at rest cannot sustain a shearing force—a force tangential to its surface—because the fluid layers simply slide over one another. Fluids tend to change shape under applied forces. Within a *rigid* object, opposing force vectors produce a scalar compressive force. There is also a scalar compressive phenomenon in a fluid. However, because the fluid is not rigid, the opposing force vectors must be applied equally in all three dimensions in order to contain the fluid. The balanced *containing forces* produce an internal pressure equal to the normal force per unit area acting on the surfaces of the quantity of fluid. Imagine enclosing a point in the fluid with a small fictitious spherical surface of radius r. A free-body diagram of such a surface and the enclosed fluid is shown in Figure 5.1. The surface vector $\vec{S}$ is an inward-pointed vector with magnitude equal to the area of a small piece of the surface. The inward-pointing force vector $\vec{F}(\vec{S})$ acts in the direction normal to that piece of surface. The **absolute fluid pressure** P at the center is the limiting value of the ratio of the force and surface-area vectors:

$$P = \lim_{r \to 0} \frac{\vec{F}(\vec{S})}{\vec{S}} \tag{5.1}$$

Pressure is a scalar quantity. Although fluid pressure varies in three dimensions, we shall be concerned primarily with its variation along the pipes in which the fluid flows.

Fluid pressure P is a potential; that is, it obeys the path law given by equation (2.28). The pressure difference between two points in the fluid is a potential difference. As we traverse a fluid path between two arbitrary points in the piping system, we observe a sequence of pressure differences whose sum is the total pressure difference between the

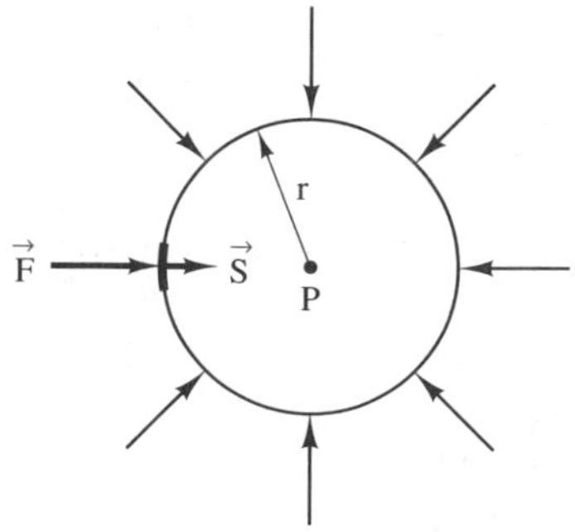

Figure 5.1 Fluid pressure.

two points. The total pressure difference we observe is the same, regardless of which path we follow between the two points.

The measurement of pressure is accomplished by means of a gage. As is typical of devices that measure potential, pressure gages measure *pressure difference* rather than absolute pressure. (Some pressure gages measure the pressure at a point in a pipe relative to the atmospheric pressure surrounding the gage; others measure the pressure drop between two points in a pipe.) Therefore, we must distinguish between absolute pressure and gage pressure. The SI unit of pressure is the pascal, abbreviated Pa (1 Pa = 1 N/m^2). Other common units of pressure are *atmospheres* and *mm of Hg*. In water systems, the expression *meters of head* is also used. (One meter of head is the pressure rise encountered in dropping vertically 1 meter in water.)

Absolute pressure is considered positive if the liquid is in a state of compression. Because liquids vaporize at low absolute pressures, negative absolute pressures do not occur in hydraulic systems. The flows in a hydraulic system depend only on pressure *differences*. Accordingly, we can designate a zero-pressure reference point arbitrarily and then think of all pressures as gage pressures relative to that reference point. We use the symbol P_0 to denote the reference pressure and the symbol $\underline{\nabla}$ to denote a physical point at which the pressure is P_0. In most problems, we assume that $P_0 = 1.01 \times 10^5$ Pa (1 atmosphere), the ambient air pressure at sea level.

If the pressure is increased at one end of a level pipe, the fluid flows toward the other end. In general, fluid particles flow in the direction of a drop in pressure. The mass flow rate constitutes a flow in the sense of equation (2.27); that is, mass is conserved in the neighborhood of each point in the system. The volume flow rate is the ratio of mass flow rate to the density of the fluid. For a liquid (which is essentially incompressible), the mass flow rate and the volume flow rate are equivalent alternative flows. It is customary to use the **volume flow rate** Q to represent flow in a hydraulic system. The SI unit for volume flow rate is m^3/s; we often use the smaller unit liter/s, where 1 liter = 0.001 m^3.

The particles of fluid in a pipe do not all flow at the same rate or in the same direction. We often visualize flow in terms of streamlines. A *streamline* is a continuous line in the fluid that has the direction of the velocity vector at each point on the line (see Figure 5.2(a)). The flow follows the streamlines, and no flow crosses a streamline. We call a bundle of streamlines a *streamtube*.

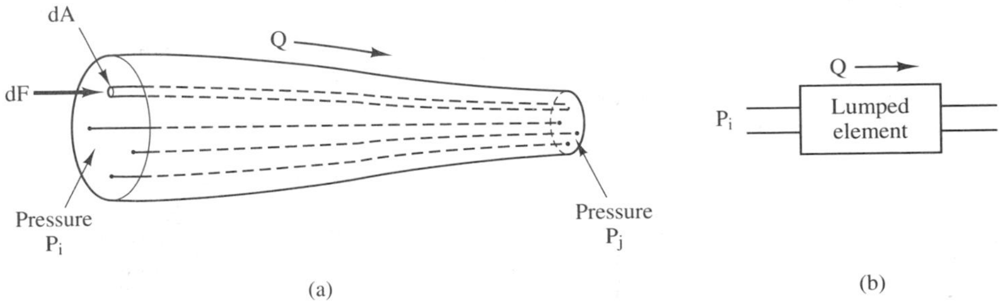

Figure 5.2 Fluid flow through a hydraulic element. (a) Streamlines in the element. (b) A lumped model of the element.

A surface that is perpendicular to a set of streamlines is a surface of equal pressure, sometimes referred to as an *equipotential surface.* (If it were not a surface of equal pressure, the pressure differences on the surface would cause flow to cross the streamlines.) If all streamlines in a pipe are parallel to the pipe, any surface perpendicular to the pipe is an equipotential surface. If the pipe bends or is not of uniform cross section, the streamlines in the pipe are not parallel, and the equipotential surfaces are not flat.

Energy and Power

Figure 5.2(a) shows the flow geometry of an element of a hydraulic system. The side boundaries of the element are streamlines. The end boundaries are equipotential surfaces, with inlet and outlet pressures $\mathrm{P_i}$ and $\mathrm{P_j}$, respectively. Let dA be the inlet *area* of an infinitesimally narrow streamtube within the element. Suppose an external force vector of infinitesimal *magnitude* dF is perpendicular to the surface dA and causes the fluid to move an infinitesimal distance dx during time increment dt. The force is balanced by the surface pressure, $\mathrm{dF = P_i\,dA}$. The incremental volume of flow across the surface during the time increment is the product $\mathrm{dV = dA\,dx}$. The energy delivered to the streamtube in order to cause the incremental flow is $\mathrm{d}E = \mathrm{dF\,dx = P_i\,dA\,dx = P_i\,dV}$. Thus, the inlet pressure $\mathrm{P_i}$ is the energy consumed per unit volume of flow into the narrow streamtube. (An analogous relation is derived for electric potential in equation (4.1).) The infinitesimal of power delivered to the narrow streamtube is $\mathrm{d}P = \mathrm{d}E/\mathrm{dt} = \mathrm{P_i\,dV/dt = P_i\,dQ}$, where dQ is the volume flow rate into the streamtube.

The pressure $\mathrm{P_i}$ appears across the whole equipotential surface at the inlet to the element. Let $\mathrm{Q} = \int \mathrm{dQ}$, the total flow across the element inlet. To produce the flow Q, an external source must deliver power

$$P_\mathrm{i} = \mathrm{P_iQ} \tag{5.2}$$

to the element, regardless of the way the flow is distributed across the equipotential surface. The fact that the flow Q is directed into the element at pressure $\mathrm{P_i}$ implies that power $\mathrm{P_iQ}$ flows *into* the element.

Because of the incompressibility of the fluid, the outlet flow rate is the same as that at the inlet. The flow out of the element is resisted by the system connected to the element. This resistance is characterized by the pressure $\mathrm{P_j}$. The power the element (or the moving fluid) transmits to the connected system is

$$P_\mathrm{j} = \mathrm{P_jQ} \tag{5.3}$$

The net power P absorbed by the element is the difference between the input power and the output power, i.e.,

$$P = \mathrm{P_{ij}Q} \tag{5.4}$$

where $\mathrm{P_{ij} = P_i - P_j}$ and Q is the volume flow rate through the element in the direction indicated by the order of the subscripts. The absorbed power is positive if the pressure drops in the direction of flow.

Figure 5.2(b) shows a lumped model of the hydraulic element. The positive direction for the flow variable Q is indicated on both the pictorial diagram of Figure 5.2(a) and the lumped model. The positive direction for Q can be chosen arbitrarily. At each end of the lumped model, there is a section of bare pipe. We view all points in the section of bare pipe as having the same pressure. Thus, bare pipe is the hydraulic analog of a perfect electrical conductor or an ideal mechanical connector. It represents a region of equal pressure—a hydraulic node. We refer to a section of bare pipe as a *node* or *point* in the lumped hydraulic circuit. We denote the pressure at the node by a symbol beside the bare pipe. On occasion, we may use a dot to indicate a point, surface, or region to which a pressure variable is assigned.

Fluid Resistance

The contact of smoothly sliding particles of fluid with the internal surfaces of a pipe or other device generates friction forces on those particles which retard the motion of the particles through the device. The slowing of fluid layers close to a stationary surface causes shearing between additional layers of fluid, as well as slowing of those layers. The overall effect is dissipation of energy in the fluid (conversion from hydraulic energy to heat) and a reduction of the volume flow rate through the device. Figure 5.3(a) illustrates this smooth-flow situation. It is characterized by nearly parallel streamlines. We refer to this smooth flow as **laminar flow.** The resistance to laminar flow is dominated by viscous friction forces in the fluid. In steady laminar flow, the pressure drop across a device is proportional to the flow Q through the device. This linear relationship motivates the definition of the ideal fluid resistor which we shall give shortly.

The fluid path in Figure 5.3(b) is obstructed drastically and abruptly; those fluid particles that encounter the obstruction must accelerate and decelerate abruptly against inertial forces in order to pass through the orifice. Inertial forces rather than viscous forces dominate the resistance to flow through the orifice, and the streamlines are not smooth and parallel downstream. This type of flow is referred to as **turbulent flow.** At high flow rates, flow becomes turbulent even for an unobstructed pipe. In steady turbulent flow, the pressure drop across a device is approximately a *quadratic* function of the flow Q through the device.

Resistance to laminar flow is proportional to the volume flow rate Q, whereas resistance to turbulent flow increases with flow rate. The dimensionless ratio of the inertial resistive force to the viscous resistive force in the fluid is called the **Reynolds number** of the flow. For flow in uniform circular pipes, the Reynolds number is

$$N_R = \frac{\rho v D}{\mu} = \frac{4\rho Q}{\pi \mu D} \tag{5.5}$$

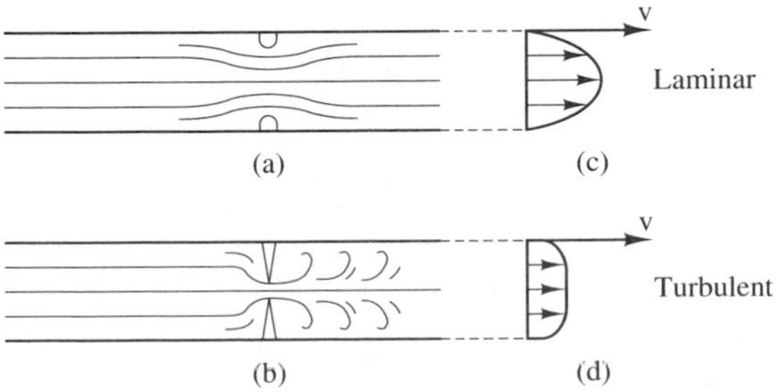

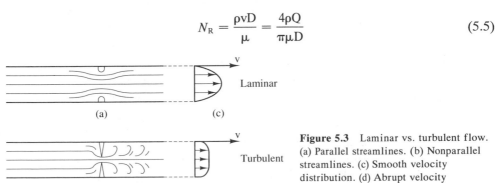

Figure 5.3 Laminar vs. turbulent flow. (a) Parallel streamlines. (b) Nonparallel streamlines. (c) Smooth velocity distribution. (d) Abrupt velocity distribution.

where D is the inside diameter of the pipe, and where ρ is the density, μ is the viscosity, and v is the velocity of the fluid; v is averaged over the cross section of the pipe. It has been found experimentally with various fluids that, for $N_R < 2{,}000$, the flow is laminar. By contrast, for $N_R > 4{,}000$, the flow is usually turbulent. The likelihood that flow will be laminar can be raised by increasing the viscosity of the fluid or the diameter of the pipe. Figures 5.3(c) and (d) show typical velocity cross sections for flow in pipes under laminar and turbulent conditions. The velocity changes more rapidly near the pipe wall in turbulent flow. As a consequence, the resistance to turbulent flow is strongly affected by the roughness of the pipe wall. The resistance to laminar flow is not affected by the roughness of the wall.

We define an **ideal fluid resistor** (or linear resistor) to be a two-terminal lumped hydraulic element such that the flow through the element is proportional to the pressure drop across the element in the direction of flow; that is,

$$P_{ij} = R_f Q \tag{5.6}$$

where the subscripts are ordered in the positive direction for Q. The proportionality factor R_f is called the **fluid resistance.** (We use the subscript f only to distinguish the fluid resistor from the electrical resistor. We often drop the subscript f and use other subscripts to distinguish one fluid resistor from another.) Comparison of equation (5.6) with equation (1.1) shows that R_f is an impedance. The SI units of fluid resistance are $N{\cdot}s/m^5$.

A section of pipe that is filled with a porous material acts like an ideal fluid resistor. Its fluid resistance is proportional to the length of the section and inversely proportional to its cross-sectional area:

$$R_f = \frac{\alpha \ell}{A} \tag{5.7}$$

The *resistivity* α of the section is a function of the porosity of the material, the mean pore size, and the viscosity of the fluid. (An analogous relation applies to an electrical resistor, as indicated in Figure 4.5.) By appropriate choice of the geometry of the pipe and the porous material, one can design a fluid resistor to match some specified resistance. Normally, we seek to minimize resistance in hydraulic systems.

An unobstructed physical pipe corresponds to extreme porosity. The resistance of a *circular* pipe to *laminar* flow can be shown to be

$$R_f = \frac{128\,\mu \ell}{\pi D^4} \tag{5.8}$$

where ℓ is the length and D is the inside diameter of the pipe. Note the strong inverse dependence of resistance on diameter.

Unfortunately, flow through pipes and valves is usually not laminar in practice. Nonetheless, even if the flow in a pipe is turbulent, we can still represent the pressure-flow relationship by equation (5.6); however, the fluid resistance will then be a function of the flow rate, $R_f(Q)$, and the equation will not be linear. The relation between pressure drop and flow under conditions of turbulent flow has been found experimentally to be approximately

$$P_{ij} = \eta_f |Q| Q \tag{5.9}$$

where

$$\eta_f = \frac{c_1 \rho \ell}{D^5} \tag{5.10}$$

and the subscripts are ordered in the positive direction for Q. The dimensionless *resistance coefficient* c_1 depends primarily on the ratio of the roughness of the wall to the diameter of the pipe.[1] It is best to measure c_1 for specific pipes. Representative values of c_1 are given in Table 5.1.

We are motivated by the nearly quadratic pressure-flow relation in turbulent flow to define an **ideal quadratic fluid resistor** to be a lumped element that satisfies equation (5.9). We call η_f the **quadratic resistor coefficient** of the element. The units of η_f are $N{\cdot}s^2/m^8$. Then the fluid resistance of the quadratic resistor is

$$R_f(Q) = \eta_f |Q| \tag{5.11}$$

Figures 5.4(a) and (b) show the symbols that we use to represent ideal linear and quadratic fluid resistors, respectively. (The shaded rectangle is intended to resemble a porous resistive material.) It is the symbols R_f and η_f that distinguish between the linear and quadratic resistors. Points i and j of each lumped model correspond to constant-pressure surfaces at the two ends of the physical hydraulic element. We shall refer to the lumped elements as the *linear resistor* R_f and the *quadratic resistor* η_f.

If we connect a linear (laminar-flow) fluid resistor in series or parallel with a quadratic (turbulent-flow) fluid resistor, the resulting resistance is neither linear nor quadratic. That is, the flow is laminar at some points and turbulent at others. Therefore, we devise

TABLE 5.1 REPRESENTATIVE c_1 VALUES FOR TURBULENT FLOW IN PIPES

Copper, steel, cast iron, or concrete	0.024
Corrugated steel	0.06

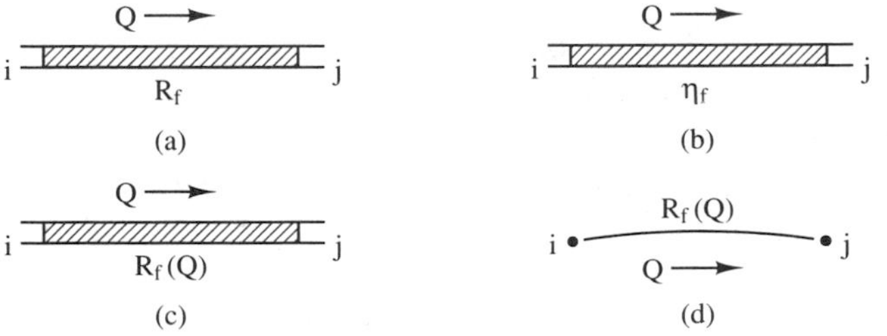

Figure 5.4 Lumped models of fluid resistors. (a) Linear resistor. (b) Quadratic resistor. (c) Flow-dependent resistance. (d) Line-graph representation of flow-dependent resistance.

[1] Fluid resistance equations are discussed in detail in chapters 3 and 21 of reference [5.4]. A good treatment of the conventional approach to hydraulics is given in reference [5.5].

the more general notation $R_f(Q)$ to denote a fluid resistance that depends on Q. The nonlinear resistance is described by the lumped-element equation

$$P_{ij} = R_f(Q)Q \tag{5.12}$$

which is comparable to the linear equation (5.6). The symbol $R_f(Q)$ has the *units* of impedance. Only if the fluid resistance is linear, however, is it correct to refer to that resistance as the impedance of the element.

The Q-dependence notation is intended as a reminder that the value of $R_f(Q)$ changes with Q. Figures 5.4(c) and (d) show a lumped representation and a line-graph representation applicable to all fluid resistors. If the resistor is linear, we drop the notation for Q dependence. The nodes of the line-graph representation correspond to constant-pressure surfaces.

Exercise 5.1: For $Q > 0$, equation (5.9) becomes $P_{ij} = \eta_f Q^2$. For $Q < 0$, it is $P_{ij} = -\eta_f Q^2$. Sketch P_{ij} vs. Q for equations (5.6) and (5.9).

As in the case of laminar flow, the quadratic resistor coefficient η_f of turbulent flow has a strong inverse dependence on the diameter of the pipe. Note the appearance of the inertial parameter ρ rather than the viscous parameter μ in equation (5.10). It may seem surprising that viscosity is not a factor in the resistance coefficient c_1. Viscosity affects the resistance indirectly, via the Reynolds number. If Q is low enough relative to μ in equation (5.5), then $N_R < 2{,}000$, the flow is laminar, and R_f is dependent on μ, as stated in equation (5.8). On the other hand, if Q is high enough relative to μ that $N_R > 4{,}000$, then the flow is turbulent, $R_f(Q) = \eta_f|Q|$, and η_f is essentially independent of μ, as shown in equation (5.10).

Fluid resistance dissipates power. According to equations (5.4) and (5.12), the power dissipated is

$$P = R_f(Q)Q^2 = \frac{P_{ij}^{\,2}}{R_f(Q)} \tag{5.13}$$

For laminar flow, $R_f(Q)$ is independent of Q; for fully turbulent flow, $R_f(Q) = \eta_f|Q|$.

Resistance of Hydraulic Events

The fluid resistance described up to now results only from friction at the pipe walls and viscous friction in the body of the fluid. If the fluid flowing in a pipe encounters a bend, an obstruction, a valve, or a contraction or expansion in the diameter of the pipe, additional turbulence is produced that can extend as far as 100 diameters down the pipe. The added turbulence causes additional friction, adds resistance to the flow, and dissipates additional power. The *additional* local resistance associated with each of these *hydraulic events* has the quadratic pressure-flow relation given in equation (5.9), with quadratic resistor coefficient

$$\eta_f = c_2\left(\frac{\rho}{D^4}\right) \tag{5.14}$$

TABLE 5.2[1] **REPRESENTATIVE VALUES OF c_2 FOR HYDRAULIC EVENTS IN PIPES**

D indicates the pipe diameter to be used in equation (5.14); D_1 and D_2 are the inlet and outlet diameters		
1. Contraction ($D = D_2$): Abrupt, $D_1/D_2 = \infty$		0.4
	Abrupt, $D_1/D_2 = 2$	0.3
	Gradual, over length D_1	0.08
2. Expansion ($D = D_1$):	Abrupt, $D_2/D_1 = \infty$	0.8
	Abrupt, $D_2/D_1 = 2$	0.6
	Gradual, over length D_2	0.4
3. Pipe bend (D = pipe diameter)		
90° pipe bend (2D < bend radius < 5D)		0.16
45° pipe bend (2D < bend radius < 5D)		0.08
90° elbow fitting		0.72
45° elbow fitting		0.32
Tee fitting (through the branch)		1.4
Tee fitting (through the run)		0.47
4. Valves, fully open (D = pipe diameter):		
	Gate type	0.12
	Globe (rotary) type	8
	Ball-check type	56
5. Restricting orifice (e.g., partially closed valve): (D = effective orifice diameter)		
	Sharp edged	2
	Round edged	1

[1] See reference [5.4] or reference [5.11].

where D is the diameter of the most restricting element of the hydraulic event, as noted in Table 5.2. We must add such turbulent-event resistances to the length-dependent resistances in the network—those described by R_f of equation (5.8) or $R_f(Q)$ of equation (5.11). The value of the dimensionless *resistance coefficient* c_2 is best determined experimentally for the specific geometry. Representative values of c_2 are shown in Table 5.2.

Steady-Flow Resistance Calculations

We illustrate the calculation of fluid resistance by an example in a familiar home setting. The geometry of the piping between the water supply inlet and a single faucet in a home is shown in Figure 5.5. We first adjust the valve to restrict the flow Q so that the Reynolds number in the pipe with the smaller diameter is $N_R = 1{,}500$. (Then flow must be laminar in both pipes, except near the elbows and tee.) We compute the flow from equation (5.5):

$$Q = \frac{\mu}{4\rho}\pi D N_R = \frac{(10^{-3}\ \text{N}\cdot\text{s/m}^2)(\pi)(0.013\ \text{m})(1{,}500)}{4(1{,}000\ \text{kg/m}^3)}$$
$$= 1.5 \times 10^{-5}\ \text{m}^3/\text{s} = 0.015\ \text{liter/s} \qquad (5.15)$$

This flow would take 67 seconds to fill a 1-liter bottle, a pretty slow rate. The calculation shows that flow in home water supply systems is usually turbulent.

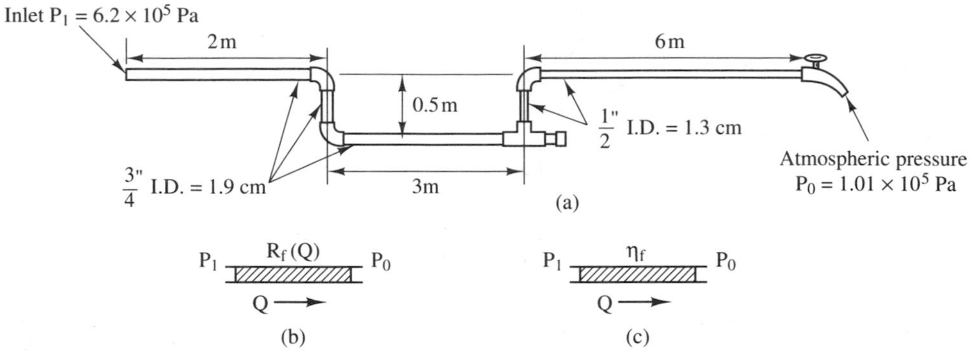

Figure 5.5 A home water supply circuit. (a) Pictorial model. (b) Whole-circuit model for low flow rate. (c) Whole-circuit model for high flow rate.

We next calculate the resistance to the flow given by equation (5.15). It has two laminar-resistance components, one for each pipe diameter. It also has five turbulent-resistance components, owing to the elbows, tee, and valve. Except for the valve, the components are

$$R_{3/4} = \frac{128\,\mu\ell}{\pi D^4} = \frac{(128)(10^{-3}\text{ N·s/m}^2)(5.5\text{ m})}{\pi\,(0.019\text{ m})^4} = 1.72 \times 10^6\text{ N·s/m}^5$$

$$R_{1/2} = \frac{128\,\mu\ell}{\pi D^4} = \frac{(128)(10^{-3}\text{ N·s/m}^2)(6.5\text{ m})}{\pi\,(0.013\text{ m})^4} = 9.27 \times 10^6\text{ N·s/m}^5$$

$$R_{3/4,\,elbows}(Q) = \frac{2c_{2a}\rho|Q|}{D^4} = \frac{2(0.72)(1000\text{ kg/m}^3)(1.5 \times 10^{-5}\text{ m}^3\text{/s})}{(0.019\text{ m})^4}$$

$$= 1.66 \times 10^5\text{ N·s/m}^5$$

$$R_{1/2,\,elbow,\,tee}(Q) = \frac{(c_{2b} + c_{2c})\rho|Q|}{D^4} = \frac{(0.72 + 1.4)(1000\text{ kg/m}^3)(1.5 \times 10^{-5}\text{ m}^3\text{/s})}{(0.013\text{ m})^4}$$

$$= 1.11 \times 10^6\text{ N·s/m}^5$$

where c_{2a}, c_{2b}, and c_{2c} represent the values of c_2 for three different hydraulic events. The sum of these resistances is

$$R_f(Q) = 1.23 \times 10^7\text{ N·s/m}^5 \tag{5.16}$$

where the notation $R_f(Q)$ acknowledges that the values of some of the resistance components depend on Q. We drop this unwieldy Q-dependence notation in some of the following calculations because Q is constant in this problem.

We cannot calculate the valve resistor coefficient η_f directly, because we do not know the valve orifice diameter D that produces the flow found in equation (5.15). Instead, we calculate R_{valve} from the pressure drop across the valve and the flow through the valve. The total pressure drop across the hydraulic circuit is $P_{10} = 5.19 \times 10^5$ Pa. The pressure drop across the sum of the resistances found in equation (5.16) is

$$R_f Q = (1.23 \times 10^7\text{ N·s/m}^5)(1.5 \times 10^{-5}\text{ m}^3\text{/s}) = 184.5\text{ Pa}$$

Therefore, the pressure drop across the valve is

$$P_{10} - R_f Q = 5.188 \times 10^5 \text{ Pa}$$

Thus, almost all of the pressure drop appears across the valve. The resistance of the valve is the ratio of the pressure drop across the valve to the flow (1.5×10^{-5} m^3/s) through the valve, i.e.,

$$R_{valve} = 3.46 \times 10^{10} \text{ N·s/m}^5 \tag{5.17}$$

The total resistance $R_f(Q)$ in the hydraulic circuit is the sum of equations (5.16) and (5.17). We portray this lumped resistance by the lumped model of Figure 5.5(b).

The resistance formula for the valve is given in equations (5.11) and (5.14). Assume that the valve has a round-edged orifice. Then the resistance formula is

$$R_{valve} = 3.46 \times 10^{10} \text{ N·s/m}^5 = \frac{c_2 \rho Q}{D^4} = \frac{1(1{,}000 \text{ kg/m}^3)(1.5 \times 10^{-5} \text{ m}^3\text{/s})}{D^4}$$

Therefore, the effective diameter of the orifice of the valve is

$$D = 0.811 \text{ mm} \tag{5.18}$$

In other words, the valve is almost turned off.

Now let us open the orifice diameter to $D = 0.5$ cm and recalculate the resistances and the flow Q. Suppose the flow is turbulent in both the large- and small-diameter pipes. (We verify this assumption after we find the flow.) Then it is appropriate to use quadratic resistors. The resistor coefficients for the two pipe sections, according to equation (5.10), are

$$\eta_{3/4} = \frac{(0.024)(1{,}000 \text{ kg/m}^3)(5.5 \text{ m})}{(0.019 \text{ m})^5} = 5.33 \times 10^{10} \text{ N·s}^2\text{/m}^8$$

$$\eta_{1/2} = \frac{(0.024)(1{,}000 \text{ kg/m}^3)(6.5 \text{ m})}{(0.013 \text{ m})^5} = 42 \times 10^{10} \text{ N·s}^2\text{/m}^8$$

The valve resistor coefficient, from equation (5.14), is

$$\eta_{valve} = \frac{(1)(1{,}000 \text{ kg/m}^3)}{(0.005 \text{ m})^4} = 160 \times 10^{10} \text{ N·s}^2\text{/m}^8$$

The resistances associated with the elbows and tee are unchanged. We get their resistor coefficients by deleting the flow factor $|Q|$ from the earlier calculations:

$$\eta_{3/4,\,elbows} = 1.1 \times 10^{10} \text{ N·s}^2\text{/m}^8, \qquad \eta_{1/2,\,elbow,\,tee} = 7.4 \times 10^{10} \text{ N·s}^2\text{/m}^8$$

The total resistor coefficient for the circuit is

$$\eta_f = 216 \times 10^{10} \text{ N·s/m}^5 \tag{5.19}$$

The lumped quadratic resistor is portrayed by the lumped model of Figure 5.5(c). The pressure-flow relationship for the lumped resistor is

$$P_{10} = \eta_f Q^2 = 216 \times 10^{10}\, Q^2 \tag{5.20}$$

The source pressure is $P_{10} = 5.19 \times 10^5$ Pa. Therefore,

$$Q = 4.9 \times 10^{-4}\ m^3/s = 0.49\ liter/s \tag{5.21}$$

This flow rate is approximately that of a wide-open water tap.

At this point, we should check our assumption of turbulence in the pipes. According to equation (5.5), the Reynolds number in the largest pipe, the one least likely to be turbulent, is $N_R = 32{,}800 \gg 4{,}000$. Thus, the flow is definitely turbulent, as assumed.

Ideal Fluid Capacitance

Tanks and reservoirs are used to store liquids. These storage devices can be treated as **fluid capacitors.** Figure 5.6(a) shows a pictorial model of a vertical-sided storage tank, perhaps the simplest example of a fluid capacitor. The pressure on the liquid surface at the top of a storage tank is the ambient air pressure P_0. The pressure difference from the tank bottom to the liquid surface is

$$P_{i0} = \rho g h \tag{5.22}$$

where h is the fluid depth (sometimes referred to as the *pressure head*), ρ is the fluid density, and g is the acceleration due to gravity. The rate of increase of the stored volume V is the inward flow Q at the tank inlet.

For a straight-sided tank, the volume of fluid stored is V = Ah. Therefore,

$$Q = A\dot{h} = \left(\frac{A}{\rho g}\right)\dot{P}_{i0} \tag{5.23}$$

We call the coefficient

$$C_f = \frac{A}{\rho g} \tag{5.24}$$

the *fluid capacitance* of the vertical-sided tank. Note that equation (5.24) applies only if the tank is neither full nor empty. The tank cannot store more fluid if it is overflowing. Neither can it supply fluid if it is empty.

Exercise 5.2: Fluid can be added to the tank only by forcing it in at the inlet against the relative tank pressure P_{i0}. The power required to insert a flow Q is $P_{i0}Q$. It might seem easier to *pour* the fluid in at the top. Show that this pouring requires the same power if the fluid must be raised from the bottom of the tank. Also, show that the fluid can be added by either route with no power consumption if the fluid is available at the top of the tank.

We define an **ideal fluid capacitor** to be a two-terminal hydraulic element that has the linear pressure-flow relation

$$Q = C_f\dot{P}_{i0} \tag{5.25}$$

The subscripts on the pressure difference are ordered in the reference direction for Q. We call the proportionality constant C_f the **fluid capacitance.** (We use the subscript f only to distinguish fluid capacitance from electrical capacitance. We often drop the subscript f

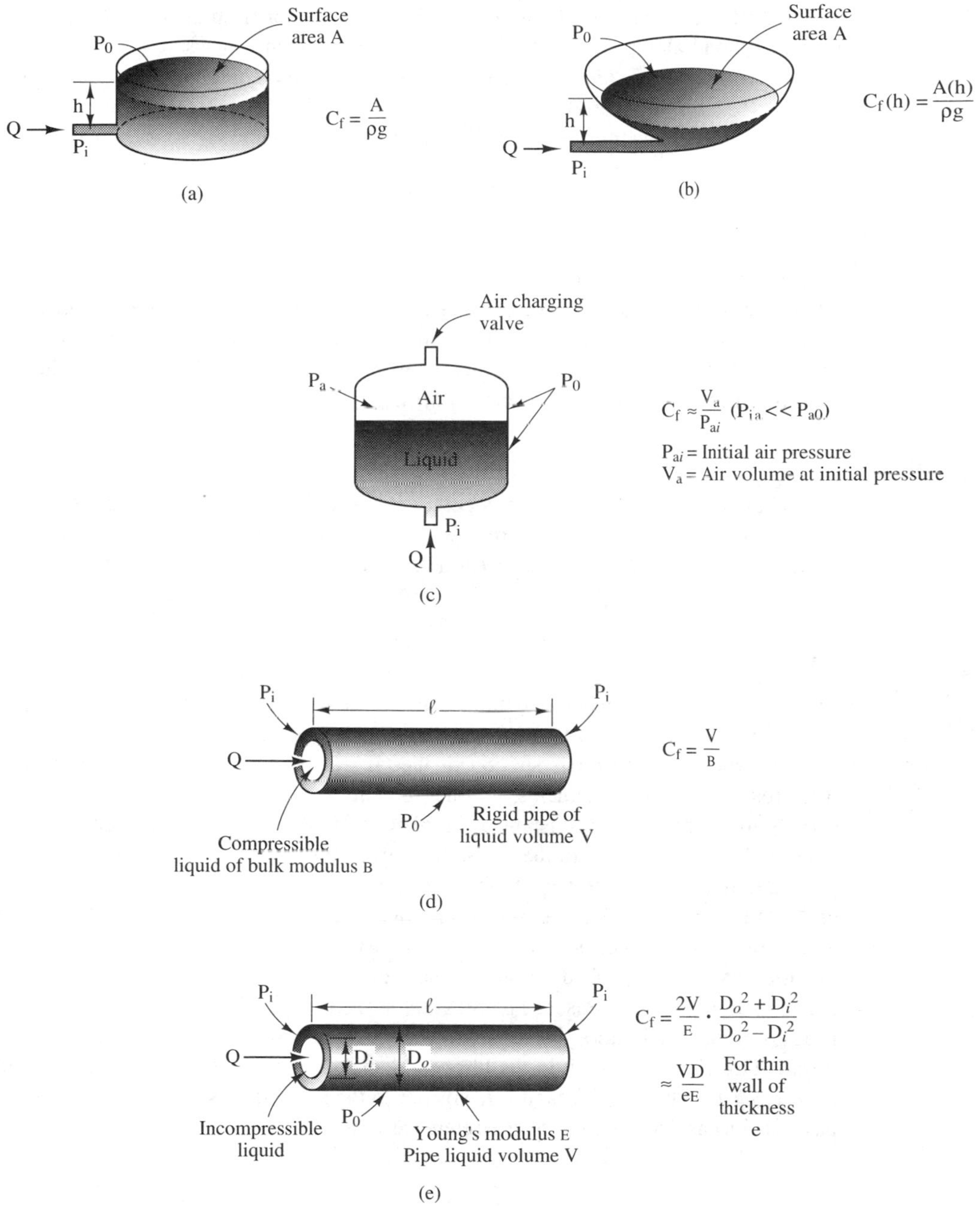

Figure 5.6 Capacitive hydraulic structures; see references [5.4] and [5.8]. The bulk modulus B of a compressible fluid is the ratio of the change in pressure to the fractional change in volume of the fluid; that is, $B = -\Delta P/(\Delta V/V)$. The bulk modulus and Young's modulus are listed for various materials in Appendix A. (a) Straight-sided tank. (b) Reservoir with sloped sides. (c) Air-pressurized accumulator tank. (d) Capacitance owing to fluid compressibility. (e) Capacitance owing to pipe compliance.

and use other subscripts to distinguish one fluid capacitor from another.) Note that one terminal of the ideal fluid capacitor is at the constant reference pressure P_0, as in the equation for the straight-sided tank. We retain the subscript zero in equation (5.25), even though $\dot{P}_0 = 0$. The SI units for fluid capacitance are m^5/N.

The fluid-capacitor equation (5.25) can also be expressed in the form

$$P_{i0} = \left(\frac{1}{sC_f}\right)Q \tag{5.26}$$

where s is the time-derivative operator and $1/sC_f$ is the impedance of the lumped element. The quantity $(1/s)Q$ should be interpreted as the integral of the flow. The integral of the flow over some time interval is the volume of fluid that enters the capacitor during that interval. If the integration begins when the capacitor is empty, the integral is the total volume of fluid stored in the fluid capacitor, i.e.,

$$V = C_f P_{i0} \tag{5.27}$$

Suppose the ideal capacitor is empty at $t = 0$. Then, by equation (5.27), $P_{i0}(0) = 0$. According to equation (5.4), the inflow Q delivers the power $P_{i0}Q$ to the capacitor. Therefore, the energy stored in an ideal fluid capacitor (at time t) is the integral

$$E = \int_0^t P_{i0}Q\,dt = \int_0^t P_{i0}C_f\dot{P}_{i0}\,dt = C_f\int_0^{P_{i0}(t)} \sigma\,d\sigma = \frac{C_f P_{i0}^2}{2} \tag{5.28}$$

Figure 5.7(a) shows the symbol that we use to represent a lumped fluid capacitor, a symbol patterned after the vertical-sided tank. The opening or port at one end of the symbol corresponds to the cross section of the tank at the level of the inlet. The pressure at this cross section is the same as the inlet pressure P_i. The dashed end of the symbol corresponds to the reference (constant-pressure) end of the physical device. In the case of an open tank, it corresponds to the whole top surface of the fluid, at the constant atmospheric pressure. When fluid enters the tank inlet, it may seem that there is no corresponding flow at the tank surface (and at the dashed end of the lumped-capacitor symbol). There is flow across the tank surface, however. As a quantity of fluid enters the inlet, a corresponding quantity of fluid is pushed across the surface, the surface level rises (slightly), and the pressure at the inlet increases. The pressure at the fluid surface remains constant. The rise in level at the fixed atmospheric pressure is the manifestation of flow at the reference terminal. That flow is explicit in the lumped-element symbol and in the line-graph symbol of Figure 5.7(b). The fluid capacitor is similar to the lumped mass of section 2.1 in that both have an inconspicuous flow at the reference terminal.

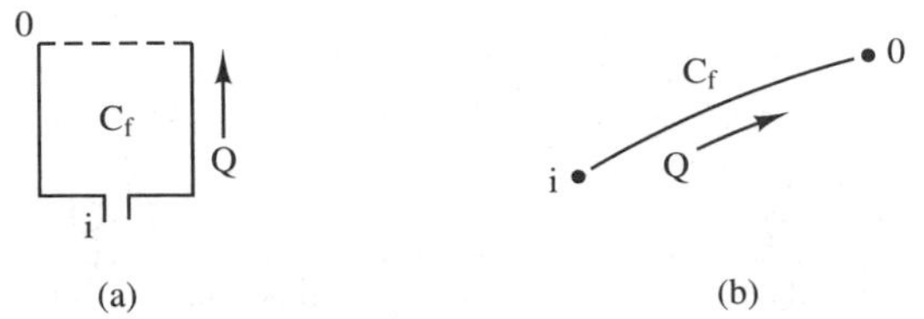

Figure 5.7 An ideal fluid capacitor. (a) Lumped model. (b) Line-graph representation.

The term *capacitance* should be distinguished from the term *capacity.* The **capacity** of a tank is the maximum volume of fluid the tank can store. The capacitance of a tank, on the other hand, indicates nothing about the tank's capacity. Rather, it measures the stored volume of fluid for a given pressure, as shown in equation (5.27). According to equation (5.25), it also measures the rate at which pressure rises as fluid enters the tank. When the tank is full (the pressure difference P_{i0} is maximum), a relatively large value of C_f implies a relatively large stored energy. Therefore, the capacitance is also an indicator of the ability of the tank to store energy.

If a reservoir does not have vertical sides, the fluid capacitance varies with the amount of stored fluid (see Figure 5.6(b)). Equation (5.24) still applies, but the area A on which it is based must be the area of the fluid *surface*. Therefore, the capacitance C_f to be used in the capacitor equation (5.25) varies with the fluid depth h, the stored volume V, and the pressure P_{i0}. The capacitance of the tank would be described better by a *plot* of V versus P_{i0}; for each value of P_{i0}, the capacitance is the slope $C_f = dV/dP_{i0}$.

Exercise 5.3: Sketch the approximate shape of the curve V vs. P_{i0} for the slope-sided reservoir of Figure 5.6(b). Is the capacitance C_f highest for a full reservoir or an empty reservoir?

Another example of a fluid capacitor is the air-pressurized *accumulator tank* of Figure 5.6(c). The fluid capacitance of an accumulator is much smaller than that of a storage reservoir. Such accumulators are used, for example, to stabilize the pressure in the hot water circulation loop of a home heating system. The tank is pressurized to the desired system pressure P_{ai}. (The pressure difference P_{ia} owing to the water in the tank is negligible compared to the system pressure and can be ignored.) Abrupt closing of a thermostatically controlled valve causes a pressure surge (owing to the momentum of the water in the pipes). The accumulator accommodates the pressure surge by allowing water to displace air in the tank.

A simple section of pipe also has fluid capacitance. Suppose we close one end of the pipe and gradually increase the fluid pressure by forcing an inflow Q at the other end. Because the incoming fluid has nowhere to go, it stretches the pipe and is, itself, compressed. From the inlet to the pipe, it appears that the pipe has accepted a small volume of fluid. We treat that fluid as if it flows into an attached container and represent that container by the lumped capacitor symbol of Figure 5.7(a). Since the external pressure on the pipe is the ambient pressure P_0, we assign the pressure P_0 to the second end of the lumped capacitor. The capacitance owing to the compressibility of the fluid is shown in Figure 5.6(d), and that owing to stretching of the pipe is given in Figure 5.6(e).

The fluid capacitances of pipes and fluids are very small indeed. Hence, they might seem of no importance. To the contrary, however, the rapid shutoff of flow in a pipe causes a pressure surge owing to the interaction between the momentum of the fluid (described in the next section) and the minute capacitance of the pipe and fluid. These pressure surges are observed as *fluid hammer.* They can be so large that they burst the pipes. Oil hammer is explored in example problem 5.5. The behavior of the cardiovascular system, treated as a network of flexible pipes, is examined in example problem 5.2.

Ideal Fluid Inertia

Because of its mass, a quantity of fluid can be accelerated only in proportion to the net force applied to it. In the context of flow in pipes, the source of force is a pressure difference, as shown in Figure 5.2. We define an **ideal fluid inertia** to be a lumped hydraulic element that obeys the formula

$$P_{ij} = I\dot{Q} \tag{5.29}$$

That is, the volume flow rate Q through the element increases in proportion to the pressure difference across the element. The subscripts on the pressure difference are ordered in the direction chosen for the flow Q. The coefficient I is called the **inertance** of the element. Figure 5.8 shows the symbol that we use to represent an ideal fluid inertia. (It is intended to resemble a rigid chunk of fluid.) Also shown is the line-graph representation of the fluid inertia. The SI units of inertance are $N{\cdot}s^2/m^5$.

The fluid-inertia relation of equation (5.29) can also be expressed in the impedance form

$$P_{ij} = (sI)Q \tag{5.30}$$

where s is the time-derivative operator. Because fluid inertia represents a tendency to resist acceleration, it may seem natural to view fluid inertia as analogous to mass. To the contrary, however, a comparison of equation (5.30) with equations (2.11) and (2.22) shows that fluid inertia is analogous to a spring rather than a mass: Inertance is analogous to the spring compliance 1/k (see Table 4.3). Fluid inertia prevents hydraulic flow from changing abruptly. Mass, described by equation (2.22), prevents velocity (potential difference) from changing abruptly. It is this contrast in the roles of potential and flow in the two cases that leads us to reject an analogy between fluid inertia and mass.

Exercise 5.4: Show that the energy stored in a fluid inertia is

$$E = \int QP_{ij}\,dt = \frac{IQ^2}{2} \tag{5.31}$$

Figure 5.9 shows the geometry of a section of a pipe filled with liquid. The mass of the *chunk* of liquid in the section of pipe is $m = \rho A\ell$, where ρ is the density of the fluid. The velocity of the fluid (averaged over the cross section) is $v = Q/A$. According to equa-

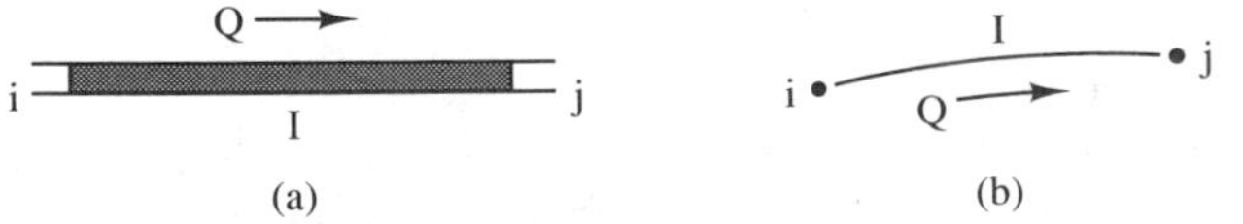

Figure 5.8 An ideal fluid inertia. (a) Lumped model. (b) Line-graph representation.

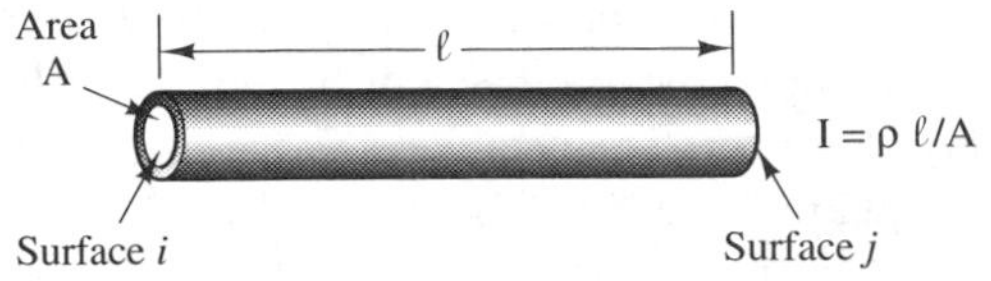

Figure 5.9 Inertance of a straight pipe.

tion (2.20), the net force on the chunk of fluid is proportional to the acceleration of the fluid; that is, $F_{ij} = m\dot{v}$, or $P_{ij}A = (\rho A\ell)\dot{Q}/A$. Hence, $P_{ij} = (\rho\ell/A)\dot{Q}$, and

$$I = \rho\ell/A \tag{5.32}$$

The inertance equation (5.32) applies to a section of pipe even if it is curved. The equation assumes that the cross sectional area of the pipe is uniform, however. For a more complicated geometry, I is more difficult to calculate.

The inertance of the home water circuit of Figure 5.5 is

$$I = \frac{4(1{,}000 \text{ kg/m}^3)}{\pi}\left(\frac{6.5 \text{ m}}{(0.019 \text{ m})^2} + \frac{5.5 \text{ m}}{(0.013 \text{ m})^2}\right) = 6.4 \times 10^7 \text{ N·s}^2/\text{m}^5 \tag{5.33}$$

The calculation treats the two sections of pipe with different diameters separately, because the fluid accelerations are different in the two sections. The sections are connected in series. In section 4.2, we found that the impedances of series elements add. Since the impedance of a fluid inertia is sI, the inertances of the two sections add as well. In Figure 5.5, we represented the home water circuit by a single fluid resistor. Figure 5.10 shows a more accurate model that includes the inertance.

According to equation (5.21), for the valve open, and with the system in steady state, $Q = 4.9 \times 10^{-4}$ m^3/s. The resistive pressure drop across the valve is $P_{30} = \eta_{valve}Q^2 = 160 \times 10^{10}Q^2 = 3.8 \times 10^5$ Pa. The pressure drop across the pipe resistor η_{pipe} is 1.4×10^5 Pa. If we close the valve, the system eventually reaches a new steady state in which $\eta_{valve} = \infty$, $Q = 0$, $P_1 = P_2$, and the pressure drop across the valve becomes the source pressure $P = 5.2 \times 10^5$ Pa. If the valve is closed quickly, however, the pressure *rises* temporarily at the valve, owing to the rapid deceleration of the fluid inertia. For a valve-closing time of 0.1 s, according to equation (5.29), the pressure drop across the inertia in the direction of flow is

$$P_{23} = I\dot{Q} \approx I\left(\frac{-Q}{0.1 \text{ s}}\right)$$

$$= \frac{(6.4 \times 10^7 \text{ N·s}^2/\text{m}^5)(-4.9 \times 10^{-4} \text{ m}^3/\text{s})}{0.1 \text{ s}} = -3.14 \times 10^5 \text{ Pa} \tag{5.34}$$

That is, the pressure *rises,* owing to deceleration, by an amount comparable to the source pressure. (To see the actual pressure versus time at the valve, we must find and solve the system equation for P_3. The differential equation is nonlinear; see problem 5.10 at the end of the chapter.) If the valve were closed slowly—say, in 1 s—the pressure rise owing to decelerating fluid inertia would be relatively small.

Note that we ignore the changes in direction of the pipe in calculating I. These changes do have an inertial effect, however. For *steady* flow Q, a 90° turn requires that we

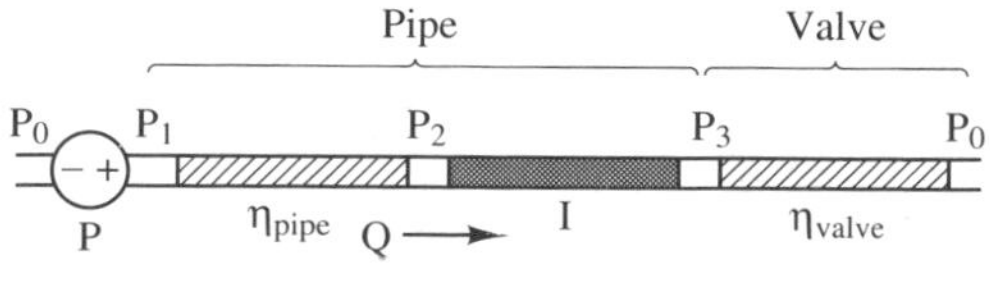

Figure 5.10 Model of home water supply circuit of Figure 5.5.

provide an *external* force in opposition to the original direction of flow to stop the forward momentum. Another external force is required in the new direction of flow. Otherwise, the turn would cause the pipe to rotate away from the new direction of flow. We see such rotation in some lawn sprinklers and in unsecured hoses.

The importance of inertance in a hydraulic network depends on the rapidity of the hydraulic source signals. Fluid inertance prevents flow from starting abruptly. Most hydraulic sources are not powerful enough to cause damaging pressure surges owing to rapid starting. Therefore, we can often ignore inertance when analyzing a hydraulic system. Rapid *decelerations* can occur, however, if valves close quickly. The **water hammer** that sometimes occurs when the valve in a washing machine or other device closes is an example of a situation in which inertance is significant. Severe water hammer can destroy a hydraulic system.

Pressure and Flow Sources

It is difficult to pump water to high elevations. To pump a liquid upward through a height h, we must generate a pressure difference $P = \rho g h$, where ρ is the density of the liquid and g is the acceleration due to gravity. Conversely, if we have a quantity of liquid at a height h above some reference level, it can provide a pressure difference equal to $\rho g h$ at the reference level. Therefore, we elevate water tanks to provide pressurized running water in buildings. Similarly, water behind dams is used to turn the hydraulic turbines that drive the electric generators in hydroelectric plants.

Figure 5.11(a) shows a lumped representation of an **ideal pressure source.** Such a source maintains a specified pressure difference P_S, regardless of the resulting flow into the attached system. The node numbers i and j at the ends of the source model correspond to constant-pressure surfaces at the inlet and outlet of a physical source. The line-graph representation of the ideal pressure source is shown in Figure 5.11(b).

A change in elevation in a pipe can be modeled as an ideal constant-pressure source if the volume flow rate through the pipe as it changes elevation is nearly zero. If the flow rate is not zero, then some pressure is lost to friction in the pipe. In this event, the lumped model shown in Figure 5.11(c) is more realistic. The effect of the pressure loss owing to friction is represented by the series resistor. The node labeled a does not exist in the physical system. The source resistance shown in the figure is analogous to the electrical source resistance discussed in section 4.2. However, unlike the electrical case, hydraulic source resistance is quadratic unless the flow is quite small.

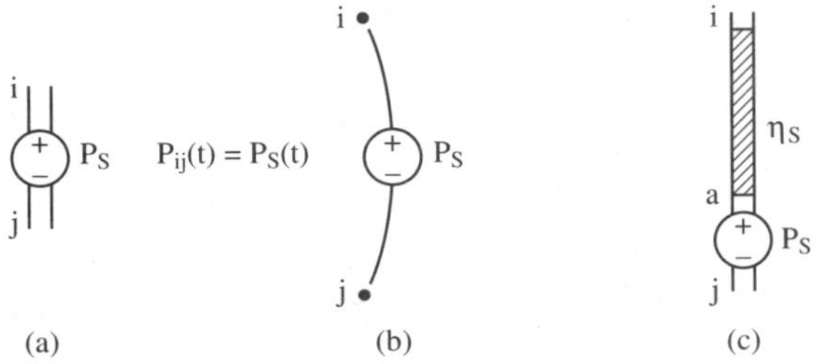

Figure 5.11 Pressure-source models. (a) Ideal lumped model. (b) Line-graph representation of the ideal lumped model. (c) Model with source resistance.

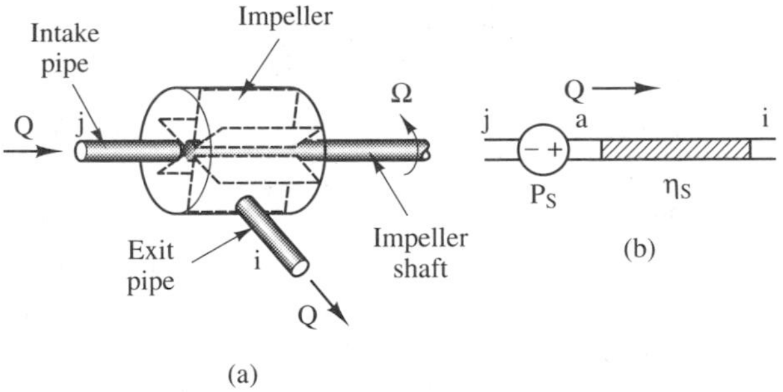

Figure 5.12 A centrifugal pump. (a) Pictorial model. (b) Lumped model.

Various types of pumps are used to move liquids in piping systems. The basic structure of one common version of *centrifugal pump* is shown in Figure 5.12(a). An electric motor rotates the impeller shaft. The impeller blades are immersed in the liquid. The centrifugal force on the liquid between the blades, owing to rotation, increases the pressure at the outer radius and decreases the pressure at the inner radius of the impeller. Hence, a centrifugal pump is basically a *pressure pump.* A surrounding housing directs the high-pressure outflow to an exit port. Replacement fluid is then sucked in through an intake port at the inner radius. Figure 5.12(b) is an appropriate lumped model for this pump. The quadratic resistance in the model arises from turbulent fluid friction.

The *gear pump* shown in Figure 5.13(a) is an example of a *positive-displacement* pump. Because the liquid cannot bypass the gears, the flow rate Q from the pump is directly proportional to the rate at which the gears are turned. Figure 5.13(b) shows another positive-displacement pump known as a *squeegee pump.* Rotation of the roller wheels forces fluid through the flexible tube. Positive-displacement pumps are basically *flow pumps.* They are typically limited to flow rates smaller than 5 liter/s.

Figures 5.14(a) and (b) show a lumped model and a line-graph representation of an **ideal fluid-flow source.** Figure 5.14(c) shows a more realistic lumped flow-source model that allows for a small amount of leakage past the gears or roller wheels. As the pressure resisting the flow increases, we would expect this leakage to increase, as the model indicates.

Section 4.2 develops rules for combining of series- or parallel-connected linear impedances. The section also shows the equivalence of Thévenin and Norton source representations. These source representations correspond to two expressions for a *straight-line* potential-flow characteristic. Since fluid flow is usually turbulent, a hydraulic

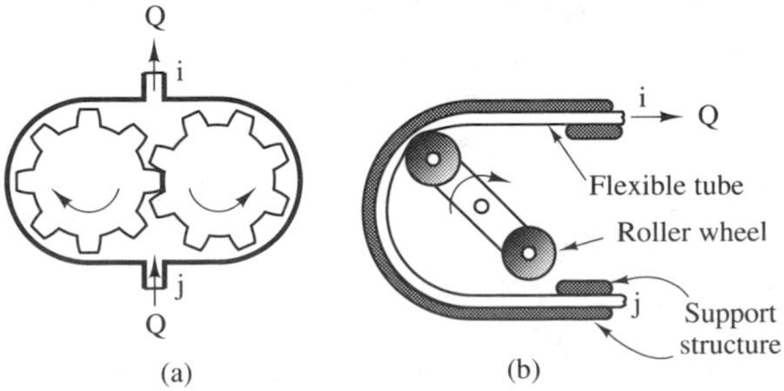

Figure 5.13 Displacement pumps. (a) A gear pump. (b) A squeegee pump.

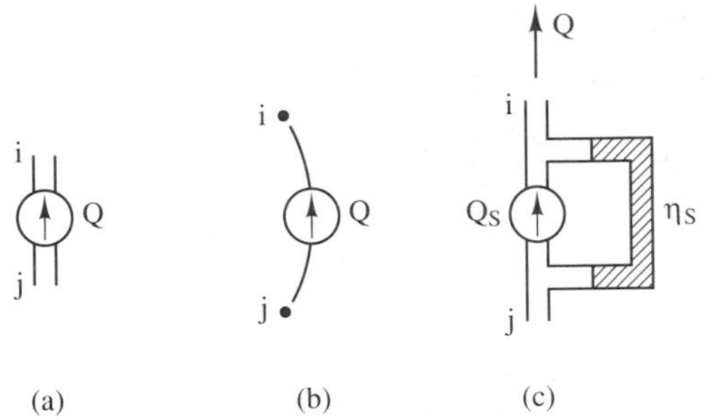

Figure 5.14 Flow-source models. (a) Ideal flow source. (b) Line-graph representation. (c) Model with source resistance.

pressure-flow characteristic is usually quadratic rather than linear. The Thévenin-Norton equivalence is not valid for quadratic pressure-flow characteristics. We examine rules for combining quadratic fluid resistors that are connected in series or parallel in problem 5.3 at the end of the chapter.

Lumped Modeling of Hydraulic Systems

The procedure for lumped modeling of a hydraulic system is essentially identical to the corresponding procedures for modeling other types of systems. The reference pressure in a hydraulic system is not zero. Nonetheless, we can assign zero potential to the reference node if we use relative (gage) pressure as the pressure variable. A fluid inertia, in contrast to mechanical inertia or mass, need not connect directly to the reference node. On the other hand, a fluid capacitor must connect directly to the reference node. That is, fluid capacitance is analogous to mass, and inertance is analogous to spring compliance, as noted in the comparison of impedances in Table 4.3. We summarize the procedure for modeling hydraulic systems here and illustrate the procedure in example problems 5.1–5.5.

THE LUMPING PROCEDURE FOR HYDRAULIC SYSTEMS

1. Draw a pictorial diagram.
2. Identify and number the equal-pressure fluid surfaces that are of significant interest. One of these surfaces must be the pressure reference, typically atmospheric pressure. Represent each numbered surface by a numbered node. Place the nodes in approximate correspondence to the layout in the pictorial diagram. Denote the reference pressure by the symbol P_0. Use the ground symbol $\underline{\nabla}$ to signify points in the pictorial diagram that are at the reference pressure.
3. Represent flow tubes (such as pipes and flow-control valves) that have significant friction or significant changes in diameter or direction by lumped fluid resistors between nodes. These lumped resistors are usually quadratic.
4. For a relatively long flow tube that might be abruptly blocked, insert a lumped fluid inertia in series with the resistor.
5. Attach a lumped fluid capacitor to each node at which *temporary* fluid storage occurs. The other end of the fluid capacitor is usually connected to the reference node. The reference node need not be isolated at a single location.

6. Insert ideal pressure sources or flow sources between appropriate nodes. In particular, a change in elevation in a pipe should be accounted for by an ideal pressure source. It may be necessary to include source resistance in order to model a source adequately.
7. Redraw the lumped network to obtain a convenient layout. In our lumped models, points connected to each other by bare pipe have the same pressure and are treated as a single node. (That is, bare pipe is the analog of a connecting line in an electrical or mechanical network.) Establish and label a positive flow direction for each branch of the network.

We demonstrate the lumping procedure by modeling the automobile cooling system of Figure 5.15. The water pump is a gear pump driven by a belt connected to the engine crankshaft. The pump forces water through the winding passages of the hot engine block. The heated water then flows through the radiator, a long folded pipe embedded in heat-radiating fins. From the radiator, the water returns to the pump.

The rate at which the radiator removes heat from the water is proportional to the temperature difference between the water and the outside air. During operation of the automobile, the water temperature rises until the amount of heat being generated by the engine equals the amount being removed at the radiator. Some of the water turns to steam. This steam increases the pressure in the system. In turn, the increase in pressure raises the boiling temperature of the water. Thus, the final pressure depends on the final temperature, which is usually over 100° C. (If we were to remove the radiator cap when it was at operating temperature, the pressure would drop suddenly, much of the hot water would flash to steam, and boiling water would spray out of the radiator.)

Our hydraulic model ignores temperature effects. (Temperature effects are examined in section 6.1.) The lumped hydraulic model will show the pressure variations around the hydraulic circuit, but will not show the absolute pressure in the circuit relative to the outside air. The pressures of most interest are the pump outlet pressure and the pump inlet pressure. These are the highest and lowest pressures in the system. We choose the inlet

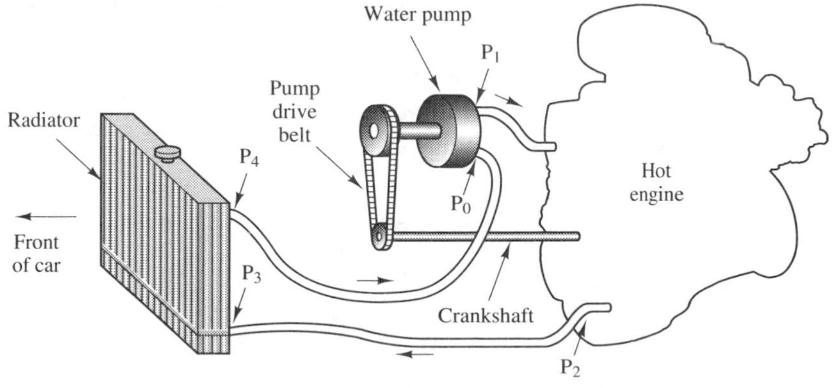

Figure 5.15 An automobile cooling system.

pressure as the reference pressure and denote it by P_0. We let P_a denote the *ambient* (surrounding) air pressure.

The flow from a gear pump is proportional to the speed at which the pump is turned. We model the pump by an ideal flow source $Q_S(t)$ that depends on the engine speed. There can be slight back-leakage of water inside the pump owing to the pressure against which the pump must push the water. We account for that back-leakage by a parallel quadratic fluid resistor with resistor coefficient η_S. We use a quadratic resistor because the flow through the convoluted passages of a pump is certain to be turbulent.

We represent the passages of the hot engine block by another quadratic fluid resistor, η_{eng}. (One might be tempted to represent the engine block as a capacitor because water is stored in the passages. However, the amount of water stored does not change with pressure. Rather, the passages are always filled, just like the passage of a pipe.) The hoses that connect the engine to the pump and to the radiator are also represented by quadratic resistors, denoted, respectively, by η_{h1} and η_{h2}. Finally, we represent the radiator and its outlet hose by quadratic resistors, denoted by η_{rad} and η_{h3}, respectively. See the lumped model in Figure 5.16.

It would also be appropriate to insert an ideal pressure source in series with *each* lumped element to account for the change in elevation between the inlet and the outlet of the corresponding physical element. For example, if the radiator outlet is higher than the inlet by h_{rad} meters, then we insert a pressure source of value $\rho g h_{rad}$ in series with η_{rad}, with the + sign oriented toward the end of the radiator model that corresponds to the lower end of the radiator. Such a pressure source is shown in Figure 5.16. The figure also shows such a pressure source for the engine, but not for the hoses and pump. We need not include a pressure source based on a difference in elevation if the pressure of that source is small relative to the resistive pressure drop across the corresponding portion of the system.

It would be appropriate to include an inertance in series with each quadratic resistor. If we are interested only in determining the flow rate Q around the main circulation path (and not interested in the pressure distribution along the path), we can lump the total inertance at a single location as shown in Figure 5.16. The inertance in the η_S branch is not important because the pump leakage is quite small. Subsequent analysis of the lumped network would show that the inertance I has little effect on behavior because the flow does not start and stop rapidly. Therefore, we can eliminate I from the model.

Finally, it is appropriate to include a lumped capacitor in each hose model. As the hose expands, owing to an increase in pressure, the amount of fluid that enters the hose

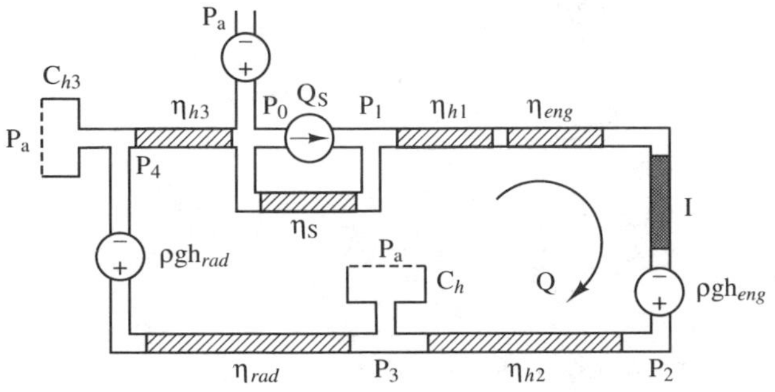

Figure 5.16 Lumped model of cooling system.

does not equal the amount that leaves. Therefore, the fluid capacitor is connected to provide an *alternate path* for the flow of fluid, as shown in Figure 5.16. We connect the other end of the capacitor to the air-pressure node Pa because the hose expands in proportion to the difference between the internal hose pressure and the outside air pressure. To which end of the hose should we assign the hose capacitance? Either end is appropriate. Or we could split the hose resistance and attach the hose capacitance at the middle. If a hose is stiff, its expansion and associated capacitance can be ignored.

The model in Figure 5.16 shows a pressure source connected between the ambient air pressure node P_a and the node P_0 that we have selected as the reference. Using this pressure source in the model amounts to assuming that we know the pressure P_0 relative to the air pressure. In general, we cannot compute P_{0a} unless we develop and use the relation between temperature and pressure. (For example, try to compute P_{0a} if C_h and C_{h3} are not included in the model.)

The procedures for drawing line-graph representations of lumped models and for formulating system equations are identical to the procedures used for mechanical and electrical networks. In line-graph representations of networks, fluid elements can be distinguished from mechanical and electrical elements by means of the unique symbols R_f, C_f, I, P, and Q or by means of their unique units.

Exercise 5.5: Draw the line-graph representation of Figure 5.16. Write the input-output system equations for the variables Q and P_{10}. (Treat the pressure P_{0a} as a known quantity.)

Example problem 5.1 develops a lumped model of a water distribution system. That model, again, exhibits the nonlinearity typical of hydraulic networks. Example problem 5.2 examines a *linear* hydraulic system: the human cardiovascular system.

The solution process for a nonlinear system equation is usually more difficult and less routine than the process developed for linear equations in sections 3.3 and 3.4. The network simplification concepts of chapter 4 apply only to *linear subnets* of nonlinear networks. Operational-model programs (such as TUTSIM) that solve the system equations numerically and network simulation programs (such as SPICE) that iterate to satisfy the node equations are able to deal with nonlinear elements. Although nonlinearity complicates analysis and design greatly, it is not an insurmountable barrier. The next section examines ways to deal with it.

Example Problem 5.1: Lumped modeling of a water distribution system.
Natural runoff in a drainage basin is trapped in a reservoir behind a dam. The water flows at at constant elevation, under control of a valve, from the reservoir to a water treatment plant. Then a centrifugal pump delivers the water to an elevated water tower. A piping system connects the water tower to two neighborhoods in the area. Sketch a pictorial model of this water collection and distribution system. Make a lumped model of the system. Discuss the hydraulic signals in the network.

Solution: A pictorial diagram is shown in Figure 5.17(a). Water is drawn from the bottom of the dam. The pressure at the bottom of the reservoir drives the water through the pipe to the treatment plant. A valve at the outlet of the dam controls the flow.

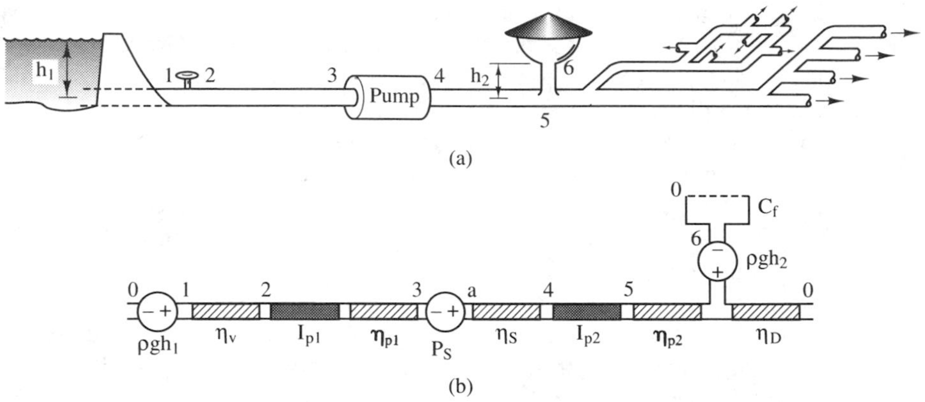

Figure 5.17 A water distribution system. (a) Pictorial model. (b) Lumped model.

A lumped model is shown in Figure 5.17(b). We represent the pressure at the dam's outlet by an ideal pressure source of magnitude $\rho g h_1$, where h_1 is the elevation of the surface of the reservoir relative to the outlet. We represent the valve by a variable resistor η_v. There is both inertance I_{p1} and fluid resistance with quadratic coefficient η_{p1} in the pipe that carries the water to the treatment plant. We use quadratic resistors throughout because the flow is probably turbulent.

The treatment process requires water transfer between a sequence of shallow ponds. From a water distribution standpoint, however, the treatment plant can be viewed as a single shallow storage tank. As long as the reservoir of treated water does not overflow or become empty, that reservoir has no effect on the flow in and out of the treatment plant. Hence, we do not include the treatment plant in the pictorial model.

Water is pumped from the treatment plant to the water tank. We presume that the base of the water tank is at the same elevation as the treatment plant. We represent the centrifugal pump by a pressure source P_S and a series resistor η_S. The water tower is represented by an ideal pressure source (corresponding to the height of the narrow column of the tower) topped by a fluid capacitor. We ignore the resistance and inertance in the short tower column, but not the resistance η_{p2} and inertance I_{p2} in the long water main that carries the water to the tower.

Water is carried from the tower by a network of large mains and smaller distribution pipes. A simplified piping network is shown. The flow in each distribution pipe depends on its fluid resistance and on how many faucets are turned on in the neighborhood at the end of the pipe. From the viewpoint of the water manager, the distribution network is a single variable resistor η_D. The value of η_D is usually a predictable function of the time of day and the weather.

All pressure sources in the lumped model are constant. Since the pump is turned off when the water tower is full, the source P_S can be turned on and off. Flow variations in the network result from variations in the valve setting at the dam outlet, from on-off cycling of the pump, from variations in the load resistances (the net effects of the faucets) in the two communities, and from variations in the level of the reservoir. The water tower keeps the inlet pressure at the various homes relatively high, regardless of the flow. (The pressure at the entrance to a home is often higher than necessary. A pressure reducer—an automatically adjusted fluid resistor—reduces the inlet pressure to a suitable level.)

Example Problem 5.2: Lumped modeling of the cardiovascular system.[2]

The human cardiovascular system is hydraulic. The right heart circulates the blood, at relatively low pressure, through the pulmonary subsystem to receive oxygen (see Figure 5.18). The left heart circulates the oxygenated blood, at high pressure, throughout the body in the following manner. The left atrium collects a quantity of blood, which arrives gradually from the pulmonary veins at low pressure, and pumps it to the left ventricle. The ventricle immediately pumps the blood into the arterial system. The aorta balloons to accept the pulse of blood. The pressurized blood in the aorta flows through branching and narrowing arteries to the capillaries and then through merging and enlarging veins to the right atrium.

The hydraulic circuit from the left ventricle to the right atrium is called the *systemic subsystem*. The heart valves are passive; each is opened and closed by the pressure difference across the valve. The systemic subsystem has high resistance and a large pressure drop enroute to the veins. A typical pressure pattern in the left ventricle is shown in Figure 5.19. It would be most appropriate to treat both ventricles as positive-displacement pumps. However, because we know the pressure pattern, we treat the left ventricle as a pressure pump. Since the arteries and veins are elastic and blood flow is laminar, *the behavior will be linear*. Typical parameters of the human cardiovascular system are shown in Table 5.3.

(a) Construct a lumped model of the systemic subsystem, and estimate its parameters.

(b) Use a network simulation tool to simulate the operation of the lumped model. Use the left-ventricle pressure pulse of Figure 5.19 as the source signal. Find the pressure pattern in the aorta. Compare the results with the typical aorta pattern of Figure 5.19.

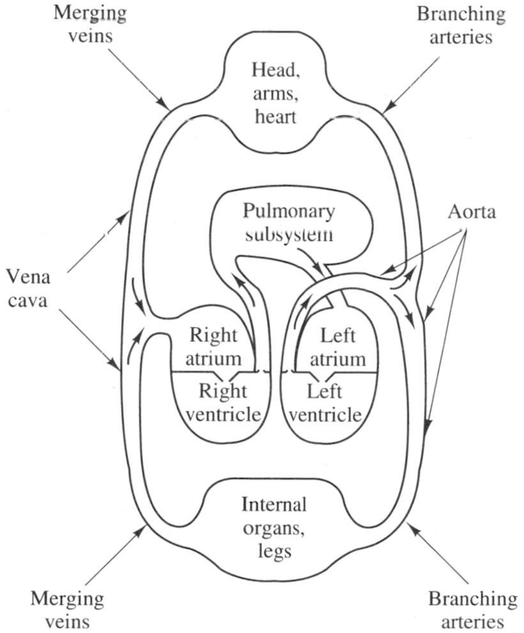

Figure 5.18 Structure of the human cardiovascular system.

[2] See reference [5.2] for a detailed explanation of cardiovascular dynamics. Lumped models for the cardiovascular system are discussed in reference [5.6].

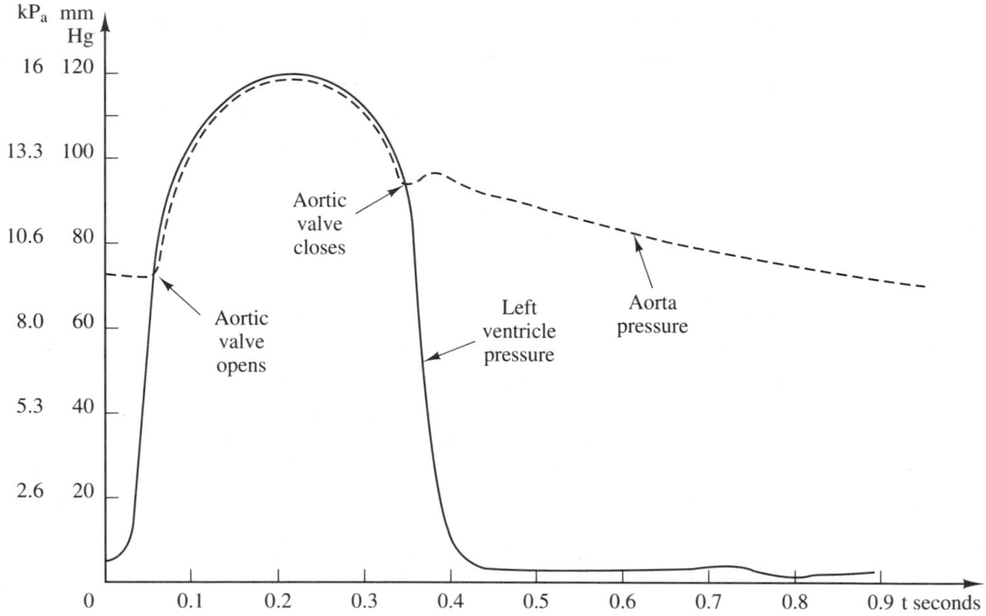

Figure 5.19 Typical human blood-pressure patterns, relative to air pressure; adapted from Figure 5.5 of reference [5.2]. © 1980 Adapted by permission of Prentice Hall, Englewood Cliffs, New Jersey.

TABLE 5.3 TYPICAL VALUES FOR THE HUMAN CARDIOVASCULAR SYSTEM (ADAPTED FROM TABLE 5-1 OF REFERENCE [5.2])

Vessel	Number	Diameter (mm)	Length (mm)	Velocity (cm/s)	Pressure (kPa)
Aorta	1	10.5	400	40	13.3
Terminal Arteries	1,800	0.6	10	<10	5.33
Arterioles	4×10^7	0.02	2	0.5	4.3
Capillaries	$>10^9$	0.008	1	<0.1	2.4
Venules	8×10^7	0.03	2	<0.3	1.33
Terminal Veins	1,800	1.5	100	1	<1
Vena Cava	1	12.5	400	20	0.33

Blood viscosity = 0.01 Pa·s   Bulk modulus of water = 2.18×10^9 Pa
Blood density = 1,060 kg/m^3 Young's modulus for aorta = 1 MPa
Ratio of wall thickness to diameter for aorta = 0.08

Solution: **(a)** The lumped model of Figure 5.20(a) is patterned after the pictorial model of Figure 5.18. The left ventrical is represented by a pressure source with the pressure pattern P(t) shown in that portion of Figure 5.19 for which the aortic valve is open. (We use a pressure source primarily because we have access to the pressure data of Figure 5.19.) The valve

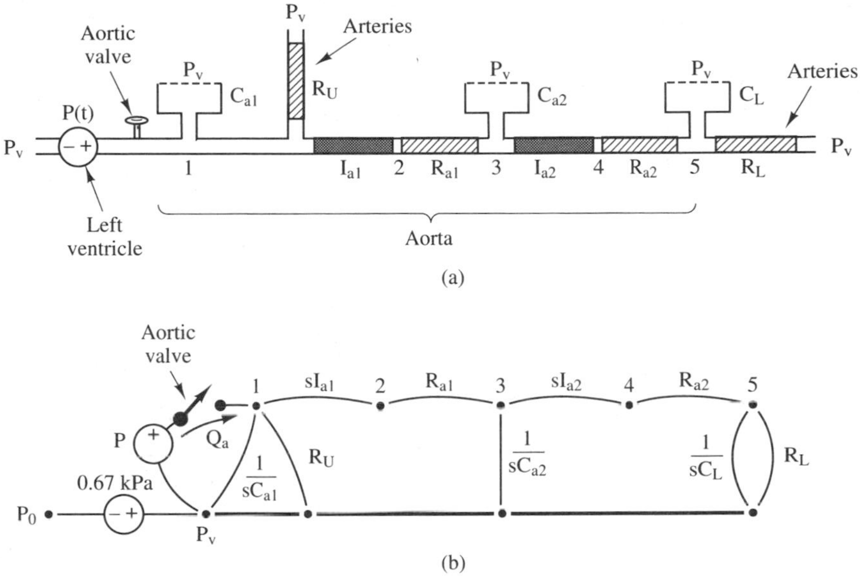

Figure 5.20 Lumped model of the systemic subsystem. (a) Lumped model. (b) Line-graph representation.

attached to the pump represents the aortic valve. The reference pressure P_v is the internal pressure of the body. We take it to be twice the pressure in the vena cava. That is, $P_v - P_0 = 0.67$ kPa, where P_0 is the ambient air pressure. We treat the flow from the vena cava to the left atrium as if it were direct, with no change in pressure.

The capacitances of the ascending and descending portions of the aorta are C_{a1} and C_{a2}. The fluid resistor R_U represents the flow resistance of the upper-body arterial tree. (Any capacitance in that arterial tree is lumped with C_{a1}.) The inertance and resistance for the ascending aorta are represented by I_{a1} and R_{a1}; for the descending aorta, they are I_{a2} and R_{a2}. We include the inertance because of the rapid closure of the aortic valve at the entrance of the aorta. The resistance and capacitance of the lower-body arterial tree are represented by R_L and C_L, respectively. Since the smaller blood vessels do not balloon in the manner of the aorta, C_L is probably not significant.

Subdividing the aorta into a larger number of segments would represent the travel of pressure waves along the aorta more accurately, but it would complicate the model further. Only testing of the behavior of the model against a real cardiovascular system can show whether the model is accurate. We compare the behavior of the model with typical measured behavior next. The model assumes that the individual is lying down; it thus ignores the effects of gravity. (The pressure associated with a 1-m column of blood is nearly as large as the average pressure produced by the left ventricle! How do you suppose your body deals with this change as you lie down and stand up? Your muscles automatically vary the resistance of the blood vessels to flow.)

We calculate the parameters of the model as follows. The dimensions and physical properties of blood and blood vessels are given in Table 5.3. The capacitance of the whole aorta (owing to its flexibility) is calculated from Figure 5.6(e):

$$C_a = \frac{VD}{E\epsilon} = \frac{\pi(0.0105\ \text{m})^2(0.4\ \text{m})}{4(10^6\ \text{N/m}^2)(0.08)} = 1.33 \times 10^{-10}\ \text{m}^5/\text{N} \tag{5.35}$$

(If we assume that the bulk modulus of blood is about the same as that of water, we find that the capacitance of the aorta owing to the compressibility of the blood is negligible compared to C_a.) The inertance of the whole aorta is calculated from equation (5.32):

$$I_a = \frac{\rho \ell}{A} = \frac{(1{,}060 \text{ kg/m}^3)(0.4 \text{ m})(4)}{\pi(0.0105 \text{ m})^2} = 4.9 \times 10^6 \text{ N·s}^2/\text{m}^5 \tag{5.36}$$

The resistance of the whole aorta, from equation (5.8), is

$$R_a = \frac{128\, \mu \ell}{\pi D^4} = \frac{(128)(0.01 \text{ N·s/m}^2)(0.4 \text{ m})}{(\pi)(0.0105 \text{ m})^4} = 1.34 \times 10^7 \text{ N·s/m}^5 \tag{5.37}$$

In the absence of detailed information about the aorta, we split this capacitance, inertance, and resistance into equal halves: $C_{a1} = C_{a2} = 2.17 \times 10^{-10}$ m^5/N, $I_{a1} = I_{a2} = 2.45 \times 10^6$ N·s^2/m^5, and $R_{a1} = R_{a2} = 6.7 \times 10^6$ N·s/m^5. We assume that the effective capacitance of the arterial tree is much smaller than that of the aorta. With little detailed information on the nature of the arterial branching, we arbitrarily let $C_L = 0.01C_{a2} = 2.17 \times 10^{-12}$ m^5/N and ignore the effect of the upper-body artery capacitance on C_{a1}.

We calculate the resistance of the whole systemic arterial tree from the data in Table 5.3 as follows. According to the formula for laminar resistance, equation (5.8), the respective resistances of individual capillaries, arterioles, and terminal arteries are $R_{cap} = 9.95 \times 10^{16}$, $R_{art} = 5.09 \times 10^{15}$, and $R_{ta} = 3.14 \times 10^{10}$, each in N·s/m^5. (Because the pressure difference across the vein tree is quite small, we ignore vein resistance). There are, on average, 25 capillaries per arteriole, 22,000 arterioles per terminal artery, and 1,800 terminal arteries associated with the single aorta. The resistance of 25 capillaries in parallel is $R_{cap}/25 = 3.98 \times 10^{15}$ N·s/m^5. Since 25 parallel capillaries are in series with each arteriole, the net resistance of the arteriole subtree is $R_{artsub} = R_{art} + R_{cap}/25 = 9.07 \times 10^{15}$ N·s/m^5. The resistance of 22,000 arteriole subtrees acting in parallel is $R_{artsub}/22{,}000 = 4.1 \times 10^{11}$ N·s/m^5. Attachment of these parallel subtrees in series with a terminal artery gives a terminal-artery subtree resistance of $R_{tasub} = R_{ta} + R_{artsub}/22{,}000 = 4.4 \times 10^{11}$ N·s/m^5. The 1,800 terminal-artery subtrees act in parallel. If they are divided equally between the upper-body and lower-body circuits, then $R_U = R_L = R_{tasub}/900 = 4.89 \times 10^8$ N·s/m^5.

(b) We used a SPICE-based program to simulate the network behavior. We simulated the behavior in two stages, to accommodate the change in network structure that occurs when the aortic valve opens or closes. The first stage begins at the instant the ventricle contracts and the aortic valve closes ($t = 0.06$ s in Figure 5.19). We estimated the initial pressures and flows at the end of the previous cardiac cycle in order to set the initial conditions on the fluid capacitors and inertias. We ran the simulation for 0.4 s, long enough to include the time at which flow from the ventricle reverses direction and closes the aortic valve.

The computed first-stage responses of the aortic pressure P_1 and flow Q_a are displayed in the left half of Figure 5.21. The pressure in the aorta must (essentially) equal the pressure in the ventricle when the aortic valve is open. At $t = 0.27$ s, Q_a becomes negative, closes the aortic valve, and ends the first stage of the simulation. Therefore, we ignore the response for $t > 0.27$ s.

We remove the ventricle pressure source from the model and use the network pressures and flows at $t = 0.27$ s as initial conditions for simulation of the second stage of the cardiac cycle. The second stage ends at the time the contracting ventricle again opens the aortic valve, $t = 0.89$ s. We show the results of the second-stage simulation in the right half of Figure 5.21. The second-stage simulated response is shifted to the right by 0.27 s in the figure.

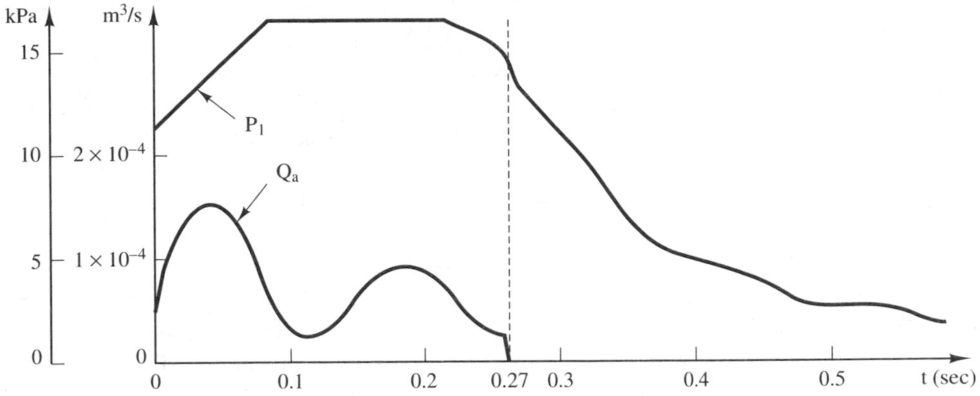

Figure 5.21 Simulated behavior of the cardiac systemic subsystem.

Let us compare the simulated aortic pressure waveform P_1 of Figure 5.21 with the typical measured aortic pressure waveform shown in Figure 5.19. (The simulated cycle is plotted earlier in time than the measured cycle by 0.06 s.) The flow reversal in the simulation closes the aortic valve after 0.27 s, as compared with the measured 0.29 s. A slight oscillation in pressure after closure of the valve appears in both plots. The physical flow reversal and pressure oscillation occur because the aortic pressure wave reflects from the points where the arteries divide. The aortic pressure in the model decays to a considerably lower value during the cycle than does the measured aortic pressure. Therefore, the beginning and ending aortic pressures from the model do not match each other.

The total volume of flow during one beat of the heart, according to the model, is about 18 ml (the area under the Q_a curve). This flow rate is smaller than typical human flow by a factor of 3.5. This fact suggests that C_a is too small by that factor. The aorta pressure P_1 in the model decays in a fashion similar to a time constant of 0.13 s, whereas the actual aorta pressure in Figure 5.19 decays about fourteen times more slowly. Since the behavior of the model is dominated by the elements R_1, R_U, and C_a, we can presume that the values of R_L and R_U are also too small.

The lumped model of Figure 5.20 is accurate enough to produce qualitatively correct behavior. Yet, we do not consider the model validated by the comparison between the measured and simulated responses. Experimental measurements of pressure along the arterial tree would help us distribute the resistances and capacitances more appropriately. A more detailed lumped model could represent tapering of the aorta and subdivision of the arteries more accurately. These refinements could be expected to produce a better match to the measured oscillations, as well as matching of the beginning and ending pressures.

5.2 NONLINEAR NETWORKS AND LINEARIZATION

Most of the physical objects discussed in chapters 2–4 act linearly enough to justify using linear analysis. But almost all physical elements are somewhat nonlinear. Springs typically become more stiff near the ends of their range of motion. The fluid capacitance of a water reservoir usually increases with stored volume. Friction phenomena tend to be nonlinear.

The coulomb friction of Figure 2.9, for example, is abruptly nonlinear. It is usually unrealistic to treat the viscous friction in a hydraulic system as if it were linear.

The analytical solution techniques of sections 3.3 and 3.4, the impedance-based equivalent-network rules developed in section 4.2, and the superposition techniques derived in section 4.3 depend fundamentally on linearity of the element equations. None of the methods should be used with any lumped element that has a nonlinear potential-flow relation. (Of course, we can apply equivalent-network transformations to the linear subnets of a nonlinear network.)

In this section, we examine four approaches to analyzing nonlinear networks: (1) Find an analytical solution to the nonlinear input-output system equation. (2) Use a network simulation tool (e.g., SPICE) to compute numerical solutions for specific parameters and source signals. (3) Solve the input-output system equation numerically, for specific parameters and source signals, by means of an operational-block simulator. (4) Replace the network by a linear network that behaves nearly the same as the nonlinear network. We call this last approach *linearization.*

Analytical Solution of a Nonlinear System Equation

The simple system of Figure 5.22(a) transfers water from a large reservoir to a smaller elevated reservoir. The system requires that we deal with nonlinear fluid resistance. A lumped model of this simple system and its line-graph representation are shown in Figures 5.22(b) and (c). We assume that the capacity of the large reservoir is infinite.

Figure 5.23 displays the pressure-flow characteristic of the pump, obtained from the manufacturer's data sheet. Since pressure loss owing to fluid friction is approximately quadratic, this pump characteristic can be approximated mathematically by the formula

$$\begin{aligned} P_{20} &= P_S - \eta_S Q^2 \\ &= 414\ \text{kPa} - (5.87\ \text{Pa}\cdot\text{s}^2/\text{liter}^2)Q^2 \end{aligned} \tag{5.38}$$

We find the parameters $P_S = 414$ kPa and $\eta_S = 5.87$ Pa·s^2/liter^2 by inserting two points from the measured curve into the quadratic equation (5.38). (If there were also some laminar flow in the system, we would add a term that is linear in Q to this quadratic form. The nature of the linearization process would not be affected.)

Exercise 5.6: Use two points from the pump characteristic of Figure 5.23 to verify (or improve) the numerical values of the parameters P_S and η_S of equation (5.38).

The pressure drop owing to friction in the pipe, according to equation (5.9), is $P_{34} = \eta_P Q^2$. For the specified pipe, the resistor coefficient is $\eta_P = c_1\rho\ell/D^5 = 7.5 \times 10^7$ N·s^2/m^8 = 75 Pa·s^2/liter^2. (The additional resistance owing to the sudden enlargement of the pipe at the reservoir is negligible compared to the friction in the pipe.) The pressure drop in the pipe owing to the rise in elevation ($h_1 = 15$ m) is $P_G = \rho g h_1 =$ 147 kPa. If the upper reservoir is filled to a level h less than 25 m, then the pressure drop across the fluid capacitor is $P_C \triangleq P_{40} = \rho g h$.

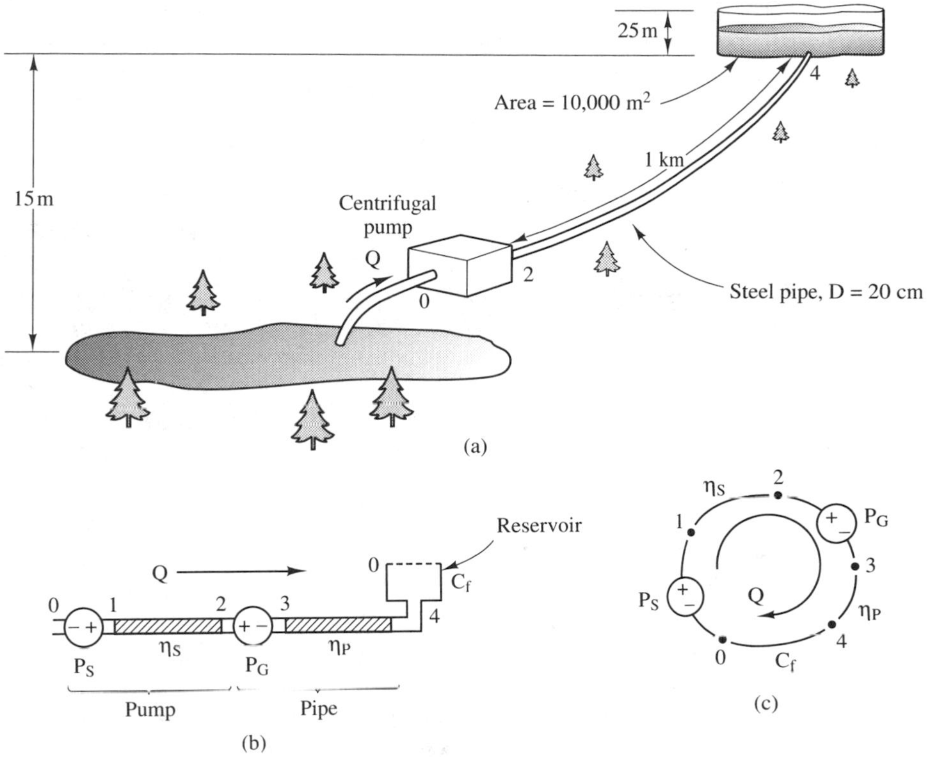

Figure 5.22 A water transfer system. (a) Pictorial model. (b) Lumped model. (c) Line-graph representation.

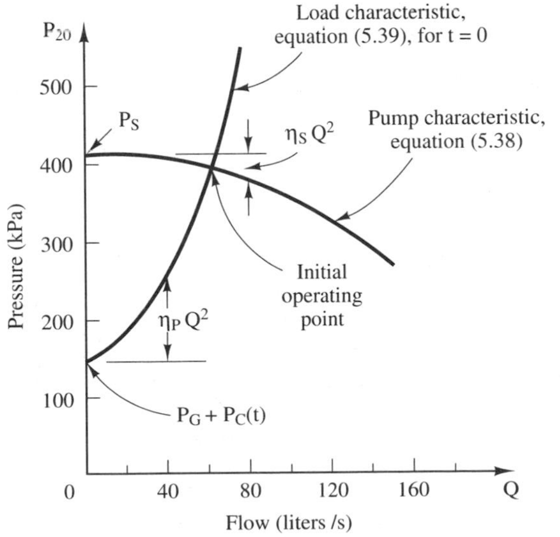

Figure 5.23 Pressure-flow characteristics.

The *load pressure* P_{20} *felt* by the pump at any instant is the sum of the gravity pressures in the pipe and reservoir and the pressure drop owing to friction in the pipe:

$$\begin{aligned} P_{20}(t) &= P_G + P_C(t) + \eta_P Q^2(t) \\ &= 147 \text{ kPa} + P_C(t) + (75 \text{ Pa}\cdot\text{s}^2/\text{liter}^2)Q^2 \end{aligned} \quad (5.39)$$

If we begin with an empty reservoir, then $P_C(0) = 0$.

We superimpose the pump characteristic given by equation (5.38) and the initial load characteristic of equation (5.39) in Figure 5.23. If we then compare this figure with the battery characteristic and load line of Figure 4.28(b) and with the steady-state operating characteristics of Figure 4.40, we see that the sets of curves are analogous. Equation (5.38) is a nonlinear source characteristic, and equation (5.39) is a nonlinear load characteristic.

The intersection of the two curves is the **initial operating point**. We find this point by solving the pair of nonlinear equations (5.38) and (5.39). The solution is

$$Q(0) = 57 \text{ liter/s} \quad \text{and} \quad P_{20}(0) = 395 \text{ kPa} \quad (5.40)$$

The graph verifies these numbers. (In this instance, solution of the equations is not difficult. It can be difficult to solve simultaneous nonlinear equations analytically, however. It is usually easier to solve them numerically by an iterative trial-and-error procedure.)

The load characteristic plotted in Figure 5.23 applies only at $t = 0$. As the upper reservoir fills, the growing reservoir pressure P_C reduces P_{24}, and the flow Q diminishes. As viewed in the figure, the growing pressure P_C raises the curve and its intercept $(P_G + P_C(t))$ and moves the intersection of the two curves to a lower value of Q. Eventually, $Q = 0$ and $P_G + P_C = P_S$.

We wish to find the flow Q(t) versus time. The loop equation for the fluid circuit is

$$-P_S + \eta_S Q^2 + P_G + \eta_P Q^2 + \left(\frac{1}{sC_f}\right)Q = 0 \quad (5.41)$$

Multiplying equation (5.41) by s, replacing s with the time derivative, and carrying out the differentiations produces

$$2(\eta_S + \eta_P)Q\dot{Q} + \frac{Q}{C_f} = 0 \quad (5.42)$$

This differential equation is first order, but nonlinear. Fortunately, dividing by Q converts it to the simple linear equation

$$\dot{Q} = \frac{-1}{2C_f(\eta_S + \eta_P)} \quad (5.43)$$

According to equation (5.24), the fluid capacitance of the reservoir is $C_f = 1.02 \text{ m}^5/\text{N}$. Therefore,

$$\dot{Q} = -6.06 \times 10^{-6} \text{ liter/s}^2 \quad (5.44)$$

We integrate equation (5.44) over the interval [0, t] and then substitute the initial flow Q(0) from equation (5.40), to obtain the solution

$$Q(t) = 57 - 6.06 \times 10^{-6} t \text{ liter/s} \tag{5.45}$$

Observe that this solution, unlike the solutions to first-order *linear* system equations, is not exponential.

When the reservoir is full, the gravity head owing to the 15-m rise of the pipe and the 25-m depth of the reservoir is 392 kPa, almost equal to the pump pressure P_S. Therefore, when the reservoir is nearly full, the flow is very slow. It takes a long time (about 80 days) to fill the reservoir completely.

We were fortunate in being able to convert the nonlinear input-output system equation (5.42) to linear form. Most nonlinear differential equations must be attacked directly. Many first-order or second-order nonlinear differential equations can be solved analytically. Each new *form* of nonlinear equation is a new challenge, however, and requires different tricks to produce a solution. Solving high-order nonlinear differential equations analytically is impractical.

Numerical Analysis of Nonlinear Networks

Numerical solution techniques are not hampered much by nonlinearity of the network elements. The drawback of these techniques, of course, is that they can be used only for specific network parameters. Numerous cases must be computed to determine the way the parameters affect the network behavior. Only then can we pick the best values for the parameters. Still, numerical solutions do give insight concerning network behavior.

In this section, we use a network simulation tool (PSPICE) to solve the nonlinear reservoir problem of the previous section. Then we change to a variable-capacitance reservoir, which makes the network more nonlinear. We use an operational-block program (TUTSIM) to find a numerical solution to the nonlinear input-output system equation for that network. Although operational-block programs typically handle a greater variety of nonlinearities than do network simulation tools, both types of numerical tools are quite versatile. We finish the section by using PSPICE to analyze the oil hammer in a cross-country oil pipeline—a nonlinear *distributed* network. We present each of these numerical solutions in the form of an example problem.

Example Problem 5.3: Simulation of nonlinear network behavior.
Consider the water transfer system and nonlinear lumped-network model shown in Figure 5.22. The water flow pattern Q(t) for this nonlinear network is given in equation (5.45). In contrast to the exponential solutions to linear system equations, this flow pattern is a straight-line function of time. The fill time of the upper reservoir is 81.6 days. Use a network simulation tool to simulate filling of the reservoir. Compare the simulation with the analytical solution.

Solution: The network simulation tool PSPICE includes a flow-controlled potential-difference source.[3] By making the potential difference of the source dependent on the flow through

[3] Dependent sources are introduced in section 2.3. A brief explanation of the use of SPICE-based simulators for nonelectrical networks is presented in Appendix C.

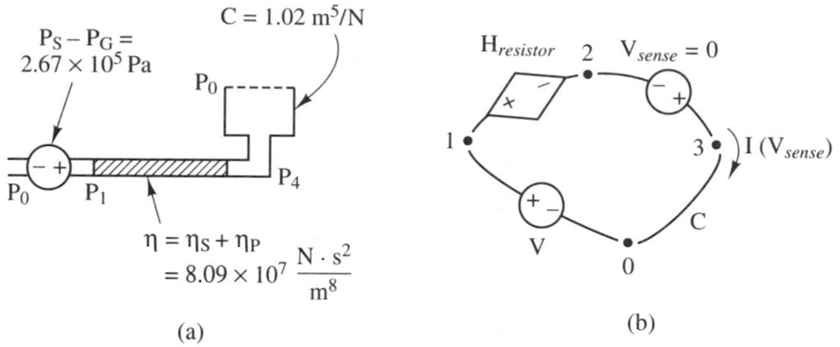

Figure 5.24 PSPICE representation of the nonlinear network of Figure 5.22. (a) Simplified lumped model. (b) PSPICE representation.

the source, we can use that source to represent a nonlinear resistance of almost any form. Figure 5.24(a) shows a simplified network that is equivalent to the lumped model in Figure 5.22(b). (The simplified version changes the order among the lumped elements, corresponding to a change in the order of the terms of the loop equation.)

Figure 5.24(b) shows a line-graph representation of the simplified network. The line graph is labeled in PSPICE notation. The symbols V and I are the electrical analogs of pressure and fluid flow. The line graph represents the quadratic resistor by a **flow-controlled pressure source** denoted $H_{resistor}$. The independent pressure source V_{sense}, with the value 0 Pa, is the PSPICE representation of a flow meter. PSPICE computes the flow, denoted $I(V_{sense})$, in the − to + direction through that source.

The pressure drop in the + to − direction across $H_{resistor}$ is denoted by the symbol $V(H_{resistor})$. PSPICE requires that we specify the pressure drop across $H_{resistor}$ as a polynomial function of the flow through $H_{resistor}$. We use the quadratic relation

$$V(H_{resistor}) = 0 + 0(I(V_{sense})) + \eta(I(V_{sense}))^2 \tag{5.46}$$

Since the pressure drop across V_{sense} is zero, equation (5.46) corresponds to $P_{13} = P_{12} = \eta Q^2$, the equation for the quadratic resistor in the network.

The capacity of the upper reservoir is 250,000 m^3. We simulate filling of the reservoir over an 80-day (6.9×10^6 s) period. The PSPICE data file for this simulation is given in Table 5.4.

The straight-line flow pattern in the simulated response shown in Figure 5.25 agrees with the analytically derived response given by equation (5.45). As one would expect, the reservoir pressure P_{40} rises more slowly as the reservoir fills.

TABLE 5.4 PSPICE SIMULATION DATA

```
Hydraulic Network of Figure 5.24
V        1   0   2.67E5:N/m2
C        3   0   1.02:m5/N      IC = 0
Hresistor    1   2    POLY(1)    Vsense    0    0 8.09E7:Ns2/m8
Vsense       3   2    0:N/m2
.TRAN     6.9E5     6.9E6     UIC
.PLOT     TRAN      I(Hresistor)     V(3)
.END
```

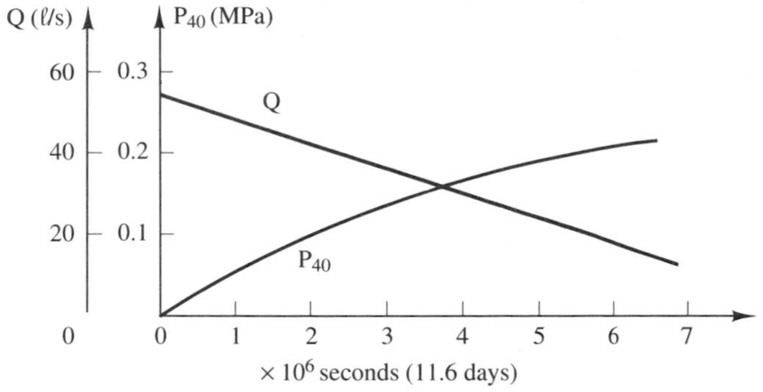

Figure 5.25 Simulated behavior of water transfer system.

Example Problem 5.4: Operational-model simulation for a nonlinear system.
Suppose the straight-sided upper reservoir in the water transfer system of Figure 5.22(a) is replaced by a natural basin that has the quadratic cross section shown in Figure 5.26. (This basin has about half the capacity of the straight-sided reservoir.) The capacitance of a reservoir is proportional to the water surface area; therefore, the capacitance of this reservoir increases as the reservoir fills.

(a) Find the capacitance of the reservoir as a function of its water level h.

(b) Find the nonlinear system equations that describe the behavior of the network. Generate an operational model of the system equations.

(c) Simulate the behavior of the network by numerical solution of the operational model.

Solution: **(a)** According to Figure 5.6(b), the capacitance of the reservoir is

$$C = \frac{A}{\rho g} = \frac{\pi r^2}{\rho g} = \frac{\pi}{\alpha \rho g} h \tag{5.47}$$

(b) The line graph for the network is shown in Figure 5.22(c). The system is characterized by the single loop equation

$$P_S - P_G = (\eta_S + \eta_P)Q^2 + \frac{V}{C} \tag{5.48}$$

where $V = (1/s)Q = \int Q$, the volume of water in the reservoir. The capacitance C varies with the reservoir water level h. That water level, in turn, is related to the water volume V. Specifically,

$$V = \int_0^V dV = \int_0^h \pi r^2\,dh = \int_0^h \pi\left(\frac{h}{\alpha}\right)dh = \frac{\pi}{2\alpha}h^2 \tag{5.49}$$

Therefore,

$$h = \sqrt{\frac{2\alpha}{\pi}}\sqrt{V} \tag{5.50}$$

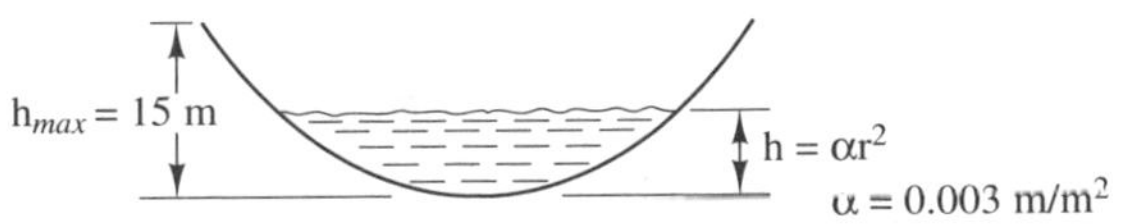

Figure 5.26 Reservoir cross section.

where V is expressed in m^3 and h is in meters. Then, from equations (5.47) and (5.50), it follows that

$$C = \frac{\sqrt{V}}{\rho g \sqrt{\alpha/(2\pi)}} \tag{5.51}$$

Substituting equation (5.51) into equation (5.48) to eliminate C and then solving for the highest order derivative yields

$$\dot{V} \equiv Q = \sqrt{\frac{(P_S - P_G) - \rho g \sqrt{\frac{\alpha}{2\pi}}\sqrt{V}}{\eta_S + \eta_P}}$$

$$= \sqrt{\frac{2.67 \times 10^5\ \text{N/m}^2 - (1{,}000\ \text{kg/m}^3)(9.8\ \text{m/s}^2)\sqrt{\frac{0.003/\text{m}}{2\pi}}\sqrt{V}}{8.09 \times 10^7\ \text{N·s}^2/\text{m}^8}}$$

$$= \sqrt{3.3 \times 10^{-3} - 2.65 \times 10^{-6}\sqrt{V}}\ \text{m}^3/\text{s} \tag{5.52}$$

The pressures and quadratic resistor coefficients that appear in equation (5.52) came from equations (5.38) and (5.39). An operational model of the nonlinear input-output system equation (5.52) with the water volume V as output is shown in Figure 5.27. The model uses two square-root blocks. The initial stored volume is $V(0) = 0$.

(c) The operational diagram of Figure 5.27 was implemented using TUTSIM. The simulated flow behavior is shown in Figure 5.28. Compare this result with the behavior of the straight-sided reservoir described by equation (5.45) and Figure 5.25.

Example Problem 5.5: Lumped model of a cross-country crude-oil pipeline.

A 50-cm (inside diameter) steel pipe with 1.3 cm wall thickness carries light crude oil at 15.6°C over the 50-km distance from the oil field to a storage area. The pump at the well head discharges the oil into the pipe at a gage pressure of 5.9 MPa. At the storage area, the oil is discharged through an open gate valve into tanks at a gage pressure of 80 kPa. The properties of the oil are as follows: specific gravity $\gamma = 0.855$, viscosity $\mu = 8.5 \times 10^{-3}$ Pa·s, and bulk modulus $B = 15 \times 10^8$ Pa. Young's modulus for the steel in the pipe is $E = 19.3 \times 10^{10}$ Pa. The velocity of sound in liquids is about 1,000 m/s.

(a) Develop a lumped model for the pumping system. Determine the parameters of the model.

(b) Determine the steady-state flow rate with the storage-tank valve fully open. Determine the corresponding pump power (at 100% pump efficiency) necessary to sustain the flow. Determine the difference in flow if there is a 300-m rise in elevation over the length of the pipeline.

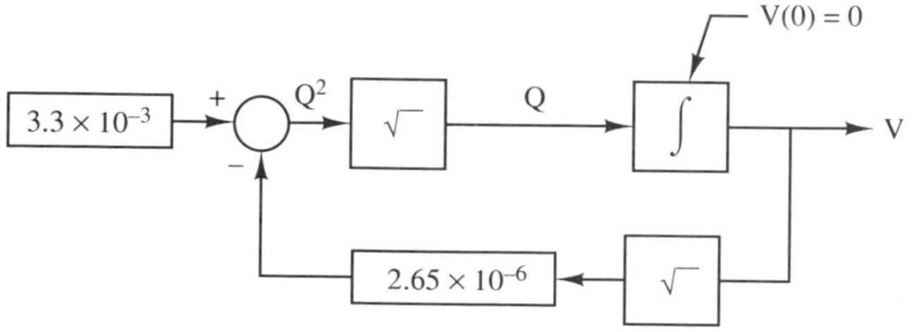

Figure 5.27 Operational diagram for equation (5.52).

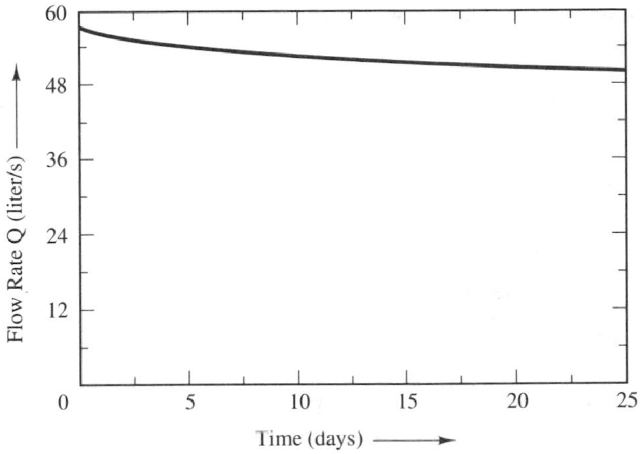

Figure 5.28 Reservoir flow pattern from Figure 5.27.

(c) Determine the kinetic energy stored in the inertance of the oil during steady flow. If abrupt valve closure were to convert that kinetic energy to potential energy (by compressing the oil and stretching the pipe), what would be the corresponding rise in pipe pressure?

(d) Closing the valve causes a pressure wave, commonly referred to as **oil hammer**, to travel along the pipe. (Severe pressure waves cause audible hammering sounds.) Suppose the valve closes in 120 seconds. Use a network simulation tool to simulate the oil hammer.

Solution: **(a)** A lumped model for the pumping system is shown in Figure 5.29(a). Since no information is given concerning the efficiency (internal resistance) of the pump, it is treated as an ideal pressure source, denoted P_p. The discharge pressure at the tank is also treated as a pressure source, denoted P_t. The pipe is divided into segments, each of which is represented by a combination of an inertia, a resistor, and a capacitor. The inertance per kilometer of pipe, according to equation (5.32), is

$$I = \frac{\rho \ell}{A} = \frac{(855 \text{ kg/m}^3)(1{,}000 \text{ m})}{\pi(0.25 \text{ m})^2} = 4.35 \times 10^6 \text{ kg/m}^4 \qquad (5.53)$$

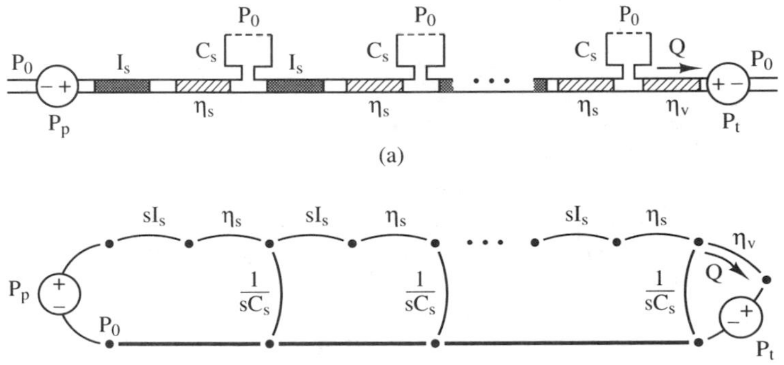

Figure 5.29 Oil pipeline. (a) Lumped model. (b) Line-graph representation.

According to Figure 5.6(d) and (e), the fluid capacitances per kilometer owing to the compressibility of the oil and the flexibility of the pipe are, respectively,

$$C_{oil} = \frac{V}{B} = \frac{(\pi)(0.25\ \text{m})^2(1{,}000\ \text{m})}{15 \times 10^8\ \text{N/m}^2} = 1.31 \times 10^{-7}\ \text{m}^5/\text{N} \tag{5.54}$$

and

$$C_{pipe} = \frac{VD}{eE} = \frac{(\pi)(0.25\ \text{m})^2(1{,}000\ \text{m})(0.5\ \text{m})}{(0.013\ \text{m})(19.3 \times 10^{10}\ \text{N/m}^2)} = 3.91 \times 10^{-8}\ \text{m}^5/\text{N} \tag{5.55}$$

The total fluid capacitance per kilometer of pipe is

$$C = C_{pipe} + C_{oil} = 1.70 \times 10^{-7}\ \text{m}^5/\text{N} \tag{5.56}$$

We assume that the flow is turbulent. (We check this assumption later.) Then, according to equation (5.10), with $c_1 = 0.024$ from Table 5.1, the quadratic resistor coefficient per kilometer is

$$\eta = \frac{c_1 \rho \ell}{D^5} = \frac{(0.024)(855\ \text{kg/m}^3)(1{,}000\ \text{m})}{(0.5\ \text{m})^5} = 6.57 \times 10^5\ \text{kg/m}^7 \tag{5.57}$$

The inertance, capacitance, and resistor coefficient of each segment of pipe depend on the number of segments used in the model. The number of segments needed, in turn, depends on the rapidity of the signals (pressures and flows) to be observed. We are to examine the pressure wave that results from closure of the valve. According to equation (2.51), the distributed model can predict the behavior of that pressure wave if the segments in the model are shorter than $\nu t_r/5$, where ν is the velocity of the wave and t_r is the *rise time* (closure time) of the valve. Since the valve's closure time is 120 s, accurate representation of the distributed pipeline requires that each segment of the model be no longer than

$$\text{segment length} = \frac{\nu t_r}{5} = \frac{(1{,}000\ \text{m/s})(120\ \text{s})}{5} = 24\ \text{km} \tag{5.58}$$

This analysis indicates that we need three segments in the model. Actually, it can be shown that the velocity of wave propagation is 1,160 m/s.[4] (This wave velocity can be observed in the simulation in part (d) of this problem.) Therefore, the segments could be as large as 25 km. We choose to use four 12.5-km segments. The parameters of the 12.5-km segments are $I_s = 5.44 \times 10^7$ kg/m^4, $C_s = 2.13 \times 10^{-6}$ m^5/N, and $\eta_s = 8.21 \times 10^6$ kg/m^7.

The valve that controls the flow at the pipe outlet is represented by a quadratic resistor. According to equation (5.14) and the value of c_2 from Table 5.2, the resistance of the open valve is insignificant relative to the resistance in each model segment. The series-parallel nature of the model is apparent in the line graph of Figure 5.29(b). We denote the flow through the valve by Q. Under normal (steady-state) operation, the flow is the same through the whole length of the pipe.

(b) In steady flow, we short-circuit the fluid inertias and open-circuit the capacitors. That is, we let $s = 0$ in the impedances sI_s and $1/sC_s$. Then the steady-state network is described by the single loop equation

$$P_p - P_t = \eta_{tot} Q^2 \tag{5.59}$$

[4] The velocity of wave propagation depends on the inertance and capacitance per unit length (see section 6.4).

where $\eta_{tot} = 4\eta_s = 3.28 \times 10^7$ kg/m^8, the total resistor coefficient for the pipeline, and $P_p - P_t = 5.9 \times 10^6$ Pa $- 8 \times 10^4$ Pa $= 5.82$ MPa. (Problem 5.3 at the end of the chapter shows that the resistor coefficients of quadratic resistors connected in series add directly.) Therefore, $Q = 0.42$ m^3/s $= 420$ liter/s (228,000 barrels/day). The power consumed by friction in the pipe is $P = (P_p - P_t)Q = 2.44 \times 10^6$ N·m/s $= 2.44$ Mw. The Reynolds number (equation (5.5)) for this flow rate is

$$N_R = \frac{4\rho Q}{\pi\mu D} = \frac{(4)(855 \text{ kg/m}^3)(0.42 \text{ m}^3\text{/s})}{(\pi)(8.5 \times 10^{-3} \text{ N·s/m}^2)(0.5 \text{ m})} = 107{,}600 \tag{5.60}$$

The flow is definitely turbulent.

The loss in pressure owing to a 300-m rise in elevation is

$$P_g = \rho gh = (855 \text{ kg/m}^3)(9.8 \text{ m/s}^2)(300 \text{ m}) = 2.5 \text{ MPa} \tag{5.61}$$

As a consequence, the pressure across η_{tot} in equation (5.59) is reduced to 3.32 MPa, and the flow is reduced to $Q = 0.318$ m^3/s. Owing to the quadratic nature of the fluid resistance, a 43% reduction in pressure causes only a 24% reduction in flow.

(c) The total kinetic energy stored in the fluid inertia, according to equation (5.31), is

$$E = \frac{IQ^2}{2} = \frac{(50)(4.35 \times 10^6 \text{ kg/m}^4)(0.42 \text{ m}^3\text{/s})^2}{2} = 19.2 \text{ MJ} \tag{5.62}$$

where I is the total inertance of the 50-km chunk of oil. Suppose this kinetic energy were converted suddenly to potential energy by stopping the flow abruptly. Hydraulic potential energy is stored in fluid capacitance—stretched pipe and compressed fluid. Let us lump all of the distributed capacitance into a single lumped fluid capacitor. According to equation (5.28), converting the kinetic energy of equation (5.62) to potential energy amounts to raising the pressure at the closed end of the pipe by the amount

$$P = \sqrt{2E/C} = \sqrt{\frac{(2)(19.2 \text{ MJ})}{(50)(1.7 \times 10^{-7}) \text{ m}^5\text{/N}}} = 2.13 \text{ MPa} \tag{5.63}$$

(d) We used PSPICE to simulate the response of the distributed network to closing of the valve. We simulated the 120-s closure of the valve by connecting an independent flow source Q_{valve} to the pipe outlet. This source had an initial value equal to the 0.42 m^3/s steady flow and reduced to zero linearly over the 120-s interval. We used a four-segment model, as determined in part (a), to represent the pipeline. Figure 5.30 shows the PSPICE model of the first segment. The quadratic fluid resistor of the segment is represented by the flow-controlled pressure source H_{r1}, in the manner described in example problem 5.3. The flow through H_{r1} in the direction of pressure drop (+ to −) is measured by the zero-pressure independent source V_{s1}. The initial value of the pressure drop across the capacitor in the final segment is $P_t = 80$ kPa.

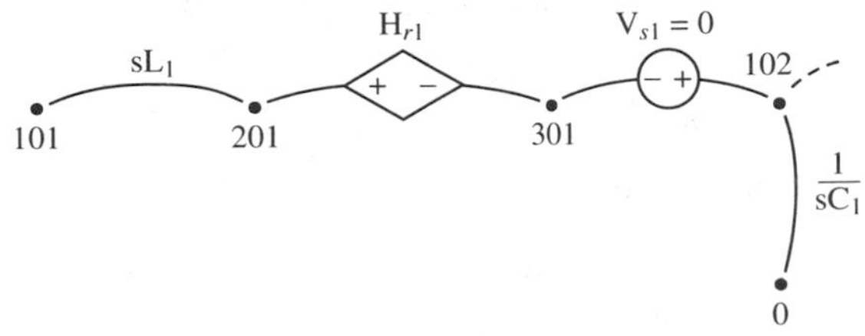

Figure 5.30 PSPICE model of first pipeline segment.

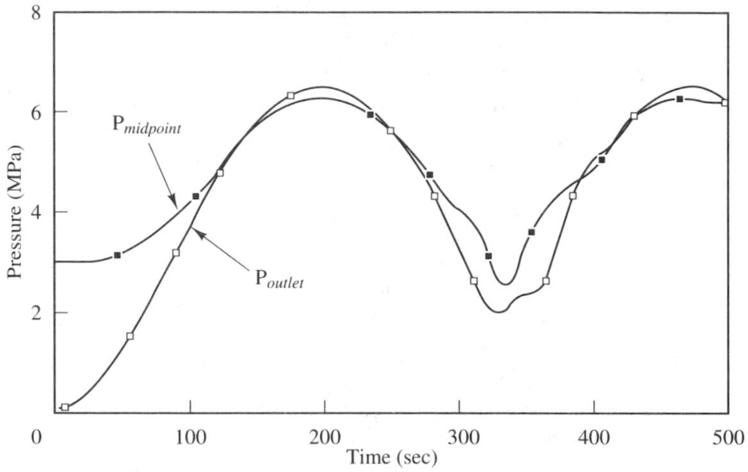

Figure 5.31 Pressure-wave behavior of a pipeline.

Figure 5.31 shows the simulated pressure waveforms at the outlet valve and at the midpoint of the pipeline. If there were no inertance, the pressures at both points would rise to the pump pressure $P_p = 5.9$ MPa by the time the flow was stopped at the valve, i.e., $t = 120$ s. The deviation of the pressure at the outlet valve from 5.9 MPa after $t = 120$ s owes partly to the limited speed (1,160 m/s) with which the fact of valve closure at the pipe outlet can be communicated back along the pipe to the pump, 50 km away. It also owes to the much lower rate at which the vast amount of kinetic energy in the moving oil can be removed. The 50-km chunk of oil continues to move toward the closed valve and becomes compressed. Once the *compressing* fluid actually ceases to accumulate at the valve (at about $t = 200$ s), the fluid has become compressed beyond the pump's 5.9-MPa pressure. Then the pressure begins to be relieved by motion of the compressed fluid back toward the pump.

As soon as the valve begins to close, a pressure wave begins to travel from the valve toward the pump. The waveforms of Figure 5.31 verify the 1,160-m/s velocity of the wave. At that velocity, the travel time of the pressure wave from the outlet valve to the midpoint of the pipeline is 21.5 s. The figure shows that the initiation of the rise in pressure at the midpoint of the simulated pipeline occurs about 22 s later than the initiation of the rise in pressure at the valve.

The strong dip in pressure at about $t = 320$ s is a result of the chunk of compressed fluid having built up reverse momentum during the release of pressure. Note that the dip arrives at the midpoint of the pipe about 20 s after it reaches the outlet valve. Then the pressure rises at the midpoint about 20 s before it rises at the outlet. Pressure dips (on the order of 4 MPa, as suggested in equation (5.63)) travel repeatedly up and down the pipe. At each location, they appear about every 6 minutes. The dips are most severe at the end of the pipe. Since the amount of flow associated with the pressure wave is low, the resistance of the pipe does not dissipate much energy. Therefore, the periodic pressure dips continue long after the valve is closed. These pressure dips constitute *oil hammer*. If we were to close the valve more quickly, the pressure dips would be more severe.

Linearization of a Nonlinear Network

In this section, we replace a nonlinear network with a linear network that behaves about the same way. Then we can use all of the linear techniques of this book to analyze the behavior of the network. The linearization process can be applied to any nonlinear phenomenon that is not infinitely abrupt. Analysis by linearization can provide insight that is not available from numerical simulations.

Consider again the water transfer system and nonlinear lumped model shown in Figure 5.22. The pump characteristic and load characteristic are superimposed to find an operating point in Figure 5.23. In Figure 5.32, we approximate the pump and load characteristics by their tangents at the initial operating point. The linear approximations are accurate for values of Q that do not vary *too far* from the point of tangency. The equation for the straight-line pump approximation can be expressed by the *linear* equation

$$\begin{aligned} P_{20} &\approx \hat{P}_S - \hat{R}_S Q \\ &= 433 \text{ kPa} - (0.669 \text{ kPa·s/liter})Q \end{aligned} \tag{5.64}$$

Equation (5.64) represents an ideal pressure pump of pressure $\hat{P}_S$ in series with a linear resistor $\hat{R}_S$. The resistance is the negative slope of equation (5.38), $\hat{R}_S = -dP_{20}/dQ$, evaluated at the initial flow $Q(0) = 57$ liter/s. We compute the pressure $\hat{P}_S$ from equation (5.64) by substituting the initial values $P_{20} = 395$ kPa and $Q = 57$ liter/s from equation (5.40). The numerical values shown in equation (5.64) can be verified graphically in Figure 5.32.

The equation of the straight-line approximation to the load (pipe) characteristic is

$$\begin{aligned} P_{20} &\approx \hat{P}_P + \hat{R}_P Q \\ &= -92.4 \text{ kPa} + (8.55 \text{ kPa·s/liter})Q \end{aligned} \tag{5.65}$$

The linear resistance is the slope of equation (5.39), $\hat{R}_P = dP_{20}/dQ$, evaluated at $Q(0) =$ 57 liter/s. We compute the pressure $\hat{P}_P$ from equation (5.65) by letting $P_{20} = 395$ kPa and $Q = 57$ liter/s.

Exercise 5.7: Use equations (5.38), (5.39), (5.64), and (5.65) to carry out a numerical evaluation of $\hat{R}_S$, $\hat{P}_S$, $\hat{R}_P$, and $\hat{P}_P$.

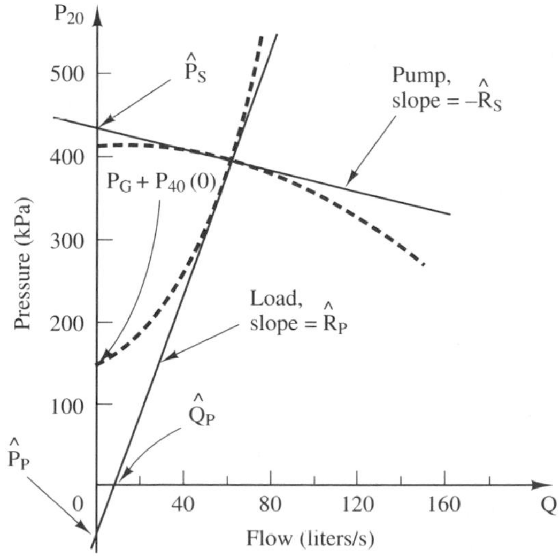

Figure 5.32 Linearized pressure-flow characteristics.

Exercise 5.8: Since equation (5.65) is linear, the Thévenin-Norton equivalence applies. Show that the Norton equivalent of equation (5.65) is described by

$$Q \approx \hat{Q}_P + \frac{P_{20}}{\hat{R}_P}$$

$$= 10.8 \text{ liter/s} + \frac{P_{20}}{8.55 \text{ kPa·s/liter}} \tag{5.66}$$

where $\hat{Q}_P$ is the Q-intercept of the straight-line load approximation in Figure 5.32.

The *linearized lumped* model that corresponds to equations (5.64)–(5.66) is shown in Figure 5.33. The fluid capacitor accounts for the rising pressure P_4 at the inlet to the reservoir. As $P_C \triangleq P_{40}$ increases, the load curve and its linear approximation in Figure 5.32 shift upward. The linear approximation remains a good approximation to the load curve at the shifted operating point for an upward shift of as much as 100 kPa. The linear loop equation is

$$-\hat{P}_S + \hat{P}_P + (\hat{R}_S + \hat{R}_P)Q + \left(\frac{1}{sC_f}\right)Q = 0$$

Rearranging this equation and replacing s by time differentiation produces the first-order linear differential equation

$$\dot{Q} + \frac{Q}{C_f(\hat{R}_S + \hat{R}_P)} = 0 \tag{5.67}$$

The solution to equation (5.67) is

$$Q(t) = Q(0)e^{-t/\tau} \tag{5.68}$$

where $\tau = C_f(\hat{R}_S + \hat{R}_P)$ and $Q(0) = 57$ liter/s.

The time constant of the linearized network is $\tau = 9.4 \times 10^6$ s = 109 days. Figure 5.34 is a plot of the exponential solution to the linearized network, equation (5.68), for a 10-day interval. It is indistinguishable from the exact straight-line solution of equation (5.45) which is plotted in Figure 5.25.

The linearized solution is accurate for $Q > 40$ liter/s. Once Q becomes smaller than 40 liter/s, the linear approximation of Figure 5.32 is no longer accurate, and we terminate the approximate solution. At that point, we can relinearize the equation about the new op-

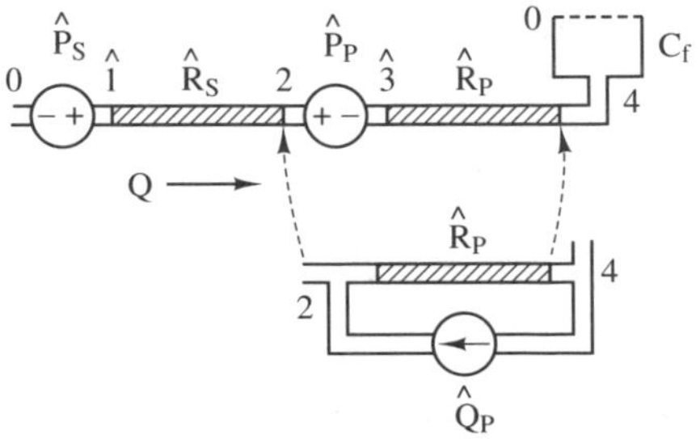

Figure 5.33 Linearized model of water transfer network.

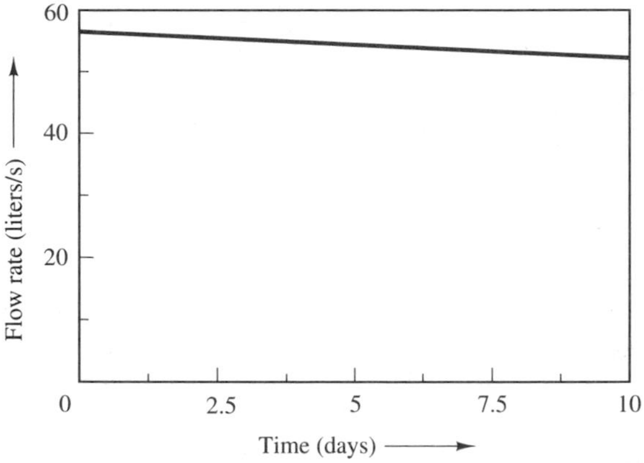

Figure 5.34 Response of the linearized network.

erating point and use the new linear approximation to compute the behavior over another interval. The process can be repeated iteratively.

This example uses the **linearization principle:** The relation between the *change* in potential drop across an element and the *change* in flow through that element is approximately linear if those changes are small enough. We can use this principle to analyze almost any nonlinear lumped network.

In general, we apply a source signal and find the initial operating point of the nonlinear network. Then we replace the network by a linear approximation that is valid for that initial operating point. Finally, we find the response of that linear approximation, beginning at the initial operating point. We shall call it the **large-signal response** to distinguish it from the small-signal response of the next paragraph. If the signals deviate from their initial values so much that the linear approximation begins to become inaccurate, we take their values at a new instant as a new initial operating point and repeat the linearization and solution process. Then the large-signal response is the concatenation of the sequence of solution segments.

Some systems are operated in such a way that the signals do not deviate much from a nominal operating point. (For example, when we drive a car on a highway, we typically keep the speed near the speed limit.) In such a situation, we first find the values of the signals at the nominal operating point of the nonlinear network. Then we use linearization to find, separately, the small deviations of the signals from that nominal operating point. We call those deviations the **small-signal response.** We add the small-signal response to the nominal operating point to get the complete response.

In the following linearization procedure, we handle the small-signal and large-signal analyses somewhat differently.

LINEARIZATION PROCEDURE

1. Find a nominal or initial network operating point at which we can linearize the network.
   - **a.** In *small-signal operation,* the sources remain nearly constant. Set the sources to typical (or average) constant values and find the steady-state operating point. Call it the *nominal operating point.*

b. In *large-signal operation,* the network signals deviate far from their initial values. Find the *initial operating point* of the network.

2. Linearize each nonlinear element in the neighborhood of its nominal or initial operating point. That is, represent each nonlinear element by a potential-flow relation of the form

$$A_{ij} - A_{ijbias} = z(T - T_{bias}) \tag{5.69}$$

where A_{ij} and T are the potential drop across the element and the flow through the element, respectively, and where $A_{ijbias} \triangleq A_{ibias} - A_{jbias}$. The quantities A_{ibias}, A_{jbias}, and T_{bias} are the **bias values** of the variables—the values at the nominal or initial operating point. Determine the extent of the deviation from each bias value for which each linear approximation is valid.

The linear equation (5.69) corresponds to a two-term Taylor series approximation of the nonlinear relation between A_{ij} and T, with $z = dA_{ij}/dT$. The equation can be represented by any of the linear models of Figure 5.35. Hence, any of those models can be substituted for the nonlinear element. In those models, the bias values correspond to constant sources.

3. Find the response to the *linearized* network.

a. *Small-signal response.* Zero the bias sources in the linearized model. (Those sources, acting simultaneously, produce the nominal operating point found in step 1.) Find the response of the remaining network to some small source signal of interest—a small step input, a small sinusoid, etc. The response of each variable represents the deviation of that variable from its bias value. Verify that the deviations from bias for the nonlinear elements do not leave the linear region. Add each small-signal response to the corresponding bias value to obtain the total response of the nonlinear network.

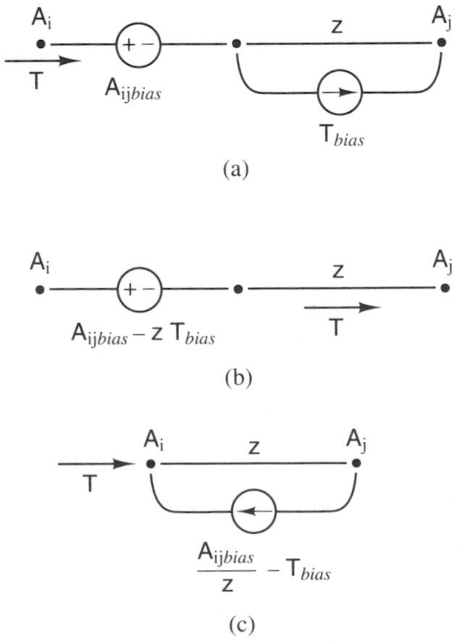

Figure 5.35 Lumped model of a linearized element; all three versions are equivalent.

b. *Large-signal response.* Find the response of the linearized network to the bias sources and the true network source acting simultaneously. The initial values were found in part 1. Superposition can be used to treat each source separately if that is helpful. When any signal for a nonlinear element approaches the boundary of its linear region, terminate the response. Take the response at the instant of termination as a new initial operating point, relinearize the nonlinear network, and calculate an additional segment of the response.

Let us review the large-signal analysis of Figure 5.22 in terms of the linearization procedure. Suppose that initially the pipe is filled with water, the upper reservoir is empty, and the pump is turned off. The network is then in the zero state. The pump is turned on abruptly to the value P_S at $t = 0$. For a brief interval, the pump accelerates the water in the pipe. Since this interval is less than 4 seconds, a negligible amount compared to the fill time of the reservoir, we ignore the fluid acceleration and do not include inertance in the model. Therefore, the condition of the network just after the water has accelerated corresponds to the initial operating point of step 1(b). Hence, we short circuit (or ignore) the inertance, but open circuit the reservoir capacitance. That is, the initial operating point corresponds to steady state with respect to the inertance, but to the initial state with respect to the reservoir.

Exercise 5.9: Determine the inertance of the pipe, and show that the inertial time constant, ignoring the capacitance of the reservoir, is $\tau = I/(\hat{R}_S + \hat{R}_P) = 3.5$ s. Why can we ignore the capacitance of the reservoir as we examine the effect of inertance?

To find the initial operating point, we set the pump pressure to the value P_S and short-circuit the fluid capacitor. The gravity head in the pipe is represented by the constant source pressure P_G (see Figure 5.22(c)). For this initial instant, the loop equation that describes the network equates the nonlinear algebraic equations (5.38) and (5.39). The curves of Figure 5.23 provide an equivalent graphical description of the network at the initial instant. We solve for the initial operating point analytically or graphically. The result is $Q \equiv Q_{bias} = 57$ liter/s, $P_{10} \equiv P_{10bias} = P_S = 414$ kPa, $P_{20} \equiv P_{20bias} = 395$ kPa, $P_{30} \equiv P_{30bias} = P_{20} - P_G = 248$ kPa, and $P_{40} \equiv P_{40bias} = 0$ kPa.

The only nonlinear elements in the network are the fluid resistors that represent friction in the pump and pipe, described by $P_{12} = \eta_S Q^2$ and $P_{34} = \eta_P Q^2$. At the initial operating point, $P_{12bias} = 19$ kPa, $P_{34bias} = 248$ kPa, and $Q_{bias} = 57$ liter/s. The small-signal impedances of these two nonlinear elements are the slopes of the curves (dP/dQ) at the initial operating point: $z_S \equiv \hat{R}_S = 2Q_{bias}\eta_S = 0.669$ kPa·s/liter and $z_P \equiv \hat{R}_P = 2Q_{bias}\eta_P = 8.55$ kPa·s/liter. The linearized equations that correspond to equation (5.69) of step 2 are $P_{12} - P_{12bias} = \hat{R}_S(Q - Q_{bias})$, or

$$P_{12} = 19 \text{ kPa} + (0.669 \text{ kPa·s/liter})(Q - 57 \text{ liter/s}) \tag{5.70}$$

and $P_{34} - P_{34bias} = \hat{R}_P(Q - Q_{bias})$, or

$$P_{34} = 248 \text{ kPa} + (8.55 \text{ kPa·s/liter})(Q - 57 \text{ liter/s}) \tag{5.71}$$

Substitute the linear model form of Figure 5.35(b) for each of the quadratic resistors of Figure 5.22(b) to produce a linearized network. Use equations (5.70) and (5.71) to determine the parameters of the substituted portions. The equation that describes the left half of that network is equation (5.64), with $\hat{P}_S = P_S - P_{12bias} + R_S Q_{bias}$. Similarly, since $P_{40}(0) = 0$, the equation that describes the right half of that network is equation (5.65), with $\hat{P}_P = P_G + P_{34bias} - R_P Q_{bias}$.

Exercise 5.10: Sketch the pressure-flow equations for the nonlinear resistors η_S and η_P; also sketch their linear approximations (as described by equations (5.70) and (5.71)). Use the sketches to determine that the linear approximations are reasonably accurate for $|Q - 57 \text{ liter/s}| < 15$ liter/s.

Exercise 5.11: Show that substitution of equation (5.70) into the pump equation $P_{20} = P_{21} + P_S$ produces the linear approximation given by equation (5.64) and that substitution of equation (5.71) into the load equation $P_{20} = P_G + P_{34} + P_C$, with $P_C = 0$, produces equation (5.65).

The linearized network is shown in Figure 5.33. The linear differential equation is equation (5.67). The solution Q(t), shown in Figure 5.34, reduces slowly from 57 liter/s to about 52 liter/s over a 10-day interval. This variation remains within the bounds of the linear region. We could use the network solution at the end of 10 days as a new operating point, linearize about that operating point, and use the new linear model to calculate another segment of the response.

5.3 SINUSOIDAL TRANSFER FUNCTIONS AND PHASOR NOTATION

Some systems are driven by steady periodic sources. Sea waves, the daily temperature cycle, the periodic contact of automobile tires with expansion joints in a highway, the oscillation of a clock pendulum, and the vibration of the electromechanical crystal in an electric watch or a computer clock are familiar examples of such sources. In many countries, the electric power network is sinusoidal. Although the step-response analyses of chapter 3 exercise a network at both the static and abruptly changing extremes, they do not expose the steady-state periodic behavior of the network. In this section, we examine the steady-state behavior of networks for sinusoidal source signals.

If a source signal is sinusoidal, then, for linear networks in a steady state, all the signals become sinusoidal at the source frequency. A sinusoidal signal has only three parameters: amplitude, phase, and frequency. If the frequency of such a signal is fixed by the source, the remaining two parameters of the signal can be represented by a single complex number that we call a **phasor**. We shall represent all network signals by phasors.

The derivative of a sinusoidal signal is also sinusoidal. When we represent sinusoidal signals by phasors, the time-derivative operator s acts as a complex multiplier. Specifically, s acts as the **complex-frequency variable** $j\omega$. As a consequence, a transfer function—an s-notation description of the relation between signals—also reduces to a complex multiplier. We call that multiplier a **sinusoidal transfer function.** This section shows how to use sinusoidal transfer functions to understand sinusoidal steady-state behavior.

We also show here that we can approximate a source signal of any waveform by a sum of sinusoids of various frequencies. High-frequency sinusoids ($s = j\omega \to \infty$) are similar to abruptly changing signals, while low-frequency sinusoids ($s \to 0$) are similar to constant signals. To find the response to a complicated source waveform, we decompose that waveform into sinusoidal components, find the response to each sinusoid by (complex) multiplication of phasors, and then add the responses. This application of superposition is similar to the superposition of step responses discussed in section 4.3. This use of the superposition principle is known as **frequency domain analysis.**

Each network variable is related to the source by an input-output transfer function (or system function). For a sinusoidal source, the system function acts like a complex number. A plot of that complex number as a function of the source frequency is called the network **frequency response**. The responses for all frequencies can be determined simultaneously.

In sum, for sinusoidal signals, the network system equations reduce to algebraic equations with complex-number signals as the variables. We show in section 5.4 that the frequency response is equivalent to the system differential equation. Just as we found the system differential equation from certain key features of the zero-state step response, we shall also find the differential equation from key features of the frequency response.

Complex Algebra

Linear networks with sinusoidal sources are analyzed most easily in terms of complex numbers. Complex numbers can be viewed as points, arrows, or vectors in the complex plane. We represent them mathematically in either rectangular or polar form:

$$Z = x + jy = r(\cos\theta + j\sin\theta) \triangleq r\angle\theta \tag{5.72}$$

We use the notations $\mathcal{Re}[Z] = x$ and $\mathcal{Im}[Z] = y$ to denote the *real part* of Z and the *imaginary part* of Z, respectively. We use the notations $\arg[Z] = \theta$ and $|Z| = r$ to denote the *angle* of Z and the *magnitude* of Z, respectively. The relations between the rectangular and polar forms are illustrated in Figure 5.36. The angle θ can be expressed in radians or in degrees.

We can add and subtract complex numbers graphically by the *parallelogram law,* shown in Figure 5.37. Mathematical addition or subtraction is most easily expressed in rectangular form:

$$\begin{aligned} Z_1 \pm Z_2 &= (x_1 + jy_1) \pm (x_2 + jy_2) \\ &= (x_1 \pm x_2) + j(y_1 \pm y_2) \end{aligned} \tag{5.73}$$

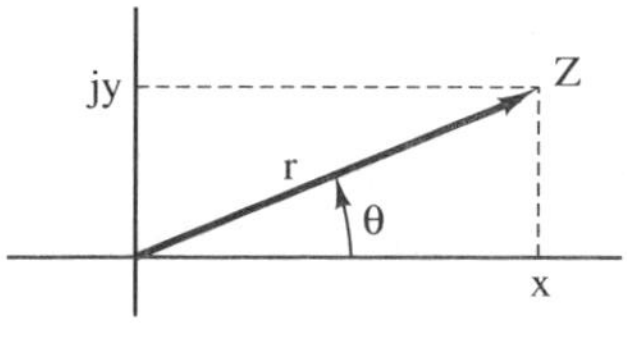

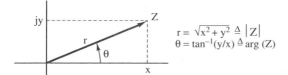

$r = \sqrt{x^2 + y^2} \triangleq |Z|$
$\theta = \tan^{-1}(y/x) \triangleq \arg(Z)$

Figure 5.36 Representation of a complex number.

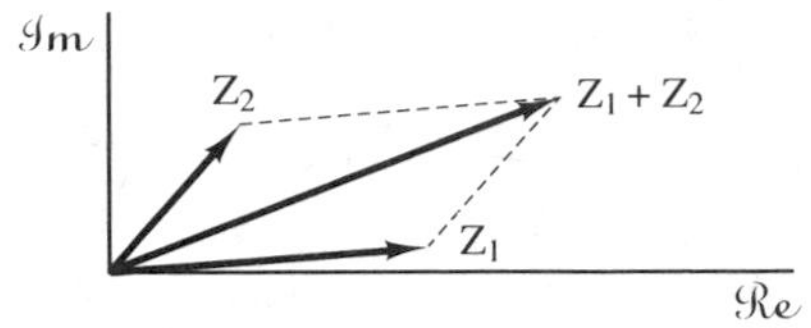

Figure 5.37 Parallelogram law for complex addition.

Multiplication is expressed most easily in polar form:

$$\begin{aligned} Z_1 Z_2 &= r_1(\cos \theta_1 + j \sin \theta_1) r_2(\cos \theta_2 + j \sin \theta_2) \\ &= r_1 r_2[\cos(\theta_1 + \theta_2) + j \sin(\theta_1 + \theta_2)] \\ &= r_1 r_2 \angle(\theta_1 + \theta_2) \end{aligned} \tag{5.74}$$

Similarly, division is expressed as

$$\frac{Z_1}{Z_2} = \frac{r_1}{r_2} \angle(\theta_1 - \theta_2) \tag{5.75}$$

According to the Euler formula,

$$e^{\pm j\theta} = \cos \theta \pm j \sin \theta \tag{5.76}$$

(To derive this formula, replace the exponential, cosine, and sine functions by their respective power series representations.) Hence, we can also represent the complex number given by equation (5.72) in the exponential form

$$Z = re^{j\theta} \tag{5.77}$$

We use the exponential notation of equation (5.77) and the *angle* notation of equation (5.72) interchangeably.

Exercise 5.12: Show that $c\mathcal{R}e[Z] = \mathcal{R}e[cZ]$, where c is a real number. Show that $\mathcal{R}e[Z_1] + \mathcal{R}e[Z_2] = \mathcal{R}e[Z_1 + Z_2]$. Hence, $\mathcal{R}e(\cdot)$ is a linear operator.

Exercise 5.13: Show that

$$e^{j\pi/2} = j1 = j = 1\angle(\pi/2) \tag{5.78}$$

Since a complex number has real and imaginary parts, an equation in which the two sides are complex is equivalent to two real-number equations. Specifically, if $x_1 + jy_1 = x_2 + jy_2$, then $(x_1 - x_2) + j(y_1 - y_2) = 0 + j0$. Therefore, $x_1 = x_2$ and $y_1 = y_2$. That is, the real parts of the two complex numbers must be equal to each other, and the imaginary parts must be equal to each other. The two real-number equations determine the real and imaginary parts of the complex-number solution to the equation.

Complex Exponential Signals

If a linear network is driven by a sinusoidal source with frequency ω, then *in a steady state,* all signals in the network become sinusoidal with the same frequency. (Just bounce periodically on the bumper of an automobile or wiggle the end of a clothesline to see that this is true). Therefore, we begin the analysis of the sinusoidal behavior of linear networks by examining simple ways to represent and manipulate sinusoidal functions. We find that complex exponential notation is the most convenient.

Any sinusoidal signal can be expressed in the form $A\cos(\omega t + \phi)$, where A is the **amplitude**, ϕ is the **phase**, and ω is the **frequency** of the signal. (Although we could, just as well, express any sinusoid in sine form, we use the cosine form throughout this book.) According to equation (5.76) and exercise 5.12, the same sinusoid (with the same parameters A, ϕ, and ω) can be expressed in the complex exponential form,

$$\begin{aligned} A\cos(\omega t + \phi) &= A\{\mathcal{Re}[e^{j(\omega t+\phi)}]\} \\ &= \mathcal{Re}[(Ae^{j\phi})e^{j\omega t}] \end{aligned} \tag{5.79}$$

This form may seem even more complicated than the original cosine form. However, we show that the operator $\mathcal{Re}[\,]$ and the exponential $e^{j\omega t}$ perform no useful functions during the manipulation of any of these equations. Instead, they are used only to interpret results. Therefore, we eventually suppress them and represent the sinusoidal signal of equation (5.79) by the complex number $Ae^{j\phi}$ alone.

What are we going to do with these complex exponential signals? We shall find which ones are solutions to the network equations. We must examine, then, the operations we perform on signals in the network equations. Suppose, for example, that a series connection of a resistor and capacitor make up part of the network. The equation which describes that part of the network is $Csv_1 = (RCs + 1)i$, where v_1 is the voltage across the RC pair and i is the current through the pair. The mathematical operations in the equation are represented by the two polynomials in s. The individual operations are (a) differentiation of a signal with respect to time (represented by s); (b) multiplication of a signal by a real number (a coefficient of the polynomial); and (c) addition of signals (terms of the polynomial). Therefore, we must examine the effects of these three operations on complex exponential signals.

Exercise 5.14: Show that

$$\frac{d}{dt}(\mathcal{Re}[Ae^{j\phi}e^{j\omega t}]) = -A\omega\sin(\omega t + \phi) = \mathcal{Re}\left[\frac{d}{dt}(Ae^{j\phi}e^{j\omega t})\right]$$

That is, show that the order of the time derivative and the $\mathcal{Re}[\,]$ operation can be interchanged.

According to exercises 5.12 and 5.14, the $\mathcal{Re}[\,]$ operations on the complex exponential signals in a linear differential equation can be deferred until after we carry out the differentiations, scalar multiplications, and additions. Therefore, it is appropriate to suppress the $\mathcal{Re}[\,]$ notation. We write $Ae^{j\phi}e^{j\omega t}$ when we really mean the real part of this signal—the cosine function of equation (5.79). We call the *complex amplitude* $Ae^{j\phi}$ a **phasor**. We *think* of $Ae^{j\phi}e^{j\omega t}$ as a sinusoid and manipulate it according to the mathematical operations specified in the system equations. Only when we have found which complex exponential is the network solution do we take its real part.

With the $\mathcal{R}e[\,]$ operation suppressed, we see more clearly the effects of the three mathematical operations—scalar multiplication, differentiation with respect to time, and addition—on complex exponential signals. They produce

$$c(Ae^{j\phi}e^{j\omega t}) = (cAe^{j\phi})e^{j\omega t} \tag{5.80}$$

$$\begin{aligned}\frac{d}{dt}(Ae^{j\phi}e^{j\omega t}) &= Ae^{j\phi}\frac{d}{dt}(e^{j\omega t})\\ &= (j\omega Ae^{j\phi})e^{j\omega t}\end{aligned} \tag{5.81}$$

$$\begin{aligned}A_1e^{j\phi_1}e^{j\omega t} + A_2e^{j\phi_2}e^{j\omega t} &= (A_1e^{j\phi 1} + A_2e^{j\phi 2})e^{j\omega t}\\ &= [(A_1 \cos \phi_1 + A_2 \cos \phi_2) + j(A_1 \sin \phi_1 + A_2 \sin \phi_2)]e^{j\omega t}\\ &= (Ae^{j\phi})e^{j\omega t}\end{aligned} \tag{5.82}$$

where

$$A = \sqrt{(A_1 \cos \phi_1 + A_2 \cos \phi_2)^2 + (A_1 \sin \phi_1 + A_2 \sin \phi_2)^2}$$

and

$$\phi = \tan^{-1}\frac{A_1 \sin \phi_1 + A_2 \sin \phi_2}{A_1 \cos \phi_1 + A_2 \cos \phi_2}$$

The relations among $A_1 \angle\phi_1$, $A_2 \angle\phi_2$, and $A \angle\phi$ become clearer if we view them in the complex plane (see Figure 5.38); these quantities are related by the parallelogram rule of Figure 5.37.

Equation (5.82) shows that the sum of two sinusoids of the same frequency ω is a third sinusoid of that same frequency ω. Equations (5.80) and (5.81) show that scalar multiplication and differentiation do not change the frequency of a sinusoid. Since linear network equations contain only these three mathematical operations, the solution to a set of linear system equations must be a sinusoid that has the same frequency as the source signal. This is the conclusion that we arrived at intuitively at the beginning of the section.

Each of the operations used in linear network equations leaves the factor $e^{j\omega t}$ unchanged in each term. We can simplify the writing of these equations by suppressing this

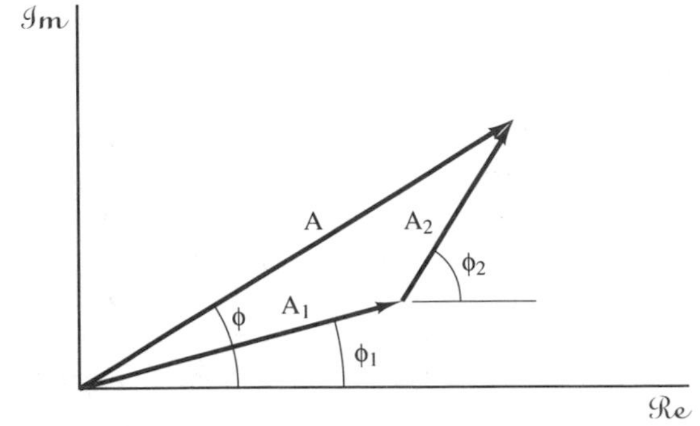

Figure 5.38 Addition of phasors.

factor. Then we express the three types of operations on complex exponential signals in the simpler notation

$$c(Ae^{j\phi}) = (cA)e^{j\phi} \tag{5.83}$$

$$s(Ae^{j\phi}) = j\omega(Ae^{j\phi}) \tag{5.84}$$

$$A_1e^{j\phi_1} + A_2e^{j\phi_2} = (A_1 \cos \phi_1 + A_2 \cos \phi_2) + j(A_1 \sin \phi_1 + A_2 \sin \phi_2) = Ae^{j\phi} \tag{5.85}$$

Note that the effect of differentiating a complex exponential signal with respect to time is to multiply the phasor representation of that signal by the factor $j\omega$. (Even though we suppress the factor $e^{j\omega t}$, we must remember the frequency ω.) For sinusoidal steady-state operation of a network, then, we interpret each time-derivative operator s in the network equations as the complex number $j\omega$. We can think of s as a **complex frequency variable.**

The complex function $e^{j\omega t} \equiv 1 \angle \omega t$ of equation (5.79) is a unit vector in the complex plane; it rotates counterclockwise in the plane with frequency ω, in rad/s. The complex amplitude in equation (5.79), the phasor $Ae^{j\phi}$, denotes a rotating vector of magnitude A. At $t = 0$, that vector is oriented at the angle ϕ, as illustrated in Figure 5.39. As time proceeds, the factor $e^{j\omega t}$ of equation (5.79) rotates the phasor at the frequency ω. The projection of that rotating phasor on the real axis is the sinusoid $A \cos(\omega t + \phi)$ of equation (5.79). The abscissa for the sinusoidal waveform can be labeled in terms of either the time variable t or the corresponding *phase variable* ωt.

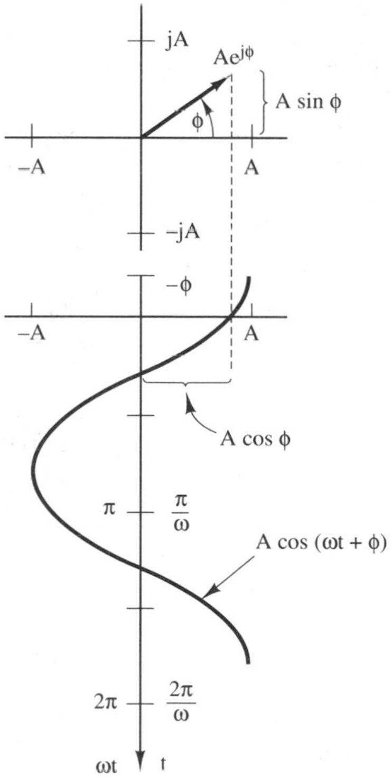

Figure 5.39 Phasor representation of a sinusoidal signal. (The phasor, shown at $t = 0$, rotates counterclockwise at ω rad/s.)

Figure 5.39 shows the phasor $Ae^{j\phi}$ and the sinusoid it generates. The rotation vector $e^{j\omega t}$ is implicit in the figure. (We suppress the factor $e^{j\omega t}$ in drawings, as well as in mathematical representations.) If we let $s = j\omega$ in the network equations, we do not need $e^{j\omega t}$, except to use its implicit rotation of phasors at the frequency ω to visualize the sinusoidal signals they represent.

A positive phase angle ϕ represents a *counterclockwise shift* of the phasor, relative to the positive real axis. A positive ϕ also implies a *shift of the cosine waveform to the left* by an amount of phase $\omega t = \phi$, a shift earlier in time of $t = \phi/\omega$. That is, the maximum of the cosine occurs earlier than it would if ϕ were zero.

Network Equations in Phasor Notation

We define the **phasor transformation** $\wp$ of a sinusoidal signal by

$$\wp[A\cos(\omega t + \phi)] \triangleq Ae^{j\phi} \tag{5.86}$$

The resulting **phasor representation** $Ae^{j\phi}$ is the snapshot, at $t = 0$, of that rotating vector which generates the sinusoid in the fashion shown in Figure 5.39. The phasor transformation is uniquely reversible. For example, $\wp^{-1}[1 + j\sqrt{3}] = \wp^{-1}[2e^{j\pi/3}] = 2\cos(\omega t + \pi/3)$. The phasor transformation merely formalizes the notational simplifications of the previous section. The specific frequency ω of the sinusoid—the rate of rotation of the phasor—is suppressed in phasor notation. But that frequency must not be forgotten: Without it, the sinusoidal signal cannot be reconstructed from the phasor. Phasor notation makes it possible to compare and manipulate sinusoidal time functions (of the same frequency) by comparing and manipulating the simpler phasors (complex numbers) that represent them.

The signal $\cos(\omega t)$ is represented by the phasor $1\angle 0$. The signal $\sin(\omega t) = \cos(\omega t - \pi/2)$ is represented by the phasor $1\angle -\pi/2$. That is, the cosine representation of $\sin(\omega t)$ **lags behind** the cosine representation of $\cos(\omega t)$ by the phase angle $\pi/2$. (The peak of the sine occurs later in time than the peak of the cosine by $\pi/2\omega$ seconds.) Such phasor comparisons are facilitated by a **phasor diagram** in the complex plane. The phasor representations of $2\cos(\omega t + \pi/4)$, $\sin(\omega t + \pi/4)$, and $-\cos(\omega t)$ are shown in the phasor diagram of Figure 5.40. The whole configuration of phasors rotates at the single frequency ω.

According to equation (5.84), differentiating a sinusoid with respect to time corresponds to multiplying its phasor representation by $j\omega$. Multiplying by ω changes the length of the phasor (and multiplies its units by rad/s). Multiplying by j ($\equiv 1\angle\pi/2$) produces a 90° counterclockwise rotation of the phasor, advancing the sinusoid in phase by $\pi/2$.

Exercise 5.15: On Figure 5.40, sketch the phasors that represent $\cos\omega t$, $\sin\omega t$, and $-\sin\omega t$.

Exercise 5.16: Use phasor addition in a phasor diagram to show that $\cos\omega t - \sin\omega t = \sqrt{2}\cos(\omega t + \pi/4)$.

Exercise 5.17: On Figure 5.40, sketch the phasors that represent the derivatives of the signals $\sin\omega t$ and $2\cos(\omega t + \pi/4)$.

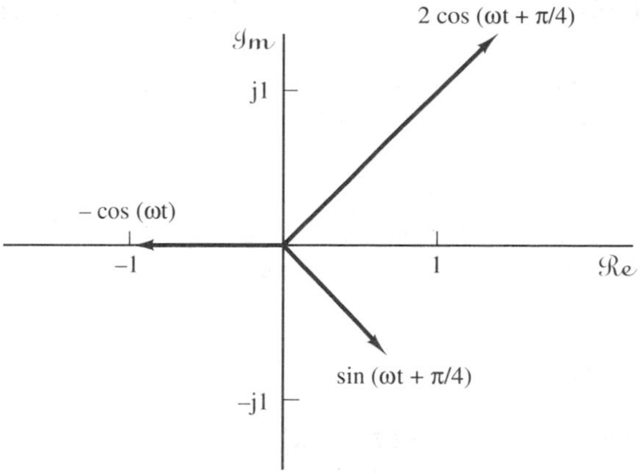

Figure 5.40 Phasor diagram for signals of frequency ω.

We are interested primarily in the *relative* magnitudes and phases among sinusoids at different locations in a network. We use the phasor representations of those sinusoids to construct a phasor diagram for the network. We can rotate the configuration of phasors in the diagram to whatever orientation provides the most convenient computational or visual geometry. Such a rotation corresponds to taking a snapshot of the rotating configuration of phasors at a time different from $t = 0$.

We distinguish the phasor representation of a signal from the corresponding sinusoidal time function by use of a bold italic uppercase symbol. (Bold italic symbols adequately distinguish phasors from those signals, such as the scalar force F, which we represent by uppercase symbols.) Thus, we denote the phasor representation of a sinusoidal velocity v(t) by the symbol ***V***, where ***V*** is a complex number. Similarly, we represent the phasor corresponding to a sinusoidal electric current i by the symbol ***I***.

The impedance of an element is the ratio $\mathsf{Z} = \mathsf{A}/\mathsf{T}$, where A is the potential drop across the element in the direction of the flow T through the element. In general, Z is a function of the time-derivative operator s. If the signals in the network are steady-state sinusoids, the signals A and T can be represented by phasors, and the substitution $s = j\omega$ can be made in the impedance function. In that case, the impedance reduces to a complex number: the ratio of the complex phasors that represent the potential drop and flow. We call that impedance the **sinusoidal impedance** (or **complex impedance**) of the element.

Each pair of signals in a linear network is related by a transfer function of the form of equation (4.47): $y_2 = G(s)y_1$. If the network is in a sinusoidal steady state, the signals can be represented by phasors, and the substitution $s = j\omega$ can be made in the transfer function. We call the complex number $G(j\omega)$ a **sinusoidal transfer function.** If the signal y_1 is the source function, then the equation is the *input-output* transfer function (or system function) for the variable y_2, and we call $G(j\omega)$ the **sinusoidal system function** (or **complex system function**). (In this text, we usually denote an *input-output* transfer function by the symbol H(s).)

The node and loop laws, equations (2.27) and (2.28), are statements about signals in lumped networks. Those laws apply to sinusoidal signals in phasor notation. The sinusoidal impedance of a linear lumped element is a proportionality between its through and across phasors. Therefore, the equations for any linear network in a sinusoidal steady state can be expressed in phasor notation. Such notation extends to all linear analysis techniques—series-parallel combination, Thévenin-Norton transformation, and superposition.

Let F(t) = cos 2t N be the source signal for the translational network of Figure 5.41. We seek the steady-state motion $v_1(t)$ of the network and the steady-state compressive forces $F_1(t)$, $F_2(t)$, and $F_3(t)$ in the branches. The sinusoidal source function F(t) can be represented, in the steady state, by the phasor $\boldsymbol{F} = 1\angle 0$ N. (The frequency $\omega = 2$ rad/s is implicit.) The sinusoidal impedances of the branches are:

$$\left.\begin{aligned} z_m &= \frac{1}{sm} = \frac{1}{(j2\ \text{rad/s})(4\ \text{kg})} = -j0.125\ \text{m/N·s} \\ z_b &= \frac{1}{b} = \frac{1}{5.66\ \text{N·s/m}} = 0.177\ \text{m/N·s} \\ z_k &= \frac{s}{k} = \frac{j2\ \text{rad/s}}{2\ \text{N/m}} = j1\ \text{m/N·s} \end{aligned}\right\} \tag{5.87}$$

The complex numbers produced by the substitution $s = j\omega = j2$ rad/s are the sinusoidal impedances of the elements. According to the node law at node 1,

$$F = \left(\frac{k}{s} + b + sm\right)v_1 \tag{5.88}$$

Therefore,

$$v_1 = \frac{s}{ms^2 + bs + k}F \triangleq H(s)F \tag{5.89}$$

where H(s) is the system function for v_1. (In this instance, H(s) can also be viewed as the input impedance of the parallel k-m-b network, as *seen* by the source.) The *sinusoidal system function* is

$$H(j2) = \frac{j2}{(4)(j2)^2 + (5.66)(j2) + 2} = 0.111\angle -51°\ \text{m/N·s} \tag{5.90}$$

In phasor notation, equation (5.89) becomes

$$\boldsymbol{V}_1 = H(j2)\boldsymbol{F} = (0.111\angle -51°\ \text{m/N·s})(1\angle 0°\ \text{N}) = 0.111\angle -51°\ \text{m/s} \tag{5.91}$$

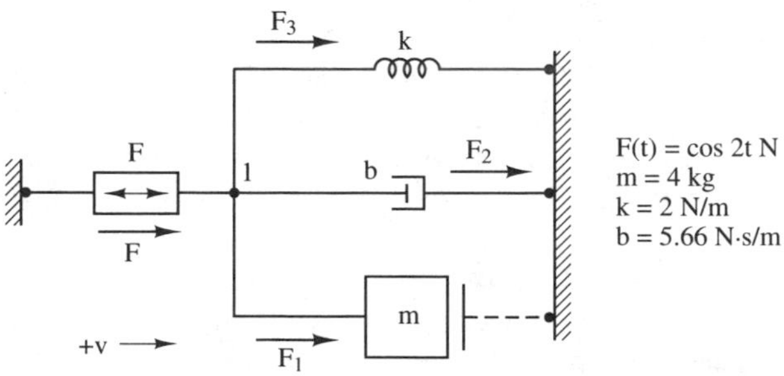

Figure 5.41 A sinusoidally driven network.

Hence, the steady-state motion of node 1 is $v_1(t) = 0.111\cos(2t - 51°)$. The phasors that represent the forces in the branches, according to equations (5.87), are

$$\left.\begin{aligned} \boldsymbol{F}_1 &= \frac{\boldsymbol{V}_1}{z_m} = \frac{(0.111 \angle -51° \text{ m/s})}{(0.125 \angle -90° \text{ N·s/m})} = 0.89 \angle 39° \text{ N} \\ \boldsymbol{F}_2 &= \frac{\boldsymbol{V}_1}{z_b} = \frac{(0.111 \angle -51° \text{ m/s})}{(0.177 \angle 0° \text{ N·s/m})} = 0.63 \angle -51° \text{ N} \\ \boldsymbol{F}_3 &= \frac{\boldsymbol{V}_1}{z_k} = \frac{(0.111 \angle -51° \text{ m/s})}{(1 \angle 90° \text{ N·s/m})} = 0.111 \angle -141° \text{ N} \end{aligned}\right\} \tag{5.92}$$

We use the rectangular-to-polar identities $j = 1\angle\pi/2 = 1\angle 90°$ and $-j = 1\angle -\pi/2 = 1\angle -90°$ to convert the rectangular forms in equations (5.87) to the polar forms needed in equations (5.92).

Exercise 5.18: Determine the steady-state signals that correspond to the phasors in equations (5.92).

It can be seen from equations (5.87) that the impedance of a spring increases with the frequency of the source sinusoid, whereas the impedance of a mass decreases with increasing frequency of the source sinusoid. Damper impedances are independent of frequency. Therefore, according to equation (5.88), if the frequency of the source in Figure 5.41 were increased, the force in the spring would be reduced and the force in the mass increased, relative to the force in the damper. As the frequency approaches infinity, the force in the spring approaches zero (an open circuit) and all of the force flows to the mass (a short circuit).

This behavior is similar to the open-circuit and short-circuit behavior of energy-storage elements for abrupt jumps in a source (see Table 4.3). *High-frequency* sinusoidal steady-state signals ($s \to \infty$) are *abruptly changing* signals. By contrast, *low-frequency* sinusoidal steady-state signals ($s \to 0$) are *nearly static* signals. Accordingly, we can open-circuit and short-circuit energy-storage elements to find the limiting steady-state sinusoidal behavior of networks as the frequency becomes very high or very low.

As the frequency of the network source increases from low frequency to high frequency, the sinusoidal system function changes in both magnitude and phase. We use the term **frequency response** of the network to refer to the variations in magnitude and phase versus the source frequency. Because the sinusoidal system function indicates the precise input-output relationship for a sinusoidal steady state, the frequency response gives a general intuitive picture of the network behavior. Section 5.4 is devoted to constructing and analyzing the frequency response.

Network Phasor Diagrams

In the sinusoidal steady-state analysis of networks, both the signals and the network elements are represented by complex numbers. Hence, we must be careful not to confuse the two entities. Phasors are parameters of the signals; sinusoidal impedances are parameters of the network. Sinusoidal impedances combine to form sinusoidal transfer functions. Thus, the sinusoidal transfer functions are also parameters of the network. A **phasor diagram** is a display of the network signal phasors. Information about the network impedances and transfer functions appears only indirectly in phasor diagrams. Specifically, it is the relative sizes and orientations of the phasors that characterize the network.

The phasor representations of the signals in the network of Figure 5.41 are displayed in Figure 5.42. Two equivalent versions of the network phasor diagram are shown. In both figures, the source signal F is treated as the reference, in the sense that it is oriented at angle 0. The open (or tail-to-tail) version of the diagram in Figure 5.42(a) emphasizes the relative sizes and orientations of the various signals. The closed (or tail-to-tip) form shown in Figure 5.42(b) provides a useful check on the zero sum of flows at a node or of potential drops around a loop.

For this particular network, the balance of flows at node 1 is demonstrated by the vector equality of the source phasor $\boldsymbol{F}$ to the tail-to-tip chain of branch phasors $\boldsymbol{F}_1$, $\boldsymbol{F}_2$, and $\boldsymbol{F}_3$ in Figure 5.42(b). Both diagrams display all essential information about the steady-state solutions. Neither diagram is unique in form. The designer should arrange the diagram to provide insight concerning the behavior of the network. Sketching of the actual sinusoids, if desired, is straightforward: Merely rotate the (tail-to-tail) phasors counterclockwise at the frequency $\omega = 2$ rad/s, and sketch the projections of each phasor on the real axis.

Exercise 5.19: Sketch the source sinusoid F(t) and the response sinusoids $F_1(t)$, $F_2(t)$, and $F_3(t)$ corresponding to Figure 5.42 on a single abcissa.

The flow $\boldsymbol{F}_2$ through the damper is in phase with the velocity $\boldsymbol{V}_1$ across the damper. Therefore, $F_2(t)$ and $v_1(t)$ reach their respective peak values simultaneously. The flow $\boldsymbol{F}_1$ through the mass leads the velocity $\boldsymbol{V}_1$ across the mass by 90°, owing to the factor $s = j\omega$ that multiplies the velocity. Because of this 90° lead, the peak of $F_1(t)$ occurs earlier than the peak of $v_1(t)$ by $t = (\pi/2)/\omega = 0.785$ s. The flow $\boldsymbol{F}_3$ in the spring lags the velocity $\boldsymbol{V}_1$ across the spring by 90°. Consequently, the peak of $F_3(t)$ occurs 0.785 s after the peak of v_1. The flows in the mass and spring oppose each other. For the specified fre-

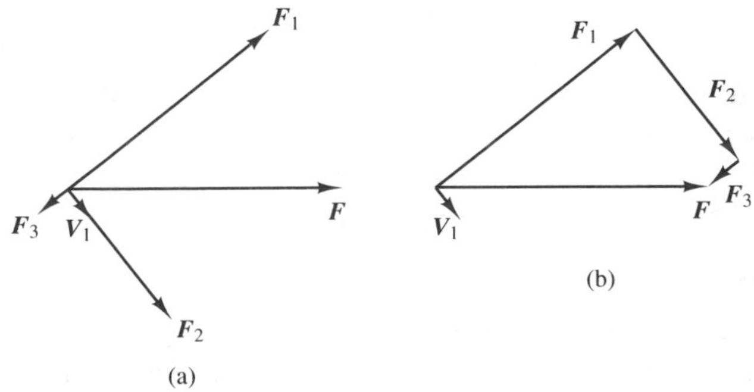

Figure 5.42 Phasor diagram for the network of Figure 5.41. (a) Open form. (b) Closed form.

quency ($\omega = 2$ rad/s), the flow in the mass dominates the flow in the spring, and the net effect of the two is the same as that of a smaller mass with no spring. (This opposition and partial cancelling of complementary energy-storage elements is typical of sinusoidal steady-state behavior.)

Let us focus attention on the source $\boldsymbol{F}$ and the velocity $\boldsymbol{V}_1$ alone. These phasors represent the input and output signals, respectively. Their magnitude ratio (0.111 m/N·s) and phase difference ($-51°$) constitute the input-to-output response of the network at the frequency $\omega = 2$ rad/s. The velocity lag is consistent with the fact that the mass-spring combination acts like a mass. (That is, for a mass, the velocity lags the force.) The lag is 51° rather than 90° because the network acts as a mass and damper rather than an ideal mass. The 51° lag corresponds to $51°(2\pi/360°) = 0.89$ rad. Therefore, the velocity $v_1(t)$ reaches its peak later than the peak of F(t) by

$$\frac{\phi}{\omega} = \frac{0.89 \text{ rad}}{2 \text{ rad/s}} = 0.446 \text{ s}$$

If the source frequency ω is increased, the magnitude of the sinusoidal impedance of each energy-storage element changes. (The arguments of those impedances remain at $\pm 90°$.) The sinusoidal system function also changes. Therefore, the magnitudes and phases of all network phasors (except the source phasor $\boldsymbol{F}$) can be expected to change. The variations with frequency of the magnitude and phase of the sinusoidal system function constitute the network frequency response. PSPICE (see Appendix C) can compute the frequency response of a network by simulating the sinusoidal steady-state behavior for various frequencies. Appendix D describes a computer program that computes the frequency response directly from a specified input-output transfer function. We explore the frequency response further in section 5.4.

Example Problem 5.6: Phasor diagrams for sinusoidal steady state.
A lumped translational network and its line graph are shown in Figure 5.43.
(a) Determine the sinusoidal source function F(t) that will cause the steady-state motion of the mass to be $v_2(t) = \sin t$ m/s. Use a sequence of computations, beginning with the phasor representation of the *specified* velocity $v_2(t)$, to find the phasors that represent $v_2(t)$, $F_4(t)$, $F_3(t)$, $F_2(t)$, $v_1(t)$, $F_1(t)$, and F(t). Then determine F(t).
(b) Sketch and label a phasor diagram that relates the phasors found in part (a). Interpret the diagram in terms of the lumped network.
(c) Determine the complex value $H(j\omega) = \boldsymbol{V}_2/\boldsymbol{F}$ of the transfer impedance between F and v_2 for $s = j1$ rad/s. Sketch a phasor diagram that displays $\boldsymbol{V}_2$ and $\boldsymbol{F}$ for the source waveform $F(t) = \cos t$ N. How does this phasor diagram relate to the phasor diagram of part (a)?
Solution: **(a)** The specified response waveform is $v_2(t) = \sin t = \cos(t - \pi/2) = \mathcal{Re}[1e^{-j\pi/2}e^{jt}]$. Since it has the frequency $\omega = 1$ rad/s, $s = j1 = 1\angle 90°$. We represent the specified waveform by the phasor

$$\boldsymbol{V}_2 = 1\angle -90° \text{ m/s} \tag{5.93}$$

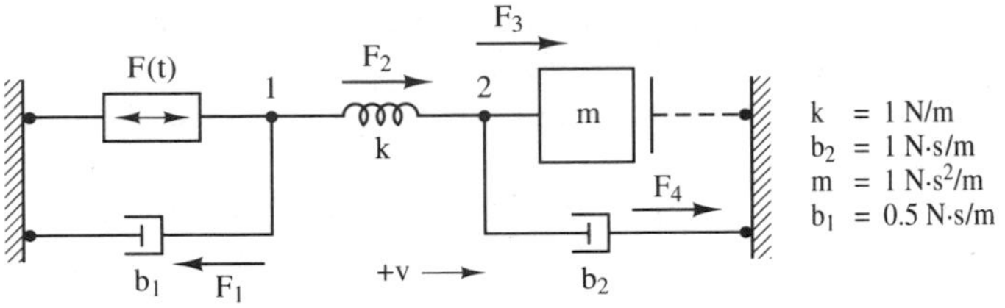

(a)

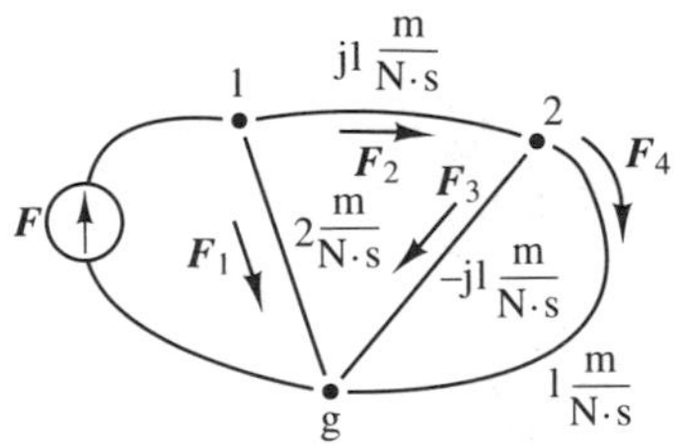

(b)

Figure 5.43 A mechanical network with a sinusoidal steady-state source. (a) Lumped model. (b) Line graph for sinusoidal steady state, labeled with phasors and complex impedances.

We facilitate comparison of relationships among network phasors by labeling the line-graph elements with their complex impedances, thus: $z_k = s/k = j1$, $z_m = 1/(sm) = -j1$, $z_{b1} = 1/b_1 = 2$, and $z_{b2} = 1/b_2 = 1$, each with units m/(N·s). The requested sequence of phasors is

$$\left.\begin{aligned}
\boldsymbol{F}_4 &= b_2\boldsymbol{V}_2 = (1\angle 0^\circ\ \text{N·s/m})(1\angle -90^\circ\ \text{m/s}) = 1\angle -90^\circ\ \text{N}\\
\boldsymbol{F}_3 &= sm\boldsymbol{V}_2 = (1\angle 90^\circ\ \text{rad/s})(1\ \text{N·s}^2/\text{m})(1\angle -90^\circ\ \text{m/s}) = 1\angle 0^\circ\ \text{N}\\
\boldsymbol{F}_2 &= \boldsymbol{F}_3 + \boldsymbol{F}_4 = 1\angle 0^\circ\ \text{N} + 1\angle -90^\circ\ \text{N} = (1 + j0\ \text{N}) + (0 - j1\ \text{N})\\
&= \sqrt{2}\angle -45^\circ\ \text{N}\\
\boldsymbol{V}_{12} &= \left(\frac{s}{k}\right)\boldsymbol{F}_2 = \frac{(1\angle 90^\circ\ \text{rad/s})}{(1\angle 0^\circ\ \text{N/m})}(\sqrt{2}\angle -45^\circ\ \text{N}) = \sqrt{2}\angle 45^\circ\ \text{m/s}\\
\boldsymbol{V}_1 &= \boldsymbol{V}_{12} + \boldsymbol{V}_2 = \sqrt{2}\angle 45^\circ\ \text{m/s} + 1\angle -90^\circ\ \text{m/s}\\
&= (1 + j1\ \text{m/s}) + (0 - j1\ \text{m/s}) = 1\angle 0^\circ\ \text{m/s}\\
\boldsymbol{F}_1 &= b_1\boldsymbol{V}_1 = (0.5\angle 0^\circ\ \text{N·s/m})(1\angle 0^\circ\ \text{m/s}) = 0.5\angle 0^\circ\ \text{N}\\
\boldsymbol{F} &= \boldsymbol{F}_1 + \boldsymbol{F}_2 = 0.5\angle 0^\circ\ \text{N} + \sqrt{2}\angle -45^\circ\ \text{N} = (0.5 + j0\ \text{N}) + (1 - j1\ \text{N})\\
&= 1.5 - j1\ \text{N} = 1.8\angle -33.7^\circ\ \text{N}
\end{aligned}\right\}\quad(5.94)$$

Since $\omega = 1$ rad/s and $33.7^\circ = 0.59$ radians, $F(t) = 1.8\cos(t - 0.59)$ N.

(b) The phasors are displayed in Figure 5.44(a). The *tail-to-tip* format illustrates the flow balance at nodes 1 and 2 and the zero sum of potential drops around the center loop of the line graph. The input (source) phasor is $\boldsymbol{F}$. The output phasor is $\boldsymbol{V}_2$. The 56.3° lag of $\boldsymbol{V}_2$ rela-

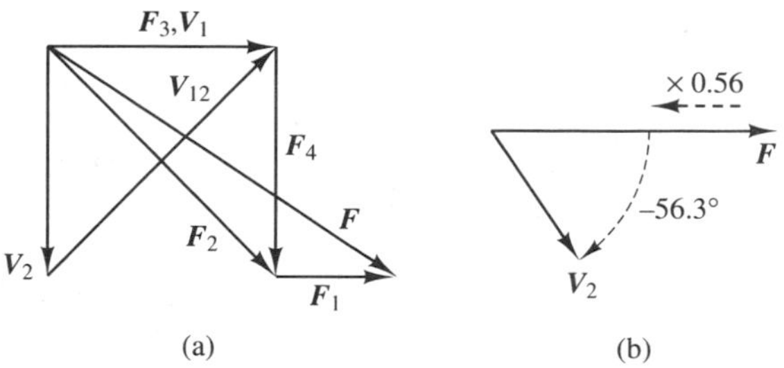

Figure 5.44 Phasor diagrams for the translational network. (a) Complete set of network phasors. (b) Input and output phasors.

tive to $\boldsymbol{F}$ implies that the peaks of the output velocity sinusoid $v_2(t)$ are delayed relative to the peaks of the input force sinusoid F(t) by 56.3°, or 15.6% of the period. Since the period is $p = 2\pi/\omega = 6.28$ s, the delay is 0.98 s.

The other phasors have similar interpretations; e.g., the 90° lead of $\boldsymbol{F}_3$ relative to $\boldsymbol{V}_2$ implies that each surge in the force $F_3(t)$ on the mass occurs one quarter-period (1.57 s) earlier than the corresponding surge in the velocity $v_2(t)$ of the mass.

(c) The transfer function that relates the input F(t) and the output $v_2(t)$ is

$$H(s) = \frac{v_2}{F} \tag{5.95}$$

We could find the functional form of H(s) from the system equation for v_2 and then substitute $s = j1$ to find the relation between the input and output phasors. On the other hand, we found the relations among the various network phasors for the frequency $s = j1$ in parts (a) and (b). Therefore, we merely take the ratio

$$\begin{aligned} H(j1) &= \frac{\boldsymbol{V}_2}{\boldsymbol{F}} = \frac{1 \angle -90° \text{ m/s}}{1.8 \angle -33.7° \text{ N}} \\ &= 0.56 \angle -56.3° \text{ m/(N·s)} \end{aligned} \tag{5.96}$$

The phasor representation for F(t) = cos t is $\boldsymbol{F} = 1 \angle 0°$ N. Then, according to equation (5.96),

$$\begin{aligned} \boldsymbol{V}_2 &= H(j1)\boldsymbol{F} = (0.56 \angle -56.3° \text{ m/(N·s)})(1 \angle 0° \text{ N}) \\ &= 0.56 \angle -56.3° \text{ m/s} \end{aligned} \tag{5.97}$$

The phasors $\boldsymbol{F}$ and $\boldsymbol{V}_2$ are shown in Figure 5.44(b). They are smaller in magnitude by the factor 0.56 and rotated in phase by the angle 33.7°, relative to their counterparts in Figure 5.44(a). The relation between $\boldsymbol{V}_2$ and $\boldsymbol{F}$ is the same in both cases because the frequency is the same; that is, $\boldsymbol{V}_2$ and $\boldsymbol{F}$ are related by the impedance relation $\boldsymbol{V}_2 = H(j1)\boldsymbol{F}$, as shown in Figures 5.44(a) and (b). For any sinusoidal input force F that has frequency $\omega = 1$ rad/s, the peaks of the velocity sinusoid v_2 will be reduced by the factor $|H(j1)|$ and delayed (in phase units) by arg(H(j1)) relative to the peaks of the applied force.

Periodic Signals as Sums of Sinusoids

A sinusoid of frequency ω is periodic with period $2\pi/\omega$. Sinusoids which have frequencies that are *integer multiples* of ω are called *harmonics* of the sinusoid of frequency ω. A sinusoid of frequency 2ω is referred to as the second harmonic, one of frequency 3ω as the third harmonic, etc. The original sinusoid is referred to as the *fundamental* sinusoid, and its frequency ω is the *fundamental frequency.* The period of the nth harmonic is smaller by the factor n than the fundamental period. The sum of a sinusoid and its harmonics is necessarily periodic, with the same period $2\pi/\omega$ as that of the fundamental sinusoid.

Figure 5.45 shows the addition of a fundamental and a third harmonic. Since the two signals do not have the same frequency, the resulting signal is periodic, but not sinusoidal. The figure demonstrates that changing the relative magnitudes or phases of the harmonic sinusoids can change the shape of the resulting periodic signal dramatically. The amplitudes and phases of various harmonics can be adjusted in order to *tailor* the combined signal to any desired shape over the fundamental period.

A sum of sinusoids that generates a particular periodic function is known as a **Fourier series expansion** of that function. The fundamental sinusoid in the expansion has the same frequency as that of the desired periodic function. The other sinusoids in the sum are harmonics (frequency multiples) of the fundamental sinusoid. Suppose that the function f(t) has period p. It can be shown that the Fourier Series expansion of f(t) is[5]

$$f(t) = a_0 + \sum_{n=1}^{\infty} \left(a_n \cos \frac{n2\pi t}{p} + b_n \sin \frac{n2\pi t}{p} \right) \tag{5.98}$$

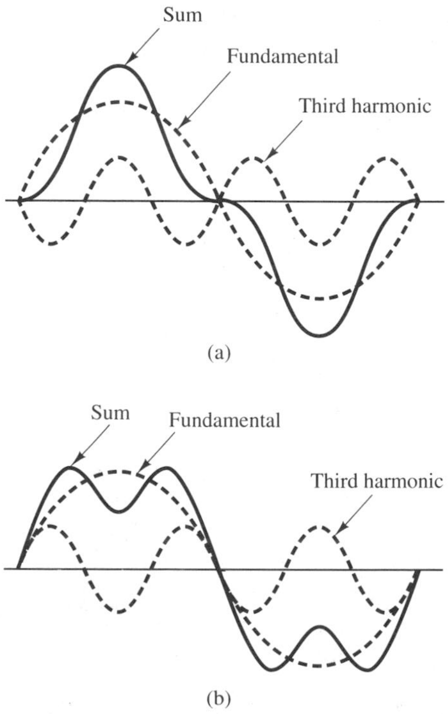

Figure 5.45 Sums of harmonic sinusoids [5.9].

[5] See reference [5.10], for example.

where

$$\left.\begin{aligned} a_0 &= \frac{1}{p}\int_{t_0}^{t_0+p} f(t)\,dt \\ a_n &= \frac{2}{p}\int_{t_0}^{t_0+p} f(t)\cos\frac{n2\pi t}{p}\,dt \\ b_n &= \frac{2}{p}\int_{t_0}^{t_0+p} f(t)\sin\frac{n2\pi t}{p}\,dt \end{aligned}\right\} \tag{5.99}$$

The integral is carried out over one full period of the signal; the symbol t_0 is an arbitrary beginning point for that period. The constant term of the expansion is the average value of the periodic function. The sine and cosine terms for each harmonic frequency can be combined into a single phase-shifted cosine. The amplitude and phase of each combined term then constitute a phasor representation of that term. If a source signal is not periodic, but is of limited duration, it can be treated for that duration as one period of a periodic signal. Hence, Fourier series expansions extend to nonperiodic signals.

Figure 5.46 constructs a *square wave* by adding harmonics of the fundamental frequency. If the source that drives a linear network is described by a periodic square wave, we can use the principle of superposition (see section 4.3) to find the steady-state solution: First, we find the harmonic components (the Fourier series representation) of the source signal. Then we find the network response for each sinusoidal sourcc component. Finally, we add the response components.

The square-wave decomposition of Figure 5.46 illustrates a general characteristic of Fourier series expansions: The higher order harmonics always have smaller amplitudes than the lower order harmonics. (If not, the series would not converge to a limit.) In section 5.4, we find that physical networks are typically *low-pass;* that is, as the frequency of the input signal increases, the amplitude of the response of each network variable decreases. In other words, high-frequency input signals have difficulty *passing through* the network. (This fact is intuitively obvious in the case of mechanical networks. If we apply a sinusoidal force of very high frequency to a mechanical network, the masses in the network are unable to respond with motions of significant amplitude.) The higher order harmonic components of a periodic source signal do not, therefore, produce significantly large response components throughout the network.

As a consequence of these two phenomena—the smaller amplitudes of the higher order components of a periodic source signal and the greater attenuation of these components by the dynamic system—we seldom need concern ourselves with more than three or four terms of a Fourier series expansion of a signal. In sum, the steady-state response of a linear network to a periodic source of arbitrary waveform can be thought of as the sum of the responses to a few sinusoidal inputs of appropriately chosen harmonic frequencies.

Example Problem 5.7: Response approximation by Fourier series.

Apply the square-wave force signal of magnitude F_c and period p shown in Figure 5.47 to the mechanical network of Figure 5.41.

(a) Find the Fourier series expansion of the source signal.

(b) Find the response v_1 of the network to each component of the expansion found in part (a) for $p = \pi$ seconds.

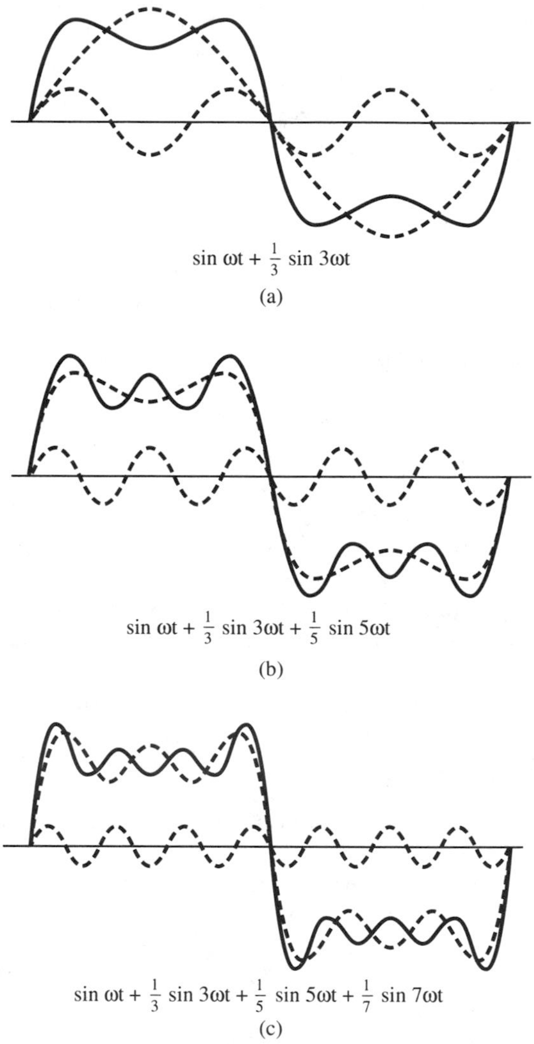

$\sin \omega t + \frac{1}{3} \sin 3\omega t$
(a)
$\sin \omega t + \frac{1}{3} \sin 3\omega t + \frac{1}{5} \sin 5\omega t$
(b)
$\sin \omega t + \frac{1}{3} \sin 3\omega t + \frac{1}{5} \sin 5\omega t + \frac{1}{7} \sin 7\omega t$
(c)

Figure 5.46 Synthesis of a square wave [5.9]. (a) Two terms. (b) Three terms. (c) Four terms.

(c) Add graphically enough response components from part (b) to sketch the response waveform.

Solution: **(a)** The fundamental frequency of the square-wave signal is $\omega = 2\pi/p$. We compute the Fourier coefficients from equation (5.99):

$$a_0 = \frac{1}{p}\int_0^p F(t)\,dt = \frac{1}{p}\int_0^{p/2} F_c\,dt = \frac{F_c}{2} \tag{5.100}$$

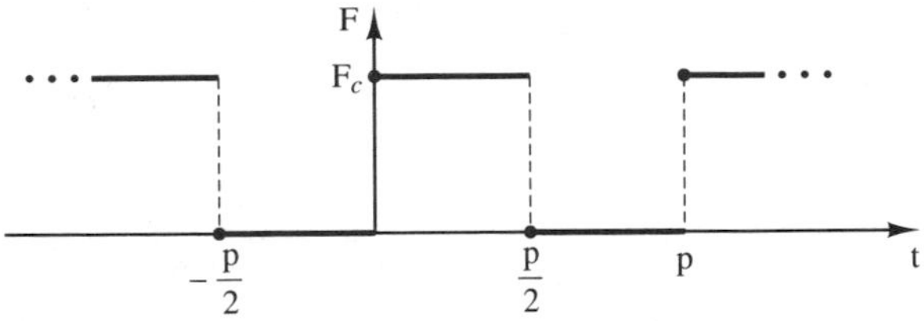

Figure 5.47 A periodic force source signal.

$$a_n = \frac{2}{p}\int_0^p F(t)\cos\frac{n2\pi t}{p}\,dt$$
$$= \frac{2}{p}\int_0^{p/2} F_c\cos\frac{n2\pi t}{p}\,dt$$
$$= \frac{2F_c}{p}\left(\frac{p}{n2\pi}\right)\left(\sin\frac{n2\pi t}{p}\right)_0^{p/2} = 0, \quad \text{for } n = 1, 2, 3, \ldots \tag{5.101}$$

$$b_n = \frac{2}{p}\int_0^p F(t)\sin\frac{n2\pi t}{p}\,dt$$
$$= \frac{2}{p}\int_0^{p/2} F_c\sin\frac{n2\pi t}{p}\,dt$$
$$= \frac{2F_c}{p}\left(\frac{p}{n2\pi}\right)\left(-\cos\frac{n2\pi t}{p}\right)_0^{p/2}$$
$$= \begin{cases} \dfrac{2F_c}{n\pi}, & \text{for } n = 1, 3, 5, \ldots \\ 0, & \text{for } n \text{ even} \end{cases} \tag{5.102}$$

Therefore, the force signal can be expressed as

$$F(t) = \frac{F_c}{2} + \frac{2F_c}{\pi}\sin\omega t + \frac{2F_c}{3\pi}\sin 3\omega t + \cdots \tag{5.103}$$

where $\omega = 2\pi/p$. The first term is the average value of F. Note the absence of cosine terms. We could have foreseen this absence because F(t) has odd symmetry about $t = 0$. Sine terms have such symmetry, whereas cosine terms do not.

(b) Suppose the input signal consisted only of the nth-harmonic term of equation (5.103). We could then represent it by the phasor $\boldsymbol{F}_n = 2F_c/(n\pi)\angle -\pi/2$, with the implicit frequency $n\omega$. Let us find the steady-state response phasor $\boldsymbol{V}_{1n}$ that would correspond to that nth-harmonic input component.

The input-output transfer function for the network, from equation (5.89), is

$$H(s) = \frac{v_1}{F} = \frac{s}{ms^2 + bs + k} \tag{5.104}$$

Since the specified period of the square-wave signal is $p = \pi$, the fundamental frequency is $\omega = 2\pi/p = 2$ rad/s. For the values of m, b, and k specified in Figure 5.41, and for $s = jn\omega = j2n$,

$$H(j2n) = \frac{j2n}{4(j2n)^2 + 5.66(j2n) + 2} = \frac{2n\angle\pi/2}{(2 - 16n^2) + j11.32n} \tag{5.105}$$

The nth-harmonic component of the response, then, is the phasor

$$\boldsymbol{V}_{1n} = \mathrm{H}(j2n)\boldsymbol{F}_n = \frac{4F_c/\pi}{(2 - 16n^2) + j11.32n} \tag{5.106}$$

The value of the sinusoidal transfer function for the constant (zero-frequency) term is $H(j0) = 0$. Therefore, v_1 is not affected by the constant component of F. (It is apparent from Figure 5.41 that there is a constant *displacement* $x_1 = F_c/2k$ corresponding to the constant component of F.)

(c) The component of the response v_1 owing to the fundamental-frequency component of F is represented by the phasor

$$\boldsymbol{V}_{11} = \frac{4F_c/\pi}{-14 + j11.32} = \frac{1.27F_c}{18\angle 141°}$$

$$= 0.0706F_c \angle -141° = 0.0706F_c \angle -2.46 \text{ rad} \tag{5.107}$$

A similar computation for the component of v_1 owing to the third-harmonic component of F produces the phasor

$$\boldsymbol{V}_{13} = 0.0087F_c \angle -167° = 0.0087F_c \angle -2.91 \text{ rad} \tag{5.108}$$

Note the strong attenuating effect of the system on the third-harmonic component. We can expect the fifth-harmonic component to be negligible. Therefore, we conclude that

$$v_1(t) \approx F_c[0.0706 \cos(2t - 2.46) + 0.0087 \cos(6t - 2.91)] \text{ m/s} \tag{5.109}$$

We show this approximation, together with the exact response, in Figure 5.48.

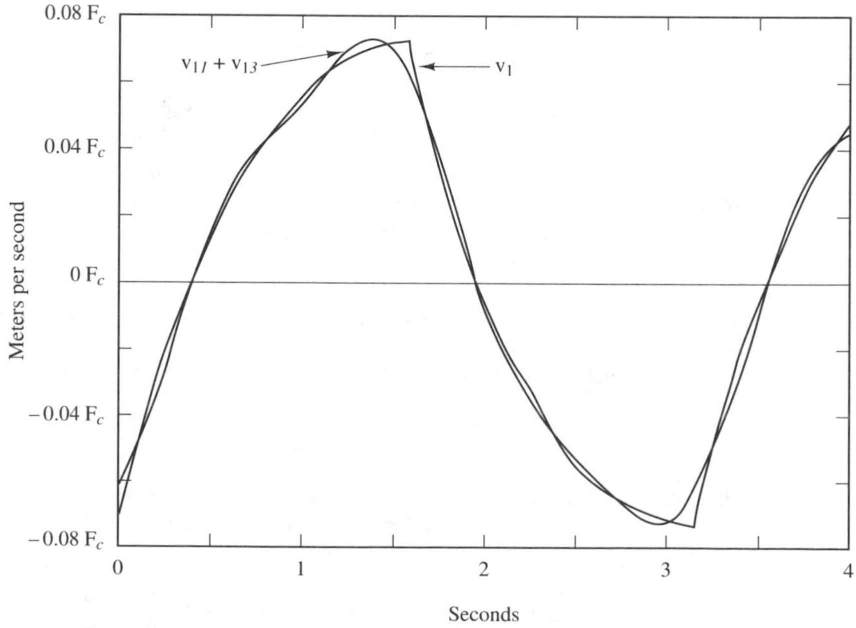

Figure 5.48 Two-term Fourier series approximation.

5.4 THE FREQUENCY RESPONSE OF A LINEAR NETWORK

Suppose we energize a linear network with a unit-amplitude sinusoidal source. Then the steady-state response of each network variable is represented by a phasor that has the same complex value as the sinusoidal system function for that variable. The **frequency response** of a variable of the network is a plot of the corresponding sinusoidal system function (amplitude and phase) as the frequency of the input sinusoid varies from zero to infinity. This section develops ways to determine and display the frequency response of a linear network. It also uses the frequency response to understand the network behavior better.

Why study the frequency response of a network variable? There are several reasons. (a) A periodic source, such as the vibration of an engine, can produce extraordinary resonant behavior. Frequency-response analysis exposes such behavior and shows how to deal with it. (b) The roots of the denominator and numerator of an input-output transfer function are called the *poles* and *zeros* of that transfer function. Those poles and zeros characterize the system behavior. They are more directly related to the frequency response of the output variable than they are to its step response. (c) We must measure the values of variables in the physical system to estimate the values of the lumped-model parameters. Accurate measurements are difficult to make if the system variables change rapidly. It is relatively easy to measure the steady-state parameters (amplitude, phase, and frequency) of sinusoidal signals. Measuring the frequency response of a system is an effective practical way to determine the system parameters.

Frequency Response

A low-frequency sinusoid is a slowly changing source signal. A high-frequency sinusoid is rapidly changing. Just as a step input exercises a network at both the slowly changing and rapidly changing extremes, the frequency response captures the behavior associated with both low and high frequencies (from zero to infinity).

Let H(s) denote the transfer function that relates the input variable y_I to the output variable y_O for a linear network. Then the response to the source signal y_I is $y_O = H(s)y_I$. For a sinusoidal input of frequency ω, s can be replaced by jω in the steady state. Also, the signals y_I and y_O can be represented by the phasors $\boldsymbol{Y}_I$ and $\boldsymbol{Y}_O$, respectively. Therefore,

$$\boldsymbol{Y}_O = H(j\omega)\boldsymbol{Y}_I \tag{5.110}$$

where H(jω) is the sinusoidal system function of section 5.3. To simplify the notation, we define the output-over-input amplitude ratio

$$M_H(\omega) \triangleq |H(j\omega)| \tag{5.111}$$

and the input-to-output phase shift

$$\phi_H(\omega) \triangleq \arg[H(j\omega)] \tag{5.112}$$

Then

$$H(j\omega) = M_H(\omega)\angle\phi_H(\omega) \tag{5.113}$$

If the source is the unit sinusoid $y_I(t) = \cos \omega t$, represented by the phasor $\boldsymbol{Y}_I = 1\angle 0°$, then the steady-state response is the sinusoid

$$y_O(t) = M_H(\omega)\cos[\omega t + \phi_H(\omega)] \tag{5.114}$$

That is, the phasor response $\boldsymbol{Y}_O$ to the *unit* sinusoid equals the complex number H(jω). We must be careful here to retain separation between the concepts of signal and transfer function. H(jω) represents the input-output transfer function for sinusoidal signals. $\boldsymbol{Y}_O$ is the phasor representation of a particular output signal. If the input-signal phasor is exactly 1, both $\boldsymbol{Y}_O$ and H(jω) are the same complex number, but their units and physical interpretations are different.

We call equation (5.113) the **frequency response** of the network. In general, the frequency response of a linear network is the output phasor, as a function of the frequency ω, for a *unit* sinusoidal input. (The input *phasor* does not change with frequency.) Because equation (5.113) is complex valued, the frequency response consists of two functions, $M_H(\omega)$ and $\phi_H(\omega)$. The term *frequency response* is sometimes used to refer to the magnitude response alone. (We shall show that the phase response can be determined from the magnitude response.)

Figures 5.49(a) and (b) show a first-order translational network. The sinusoidal system function for F is

$$H(s) \triangleq \frac{F}{v} = \frac{ms}{1 + (m/b)s} \text{ kg/s} \tag{5.115}$$

where $\tau = m/b$ is the time constant of the network.

To find the frequency response of the variable F to a *unit* sinusoidal velocity-source input, we merely replace s by jω in equation (5.115) to obtain

$$\boldsymbol{F}(j\omega) = H(j\omega) = \frac{j\omega m}{1 + j\omega m/b} \text{ N} \tag{5.116}$$

Note that we speak of the frequency response of F, for which the units are newtons, yet we compute H(jω), for which the units are kg/s. Actually, to get $\boldsymbol{F}$, we multiply H(jω) by the implicit unit-valued phasor $\boldsymbol{V}(j\omega)$, for which the units are m/s. It is only the numerical values of $\boldsymbol{F}(j\omega)$ and H(jω) that are equal. We shall not continue to carry the physical units for H, because it is the *shape* of the frequency response that explains the behavior of the network.

According to the rules of complex algebra, the magnitude and phase of equation (5.116) are, respectively,

$$M_H(\omega) = \frac{\omega m}{\sqrt{1 + \omega^2 m^2/b^2}} \tag{5.117}$$

and

$$\phi_H(\omega) = 90^\circ - \tan^{-1}\left(\frac{\omega m}{b}\right) \tag{5.118}$$

We have normalized the constant term (rather than the coefficient of s) in the denominator of equation (5.115) in order to ease the computation of equations (5.117) and (5.118). Calculating $\phi_H(\omega)$ is relatively simple. Calculating $M_H(\omega)$ is more complicated. Equations (5.117) and (5.118) are plotted against frequency in Figures 5.49(c) and (d) for particular values of m and b.

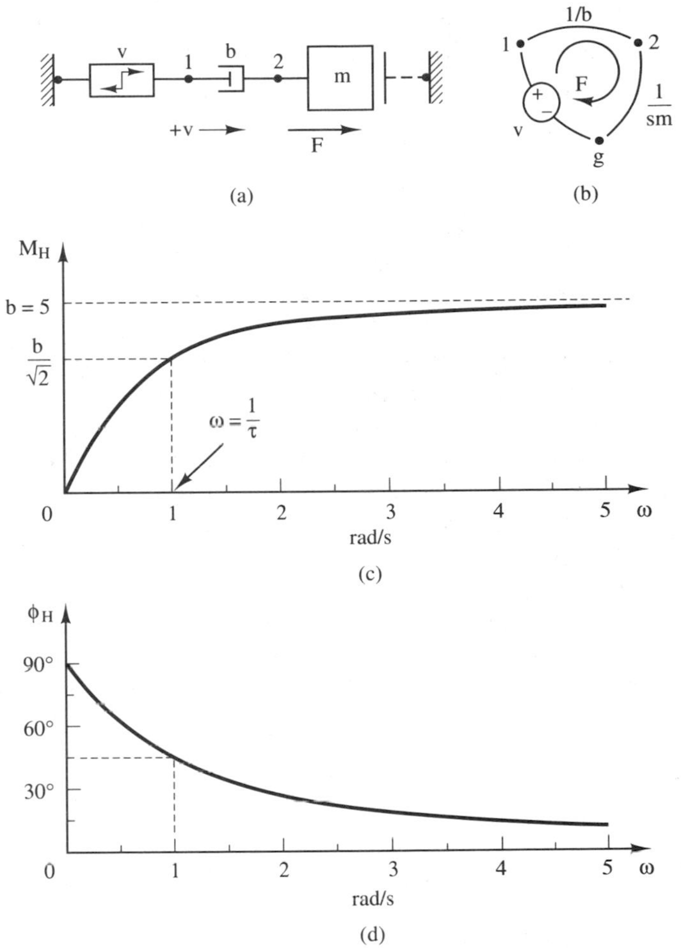

Figure 5.49 A first-order translational network and its frequency response; $m = 5$ kg, $b = 5$ N·s/m, $\tau = m/b = 1$ s. (a) Lumped model. (b) Line-graph representation. (c) Magnitude response. (d) Phase response.

The network described by this pair of frequency-response plots can be thought of as a **high-pass** network because low-frequency signals (those that change slowly relative to the time constant of the network) are severely attenuated as they *pass* from input to output. This attenuation is clearly visible in the magnitude plot. We consider the signal to be blocked by the network if $\omega < 1/\tau$.

Figure 5.41 shows a second-order translational network. The input-output transfer function for the velocity v_1 of that network is given in equation (5.89). It has natural frequency $\omega_n = \sqrt{k/m}$. We renormalize that transfer function to the form

$$H(s) = \frac{s/k}{1 + (b/k)s + (m/k)s^2} \tag{5.119}$$

(Normalizing the constant term of the denominator to 1 produces a fixed standard against which to compare the sizes of the other terms.) For $s = j\omega$, equation (5.119) is

$$H(j\omega) = \frac{j\omega/k}{(1 - \omega^2 m/k) + j\omega b/k} \tag{5.120}$$

The magnitude $M_H(\omega)$ and phase $\phi_H(\omega)$ of equation (5.120) are plotted in Figure 5.50 for specific values of m, k, and b. The damping ratio is $\zeta = 0.1$. Because the network is *lightly damped,* the frequency response is highly sensitive to changes in ω if ω is in the neighborhood of the natural frequency ω_n. This sensitivity is manifest as a sharp peak in the magnitude plot and as a steep negative slope in the phase plot. The markedly stronger response near the natural frequency is a phenomenon known as **resonance**.

A network with the frequency response shown in Figure 5.50 can be thought of as a **band-pass** network because it strongly accentuates frequencies in the narrow band near the resonant (or peak-magnitude) frequency; alternatively, we can view the other frequencies as severely attenuated. (It is customary, when speaking of the **filtering** characteristics of a network, to speak of *frequencies* when we really mean sinusoidal signals that possess those frequencies.)

The frequency-response plots of Figure 5.49 and 5.50 highlight the signal-filtering and resonant characteristics of the two networks. The plots even give some indication of the values of the system parameters (the time constant or natural frequency). In the next

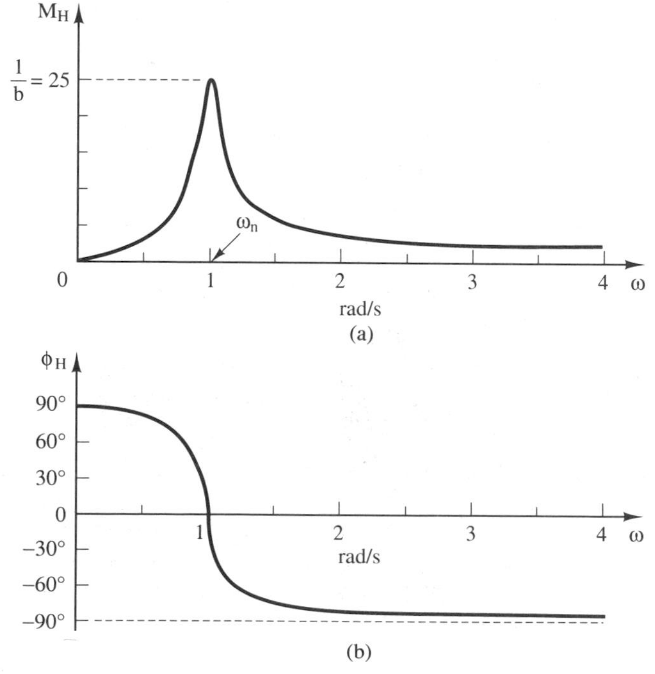

Figure 5.50 Frequency response of the second-order transfer function of equation (5.119); $m = 0.2$ kg, $k = 0.2$ N/m, $b = 0.04$ N·s/m, $\omega_n = \sqrt{k/m} = 1$ rad/s, $\zeta = 0.1$. (a) Magnitude response. (b) Phase response.

section, we redisplay these same frequency-response data in a manner that pinpoints more clearly the distinguishing features of the network.

Logarithmic Scales

H.W. Bode (see reference [5.1]) noted that the key features of a dynamic system are exposed more clearly if the frequency-response data are plotted against logarithmic scales. We use log scales for the frequency ω and magnitude M_H, but a linear scale (in degrees) for the phase ϕ_H. The logarithm converts multiplication and division of factors to addition and subtraction of their logarithms. Equal distances on a log scale represent equal ratios of the variable plotted.

A log scale is shown in Figure 5.51. Distance along the scale is proportional to the log of the variable. The scale is labeled above the line with values of the variable itself. It is convenient to use this direct form of labeling for the logarithmic frequency axis of frequency-response plots. The distance between the values 1 and 2 equals the distance between the values 2 and 4 and the distance between the values 5 and 10. Each such distance on the log scale represents a doubling of value and is called an **octave** change in value. (The term *octave* originates in a musical scale, which exhibits a doubling of frequency over a span of eight notes. We use the term for the frequency axis.) The distance between the values 0.1 and 1 equals the distance between the values 1 and 10. Each such distance represents an increase of a factor of 10 and is called a **decade** change in value. The fact that the distance between the values 4 and 5 (or 0.4 and 0.5) is very nearly one-third of an octave makes it easy to sketch a log scale on uniformly lined paper: 3 lines cover one octave; 10 lines cover one decade.

Figure 5.51 also labels the same log scale with base-10 log values. The log of the variable is not much used as an axis label. However, the quantity

$$\mathrm{Lm}(H) \triangleq 20 \log_{10} M_H, \tag{5.121}$$

is often used for labeling the magnitude-response axis of frequency-response plots. It is referred to as the **log-magnitude** of H and is expressed in **decibels** (abbreviated dB). Figure 5.51 shows this decibel labeling. (The term *decibel* was first defined as a measure of the ratio of output power to input power. Equation (5.121) is fully equivalent to the original definition only if M_H^2 is a power ratio.) The notations M_H and Lm(H) pertain only to a system in sinusoidal steady state ($s = j\omega$).

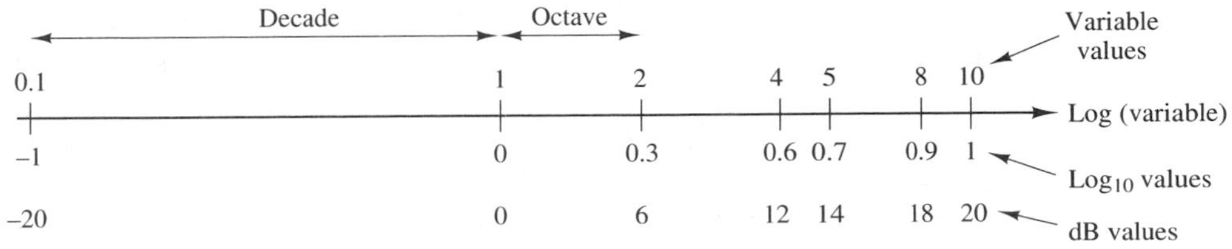

Figure 5.51 Logarithmic scales.

Bode Diagrams

The logarithm of the magnitude-response function (5.117) is

$$\log M_H = \log \omega m - \log \sqrt{1 + \frac{\omega^2 m^2}{b^2}}$$

$$= \log m + \log \omega - \tfrac{1}{2} \log\left(1 + \frac{\omega^2 m^2}{b^2}\right) \tag{5.122}$$

For low frequencies ($\omega \ll b/m$, where m/b is the time constant τ), equation (5.122) is approximately

$$\log M_H \approx \log m + \log \omega - \tfrac{1}{2} \log(1)$$

$$= \log m + \log \omega \tag{5.123}$$

a straight line on log-scaled axes. For high frequencies ($\omega \gg b/m$), equation (5.122) becomes

$$\log M_H \approx \log m + \log \omega - \tfrac{1}{2} \log\left(\frac{\omega^2 m^2}{b^2}\right)$$

$$= \log m + \log \omega - \log\left(\frac{\omega m}{b}\right)$$

$$= \log b \tag{5.124}$$

again a straight line on log-scaled axes. The magnitude function is displayed against log-scaled axes (for specific values of m and b) in Figure 5.52(a). The low-frequency and high-frequency asymptotes are also displayed. We call the frequency $\omega = b/m = 1/\tau$ at which the two asymptotes intersect a **corner frequency** of the log-log plot. The low-frequency and high-frequency terms of the square root are equal at the corner frequency. According to equation (5.115), the corner frequency $\omega = b/m$ is the negative of the network pole.

The phase function ϕ_H in equation (5.118) is constant at the value 90° for low frequencies and constant at the value 0° for high frequencies. In these two regions, ϕ_H can be viewed as linear in log ω. The phase response is displayed against a log(ω) axis in Figure 5.52(b). Also displayed are the low-frequency and high-frequency asymptotes. These two asymptotes are not accurate representations of the true response in the two-decade interval centered at the corner frequency (of the magnitude response). However, the straight line that intersects the low-frequency and high-frequency asymptotes exactly one decade above and below the corner frequency does provide a good approximation to ϕ_H.

We usually display the frequency response of a linear network as a pair of plots: the log-magnitude plot, Lm(H) vs. log ω; and the phase plot, ϕ_H vs log ω. These plots are referred to as **Bode diagrams**, in recognition of the early frequency-response work of H.W. Bode. We usually label the log ω-axis with values of ω rather than with values of log ω, in order to ease interpretation of the plot. For the same reason, we sometimes label the Lm(H) axis with values of M_H, instead of or in addition to the dB scale.

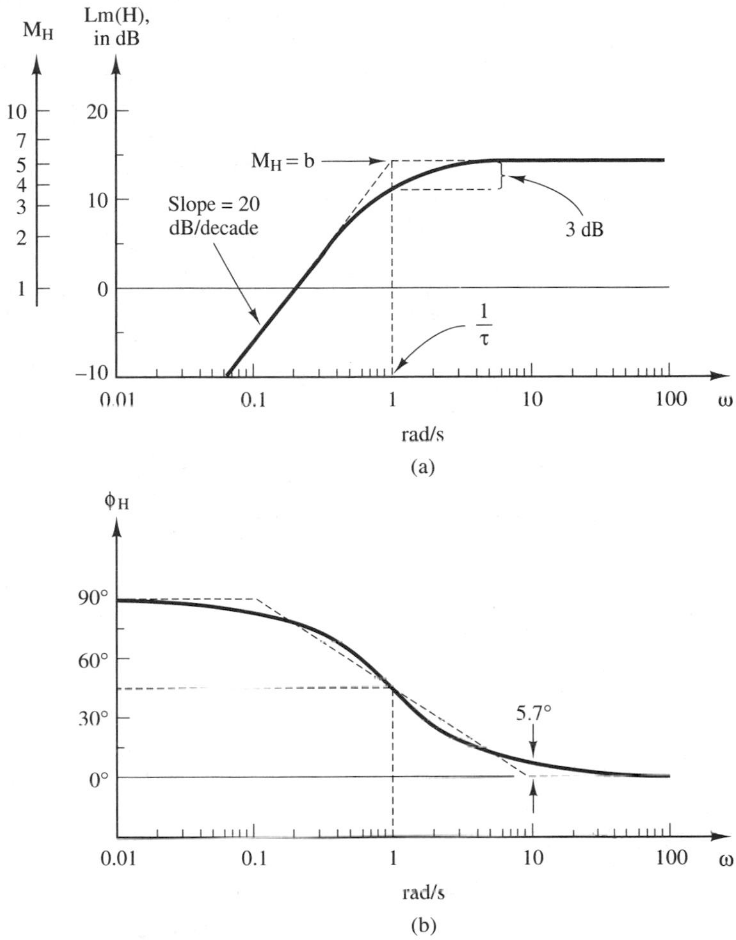

Figure 5.52 Bode diagrams for the system equation (5.115); m = 5 kg, b = 5 N·s/m, $\tau = m/b = 1$ s. (a) Magnitude diagram. (b) Phase diagram.

The Bode diagrams for the second-order transfer function (5.119) are shown, for specific values of m, k, and b, in Figure 5.53. In this example, the corner frequency is the natural frequency $\omega_n = \sqrt{k/m}$ of the pole pair. The deviation of the log-magnitude plot from the asymptotic approximation is greater than in the first-order example of Figure 5.52, because the pole pair is lightly damped (with a tendency to oscillate). Associated with this greater deviation from the asymptotes in the Lm plot is a more rapid change of phase with frequency in the phase plot ϕ_H. That is, the phase switches value over a frequency interval much shorter than two decades.

The foregoing examples demonstrate several universal properties of Bode diagrams of linear networks: (1) The log-magnitude plots can be approximated by straight lines

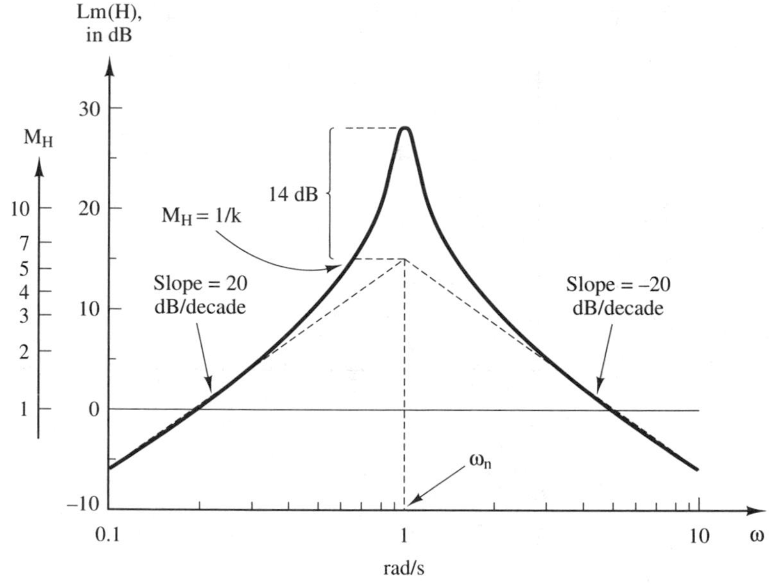

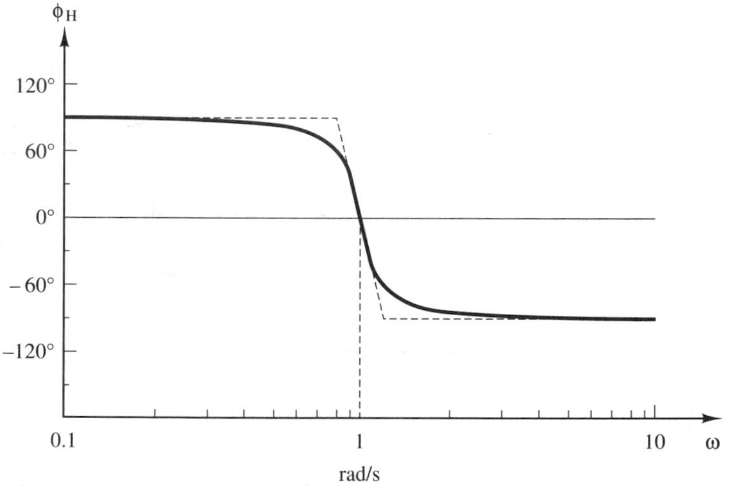

Figure 5.53 Bode diagrams for the system equation (5.119); m = 0.2 kg, k = 0.2 N/m, b = 0.04 N·s/m, $\omega_n = \sqrt{k/m} = 1$ rad/s, $\zeta = 0.1$. (a) Magnitude diagram. (b) Phase diagram.

whose slopes are integral multiples of 20 dB/decade (or 6 dB/octave). (2) The straight-line approximations to the log-magnitude plots intersect at corner frequencies that correspond to the poles and zeros of the network—namely, frequencies equal to the negatives of the first-order poles and zeros or equal to the natural frequencies (ω_n) of the complex conjugate pole pairs and zero pairs. (3) The straight lines are quite accurate approximations to the log-magnitude plots, except in the neighborhoods of the corner frequencies. (4) The phase plots are characterized by regions of constant phase—that is, regions in which the phase is an integral multiple of 90°; phase changes occur only in the two-decade neighborhood of each corner frequency. The phase changes associated with a lightly damped pole pair or zero pair occur over a much smaller frequency interval.

Constructing Bode Diagrams

A transfer function H(s) contains products and quotients of pole and zero factors. (We treat complex pole pairs and zero pairs as quadratic factors in order to minimize use of complex numbers.) The Bode phase plot ϕ_H is the sum of the phase plots associated with the individual factors. (See equation (5.118), for example.) Because the logarithm converts multiplications to additions, the log-magnitude plot Lm(H) is the sum of the log-magnitude plots for the individual factors. In this section, we first define a standard normalized form for the transfer function. Then we determine the forms of the Bode plots for the various types of factors of the transfer function. Finally, we present a procedure for constructing the Bode plots for an arbitrary transfer function.

First, we factor the numerator and denominator of the transfer function H(s). We normalize each pole and zero factor to set the constant term equal to 1. We accumulate all the constant factors into a single constant A *in the numerator*. We call this normalized form the *Bode form* of H(s). (The factor s, which corresponds to a pole or zero at the origin, contains no constant term and cannot be normalized.) If there is a pair of complex conjugate factors, we multiply those factors to produce the conventional form $s^2 + 2\zeta\omega_n s + \omega_n^2$ and then normalize it to the form $1 + 2\zeta(s/\omega_n) + (s/\omega_n)^2$. This quadratic form avoids the use of complex numbers.

For example, suppose the numerator and denominator are factored to produce

$$H(s) = \frac{2(s + 3)}{3s(s + 5)(s + 8)} \tag{5.125}$$

We factor out the constant terms in the numerator and denominator to produce

$$H(s) = \frac{2(3)(1 + s/3)}{3(5)(8)s(1 + s/5)(1 + s/8)} = \frac{0.05(1 + s/3)}{s(1 + s/5)(1 + s/8)} \tag{5.126}$$

This is the form of normalization that we used in the two previous examples. The constant factor for equation (5.115) is $A = m$; for equation (5.119), the constant factor is $A = 1/k$.

We now find the Bode diagrams for each type of factor of H(s). We must remember that the Bode diagrams assume sinusoidal steady-state operation with $s = j\omega$. The Bode diagrams for the constant **factor A** have the values

$$\mathrm{Lm}(A) = 20 \log(A) \quad \text{and} \quad \phi(A) = 0° \tag{5.127}$$

for all frequencies. The effect of the constant factor is to raise or lower the total Lm(H) curve by a fixed amount. The constant factor does not affect ϕ_H.

The **factor 1/s** represents a pole of H(s) at s = 0. According to equations (5.121) and (5.111), the log magnitude of the factor 1/s is

$$\text{Lm}\left(\frac{1}{s}\right) = 20 \log\left|\frac{1}{(j\omega)}\right| = 20 \log\left(\frac{1}{\omega}\right) = -20 \log \omega \text{ dB} \tag{5.128}$$

This is a straight line of slope −20 dB/decade on the log ω scale. The line passes through 0 dB at ω = 1 rad/s. The phase function for the factor 1/s, from equation (5.112), is the constant

$$\phi = \arg\left(\frac{1}{j\omega}\right) = -90° \tag{5.129}$$

The **factor s** represents a zero of H(s) at s = 0. The log-magnitude plot for the factor s is

$$\text{Lm}(s) = 20 \log \omega \tag{5.130}$$

Hence, the plot for a zero at the origin differs from the plot of equation (5.128), for a pole at the origin, only in sign. The phase function for the factor s is the constant

$$\phi = \arg(j\omega) = 90° \tag{5.131}$$

Thus, moving the factor s from the denominator to the numerator of H(s) merely changes the signs of the magnitude and phase curves for the factor.

All **other first-order factors** have the form (1 + τs). If this factor is in the denominator, then its magnitude-plot satisfies

$$\text{Lm}\left(\frac{1}{1 + \tau s}\right) = 20 \log\left|\frac{1}{1 + j\omega\tau}\right| = -10 \log(1 + \omega^2\tau^2) \tag{5.132}$$

This quantity is plotted in Figure 5.54(a) against a log ω scale that is *labeled* in terms of the normalized quantity ωτ. It has a corner frequency at ω = 1/τ, a value of zero for low frequencies, a −20-dB/decade slope at high frequencies, and a 3-dB maximum deviation from those two asymptotes. If the factor is in the numerator instead of the denominator, the sign of the Lm function is reversed, and the high-frequency slope is 20 dB/decade, as displayed in Figure 5.54(a). If the factor is in the denominator, the coefficient τ is the time constant corresponding to the pole. (*There is no time constant associated with a zero* of H(s).)

The phase plot for the denominator factor $\left(\frac{1}{1 + \tau s}\right)$ satisfies

$$\phi = \arg\left(\frac{1}{1 + j\omega\tau}\right) = -\arg(1 + j\omega\tau) \tag{5.133}$$

The phase is zero for low frequencies and −90° for high frequencies. It can be approximated in the two-decade neighborhood of ω = 1/τ by a straight line that passes through −45° at the corner frequency of the log-magnitude plot and that has −45°/decade slope, as

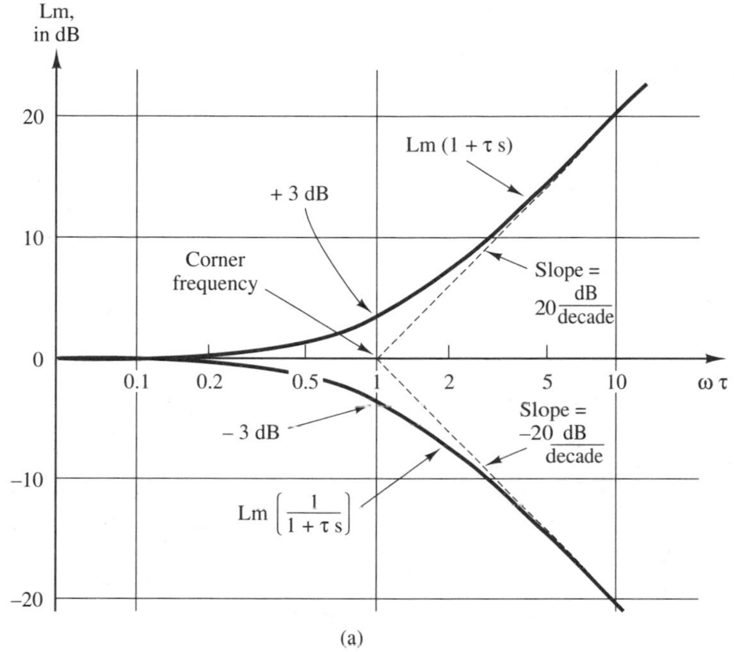

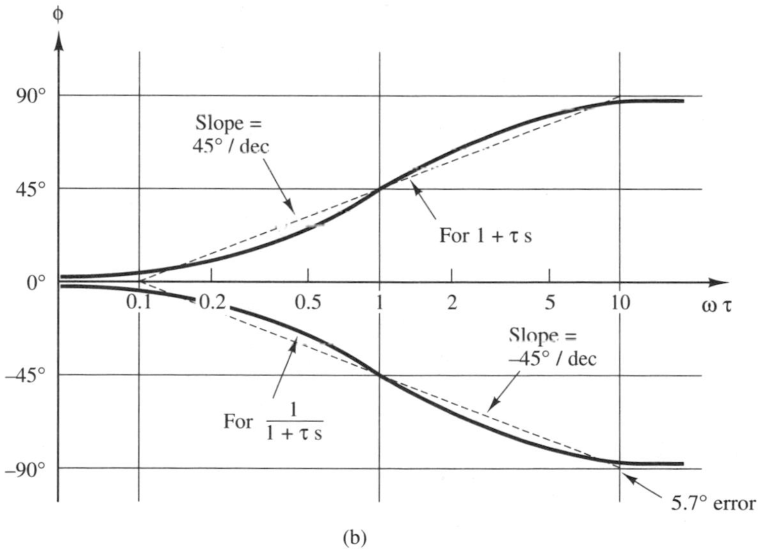

Figure 5.54 Bode diagrams for the factors $(1 + \tau s)$ and $1/(1 + \tau s)$; dashed lines show straight-line approximations. (The frequency axis is normalized; the corner frequency is $\omega = 1/\tau$.) (a) Magnitude diagrams. (b) Phase diagrams.

illustrated in Figure 5.54(b). If the factor $(1 + \tau s)$ is in the numerator instead of the denominator of H(s), the sign of the phase plot is reversed, and the midfrequency slope is 45°/decade.

A factor associated with a **complex conjugate pole pair** has the form $1/[1 + 2\zeta(s/\omega_n) + (s/\omega_n^2)]$, with log magnitude

$$-20 \log \left| 1 - \frac{\omega^2}{\omega_n^2} + j\frac{2\zeta\omega}{\omega_n} \right| \tag{5.134}$$

This function is plotted in Figure 5.55(a) for various values of damping ratio ζ. Note that the *shape* of the plot is determined by ζ alone. ω_n specifies the frequency above which the system is unresponsive.

The low-frequency asymptote for the conjugate pole pair $(\omega \ll \omega_n)$ is constant at 0 dB. The high-frequency asymptote $(\omega \gg \omega_n)$ has a −40-dB/decade slope (because there are two poles). The corner frequency at which the two asymptotes intersect is the natural frequency ω_n. The maximum deviation from the straight-line approximation is 6 dB if $\zeta = 1$. For $\zeta < 0.707$, the log-magnitude plot has a peak in the neighborhood of $\omega = \omega_n$. If ζ is less than 0.3, the magnitude of the peak is large (implying resonance), and the pole pair can cause a large response in the total transfer function.

The phase response for the quadratic pole pair is plotted in Figure 5.55(b). It has the asymptotic characteristics of two individual poles; namely, 0° for low frequencies $(\omega \ll \omega_n)$ and −180° for high frequencies $(\omega \gg \omega_n)$. It can be represented in the neighborhood of the log-magnitude corner frequency $(\omega = \omega_n)$ by a straight line that passes through −90° at the corner frequency. That straight line has −90°/decade slope if $\zeta = 1$; for smaller values of ζ, the slope is steeper, approaching $-\infty$ as $\zeta \to 0$.

The Bode plots for complex conjugate zeros are the negatives of the plots for conjugate poles with the same locations in the s-plane. Thus, Bode plots for both the pole and the zero factors can be obtained from Figure 5.55. Complex conjugate zeros are uncommon. The symbols ζ and ω_n were defined in relation to characteristic roots—roots of the *denominator* of H(s). Roots of the numerator are not directly correlated with the form of network response.

The normalized second-order frequency response forms shown in Figure 5.55 are comparable to the normalized zero-state step-response forms of Figure 3.52. The *shape* of the frequency response is determined by ζ and is a measure of the *degree of stability* of a second-order network. The **corner frequency** (or *cutoff frequency*) ω_n is a measure of the *responsiveness* of the network.

Similarly, the frequency-response forms for a first-order denominator factor (see Figure 5.54) are comparable to the exponential step-response forms for first-order networks (see section 3.3). The corner frequency $1/\tau$ is a measure of responsiveness for a first-order network.

BODE-DIAGRAM CONSTRUCTION PROCEDURE

1. Factor the numerator and denominator polynomials. Typically, all poles and zeros lie in the left half-plane.
2. Normalize to the value 1 the constant terms of the pole and zero factors (except for poles and zeros at the origin). Accumulate all the constant factors into a

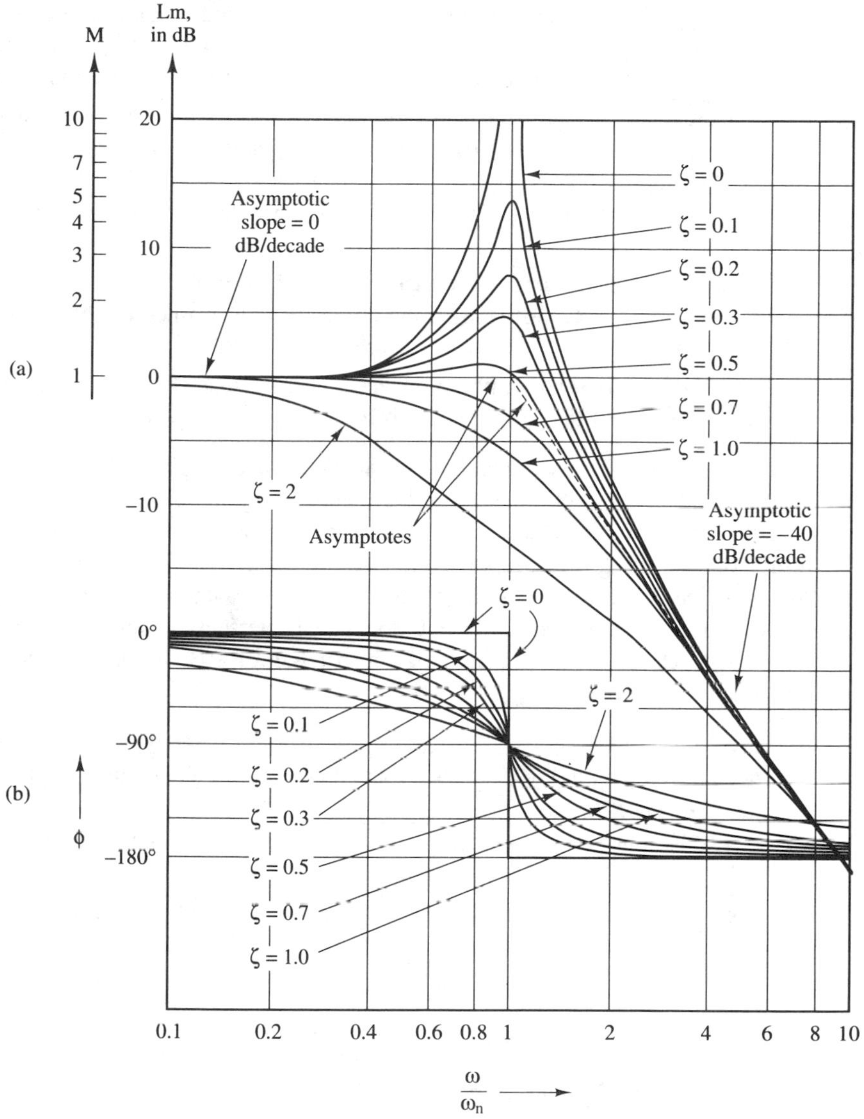

Figure 5.55 Bode diagrams for the factor $1/[1 + 2\zeta(s/\omega_n) + (s/\omega_n)^2]$. The frequency axis is normalized; the corner frequency of the straight-line Lm approximation is $\omega = \omega_n$. (a) Magnitude diagrams. (b) Phase diagrams.

single constant multiplier A in the numerator. That multiplier has a constant log magnitude equal to 20 log(A), in decibels. Plot that constant value vs. log ω.

3. Plot the straight line corresponding to the log magnitude of any pole(s) or zero(s) which lies at the origin; the straight line should cross the 0-dB axis at $\omega = 1$ rad/s.

4. Plot the straight-line approximation to the log-magnitude curve for each first-order or second-order factor (see Figure 5.54 or 5.55) at the location on the log ω-axis that is determined by the corner frequency for that factor.
5. Add the straight-line Lm plots for the factors to obtain the asymptotic Lm(H) curve. (Adding the dB values corresponds to multiplying the factors.) The asymptotic curve is inaccurate in the neighborhoods of the corner frequencies. The asymptotic Lm(H) curve can be corrected for each term, as implied by Figures 5.54 and 5.55.
6. Plot the phase curve for each first-order or second-order factor (see Figure 5.54 or 5.55) at the position on the log ω-axis determined by the corner frequency for that factor.
7. Add the individual phase curves and then raise (lower) the sum by 90° for each zero (pole) that lies at the origin, to produce the ϕ_H curve.

This construction procedure is illustrated in example problem 5.8. The Bode diagrams can be plotted on linear, semilog, or log-log graph paper, depending on the labeling selected for the axes. The observations made earlier concerning the slopes and intersections of the straight-line approximations assist in the plotting process. Approximate straight-line Bode plots can be sketched quickly, without reference to the normalized curves, once the observations about slopes and intersections are understood.

As a check on the accuracy of construction of a Bode diagram, note the following facts about Lm(H) and ϕ_H:

a. The low-frequency asymptote of Lm(H) should have a slope equal to $-20(n_p - n_z)$ dB/decade, where n_p and n_z are the numbers of poles and zeros *at the origin;*
b. The low-frequency asymptote (or its extension) should have the value 20 log(A) at $\omega = 1$ rad/s;
c. The high-frequency asymptote should have a slope equal to $-20(N_p - N_z)$ dB/decade, where N_p and N_z are the numbers of poles and zeros in H(s);
d. All other asymptotes should have slopes that are integral multiples of 20 dB/decade;
e. The low-frequency phase should equal $-90°\ (n_p - n_z)$;
f. The high-frequency phase should equal $-90°\ (N_p - N_z)$.

Computer programs can produce accurate Bode diagrams for a specified H(s). The program CC (see appendix D), for example, provides convenient procedures for describing the transfer function H(s) and permits the user to choose the form of presentation for the computed frequency response. PSPICE (see appendix C) can generate the frequency response directly from a specification of the network elements and the interconnection structure, without the necessity of determining the system equations or the transfer function.

The observations concerning the slopes and intersections of the straight-line approximations serve as guides for *interpreting* computer-generated Bode diagrams. We use the straight-line approximations to determine the system poles and zeros in a later section.

Let us apply the construction procedure to the transfer function of equation (5.115) for m = 5 kg and b = 5 N·s/m. The transfer function is already factored and normalized. First, we find the log-magnitude plot. The magnitude response corresponding to the constant m = 5 in the numerator is a horizontal line at the level Lm(5) = 20 log 5 = 14 dB. We superimpose that line on Figure 5.52(a). The s in the numerator of H(s) indicates a zero at the origin. We represent the corresponding component of the magnitude response by a straight line with a 20-dB/decade slope. This line should pass through 0 dB at $\omega = 1$ rad/s. We superimpose this line also on Figure 5.52(a).

The factor $1 + (m/b)s = 1 + s$ in the denominator has the magnitude response shown in the bottom half of Figure 5.54(a), with $\tau = m/b = 1$ s. The asymptotic approximation to the magnitude response for this factor has two parts: a horizontal line at the 0-dB level on the left and a line with a −20 dB/decade slope on the right. The sloping line intersects the 0-dB horizontal line at $\omega\tau = 1$. Thus, the corner frequency for this factor is $\omega = 1/\tau = 1$ rad/s. We superimpose this straight-line approximation on Figure 5.52(a). We add the straight-line approximations for the factors 5, s, and $1/(1 + s)$ to obtain the total straight-line approximation for $H(j\omega)$ shown in Figure 5.52(a). Figure 5.54(a) shows that we should make a −3 dB correction to the straight-line approximation at the corner frequency $\omega = 1$, owing to the pole at $s = -1$. The corrections at neighboring points can be sketched by eye or transferred from Figure 5.54(a).

Next we find the phase plot. The phase shift owing to the constant factor is 0°. Hence, we ignore it. The phase shift owing to the factor s in the numerator of H(s) is 90°. We superimpose the 90° line on Figure 5.52(b). The phase plot for the factor $1/(1 + s)$, according to Figure 5.54(b), can be approximated by three straight-line segments. The center segment passes through −45° at $\omega = 1/\tau = 1$ rad/s. That segment has a −45°/decade slope and extends one decade above and below the corner frequency $\omega = 1$, from 0.1 rad/s to 10 rad/s. The low-frequency line segment ($\omega \leq 0.1$ rad/s) is horizontal, with value 0°. The high-frequency line segment ($\omega \geq 10$ rad/s) is horizontal, with value −90°. We superimpose this three-segment straight-line approximation on Figure 5.52(b) and then raise it by 90° to account for the 90° phase shift owing to the zero at the origin. This straight-line approximation differs from the exact phase curve by no more than 5.7°, as shown in Figure 5.54(b).

Exercise 5.20: Apply the Bode plot construction procedure to equation (5.119) for m = 0.2 kg, k = 0.2 N/m, and b = 0.04 N·s/m to verify Figure 5.53.

Exercise 5.21: Apply the checks on the accuracy of construction mentioned on page 390 to the Bode diagrams of Figures 5.52 and 5.53.

Example Problem 5.8: Bode diagram construction.

(a) Sketch the asymptotic straight-line approximations to the Bode diagrams for the transfer function

$$H(s) = \frac{1{,}000s + 2{,}000}{2s^4 + 25s^3 + 812s^2 + 400s} \tag{5.135}$$

Also, sketch the corrections to the straight-line approximations in the neighborhoods of the corner frequencies.

(b) Compare your sketch from part (a) with a computer-generated Bode diagram.

Solution: **(a)** In some instances, a transfer function can be obtained directly in factored form. In this case, however, we must first factor the numerator and denominator polynomials. We also must normalize the constant term of each factor to 1. The factorization can be done by hand or by computer. Computer programs (such as CC) that analyze transfer functions can do both tasks. The factored and normalized forms are

$$\begin{aligned} H(s) &= \frac{500(s+2)}{s(s+0.5)(s^2+12s+400)} \\ &= \frac{5(1+s/2)}{s(1+s/0.5)[1+0.6(s/20)+(s/20)^2]} \end{aligned} \tag{5.136}$$

For the magnitude diagram, we use semilog paper with four decades of logarithms on the abscissa. We use a linear dB ordinate. (Rectangular or log-log paper could also be used.) We plot the straight-line approximation for each factor (in dB) against a log ω abscissa, as shown in Figure 5.56. For example, the numerator factor $(1 + s/2)$ has a magnitude plot of the form shown in the top half of Figure 5.54(a), with the corner frequency placed at $\omega = 2$. We then add the straight-line approximations for the factors to obtain the total log-magnitude approximation shown in Figure 5.56. (Adding the dB values is equivalent to adding the logarithms or multiplying the factors.) The (extended) low-frequency asymptote of the "total approximation" intersects the line $\omega = 1$ at $\text{Lm}(H) = 14$ dB (or $M_H(1) = 5$).

An alternative approach to plotting the straight-line magnitude approximation is the following: (1) Set the slope of the low-frequency asymptote to -20 dB per decade (because the pole at the origin is of order 1). (2) Set the vertical placement of the low-frequency asymptote so that it intersects $\omega = 1$ at the dB value of the constant factor, $\text{Lm}(5) = 14$ dB, and draw the low-frequency asymptote. (3) Since the next larger pole or zero value is a pole at $s = 0.5$, make the slope 20 dB more negative (i.e., -40 dB) beginning at $\omega = 0.5$. Similarly, make the slope 20 dB/decade more positive beginning at $\omega = 2$ (because of the zero at

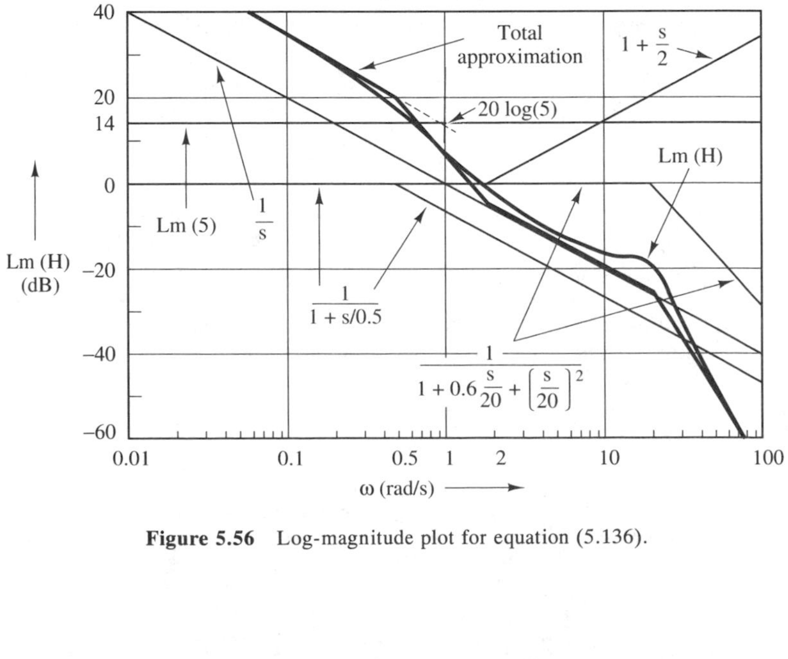

Figure 5.56 Log-magnitude plot for equation (5.136).

$s = 2$); and make the slope 40 dB/decade more negative beginning at $\omega = 20$ (because of the quadratic pole pair with natural frequency $\omega_n = 20$).

The corrections to the straight-line approximations can be read from Figures 5.54 and 5.55. There is a -3-dB correction at the corner frequency $\omega = 0.5$ and a $+3$-dB correction at the corner frequency $\omega = 2$. The two correction curves extend approximately two octaves to each side of the corner frequencies and can be sketched sufficiently accurately by eye. (There is some cancellation of the corrections between the corner frequencies because they are close to each other.)

The damping ratio associated with the quadratic pole pair is $\zeta = 0.3$. According to the corresponding curve of Figure 5.55, the correction at the corner frequency $\omega = \omega_n = 20$ rad/s is $+4$ dB. There is a $+4.5$-dB correction at the resonant (peak) frequency $\omega = 0.9\omega_n = 18$ rad/s. The correction curve extends approximately two octaves to each side of the corner frequency and can be sketched by eye. The straight-line approximation is a good approximation to the true log-magnitude curve in this example.

For the phase diagram, we use semilog paper with four decades of frequency and a linear phase ordinate. (Rectangular paper could also be used.) We plot the straight-line approximation for each factor against a log ω abscissa, as shown in Figure 5.57. For example, the numerator factor $(1 + s/2)$ has a phase plot of the form shown in the top half of Figure 5.54(b), with a $+45°$ phase shift at the corner frequency, $\omega = 2$. We determine from Figure 5.55(b) that a second-order factor produces a $-90°$ phase shift at the natural frequency, $\omega = 20$; for $\zeta = 0.3$, the phase in the neighborhood of $\omega = \omega_n$ has a slope of about $-90°$/octave.

Approximate phase corrections for each of the factors are also displayed in Figure 5.55(b) The largest phase correction for a single pole or zero is 5.7°. This correction occurs at the two *phase-corner* frequencies, one decade above and one decade below the magnitude-corner frequency. Corrections can be as large as 11° for the quadratic pole pair. We add the straight-line approximations for the factors to obtain the total phase approximation shown in Figure 5.57. The corrections can be sketched by eye. The straight-line approximation is a good approximation to the true phase curve.

The low-frequency total-phase asymptote is level at $-90°$, and the low-frequency magnitude asymptote has a -20 dB/decade slope; both attributes correspond to the single pole at the origin. The high-frequency total-phase asymptote is level at $-270°$, and the high-frequency magnitude asymptote has a slope of -60 dB/decade; again, both attributes correspond to $(N_p - N_z)$, the excess of poles over zeros.

(b) The computer program CC (see appendix D) was used to generate the Bode diagrams for the transfer function of equation (5.136). The diagrams are shown in Figure 5.58. The sketches deviate little from the computed diagrams.

Filtering

An input-output transfer function is a ratio of polynomials in s. For a physical system, the order of the denominator polynomial is typically equal to or greater than the order of the numerator polynomial; that is, it would be unusual for the system equation to have higher order derivatives of the dependent variable than of the independent variable. At high frequencies, therefore, M_H is quite small, and the slope of the Lm(H) curve is negative (at -20, -40, or even -60 dB/decade). Physical systems are usually **low-pass** systems. That is, they respond very little to high-frequency sources. In a sense, such systems act as

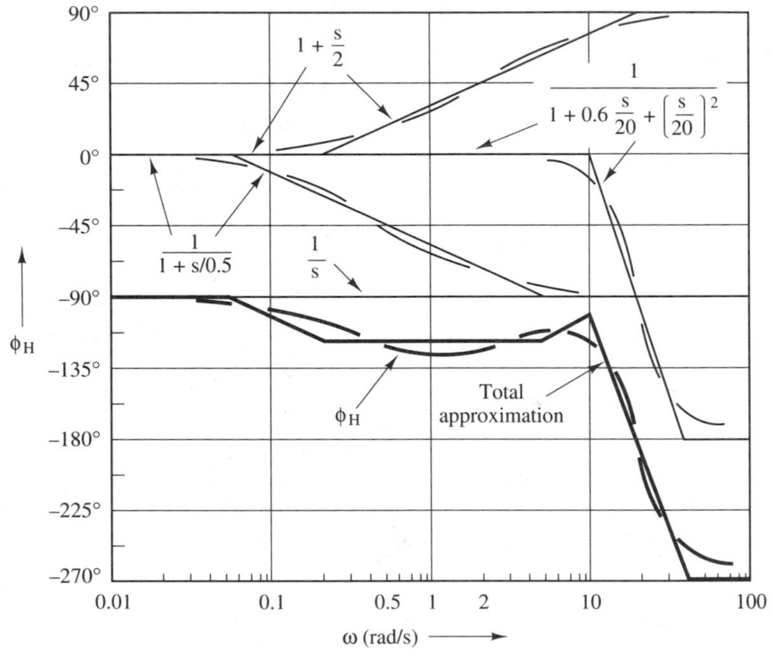

Figure 5.57 Phase plot for equation (5.136)

filters for source signals, selectively filtering out high frequencies relative to low frequencies. For example, the human arterial system filters out the pulsing of the heart, leaving essentially a steady flow of blood in the extremities of the body (see example problem 5.2). Some systems are designed specifically to filter out certain bands of source frequencies and to enhance others. Mechanical crystal oscillators, which serve as clocks

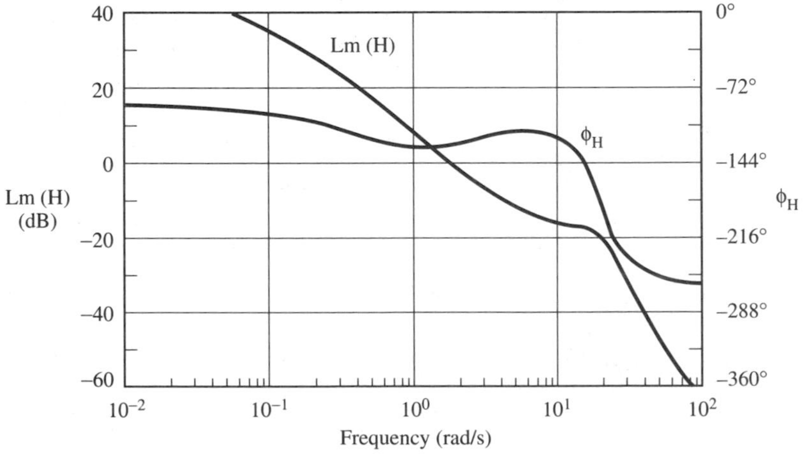

Figure 5.58 Computer-generated Bode plots for equation (5.136).

for digital devices, are designed to resonate strongly at a specific frequency and to discriminate against (i.e., filter out) other frequencies.

A high-fidelity music system is designed to reproduce, at the ear of the listener, the same acoustical pressure variations that would occur at a live performance. We can think of the musical signal as the sum of a large number of sinusoidal signals differing in frequency, amplitude, and phase. The original sound waveform is modified by each piece of equipment in the recording and reconstruction process. In order for the *system* to possess perfect fidelity, the reproduced signal must have the same waveform as the original, perhaps increased in magnitude (amplified). That is, we want the frequency response of the whole music system to be $H(j\omega) = A$, where A is the amplification factor. But physical systems cannot pass all frequencies equally well. Fortunately, the human ear is, itself, a dynamic system. In most people, the ear does not respond significantly to frequencies lower than 50 Hz or higher than 16 kHz. The frequency response of the ear is at its peak level over the range from about 300 Hz to 5 kHz. (See problem 5.24 at the end of the chapter.) The voice telephone system passes frequencies only in the range from 300 Hz to 3 kHz. The better high-fidelity music systems seek a roughly flat frequency response in the 50-Hz to 16-kHz range. (Although no musical instruments produce fundamental tones in the upper part of this range, they do produce harmonics (overtones) in that part of the range, and these overtones affect the nature of the sound.)

The listening room has a strong effect on the way a loudspeaker sounds; indeed, it is an extension of the loudspeaker. Because the designer has no control over listening-room acoustics, some high-fidelity sound systems include a sound equalizer—a set of electrical filters that increase or decrease the frequency response over selected frequency ranges. The listener can then adjust the frequency response so the music sounds best.

It is not necessary that each piece of equipment in the system have a flat frequency response in order for the system as a whole to have a flat response (over the audible range). Engineers take advantage of this fact to improve the quality of sound. Every piece of electrical equipment introduces a small amount of noise into the signal as it passes through. (Consider, for example, the hissing background noise that occurs when the volume of a stereo system is turned up and the music is quiet.) By using an electrical filter to strengthen certain frequency ranges during recording, the signals are made much stronger than the noise generated by the recorder in that range. Then, during playback, the same frequency range is deemphasized; the signals are returned to their original level, and the noise introduced by the recorder becomes negligible. This emphasis and deemphasis of signals follows industrywide *equalization standards* (such as Dolby), so that users will be able to set their playback equipment to reconstruct the signals properly.

Resonant Networks

A lightly damped second-order network resonates for source frequencies near its natural frequency. That is, the amplitudes of all the network variables increase dramatically for that frequency range. The lower the damping, the more pronounced will be the resonance. A higher order network also resonates if its characteristic roots include a lightly damped pair of poles. A network with multiple pole pairs can resonate in more than one frequency band.

The frequency response of a network deviates drastically from the asymptotic straight-line approximation in the neighborhood of a resonant frequency. In this section, we examine that deviation in detail. We show that the maximum deviation is a precise indicator of the damping ratio ζ associated with the resonant pole pair.

The magnitude frequency response M_H of a resonant network has a large, narrow peak. We call the frequency at the peak the **resonant frequency** and denote it ω_r. We call the peak magnitude the **resonant peak** and denote it M_r.

Suppose the input-output transfer function H(s) for a variable y of a second-order network is the one described in normalized form in Figure 5.55 (for a particular value of ζ). Then the system equation for y is

$$(s^2 + 2\zeta\omega_n s + \omega_n^2)y = \omega_n^2 u \tag{5.137}$$

Since the right side is strictly proportional to the source function u, y is a prototype variable for the network (see section 3.4). For the particular right-side coefficient ω_n^2, the final value y_f for the *unit-step input* $u(t) = u_s(t)$ is 1. According to Figure 5.55, the low-frequency response M_H is also equal to 1. We shall refer to such a prototype equation as a *second-order unit-prototype* system equation. The corresponding transfer function H(s) is a second-order unit-prototype transfer function.

Figure 5.55 shows that the magnitude frequency response M_H of a second-order unit-prototype transfer function exhibits resonance if $\zeta \leq 0.707$. The resonant peak M_r *for the unit-prototype* transfer function is the factor by which the resonant peak rises above the low-frequency value of 1. The peak value M_r is also the factor by which the resonant peak rises above the corner point of the asymptotic approximation. This second interpretation is the more useful of the two.

The quantities M_r and ω_r are key features of the unit-prototype response. The resonant peak is a measure of the *degree of stability* (or response *shape*) of the second-order network. The resonant frequency ω_r is a measure of the network *responsiveness.* (The network responds well for frequencies as high as ω_r; the frequency response drops off rapidly for higher frequencies.) These key features are usually relatively easy to measure. There is a unique relationship between the characteristic roots ζ and ω_n of the unit-prototype transfer function and the key features of its resonant peak—M_r and ω_r.

Exercise 5.22: Find the maximum of $M_H \equiv |H(j\omega)|$ for the unit-prototype transfer function of Figure 5.55. That is, equate to zero the derivative of M_H with respect to ω/ω_n to show that:
(a) The resonant frequency is

$$\omega_r = \omega_n\sqrt{1 - 2\zeta^2} \tag{5.138}$$

(b) The magnitude of the resonant peak is

$$M_r \triangleq M_H(\omega_r) = \frac{1}{2\zeta\sqrt{1 - \zeta^2}} \tag{5.139}$$

Thus, the resonant peak is uniquely related to ζ. For values of $\zeta < 0.3$, $M_r \approx 1/2\zeta$ in equation (5.139). Figure 5.55 shows that all *highly resonant* pole pairs have damping ratios smaller than 0.3. We conclude that for any highly resonant second-order unit-prototype transfer function,

$$M_r \approx \frac{1}{2\zeta} \tag{5.140}$$

According to equation (5.138), ζ is not a strong factor in determining ω_n. For a highly resonant transfer function, ω_r is nearly equal to (but slightly smaller than) ω_n.

We view M_r and ω_r as measurable key features of the second-order unit-prototype frequency response. From a measured value of M_r, we use equation (5.139) or equation (5.140) to determine ζ. Then we use the calculated value of ζ together with a measured value of ω_r in equation (5.138) to determine ω_n.

All of the preceding discussion is based on the second-order unit-prototype transfer function. Other second-order prototype transfer functions differ from the unit-prototype transfer function only by having a right-side coefficient different from the coefficient ω_n^2 of equation (5.137). This slight difference in the transfer function merely increases the frequency response by a constant factor. That is, it raises or lowers the whole Lm(H) in Figure 5.55(a). Raising or lowering the Lm plot changes the peak value M_r, but does not change the factor by which M_r rises above the corner point of the asymptotic straight-line approximation. (The corner point has the same level as the low-frequency value of M_H.) Generally speaking, it is not the value of M_r, but rather, the factor $\hat{M}_r$ by which M_r rises above the corner point, that we should use in equation (5.140) to determine ζ. Intuitively speaking, $\hat{M}_r$ is the value of M_r *relative to* the value of the asymptotic approximation at the resonant frequency.

Suppose the input-output transfer function H(s) for a network has other numerator and denominator factors, in addition to the conjugate pole pair. Then, to obtain the Lm plot for H(s), we add the Lm plots for the additional factors to the unit-prototype Lm plot. Adding the Lm plots for the additional factors changes both the peak value M_r and the peak frequency ω_r (see Figure 5.55(a)). Thus, the relations between (M_r, ω_r) and (ζ, ω_n) given by equations (5.138)–(5.140) do not apply directly to a network of order higher than two. Rather, they apply only to the second-order component of such a network.

In general, if we wish to use measured values of M_r and ω_r to determine ζ and ω_n, we must first subtract from the measured frequency response the response components for all factors other than the second-order denominator factor. Then we find the values of M_r and ω_r pertaining to the second-order factor alone. We use these values to determine ζ and ω_n for the resonant pole pair. Example problem 5.9 illustrates how to determine the factors of the response that we must subtract from the measured response curve.

In practice, if the resonance is strong (the resonant pole pair is lightly damped), and if there is not another resonant peak at a neighboring frequency, we need not actually subtract these other factors from M_H. We can get a good estimate of ζ and ω_n from directly measured values of M_r and ω_r. Even though the response components owing to factors other than the unit-prototype factor affect the peak value and peak frequency, the factor $\hat{M}_r$ by which the total resonant peak M_r rises above the corner point of the total asymptotic straight-line approximation still nearly equals the amount of rise at the peak owing to the resonant pole pair alone. Furthermore, the additional components of the frequency response usually do not move the resonant frequency far from the value owing to the second-order component alone.

Consider the resonant response in Figure 5.53, for example. The transfer function (5.119) from which that response is derived is

$$H(s) = \frac{5s}{1 + 0.2s + s^2} \tag{5.141}$$

The coefficients in the quadratic denominator factor show that $\omega_n = 1$ rad/s and $\zeta = 0.1$ for that pole pair. The frequency of the peak in Figure 5.53 is exactly ω_n. However, that is not the resonant frequency ω_r of the pole pair. If we remove the portions of the response owing to the factors 5 and s in the numerator of H(s), we obtain the $\zeta = 0.1$ curve of Figure 5.55. According to that figure, $M_r = 5$ and $\omega_r \approx \omega_n \approx 1$ rad/s. We cannot easily distinguish between the frequency ω_r of the resonant peak and the corner frequency ω_n in the plot of Figure 5.55(a) either. However, according to equation (5.138), $\omega_r = 0.99\omega_n = 0.99$ rad/s.

Note that the frequency response in Figure 5.53 shows a strong resonance. Let us apply equations (5.140) and (5.138) directly to the measured peak. The peak value is $M_r = 25$ (28 dB). The peak rises above the corner point of the asymptotic approximation by the factor $\hat{M}_r = 5$. (That is, Lm(5) = 14 dB, or $\hat{M}_r = 10^{14/20} = 5$.) Then, according to equation (5.140), $\zeta \approx 0.1$. This value is identical to the true value. (If we were to use the more accurate equation (5.139), the two values would not be quite identical.) Again, from Figure 5.53, we read the resonant frequency as $\omega_r = 1$. The true value is $\omega_r = 0.99$. The error is only 1%.

Many systems are designed to resonate at a particular frequency so that they can respond well to that frequency. The tuner of a television set is an example. By adjusting the tuner to the frequency of a particular channel—perhaps by adjusting the capacitance of a variable capacitor—we can make the circuit accept (pass) the TV signal at that frequency and discriminate against (not pass) signals at frequencies of other channels. The narrowness of the resonant peak determines the accuracy with which the tuner can discriminate between different frequencies. Thus, the width of the peak is a measure of the selectivity of the resonant tuner circuit.

We define the **bandwidth** $\Delta\omega$ of a resonant transfer function to be the width of the resonant peak at the **half-power level.** That is, $\Delta\omega$ is the difference between the frequencies at which M_H^2 is reduced from its peak value M_r^2 by the factor 0.5 (or −3 dB). We denote these frequencies by ω_u and ω_ℓ (See Figure 5.59). We use linear scales in that figure because they show directly the quantities that we wish to measure. There is no advantage to using logarithmic scales for the small resonant interval.

The resonance curves in Figure 5.55 have various bandwidths. The frequency axis in that figure is normalized relative to ω_n. Therefore, for a given level of damping ζ, the bandwidth is proportional to ω_n. It is the ratio of the bandwidth to ω_n that determines the **selectivity** of the resonance—the accuracy with which it discriminates between frequencies. For example, if the bandwidth is 1 kHz and the natural frequency is 10 kHz, then $\Delta\omega$ is a 10% change in frequency. On the other hand, if the bandwidth is 1 kHz but the natural

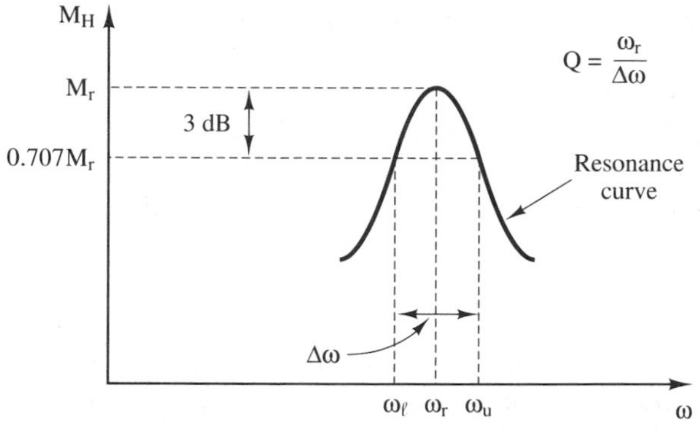

Figure 5.59 A resonant peak.

frequency is 1 MHz, then $\Delta\omega$ is a 0.1% change in frequency. The latter case is much more selective.

The resonant frequency ω_r is nearly equal to ω_n. Since it is easier to measure ω_r (the frequency at the peak) than ω_n (the corner-point frequency), we define the **quality** Q of the resonance by

$$Q \triangleq \frac{\omega_r}{\Delta\omega} \tag{5.142}$$

A large value of Q corresponds to a narrow bandwidth and high selectivity. By setting $M_H(\omega) = M_r/\sqrt{2}$ and solving for ω_u and ω_ℓ, it can be shown that, for a second-order prototype transfer function,

$$\Delta\omega \triangleq \omega_u - \omega_\ell \approx 2\zeta\omega_r \tag{5.143}$$

Therefore,

$$Q \approx \frac{1}{2\zeta} \tag{5.144}$$

By comparing equation (5.144) with equation (5.140), we see that $Q \approx M_r$. Thus, measuring $\Delta\omega$ and ω_r directly and then computing Q provides an alternative method for determining ζ. Even for a higher order system, we need to test the system only for frequencies in the resonant range to determine Q. The calculation of Q does not require that we subtract from the peak value the contribution owing to other factors of the transfer function (or that we find the amount $\hat{M}_r$ by which M_r rises above the corner point).

Resonance Removal

Every piece of rotating machinery (electric motor, hydraulic turbine, internal combustion engine, washing-machine tub) introduces slight translational vibrations (forces or motions) that are sinusoidal in time, because of slight imbalances in the rotating parts. Every mechanical system acts as some interconnection of lumped springs and masses. There is a resonant frequency associated with each spring-mass pair. If a piece of rotating machinery vibrates the mechanical system at one of its resonant frequencies, we can expect the system to exhibit a large-amplitude response. Hence, machine vibrations can cause irritating noise, uncomfortable vibratory motions, and perhaps damaging forces or motions in neighboring equipment (or in the rotating machinery itself).

We can *avoid* a vibration problem by balancing the rotating machine carefully to eliminate the source. But perfect balance is impossible, and machine wear can introduce additional imbalance. One approach to *eliminating* or *reducing* a vibration problem is to change the mass or stiffness of the affected mechanical system in order to change the resonant frequency. If the resonant frequency is not close to the vibration frequency, the vibration will not be magnified by the system.

We examine here two ways to *reduce* the effect of vibrations if the resonant frequency cannot be changed. One way is to increase the damping associated with the resonance by attaching a damping device. (For example, we could hold the piece of equipment tightly with our hand, making it more difficult for the system to vibrate.) According to Figure 5.55, if the damping ratio can be increased to $\zeta = 0.3$ or more, the resonance will not be severe, and vibration near the natural frequency will not be greatly magnified. We refer to this approach as **vibration absorption**, because the damper absorbs the vibration energy. A second approach is to attach to the equipment an additional resonant spring-mass combination that is tuned to the vibration frequency. The intent is that the resonant attachment cancel the vibration. Hence, we refer to this approach as **vibration cancellation**.

Figures 5.60(a) and (b) model a piece of equipment resting on a vertically vibrating floor. (The lumped model ignores the gravity force and average spring force, which balance each other.) The spring in the lumped model is attributed primarily to the compliance of the supporting feet. The frequency of vibration is near the resonant frequency of the spring-mass pair. We attach the damper b to attenuate the vibration of the piece of equipment. Figure 5.60(c) shows a line-graph representation of the network. We must avoid a buildup of the machine motion v_2 owing to the floor vibration $v_1 = v$. Therefore, we examine the frequency response of $H_a(s) = v_2/v$, where the subscript a refers to the *absorp-*

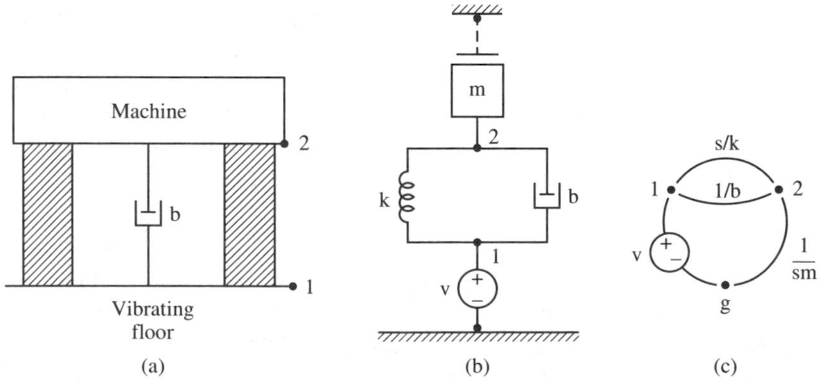

Figure 5.60 Absorption of vibration. (a) Pictorial model. (b) Lumped model. (c) Line-graph representation.

tion method of vibration management. The impedance of the parallel spring-damper pair is s/(bs + k). The transfer function v_2/v is the impedance ratio

$$H_a(s) = \frac{\dfrac{1}{sm}}{\dfrac{1}{sm} + \dfrac{s}{bs + k}}$$

$$= \frac{sb/k + 1}{s^2m/k + sb/k + 1} \quad (5.145)$$

The normalized transfer function is

$$H_a(s) = \frac{1 + 2\zeta(s/\omega_n)}{1 + 2\zeta(s/\omega_n) + (s/\omega_n)^2} \quad (5.146)$$

where $\omega_n = \sqrt{k/m}$ and $\zeta = b/(2\sqrt{km})$. The magnitude-response $M_{Ha}(\omega)$ for equation (5.146) is plotted against ω in Figure 5.61, for several values of ζ. (As in Figure 5.61, we use linear scales to examine the resonance.) Since the frequency of the floor vibration is close to the resonant frequency, $\omega \approx \omega_n$. We select the damping ratio ζ by choosing b (relative to the product km). It is apparent that the use of an appropriate damper eliminates the resonant buildup of vibration. In some cases, the vibration damper and the compliant support might be combined in appropriately designed rubber feet.

In Figure 5.62, a machine of mass m and support stiffness k is vibrated vertically at its resonant frequency $\omega_n = \sqrt{k/m}$. A vibration canceller, consisting of the mass-spring combination (m_2, k_2) is attached to the machine. According to the lumped model and line

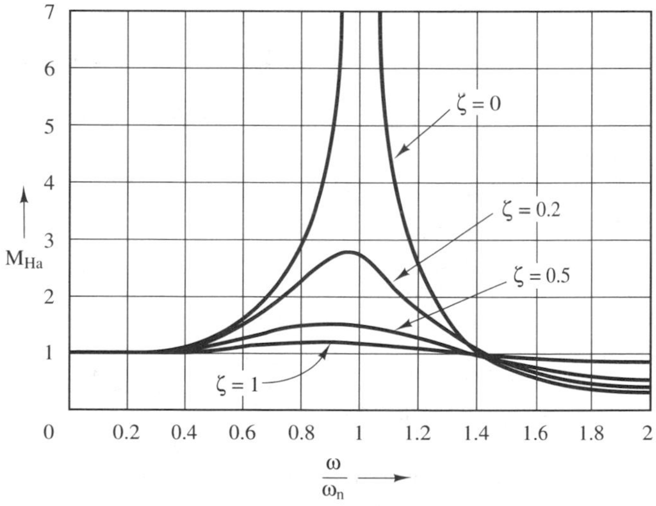

Figure 5.61 Vibration-absorber response. Adapted from [5.7]. © 1992. Adapted by permission of Prentice Hall, Englewood Cliffs, New Jersey.

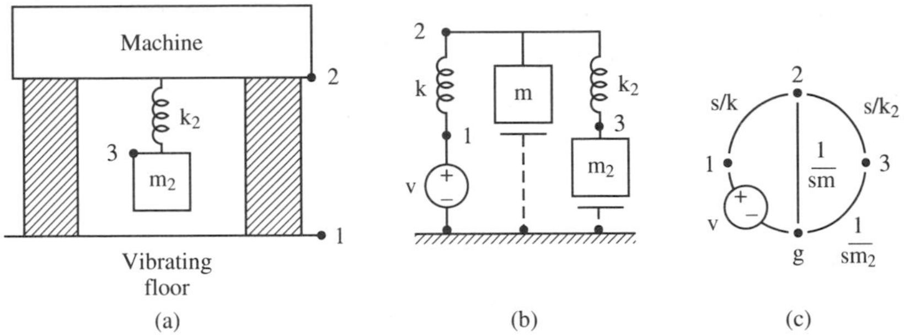

Figure 5.62 A vibration canceller. (a) Pictorial model. (b) Lumped model. (c) Line-graph representation.

graph of the figure (which ignore the balanced gravity and spring forces), the series combination of m_2 and k_2 has impedance $(s^2m_2 + k_2)/(sm_2k_2)$. This impedance can be combined with that of the parallel mass m to produce

$$z = \frac{s^2m_2 + k_2}{s(s^2mm_2 + mk + m_2k_2)} \tag{5.147}$$

Then the transfer function $H_c = v_2/v$ that describes the motion v_2 of the machine mass m is the impedance ratio

$$\begin{aligned} H_c(s) &= \frac{z}{z + s/k} \\ &= \frac{k(s^2m_2 + k_2)}{s^4mm_2 + s^2(m_2k_2 + mk_2 + km_2) + kk_2} \end{aligned} \tag{5.148}$$

The subscript c denotes the *cancellation* method of vibration management. The transfer function has a zero at the resonant frequency $\omega_{n2} = \sqrt{k_2/m_2}$ of the attached spring-mass pair. We tune the spring-mass pair (by appropriate choice of k_2 and m_2) to the vibration frequency $s = j\omega_n = j\sqrt{k/m}$. Then the frequency response has zero magnitude *at the vibration frequency* and produces complete cancellation!

The magnitude-response $M_{Hc}(\omega)$ is shown in Figure 5.63 for two different values of the vibration-canceller mass m_2. Attaching the tuned vibration canceller produces zero response at the tuned frequency. But it also produces a pair of resonances, one on each side of the zero. The attached mass m_2 cannot be too small relative to the mass of the machine, or a very slight change in the vibration frequency will excite one of these side resonances and defeat the canceller. If we attach a vibration-absorbing damper in addition to the canceller, the two side resonances will be much less severe.

Exercise 5.23: Determine how to tune the vibration absorber if the vibration frequency is not quite equal to the resonant frequency of the machine.

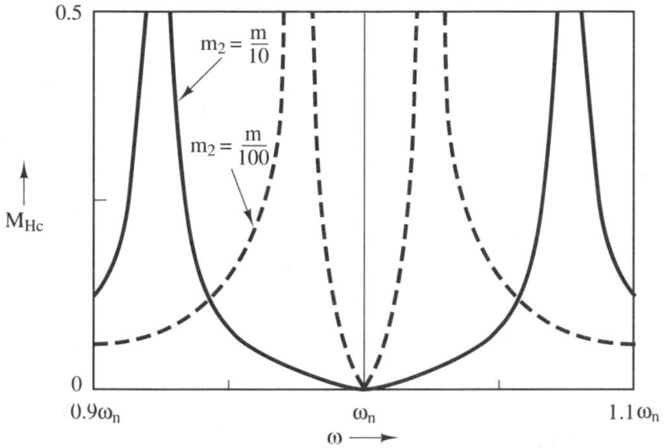

Figure 5.63 Vibration-canceller response.

Experimental Determination of Transfer Functions

How can we assure that the lumping process produces an accurate lumped model of a physical system? Even if the form of the network is accurate, the values of the element parameters can be hard to determine accurately. For some physical systems, it is easier to measure the steady-state sinusoidal response than it is to measure the individual element parameters.

For example, a compact electric circuit, built of elements of known value, is affected by *parasitic* capacitances between the various conductors in the circuit (owing to the particular physical layout). The values of the parasitic capacitances cannot be calculated accurately. Yet, the true frequency response of the circuit includes the effects of parasitic capacitance, and that true frequency response is easy to measure.

The asymptotic slope and corner-frequency properties of the Lm(H) curves embrace the key features of the transfer function. The slopes and corner frequencies can be obtained from a measured frequency response. The experimentally discovered slopes and corner frequencies determine directly an accurate transfer function for the physical system. That transfer function, in turn, specifies completely the system equation (see section 4.2) and the dynamic behavior of the system (see section 3.4). In this section, we describe a procedure for finding the transfer function that corresponds to a measured frequency response.

TRANSFER FUNCTION CONSTRUCTION PROCEDURE

1. We derive the transfer function from a measured log-magnitude frequency response. Draw asymptotes to the measured log-magnitude curve. These asymptotes must have slopes that are integral multiples of 20 dB/decade. The factors in the transfer function can be determined from these asymptotes.
2. If the slope of the straight-line approximation to the log-magnitude curve changes by 20 dB/decade at $\omega = \omega_1$, there is a factor $(1 + s/\omega_1)$ in the transfer function. If the slope becomes more negative with increasing frequency at $\omega = \omega_1$, the factor is a pole (in the denominator); if it becomes less negative, the factor is a zero (in the numerator).

3. If the slope changes by 40 dB/decade at $\omega = \omega_2$, then there is either a resonant factor $[1 + 2\zeta(s/\omega_2) + (s/\omega_2)^2]$ or a nonresonant factor $(1 + s/\omega_2)^2$ in the transfer function. The factor is in the denominator if the slope becomes more negative with increasing frequency; it is in the numerator if the slope becomes less negative with increasing frequency. If the deviation of the response from the corner point ω_2 is more than 6 dB, use the resonant factor. Otherwise, use the nonresonant factor. For the resonant case, measure the bandwidth $\Delta\omega$ and compute $\zeta \approx \Delta\omega/2\omega_2$. Alternatively, measure the factor $\hat{M}_r$ by which the resonant peak M_r rises above the corner point, and compute $\zeta \approx 1/2\hat{M}_r$.
4. If the slope of the *low-frequency asymptote* is 20n dB/decade, there is a factor s^n in the numerator of the transfer function; if the slope is $-20n$ dB/decade, the factor s^n is in the denominator.
5. The constant factor in the numerator of the transfer function is the magnitude (not log magnitude) of the low-frequency *asymptote* (not the curve itself) at $\omega = 1$.

It is fortunate that the foregoing procedure uses only the magnitude plot; it is easier to make accurate amplitude measurements than accurate phase measurements. Once we know the transfer function, we can use it to determine the phase plot. Only if the correct transfer function includes propagation delays (see section 6.4) or zeros in the right half of the s-plane is the phase curve needed to determine the transfer function. The measured phase data can be compared with the phase plot derived from the transfer function to verify that there are no propagation delays and no zeros in the right half-plane.

If the physical system is somewhat nonlinear, the 20-dB/decade slopes cannot match the measured data accurately. In that case, the procedure provides a lumped linear model that *approximates* the system behavior.

Example Problem 5.9: Experimental determination of a transfer function.
A feedback-controlled single-link robotic arm is shown in Figure 5.64(a). The shaft rotation angle θ_d at the input represents the desired angle of the arm. The actual arm angle is θ. We seek the transfer function $H(s) = \theta/\theta_d$. The magnitude-response Lm(H) of the system was measured experimentally. A motorized drive was used to rotate the input shaft sinusoidally with a fixed amplitude and an adjustable frequency. The output amplitude was measured for various input frequencies.

The measured frequency-response data are plotted in Lm(H)-vs.-log ω form in Figure 5.64(b). Because of the apparent resonant behavior near $\omega = 60$ rad/s, careful measurements were made of the quality of the resonance. Specifically, the frequency of the input was adjusted to obtain the maximum response value, and the data were recorded. Then the frequency was adjusted (both higher and lower than the resonant frequency) so as to produce a 3-dB reduction in output, and the data were recorded again. These data are listed in Figure 5.64(b). Determine the transfer function of the system.

Solution: We fit straight-line approximations to the plotted log-magnitude data using slopes that are multiples of 20 dB/decade in Figure 5.64(b). The slope of the low-frequency asymptote is zero. Therefore, the transfer function has no pole or zero at the origin. Since the low-frequency asymptote has the value 0 dB at $\omega = 1$, the constant factor in the transfer function is 1.

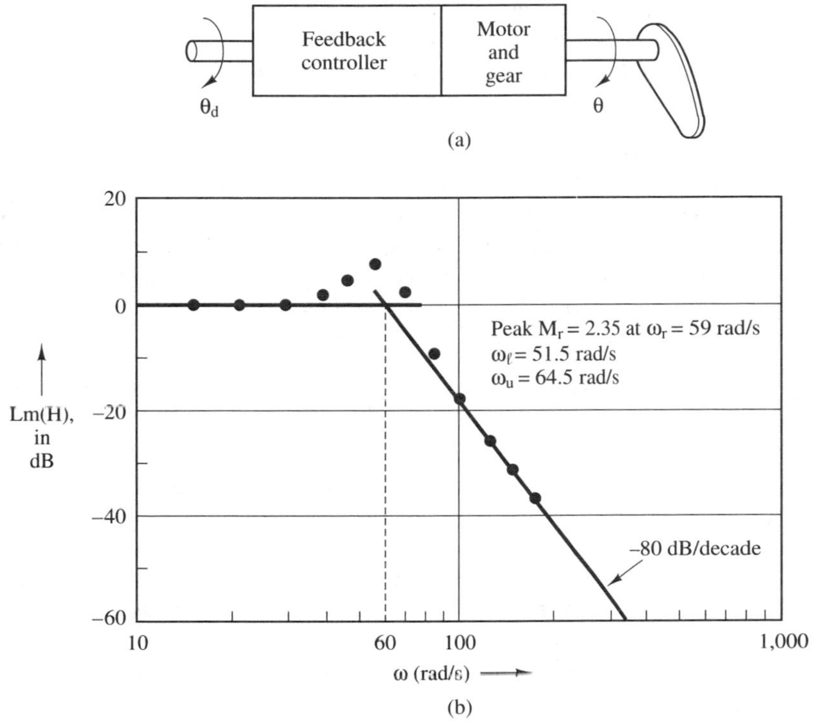

Figure 5.64 Experimental frequency response data for a robotic arm. (a) Pictorial model of the system. (b) Experimental frequency-response data.

The -80-dB/decade slope of the high-frequency asymptote implies that there are four more poles than zeros. The only obvious change in slope is the negative-going change near the resonant frequency. It is appropriate, therefore, to assume that there are four poles and no zeros. The corner frequency is $\omega = 60$ rad/s. Apparently, all corner frequencies are superimposed on each other.

The frequency response looks much like that of a resonant pole pair, but the high-frequency slope is twice that of a pole pair. We conclude that the system includes a resonant pole pair and two first-order poles, all with identical corner frequencies. We place the two first-order poles at $s = -60$ rad/s. Then we let $\omega_n = 60$ rad/s for the resonant pole pair. The measured 3-dB bandwidth is $\Delta\omega = 64.5 - 51.5 = 13$ rad/s. Then $Q = \omega_r/\Delta\omega = 59/13 = 4.54$, and the damping ratio for the pole pair must be $\zeta = 1/(2Q) = 0.11$. Hence, the transfer function, expressed in the normalized form used for Bode diagrams, must be

$$H(s) = \frac{1}{[1 + (s/60)]^2[1 + 2(0.11)(s/60) + (s/60)^2]}$$

$$= \frac{1.3 \times 10^7}{(s + 60)^2(s^2 + 13.2s + 3{,}600)} \tag{5.149}$$

Alternatively, we can compute ζ using equation (5.140). The measured peak value is $M_r = 2.35 = 7.4$ dB. We must subtract out the contribution owing to the other factors of the transfer function. The constant term is 1 (0 dB). The level of the corner point of the asymptotic approximation is also 0 dB. There are two −3-dB corrections at $\omega = 60$ owing to the two nonresonant poles. Therefore, the value of M_r attributable to the resonant pole pair is 7.4 dB − (−6 dB) = 13.4 dB, or $M_r = 10^{13.4/20} = 4.68$. It follows that $\zeta = 1/[(2)(4.68)] = 0.107$. The difference between the two values of ζ is slight.

We placed the two first-order poles at $s = -60$ in equation (5.149) because the corner frequencies associated with the poles are so close together that we cannot distinguish them. Suppose, however, that we were to assume instead an overdamped pole pair in the neighborhood of $s = -60$. As the damping of a pole pair is increased, the geometric mean of the pole positions remains constant. We choose poles at $s = -40$ and $s = -90$, with geometric mean at $s = -60$. Then the transfer function must be

$$H(s) = \frac{1}{[1 + (s/40)][1 + (s/90)][1 + 2(0.11)(s/60) + (s/60)^2]}$$
$$= \frac{1.3 \times 10^7}{(s + 40)(s + 90)(s^2 + 13.2s + 3{,}600)} \tag{5.150}$$

On the other hand, one might propose that the system acts like two identical resonant pole pairs with the same resonant frequency $\omega_n = 60$. If so, then the measured peak value $M_r = 2.35$ is the product of the peaks owing to the two pole pairs. Therefore, we use $M_r = \sqrt{2.35} = 1.53$ for each pair. Since this value does not correspond to strong resonance, we use equation (5.139) to determine that $\zeta = 0.35$ for each pair. The corresponding transfer function is

$$H(s) = \frac{1}{[1 + 2(0.35)(s/60) + (s/60)^2]^2}$$
$$= \frac{1.3 \times 10^7}{(s^2 + 42s + 3{,}600)^2} \tag{5.151}$$

The log-magnitude plots associated with equations (5.149)–(5.151) are superimposed on the measured data in Figure 5.65(a). (The plots were computed by means of the computer program CC.) The three plots have the same straight-line approximation. They also have about the same peak value. Because the bandwidth of the resonant peak for equation (5.151) is much too large, we conclude that the double pole pair does not fit the data accurately.

The unit step responses of the two accurate transfer functions are compared in Figure 5.65(b). (These responses were also computed from the transfer functions by means of the computer program CC.)

As the step responses of the preceding example demonstrate, if two sets of transfer functions represent a measured frequency response equally well, either transfer function can be considered a good representation of the system. The example also demonstrates that the straight-line approximation to a Bode plot is not a sufficient characterization of the frequency response in the neighborhood of resonance.

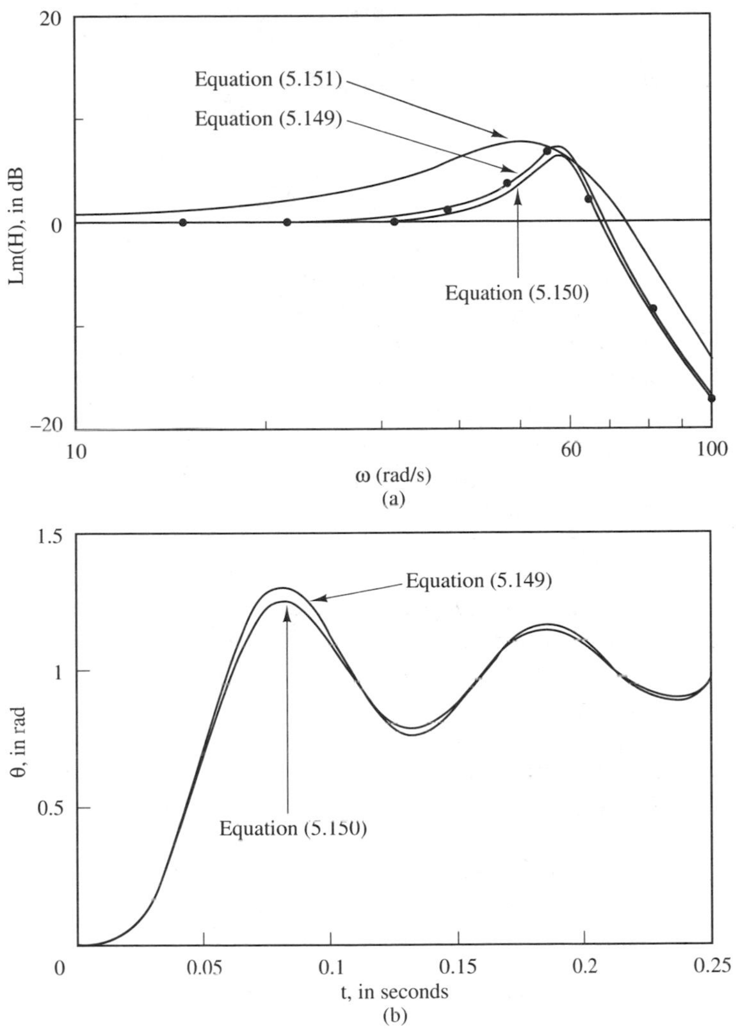

Figure 5.65 Computed behaviors for system equations (5.149)–(5.151). (a) Frequency responses. (b) Step responses.

Frequency Response, Time Response, and Transfer Functions

We begin analyzing a physical system by generating a lumped model of the system (see section 2.2). The lumped model can represent the physical system quite well, but not perfectly. We obtain the input-output system equation for any single network variable from the lumped model. That system equation describes completely the behavior of that variable in the lumped network. The input-output transfer function (or system function) for that variable is *completely equivalent* to the (differential) system equation for that variable.

The **key features of the input-output transfer function** are the poles and zeros and the constant factor. The poles alone show the form of behavior of the system—the time constants, natural frequencies, and damping ratios. The time constants and natural frequencies are indicators of network responsiveness. The damping ratios are indicators of the degree of network stability.

We may choose to focus attention on a prototype variable of the network as representative of the behaviors of all variables in the network. A prototype variable is one for which the transfer function has no zeros. Hence, prototype variables have simpler input-output transfer functions than do other variables, yet they exhibit behaviors which are representative of all variables of the network.

For a linear network, the measured frequency response of a network variable is completely equivalent to the transfer function for that variable (to within the accuracy of the measurements and the accuracy of the lumping process). The **key features of the frequency response** of a network variable are the corner frequencies (including the resonant frequencies and the reciprocals of the time constants), the values of the resonant peaks (or the corresponding values of Q), and the value of the low-frequency asymptote at $\omega = 1$ rad/s (which determines the scale factor). These key features serve as measures of the network responsiveness and the degree of network stability. Moreover, the key features alone are completely equivalent to the transfer function.

We usually experience the actual behaviors of physical systems in the time domain. The **key features of the zero-state step response** of a network variable are the final value of the variable (which determines the input-to-output multiplier), the percent overshoot, the peak time, the rise time, the settling time, and the period of oscillation (if there is any oscillation). These key features serve as measures of the responsiveness (peak time, rise time, settling time, or oscillation period) and the degree of stability (percent overshoot) of the network.

The key features of the step-response are completely equivalent to the input-output system equation for the variable if the network is first order or second order and linear. For these networks, it is apparent that the transfer function, the step response, and the frequency response provide precisely equivalent information. From any one of the three, we can determine the other two. We shall find in section 5.5 that this equivalence extends to linear lumped networks of any order—that the transfer function can be constructed by transforming the network impulse response.

This equivalence between the mathematical model, the step response, and the frequency response provides three different ways to measure the parameters of a network model: (1) We can make measurements on the physical system to determine the values of the parameters of the lumped elements and then use those values in the mathematical model. (2) We can measure the frequency response and then use the key features of that response to determine the transfer function of the network (as we have done in this section). Or (3) we can measure the step response and then use the key features of that response (or the transformation of section 5.5) to determine the transfer function of the network. If the three approaches do not produce the same transfer function, the differences are explained by errors in measurement and by inadequacy of the linear lumped-model approximation. (The model may need finer subdivision, or the system may be somewhat nonlinear.)

5.5 LAPLACE TRANSFORMS

We distinguish between a system and the signals in that system. Yet we perceive a system primarily through its behavior. Therefore, our image of a system tends to consist of representative signals. For a linear network, the unit-step response (see section 3.4) is such a system-characterizing signal. We should view the unit-step response as a description of the network, not as a signal. The unit-impulse response is another description of the network, equivalent to the unit-step response. (In section 4.3, we show that the unit-step response is the integral of the unit-impulse response.)

The input-output transfer function is an accurate description of a network. It describes, in terms of time-derivatives s, what the network does to signals. We can determine from that transfer function—or the equivalent differential equation—the behavior of the network for any source waveform and any initial state. The transfer function for a network seems clearly distinguishable from the signals in the network. In this section, however, the distinction between transfer functions and behavior-characterizing signals becomes blurred. We find that the input-output transfer function is merely a mathematical rearrangement (or transformation) of the unit-impulse response; the two entities are equivalent.

We can view a transfer function, then, as a behavior-characterizing signal. We can transform that transfer function into a function of time—the unit-impulse response. This response is the **time-domain interpretation** of the transfer function. Alternatively, we can view s as a complex frequency variable. Then the transfer function is the complex multiplier that converts a sinusoidal source phasor into a response phasor (see section 5.3). This phasor, as a function of frequency is the **frequency-domain interpretation** (or frequency response) of the transfer function. In this section, we use the **Laplace transformation** to unify the two interpretations of the transfer function. In the process, we find that the frequency-domain interpretation extends to all signal waveforms, not just sinusoids.

The Laplace Transformation

The Laplace transformation converts signals (functions of time) into functions of a *complex frequency variable* $s = \sigma + j\omega$. There is a unique (one-to-one) correspondence between a signal and its Laplace transform. We can compare the transformation process with a distorting mirror: It produces an image that preserves all the information from the original function. We can retrieve the original function by inverse transformation.

Like the logarithm and the phasor transformation, the Laplace transformation produces images that have some more convenient properties than do the original signals. In particular, differentiating a signal with respect to time corresponds to multiplying its Laplace transform by the complex frequency variable s; convolving two time functions corresponds to multiplying their Laplace transforms. It is these simplifications of time-domain operations that make the Laplace transform useful.

The **one-sided Laplace transformation** $\mathscr{L}$ is an integral operator that converts a signal f(t) to a complex-valued function **F**(s) of a complex variable s according to[6]

$$\mathscr{L}[f(t)] \equiv \mathbf{F}(s) \triangleq \int_{0^-}^{\infty} f(t)e^{-st}\,dt \tag{5.152}$$

We refer to the transformed function **F**(s) as the **Laplace transform** of the signal f(t). We picture the lower limit 0^- of the integral as a *specific* instant prior, but infinitesimally close, to $t = 0$.

It is customary to use a lowercase symbol (e.g., f) to represent a signal waveform and an uppercase symbol (e.g., F) to represent its Laplace transform. However, since we use both uppercase and lowercase symbols to represent signals in various applications, we set the transform symbol in boldface uppercase to assure that we can distinguish it from the signal waveform in all cases.

We shall use the Laplace transformation to transform network signals. The behavior of a network for $t \geq 0$ depends only on the source for $t \geq 0$ and on the prior state (at $t = 0^-$). Hence, it does not matter that the Laplace transformation ignores f(t) for $t < 0^-$.

Some authors use the initial instant $t = 0^+$ as the lower limit of the one-sided Laplace transform. We use 0^- as the lower limit in order that the integration interval clearly include the abrupt changes that occur in some network signals at $t = 0$. This infinitesimally small difference in the lower limit does not affect the integral, unless the signal f(t) includes an impulse at $t = 0$. We can *think* of the lower limit as zero.

The process of finding the time function f(t) that corresponds to a particular Laplace transform **F**(s) is called inverse Laplace transformation and is denoted by $\mathscr{L}^{-1}$. We also call f(t) the **inverse Laplace transform** of **F**(s). Since the one-sided Laplace transformation ignores $t < 0^-$, **F**(s) contains no information about f(t) for $t < 0^-$. Therefore, inverse Laplace transformation cannot reconstruct f(t) for $t < 0^-$. We treat all signals as if they are defined only for $t \geq 0^-$. Then there is a one-to-one relation between f(t) and **F**(s).

To illustrate the Laplace transformation, we find the Laplace transform of the decaying exponential

$$f(t) = e^{-\alpha t}, \qquad t \geq 0^- \tag{5.153}$$

The transform is

$$\begin{aligned} F(s) &= \int_{0^-}^{\infty} e^{-\alpha t}e^{-st}\,dt \\ &= \int_{0^-}^{\infty} e^{-(s+\alpha)t}\,dt \\ &= \left.\frac{e^{-(s+\alpha)t}}{-(s+\alpha)}\right|_{0^-}^{\infty} \end{aligned}$$

[6] Although we are concerned about time signals, there is nothing in the defining equation (5.152) that requires f(t) to be a function of time. The transformation is defined for functions of any variable t.

$$= \left.\frac{e^{-(\sigma+\alpha)t}e^{-j\omega t}}{-(s+\alpha)}\right|_{0^-}^{\infty}$$

$$= \frac{1}{s+\alpha} \quad \text{for } \mathcal{R}e[s] > -\alpha \tag{5.154}$$

We must require that $\sigma > -\alpha$, where σ is the real part of s, in order that the factor with the real exponent converge to zero at the upper limit. (The magnitude of the factor with the imaginary exponent remains 1 for all t.) Therefore, the Laplace transform of the decaying exponential is defined only for $\mathcal{R}e[s] > -\alpha$. This restriction on the domain of **F** in the s-plane is comparable to the restriction $t \geq 0^-$ on the domain of f.

The significant features of the complex frequency function $1/(s + \alpha)$ are the existence of a single pole and the location of that pole, $s = -\alpha$ rad/s. (The pole defines the left boundary of that region of the complex s-plane over which the transform $1/(s + \alpha)$ is defined.) The significant features of the corresponding time function are the fact of decay and the rate of decay, with time constant $\tau = 1/\alpha$ seconds. We can thus see parallels between the features of f(t) and **F**(s).

None of the values of **F**(s) can be determined without using all of f(t) for $t \geq 0^-$. The values of the network signal f(t) represent the variations of a particular physical quantity. It is conceptually difficult, however, to ascribe physical meaning to the value of **F**(s) at the complex frequency s. It is easier to think of the whole function **F** as representing, in a different form, the whole waveform of the physical network quantity. We speak of f(t) as the **time-domain representation** of the signal and **F**(s) as the **frequency-domain representation**—the representation in terms of the complex frequency variable s.

As a second transformation example, let $f(t) = \delta(t)$, the unit impulse function. According to section 4.3, $\delta(t)$ is a unit area pulse of very short duration. It acts at $t = 0$, barely within the lower limit of the Laplace integral, and has the value zero at $t = 0^-$. (Because we use 0^- as the lower limit of the defining integral, it does not matter whether the impulse straddles $t = 0$ or begins to rise at $t = 0$.) The impulse is nonzero only for $t \approx 0$, where $e^{-st} \approx 1$. Therefore, the Laplace transform is

$$\mathbf{\Delta}(s) = \int_{0^-}^{\infty} \delta(t)e^{-st}\,dt = \int_{0^-}^{\infty} \delta(t)\,(1)\,dt = 1 \tag{5.155}$$

Exercise 5.24: Use equations (5.153) and (5.154), with $\alpha = 0$, to show that the Laplace transform of the unit step function $u_s(t)$ is $\mathbf{U}_s(s) = 1/s$. Note that $u_s(t)$, by the definition in section 3.3, has a jump at $t = 0$, not at $t = 0^-$.

It is not necessary to derive the Laplace transform for each signal we use in the study of networks. Laplace transform tables give the transforms of many signals, including the kinds of signals that appear in a linear network driven by a step function or a sinusoid. Table 5.5 gives the transforms of a few signal waveforms.

Exercise 5.25: Verify the Laplace transform **F**(s) given in Table 5.5 for $f(t) = \cos(\omega_0 t) = (e^{j\omega_0 t} + e^{-j\omega_0 t})/2$. Where are the poles of **F**(s)?

TABLE 5.5 LAPLACE TRANSFORM PAIRS

	$f(t) = \mathcal{L}^{-1}[\mathbf{F}(s)],\ t \ge 0^-$		$\mathbf{F}(s) = \mathcal{L}[f(t)]$
1	unit impulse $\delta(t)$		1
2	unit step $u_s(t)$	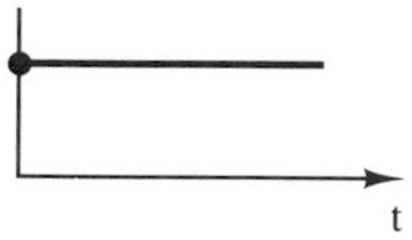	$\frac{1}{s}$
3	$t^n,\ n = 1, 2, \ldots$		$\frac{n!}{s^{n+1}}$
4	$e^{-\alpha t}$	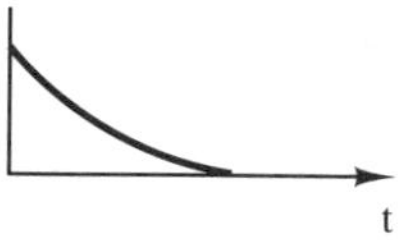	$\frac{1}{s + \alpha}$
5	$t^n e^{-\alpha t},\ n = 1, 2, \ldots$	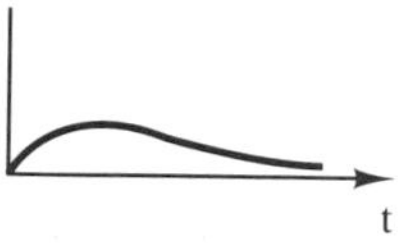	$\frac{n!}{(s + \alpha)^{n+1}}$
6	$\sin(\omega_0 t)$	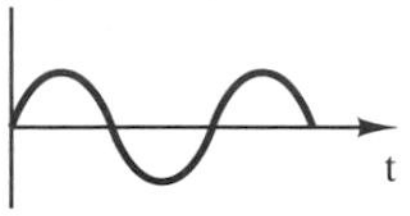	$\frac{\omega_0}{s^2 + \omega_0^2}$
7	$\cos(\omega_0 t)$	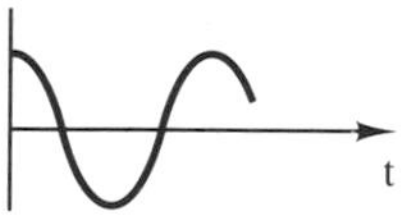	$\frac{s}{s^2 + \omega_0^2}$
8	$e^{-\alpha t}\sin(\omega_d t)$	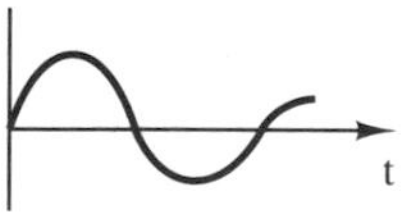	$\frac{\omega_d}{(s + \alpha)^2 + \omega_d^2}$
9	$e^{-\alpha t}\cos(\omega_d t)$	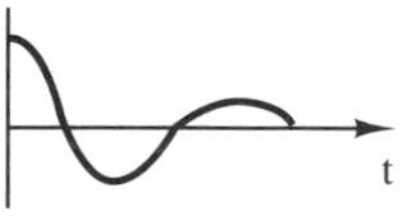	$\frac{s + \alpha}{(s + \alpha)^2 + \omega_d^2}$

In Laplace-transform analysis, we treat the signal f(t) as if it is not defined for $t < 0^-$. If we wish to extend the definition of f(t) to negative t, we must define $f(t) = 0$ for $t < 0^-$. If $f(0^-)$ is not zero for a particular functional form f, then setting $f(t) = 0$ for $t < 0^-$ produces a jump of size $f(0^-)$ at $t = 0^-$, just prior to $t = 0$. We see such jumps in the signals in entries 4, 7, and 9 of Table 5.5. The unit step in entry 2 does not have such a jump. Rather, it jumps precisely at $t = 0$. The unit impulse in entry 1 lies totally between 0^- and 0^+. Therefore, it has no infinite jump at $t = 0^-$.

Since the Laplace integral is improper (upper limit of infinity), there are functions f(t) for which the integral does not converge. In general, a function is Laplace transformable (i.e., there is a region of the s-plane for which the transformation integral converges) if it grows no faster than an exponential as $t \to \infty$. All signals of interest to us are transformable.

Transform Properties

A number of useful properties of the Laplace transformation $\mathscr{L}$ are summarized in Table 5.6. The first two properties in the table indicate that $\mathscr{L}$ operates linearly on signals. (These two properties follow from the fact that sums and scalar multipliers can be taken outside the integral sign.) We shall use the linearity of $\mathscr{L}$ in solving network system equations.

TABLE 5.6 PROPERTIES OF THE LAPLACE TRANSFORMATION $\mathscr{L}$

1	Magnification	$\mathscr{L}[af(t)] = a\mathbf{F}(s)$
2	Addition	$\mathscr{L}[f_1(t) + f_2(t)] = \mathbf{F}_1(s) + \mathbf{F}_2(s)$
3	Derivative	$\mathscr{L}[\dot{f}(t)] = s\mathbf{F}(s) - f(0^-)$
4	Derivatives	$\mathscr{L}[\ddot{f}(t)] = s^2\mathbf{F}(s) - sf(0^-) - \dot{f}(0^-)$
5	Integral	$\mathscr{L}\left[\int_{0^-}^{t} f(t)\,dt\right] = \dfrac{\mathbf{F}(s)}{s}$
6	Convolution	$\mathscr{L}\left[\int_{0}^{t} f_1(\lambda)f_2(t-\lambda)\,d\lambda\right] = \mathbf{F}_1(s)\mathbf{F}_2(s)$
7	Initial value	$f(0^+) = \lim_{t\to 0^+} f(t) = \lim_{s\to\infty} s\mathbf{F}(s)$
8	Final value	$f(\infty) = \lim_{t\to\infty} f(t) = \lim_{s\to 0} s\mathbf{F}(s)$ if finite
9	Definite integral	$\int_{0}^{\infty} f(t)\,dt = \lim_{s\to 0} \mathbf{F}(s)$ if finite
10	Exponential decay	$\mathscr{L}[e^{-\alpha t}f(t)] = \mathbf{F}(s+\alpha)$
11	Delay	$\mathscr{L}[f(t-t_0)u_s(t-t_0)] = e^{-t_0 s}\mathbf{F}(s)$ for $t_0 \geq 0$
12	Time multiplication	$\mathscr{L}[tf(t)] = -\dfrac{d\mathbf{F}(s)}{ds}$
13	Time division	$\mathscr{L}\left[\dfrac{f(t)}{t}\right] = \int_{s}^{\infty} \mathbf{F}(s)\,ds$
14	Time scaling	$\mathscr{L}[f(at)] = \dfrac{\mathbf{F}(s/a)}{a}$

We use integration by parts to derive property 3, the derivative property of $\mathscr{L}$:

$$\mathscr{L}[\dot{f}(t)] = \int_{0^-}^{\infty} \dot{f}(t)e^{-st}\,dt$$

$$= e^{-st}f(t)\,\Big|_{0^-}^{\infty} - \int_{0^-}^{\infty} f(t)\,(-s)e^{-st}\,dt$$

$$= s\mathbf{F}(s) - f(0^-) \tag{5.156}$$

When we Laplace transform the equation for an energy-storage element, the derivative property *automatically incorporates the prior state* of the element (at $t = 0^-$). When we Laplace transform the system equation for a network, the derivative property automatically incorporates the whole prior network state. As a consequence, we can find the complete behavior of the network without having to examine the initial conditions (at $t = 0^+$)—a considerable simplification of the solution process! We demonstrate this process shortly.

The derivative property of $\mathscr{L}$ can be interpreted intuitively as follows. The multiplier s acts precisely like the time-derivative operator, as in previous chapters, but now in the domain of Laplace-transformed signals. Multiplication of $\mathbf{F}(s)$ by s produces the transform of the true derivative of the signal that is represented by $\mathbf{F}(s)$. That signal is zero for $t < 0^-$. If $f(0^-) > 0$ for a particular functional form f, then the signal represented by $\mathbf{F}(s)$ has a jump of size $f(0^-)$ at $t = 0^-$; it is not quite identical to f(t). Its derivative is $\dot{f}(t) + f(0^-)\,\delta(t)$; it includes an impulse of size $f(0^-)$ at $t = 0^-$. Therefore, $s\mathbf{F}(s)$ includes, in addition to the transform of $\dot{f}(t)$, the transform $f(0^-)$ of that impulse. (The whole impulse appears to be included within the interval $(0^-, \infty)$. To find the transform of $\dot{f}(t)$, which does not include an impulse, we must subtract the quantity $f(0^-)$ from $s\mathbf{F}(s)$.

Entries 4, 7, and 9 of Table 5.5 are functions that have a nonzero value at $t = 0^-$. For these functions, multiplying the transform by s does not, by itself, produce the transform of the derivative function. In addition, we must subtract the value at $t = 0^-$, as indicated by equation (5.156). The unit step function (entry 2), on the other hand, includes a jump at $t = 0$, so its derivative is precisely a unit impulse at $t = 0$. The subtraction in property 3 of Table 5.6 has no effect for the unit step function, because $u_s(0^-) = 0$.

The response of a network is the convolution of the source waveform and the unit impulse response (see section 4.3). Unless the source waveform is quite simple, performing a convolution analytically is very complicated. However, property 6 of Table 5.6 shows that convolving two signals corresponds to multiplying their Laplace transforms. Also, we shall find shortly that the Laplace transform of the unit impulse response is just the input-output transfer function for the network variable, a quantity that we already know. Therefore, it is quite easy to find analytically the Laplace transform of the network response for any source signal whose Laplace transform we can find (or look up).

Since $\mathbf{F}(s)$ contains all information about f(t) for $t \geq 0^-$, it is possible to find some features of the signal f(t) from the transform $\mathbf{F}(s)$ without performing an inverse Laplace transformation. Properties 7–9 of Table 5.6 provide three of these features, namely, the initial value ($t \to 0^+$), the final value ($t \to \infty$), and the area under the waveform. The remaining properties in the table show the effect on the transform of various operations on the signal waveform—multiplying it by t, by 1/t, or by the decay factor $e^{-\alpha t}$, changing the time scale, and delaying the time variable. Note that an exponential factor in the Laplace transform (property 11) corresponds to a time delay. It is the signal represented by the transform **F**(s), with value zero for $t < 0^-$, that is delayed. Therefore, the delayed signal is zero for $t < t_0$.

Inversion by Partial-Fraction Expansion

The usual approach to finding inverse transforms is to use a table of transform pairs. Table 5.5 demonstrates that transforms of typical network signals are ratios of polynomials in s. (Signal transforms look like transfer functions; we observe shortly that transfer functions can be interpreted as transformed signals.) In this section, we show that a ratio of polynomials in s can be decomposed into a sum of *simple* polynomial fractions, a process referred to as **partial fraction expansion.** As a consequence, the inversion process can be accomplished with a brief table of transforms. In fact, partial fraction expansion and inverse Laplace transformation can be carried out by a computer program (see appendix D).

We examine partial fraction expansion by means of an example. Suppose that the transform **F**(s) can be expressed as a sum of simple terms in the form

$$\mathbf{F}(s) = \frac{s+2}{s(s+1)(s+3)^2(s+1+j2)(s+1-j2)}$$

$$= \frac{a_1}{s} + \frac{a_2}{s+1} + \frac{a_3}{s+3} + \frac{a_4}{(s+3)^2} + \frac{a_5}{s+1+j2} + \frac{a_6}{s+1-j2} \quad (5.157)$$

We call the roots of the first-order factors of the denominator of **F**(s) the *first-order poles* of **F**(s). We call the root $s = -3$ of the double factor a *second-order pole* of **F**(s). In the expanded form, each first-order pole of **F**(s) appears in the denominator of a separate term, and multiple poles appear in multiple terms. We must determine the numerator coefficients.

There is a straightforward method for finding the coefficients a_i. We multiply equation (5.157) by the full denominator of **F**(s) to leave the polynomial (s + 2), with known coefficients, on the left. On the right, we group the coefficients of like powers of s. We then equate the right-side coefficient of each power of s to the known coefficient for the same power of s on the left side and solve the resulting six equations for the six unknowns.

The coefficients a_i can be found much more efficiently by taking advantage of the special structure of the equation. Focus attention on the first term, a_1/s. We multiply equation (5.157) by the denominator s of that term and then let $s = 0$ (the value of the pole for that term). This procedure reduces the first term on the right to a_1 and the other terms on the right to zero. Therefore,

$$a_1 = s\mathbf{F}(s)\big|_{s=0} = \left.\frac{s+2}{(s+1)(s+3)^2(s+1+j2)(s+1-j2)}\right|_{s=0}$$

$$= \frac{2}{(1)(3)^2(1+j2)(1-j2)} = \frac{2}{45} \quad (5.158)$$

We can generate the coefficient a_1 by inspection of **F**(s) alone: We merely remove the denominator factor s from **F**(s) and let $s = 0$.

The same procedure produces the coefficient of any first-order pole. We cancel the pole from **F**(s) and set s equal to the value of that pole. Thus, for a term of the form $a_i/(s + p_i)$,

$$a_i = (s + p_i)\mathbf{F}(s)\big|_{s=-p_i} \tag{5.159}$$

By this procedure, we find the coefficients

$$\begin{aligned} a_2 &= (s + 1)\mathbf{F}(s)\big|_{s=-1} = -\frac{1}{16} \\ a_5 &= (s + 1 + j2)\mathbf{F}(s)\big|_{s=-1-j2} = \frac{(4 - j3)}{320} \\ a_6 &= (s + 1 - j2)\mathbf{F}(s)\big|_{s=-1+j2} = \frac{(4 + j3)}{320} \end{aligned} \tag{5.160}$$

This example demonstrates that the coefficients of complex conjugate poles are themselves complex conjugates. Therefore, we need calculate only one of them.

This simple method for finding coefficients also works for a_4, the coefficient of the highest order term associated with the multiple pole. We merely multiply equation (5.157) by the highest power of the multiple pole, $(s + 3)^2$, and then let s take on the value at the pole, thereby isolating a_4 on the right side. That is,

$$a_4 = (s + 3)^2\mathbf{F}(s)\big|_{s=-3} = -\frac{1}{48} \tag{5.161}$$

The method does not work, however, for the lower order terms associated with a multiple pole. (Try it for a_3.) Instead, we must multiply the equation by the highest power of the pole and then differentiate the resulting equation with respect to s enough times to eliminate s from the term of interest. Finally, we let s take on its value at the pole to eliminate the other terms on the right side. For the coefficient a_3, this procedure gives

$$a_3 = \frac{d}{ds}[(s + 3)^2\mathbf{F}(s)]\big|_{s=-3} = -\frac{1}{144} \tag{5.162}$$

If the pole were of order higher than 2, additional differentiations would be required to isolate its other coefficients.

Exercise 5.26: Verify the computed values of a_1 through a_6.

By properties 1 and 2 of Table 5.6, the inverse transform of a linear combination of terms is the same linear combination of the individual inverse transforms. We find the inverse Laplace transform of the transformed signal **F**(s) of equation (5.157), term by term, by referring to Table 5.5. For that equation, we have found that $a_1 = 2/45$, $a_2 = -1/16$,

$a_3 = -1/144$, $a_4 = -1/48$, $a_5 = (4 - j3)/320$, and $a_6 = (4 + j3)/320$. Therefore, according to the table,

$$f(t) = \frac{2}{45} - \left(\frac{1}{16}\right)e^{-t} - \left(\frac{1}{144}\right)e^{-3t} - \left(\frac{1}{48}\right)te^{-3t} + \left[\frac{(4 - j3)}{320}\right]e^{-t}e^{-j2t}$$
$$+ \left[\frac{(4 + j3)}{320}\right]e^{-t}e^{j2t} \qquad \text{for } t \geq 0 \tag{5.163}$$

The last two terms of equation (5.163) can be combined to form

$$e^{-t}\left[\frac{4}{320}(e^{j2t} + e^{-j2t}) + j\frac{3}{320}(e^{j2t} - e^{-j2t})\right] = \frac{1}{160}e^{-t}[(4\cos 2t - 3\sin 2t)]$$
$$= \frac{5}{160}e^{-t}\cos(2t + 0.644) \tag{5.164}$$

We use the phasor-addition process of section 5.3 to combine the cosine and sine terms into a single phase-shifted cosine.

Because calculations with complex numbers are complicated, we might prefer to combine complex conjugate terms to avoid such calculations. For example, combining the last two terms of equation (5.157) produces

$$\frac{cs + d}{(s + 1)^2 + 2^2} \tag{5.165}$$

where the real coefficients c and d replace the complex conjugate coefficients a_5 and a_6. Once coefficients are found for the other terms (those not related to sinusoidal signals), we move them to the other (known) side of the equation. Then we substitute two convenient values of s (other than the values at the poles) to produce two linear equations in c and d and solve those equations for c and d.

For example, suppose we have found a_1 through a_4. We substitute those known coefficients into equation (5.157). Then we combine the last two terms of the equation into the form (5.165) and substitute $s = 1$ to produce the equation $c + d = 1/80$. Substituting $s = 2$ produces $2c + d = 3/80$. It follows that $c = 1/40$ and $d = -1/80$ in the term (5.165).

Exercise 5.27: Use the foregoing procedure to verify the values of c and d.

Now we must find the inverse transform of the term (5.165). Compare that term with the sinusoidal forms in Table 5.5. Note that $\alpha = 1$ rad/s and $\omega_d = 2$ rad/s. According to the table, the transform of $e^{-\alpha t}\cos(\omega_d t)$ has $(s + \alpha)$ in the numerator. The transform of $e^{-\alpha t}\sin(\omega_d t)$ has ω_d in the numerator. We need a combination of these two forms. Let us rearrange the numerator of the conjugate-pole factor (5.165) to the form

$$cs + d = \frac{1}{40}s - \frac{1}{80} = c_1(s + \alpha) + c_2(\omega_d) = c_1(s + 1) + c_2(2) \tag{5.166}$$

We equate coefficients on like powers of s to find $c_1 = 1/40$ and $c_2 = -3/160$. Then, according to Table 5.5, the sinusoidal portion of the inverse transform (or signal) f(t) is

$$e^{-t}\left[\frac{1}{40}\cos 2t - \frac{3}{160}\sin 2t\right] \tag{5.167}$$

The result is the same as equation (5.164).

Once a coefficient of a term of the expansion has been found, it is sometimes useful to subtract that term from both sides of the equation to reduce the size of the expansion problem. In problem 5.33(iii), for example, the coefficient a_1 of the expansion term a_1/s is easy to find. Subtracting that term from both sides of the expansion equation reduces the expansion to a quadratic form similar to (5.165).

There is another useful shortcut in determining the coefficients of a partial fraction expansion. Suppose we normalize **F**(s) so that the coefficient of the highest power of s is 1 in both the numerator and the denominator (with no constant factor). It can be shown that that *normalized transform* has the following *constant-sum property:*

- If the order of the denominator is one higher than the order of the numerator, the sum of the coefficients of the *first-order terms* is 1.
- If the order of the denominator is two or more higher than the order of the numerator, the sum of the coefficients of the *first-order terms* is 0.

For the example of equation (5.157), the constant-sum property requires that

$$a_1 + a_2 + a_3 + a_5 + a_6 = 0 \tag{5.168}$$

Since the calculation of a_3 in equation (5.162) is complicated, we can avoid that calculation by using equation (5.168) to calculate a_3. Alternatively, we can use the constant-sum property as a check on our calculations.

A ratio of polynomials can be expanded in partial fractions only if the denominator polynomial is of higher order than the numerator polynomial. The transforms of most signals associated with networks have a higher order denominator. If a particular transform does not, we divide the denominator into the numerator until the remainder has a higher order denominator. Then we expand the remainder in partial fractions. We illustrate this process in example problem 5.11.

In sum, the procedure for finding the inverse Laplace transform is as follows.

PROCEDURE FOR FINDING A SIGNAL WAVEFORM f(t) FROM ITS TRANSFORM F(s):

1. Divide the denominator of **F**(s) into the numerator to obtain a remainder with numerator of lower order than the denominator.
2. Factor the denominator of the remainder.
3. Expand the remainder into a sum of partial fractions.
4. Find the coefficients of the partial fraction expansion. Add any terms obtained by the division in part (1) to the partial fraction expansion. The result is an easily inverted expression for the transformed signal **F**(s).
5. Use a table of Laplace transforms to find the inverse transform of each term of the expansion. Combine terms as necessary to put the expression in a convenient form.

Factoring the denominator of $\mathbf{F}(s)$ can be difficult. Finding the coefficients of the partial fraction expansion is straightforward, but often complicated. Fortunately, for a specific function $\mathbf{F}(s)$, these are easily automated tasks. The computer program CC (see appendix D), for example, performs each task with little effort on the part of the user.

Transformation and Solution of Network Equations

Consider the translational mechanical network shown in Figure 5.41. The input-output transfer function H(s) for v_1, according to equation (5.89), is

$$v_1 = \frac{s}{ms^2 + bs + k} F \triangleq H(s)F \tag{5.169}$$

The system equation for v_1 is

$$\ddot{v}_1 + \left(\frac{b}{m}\right)\dot{v}_1 + \left(\frac{k}{m}\right)v_1 = \frac{\dot{F}}{m} \tag{5.170}$$

Suppose that the network is in motion at $t = 0^-$, with known energy states and a known force source. The source suddenly changes to the value $F(t) = F_c$ at $t = 0$. We seek the response $v_1(t)$.

The two sides of equation (5.170) are identical functions of time. Therefore, the Laplace transforms of the two sides are equal. Since the Laplace transformation is linear (properties 1 and 2 of Table 5.6), and since the coefficients of the differential equation are constants, the Laplace transform can be applied separately to the individual terms of each side. The result is

$$[s^2\mathbf{V}_1 - sv_1(0^-) - \dot{v}_1(0^-)] + \left(\frac{b}{m}\right)[s\mathbf{V}_1 - v_1(0^-)] + \left(\frac{k}{m}\right)\mathbf{V}_1 = \frac{[s\mathbf{F} - F(0^-)]}{m} \tag{5.171}$$

where the derivative properties of the Laplace transformation (properties 3 and 4 of Table 5.6) introduce the prior values $v_1(0^-)$, $\dot{v}_1(0^-)$, and $F(0^-)$ into the equation.

Intuitively, $v_1(0^-)$ is a measure of the energy stored in the mass at the prior instant ($t = 0^-$), and $F(0^-)$ is the force applied by the source at the prior instant. If the problem is fully defined, both quantities are known. What does $\dot{v}_1(0^-)$ represent? At $t = 0^-$, the node equation at node 1 is

$$F(0^-) = m\dot{v}_1(0^-) + bv_1(0^-) + kx_1(0^-) \tag{5.172}$$

Since the prior force $F(0^-)$ applied by the source is known and the prior node velocity $v_1(0^-)$ is known, finding the prior acceleration $\dot{v}_1(0^-)$ is equivalent to finding the prior node displacement—a measure of energy stored in the spring. Thus, the set of prior conditions required in equation (5.171) is equivalent to the prior energy state of the network.

We solve the transformed equation (5.171) for $\mathbf{V}_1(s)$ by a straightforward algebraic manipulation:

$$\mathbf{V}_1(s) = \frac{s/m}{s^2 + (b/m)s + (k/m)}\mathbf{F}(s) + \frac{sv_1(0^-) + [(b/m)v_1(0^-) + \dot{v}_1(0^-) - (1/m)F(0^-)]}{s^2 + (b/m)s + (k/m)} \tag{5.173}$$

According to equation (5.172), the quantity in square brackets is $-(k/m)x_1(0^-)$. Hence, the response transform $\mathbf{V}_1$ depends on the source-signal transform **F** and on the prior energy states $v_1(0^-)$ and $x_1(0^-)$.

Since the source has the constant value F_c for $t \geq 0$, the transform of the source function is $\mathbf{F}(s) = F_c/s$, the transform of a step of size F_c. Although we could find the signal waveform $v_1(t)$ as a function of the network parameters m, k, and b, the source parameter F_c, and the energy states $v_1(0^-)$ and $x_1(0^-)$, the process of factoring the denominator, expanding into partial fractions, and inverting the terms of the expanded transform would be tedious. Instead, we complete the solution process for specific numbers: m = 2 kg, b = 4 N·s/m, k = 10 N/m, $F(0^-) = 0$, $v_1(0^-) = 0.1$ m/s, $x_1(0^-) = 0.2$ m, and $F_c = 1$ N. Then

$$\mathbf{V}_1(s) = \frac{(s/2)}{s^2 + 2s + 5}\,\frac{1}{s} + \frac{(0.1)s + [-1]}{s^2 + 2s + 5}$$

$$= \frac{0.6s - 1}{(s + 1)^2 + 2^2} \tag{5.174}$$

The denominator polynomial shows that $\alpha = 1$ rad/s and $\omega_d = 2$ rad/s. Accordingly, we rewrite the numerator of equation (5.174) in the form $0.6s - 1 = a(s + \alpha) + b(\omega_d) = 0.6(s + 1) - 0.8(2)$. We then recognize from Table 5.5 that the inverse Laplace transform of equation (5.174) is the signal

$$v_1(t) = e^{-t}[0.6 \cos 2t - 0.8 \sin 2t]$$

$$= e^{-t}\cos(2t + 0.93) \text{ m/s} \tag{5.175}$$

One advantage of the Laplace transform method for solving network equations is the fact that it incorporates the prior state automatically; we do not have to find the initial conditions (at $t = 0^+$), as we did in section 3.4. (If we had used 0^+ as the lower limit of the Laplace transform, the solution would have been expressed in terms of the initial conditions.) If we *want* to find the initial value of $v_1(t)$, we use property 7 of Table 5.6:

$$v_1(0^+) = \lim_{s\to\infty} s\mathbf{V}_1(s) = \lim_{s\to\infty} \frac{s(0.6s - 1)}{s^2 + 2s + 5}$$

$$= \lim_{s\to\infty} \frac{0.6 - 1/s}{1 + 2/s + 5/s^2} = 0.6 \text{ m/s} \tag{5.176}$$

Example Problem 5.10: Laplace transform analysis of system behavior.

The capacitors in the circuit of Figure 5.66 have been charged to 2 volts each by holding $v = 4$ V. Hence, $v_{12} = v_{23} = 2$ V at $t = 0^-$. At $t = 0$, we close the switch and change the source voltage to 2 V.

(a) Find the input-output transfer function $H(s) = v_3/v$ for $t \geq 0$.

(b) Use Laplace transforms to determine $\mathbf{V}_3(s)$, the Laplace transform of the voltage at node 3.

(c) Use partial fraction expansion to find the time response $v_3(t)$ from the transform $\mathbf{V}_3(s)$ found in part (b).

(d) Use a computer program to inverse transform $\mathbf{V}_3(s)$ from part (b) and produce $v_3(t)$.

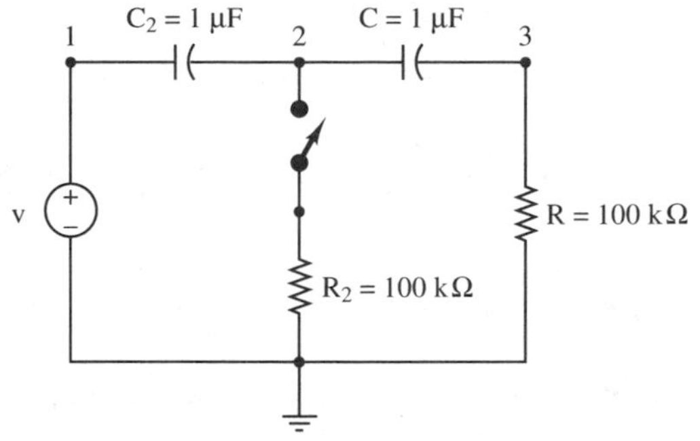

Figure 5.66 An electric circuit.

Solution: **(a)** The signal v_3 is related to v_2 by the impedance ratio

$$v_3 = \frac{R}{R + (1/sC)} v_2 = \frac{RCs}{RCs + 1} v_2 \tag{5.177}$$

The effective impedance between node 2 and ground, as *seen* by the left branch, is

$$z = \frac{R_2(R + 1/sC)}{R_2 + (R + 1/sC)} = \frac{R_2(RCs + 1)}{(R_2 + R)Cs + 1} \tag{5.178}$$

Then v_2 is related to v by the impedance ratio

$$v_2 = \frac{z}{z + 1/sC_2} v_1 = \frac{R_2C_2s(RCs + 1)}{R_2C_2RCs^2 + (R_2C_2 + RC + R_2C)s + 1} v \tag{5.179}$$

We combine equations (5.177) and (5.179)—that is, we multiply the two transfer functions—to obtain the input-output transfer function

$$\frac{v_3}{v} = \frac{RCR_2C_2s^2}{RCR_2C_2s^2 + (R_2C_2 + RC + R_2C)s + 1} = H(s) \tag{5.180}$$

For the specific parameter values $R_2 = R = 100\ k\Omega$ and $C_2 = C = 1\ \mu F$,

$$H(s) = \frac{0.01s^2}{0.01s^2 + 0.3s + 1} = \frac{s^2}{s^2 + 30s + 100} \tag{5.181}$$

Since the signals are expressed as functions of time (v and v_3), the symbol s in the transfer function should be interpreted as the time-derivative operator. We could, as well, express the signals in the Laplace transform notation ($\mathbf{V}$ and $\mathbf{V}_3$). Then the same symbol s in the transfer function would be interpreted as the complex frequency variable.

(b) We rearrange the transfer function expression to form the system differential equation for v_3,

$$\ddot{v}_3 + 30\dot{v}_3 + 100v_3 = \ddot{v} \tag{5.182}$$

Next, we Laplace transform equation (5.182), term by term and insert the prior conditions as indicated in properties 3 and 4 of Table 5.6:

$$[s^2\mathbf{V}_3 - sv_3(0^-) - \dot{v}_3(0^-)] + 30[s\mathbf{V}_3 - v_3(0^-)] + 100\mathbf{V}_3$$
$$= [s^2\mathbf{V} - sv(0^-) - \dot{v}(0^-)] \tag{5.183}$$

or

$$(s^2 + 30s + 100)\mathbf{V}_3 = s^2\mathbf{V} + s[v_3(0^-) - v(0^-)] + \dot{v}_3(0^-) - \dot{v}(0^-) + 30v_3(0^-) \quad (5.184)$$

The prior conditions, according to the problem statement, are $v(0^-) = 4$ V, $v_3(0^-) = 0$ V, and $\dot{v}_1(0^-) = \dot{v}_3(0^-) = 0$. The Laplace transform of the source signal, which is constant at 2 V for $t \geq 0$, is

$$\mathbf{V}(s) = \int_{0^-}^{\infty} v(t)e^{-st}\,dt = \int_{0}^{\infty} (2)e^{-st}\,dt = \frac{2}{s} \quad (5.185)$$

This is the transform of a 2-V step function. By substituting the source-signal transform and the prior values into equation (5.184) and solving for $\mathbf{V}_3(s)$, we get

$$\mathbf{V}_3(s) = \frac{-2s}{s^2 + 30s + 100} \quad (5.186)$$

It might seem that the step function in equation (5.185) should be negative, since the source jumps from 4 V to 2 V. However, that would be the case only if we were to treat the source as the sum of a 4-V constant and a negative 2-V step. Then we would have to superimpose the solution for a constant 4-V source. It is the actual waveform v(t) for $t \geq 0^-$ that defines the source. The previous 4-V value is accounted for by $v(0^-)$. The combined effect of $v(0^-) = 4$ V and $\mathbf{V}(s) = 2/s$ is to produce the coefficient -2 in equation (5.186).

(c) The denominator of equation (5.186) is of lower order than the numerator and factors into the product $(s + 3.82)(s + 26.2)$. Therefore, equation (5.186) can be expanded in the form

$$\mathbf{V}_3(s) = \frac{a_1}{s + 3.82} + \frac{a_2}{(s + 26.2)} \quad (5.187)$$

The two coefficients of the expansion, by equation (5.159), are

$$a_1 = [(s + 3.82)\mathbf{V}_3(s)]\big|_{s=-3.82} = \left.\frac{-2s}{s + 26.2}\right|_{s=-3.82} = 0.34$$

$$a_2 = [(s + 26.2)\mathbf{V}_3(s)]\big|_{s=-26.2} = \left.\frac{-2s}{s + 3.82}\right|_{s=-26.2} = -2.34 \quad (5.188)$$

According to transform 4 of Table 5.5,

$$v_3(t) = 0.34e^{-3.82t} - 2.34e^{-26.2t} \quad (5.189)$$

(d) We used the inverse Laplace transform feature of the computer program CC (described in appendix D) to find $v_3(t)$ from the transform in equation (5.186). The interactive CC session is shown in Table 5.7, with G1 $\triangleq \mathbf{V}_3(s)$. The computer program gives the algebraic expression for the response. At the end of the session, the CC *time* command was used to plot the waveform $v_3(t)$ shown in Figure 5.67.

The Laplace transform method can be used only for **linear differential equations with constant coefficients.** (We use linearity and time invariance to take the Laplace transformation inside the terms of the differential equation.) We can take Laplace transforms of the network equations at any stage in their development. We can even write the

TABLE 5.7 CC PROGRAM SESSION

```
CC > G1 = (−2 * s)/(s^2 + 30 * s + 100)
CC > ilt, G1
    G1(t) = −2.341641 * exp(−26.18034t) + .3416408 * exp(−3.81966t)
                for t > 0
          = 0   for t < 0
CC > time, G1      (open-loop impulse option)
```

node and loop equations directly in terms of transformed variables if we wish. The process of eliminating variables can be carried out as well in one notation as in the other.

In section 4.2, we defined the transfer function (equation (4.48)) and derived input-output transfer functions H(s) for network differential equations. In those equations, the s-notation indicates the time derivative. The transfer function $H(s) = v_1(t)/F(t)$ in equation (5.169) represents a differential equation in this manner. In section 5.3, we used the same transfer function H(s), with s interpreted as the complex frequency variable $j\omega$, to compute the sinusoidal transfer function $H(j\omega)$. That sinusoidal transfer function acts as a simple multiplier to convert the phasor $\boldsymbol{F}$ into the phasor $\boldsymbol{V}_1$.

We can also obtain the same transfer function H(s) from the Laplace-transformed differential equation (5.171) by setting the prior conditions to zero and then taking the ratio $\mathbf{V}_1(s)/\mathbf{F}(s)$. If we derive the transfer function in this manner, the symbol s denotes the more general complex frequency variable of the Laplace transformation. In this more general interpretation, the transfer function again acts as a simple multiplier, but on the transform of *any signal waveform,* not just a sinusoid.

It is appropriate to define the transfer function directly in terms of Laplace-transformed signals. The transfer function from $\mathbf{Y}_1$ to $\mathbf{Y}_2$ is

$$G(s) \triangleq \left.\frac{\mathbf{Y}_2(s)}{\mathbf{Y}_1(s)}\right|_{\text{p.c.}=0} \tag{5.190}$$

where $\mathbf{Y}_1(s)$ and $\mathbf{Y}_2(s)$ are the Laplace transforms of the signals $y_1(t)$ and $y_2(t)$ and the notation p.c. = 0 means that all prior conditions are set to zero. This transform-domain definition is equivalent to the signal-domain definition given in equation (4.48). However,

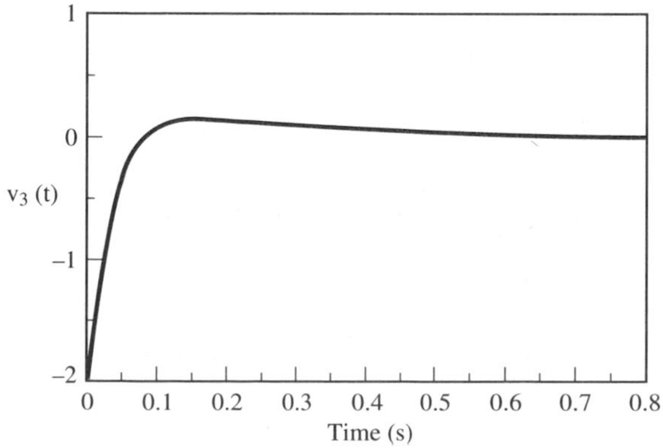

Figure 5.67 Network response.

since equation (5.190) is in the transform domain, we interpret the variable s as the complex frequency variable rather than the time derivative. In a similar fashion, we define the input-output transfer function (or system function) of a network as the ratio

$$\mathbf{H}(s) \triangleq \left.\frac{\mathbf{Y}_O(s)}{\mathbf{Y}_I(s)}\right|_{p.c.=0} \tag{5.191}$$

where $\mathbf{Y}_I(s)$ and $\mathbf{Y}_O(s)$ are the Laplace transforms of the network input signal and output signal, respectively.

According to the time-domain definition of the transfer function in equation (4.48), the input and output signals for a zero-state network are related by the differential equation

$$y_O(t) = H(s)y_I(t) \tag{5.192}$$

where s is the time derivative. By equation (4.114), the response is the convolution

$$y_O(t) = h(t) * y_I(t) \tag{5.193}$$

where h(t) is the unit impulse response of the network. The Laplace transform of equation (5.193), by property 6 of Table 5.6, is

$$\mathbf{Y}_O(s) = \mathbf{H}(s)\mathbf{Y}_I(s) \tag{5.194}$$

where $\mathbf{H}(s)$ is the Laplace transform of the impulse response h(t) and s is the complex frequency variable.

It is apparent that $\mathbf{H}(s)$ of equation (5.194) and H(s) of equation (5.191) are the same function. Therefore, we can **interpret the time-domain input-output transfer function** H(s) of equation (5.192) directly **as the Laplace transform of the unit impulse response** h(t).

Example Problem 5.11: The transfer function and the impulse response.
The input-output transfer function H(s) for node 3 of the electrical network of Figure 5.66 is given in equation (5.181). Inverse-transform H(s) to find the unit impulse response h(t) at node 3.

Solution: The unit impulse response h(t) is the inverse Laplace transform of the transfer function H(s) in equation (5.181). To inverse-transform H(s), we first divide the denominator polynomial into the numerator polynomial to obtain a remainder in which the numerator has lower order than the denominator. The result is

$$H(s) = 1 - \frac{30s + 100}{s^2 + 30s + 100} \triangleq 1 - G(s) \tag{5.195}$$

Then we expand the remainder G(s) into partial fractions:

$$G(s) = \frac{b_1}{s + 3.82} + \frac{b}{s + 26.2} \tag{5.196}$$

The coefficients are

$$b_1 = [(s + 3.82)G(s)]|_{s=-3.82} = \left.\frac{30s + 100}{s + 26.2}\right|_{s=-3.82} = -0.65$$

$$b_2 = [(s + 26.2)G(s)]|_{s=-26.2} = \left.\frac{30s + 100}{s + 3.82}\right|_{s=-26.2} = 30.65 \tag{5.197}$$

Therefore,

$$H(s) \equiv 1 - G(s) = 1 + \frac{0.65}{s + 3.82} + \frac{-30.65}{s + 26.2} \tag{5.198}$$

The inverse Laplace transform of equation (5.198), according to entries 1 and 4 of Table 5.5, is the unit impulse response

$$h(t) = \delta(t) + 0.65\, e^{-3.82t} - 30.65\, e^{-26.2t} \tag{5.199}$$

The system function H(s) for a linear network has two interpretations. In the frequency domain, H(s) is a multiplier that produces the response—by multiplying the source-signal transform, as in equation (5.194). In the time domain, we use a representative response signal—the impulse response h(t)—to characterize the system. The system function H(s) is the Laplace transform of that characteristic response.

5.6 SUMMARY

Section 5.1 examines hydraulic systems. The volume flow rate Q is a flow, the pressure P a potential. Derivatives and integrals of Q and P also act as flows and potentials. Lumped models of hydraulic systems include three types of passive elements and two types of active sources. The ideal fluid capacitor is described by $P_{i0} = (1/sC_f)Q$. Fluid capacitors must connect to the reference node at pressure P_0. Fluid capacitance can depend on the pressure drop P_{i0}; hence, the capacitor equation can be nonlinear. The ideal fluid inertia is described by $P_{ij} = (sI)Q$. The ideal fluid resistor is described, under most practical conditions, by $P_{ij} = \eta_f Q^2$, where η_f is the *quadratic resistor coefficient.* The ideal fluid-flow source has a specified flow rate Q(t). The ideal pressure source has a specified pressure difference $P_{ij}(t)$. We orient the flow variable for each element arbitrarily. The sign convention for element equations is to order the subscripts on the pressure variable in the direction selected for the flow variable. Then positive flow is in the direction of pressure drop in passive elements. The power absorbed by an ideal hydraulic element is $P_{ij}Q$, where P_{ij} is the pressure drop in the direction of the flow Q.

Unlike the lumped models in earlier chapters, lumped models of hydraulic systems are usually nonlinear. Nonlinearity can be handled by linearizing the system equations or by simulating the nonlinear behavior. Linearizing the system equations allows the use of all the linear analysis techniques.

A linear network is much simpler to analyze if it is operated in a sinusoidal steady state. Then the signals in the network can be represented by phasors, and the network itself acts like a complex multiplier that converts input-signal phasors to output-signal phasors. The complex multiplier is the network transfer function H(s), where $s = j\omega$, and where ω is the frequency of operation of the network. We use a network phasor diagram to display explicitly the signals in the network and the relations among those signals.

Some networks exhibit resonant behavior. Resonant networks have poor stability. If a network is designed to resonate, selectivity is a better descriptor of behavior than is stability. The selectivity of a resonant network is measured by the quality of the resonance, customarily denoted Q.

We usually experience the behavior of a physical system in the time domain, as a variation in some quantity with time. Chapter 3 uses the unit-step response of a lumped linear network to provide insight about the network behavior. We refer to that approach as *time-domain analysis.*

This chapter, in contrast, uses the frequency response of a lumped linear network to provide insight concerning the network behavior. We refer to this approach as *frequency-domain analysis.* Every signal acts like a sum of frequency components. Networks act like filters that operate selectively on those frequency components. The network filter characteristic—the shape of its frequency response—fully determines the network behavior.

The input-output transfer function H(s) of chapter 4 is a shorthand notation for a differential equation. The symbol s in that transfer function denotes differentiation with respect to time. Hence, the transfer function shows what the network does to signals. Indeed, the transfer function fully determines the unit-step response. In section 5.4, we found that the same input-output transfer function, with s interpreted as a complex frequency variable, determines the frequency response of the network. In fact, the transfer function is equivalent to that frequency response. Finally, we discovered in section 5.5 that the input-output transfer function is the Laplace transform of the unit-impulse response, which, in turn, is the derivative of the unit-step response.

So the circle is complete: The transfer function, the frequency response, and the unit-step response are equivalent representations of the lumped linear network. From any one of these three network characterizations, we can find the other two.

The three network characterizations differ only in the kinds of insight they provide concerning the network. The poles of the transfer function indicate the oscillation frequencies and decay time constants of the component signals from which the network response waveforms are fashioned. The unit-step response illustrates, via its rise time and overshoot, the responsiveness and degree of stability of the network. The frequency response discloses the signal rapidity that the network can handle, another measure of network responsiveness. The frequency response also highlights any network resonance and shows the selectivity of that resonance.

To analyze the behavior of a linear network for a specific source waveform and a specific prior state, we can use either a time-domain approach or a frequency-domain approach.

Time-domain analysis: Here, we represent signals by time functions and the network by a differential equation. We use the network or the differential equation to find the initial state ($t = 0^+$) from the prior state ($t = 0^-$). We solve the differential equation for the specified source waveform and substitute the initial conditions to produce the response waveform $y_0(t)$ directly. We then note any distinguishing features of the response waveform.

Frequency-domain analysis: In this approach, we represent signals by their Laplace transforms, functions of the complex frequency variable s. We represent the network by the Laplace-transformed differential equation, also a function of the complex frequency variable s. The transformed differential equation automatically incorporates the prior state (at $t = 0^-$); it is not necessary to find the initial conditions (at $t = 0^+$). If we are interested in the time-domain behavior of the network, we either inverse-transform the response $\mathbf{Y}_O$ to produce $y_O(t)$ or infer significant characteristics of $y_O(t)$ directly from $\mathbf{Y}_O$.

5.7 PROBLEMS

5.1 For one of the lumped models of Figure P5.1:

(a) Label the system pressures and flows.

(b) Find a set of system equations in terms of the element parameters.

(c) Determine the flow in the load resistor (subscripted L) in terms of the element parameters.

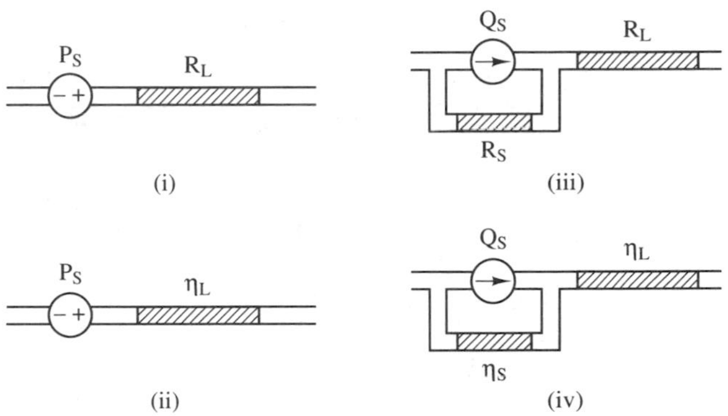

Figure P5.1

5.2 For one of the lumped models of Figure P5.2:

(a) Find the input-output system equation for Q in terms of the element parameters.

(b) The system is unenergized at $t = 0^-$. The pump is turned on (switched from 0 to a constant value denoted by subscript c) abruptly at $t = 0$. Find the initial output flow Q_i (at $t = 0^+$). That is, solve the system equation for $s \to \infty$.

(c) Suppose the pump is turned on as noted in (b). Find the final output flow Q_f (as $t \to \infty$). That is, solve the system equation for $s = 0$.

(d) Optional: Solve the system equation for Q(t), either analytically or numerically (by means of an operational-model simulator or a network simulator). In the case of numerical simulation, select specific numerical values for the parameters.

5.3 The rules of Section 4.2 for combining series- and parallel-connected elements are for linear elements only. Appropriate rules for combining series and parallel elements must be derived anew for each form of nonlinearity. Suppose that η_1 and η_2 are the quadratic resistor coefficients of two ideal quadratic fluid resistors.

(a) Show that if these resistors are connected in parallel, as in Figure P5.3(a), the effective quadratic resistor coefficient η_{par} of the parallel combination satisfies

$$\frac{1}{\sqrt{\eta_{par}}} = \frac{1}{\sqrt{\eta_1}} + \frac{1}{\sqrt{\eta_2}}$$

(b) Show that if these resistors are connected in series, as in Figure P5.3(b), the effective quadratic resistor coefficient η_{ser} of the series combination satisfies

$$\eta_{ser} = \eta_1 + \eta_2$$

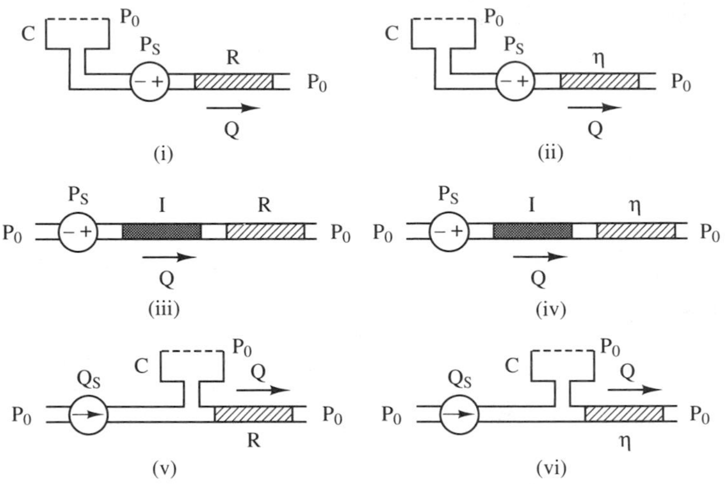

Figure P5.2

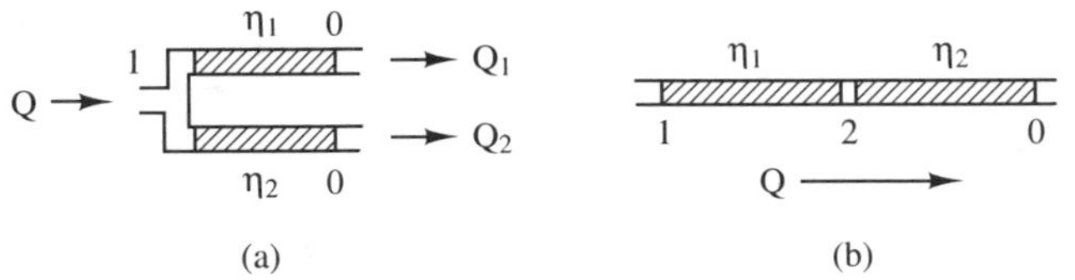

Figure P5.3

5.4 Derive a simple lumped model that describes the main features for one of the following situations or systems. State your assumptions.

(i) A centrifugal pump and hose are used to empty a swimming pool. The water is discharged into a drain at the level of the top edge of the pool.

(ii) A centrifugal pump is used to circulate the water in a closed hot-water heating system for a building. The water circulates in copper pipe. The pump forces the water directly through a coil in the heater at basement level. The water then rises two stories to the second floor and travels around the perimeter of the building at that level. Finally, the water returns to the pump at basement level.

(iii) A siphon removes water from a fish tank to a neighboring filter tank. The siphoned water settles through a filter, then is pumped by a centrifugal pump back into the fish tank.

5.5 The pressure-flow characteristic of a fully-open lever-operated bathroom water valve is shown in Figure P5.5. The valve is connected to a water main via 14 m of 1.3 cm (inside diameter) copper pipe. The water line has 3 elbows. A pressure reducer at the water main holds the inlet pressure constant at 480 kPa (relative to atmospheric pressure P_0).

(a) Make a lumped model of the system. Denote the pressure at the valve inlet by P_1.

(b) Write the equation that relates P_{10} to the flow Q as required by the combined pressure source and water line.

(c) Write the equation that relates P_{10} to Q as required by the valve. Use the data in Figure P5.5 to find the quadratic resistor coefficient for the fully-open valve.

(d) Plot the pressure-flow characteristic found in (b). Superimpose it on the pressure-flow characteristic of the open valve (Figure P5.5). Determine graphically the value of the flow for the fully-open condition.

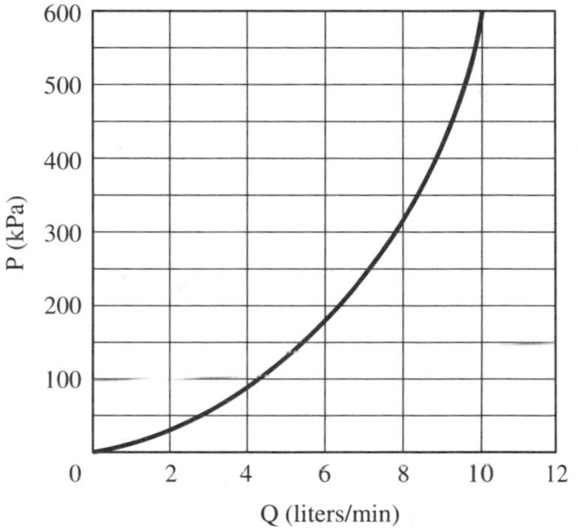

Figure P5.5

5.6 For one of the lumped networks of Figure P5.6, find the values of all flows and pressures (relative to the ambient pressure P_0) in terms of the network parameters. (Hint: Use the rules from problem 5.3 for combining series and parallel elements to simplify the network.)

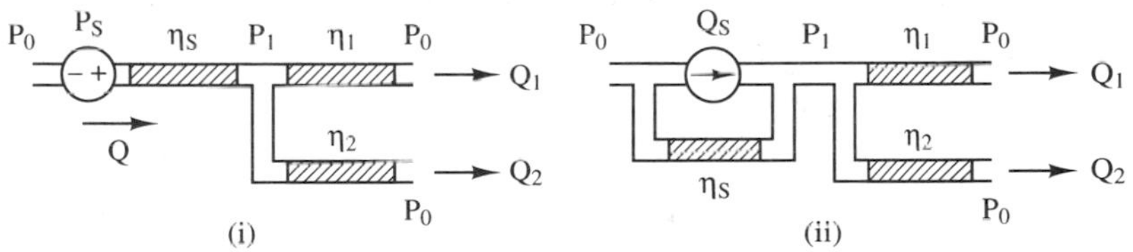

Figure P5.6

5.7 A simplified cold-water portion of a home water piping system is shown in Figure P5.7. The only elbows and tees are those shown in the figure. All points in the system are at the same elevation. All pipe diameters are 1.3 cm. The inlet pressure is 500 kPa (relative to atmospheric pressure). When the toilet is flushed or the washer turns on, the shower flow is affected. The rules for combining series and parallel elements that are found in problem 5.3 can simplify analysis of the interactions.

(a) Construct a lumped model of the system. Assume that all flow is turbulent. Find the quadratic resistor coefficient of each portion of the system. Treat each terminal device as a valve which has quadratic resistor coefficient $\eta_v = 3 \times 10^8\ \text{N·s}^2/\text{m}^8$ when fully open.

(b) Find the effective resistor coefficient of the network, as seen at the inlet, if only the shower is turned on (with fully open valve). Find the shower flow.

(c) Find the effective resistor coefficient of the network, as seen at the inlet, if both the shower and the washer are turned on. Find the shower flow. What happens if the washer is suddenly turned on while someone is using the shower?

(d) Find the effective resistor coefficient of the network, as seen at the inlet, if all three items (shower, washer, and toilet) are turned on. Find the shower flow.

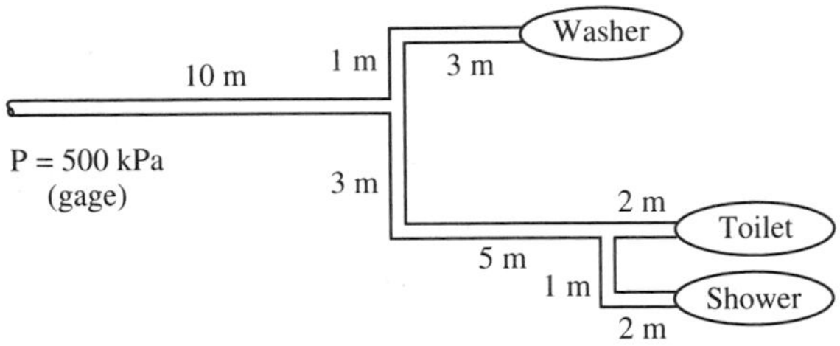

Figure P5.7

5.8 The two tanks shown in Figure P5.8 are connected by means of a pipe of diameter D. When the outlet valve is closed, the water levels in the tanks equalize. When the valve is opened, water is released at atmospheric pressure, the pipe flows are unequal, and the levels drop unequally.

(a) Make a lumped model of the system using ideal fluid elements. Assume flow is turbulent and inertance is negligible. Represent the initial fluid levels in the capacitors by pressure sources.

(b) Suppose $\ell_1 = 25$ m, $\ell_2 = 7$ m, $A_1 = 11$ m^2, $A_2 = 7$ m^2, $h_1(0) = 3$ m, $h_2(0) = 8$ m, $h_3 = 7$ m, $h_4 = 2$ m, D = 5 cm, and the valve is a gate valve. Estimate the values of the elements of the model.

(c) Use the model to find the initial values of the flows after the valve is opened. Hint: Let $s \to \infty$. Although the equations are nonlinear, they can be solved analytically.

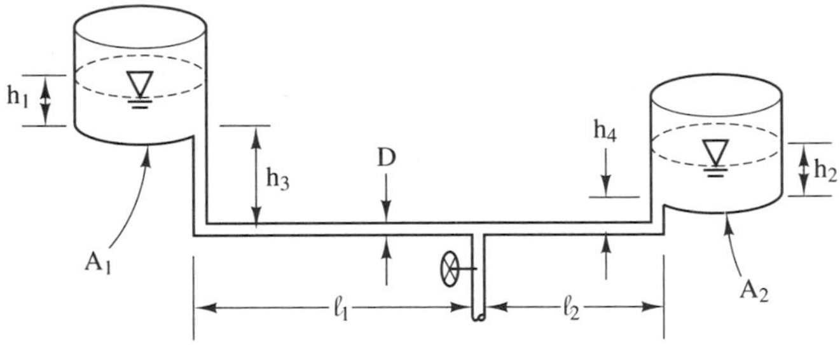

Figure P5.8

5.9 The lumped network of Figure P5.9 represents the system of Figure 5.22(a) with the rise in elevation removed and the pipe inertance included. The values of the network parameters are $P_S = 414$ kPa, $I = 3.18 \times 10^7$ N·s^2/m^5, $\eta = 8.09 \times 10^7$ N·s^2/m^8, and $C = 1.02$ m^5/N.

(a) Find the input-output system equation for the flow Q.

(b) Presume the capacitor is initially unenergized (the reservoir is empty) and that the pump is turned on at t = 0. Solve the system equation numerically (using an operational-model simulator or a network simulator).

(c) The flow Q(t) reaches its maximum very quickly then diminishes extremely slowly. Demonstrate that you can ignore the capacitance during the initial increase in flow. Demonstrate that you can ignore the inertance during the later decrease in flow.

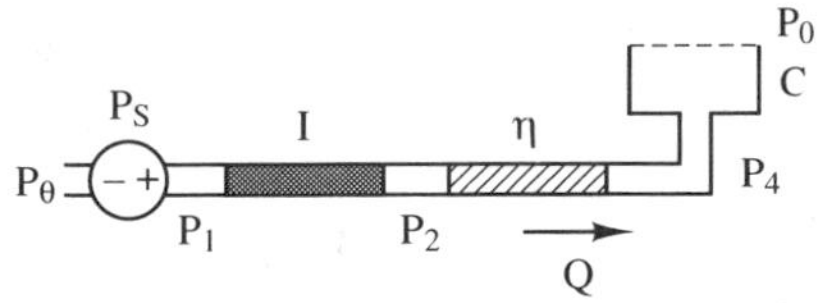

Figure P5.9

5.10 In this problem we examine the water hammer owing to rapid shutoff of flow. Figure 5.10 gives a simple lumped model of the home water circuit of Figure 5.5. The parameter values are $P = 519$ kPa, $\eta_{valve} = 160 \times 10^{10}$ N·s²/m⁸, $\eta_{pipe} = 56 \times 10^{10}$ N·s²/m⁸, and $I = 6.4 \times 10^7$ N·s²/m⁵. The steady-state flow is $Q_{steady} = 4.9 \times 10^{-4}$ m³/s.

(a) Suppose we close the valve over the time interval [0,T] in such a manner that $\dot{Q}(t) = -Q_{steady}/T$; that is, the flow is reduced to zero at a uniform rate. To model this shutoff process, we replace the valve by a flow source $Q_S(t)$. Let $Q_S(0) = Q_{steady}$. Determine the waveform for $Q_S(t)$.

(b) Solve numerically for the pressure $P_{30}(t)$ across the valve during shutoff. Use either an operational-model simulator or a network simulator. Find the maximum pressure for $T = 1$ s, for $T = 0.1$ s, and for $T = 0.01$ s. Compare the results to the estimate found in equation (5.34).

5.11 Carry out one of the following operations on sinusoids by converting each sinusoid to its phasor representation, operating on the phasors, then converting the resultant phasor to a sinusoid. Illustrate graphically the component phasors and the resultant phasor.

(i) $2\cos\omega t + 3\cos(\omega t + \pi/3)$

(ii) $\cos(\omega t + \pi/4) - \sin\omega t$

(iii) $2\sin\omega t - 3\cos(\omega t + \pi/4)$

(iv) $e^{-2t}\cos\omega t + 2e^{-2t}\cos(\omega t + \pi/3)$

(v) $\dfrac{d}{dt}\left[3\cos\left(\dfrac{\pi}{2}t + \dfrac{\pi}{4}\right)\right]$

(vi) $\dfrac{d}{dt}\left[\sin\left(\dfrac{\pi}{4}t - \dfrac{\pi}{4}\right)\right]$

5.12 Each of the networks of Figure P5.12 is operating in sinusoidal steady state. The source signal is given. For one of these networks:

(a) Find the phasor representation of each node and branch variable, including the source. Note that the network simplification techniques of Section 4.2 apply to sinusoidal impedances.

(b) Sketch the phasor diagram for the network.

5.13 An electrical network is shown in Figure P5.13.

(a) The steady-state current through the load resistor is to be $i_2(t) = \cos 5000t$ amps. Find the phasors which represent i_2, v_2, i_1, i, and v_S by a sequence of computations that begins with the phasor for the known current i_2.

(b) Sketch and label a phasor diagram which relates these quantities. Find the sinusoidal source function $v_S(t)$.

(c) Find the complex value of the input-output transfer function $H(j5000) = \boldsymbol{V}_2/\boldsymbol{V}_S$.

(d) Sketch a phasor diagram which relates $\boldsymbol{V}_2$ and $\boldsymbol{V}_S$ for the source waveform $v_S(t) = 20\sin 5000t$ volts.

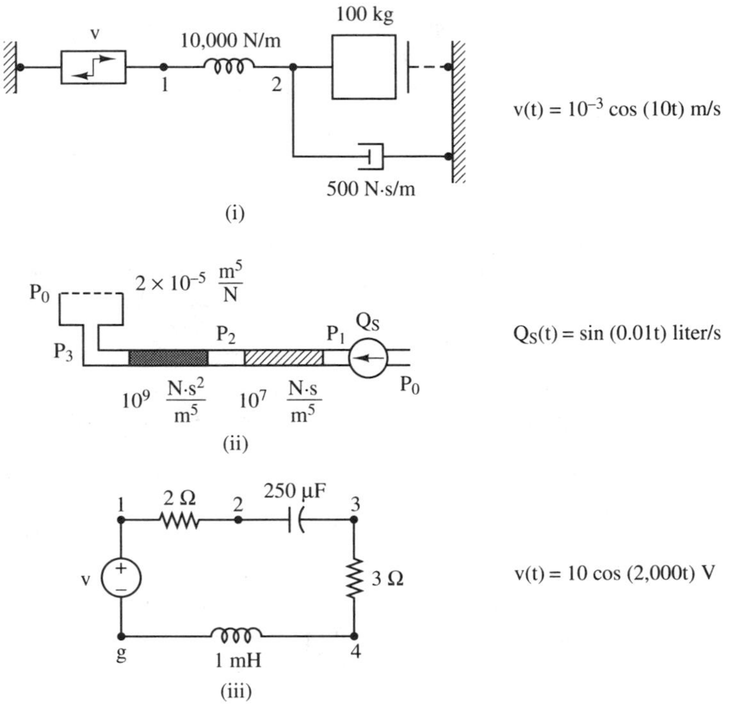

Figure P5.12

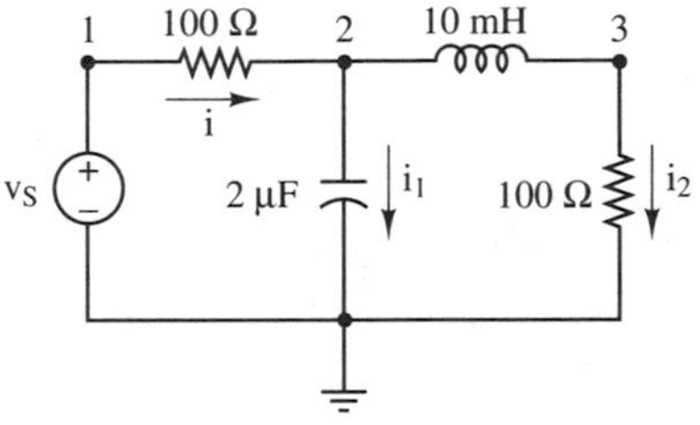

Figure P5.13

5.14 Each of the networks of Figure P5.14 is operating in sinusoidal steady state. The specified sinusoid is the desired *output* signal. For one of these networks:

(a) Find the phasors which represent the node potentials and branch flows.

(b) Sketch the network phasor diagram.

(c) Find the sinusoidal source signal.

(d) Find the complex value of the input-output transfer function for the specified frequency.

5.15 The voltage signal at an ac electric power outlet is $v(t) = v_m \cos(\omega t)$ volts, where $\omega = 2\pi f$ and $f = 60$ Hz. A resistor of resistance R (perhaps a light bulb) is connected to the outlet.

(a) Show that the power delivered to the resistor is

$$P(t) = \frac{v_m^{\,2}}{2R}(1 + \cos 2\omega t)$$

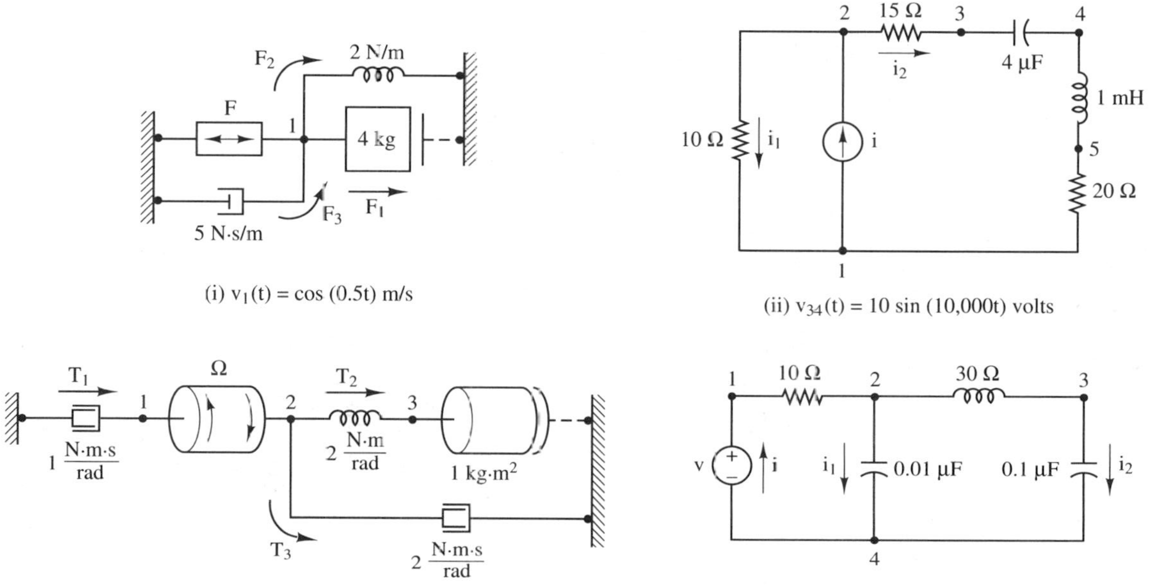

(i) $v_1(t) = \cos(0.5t)$ m/s

(ii) $v_{34}(t) = 10 \sin(10{,}000t)$ volts

(iii) $\Omega_3(t) = 10 \sin t$ rad/s

(iv) $i_2(t) = 5 \cos(10^6 t)$ A

Figure P5.14

(b) It is customary to rate the electric power outlet at v_{eff} volts, where $v_{eff} = v_m/\sqrt{2}$. Then $i_{eff} = i_m/\sqrt{2}$. Show that the *average* power delivered to the resistor is $v_{eff}^2/R = i_{eff}^2 R$.

5.16 The voltage signal at an ac electric power outlet is $v(t) = v_m \cos(\omega t)$ volts. An inductor with impedance sL is connected to the outlet.

(a) Show that the current in the inductor is

$$i(t) = \frac{v_m}{\omega L} \sin(\omega t) = \frac{v_m}{\omega L} \cos(\omega t - \pi/2)$$

(b) Show that the power delivered to the inductor is

$$P(t) = \frac{{v_m}^2}{2\omega L} \sin 2\omega t$$

(c) Show that the *average* power delivered to the inductor is zero.

5.17 The voltage signal at an ac electric power outlet is $v(t) = v_m \cos(\omega t)$ volts. A capacitor with impedance 1/sC is connected to the outlet.

(a) Show that the current in the capacitor is

$$i(t) = -\frac{v_m}{1/\omega C} \sin(\omega t) = \frac{v_m}{1/\omega C} \cos(\omega t + \pi/2)$$

(b) Show that the power delivered to the capacitor is

$$P(t) = -\frac{{v_m}^2}{2/\omega C} \sin 2\omega t$$

(c) Show that the *average* power delivered to the capacitor is zero.

5.18 The circuit in Figure P5.18 represents a 60 Hz ac electric power system. The sinusoidal impedances of the circuit elements are shown. The power company adjusts the voltage level at the generator (node 1) to maintain the *effective value* of v_2 at 120 volts. (See the definition of effective ac voltage in problem 5.15. Also note the current relations and power consumption relations in problems 5.15–5.17.)

(a) Find the power consumed by the resistive load (2 Ω).

(b) Find the currents in the two parallel loads and the resultant current i in the power line. (Hint: Use phasor representations.) Use the power-line current to determine the power consumed in the power-line resistance (0.25 Ω).

(c) Attach the capacitor in parallel with the electric motor as shown in the Figure P5.18. Show that the inductive (j2 Ω) and capacitive (−j2 Ω) impedances cancel. (That is, the parallel combination has infinite impedance.)

(d) Find the power line current and the power consumption in the line with the capacitor in place. How much power is saved by attaching the capacitor?

5.19 For one of the following transfer functions, find:

(a) M_G vs. ω and ϕ_G vs. ω

(b) Lm(G) vs. log ω and ϕ_G vs. log ω

(i) $G(s) = 5$ **(iv)** $G(s) = 1 + \frac{s}{2}$

(ii) $G(s) = 5s$ **(v)** $G(s) = \frac{1}{(s + 2)}$

(iii) $G(s) = \frac{1}{s}$ **(vi)** $G(s) = \frac{s}{(1 + 0.5\,s)}$

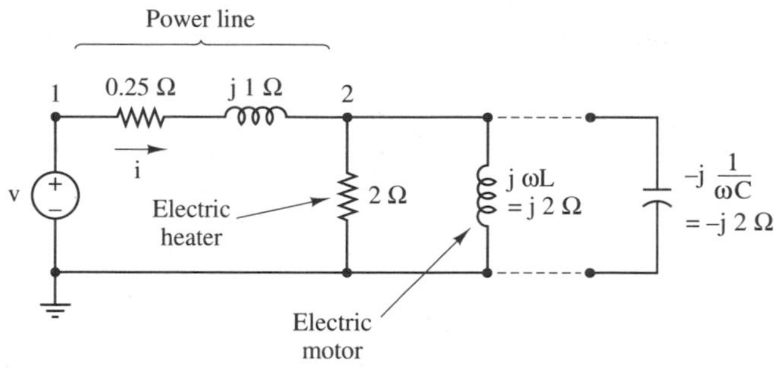

Figure P5.18

5.20 For one of the following transfer functions:

(a) Use a transfer function analysis program (e.g., CC) to generate the Bode diagrams ($Lm(G)$ and ϕ_G vs. log ω).

(b) Draw the straight-line approximations to the log-magnitude curve and determine the frequencies at which they intersect. (Be sure to use slopes that are multiples of 20 dB/decade.) Relate the measured intersection frequencies to the poles and zeros of the transfer function. What other features of the straight-line approximations are characteristic of that transfer function?

(i) $G(s) = s + 2$

(ii) $G(s) = \dfrac{1}{(1 + 0.5\,s)}$

(iii) $G(s) = \dfrac{1}{(s + 2)(s + 20)}$

(iv) $G(s) = \dfrac{1}{(s + 2)^2}$

(v) $G(s) = \dfrac{1}{(s + 2)(s + 4)}$

(vi) $G(s) = \dfrac{s}{(s + 2)(s + 20)}$

(vii) $G(s) = \dfrac{1}{s(s + 2)(s + 20)}$

(viii) $G(s) = \dfrac{(s + 10)}{s(s + 2)(s + 20)}$

5.21 For one of the following conjugate-pole transfer functions:

(a) Use a transfer function analysis program to generate the Bode diagrams ($Lm(G)$ and ϕ_G vs. log ω).

(b) Draw the straight-line approximations to the log-magnitude curve and find the corner point (intersection). What is the natural frequency ω_n for the conjugate-pole pair? Use $\hat{M}_r$, the resonant rise above the corner point, to determine the damping ratio ζ for the pole pair.

(c) Use the computer program to plot M_G vs. ω in the neighborhood of resonance.

(d) Use the M_G vs. ω plot to determine the resonant frequency ω_r and the −3 dB bandwidth. (A numerator factor s slightly shifts the frequency of the peak away from the resonant frequency ω_r of the pole pair. However, the frequency at the peak is still *approximately* ω_r.) Use ω_r and the bandwidth to determine the quality Q of the resonance and the damping ratio ζ of the pole pair. Can you use these resonance data to determine ω_n for the pole pair?

(i) $G(s) = \dfrac{1}{1 + 0.02s + 0.01s^2}$

(ii) $G(s) = \dfrac{1}{s^2 + 2s + 100}$

(iii) $G(s) = \dfrac{s}{s^2 + 4s + 100}$

(iv) $G(s) = \dfrac{s}{1 + 0.03s + 0.01s^2}$

(v) $G(s) = \dfrac{1}{s^2 + 6s + 900}$ (vi) $G(s) = \dfrac{s}{s^2 + 0.3s + 9}$

5.22 For one of the following transfer functions:

(a) Sketch the straight-line approximations to the Bode diagrams (Lm(G) and ϕ_G vs. log ω);

(b) Sketch the corrections to the straight-line approximations in the neighborhoods of the corner points.

(i) $G(s) = \dfrac{10s}{(1 + 0.5s)(1 + 0.1s)}$ (iv) $G(s) = \dfrac{s + 3}{s(s + 30)(s + 120)}$

(ii) $G(s) = \dfrac{5(s + 10)}{(s + 1)(s + 4)(s + 40)}$ (v) $G(s) = \dfrac{s + 15}{s(s^2 + 6s + 900)}$

(iii) $G(s) = \dfrac{5s}{(s^2 + 0.3s + 9)(s + 60)}$ (vi) $G(s) = \dfrac{s + 50}{s(s + 1000)^2}$

5.23 For one of the following transfer functions:

(a) Use a transfer function analysis program to generate a diagram of Lm(G) vs. log ω;

(b) View the magnitude curve generated in (a) as measured data, and use it to determine the transfer function.

(i) $G(s) = \dfrac{s + 2}{s^2 + 4s + 400}$ (iv) $G(s) = \dfrac{1}{(s^2 + 400s + 400)(s + 10)}$

(ii) $G(s) = \dfrac{s + 200}{s^2 + 20s + 400}$ (v) $G(s) = \dfrac{1}{(s^2 + 30s + 200)(s + 2)}$

(iii) $G(s) = \dfrac{s + 40}{s^2 + 20s + 400}$ (vii) $G(s) = \dfrac{(s + 80)}{(s^2 + 30s + 200)(s + 160)}$

5.24 Measured equal-loudness contours for the human ear are shown in Figure P5.24.[7] The frequencies are recorded in Hz rather than rad/s. In order for a 100 Hz tone to be perceived at the same loudness as a 1000 Hz tone, the 100 Hz tone must be about 30 dB stronger. Therefore, the ear response is about 30 dB weaker at 100 Hz than at 1000 Hz. Find a transfer function which has essentially the same frequency response as the third curve from the bottom. (What might cause the ear to have such a response? Can you explain why the response curve is different for loud sounds than for quiet sounds?)

5.25 A series connection of an ideal (resistance-less) inductor and a capacitor has zero impedance at the resonant frequency. A parallel connection of an ideal inductor and a capacitor has infinite impedance at resonance. These properties are used to design frequency-selective circuits for radio receivers. Such a frequency-selective circuit is shown in Figure P5.25. The transfer ratio $H(s) = v_{out}/i_{in}$ determines the strength of the received signal.

(a) Find the resonant frequency, the −3 dB bandwidth, and the Q of the circuit. Sketch M_H vs. ω in the neighborhood of resonance.

(b) Determine the fraction by which we must change the value of the variable capacitor in order to move the −3 dB pass band so that it does not overlap with the original −3 dB pass band.

(c) The circuit selectivity is ultimately limited by the resistance in the coil. Determine the change in Q if R_L is halved. What effect does this change in Q have on the bandwidth?

[7] From Figure 135 of Fletcher [5.3].

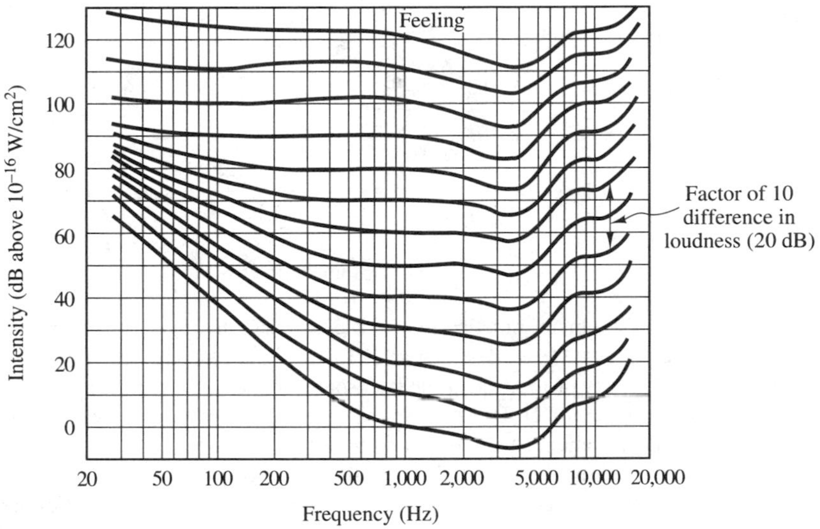

Figure P5.24

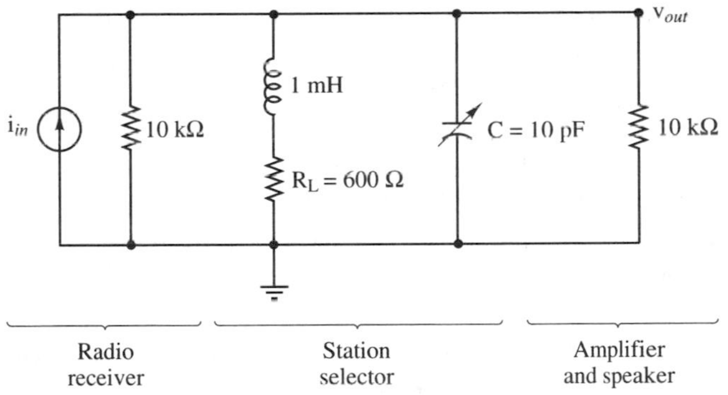

Figure P5.25

5.26 A system for measuring tide level is shown in Figure P5.26.[8] The inlet pressure P_2 depends on the water height h_2. View $h_2(t)$ as the sum of three components, an average value (3 m), a slow sinusoid of period 12 hours (owing to the tide), and a fast sinusoid of period 12 s (owing to waves). The average value of h_1 is the same as the average value of h_2. The wave amplitude is typically 1.8 m. The tide amplitude is typically 2.7 m. We want to choose values of A, D, and ℓ so that the variation in tank level h_1 owing to waves is less than 1% of the actual wave variation; and so that the variation in h_1 owing to the tide is in error by no more than 1% of the true tidal variation. The initial design parameters are $\ell = 6$ m, D = 5 cm, and A = 0.1 m^2.

(a) Make a lumped model of the system. Assume flow in the pipe is laminar. Find the transfer function $H(s) = h_1/h_2$.

[8] Adapted from problem 15.2 of Shearer [5.8]

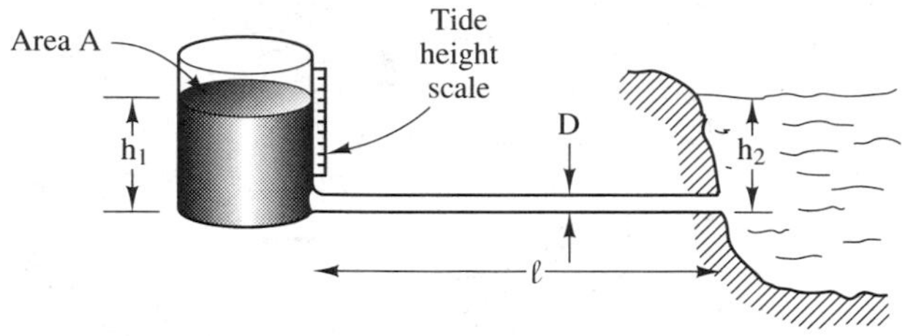

Figure P5.26

(b) Use the transfer function to evaluate the performance of the system with the initial design parameters. (Hint: The system is linear. By the superposition principle, we can examine the effects of the system on each source-signal component separately. Find the frequency response Lm(H). Use it to assess the extent to which the two sinusoidal signal components are attentuated while traversing the system.)

(c) By changing ℓ, D, or A we can adjust the corner frequencies of Lm(H). For the given values of ℓ and D, choose the value of A to satisfy the system performance specifications.

5.27 Suppose G(s) is the input-output transfer function for a unit-prototype second-order system. Correlate the features of the frequency response and the features of the step response with the locations of the poles. Specifically:

(a) Make a table that compares the values of the resonant peak M_r, the quality of resonance Q, and the fractional overshoot f_{OS}, as a function of the damping ratio ζ.

(b) Let $\zeta = 0.2$ and make a table that compares the values of the resonant frequency ω_r and the time-of-peak t_p as a function of the natural frequency ω_n.

5.28 Use the defining integral to find the Laplace transform of one of the following functions.
(i) $f(t) = \cos 30t$ **(ii)** $f(t) = \cosh 2t$ **(iii)** $f(t) = \delta(t - t_0)$

5.29 Use the defining integral and integration by parts to find the Laplace transform of one of the following functions.
(i) $f(t) = t$ **(ii)** $f(t) = te^{-2t}$ **(iii)** $f(t) = t \cos \omega t$

5.30 Find the Laplace transform of one of the following functions without using entries 5, 8, or 9 of Table 5.5. Use the properties in Table 5.6 as needed.

(i) $f(t) = e^{-4t}\cos(3000t)$
(ii) $f(t) = t \cos(50t)$
(iii) $f(t) = te^{-t}$
(iv) $f(t) = \cos(40t + 0.9)$
(v) $f(t) = \begin{cases} 0, & t < 1 \\ e^{-2(t-1)}, & t \geq 1 \end{cases}$
(vi) $f(t) = \begin{cases} 0, & t < 0 \\ e^{-2(t-1)}, & t \geq 0 \end{cases}$
(vii) $f(t) = e^{3t}\sin(2t)$
(viii) $f(t) = t \sin(30t)$
(ix) $f(t) = e^{-2t+3}$
(x) $f(t) = \cos(\omega_0 t + \phi)$
(xi) $f(t) = \sinh(2t)$

5.31 Find the Laplace transform of one of the waveforms shown in Figure P5.31.

5.32 Prove one of the following transform properties of Table 5.6.
(i) Property 10 **(iii)** Property 12
(ii) Property 11 **(iv)** Property 14

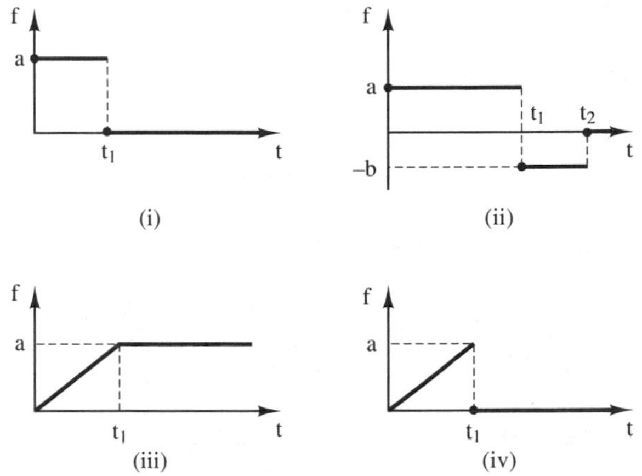

Figure P5.31

5.33 Use partial fraction expansion together with Tables 5.5 and 5.6 to find the inverse Laplace transform of one of the following functions.

(i) $F(s) = \dfrac{s + 2}{(s + 1)(s + 3)}$ **(ii)** $F(s) = \dfrac{s + 2}{(s + 4)^2 + 36}$

(iii) $F(s) = \dfrac{1}{s(s^2 + 100)}$ **(iv)** $F(s) = \dfrac{1}{s(s + 2)(s + 3)}$

(v) $F(s) = \dfrac{7s}{(s + 2)(s + 30)}$ **(vi)** $F(s) = \dfrac{2e^{-s}}{(s + 5)}$

5.34 Use Laplace transforms to solve one of the following input-output system equations. The source signal is the unit step, $v = u_s(t)$ m/s. All prior values (at $t = 0^-$) are zero. (These equations are derived from equations (3.92) and (3.93).)

(i) $\ddot{v}_3 + 15\dot{v}_3 + 36v_3 = 36v$

(ii) $\ddot{v}_3 + 6\dot{v}_3 + 36v_3 = 36v$

(iii) $\ddot{v}_2 + 15\dot{v}_2 + 36v_2 = 36\dot{v}$

(iv) $\ddot{v}_2 + 6\dot{v}_2 + 36v_2 = 36\dot{v}$

5.35 For one of the following input-output system equations:

(a) Find the Laplace transform $\mathbf{Y}(s)$.

(b) Use the transform from part (a) to find the initial value y_i (at $t = 0^+$) and the final value y_f ($t \to \infty$).

(c) Use the transform from part (a) to find $y(t)$ for $t \geq 0$.

(i) $\ddot{y} + 2\dot{y} + 2y = 1,\ y(0^-) = 1,\ \dot{y}(0^-) = 0$

(ii) $\ddot{y} + 5\dot{y} + 6y = 2,\ y(0^-) = 1,\ \dot{y}(0^-) = 0$

(iii) $\ddot{y} + 4\dot{y} + 13y = 1,\ y(0^-) = 1,\ \dot{y}(0^-) = 0$

(iv) $\ddot{y} + 7\dot{y} + 12y = t,\ y(0^-) = 1,\ \dot{y}(0^-) = 0$

(v) $\ddot{y} + 5\dot{y} + 6y = 2,\ y(0^-) = 0,\ \dot{y}(0^-) = 1$

5.36 The signal v_2 in an electric circuit satisfies the input-output system equation $\ddot{v}_2 + 3\dot{v}_2 + 2v_2 = v$, $t \geq 0$, where v is the source voltage. Let $\mathbf{V}_2$ and $\mathbf{V}$ denote the Laplace transforms of v_2 and v, respectively.

(a) Express $\mathbf{V}_2$ in terms of $\mathbf{V}$ and the prior values v_{2p} and $\dot{v}_{2p}$ at $t = 0^-$.

(b) Suppose the source signal has the constant value v_c. Inverse transform the expression found in (a) to find $v_2(t)$.

5.37 We can write the *zero-state* system equations for a network directly in terms of Laplace transforms of the variables. For one of the networks of Figure P5.37:

(a) View the unspecified variable of the source as the output signal. Find the input-output transfer function for that variable, directly in Laplace transform notation, in terms of the lumped-element parameters.

(b) Use the transfer function from part (a) to find the zero-state unit-impulse response and the zero-state unit-step response of the unspecified source variable.

(c) Suppose the energy storage element of the network contains one joule of prior stored energy (at $t = 0^-$). Use the transfer function from part (a) together with the derivative property of Laplace transforms to find the zero-input response of the unspecified variable of the source to that stored energy.

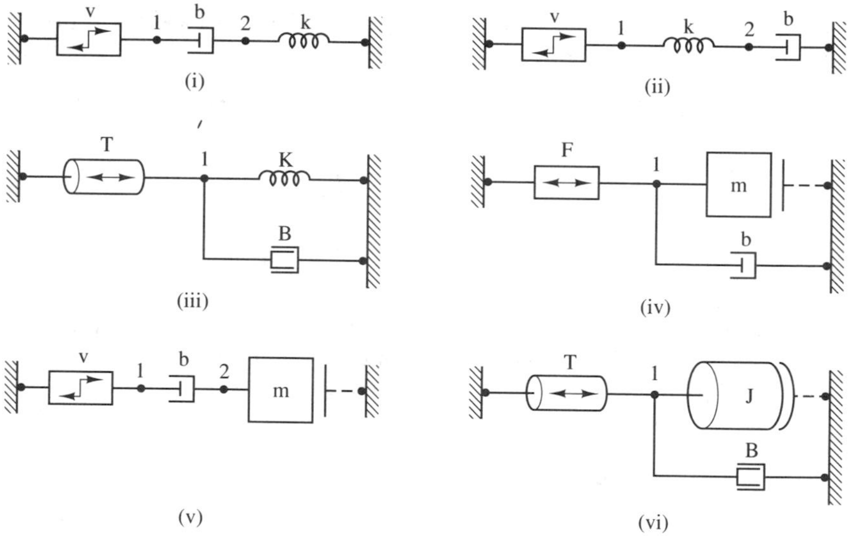

Figure P5.37

5.8 REFERENCES

5.1 Bode, H.W., Relations Between Attenuation and Phase in Feedback Amplifier Design. *Bell System Tech. J.* 19 (July 1940): pp 421–54.

5.2 Cromwell, L., Weibell, F. J., Pfeiffer, E. A., and Usselman, L. B., *Biomedical Instrumentation And Measurements.* Englewood Cliffs, NJ: Prentice Hall, 1973.

5.3 Fletcher, Harvey, *Speech and Hearing in Communication.* Princeton NJ: D. Van Nostrand, 1953.

5.4 V. L. Streeter, ed., *Handbook of Fluid Dynamics.* New York: McGraw-Hill, 1961.

5.5 Hwang, Ned H. C., and Hita, Carlos E., *Fundamentals of Hydraulic Engineering Systems,* Englewood Cliffs, NJ: Prentice Hall, 1987.

5.6 Li, John K-J., *Arterial System Dynamics.* New York: New York University Press, 1987.

5.7 Ogata, Katsuhiko, *System Dynamics.* Englewood Cliffs, New Jersey: Prentice-Hall, 1978.

5.8 Shearer, J. L., Murphy, A. T., and Richardson, H. H., *Introduction to System Dynamics,* Reading, MA: Addison-Wesley, 1967.

5.9 Skilling, H. H., *Electrical Engineering Circuits.* New York: Wiley, 1958.

5.10 Wylie, C. R., *Advanced Engineering Mathematics.* New York: McGraw-Hill, 1975.

5.11 Yeaple, Franklin D. (ed.), *Hydraulic and Pneumatic Power and Control.* New York: McGraw-Hill, 1966.

CHAPTER

6

Coupling Devices

The source signal that energizes a system typically originates in the sensing of a potential or flow in some other system. Usually, the sensed signal is relatively weak, so it must be conditioned in various ways before it can energize the system. *Signal-conditioning* operations adjust signal levels, power levels, energy forms, and signal locations to match the demands of the load. The various signal-conditioning functions can, in principle, be carried out in any order.

A race car driver, for example, uses foot motions that are sensed by an accelerator pedal to tell the car when and by how much to accelerate. Although the pedal-deflection signals are relatively weak, they control the engine fuel supply, which in turn determines the engine torque. The pedal-controlled engine, then, *strengthens* the pedal-deflection signals. The transmission and drive wheels convert the powerful torque signals to the force signals that propel the car.

We can view the sensed subsystem (race car driver) as a two-terminal device in which one terminal (the foot) moves relative to the other (the ground). We can also treat the energized load (race car) as a two-terminal device in which one terminal can be moved relative to ground. The sensor (accelerator pedal) is a *two-port* (four-terminal) device, with an input signal (foot motion) and an output signal (motion of the pedal linkage). The signal strengthener (engine) and the other signal conditioners (transmission and wheels) are also two-port devices. We must understand each of these two-port devices in order to design the system.

A tape deck, amplifier, and loudspeaker form a sound system. The tape transport system moves a prerecorded magnetic tape across a gap in the tape head. The tape transport system is the sensed system, and the tape head is the sensor. The tape head converts the changing magnetic field in the head gap (owing to motion of the recorded tape) to a proportional electric voltage. However, the voltage signal is too weak to evoke audible sound from a loudspeaker, so we use the amplifier to strengthen the voltage signal. The signal that emerges from the amplifier is powerful enough to drive the loudspeaker. The loudspeaker response waveform is a replica of the sensed voltage waveform.

We use the term *transducer* to mean a device, such as an electric motor, that converts source energy to the *form* needed by a load. We use the term *transformer* to mean a device, such as a gear, that adjusts source signals to the *level* needed by the load. Transformers and transducers are passive; that is, the signal power at their output port is no greater than at their input port. We call both of these passive devices *converters*. We examine converters in section 6.2.

Independent source signals typically originate in the measurement of some quantity in another system. The sensor that makes the measurement produces a very weak signal, too weak to energize a load directly. Therefore, we insert a *power amplifier* between the sensor and the load to strengthen the measured signal. A power amplifier is a valvelike device in which a low-power signal controls a high-power signal. The output signal power of an amplifier is greater than the input signal power. Section 6.3 explores power amplifiers.

Sometimes a source and load are physically separated, and it is necessary to use a cable, pipe, pushrod, transmission line, or other type of device to transport signals from the source to the load. We call these devices *transporters*. A transporter is inherently distributed; hence, it introduces *transport delay,* which reduces the stability of the system. We examine transporters in section 6.4.

Sources and loads are single-port subsystems. A *linear* load can be represented by a single equivalent impedance. A linear source can be represented by an ideal source and an equivalent impedance. We found in section 4.2 that we can represent each half of a linear network by one of two simple equivalent subnets, each containing a single impedance and an ideal source. In section 6.5, we partition a linear network into three sequential subnets, the middle subnet of which has four terminals, i.e., two ports. A two-port subnet can also be represented by a simple equivalent subnet. We shall find that there are six different equivalent representations. We examine these representations in section 6.5.

Models for two-port devices permit the analysis and design of complete systems—for example, a water distribution system, a sound reproduction system, or the drive train of an automobile. External **feedback** can be used in any of these systems to increase its responsiveness and to decrease its sensitivity to changes or uncertainties in the system parameters. However, the careless use of feedback can cause a system to become unstable. We examine the benefits and stability problems inherent in feedback in section 6.6.

We begin the chapter by developing lumped-modeling procedures for two types of systems that are quite distinct from the mechanical, electrical, and hydraulic systems of the previous chapters. The first type is thermal systems. Thermal flow—that is, heat flow—is thermal power; hence, we must treat power-related issues differently for thermal systems. The second type is systems of populations, such as people, manufactured items, or dollars. When we model population systems, we extend the concepts of potential and flow far beyond the usual technical realm, into ecology, economics, demographics, and similar fields.

6.1 LUMPED MODELS FOR DIVERSE SYSTEMS

The lumping process used for mechanical, electrical, and hydraulic systems in the previous chapters can be extended to other types of systems as well. When we apply the process to thermal systems and population systems, we find structural differences that have major impacts on their behaviors.

We examine thermal systems first. In a thermal system, the flow variable is heat flow, with units of power flow. Hence, the relationships among potential, flow, and power that are common to the previous systems we have examined do not apply to thermal systems. There is no thermal element that maintains flow in the manner of the electrical inductor, the mechanical spring, and the hydraulic inertance. Therefore, there are no natural oscillations of thermal origin, and the response behaviors of thermal systems are always overdamped.

The second class of systems examined in this section is population systems. In a population system, we are interested in the number of objects in each of several domains—geographical regions or spheres of association. Bacteria, rabbits, people, dollars, and manufactured items are examples of such objects. These objects migrate from domain to domain according to natural laws or deliberate policies. Population systems exhibit characteristics that are not typical of all physical systems. One such characteristic is *directionality:* Migrations from domain to domain tend to favor one direction over another. Many population systems exhibit *exponential growth,* another nontypical characteristic. Exponential growth cannot be sustained, of course. Hence, population systems that grow exponentially always undergo nonlinear changes that limit that growth.

Thermal Variables: Temperature and Heat Flow

The temperature of an object is a measure of the mean kinetic energy of the molecules of the object. Since a molecule cannot have negative kinetic energy, the absolute temperature of an object cannot be negative. Therefore, positivity of temperature and the zero value of temperature have well-accepted definitions. The SI unit of temperature is the kelvin (abbreviated K). However, the Celsius scale is often used instead of the kelvin scale. The Celsius scale is a *relative* scale that has units of the same *size* as the kelvin scale. That is, a 1°C change in temperature is a change of 1 K.

We denote the temperatures of points in a system by the symbols θ_i. **Temperature is a potential;** it obeys the compatibility condition (or **path law**) given by equation (2.28). Derivatives and integrals of temperature are also potentials.

We measure temperature in a variety of ways. The most familiar device for measuring temperature is a thermometer. Most thermometers convert the thermal expansion and contraction of a liquid or solid into a visible mechanical displacement. In a resistance thermometer, the thermally induced change in electrical resistance of an object is measured electrically. A pair of dissimilar metal wires, twisted together at the ends, generates a voltage between the two junctions that depends on the temperature difference between the junctions. We use this phenomenon in a temperature-measuring device called a *thermocouple.* At very high temperatures, objects emit light of a color that depends on temperature. A colorimeter uses that color to measure the temperature.

If two objects of different temperature are brought in contact with each other, the temperature of the hotter object decreases and that of the cooler object increases. These changes in temperature are explained by a transfer of heat energy from the hotter object to the cooler object. Let us represent the quantity of heat in an object by the symbol H. Let $q \triangleq dH/dt$ denote the rate of flow of heat into the object.

According to the first law of thermodynamics, **heat flows obey the node law,** equation (2.27); that is, energy is conserved. Therefore, q is a flow in the mathematical sense discussed in section 2.2. It follows that its integral H must also be a flow.

The flow of heat between two points in a system is related to the temperature difference between the points in the much the same way that a flow of liquid is related to pressure difference or an electric current is related to voltage difference. There are fundamental differences between thermal systems and the systems studied in previous chapters, however. In each of the previously examined energy domains, the power delivered to an object is the product of the potential drop across the object and the flow through the object. But heat H is, itself, thermal energy; hence, **heat flow q is a flow of thermal power.**

We shall see shortly that physical materials exhibit only two thermal characteristics: a resistance to the flow of heat and a capability to store heat (i.e., to accumulate the flow of heat). These two characteristics are analogous to the resistive and capacitive phenomena of other energy domains. Physical materials do not exhibit a tendency to *sustain* heat flow. That is, there is no heat analog of fluid inertance, electrical inductance, or mechanical compliance.

The amount of heat H delivered to an object is treated as positive if it causes the temperature of the object to rise. The SI unit of heat is the joule (denoted J). It is defined as the amount of heat necessary to raise the temperature of 1 kg of water by 1 K (or 1°C). Another unit of heat in common use is the kilocalorie (abbreviated kcal); 1 kcal = 4,187 J. The SI unit for the heat flow rate q is the watt (abbreviated W), where 1 W = 1 J/s.

It is more difficult to measure the *flow* of heat through a portion of a system than it is to measure the temperature at a point in the system. We usually measure a heat flow rate by inserting an object of known thermal behavior in the flow path and measuring the temperature drop across the object.

Thermal Resistance

There are three different mechanisms for heat transfer: conduction, convection, and radiation. In each, heat flows from hot objects to cold objects, and the rate of heat flow increases with the temperature difference. We represent the relation between temperature difference and flow by a thermal resistance.

The flow of heat through a solid occurs by **conduction**, in which kinetic energy is passed from molecule to molecule. According to Fourier's law, the conducted heat flux (heat flow per unit area) is oriented in the direction of the greatest rate of decrease in temperature and is proportional to that rate of decrease. (The proportionality factor k_t is known as the *thermal conductivity* of the material.) In the object shown in Figure 6.1, heat moves from high temperature to low temperature. We portray the paths of heat flow by *streamlines.* No flow crosses a streamline. A surface that is perpendicular to a set of streamlines is a surface of equal temperature, called an *isothermal surface* (or **isotherm**). If a surface perpendicular to the streamlines were not an isotherm, the temperature differences on the surface would cause flow to cross the streamlines.

It can be demonstrated experimentally that heat is conducted from an isotherm at temperature θ_i to another isotherm at temperature θ_j at a rate proportional to the temperature difference $\theta_i - \theta_j$. We represent this conductive heat flow relation in the form

$$\theta_{ij} = R_t q \tag{6.1}$$

where θ_{ij} denotes the temperature difference $\theta_i - \theta_j$ and the proportionality factor R_t is the **thermal resistance** of the object between the isotherms. Thermal resistance is an impedance—a ratio of potential (temperature) difference to flow. Its units are °C/W = °C·s/J. (We use the subscript t to distinguish thermal resistance from electrical resistance and fluid resistance. We omit this subscript if the context is clear.)

Equation (6.1) assumes that all of the heat that leaves isotherm i arrives at isotherm j; i.e., no heat is stored or converted to work. Unlike electrical and hydraulic resistors, thermal resistors do not dissipate energy; they merely accommodate the transport of energy from higher to lower temperatures.

We define an **ideal thermal resistor** to be a lumped element that obeys the linear temperature-flow relation of equation (6.1). We represent a thermal resistor by the symbol shown in Figure 6.2. The two nodes represent isothermal surfaces of temperatures θ_i and

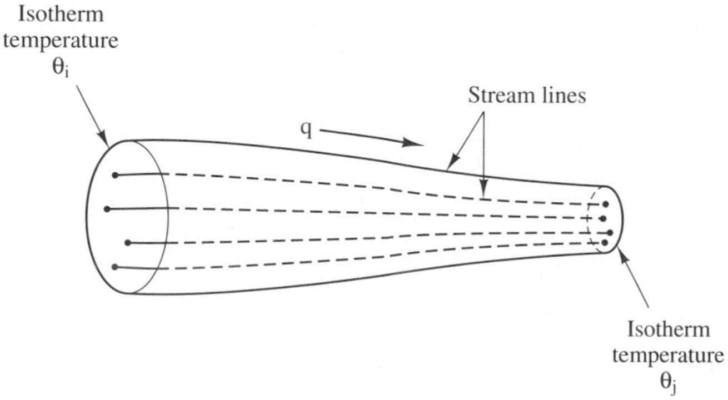

Figure 6.1 Conductive heat flow in a solid object.

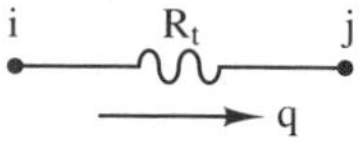

Figure 6.2 An ideal thermal resistor.

θ_j. The arrow beside the lumped element designates the positive direction chosen for the flow q. The sign convention for the resistor equation (6.1) is to order the subscripts on the temperature difference in the direction of q.

Although we have used the mechanism for heat conduction to introduce thermal resistors, we shall find shortly that heat convection also obeys the ideal resistor equation (6.1). Heat radiation, too, can be represented by equation (6.1), although the resistance varies with temperature. Hence, we use thermal resistors to represent all three heat transfer mechanisms.

A homogeneous object with parallel planar isotherms is shown in Figure 6.3(a). It follows from Fourier's law that the thermal resistance of the object is proportional to the length (or thickness) of the object and inversely proportional to the cross-sectional area, as indicated in the diagram. The proportionality factor is the *thermal resistivity* of the material (the inverse of the *thermal conductivity* k_t).[1] The thermal resistances of objects with other configurations are also shown in the figure. Heat flow is proportional to the temperature difference across the object for each configuration.

It is apparent from the formula for the resistance of the object of Figure 6.3(a) that thermal resistors follow the rules derived in section 4.2 for combining series and parallel impedances. Resistances add for objects in series, while conductances (inverse resistances) add for objects in parallel. When two solid objects are connected in series, however, the thermal contact is often imperfect. The thermal resistance of very small air spaces can be appreciable. We account for such air spaces by an additional **contact resistance.** Contact resistance is best measured experimentally because the dimensions of the air spaces are quite uncertain.

We denote the thermal conductivity of a material by k_t. We call a material that has a very high thermal conductivity a *conductor*. Materials that are good electrical conductors are usually good thermal conductors. An *insulator* is a material that has a very low thermal conductivity. The best thermal insulator is dry air. For this reason, most commercial insulating materials suspend small pockets of air in a supporting medium.

Let us demonstrate these thermal conduction concepts by an example. Suppose that the dimensions of a particular mobile home are 8 m $\times$ 3 m $\times$ 2.2 m and that the exterior walls, floor, and ceiling contain a 2.5-cm-thick layer of polyethylene foam. That foam provides the only significant resistance to heat flow out of the home. Let the outside temperature be 0°C (32°F) and the inside temperature be 21°C (70°F). Let us calculate the heat loss to the outdoor air owing to conduction.

Suppose it is windy outside. Then the temperature of the outside surface will be nearly the ambient air temperature. Assume that the mobile home is empty of carpets or furnishings and that a home heater blows the warm air against the inside surfaces. Then the temperature of the inside surface will be nearly the same as the room temperature. We represent the whole outside surface by node 1 and the inside surface by node 2. The heat flow q through the exterior of the mobile home is related to the temperature difference θ_{21} by equation (6.1). We must calculate the resistance R_t of the exterior of the mobile home.

[1] The thermal conductivities of various materials are given in appendix A.

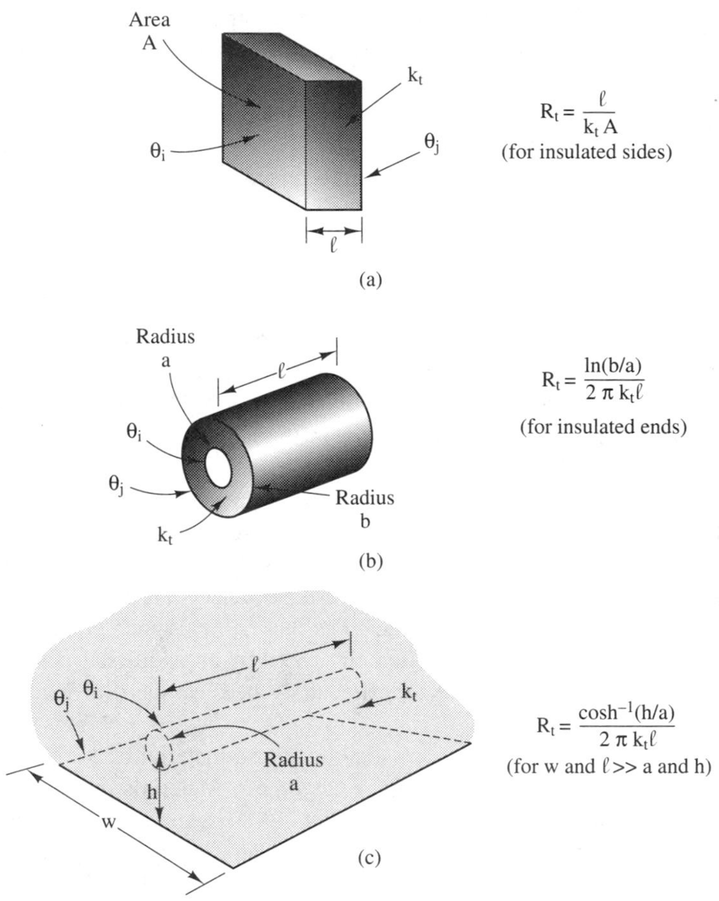

Figure 6.3 Thermal resistance to conduction between two *perfect* conductors (isotherms) through a material of conductivity k_t; compare with the electrical configurations of Figure 4.11. (a) Parallel planar conductors. (b) Coaxial cylindrical conductors. (c) Planar conductor and parallel rod-shaped conductor. (Continued)

The formula for the resistance of the exterior is given in Figure 6.3(a). (The six perpendicular surfaces through which the heat passes act as one surface—i.e., as resistors in parallel.) The length of the heat-flow streamlines is the wall thickness, 0.025 m. The total area through which the heat flows is 96.4 m^2. According to appendix A, the thermal conductivity of polyethylene is 0.026 W/m°C (about the same as for still air). The resistance of the exterior, then, is

$$R_t = \frac{0.025 \text{ m}}{(0.026 \text{ W/m°C})(96.4 \text{ m}^2)} = 0.01\text{°C/W} \tag{6.2}$$

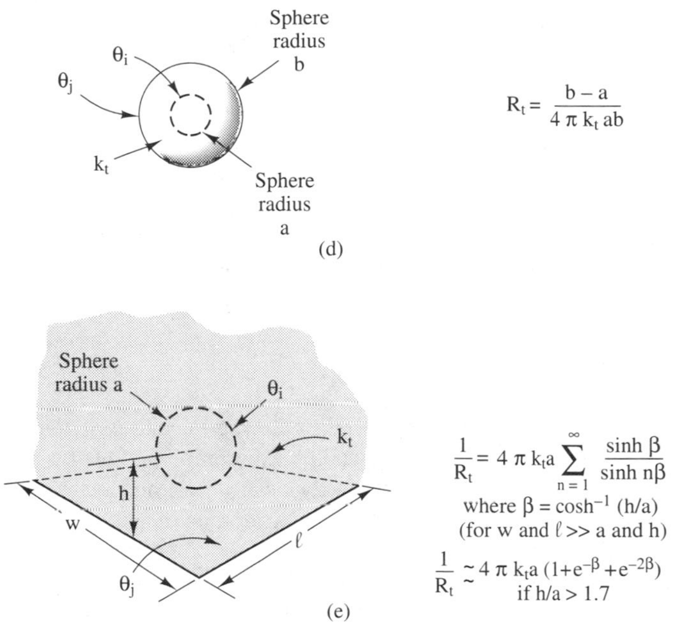

Figure 6.3 (Continued) (d) Concentric spherical conductors. (e) Spherical conductor next to a planar conductor.

Therefore, the heat loss is

$$q = \frac{\theta_{21}}{R_t} = \frac{21°\text{C}}{0.01°\text{C/W}} = 2.1 \text{ kW} \tag{6.3}$$

This is about the amount of heat delivered by two small electric space heaters.

Convection and Radiation

The heat conduction process just described also applies to *stationary* liquids and gases. Unless some mechanism (such as suspension of air pockets in polyethylene foam) holds the fluid stationary, however, conductive heating of the fluid next to a hot object causes the fluid to expand and then to rise owing to buoyant forces. The displaced fluid is replaced by cooler fluid, which, in turn, heats and rises. The heated fluid eventually comes in contact with a cooler object, loses heat, and falls.

The transfer of heat from a hot object to a cooler object by circulation of a fluid is called **convection**. Convection is the mechanism for heat transfer between the air in a room and the walls and ceiling of the room and between the outside walls and roof of a building and the surrounding air. It is also the mechanism by which the heat in the air of a room is transferred to the coils of an air conditioner. Convective heat transfer is more effective than conductive heat transfer; i.e., the effective thermal resistance of the fluid is lower because the fluid circulates. The convection can be *natural* (or free), as just described for the walls and ceiling, or *forced,* as when it is blown by a fan across the coils of an air conditioner. Increasing the fluid flow rate decreases the effective thermal resistance between the bulk of the moving fluid and the surface to or from which the heat transfers.

The primary resistance to convective heat transfer lies in the *thermal boundary layer,* a thin, near-stationary layer of fluid adjacent to the solid surface. Convective heat transfer obeys the linear temperature-flow relation of equation (6.1) if the fluid flows at a constant rate. One of the variables θ_j and θ_i in that equation denotes the surface temperature of the solid object. The other variable denotes the temperature in the moving fluid far beyond the boundary layer. The convective thermal resistance is inversely proportional to the area of the fluid-bathed surface through which the heat flows.

We represent the thermal resistance for convective heat transfer by

$$R_t = \frac{1}{h_t A} \tag{6.4}$$

where A is the surface area exposed to convection and h_t is the **heat transfer coefficient.** The heat transfer coefficient depends on the properties of the heat transfer fluid (density, viscosity, thermal conductivity, and specific heat) and on the velocity of the fluid. It is best determined experimentally for specific situations.[2]

To demonstrate these heat convection concepts, let us find the heat loss through the window of a home. Suppose the area of the window is 2.6 m^2 and the glass is 5 mm thick. Let the inside temperature be $\theta_1 = 21°C$ and the outside temperature be $\theta_2 = 0°C$.

The room air circulates naturally against the inside surface of the window. Hence, heat transfers to the glass by convection and is conducted through the glass to the outside surface. At that surface, the heat is transferred to the outside air by convection—natural circulation and wind. Thus, we can represent the heat transfer by three resistances connected in series. We must calculate these three resistances.

According to appendix A, the thermal conductivity of glass is 0.75 W/m°C. The heat transfer coefficient for natural air convection varies from 5 to 25 $W/m^2°C$. The coefficient for forced air convection varies from 10 to 200 $W/m^2°C$. Let us assume that the air is relatively undisturbed inside and use $h_{tin} = 5$. Because of the presence of winds outside, we use $h_{tout} = 50$.

The convection resistance given by equation (6.4) and the conduction resistance shown in Figure 6.3(a) are both inversely proportional to the cross-sectional area A of the window. We can express the total thermal resistance of the window as the sum of three components:

$$\begin{aligned} R_t &= \frac{1}{A}\left(\frac{1}{h_{tin}} + \frac{\ell}{k_t} + \frac{1}{h_{tout}}\right) \\ &= \frac{1}{A}\left(\frac{1}{5\ W/m^2°C} + \frac{0.005\ m}{0.75\ W/m°C} + \frac{1}{50\ W/m^2°C}\right) \\ &= \frac{1}{A}(0.2 + 0.0067 + 0.02) m^2°C/W \\ &= \frac{0.227 m^2°C/W}{A} = 0.087°C/W \end{aligned} \tag{6.5}$$

[2] Approximate heat transfer coefficients for water and air are given in appendix A.

Then we use equation (6.1) to find the heat loss:

$$q = \frac{\theta_{12}}{R_t} = \frac{21°C}{0.087°C/W} = 240 \text{ W} \tag{6.6}$$

Note that the most significant resistance to heat loss through the window lies in the inside convection boundary layer owing to the relatively still air in the room. If the air is circulated in the room, the heat loss will be increased greatly.

Exercise 6.1: It is apparent from equation (6.5) that we can represent the series of three resistances by a single equivalent *conductive* resistance. Show that the *effective* thermal conductivity of the window, in terms of the 5-mm thickness of glass, is $k_t = 0.022$ W/m°C, about the same as still air of the same thickness.

According to the Stefan-Boltzmann law, every object radiates electromagnetic energy in proportion to the fourth power of its absolute temperature. The specific amount of energy radiated from the surface of the object depends on the effective surface area and the nature of the surface. Every object also absorbs part of the radiated energy that strikes it. The net energy transfer from a warm body at temperature θ_i to a cooler body at temperature θ_j is

$$q = \xi(\theta_i^4 - \theta_j^4) \tag{6.7}$$

where the temperatures are expressed in K and the *radiation coefficient* ξ depends on the shape, surface area, and surface characteristics of the two objects.

The radiaton coefficient ξ is quite small. For example, if a perfectly radiating (black) spherical object of temperature $\theta_i = 300$ K (room temperature) radiates into a completely absorbing environment (of temperature $\theta_j = 0$ K), the radiation coefficient is $\xi = \sigma A$, where $\sigma = 5.67 \times 10^{-8}$ W/m^{2}K^4 (the Stefan-Boltzman constant) and A is the area of the sphere. If the area of the sphere is 1 m^2, then the radiated power is $q = 459$ W. However, 300 K is a relatively large temperature difference, compared with most temperature differences observed on the earth. Unless the temperature *difference* is unusually large, radiative heat transfer between objects on the earth is usually negligible, compared with other heat flows in the system.

The temperature-flow relation (6.7) is very nonlinear. If radiative heat transfer is not negligible, it may be practical to linearize equation (6.7) in the manner described for quadratic fluid resistors in section 5.2. However, we must recognize that the resulting thermal resistance changes rapidly as a function of the two temperatures.

The sun is extremely hot relative to temperatures of objects on the earth. Consequently, we can treat the sun as a constant heat source. The **solar constant,** 1.353 kW/m^2, is the amount of solar radiation received outside the earth's atmosphere on a surface normal to the incident radiation. The average radiation actually received at the earth's surface *through a cloudless atmosphere* varies from 0 to 422 W/m^2, depending on the latitude and season (surface orientation and length of day). The atmospheric transparency, cloud cover, shading from trees, and color and surface texture of objects on the earth affect the actual transfer of heat to those objects.

Thermal Capacitance

The ability of an object to store heat is indicated by its specific heat c_p.[3] Specific heat has the units J/kg°C. Let the subscript i refer to a homogeneous object with temperature θ_i and mass m (in kg). If the temperature of the object is raised by $\delta\theta_i$ (in °C), then the heat stored in the object (in joules) increases by $\delta H = mc_p\,\delta\theta_i$. This heat-storage relation is equivalent to the formula

$$q = mc_p\dot{\theta}_i \tag{6.8}$$

where q is the rate of flow of heat *into* the object.

Equation (6.8) is not quite analogous to the element equations of the previous chapters, because it is not expressed in terms of a potential *difference*. However, let us designate a reference point of constant temperature θ_0 for the system in which the object is imbedded. (For example, we could choose a point at the ambient temperature or a fictitious point at zero temperature.) Then, since θ_0 is constant, $\delta\theta_{i0} = \delta\theta_i$. If we now substitute the potential difference θ_{i0} for θ_i in equation (6.8), we can restate the equation in impedance notation as

$$\theta_{i0} = \left(\frac{1}{sC_t}\right)q \tag{6.9}$$

where s is the time-derivative operator and

$$C_t = mc_p \tag{6.10}$$

is the **thermal capacitance** of the object.

According to equation (6.8), we can continue removing heat from an object as long as we continue lowering its temperature. Since we cannot lower the temperature of an object below absolute zero, we can view the *absolute* temperature of an object as specifying the absolute amount of heat stored in the object. Specifically, we integrate equation (6.8) with respect to time to produce

$$H = \int q\,dt = mc_p\theta_i \tag{6.11}$$

Then $mc_p\theta_{i0}$ is the amount of heat necessary to raise the temperature of the object from the reference temperature θ_0 to temperature θ_i; we shall speak of it as the amount of stored heat *relative to the heat stored at the reference temperature.*

Equation (6.8) assumes that the temperature in the object is uniform, i.e., that all points of the object experience essentially identical temperature variations. If the object is not a good thermal conductor, we can subdivide it into small equal-temperature chunks. Then equation (6.8) relates to a single chunk. Highly conductive objects and stirred fluids behave very much in accordance with this equation.

[3] Values of c_p for various materials are listed in appendix A. The subscript p implies constant pressure during heat transfer. Pressure is an important variable during heat transfer to and from compressible gases.

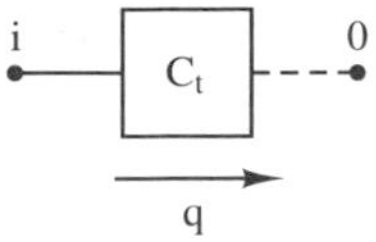

Figure 6.4 An ideal thermal capacitor.

We define an **ideal thermal capacitor** to be a lumped element that obeys the linear temperature-flow relation of equation (6.9). We represent the ideal thermal capacitor by the symbol shown in Figure 6.4. The dashed line associates the capacitor with the reference node. We use an arrow beside the lumped element to designate a positive direction for the flow q through the capacitor. The sign convention for the capacitor equation (6.9) is to order the subscripts on the temperature difference in the direction of q. If we orient q toward the reference node, then a positive value of q corresponds to a rise in θ_i.

Mathematically, the lumped capacitor behaves as if the flow q leaves node i and enters the reference node. In actuality, there is no flow of heat into the reference node. Yet, according to the node method for writing network equations (see section 2.2), the flows departing toward the reference node from adjacent nodes must sum to zero. Therefore, the fictitious heat flows into the reference node sum to zero. Intuitively speaking, this means that the heat generated equals the heat entering storage at each instant.

The thermal capacitance of an object is proportional to its specific heat and its mass (or specific gravity). (The specific heats and specific gravities of various materials are listed in appendix A.) The material with the greatest specific heat is water, which is also the material with the greatest product of specific heat and specific gravity. This fact explains the importance of the oceans of the earth in stabilizing and changing temperature.

Ideal Thermal Sources

Suppose a thermostatically controlled burner maintains a constant water temperature in a boiler, and the boiler water is circulated through the heat exchanger for a room. Let us view the water in the boiler and heat exchanger as node i (at temperature θ_i). We can treat the heat exchanger as a constant-temperature source. When the room draws heat from the heat exchanger, the thermostat turns on the burner at the boiler to keep the temperature of the water from falling. The burner delivers heat to the boiler at a rate q equal to the rate at which the room draws heat from the heat exchanger. We can treat that heat as if it comes from the reference node (at temperature θ_0).

This example guides our definition of a temperature source. We define an **ideal independent temperature source** to be a thermal source that maintains a specified temperature difference $\theta_{i0} = \theta(t)$ relative to the reference node, regardless of the thermal system that is connected to the source. (Hence, one terminal of a temperature source must be the reference node.) The power delivered by the source to the attached system is the heat flow q through the source in the direction from the reference node to node i. Figure 6.5(a) shows the symbol that we use to represent an ideal temperature source.

Since the reference temperature θ_0 can be chosen arbitrarily, delivery of power by a thermal source does not require that the potential (temperature) rise across the source in the direction of the heat flow. This behavior is in contrast to the behaviors of mechanical, electrical, and hydraulic systems. We shall find that the reference temperature does not play any direct role in heat flow calculations.

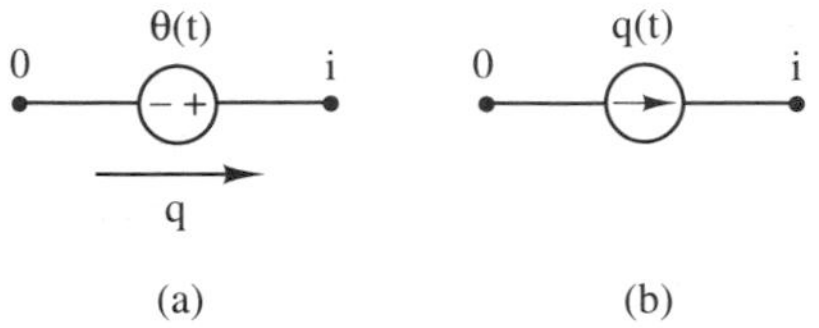

Figure 6.5 Independent thermal sources. (a) Temperature source. (b) Heat-flow source.

We define an **ideal independent heat-flow source** to be a thermal source that maintains a specified flow of heat q(t) *from the reference node* to node i, regardless of the thermal system which is connected to node i. We represent an ideal heat-flow source by the symbol shown in Figure 6.5(b). An electrical space heater is an example of a heat-flow source. The electric current heats a resistor in the heater. From the electrical viewpoint, the energy dissappears. (We say it is dissipated.) It reappears as heat in a thermal system. The rate of heat generation is controlled by the electric current. That is, the electrical power (i^2R) dissipated in the heater resistance R equals the heat flow q delivered by the heat-flow source to the attached thermal system.

Lumped Modeling of Thermal Systems

The process for lumped modeling of a thermal system parallels the processes for the other energy domains. The reference temperature in a thermal system need not be zero. Yet we can assign a temperature of zero to the reference node if we use relative temperature as the potential. We connect all thermal capacitors directly to the reference-temperature node. Thermal capacitance is analogous to mass and fluid capacitance, as noted in the comparisons of impedance in Table 4.3. We summarize the thermal modeling process here and illustrate the process in example problems 6.1 and 6.2.

THE LUMPING PROCEDURE FOR THERMAL SYSTEMS

1. Draw a pictorial diagram.
2. Identify good conductors, isothermal objects, or isothermal surfaces that have temperatures of significant interest or that have essentially constant temperature relative to their surroundings. Use a numbered node to represent each such item identified. One of these nodes must be the reference node, typically at ambient temperature. Place the nodes in approximate correspondence to the layout in the pictorial diagram. Denote the reference temperature by the symbol θ_0.
3. Insert lumped thermal resistors between nodes that are connected by objects with significant thermal conductance. Insulating materials are often a focus of interest because thermal systems are usually designed to conserve energy.
4. Attach lumped thermal capacitors to those nodes which represent objects of high mass and/or high specific heat. Thermal capacitors must be connected to the reference node, which need not be isolated at a single location.
5. Insert ideal temperature sources or heat-flow sources between appropriate nodes and the reference node. (All thermal sources must connect directly to the reference node.)

6. Redraw the lumped network to obtain a convenient layout. Treat all points connected by solid lines as a single node.
7. Establish and label a positive heat-flow direction for each branch of the network.

The procedures for drawing line-graph representations of lumped models and for formulating system equations for thermal systems are identical to the procedures used for mechanical, electrical, and hydraulic networks. In line-graph representations of networks, thermal elements can be distinguished from other kinds of elements by the unique symbols R_t, C_t, θ, and q or by means of their unique units.

Example Problem 6.1: Insulation of a hot water tank.
A cylindrical water heater of diameter D = 0.6 m and length L = 1.3 m has a 1.3-cm thickness of polyethylene foam insulation in its exterior shell. The water in the heater is maintained at 50°C. The room in which the heater resides is maintained at 21°C.
(a) Determine the rate at which heat transfers from the heater to the room.
(b) It is proposed to enclose the heater in a 5-cm-thick glass wool insulation jacket to reduce the heat loss. Determine whether or not it is worthwhile to add the insulation jacket.
Solution: **(a)** Let us choose the interior of the heater tank as node 0 (the reference node) of a lumped model, with temperature 50°C. We treat the exterior surface of the heater tank as a node 1 of the model, with temperature 21°C. We connect a 29°C constant-temperature source between the nodes (with the + symbol at node 0) to maintain $\theta_{10} = -29°C$. The resistance of the insulation between the two surfaces is determined from the formula in Figure 6.3(a). The streamline length ℓ is the 0.013-m thickness of the insulation between the two surfaces. The area of the isothermal surfaces represented by the nodes is the total area of the cylinder and its two ends, i.e.,

$$A = 2\left(\frac{\pi D^2}{4}\right) + \pi DL = \frac{\pi(0.6 \text{ m})^2}{2} + \pi(0.6 \text{ m})(1.3 \text{ m}) = 3.01 \text{ m}^2 \tag{6.12}$$

The thermal resistance of the insulation is

$$R = \frac{\ell}{k_t A} = \frac{0.013 \text{ m}}{(0.026 \text{ W/m°C})(3.01 \text{ m}^2)} = 0.166\text{°C/W} \tag{6.13}$$

Therefore, we attach a thermal resistor of size R between nodes 1 and 0. The heat loss is

$$q = \frac{\theta_{01}}{R} = \frac{29\text{°C}}{0.166\text{°C/W}} = 175 \text{ W} \tag{6.14}$$

(b) With the added insulation jacket, there are three isothermal surfaces in the system: The interior surface of the tank, the exterior surface of the tank, and the exterior surface of the insulation jacket. We treat each as a node of the lumped model. The thermal resistance of the added insulation jacket is in series with the thermal resistance of the tank shell.

The cross-sectional area of the insulation jacket (the area through which the heat flows) is essentially the same as that found in equation (6.12). The thermal conductivity of glass wool, from appendix A, is 0.04 W/m°C. Therefore, the resistance of the insulation jacket is

$$R_{jacket} = \frac{\ell}{k_t A} = \frac{0.05 \text{ m}}{(0.04 \text{ W/m°C})(3.01 \text{ m}^2)} = 0.415\text{°C/W} \tag{6.15}$$

The total resistance of the two insulation layers in series is the sum of equations (6.13) and (6.15), i.e., $R_{total} = 0.581°C/W$. The heat loss with the jacket in place is

$$q = \frac{\theta_{01}}{R_{total}} = \frac{29°C}{0.581°C/W} = 50 \text{ W} \tag{6.16}$$

To determine whether it is worthwhile to add the insulation jacket, we must compare the cost of adding the jacket with the savings owing to reduced heat loss. That comparison depends on the current cost of heating the water and the current cost of labor and materials to add the insulation jacket. The total heat saved in one year is (75 W) (24 hours/day) (365 days/yr) = 657 kW/yr. Suppose the water heater is electrically heated. Then if the cost of electrical power were \$0.10/kWhr, the savings would be \$65.70 per year.

Example Problem 6.2: A distributed baseboard-heater model.
Figure 6.6 shows the layout of two rooms of a home. The rooms are identical in size and construction. They also have the same resistance to heat loss to the outside air. The rooms are 2.3 m in height. The ceiling is insulated with 15 cm of glass wool. The walls are insulated with 5 cm of glass wool. The home has a hot-water heating system, one circuit of which heats the two rooms in the figure. Other rooms of the home are heated by separate heat-exchanger circuits.

The heat exchanger for the two rooms runs along the baseboard of the outside walls. It consists of a copper pipe with inside diameter 1.9 cm and wall thickness 1 mm. In each room, there are 120 5-cm × 10-cm × 0.5-mm aluminum heat-exchanger fins attached to the copper pipe every 0.6 cm. Water from a 75°C constant-temperature boiler is circulated at a flow rate of Q = 0.075 liter/s through the pipe of the heat exchanger. A thermostat in room 1 turns on the boiler-water circulator (pump) whenever the temperature falls below the thermostat setting. It turns off the circulator when the temperature at the thermostat reaches the set value.

When the circulator turns on, it takes about 60 s for the hot boiler water to travel from the inlet to the outlet of the heat exchanger. (As a result, heat reaches the second room 30 s

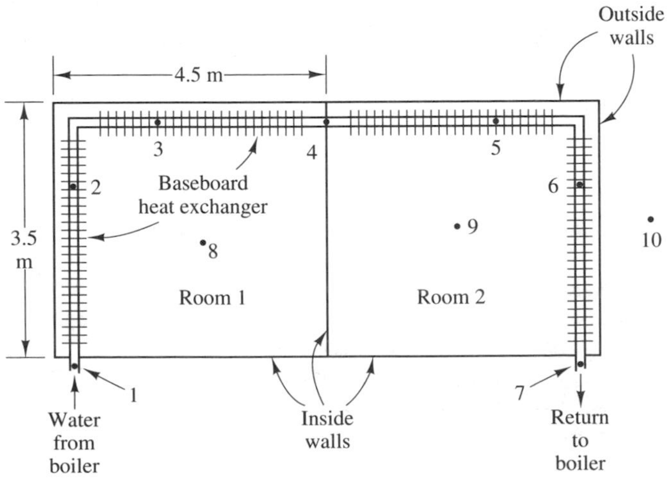

Figure 6.6 Baseboard hot-water heating system.

after it reaches the first room.) In each segment of the heat exchanger, heat stored in the hot water flows by convection to the pipe (and the heat exchanger fins) and raises the temperature of the fins to the water temperature. (There is a slight lowering of water temperature for that segment in the process.) The time constant for the transfer of heat from the water to the fins is about 2 s. (We derive this time constant in problem 6.6 at the end of the chapter.) Hence, about 65 s after the thermostat setting is increased, the last segment of the heat exchanger reaches the temperature of the water in that segment, about 75°C.

(a) Make a lumped thermal model of the rooms and heating system. Determine the values of the parameters of the lumped model.

(b) Suppose the outdoor temperature is 0°C and the temperature in the two rooms (owing to the thermostat setting) is 15.5°C (60°F). If we change the thermostat setting to 21°C (70°F), how long does it take for room 1 to reach that temperature and shut off the circulator pump?

(c) At what rate must heat be generated to maintain the rooms at the temperature 21°C? What is the temperature of the water that returns to the boiler? What is the temperature in room 2 at the time the circulator is turned off? What can we do to correct the imbalance in room temperatures?

Solution: **(a)** Suppose the air mixes throughout each room, but not from room to room. Then we represent each room by a single node (nodes 8 and 9 of Figures 6.6 and 6.7) with a single temperature. We represent the capability of the air in each room to store heat by a single thermal capacitor (C_{a1} and C_{a2} of Figure 6.7). We represent the outside air by a single node as well (node 10). Note that node 0 is a reference node whose temperature we set arbitrarily to 0°C. We use the ideal temperature source θ_{amb} to represent the ambient temperature—0°C for this example. We represent the heat transfers from the two rooms to the outside by the two resistors R_{O1} and R_{O2}.

We represent the heat source—the constant-temperature boiler water—by an ideal temperature source, as shown in Figure 6.7. The source temperature is $\theta_B = 75$°C. As the boiler water travels through the heat exchanger, it loses heat to the surrounding air, and the water temperature drops. Since the temperature of the heat exchanger (and water) varies continuously along its length, we subdivide the heat exchanger into segments. We use three equal-length segments for each room. Nodes 1 and 7 of Figure 6.7 correspond to the boiler-water entrance and exit points. Node 4 corresponds to the point where the heat exchanger crosses from room 1 to room 2. The other nodes are equally spaced along the heat exchanger, nodes 2 and 3 in room 1 and nodes 5 and 6 in room 2. Associate with each of nodes 1–6 the 2.67-m segment of the heat exchanger between that node and the next node.

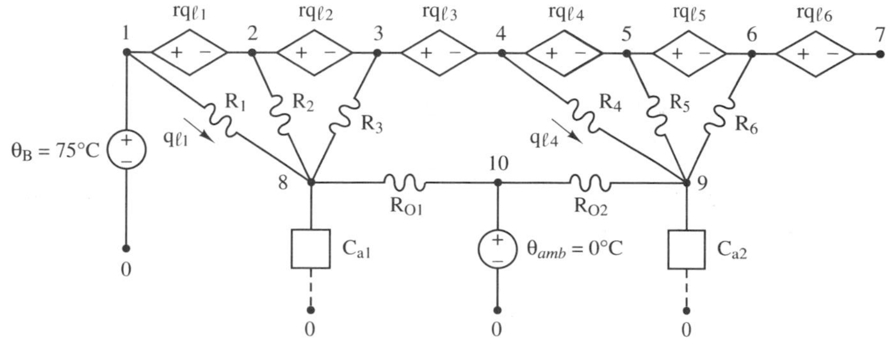

Figure 6.7 Lumped model of the heating system.

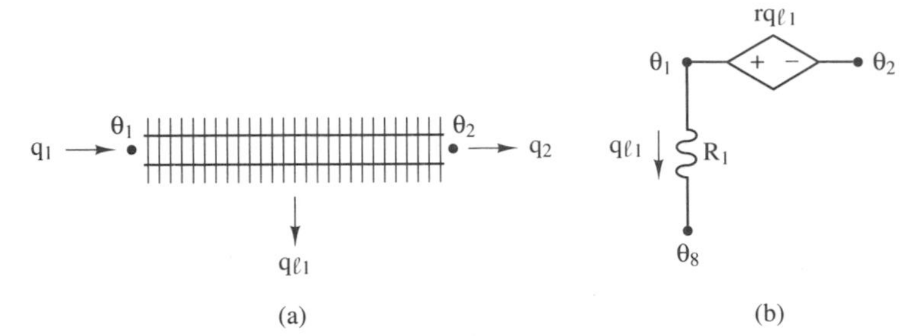

Figure 6.8 Segment 1 of the heat exchanger. (a) Pictorial model. (b) Lumped model.

Figure 6.8(a) shows an isolated view of segment 1 of the heat exchanger. Since we have chosen to associate the segment with node 1, we view θ_1 as the temperature of the whole segment. The rate of heat transfer from segment 1 of the heat exchanger to the room (represented by node 8) is $q_{\ell 1} = \theta_{18}/R_1$, where R_1 is the thermal resistance to that heat transfer. Therefore, we attach a thermal resistor to node 1, as shown in Figure 6.8(b).

The hot water that enters the left end of the segment in Figure 6.8(a) carries stored heat at the rate $q_1 = c_p\rho Q\theta_1$. The water that leaves the other end of the segment (at temperature θ_2) carries stored heat at the rate $q_2 = c_p\rho Q\theta_2$. The difference in these two heat flows is $q_1 - q_2 = c_p\rho Q\theta_{12}$. By conservation of energy (the node law), the difference in these two heat flows must equal $q_{\ell 1}$. Therefore, the temperature drop in the water and heat exchanger owing to transferring of heat to the room air is

$$\theta_{12} = \frac{q_{\ell 1}}{c_p\rho Q} = rq_{\ell 1} \qquad (6.17)$$

where $r \triangleq 1/(c_p\rho Q)$.

The temperature drop θ_{12} is proportional to a heat flow, but not the flow from node 1 to node 2. Therefore, we use the *dependent-source notation* that we developed at the end of section 2.2 to represent it. Figure 6.8(b) uses a dependent temperature-difference source to relate θ_2 to θ_1. The proportionality factor r has the units of thermal resistance.

We show the lumped model for the whole heating circuit in Figure 6.7. Because the resistor that accounts for heat removal from a segment is attached at the input node of the segment, the model exaggerates the heat removal. If we were to attach the resistor to the output node of the segment, the model would underestimate the heat removal. The two attachment locations should produce nearly the same results if we use enough segments.

Observe that the heat flow $c_p\rho Q\theta_7$ owing to the residual heat content of the water (after it passes through the heat exchanger) does not appear in the model. That is, no heat passes through node 7. The heat injected into the model by the temperature source θ_B just matches the total heat transferred out of the water stream during its traversal of the heat exchanger.

If we wish to include the residual heat content of the water in the model, we can do so by attaching a thermal *load* resistor of value $1/(c_p\rho Q)$ between node 7 and the reference node. Then the heat inflow from θ_B and the heat outflow from node 7 will both increase by the amount $c_p\rho Q\theta_7$.)

The thermal capacitance of a single room is $C_{ai} = c_p\rho V$, where $V = (4.5\text{ m})(3.5\text{ m}) \times (2.3\text{ m}) = 36.2\text{ m}^3$, the volume of air in the room. From appendix A, the specific heat of air is 1 kJ/kg°C and the density of air is about 1.3 kg/m^3. Therefore, $C_{ai} = 47.1$ kJ/°C.

Assume that the primary resistance to heat loss through the walls and ceiling of a room is the conduction resistance of the glass wool insulation. Therefore, we ignore the convection resistance at the inside and outside surfaces. The thermal resistance to heat loss through the ceiling of a single room, according to Figure 6.3(a), is $R_c = (15\text{ cm})/[(0.04\text{ W/m°C})(15.75\text{ m}^2)] = 0.238°\text{C/W}$. The resistance to heat loss through the walls is $R_w = (5\text{ cm})/[(0.04\text{ W/m°C})(18.4\text{ m}^2)] = 0.0679°\text{C/W}$. Since these two resistances act in parallel, the total effective thermal resistance between each room (node 8 and node 9) and the outside (node 10) is $R_{Oi} = R_wR_c/(R_w + R_c) = 0.0528°\text{C/W}$.

The resistance R_i represents the heat transfer $q_{\ell i}$ from node i to node 8 or node 9 (the air in room 1 or room 2). It has three components: one owing to forced convection (pumped water flow) at the inside wall of the pipe, a second owing to conduction through the copper pipe and aluminum fins, and a third owing to natural air convection from the fins. The thermal resistance of the highly conductive copper and aluminum is negligible. The thermal resistance owing to forced water convection is much less than that owing to natural air convection. Therefore, we compute R_i on the basis of the natural air convection alone.

The thermal resistance R_i associated with natural air convection from the fins of each segment of the coil is $R_i = 1/h_tA$, where h_t is the heat transfer coefficient and A is the exposed area of the fins on the segment. Since each fin has two sides, the exposed area of each is $2(5\text{ cm})(10\text{ cm}) = 100\text{ cm}^2$. Therefore, $A = 0.4\text{ m}^2$ for the 40 fins along a 2.67-m segment. From appendix A, we select $h_t = 12\text{ W/m}^2\text{°C}$. Then $R_i = 0.208°\text{C/W}$ for each segment of the heat exchanger.

Finally, the *transfer resistance* r of each dependent source in Figure 6.7 is $r = 1/[(4.2\text{ kJ/kg°C})(1{,}000\text{ kg/m}^3)(7.5 \times 10^{-5}\text{ m}^3\text{/s})] = 3.17°\text{C/kW}$.

(b) We used PSPICE to simulate the operation of the model shown in Figure 6.7 with the element values derived in part (a). We assumed that the heat exchanger and rooms began at the specified 15.5°C temperature. The temperature variations in the two rooms are shown in Figure 6.9. The model does not account for the 30 s delay in arrival of hot water to room 2,

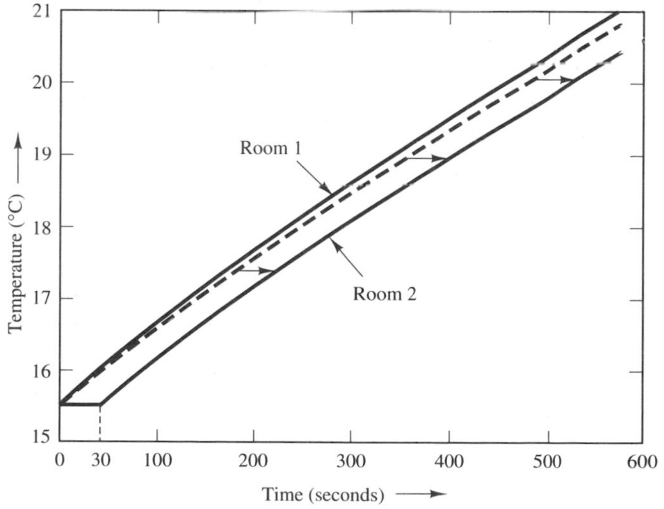

Figure 6.9 Dynamic response of the heater model of Figure 6.7.

relative to room 1. Therefore, we manually delayed the temperature plot for room 2 by 30 s. It takes 9.5 minutes for room 1 to reach the 21°C level.

(c) The heat lost from room 1 is the flow through R_{O1}; that from room 2 is the flow through R_{O2}. If we maintain the rooms at 21°C, the total heat loss is 2(21°C)/(0.0528°C/W) = 795 W. (The heater supplies heat to the rooms at about twice that rate, but intermittently.)

Once we known the room temperatures (at nodes 8 and 9), we can use the model in Figure 6.7 to calculate the temperatures at the nodes one at a time. The flow through R_1 is $q_{\ell 1} = \theta_{18}/R_1 = (75°C - 21°C)/0.208°C/W = 260$ W. Then, according to the first dependent source, $\theta_{12} = (3.17°C/kW)(260W) = 0.824°C$, or $\theta_2 = 74.2°C$. We use similar calculations to find, in sequence, the temperatures in the remaining nodes and determine that the return water temperature is $\theta_7 = 70.2°C$.

The temperature of room 2 at the time the thermostat shuts off the circulator, according to Figure 6.9, is 20.3°C. To balance the temperatures of the two rooms, remove some fins from the exchanger in room 1 or cover them to reduce the rate of heat transfer.

Exercise 6.2: Estimate the resistance to heat loss from a single room to the outside owing to convection at the inside and outside surfaces. Is it negligible compared with the conduction resistance, as assumed?

Population Variables

We apply the term *population* to many different dynamic entities. Examples are populations of bacteria, rabbits, people, dollars, and manufactured items. Because population systems exhibit growth and decay in numbers of items, lumped networks are appropriate models for such systems.

We use network nodes to represent *domains* (geographical regions or spheres of association of objects, individuals, or organizations). We call the number N of objects in a domain the **level** (or population of objects) at the node. (Imagine *storage elements* that hold the objects at each node.) The level varies from node to node. Thus, level is an *across variable.* As we move our attention from node to node, we find continuity of level. That is, no matter which path we follow in passing from node i to node j, we encounter the same difference in level between nodes i and j. We conclude that the path law of equation (2.28) applies to levels; hence **level is a potential.**

A change in level at a node can occur by three mechanisms: (1) An object can move between the node and another node. (2) A new object can be created at the node; we treat the phenomenon as a *birth*—an arrival from a **reservoir** of objects. (3) An object can be eliminated at the node; we treat this phenomenon as a *death* and return the object to the reservoir. We **treat a change in level,** then, **as a flow of objects.** We consider the net flow into a node as a flow into a storage element at the node. (This flow changes the storage level at the node.) Then the net flow into each node is zero.[4]

Node i of a population network is shown in Figure 6.10. Node 0 represents the reservoir. We connect node i to the reservoir by a storage element. The arrow labeled r_+ denotes a rate of flow of objects into the node. The arrow r_- denotes a rate of flow of

[4] The use of *level* and *rate of change of level* as the two concepts of importance in modeling dynamic systems was introduced by Forrester [6.2].

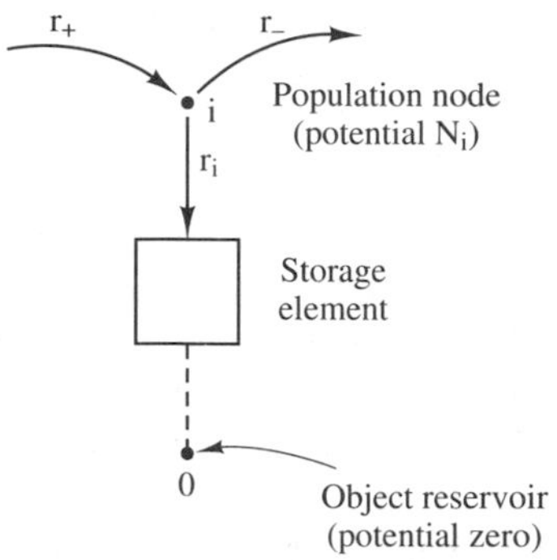

Figure 6.10 A node in a population network.

objects out of the node. Each of these flows arises from one of the three mechanisms described in the preceeding paragraph. The symbol r_i represents the *net flow* of objects into the node (and into the storage element at the node). It is apparent that the node law, equation (2.27), is satisfied at each node; that is, $r_+ - r_- - r_i = 0$.

We also want the node law to be satisfied at the reservoir node. Intuitively, the net flow of objects from the reservoir (births minus deaths) equals the net flow of objects into the set of node storage elements. In order that the net flow of objects from the reservoir be zero, we let the net flow into each node pass through the node storage element into the reservoir. The fictitious flow through the storage element is indicated in Figure 6.10 by a dashed line. It is apparent in the figure that the object flows r_+ and r_- and the rate of change r_i of the level at node i are *through variables.*

Since r_i represents the flow into the storage element (out of node i), it represents the rate of *increase* of level N_i at node i; that is,

$$r_i = \dot{N}_i \tag{6.18}$$

An equivalent statement in impedance notation is

$$N_i = \left(\frac{1}{s}\right) r_i \tag{6.19}$$

where s denotes the time-derivative operator. We **assign zero potential** (zero level) **to the reservoir** at node 0. Then equation (6.19) follows the rules used for lumped elements throughout this book. That is,

$$N_{i0} = \left(\frac{1}{s}\right) r_i \tag{6.20}$$

where the subscripts on the potential difference N_{i0} are ordered in the direction of the flow r_i. The storage element is then analogous to a thermal capacitor with *unit-valued* capacitance. Compare equation (6.20) with equation (6.9).

No alphanumeric symbol has been assigned to the storage element, because all population storage elements have *unit value.* Nor is it essential that we show the storage elements explicitly in the network diagram. We could merely recognize that **the sum of**

***explicit* flows *into* a node is the rate of increase in level at the node.** For the node in Figure 6.10, this statement becomes

$$r_+ - r_- = \dot{N}_i \tag{6.21}$$

Populations of discrete objects (people, dollars, etc.) change by one item at a time. Furthermore, the changes occur at discrete instants. Thus, rates of change and derivatives with respect to time may not seem appropriate for describing changes in such a system. Modeling and simulation of small populations is not meaningful, unless we account for the random activities of individual objects in the population. We shall not attempt to address such discrete-event behavior here.

If the populations are not small, we can treat them as if they change continuously with time. For many population systems, avoiding small populations merely means observing the system over a sufficiently long time scale relative to the rates of change of population in the system. We shall assume that the node populations in the system are sufficiently large that they appear nearly as smooth functions of time.

Modeling of Population Systems

Figure 6.11 shows a model of a rabbit population. The three numbered nodes represent geographical regions; N_i is the number of rabbits in the ith region. Node 0 is the reference (reservoir) node. The branches connecting the numbered nodes represent rabbit migration paths. Associated with each such path is a *migration rate* (or transfer rate). Rabbits tend to migrate from one region to another because the environment (food supply, weather, number of predators, etc.) appears better. Migration occurs in both directions, based on the perceptions of individual rabbits.

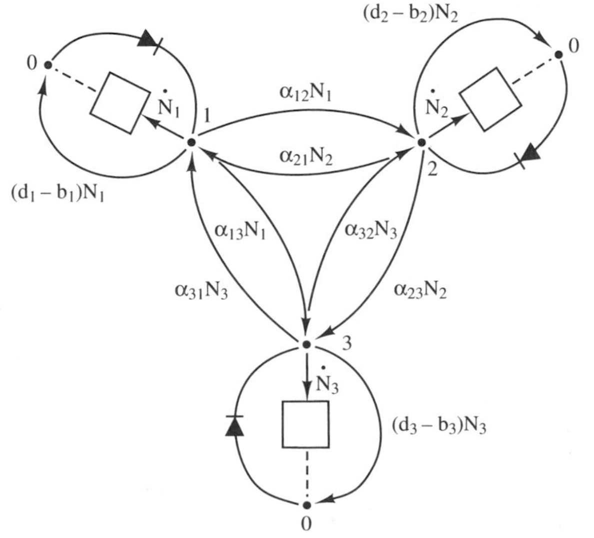

Figure 6.11 A rabbit-population network.

The rate of migration out of node i into node j is in proportion to N_i, the number of rabbits at node i. Specifically, for migration from region 1 to region 2,

$$\dot{N}_2 = \alpha_{12} N_1 = -\dot{N}_1 \tag{6.22}$$

where the rate of increase $\dot{N}_2$ at node 2 owes only to migration from node 1, $-\dot{N}_1$ is the corresponding rate of decrease at node 1, and α_{12} is the **coefficient of migration from node 1 to node 2;** α_{12} is determined by (and is a measure of) the perceptions of rabbits in region 1. The units of α_{12} are (rabbits transferred per year)/(rabbit population at node 1) = year^{-1}. Intuitively, we can think of α_{12} as the **yearly fractional migration** of rabbits. We sometimes express such a fractional quantity as a percentage, say, 2% per year.

In a population network, we **use a directed branch**—a branch with an arrowhead—**to imply directionality of flow.** We label each branch with a symbolic description of the actual amount of flow, as shown in Figure 6.11. Note that the migration coefficient multiplies the level at the *tail end* of the directed branch.

Since N_1 is a potential and $\dot{N}_2$ is a flow, the migration coefficient α_{12} has the units of admittance (inverse impedance), a ratio of flow to potential. We can think of $1/\alpha_{12}$ as a **directional resistance to flow.** It resembles resistance in the physical systems of previous chapters, but it applies only to flow directed from node 1 to node 2.

In section 2.2, we introduced the dependent-source symbol shown in Figure 6.12(a) to represent relationships like equation (6.22). The simpler line-graph notation in Figure 6.12(b) represents the same relationship. We shall use the dependent-source symbol extensively throughout this chapter to represent dependent flows.

Rabbit migrations from region 2 to region 1 are characterized by the migration coefficient α_{21} and the equations

$$\dot{N}_1 = \alpha_{21} N_2 = -\dot{N}_2 \tag{6.23}$$

The factor $1/\alpha_{21}$ is the resistance to flow from node 2 to node 1. The rates of change $\dot{N}_1$ and $-\dot{N}_2$ in equation (6.23) are not the same as in equation (6.22). In equation (6.22), they are *components of change* owing only to flows from node 1 to node 2. In equation (6.23), they are *components of change* owing only to flows from node 2 to node 1. Comparable equation pairs pertain to the flows between nodes 1 and 3 and between nodes 2 and 3.

The two branches between nodes 1 and 2 are similar to the parallel resistors of previous sections, except that these resistances are directional. The left equality of (6.22) and the right equality of (6.23), taken together, describe the effect of the two opposing flows

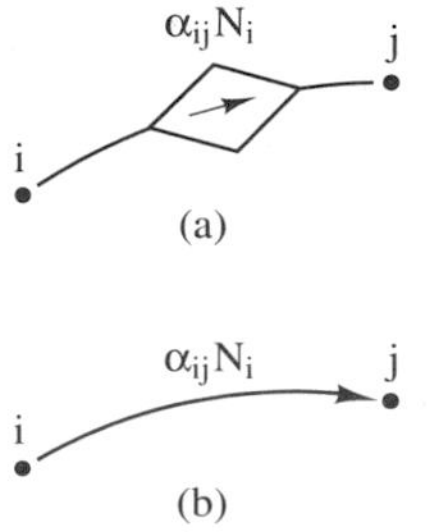

Figure 6.12 Equivalent notations for dependent flows; both notations denote r_{ij}, the directed flow from node i to node j. (a) Dependent flow-source symbol. (b) Simpler dependent flow-source notation.

on the change in level at node 2. That is, the *net* inflow at node 2 (owing only to flows between nodes 1 and 2) is

$$\dot{N}_2 = \alpha_{12}N_1 - \alpha_{21}N_2 \tag{6.24}$$

Of course, the net inflow $\dot{N}_2$ at node 2 corresponds to (and is equal to) the net outflow $-\dot{N}_1$ at node 1.

If the migration coefficients α_{12} and α_{21} are equal (i.e., the resistances to flow are the same for both directions), then we can write equation (6.24) as

$$\dot{N}_2 = \alpha(N_1 - N_2) = \alpha N_{12} = -\dot{N}_1 \tag{6.25}$$

That is, for equal flow resistance in both directions, equations (6.22) and (6.23), taken together, correspond to a linear resistor (with resistance $1/\alpha$) between nodes 1 and 2. Then the potential difference N_{12} is proportional to the flow $\dot{N}_2 \equiv -\dot{N}_1$. Compare equation (6.25) with equation (6.1).

Like most populations, the rabbit populations N_i at the nodes of Figure 6.11 cannot be negative. Hence, the flow-rate equations must be supplemented by the nonlinear nonnegativity equations

$$N_i \geq 0 \tag{6.26}$$

To represent this nonnegativity requirement graphically, we insert the ideal one-way-flow element shown in Figure 6.13 in parallel with the population storage element at node i. If $N_i > 0$, the element behaves as an open circuit and has no effect on the behavior of the network. If $N_i \not> 0$, the element behaves as a short circuit; that is, it does not permit N_i to drop *below* the zero potential level of the reservoir.[5]

The node populations are changed by births and deaths, as well as by migrations. Since births and deaths at node 1 are both proportional to N_1, we represent the birth and death flow paths at node 1 (in rabbits per year) by a single branch directed from node 1 to the reference node. That is, we maintain our convention that **the population at the tail of the directed branch determines the flow.** The migration coefficient for the flow is $d_1 - b_1$, the death rate minus the birth rate. Therefore, the flow *out* of node 1 (owing to deaths and births alone) is

$$-\dot{N}_1 = (d_1 - b_1)N_1 \tag{6.27}$$

If the death rate is larger than the birth rate, the coefficient in equation (6.27) is positive and the flow is directed out of the node, as indicated in the diagram. If the birth rate is larger than the death rate, the coefficient is negative, reversing the direction of the flow.

How good is this population model? How well does it characterize population flows? As with any model, we must compare the behavior of the model with actual popu-

Figure 6.13 Ideal one-way-flow element that prevents $N_i < 0$.

[5] One-way-flow elements are used in a number of physical contexts. The electrical version is called a *diode.* A *check valve* is a hydraulic version. A *pawl*-and-*rachet* is a translational version. It is the diode symbol, borrowed from electrical notation, that we have used in Figure 6.13.

lation systems in order to answer these questions. The rabbit population model of Figure 6.11 is simulated in example problem 6.3. In that example, we find that rabbit **populations grow exponentially if the birth rate is greater than the death rate.** This system behavior is intuitively reasonable.

In the physical world, however, exponential growth eventually must cease. Therefore, population-limiting mechanisms must be incorporated into the migration coefficients. Although births, deaths, and migrations occur in proportion to the node populations, the migration coefficients decrease as the levels increase. Hence, equations (6.22), (6.23), and (6.27) are nonlinear.[6]

None of the systems of previous chapters exhibited exponential growth. In most of the dynamic systems treated in this book, fixed natural properties of materials tie levels to rates of change of levels. For example, $\dot{v} = F/m$ for a mass; v is the level and F/m is the rate of change of level.

In population systems, however, the **rates are typically determined by available resources or selected by policy.** For example, by choosing to harvest a certain number of deer in a region each year, we control the growth rate of the deer population in that region. Similarly, if the rabbit population in an area becomes large, a number of factors act to reduce the growth of rabbits: Because the vegetation becomes sparse, the rabbits find it less attractive; weak rabbits die more easily; rabbit predators thrive and eat more rabbits; and farmers take stern measures to reduce the population. Thus, the policies or resources that determine the rates of change of levels are, in turn, *changed* by the levels.

Example problem 6.3 demonstrates the population-limiting effect of reducing the net birth rate (birth rate minus death rate) as population levels increase. These diminishing coefficients produce the so-called S-shaped curves that are typical of population growths. Without such nonlinear limiting mechanisms, all populations (bacteria, people, pollution, wealth, etc.) would grow exponentially, without bound. The difficulty in modeling populations is in determining accurately the manner in which the transfer coefficients decrease with increasing population.

THE LUMPING PROCEDURE FOR POPULATION SYSTEMS

1. Draw a pictorial diagram. If the system deals with physical regions, the diagram is essentially a map. If it deals with spheres of influence, the diagram can represent associations rather than physical regions.
2. Identify physical regions (or spheres of influence) in the system that are significant foci of population or in which the populations are of particular interest. Represent each region by a numbered node. One of these nodes must represent the object reservoir, *assigned the potential zero.* (Adding population to the reservoir does not change its apparent level.) Place the nodes in approximate correspondence with the layout in the pictorial diagram.
3. Connect a lumped storage element (of unit value) between each population node and the reservoir node. The reservoir need not be isolated at a single location.

[6] We must be cautious in our use of the impedance concept for these nonlinear equations. We cannot use the standard series and parallel combination rules of section 4.2 for nonlinear lumped elements.

Designate by r_i (or $\dot{N}_i$) the flow through the storage element from node i to the reservoir. This flow is the rate of increase of population at node i; it can be positive or negative.

4. For each node that can experience deaths and births, insert a death-birth path from that node to the reservoir, oriented toward the reservoir. Label each such path with its *death rate minus birth rate* flow, $r_{i0} \triangleq (d_i - b_i)N_i$. If the birth rate is greater than the death rate, then r_{i0} will be negative.
5. Insert directed migration paths between appropriate population nodes. Let $r_{ij} \triangleq \alpha_{ij}N_i$ denote the directional flow from node i to node j, where α_{ij} is the coefficient of migration from node i to node j. Typically, r_{ij} is a nonlinear function of the node populations.
6. Determine the manner in which the migration coefficients α_{ij} and death-birth coefficients $(d_i - b_i)$ depend on the node populations. The coefficients can depend on factors (and populations) that are outside the domain of inquiry. (For example, rabbit deaths can depend on fox populations in another network.)
7. Redraw the lumped network to obtain a convenient layout. Attach a *one-way-flow* element from the zero-potential reservoir to each node at which the population must remain nonnegative. Orient the one-way-flow elements away from the reservoir.

The node and loop procedures for generating system equations for population systems are identical to the procedures used for mechanical, electrical, hydraulic, and thermal systems. The system equations are nonlinear, owing to the nonnegativity of populations, to the variability of the coefficients α_{ij} with populations, to policy changes, and to other environmental factors. The behaviors of these nonlinear networks can be determined numerically by SPICE-based simulation programs.

Forrester has developed a dynamic model of the world that represents the interactions among major world subnets (human population, capital investment, natural resources, the fraction of capital devoted to agriculture, and pollution).[7] Each of these subnets can be treated as a population network in the sense just described. Growth in one subnet tends to promote or inhibit growth in the other subnets. There has been considerable controversy over the values to use for the migration coefficients α_{ij} and over the manner in which those coefficients are assumed to depend on the various subnet populations. It is difficult to get the data necessary to compare a simulation with the actual *running* of the world.

Example Problem 6.3: Computer simulation of population behavior.
Suppose that the initial population in each region of the three-region rabbit population model shown in Figure 6.11 is 10 rabbits. Let the migration coefficients be $\alpha_{12} = 0.05$, $\alpha_{21} = 0.02$, $\alpha_{23} = 0.05$, $\alpha_{32} = 0.02$, $\alpha_{13} = 0.02$, and $\alpha_{31} = 0$, regardless of the populations in each region. Each coefficient has the units fraction of rabbits per year (or year^{-1}).

Assume that the net birth rate in a region decreases with the population of that region according to the formula

$$b_i - d_i = b_{si}\left(1 - \frac{N_i}{500}\right) \tag{6.28}$$

[7] See reference [6.3].

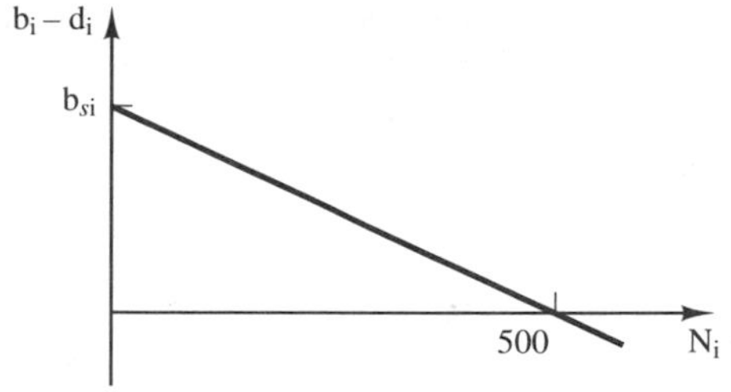

Figure 6.14 Net birth rate versus population.

where b_{si} is the net birth rate *for small populations.* This diminishing net birth rate is plotted in Figure 6.14. Then the net birth flow into node i is

$$-r_{i0} = b_{si}\left(N_i - \frac{N_i^2}{500}\right) \tag{6.29}$$

If the population N_i becomes higher than 500, the death rate exceeds the birth rate for that region. Let $b_{s1} = 0.2$, $b_{s2} = 0.3$, and $b_{s3} = 0.35$, each with the units fraction of rabbits per year.

Simulate the operation of the model. That is, determine the population patterns predicted by the model.

Solution: We used a SPICE-based program to simulate the operation of the model. The program has a dependent flow-source element like that of Figure 6.12. It also has a diode (one-way-flow) element like that of Figure 6.13.

The population patterns for the three regions, as predicted by the lumped model, are shown in Figure 6.15. Note the exponential growth for low population levels and the

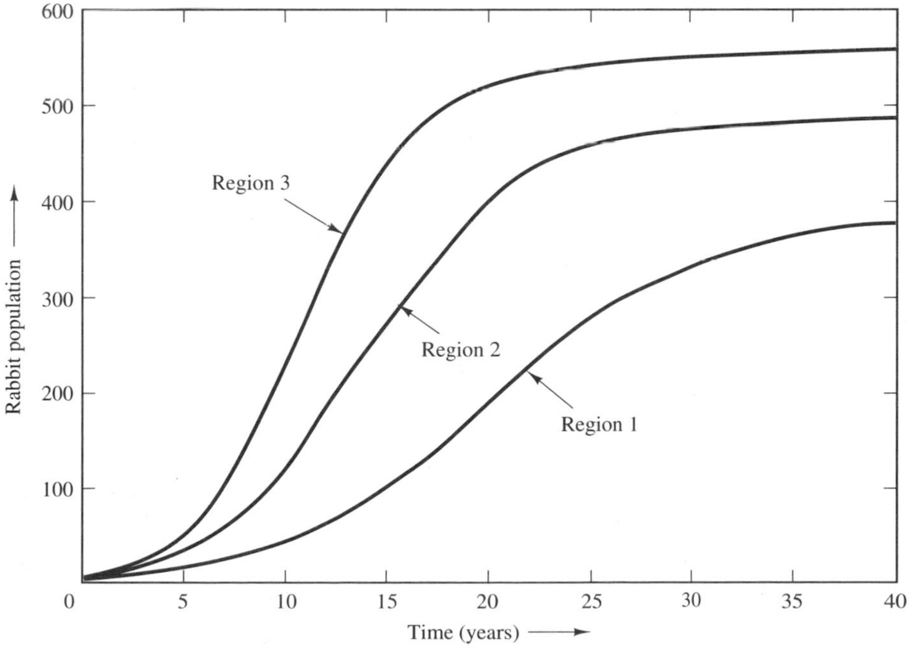

Figure 6.15 Rabbit population patterns.

diminishing growth rates for higher populations. Although the net birth rate for region 3 is not much greater than that for region 2, the initial exponential growth is much higher because the migration coefficients are biased to favor region 3. The migration rates are biased to produce the highest steady-state population in region 3 and the lowest in region 1.

6.2 TWO-PORT MODELS FOR CONVERTERS

We now introduce two-port (four-terminal) devices. We use these devices to connect sources to loads and to make them compatible with each other. In this section, we examine two types of two-port devices: *transformers* and *transducers.* We use a transducer (say, an electric motor) to convert available source energy to the *form* needed by a load. We use a transformer (say, a gear) to adjust a source signal to the *level* needed at a load. We call both types of devices *converters.*

Transformers and transducers are passive devices. Most are linear. In this section, we develop ideal lumped models for these converters and establish sign conventions for the converter equations. Then we use the ideal lumped models and equations in models of physical converters. This section also examines the way a converter transforms the source and load equations.

Transformers and Transducers

A **transformer** magnifies one variable at a port at the expense of the other variable at that port. For example, a gear in a rotational system increases the shaft torsion by *gearing down* the shaft velocity, an electrical transformer increases the port voltage by decreasing the port current, and a lever trades off velocity and force. Transformers are used to change signal levels and impedances.

A selection of transformers and the corresponding *ideal* equations is illustrated in Figure 6.16. For each device, the *ideal lumped model* should be understood to correspond exactly in behavior to the ideal equations. The *pictorial model* of the transformer shows how the physical transformer works. Pictorial models can include any desired level of detail. The ideal-transformer model shows little detail, but does retain the physical nature of each variable.

Each end of a transformer is represented by a pair of nodes that forms a port. Thus, a transformer is a two-port device. For some transformers, such as the lever and the hydraulic intensifier, one node of each port is a reference node—say, ground or ambient pressure—that is not noted explicitly in the ideal lumped model.

It is convenient to refer to the input (or source) port of a transformer as the **primary port** and the output (or load) port as the **secondary port.** However, a transformer is bidirectional: Either port can be used as the primary port. If sources are applied to both ports, the stronger source delivers power to the weaker source through the transformer. A transformer is passive. An *ideal* transformer neither consumes nor stores energy: The power leaving the secondary port equals the power entering the primary port.

If a source moves end a of the lever of Figure 6.16(a) to the right, then end b must move to the left. The arrows on the pictorial diagram of the lever make explicit the *rela-*

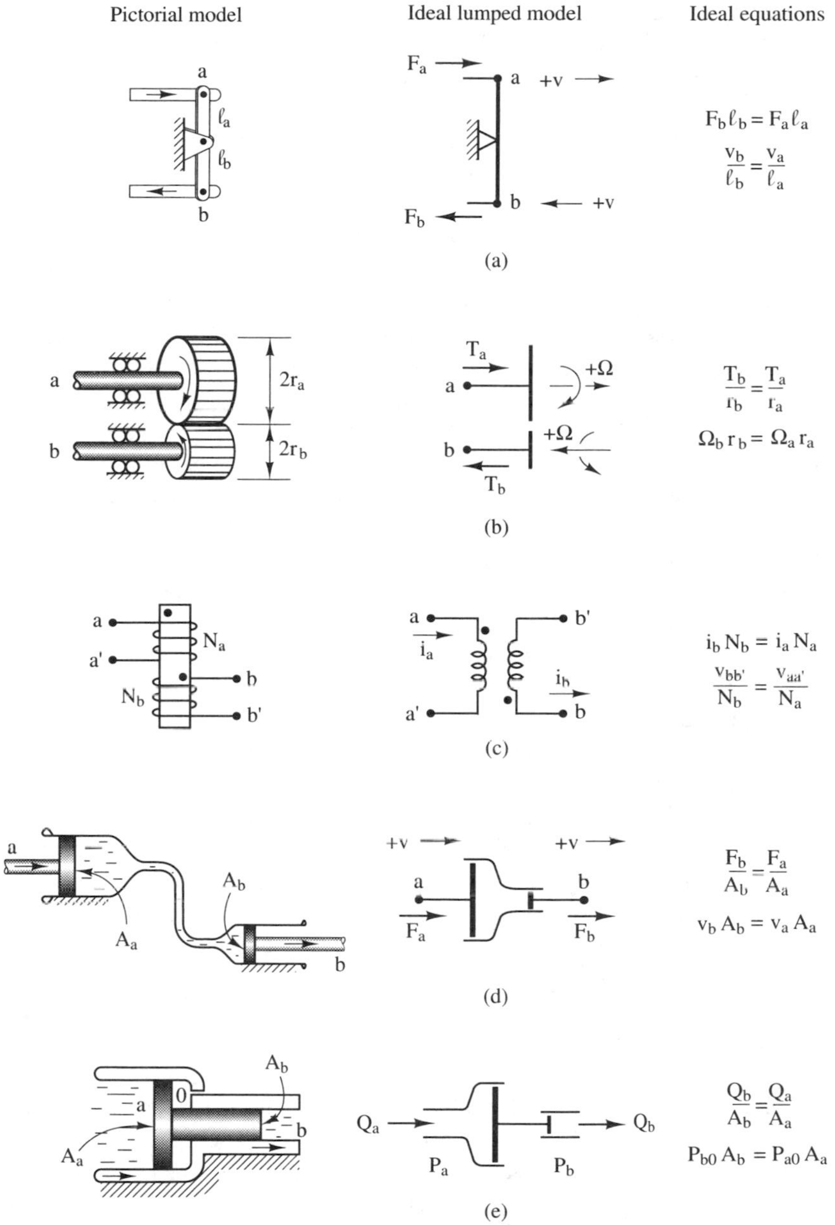

Figure 6.16 Ideal transformers. The arrows and dots on the pictorial models show the relative polarities of the two ends. The equations assume that the directions of positive motion and the orientations of the flow variables are chosen to produce positive port variables when power flows from the source to the load. The electrical transformer equations apply only for signal frequencies in the pass band. (a) A lever. (b) A gear. (c) An electrical transformer. (d) A hydraulic lever. (e) A hydraulic intensifier.

tive directions of motion of the two ends. In the electrical transformer of Figure 6.16(c), if the dotted terminal of the left winding is the terminal with higher voltage, then the dotted terminal of the right winding must also be the terminal with higher voltage. We call these arrows and dots **polarity marks.** Similar polarity marks (arrows or dots) appear on the pictorial diagram of each transformer in the figure. We shall use the polarity marks to determine the signs on the transformer equations.

If the pictorial diagram of a transformer is sufficiently detailed to show its manner of operation, then the relative polarities of the two ends are apparent, and the polarity marks are redundant. The manner of operation might not be visible in the physical device, however. Manufacturers sometimes inscribe polarity marks on each transformer so that the user need not test it to find its polarity. For many of the transformers of Figure 6.16, the manner of operation (and the corresponding polarity) can be seen in the ideal-transformer symbol. In the electrical transformer, it cannot; therefore, we attach polarity dots to the ideal electrical transformer symbol.

The physical parameters of each transformer can be grouped in the ideal-transformer equations to form a ratio—a single number. The ideal-transformer symbol, then, is a *two-port lumped model* with a single parameter. We call that parameter the *transformer parameter,* the *converter parameter,* or the *converter constant.* In specific devices, it may also have another, more familiar name. For example, the converter parameter for a lever is the *mechanical advantage,* and the parameter for a gear is the *gear ratio.*

There are a number of variations of each transformer. Levers, for example, come with the fulcrum in the middle or on the end. They also come with end linkages that transfer lever motions to adjacent locations. A bicycle chain with its pair of sprockets is essentially a gear. Belt and pulley versions are also common. Other gear geometries or belt-and-pulley geometries enable the input and output shafts to rotate with different axes. Gears with continuously adjustable gear ratios are available. Different implementations lead to different deviations from ideal behavior.

A **transducer** is an energy converter. It changes signals (and the associated power) from one energy form to another. For example, an electric generator transforms rotational power into electrical power. Similarly, the wheels of an automobile convert the rotational torque of the axle into the translational force needed to accelerate the automobile. A hydraulic cylinder and piston can be used as a hydraulic motor, which transforms hydraulic power into translational power.

A selection of transducers and the corresponding ideal equations are illustrated in Figure 6.17. The pictorial model of each transducer shows how the physical transducer works. The physical nature of each variable is carried over to the ideal lumped model. That model should be understood to correspond exactly in behavior to the ideal equations.

Each transducer is a two-port device. Each port is represented by two nodes. In some instances, one node of a port is an implicit reference node. In Figure 6.17(a), for example, the mechanical ground for the piston and the ambient pressure node for the hydraulic cylinder are implicit. Those reference nodes remain implicit in the ideal lumped models.

It is convenient to refer to the input (or source) port of a transducer as the *primary port* and the output (or load) port as the *secondary port.* Most transducers are bidirec-

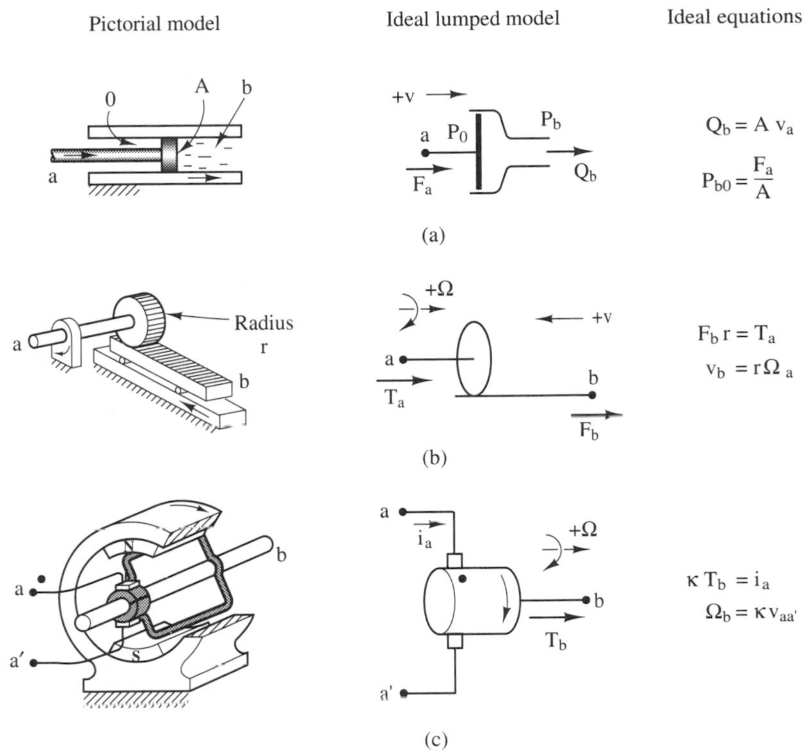

Figure 6.17 Ideal transducers. The arrows and dots on the pictorial models show the relative polarities of the two ends. The equations assume that the directions of positive motion and the orientations of the flow variables are chosen to produce positive port variables when power flows from the source to the load. (a) A hydraulic cylinder. (b) A wheel (or rack and pinion). (c) A rotary induction machine (or electric motor-generator). (Continued)

tional. That is, either port can be used as the primary port. If a translational source moves the piston in the hydraulic cylinder of Figure 6.17(a), the device acts as a pump. If a hydraulic source moves the fluid, the device acts as a hydraulic linear (translational) motor. If the induction machine of Figure 6.17(d) is energized by an electrical source, the machine acts as an electric linear motor. If it is energized by a translational source, the machine acts as an electric generator. Figures 6.17(e) and (c) show rotational devices that are analogous to the hydraulic cylinder and the linear induction machine. Dissipative heaters, such as the electric heater of Figure 6.17(f), are *not* bidirectional. They cannot convert heat to energy of another form.

Like transformers, transducers are passive. They do not supply or store energy. The power leaving the secondary port of an ideal transducer equals the power entering the primary port.

We use polarity marks—arrows or dots—to note the relative orientations of the motions and physical flows at the two ports of a transducer. For example, motion of the piston of Figure 6.17(a) to the right causes the fluid to flow to the right. Polarity marks help

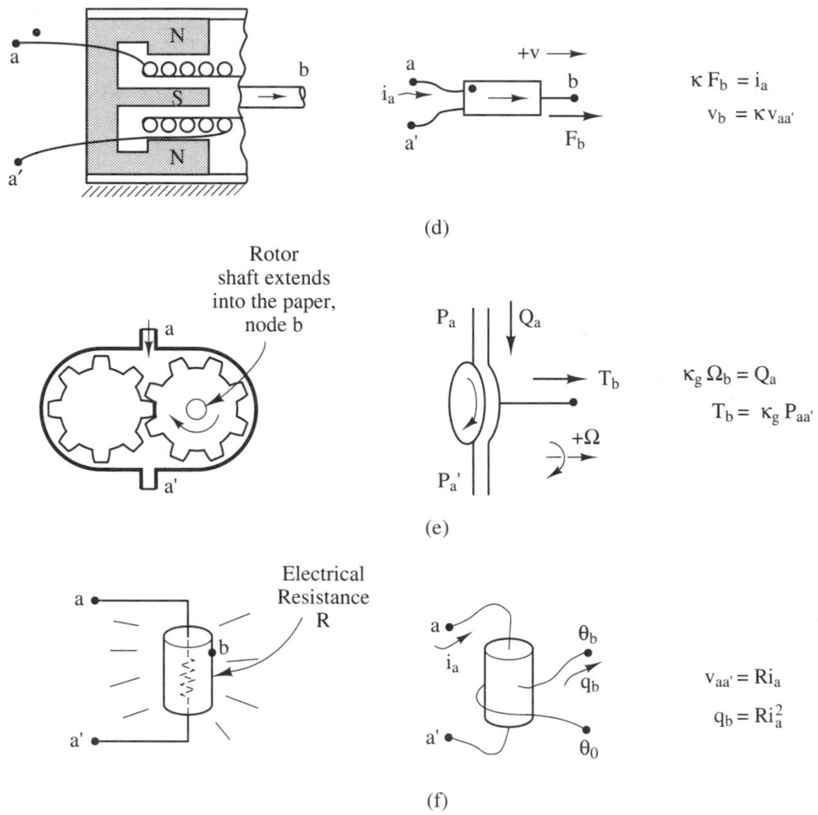

Figure 6.17 (Continued) (d) A linear induction machine. (e) A positive-displacement machine (or hydraulic motor-pump). (f) An electric heater.

us choose the correct signs for the ideal-transducer equations. We find the polarity marks and determine the signs in the equations in later sections.

The ideal-transducer symbol is a *two-port lumped model.* The physical parameters of each transducer can be grouped to form a single lumped-model parameter in the ideal transducer equations. We call that parameter the *transducer parameter,* the *converter parameter,* or the *converter constant.* In a specific device, it may also have other, more familiar names. For example, if a rotary induction machine is used as a motor, its transducer parameter is called the torque constant. If it is used as a generator, the same transducer parameter is called the generator constant.

There are numerous variations of each transducer. Electric motors, for example, come with permanent magnets or with magnetic field windings. They can be designed to operate on ac power or dc power. A displacement pump can be implemented not only by a pair of gears, but also by squeezing a flexible tube or by means of reciprocating pistons. Different implementations lead to different deviations from ideal behavior. We model the effects of such deviations in a later section.

The Ideal Lever—a Direct Transformer

A lever is used to scale the velocity or force at a point in a system. We use the lever here to illustrate features common to many two-port devices. An ideal lumped model of a lever is embedded in a lumped translational network in Figure 6.18. (Node 1 is identical to node a; we separate the two nodes only to permit the lever force F_a to appear explicitly in the lumped model.) The fulcrum of the lever is supported by a housing fastened to ground. We denote the primary (source) side of the lever by subscript a and the secondary side by subscript b. The ideal-lever model presumes that the lever is massless, frictionless, and infinitely stiff.

To analyze a translational network, we must choose the +v direction for each node and the orientation of the flow variable for each branch. These choices establish positivity for each node velocity and each branch force. In section 2.1, we decided to use a single +v direction so that we could express relative motions between points as velocity *differences*. Since the lever equations do not depend on the *relative* motions of the two ends, we can choose the +v directions independently on the two sides of the lever.

Intuitively, we can see from Figure 6.18(a) that node a will move to the right and node b will move to the left. (This observation about the relative motions of the two ends corresponds to visualizing polarity arrows like those in Figure 6.16(a).) We choose the +v directions at the two ends of the lever to agree with these observable directions of motion, as shown in the figure. Then positive motion of the primary end will produce positive motion of the secondary end. We also see that the scalar forces at the two ends of the lever will be compressive. Therefore, we orient the flow variables F_a and F_b in the respective +v directions so that both variables represent *compressive* forces.

With these choices for positivity for the four port variables, we know that a positive value of v_a will make v_b, F_a, and F_b positive. Then power must transfer from the source to the lever input (port a) because F_a flows *through the source* in the direction of velocity

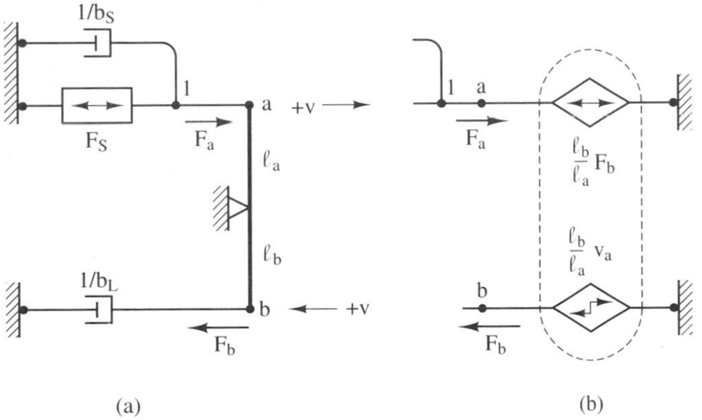

Figure 6.18 An ideal lever in a translational network. (a) Lumped model. (b) Dependent-source representation of the lumped model of the lever.

rise. Similarly, power must transfer from the lever output (port b) to the load because F_b flows *through the load* in the direction of velocity drop.

It follows that F_a flows through port a of the lever in the direction of velocity drop and that F_b flows through port b of the lever in the direction of velocity rise. However, it is difficult to discern directly the direction of velocity rise or drop across each port because the reference nodes for the ports do not appear explicitly in Figure 6.18(a). We shall make those reference nodes explicit shortly.

The lengths of the two sides of the lever are ℓ_a and ℓ_b. The compressive forces F_a and F_b correspond to *competing* force *vectors* acting on the two ends of the lever. The moment of force about the fulcrum must be zero; hence,

$$F_a\ell_a = F_b\ell_b \Longrightarrow F_b = \left(\frac{\ell_a}{\ell_b}\right)F_a \tag{6.30}$$

Since the angular velocities of the two sides of the lever must be identical, as *seen* at the fulcrum,

$$\frac{v_a}{\ell_a} = \frac{v_b}{\ell_b} \Longrightarrow v_b = \left(\frac{\ell_b}{\ell_a}\right)v_a \tag{6.31}$$

Equations (6.30) and (6.31) agree with the ideal-lever equations of Figure 6.16(a).

If we were to reverse the orientation of F_a or F_b, we would have to attach a negative sign to that variable in equation (6.30). If we were to reverse the $+v$ direction at one end of the lever, we would have to attach negative signs to both the velocity variable and the flow variable associated with that end of the lever, thereby changing the signs in both equation (6.30) and (6.31).

A lever is a *direct* converter in the sense that it relates a potential difference at one port to a potential difference at the other port; hence, it also relates a flow at one port to a flow at the other port. In other words, **a direct converter** is one that **preserves the roles of potential difference and flow.** This preservation of roles is important. The hydraulic cylinder, which we shall examine shortly, reverses these roles and, as a result, affects networks in a dramatically different way.

The ideal-lever model in Figure 6.18(a) is a *lumped* model. That is, it ignores the internal behavior of the lever and considers motions and forces only at its ends. We should view the lumped model as a *physical* model. Compare it to the pictorial model in Figure 6.16(a). The lumped model retains the motions of the nodes and the compressive or tensile natures of the scalar forces.

Equations (6.30)–(6.31) constitute a mathematical model for the lever. By defining velocity and scalar-force variables, we have converted to equations the physical laws that relate the observable physical quantities. Each equation relates a variable at one port to a variable at the other port. At the end of section 2.2 (in Figure 2.38), we introduced the *ideal dependent source* to portray the dependency of one network variable on another. We now use dependent sources to depict equations (6.30)–(6.31).

Motions at the primary port of the lever in Figure 6.18(a) are measured relative to the implicit stationary ground. We display that ground node explicitly in the upper half of

Figure 6.18(b). Let us use equation (6.30) to describe the behavior of the lever as observed at its primary port. The variable of that equation that pertains to the primary port is F_a. The equation shows that F_a is proportional to F_b. Therefore, we use the diamond-shaped **dependent flow source** symbol of Figure 6.18(b) to represent F_a and state the specific dependency on F_b beside that symbol.

Similarly, in the lower half of Figure 6.18(b), we display explicitly the ground node for the secondary port, and we use the remaining equation, (6.31), to describe the behavior of the lever at that port. Since the variable of equation (6.31) that pertains to the secondary port is the velocity variable v_b, we represent v_b by a **dependent velocity source** symbol and state beside the symbol the manner in which it depends on v_a, as shown in the figure.

The dotted line surrounding the pair of dependent sources in Figure 6.18(b) defines the extent of the dependent-source model of the lever. We still view the model as physical, because the physical natures of the variables are still exposed. The mathematical equations (6.30)–(6.31) that describe the relations among the variables are inherent in the model. It should be apparent that we could, instead, use a dependent *velocity* source in the upper half of Figure 6.18(b) to represent the dependency of v_a on v_b and use a dependent flow source in the lower half to represent the dependency of F_b on F_a.

Figure 6.19(a) shows a line-graph representation of the lumped model of Figure 6.18(b). This model is mathematical. The physical natures of the variables have been

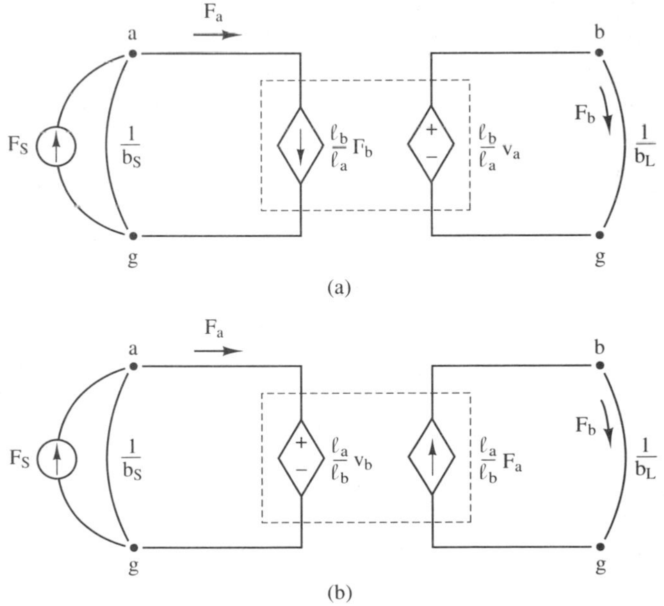

Figure 6.19 Line-graph models for the ideal lever of Figure 6.18. (a) Equation (6.30) describes the behavior at the primary side. (b) Equation (6.30) describes the behavior at the secondary side.

eliminated. (The line-graph nodes do not move. The line-graph flows are positive or negative, not compressive or tensile.) We can obtain this dependent-source line-graph representation directly from the lever equations (6.30)–(6.31), without first drawing the physical model in Figure 6.18(b). We merely solve equation (6.30) for F_a to define the dependent flow source and solve equation (6.31) for v_b to define the dependent velocity source.

Figure 6.19(b) shows an alternative dependent-source line-graph representation of the lever. Once again, we obtain it directly from equations (6.30)–(6.31). We merely solve equation (6.31) for v_a to define the dependent velocity source and solve equation (6.30) for F_b to define the dependent flow source. The dashed line that encloses each pair of dependent sources is used only to associate the pair of sources conceptually. It is not essential.

The dependent-source line-graph models simplify the writing of system equations for networks that contain converters. We just apply the loop or node method separately to each detached subgraph. The dependent sources in the subgraphs incorporate the relations between subgraphs. Network simulation programs (such as SPICE, appendix C) include models for ideal dependent sources. Hence, network simulation programs can be used to simulate networks that contain lumped models of two-port devices.

Exercise 6.3: Reverse the +v direction for one end of the lever of Figure 6.18(a). Show that: (a) the signs reverse in the lever equations (6.30) and (6.31); (b) each dependent source in Figure 6.19 either reverses direction or changes sign.

Exercise 6.4: Write the system equations for the line graph of Figure 6.19(a). Solve them for the load force F_b, in terms of the source signal and the lumped element parameters.

The Ideal Hydraulic Cylinder—a Gyrating Transducer

A hydraulic cylinder relates the pressure and flow of a hydraulic system to the compressive force and velocity of a translational system. It is used, for example, in the hydraulic brake system of an automobile. In some contexts, it is called a *hydraulic ram.* Figure 6.20(a) shows an ideal hydraulic cylinder used as a hydraulic motor in a hydromechanical network. The primary side, with subscript a, is hydraulic and is driven by a pressure source. The reference pressure is the ambient pressure, denoted P_0. The secondary side, with subscript b, is translational, with a dissipative load. The translational reference is the ground.

Suppose the hydraulic source applies pressure to port a, 0. It is obvious from the lumped model that the fluid at the input and the piston at the output both move to the right. (This observation corresponds to visualizing polarity arrows in Figure 6.17(a).) Positivity of pressure was defined for all hydraulic systems in section 5.1. Intuitively, we can see that $P_{a0} > 0$ and that the scalar force in the shaft at node b is compressive. We orient the hydraulic flow variable Q_a to the right, the +v direction on the mechanical side to the right, and the scalar-force variable F_b in the +v direction. Then F_b is *compressive* force, and all four port variables must be positive.

The volume flow rate in the cylinder is $Q_a = Av_b$, where A is the cross-sectional area of the piston. The compressive force in the piston shaft is $F_b = AP_{a0}$. Therefore, the equations that describe the ideal cylinder and piston are

$$v_b = \left(\frac{1}{A}\right)Q_a \tag{6.32}$$

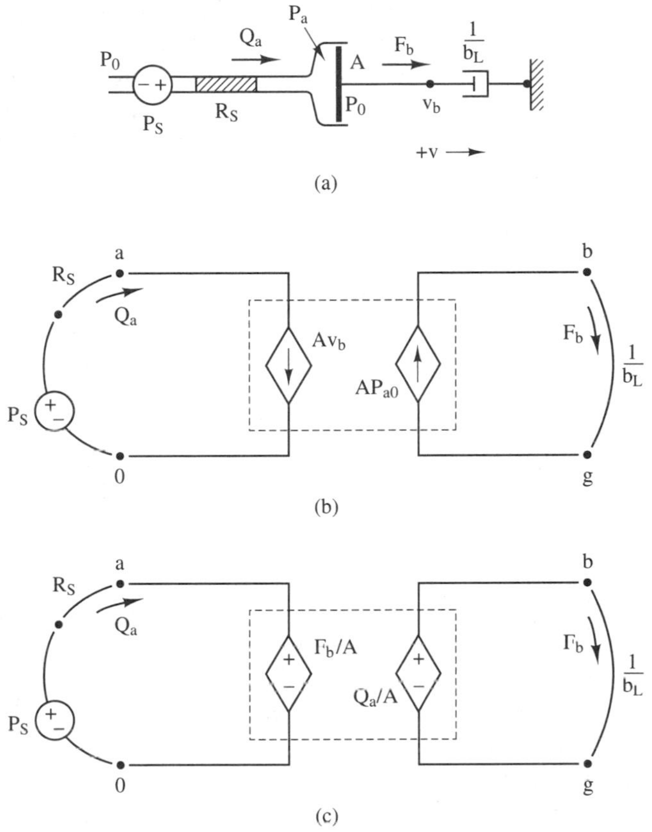

Figure 6.20 An ideal hydraulic cylinder in a hydromechanical network. (a) Lumped model. (b) Line-graph model. (c) Alternative line-graph model.

$$F_b = AP_{a0} \tag{6.33}$$

Equations (6.32) and (6.33) agree with the ideal hydraulic cylinder equations of Figure 6.17(a). Note the reversal of roles of the subscripts a and b in Figure 6.20, as compared with Figure 6.17(a). We have used the hydraulic port as the primary port. Also, note the reversal of the $+v$ direction and both flow-variable orientations in Figure 6.17(a), as compared with the hydraulic cylinder of Figure 6.20. Although the polarity arrows in Figure 6.17(a) oppose the directions of motion in Figure 6.20, reversal of the arrows does not change the *relative* polarities of the ports. Reversal of the $+v$ direction changes the signs of F_b and v_b. Reversal of the flow-variable orientations changes the signs of F_b and Q_a. The net effect is to keep the same signs on the ideal equations.

Most converters are *direct converters,* relating flow to flow and potential difference to potential difference (see Figures 6.16 and 6.17). The hydraulic cylinder, on the other hand, relates potential difference (pressure P_{a0}) to flow (compressive force F_b) and flow (volume flow rate Q_a) to potential difference (velocity v_b). We call a converter that reverses the roles of potential and flow a **gyrating converter.** We shall find that a gyrating converter changes the source and load equations much differently than does a direct converter.

Figures 6.20(b) and (c) show two dependent-source line-graph representations of the ideal hydraulic cylinder. The dashed lines function only to unite pairs of dependent sources conceptually. Compare the dependent-source representations for this gyrating converter with the dependent-source representations for the direct converter of Figure 6.19. Dependent-source models are convenient for simulation with SPICE and facilitate the writing of system equations by the loop and node methods.

Converter Polarity Marks

The behavior of an ideal converter is described by a pair of simple linear algebraic equations. Each secondary-port variable is proportional to a particular primary-port variable.[8] The *signs* on the proportionality constants depend on the orientations that we choose for the flow variables and the directions of positive motion. This section defines *polarity marks* that show the relative orientations of the physical variables at the two ports of a converter.

We first summarize, for each energy domain, the definition of positivity for each type of variable and the effects of those definitions on the expressions for power transfer. Then we discuss how to determine and interpret the converter polarity marks for each energy domain. In the next section we use these polarity marks to find orientations for the flow variables and directions of positive motion that make the signs of both converter equations positive.

Electric Circuits: In a lumped electric circuit, the voltage v at a node is the work required to transport a positive charge to that node from a reference node (section 4.1). Positivity of electric potential v, then, has been defined once and for all. We need only specify positivity for the electric currents. We use an arrow to specify a positive direction for the electric current i in a specific branch. The power delivered to an object between two electrical nodes is the product of the current passing through that object and the voltage drop between the nodes in the direction of that current. The electrical nodes are explicit at the electrical ports of Figures 6.16 and 6.17.

Hydraulic Circuits: The absolute pressure P at a node of a hydraulic network is defined in section 5.1. Thus, positivity of hydraulic potential P has been defined once and for all. We need only specify positivity for the volume flow rates. We use an arrow to specify a positive direction for the hydraulic flow Q in a specific branch. The power delivered to an object between two hydraulic nodes is the product of the flow Q through that object and the pressure drop between the nodes in the direction of that flow.

We usually express the pressure at a node relative to the pressure at a reference node, usually at ambient pressure. Ambient-pressure nodes do not appear explicitly at the hydraulic ports of Figures 6.16(e) and 6.17(a). For each of these ports we can *visualize* an explicit ambient-pressure node at some point in the ambient-pressure region.

[8] There are exceptions. The electric heater in Figure 6.17(f), for example, is nonlinear and unidirectional.

Translational Networks: We measure motions in a translational network relative to fixed ground nodes. We can assign locations for ground nodes arbitrarily. The ground node is an implicit second node for each translational port of Figures 6.16 and 6.17.

We choose a single +v direction to define positivity of motion v for all translational nodes that are connected to each other by lumped-element branches. We use an arrow to assign a positive direction for the scalar force F in each network branch. Unlike for electrical and hydraulic networks, for translational networks we must assign positivities for both the node velocity variables and the branch scalar-force variables. (The physical interpretations of the scalar-force variables—compressive or tensile—depend on the +v directions.) The power delivered to an object between two translational nodes is the product of the flow (scalar force) through that object and the velocity drop between the nodes in the direction of that flow.

Rotational Networks: Rotational networks are precise analogs of translational networks. The right-hand rule relates the two cases. We measure motions in a rotational network relative to a nonrotating ground node. We must assign positivities for both the node velocities Ω and the branch torsions T. We assign a single $+\Omega$ direction to define positivity of rotation for all nodes that are connected to each other by lumped-element branches. For each network branch, we use an arrow to assign a positive *linear* direction for the torsion T. The power delivered to an object between two rotational nodes is the product of the flow (torsion) through that object and the velocity drop between the nodes in the direction of that flow.

Thermal Networks: The temperature θ of a node in a thermal network represents the mean kinetic energy of a collection of molecules. Positivity of thermal potential θ, then, has been defined once and for all. We usually express temperature at a node relative to the absolute temperature of a reference object such as ice. We use an arrow to specify a positive direction for the heat flow q in a specific network branch. In a thermal system, power is not the product of potential-difference and flow. Rather, heat flow q is thermal power. Hence, thermal power q is delivered from a node (object) at one temperature to another node (object) at a different temperature. In other energy domains, the power delivered to a resistive object is dissipated (becomes thermal). In the thermal domain, power flows *through* thermal resistors and is accumulated in thermal capacitors.

In each energy domain except thermal, the power delivered to an object can be expressed as the product of the flow through the object and the potential drop across the object in the direction of that flow. We shall define the positivities of the converter port variables so that positive values of those variables correspond to power flow through the converter from source to load.

Polarity marks on a physical converter show the *relative* orientations of the physical quantities at the two ports of that converter. To find the polarity of a particular converter, we apply a constant source to one port, attach a **resistive load** to the other port, and then observe and mark the direction of steady-state motion or physical flow at each port.[9] Figures 6.18(a) and 6.20(a) show lumped-model examples of such polarity measurement structures.

For this source and load structure, power must flow out of the source into the primary port of the converter. Therefore, the potential must drop across the primary port in the direction of flow *through the converter,* and the potential must be higher at that node by which positive flow enters the primary port. Power must flow out of the secondary port into the load. Thus, the secondary port acts as a source to the resistive load, and the potential must rise across the secondary port in the direction of positive flow *through the converter.* The potential must be higher, then, at that node from which that flow emerges from the port into the load. (The potential must drop across the load in the direction of positive flow through the load.)

We use an arrow to mark a direction of physical motion or flow at a port. We use a dot to mark the higher potential node of a port. These two polarity marks are redundant. We can determine one from the other. Positive flow enters the dotted node of the primary port. Positive flow emerges (into the load) from the dotted node of the secondary port. Let us examine these quantities in the context of each energy domain.

We shall use an arrow to mark the direction of actual fluid flow at a **hydraulic port.** If the port is the primary, we denote that entering flow by Q_a. The pressure P_a at the entrance is higher than the reference pressure, and the input power is $P_{a0}Q_a$. If the port is the secondary, we denote that emerging flow by Q_b. The exit pressure P_b is higher than the reference pressure, and the output power is $P_{b0}Q_b$. In Figures 6.16(e) and 6.17(a), the reference pressure is the ambient pressure. In Figure 6.17(e), the reference pressure is the outlet pressure $P_0 = P_{a'}$.

We shall use a dot to mark the higher voltage node of an **electrical port.** (Sometimes + is used instead of a dot.) If the port is the primary, we denote the dotted node by the subscript a and denote the entering current by i_a. Then the input power is $v_{aa'}i_a$. If the port is the secondary, we denote the dotted node by the subscript b and denote the emerging current by i_b. Then the output power is $v_{bb'}i_b$.

We shall also use a dot to mark the higher temperature node of a **thermal port.** If it is the primary port, we denote the heat flow that enters the dotted node by q_a. It is the input power, and it is positive. If the port is the secondary, we denote the heat flow that emerges from the dotted node by q_b. It is the output power, and it is positive.

At a **translational port** we shall use an arrow to mark the direction of observed motion and assign that direction as the +v direction. Then the velocity at the explicit node of the port is positive, and we treat that node as a dotted node. If it is the primary port, the scalar-force variable oriented into that node from the source is positive. If it is the secondary port, the scalar-force variable oriented from that node into the load is positive. See the orientations in Figure 6.18, for example.

At a **rotational port** we shall mark the direction of observed rotation with a rotation arrow and assign that direction as the $+\Omega$ direction. (The right-hand rule assigns a linear

[9] For the thermal secondary port in Figure 6.17(f), we measure temperatures to determine the direction of heat flow through the resistive load. Heat flows from high temperature to low temperature.

$+\Omega$ direction to that rotation.) Then the velocity at the explicit node of the port is positive, and we treat that node as a dotted node. If it is the primary port, the torsion variable oriented into that node from the source is positive. If it is the secondary port, the torsion variable oriented from that node into the load is positive.

Polarity marks (arrows and dots) are shown on the pictorial models of Figures 6.16 and 6.17. For most of these converters, a viewer can see, from the way the converter works, how to place the two polarity marks. We show polarity marks on the ideal-converter symbol only if the relative orientations of those marks are not obvious from the structure of the symbol itself. Since the manner of operation is usually not visible in a physical transformer or transducer, manufacturers sometimes inscribe polarity marks on the physical device so that the user need not test its polarity.

Sign Conventions for Ideal-Converter Equations

The purpose of this section is to set sign conventions for converter equations. The polarity marks guide orienting of the port variables to make the signs of the proportionality coefficients positive.

The polarity marks of every converter correspond to a pair of *linear arrows.* Translational ports and hydraulic ports are marked directly with such arrows. At a rotational port, we use the right-hand rule to associate a linear arrow with the rotational arrow. At an electrical or thermal *primary* port, we orient the arrow to *enter* the dotted terminal. At an electrical or thermal *secondary* port, we orient the arrow to *emerge* from the dotted terminal. Reversing the pair of polarity arrows does not change the *relative* polarities of the two ports.

We orient the *flow variable* at each port to match the linear polarity arrow associated with the port. For a mechanical port, we orient the +v direction or the linear $+\Omega$ direction to match the polarity arrow. Then, all four port variables must have the same sign, and the proportionality coefficients in the converter equations must be positive. We used essentially this procedure to determine the signs on the lever equations (6.30) and (6.31) and the hydraulic cylinder equations (6.32) and (6.33). A summary of the procedure follows:

PROCEDURE TO PRODUCE POSITIVE CONVERTER-EQUATION COEFFICIENTS

1. Convert polarity dots and rotation arrows to linear arrows. Replace a primary-port dot by an arrow entering the primary port at the dotted node. Replace a secondary-port dot by an arrow emerging from the secondary port at the dotted node. Use the right-hand rule to replace a rotation arrow by a linear arrow.
2. Orient all port flow variables and directions of positive motion to match (or oppose) the polarity arrows. Then the signs on the constants in the converter equations are positive.

Exercise 6.5: Show that all four port variables must have the same sign if the port flow variables and directions of positive motion are oriented to match the linear polarity arrows.

Let us use this procedure to orient the variables for the ideal electrical transformer equations. We draw the lumped model, with its polarity dots, as shown in Figure 6.16(c). The dots indicate the higher voltage terminals. We label these dotted terminals a and b. We orient the primary current i_a to enter the dotted primary terminal and the secondary current i_b to emerge from the dotted secondary terminal. Then the equations that relate i_b to i_a and $v_{bb'}$ to $v_{aa'}$ must use positive signs, as shown in Figure 6.16(c).

Now suppose, instead, that we orient the flows and label the terminals arbitrarily at the two ports of the electrical transformer. Then we can still use the equations of Figure 6.16(c) if we replace i_a by the current into the dot $v_{aa'}$, i_b by the current out of the dot, and $v_{bb'}$ by the voltage drops from dotted to undotted terminals.

As a second example, let us orient the variables for the ideal displacement machine of Figure 6.17(e). We orient the primary flow variable Q_a to agree with the polarity arrow at the hydraulic port. Suppose the source produces positive Q_a. Since the source delivers power to the hydraulic port, $P_{aa'}$ must also be positive. We orient the $+\Omega$ direction for port b to agree with the rotation arrow marked on the gear so that Ω_b will be positive. Then we orient the torsion T_b at the secondary in the linear $+\Omega$ direction, out of node b toward the load, in order that T_b be positive and the device deliver power to the load. We can change any one of these variable orientations if we attach a negative sign to the corresponding variable in the equations.

For all the ideal converters of Figures 6.16 and 6.17, the directions of positive motion at the mechanical ports and the flow variables at both ports are oriented to agree with the polarity arrows. The flow variables are oriented as if port a were the primary and port b were the secondary. The transducers of Figure 6.17 can be used in either orientation. If a transducer is used in the reversed orientation, with port b as the primary, the same equations still apply (see example problem 6.4).

Lumped Modeling of Systems with Converters

The following two example problems demonstrate how to use an ideal converter as part of the lumped model for a physical system. They illustrate that most physical converters are not ideal. In general, we must attach a number of energy-storage elements and energy-dissipating elements to an ideal-converter model to represent the behavior of a physical converter adequately.

Example Problem 6.4. Lumped modeling of a hydraulic lift.
Figure 6.21(a) shows a hydraulic lift in an auto repair shop. The three-way valve has three operating positions. The two extreme valve positions connect the hose at the valve outlet to either the positive-displacement pump or the drain. The intermediate valve position closes the outlet to the hose.

Initially, the valve connects the outlet to the pump. At the flip of a switch, an electric motor begins to turn the pump shaft. The shaft reaches its final speed almost instantly. The pump extracts hydraulic oil from a reservoir (called a *sump*), forces it through the valve and hose into the lift cylinder, and raises the piston and automobile. When the lift reaches an appropriate height, the three-way valve is moved to its intermediate (closed) position, and the motor is turned off.

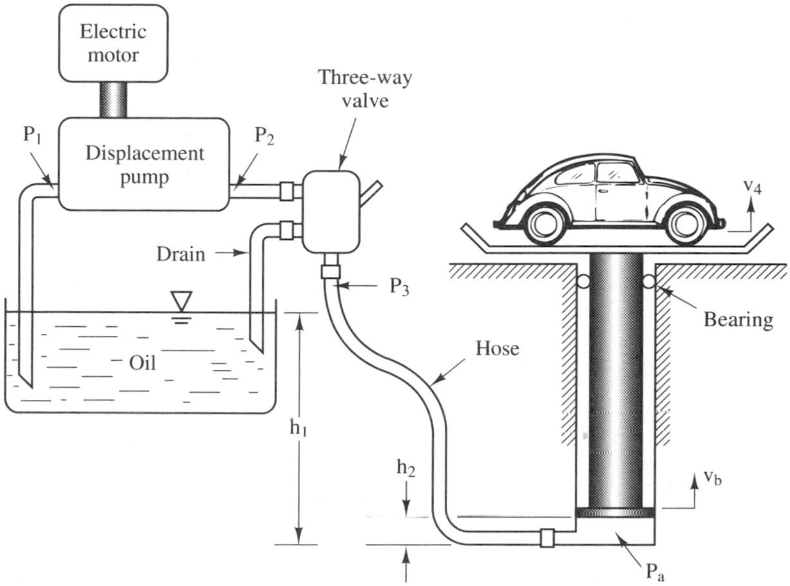

Figure 6.21 A hydraulic lift. (a) Pictorial model. (Continued)

When it is time to lower the lift, the valve is shifted to connect the hose to the drain line. Then the pressurized oil in the cylinder flows back to the oil reservoir, and the automobile sinks slowly to the floor. In practice, additional safety mechanisms are included in the system to prevent overextension of the piston or accidental draining of the cylinder.

(a) Construct a lumped model of the system. Define and label the potentials and flows. Write the ideal hydraulic cylinder equations in terms of those potentials and flows. Find a dependent-source representation for the hydraulic cylinder that is equivalent to the ideal lumped hydraulic cylinder.

(b) Draw a line-graph representation for the lumped model that applies during the lifting interval. Use the line graph to explain intuitively the behavior of the network. Explain the effect of each lumped element.

Solution: **(a)** A lumped model of the system is shown in Figure 6.21(b). An ideal positive-displacement pump is a flow pump (see the Figure 5.13(a)). The back pressure at the pump outlet causes some fluid to leak backwards through the pump mechanism. For that reason, we represent the pump by a flow source in parallel with a quadratic resistor. The pump reaches maximum speed abruptly, relative to the system time constants. Therefore, we give the source flow Q_S the constant value Q_c.

We treat the hydraulic port of the cylinder as the primary port. (This hydraulic cylinder is reversed relative to that of Figure 6.17(a).) We denote the hydraulic port variables by the subscript a and the translational port variables by the subscript b. It is apparent that the piston and the hydraulic fluid both move upward during the lifting process. Hence, we *visualize* upward-pointing arrows as polarity arrows at the two ends of the hydraulic cylinder. We orient the +v direction and the secondary flow F_b upward to match the secondary polarity arrow. We orient the primary flow Q_a upward to match the primary polarity arrow. Then the equations for the hydraulic cylinder have positive coefficients.

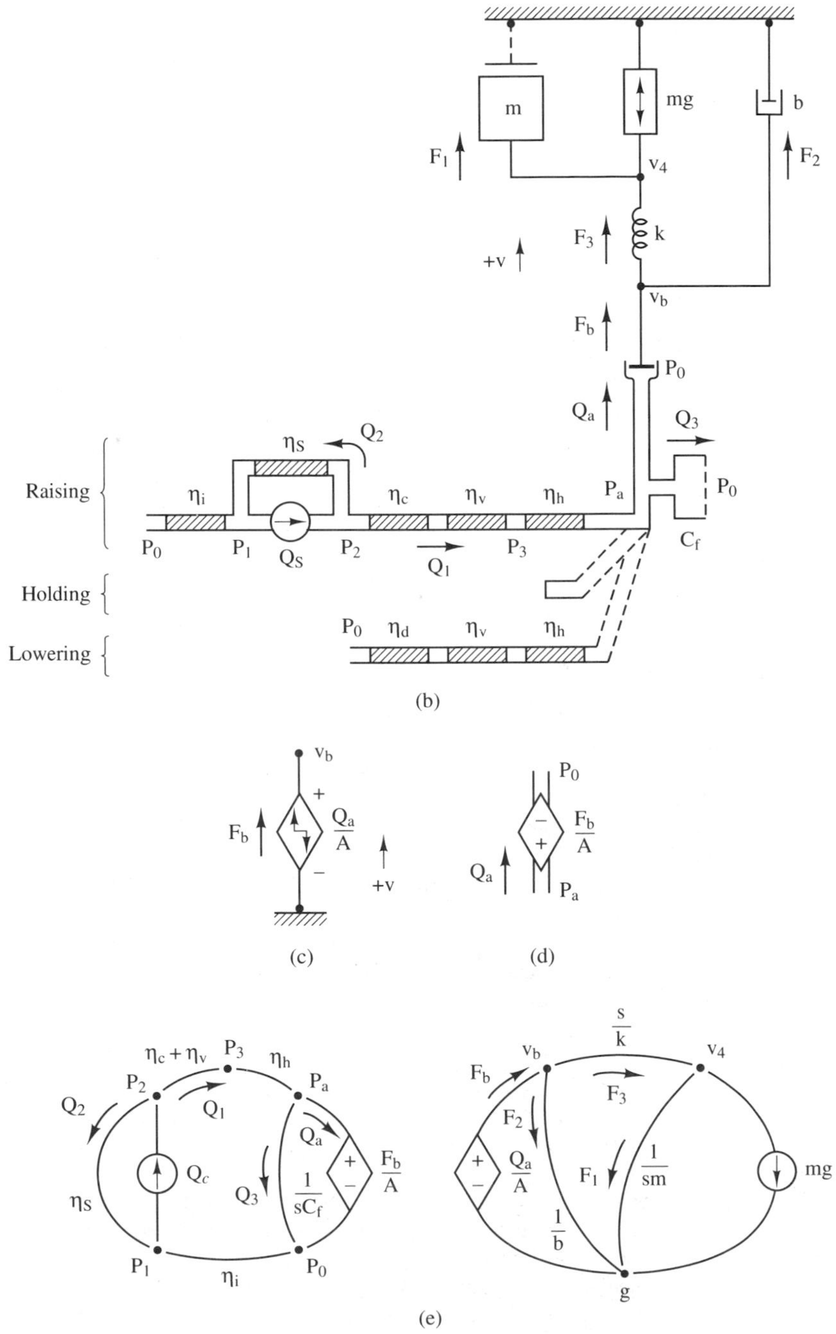

Figure 6.21 (Continued) (b) Lumped model. (c) Dependent-source model for the mechanical port. (d) Dependent-source model for the hydraulic port. (e) Line-graph representation.

The piston lifts the combined weight mg of the automobile and the support structure. We represent that weight by a scalar-force source. The inertial behavior of the load is characterized by the lumped mass m. The damper accounts for the mechanical friction in the piston oil seal and in the shaft bearings. (We treat the friction as linear because those friction points are lubricated.) Node 4 represents the velocity of the automobile. The spring between node 4 and the piston (node b) represents compliance in the support structure, primarily bending in the automobile platform.

We could account for the gravity-induced pressure owing to the oil in the cylinder and in the storage sump by inserting pressure sources ρgh_1 and ρgh_2 into the model. The heights h_1 and h_2 vary in proportion to the height x_4 of the automobile. Since the weight of the steel is much greater than the weight of the oil, we ignore these pressure sources. We could also attach a lumped mass m_p and a gravity-force source m_pg to node b to account for the mass of the piston. Instead, we assume that $m_p \ll m$.

The oil in the transfer hose and cylinder is under considerable pressure. We include the fluid capacitor C_f to account for stretching of the hose and the compressibility of the oil. We assume that all flows are turbulent. Therefore, we use quadratic resistors to represent hydraulic resistances in the intake line (η_i), the pump (η_S), the connector line (η_c), the valve (η_v), the transfer hose (η_h), and the drain line (η_d).

One might wonder whether η_h should be placed above or below the capacitor C_f. The simulated behavior at node 4 is probably about the same for either ordering. If the ordering were to make a significant difference, we would subdivide η_h and C_f and alternate resistors and capacitors to account for distributed behavior.

Note that the hydraulic structure is different during raising, holding, and lowering of the automobile. When the valve is moved abruptly to its closed position, there is a brief transient oscillation to achieve static balance among the stored energies in the compressed fluid (C_f), the bending structure (k), and the mass (m).

We have oriented the hydraulic cylinder port variables to make the equation coefficients positive. The equations for the hydraulic cylinder are $v_b = Q_a/A$ and $F_b = AP_{a0}$. These two equations correspond to the dependent velocity source of Figure 6.21(c) and the dependent pressure source of Figure 6.21(d). These two dependent sources could be inserted into the lumped model of Figure 6.21(b) in place of the ideal hydraulic cylinder symbol.

(b) The line-graph representation of Figure 6.21(b) shown in Figure 6.21(e) represents the ideal hydraulic cylinder by dependent sources that are mathematically equivalent to Figures 6.21(c) and (d). Of course, we can obtain these line-graph dependent-source models directly from the equations for the hydraulic cylinder without drawing Figures 6.21(c) and (d).

The network is third order. In general, its behavior can be complicated. Analysis is difficult because of the quadratic resistances. For specific values of the parameters, we could use a network simulation package to compute the behavior.

The system is intended to lift the weight of the automobile. The ideal behavior, then, would seem to be that of a constant-flow pump acting on an ideal hydraulic cylinder to lift the car weight. To observe this behavior in the lumped model, we delete (short-circuit) k, η_i, η_c, η_v, and η_h. Also, we delete (open-circuit) C_f, m, η_S, and b. Then the flow $Q_a = Q_c$ converts to $v_4 = v_b = Q_c/A$. According to this simplified model, the velocity continues to increase until the pump is shut off.

The hydraulic capacitance C_f prevents the abrupt source flow from producing an abrupt rise in pressure P_{a0} and, hence, slows the application of force to the automobile. The initial flow enters the capacitor (compresses the fluid and stretches the hose) instead of moving the piston. The mass m also prevents an abrupt jump in v_4. Hence, node 4 approaches its steady-state value gradually. The resistances reduce somewhat the steady-state pressure P_{a0}, flow Q_a, force F_b, and velocity v_4.

The spring (compliance) at the load interacts with the mass to produce a slight oscillation. The car tends to bounce up and down at a relatively rapid rate (about 2 Hz). As we discovered in sections 5.3 and 5.4, to excite a 2-Hz oscillation, the source must change as rapidly as a 2-Hz sinusoid. The mechanical signal that arrives at the load from the hydraulic subnet is more gradual than that because of the capacitor C_f. At shutoff, however, the flow is stopped abruptly. Hence, a brief decaying oscillation appears immediately after shutoff.

Exercise 6.6: The dependent sources of Figures 6.21(c) and (d) are both potential-difference sources. One is in the hydraulic domain, the other in the translational domain. Draw a pair of dependent flow sources that is equivalent in behavior to the pair of dependent potential-difference sources.

Electric Motor-Generators

An electric motor takes advantage of the force that a magnetic field exerts on a current-carrying wire. An electric generator takes advantage of the voltage that a magnetic field generates in a moving conductor. In the particular dc motor-generator illustrated in Figure 6.22, a stationary permanent magnet generates the magnetic field. A copper coil (called a winding) is wrapped on a steel support structure so that it can rotate on a shaft within the field. The ends of the winding are fastened to a pair of copper bars at the end of the shaft. When the winding is in a favorable orientation relative to the magnetic field, the bars contact a pair of carbon brushes connected to a pair of electrical terminals. Additional windings and bars are oriented at different angles around the rotating structure in order that one winding always be in contact with the terminals. The set of contactor bars is called the *commutator*. The whole rotating structure is called the *rotor* or the *armature*.

Let us describe the operation of the device as a motor. An electrical source produces a current in the winding. The magnetic field associated with the current in the winding reacts with the field of the permanent magnet to produce a torque on the armature shaft. The torque causes rotation of the armature and its attached load. The directions of the current and the torque (or motion) show the polarity of the motor.

According to Lenz's law, the subsequent motion of the armature winding in the magnetic field induces a voltage in the winding—the *generated voltage* or *back electromotive*

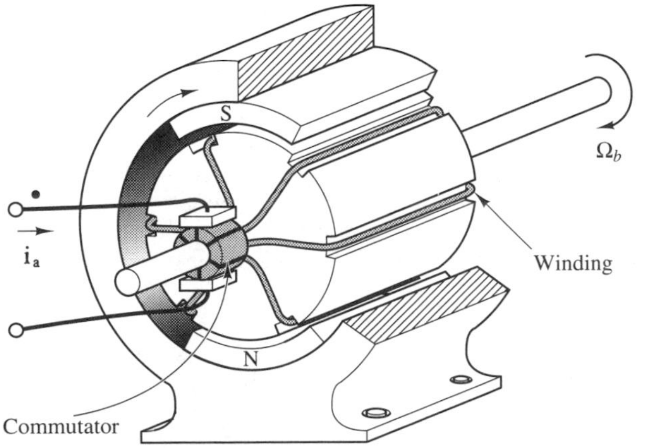

Figure 6.22 Pictorial model of a dc motor-generator. Adapted from Figure 2.24 on page 80 of Cannon, Robert H., *Dynamics of Physical Systems,* McGraw-Hill Book Co., New York, 1967.

force (emf)—in the direction that opposes the current. The faster the armature rotates, the greater is the voltage generated.

Intuitively speaking, an applied voltage causes current to flow through the resistance of the armature wire, the current causes a proportional torque on the motor shaft, and the torque causes increasing rotational velocity of the shaft and mechanical load. The rotation of the winding in the magnetic field generates a proportional back emf that opposes the applied voltage. The velocity and the voltage generated increase, and the *net* armature voltage and current decrease, until the motor achieves a stable velocity, torque, generated voltage, and current.

The device can also be operated as a generator if we replace the electrical source by an electrical load and use a mechanical source to rotate the armature. Intuitively, the applied torque causes increasing rotational velocity of the armature winding, rotation of the armature in the magnetic field generates a voltage proportional to velocity in the winding, and the voltage produces a current in the electrical load. As that current flows through the armature winding, it reacts with the magnetic field to produce a proportional torque that opposes the rotation of the mechanical source. The velocity, voltage, and current increase, and the *net* torque decreases, until the generator achieves a stable velocity, torque, voltage, and current.

Example Problem 6.5: Models for a dc motor-generator.

(a) Derive the equations for the *ideal* motor-generator (see Figure 6.17(c)). Represent them by an equivalent line graph.

(b) Develop a *realistic* lumped model for the *physical* motor-generator.

Solution: (a) We treat the device as a motor. Let the subscripts a and b denote the primary and secondary ports, respectively. We orient i_a into the dotted primary terminal and orient $+\Omega$ to agree with the rotation arrow marked on the motor case. Then, $i_a > 0$ produces $\Omega_b > 0$. Figure 6.23(a) shows an ideal-motor symbol with its polarity marks oriented to agree with the pictorial diagram of Figure 6.22. We define the primary-port node symbols (a and a′) and the secondary flow T_b to be consistent with the sign conventions of the previous section. That is, i_a should be oriented from a to a′, and T_b should be oriented in the linear $+\Omega$ direction. Then the converter constants must be positive.

The motor produces a torsion T_b in the armature shaft in proportion to the armature current i_a. If we denote the proportionality coefficient (or *torque constant*) by K_T, then

$$T_b = K_T i_a \tag{6.34}$$

(It can be shown that $K_T = 2NB\ell r$, where N is the number of *turns* in the winding, B is the magnetic flux density in webers/m^2 at the surface of the magnet, ℓ is the length of the winding perpendicular to the magnetic field, and r is the radius of the winding.)

When the armature rotates, the motor generates a back emf $v_{aa'}$ in proportion to the rotation velocity Ω_b. If we denote the proportionality coefficient (or *generator constant*) by K_G, then

$$v_{aa'} = K_G \Omega_b \tag{6.35}$$

The electrical source applies the power $v_{aa'} i_a = K_G \Omega_b i_a$ to the ideal transducer. The transducer, in turn, delivers the power $T_b \Omega_b = K_T i_a \Omega_b$ to the mechanical load. Since the

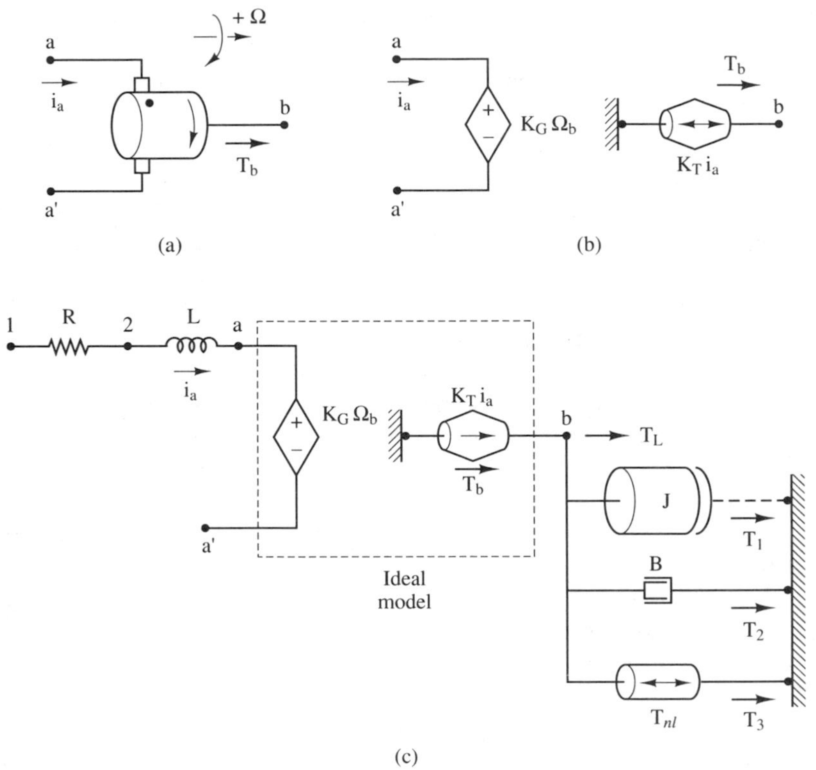

Figure 6.23 Lumped models of a motor-generator. (a) Ideal lumped model. (b) Dependent-source representation. (c) Lumped model with degrading elements.

ideal transducer is lossless, the electrical power entering the ideal-motor primary equals the mechanical power leaving the secondary. Therefore,

$$K_T = K_G \tag{6.36}$$

(This equality can be masked if the constants are not both expressed in SI units. Can you show that the units on the two sides of equation (6.36) agree?) We denote the parameter in equation (6.36) by $1/\kappa$ in Figure 6.17(c).

Figure 6.23(b) shows a dependent-source representation of the ideal motor-generator. This representation is identical in behavior to the ideal-device symbol and explicitly incorporates the ideal motor-generator equations.

(b) The armature winding is a coil of wire with resistance R and inductance L. (The coil has little capacitance, since its geometry has little similarity to parallel-plate geometry.) The rotation of the armature induces the voltage $K_G \Omega_b$ in the coil. The generated voltage, the resistance, and the inductance are distributed throughout the coil. The voltage drop between the physical armature terminals is the sum of the drops owing to resistance, inductance, and generated voltage. Therefore, we place the lumped resistor R and inductor L in series with the ideal-motor armature, which lies between nodes a and a′, as shown on the left side of Figure 6.23(c). Nodes 1 and a′ represent the physical terminals. Nodes 2 and a are fictitious nodes; they do not represent points with distinguishable voltages in the physical motor.

The motor shaft acts as a source of torsion T_L to the mechanical load. Not all of the torque T_b generated reaches the load, however. Some of it is expended on viscous friction in the lubricated bearings and in the cylinder of air between the armature and the case. We represent the viscous friction by the damper B in Figure 6.23(c). Some of the torque generated accelerates the rotor inertia, represented by J. There is also a small amount of coulomb friction, represented by a nonlinear torsion source T_{nl}, of magnitude $T_{coulomb}$. Coulomb friction always opposes the direction of motion. Thus, $T_{nl} = \text{sign}(\Omega_b)T_{coulomb}$, where

$$\text{sign}(\Omega_b) \triangleq \begin{cases} 1 & \text{for } \Omega_b > 0 \\ 0 & \text{for } \Omega_b = 0 \\ -1 & \text{for } \Omega_b < 0 \end{cases} \tag{6.37}$$

The load is fastened to the rigid shaft and rotates at the shaft velocity Ω_b. Therefore, we must connect the inertia and the friction source in parallel with the load. We model the mechanical portion of the motor, then, as shown on the right side of Figure 6.23(c). The flows T_1, T_2, and T_3 each represent an accumulation of small torques that are distributed throughout the armature. Only the load torsion T_L can be measured directly.

To complete the lumped model of the motor-generator, we must determine the values of the motor parameters. Table 6.1 shows the lumped-parameter values published by the manufacturer for a particular precision dc motor-generator. The following are appropriate procedures for measuring these parameter values:

- *To measure K_T:* Block the motor shaft (to eliminate rotation and the effects of inertia and friction). Apply a constant voltage to the armature terminals (1 and a′). Use an ammeter to measure the armature current i_a. To measure the steady-state load torque T_L, attach an arm perpendicular to the load shaft. Support the arm on a spring scale at a distance ℓ from the axis of rotation. The torque T_L is the product of ℓ and the force measured by the scale. Then $K_T = T_L/i_a$.
- *Alternatively, to measure K_G:* Open-circuit the armature winding (to eliminate the effects of resistance and inductance). Use another motor (perhaps an electric drill) to rotate the shaft at a constant speed. Use a voltmeter to measure the voltage $v_{1a'}$ at the armature terminals. Use a stroboscope to measure the shaft speed Ω_b. (A stroboscope is a light that flashes periodically at a controlled rate. When the rotating shaft is illuminated once per revolution, it appears to be stationary.) Then $K_G = v_{1a'}/\Omega_b$.

TABLE 6.1 MOTOR CONSTANTS FOR A KOLLMORGEN TT-2031 MOTOR-GENERATOR

$K_T = 0.052$ N·m/A	$R = 1.7$ ohms
$K_G = 0.052$ V·s/rad	$L = 1.55$ mH
$J = 1.04 \times 10^{-4}$ N·m·s²/rad	$\tau_M = 67.4$ ms
$B = 3.82 \times 10^{-5}$ N·m·s/rad	$\tau_E = 0.92$ ms
$T_{coulomb} = 0.025$ N·m	

- *To measure the resistance R and inductance L:* Block the rotor to eliminate rotation and the back emf $K_G\Omega_b$ in the armature winding. Apply a sinusoidal voltage $v_{1a'}$ of known frequency, say, $\omega = 2\pi(1{,}000 \text{ Hz})$, to the armature. Then i_a is sinusoidal with the same frequency. Use an oscilloscope to measure the amplitudes and the relative phase of $v_{1a'}$ and i_a. Then use these quantities to form the phasors $\boldsymbol{V}_{1a'}$ and $\boldsymbol{I}_a$. Compute R and L from the identity $R + j\omega L = \boldsymbol{V}_{ia'}/\boldsymbol{I}_a$.
- *To measure the friction torque $T_{coulomb}$:* Increase the armature current slowly, beginning at zero. Measure the current i_a at which the armature shaft just begins to turn. Use the relation $T_{coulomb} = K_T i_a$ to determine the coulomb friction torque. (We assume that the static and dynamic coulomb friction torques are essentially the same.)
- *To measure the viscous friction constant B:* Apply a substantial constant current i_a to the armature, and use the stroboscope to measure the steady-state velocity Ω_b of the unloaded motor. Because the speed is constant, J has no effect. The total torque generated is $K_T i_a$. Subtract the coulomb friction torque to compute $B = (K_T i_a - T_{coulomb})/\Omega_b$.
- *To measure the inertia J:* Remove the rotor from the motor case, and suspend it on a wire of known (or measured) rotational stiffness K. Give the rotor an initial twist, and measure the period p of the lightly damped oscillation. Calculate J from the resonant-period formula $\sqrt{J/K} = p/2\pi$. Alternatively, measure the dimensions of the rotor and weigh it to find its average density. Then use the inertia formulas of Figure 3.9 to compute J.

To test the adequacy of the two-port lumped model, we must measure the behavior of the motor-generator under a variety of steady-state and dynamic circumstances and compare that behavior to the behavior predicted by the model. In steady-state operation, the inductance and inertia of the model have no effect. We can display the full range of *steady-state motor* operating conditions as a set of torque-speed curves—curves of T_L vs. Ω_b, with either i_a or $v_{1a'}$ as a parameter. Figure 6.24 shows such curves for the motor parameters of Table 6.1.

Exercise 6.7: Torque-speed curves show the steady-state *motor* behavior of the motor-generator. What plots would you use to characterize the steady-state *generator* behavior? Sketch those plots for the parameters of Table 6.1.

To compare the actual steady-state *motor* behavior with the behavior predicted by the model, we apply various values of voltage $v_{1a'}$ to the armature winding and measure steady-state values for $v_{1a'}$, i_a, T_L, and Ω_b for various friction loads. For each voltage and load, the measured data can be superimposed as a data point on the torque-speed plot of Figure 6.24. The degree to which the measured data points agree with the model-derived curves of the figure is an indication of the accuracy of the model.

Exercise 6.8: The friction load can be either viscous or coulomb. Superimpose the load lines for a viscous load and a coulomb load on the curves of Figure 6.24.

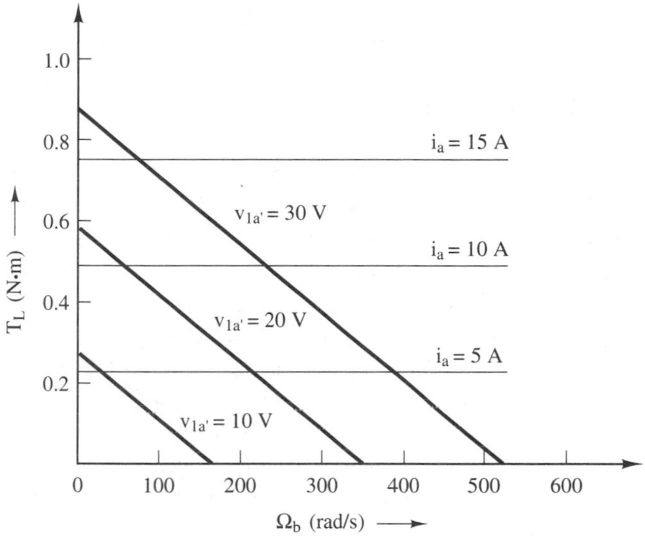

Figure 6.24 Model-derived torque-speed curves for the Kollmorgen TT-2031 motor.

To obtain the torque-speed measurements, we must measure the load torque T_L while the rotor is turning. Typically, we use a coulomb-friction load. We fasten a lever arm to the motor shaft by means of a leather friction brake with an adjustable grasp force. We measure the friction torque by supporting the arm on a spring scale.

Exercise 6.9: How would you measure the actual steady-state *generator* behavior of the motor-generator?

If we block the rotor and apply a step of voltage to the armature, the time for the current to rise to 63% of its final value is called the **electrical time constant** τ_E of the motor. If we apply a step of voltage to the *unloaded* motor, the current reaches its steady-state value before the rotor rotates appreciably. The time for the rotor of the unloaded motor to reach 63% of its final velocity is called the **mechanical time constant** τ_M of the motor. Both the electrical and the mechanical time constants are measures of the dynamic behavior of the device when operated as a motor. (The two time constants correspond to the poles of the transfer function, $\Omega_b/v_{1a'}$, of the *unloaded* motor.) Table 6.1 includes *measured values* for the electrical and mechanical time constants. Note that the value of the mechanical time constant is much greater than that of the electrical time constant. To test the adequacy of the lumped motor-generator model in *dynamic* operation, we can compare the measured values for τ_E and τ_M with the values predicted by the model.

Electrical Transformers

Most converters are low-pass devices: They work well only for source signals that do not vary rapidly. The reduction in response at higher frequencies owes to energy storage in the physical converter. The lumped inertia and inductor represent such energy storage in the motor-generator model of Figure 6.23. It is inappropriate to use *ideal* converter models—models that do not account for energy storage—if the source signals vary rapidly.

Electrical transformers, in contrast to other converters, are *band-pass* devices: They are limited in response at both low and high frequencies. Indeed, they cannot respond at all if the input signal is not changing. In this section, we derive an accurate lumped model for electrical transformers. Then we show the circumstances under which the ideal-transformer model of Figure 6.16(c) predicts the transformer behavior adequately.

An electrical transformer is constructed by winding a pair of coils so that they are magnetically coupled, as shown in the pictorial model of Figure 6.25(a). We denote the secondary current by i_2 rather than i_b because i_2 is oriented to oppose the current i_b of Figure 6.16(c). Both currents in Figure 6.25 are oriented toward the dotted terminals, as if both were delivering power to the transformer.

The primary-side current i_a generates magnetic flux linkages with the secondary coil. Hence, increases in i_a cause increases in the flux linkages. By Faraday's law, a voltage is induced in the secondary coil in proportion to the *rate of increase* of these flux linkages. Typically, this secondary voltage causes a current to flow in the secondary. The induced secondary voltage is oriented so that the secondary current will generate *opposing* flux linkages (Lenz's law). These opposing flux linkages, in turn, induce a voltage drop

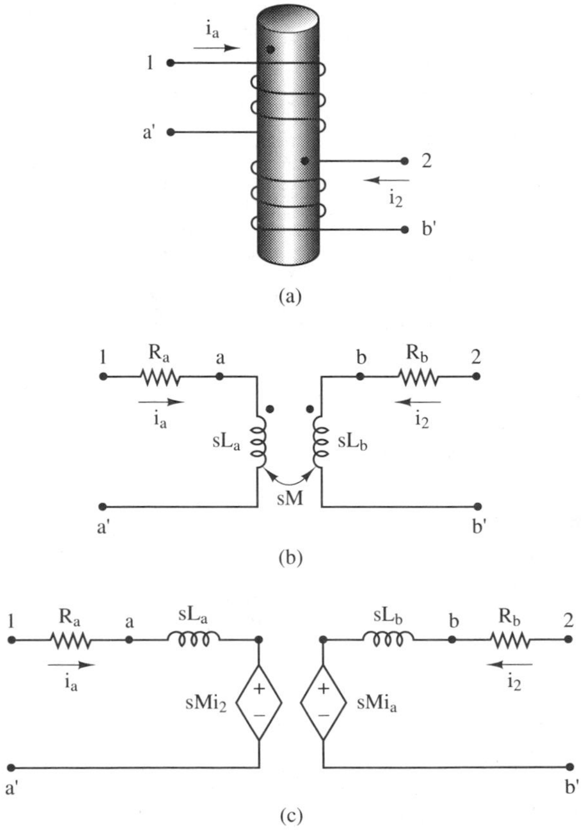

Figure 6.25 Electrical transformer models; M is the mutual inductance. (a) Pictorial model. (b) Lumped model. (c) Circuit model.

in the primary coil in the direction of i_a. Hence, power flows from the source into the transformer.

The **mutual inductance** of the transformer is the proportionality factor that relates the rate of change of current in one coil to the induced voltage in the other coil. It can be shown that the mutual inductance has the same value M in both directions, primary to secondary and secondary to primary. The dots show the orientations of the induced voltages. If di/dt is positive in one coil, then the induced voltage in the other coil is higher at the dotted terminal.

Figure 6.25(b) shows a realistic lumped model for an electrical transformer. The model shows explicitly the resistance and inductance in each coil. We position the inductors intentionally between the nodes labeled a and b. As this positioning suggests, we shall show that the self-inductances are inherent to the *ideal* lumped model, whereas the resistances are not. That is, an ideal electrical transformer is a pair of ideal inductors that interact with each other magnetically. We use the two-headed arrow in Figure 6.25(b) to imply sufficient closeness of the inductors to produce *mutual coupling* between them.

The voltage drop $v_{aa'}$ has two components. Owing to the *self*-inductance L_a of the primary coil, there is a voltage drop $L_a di_a/dt$ in the direction of i_a, from the dotted terminal to the undotted terminal. Owing to the *mutual* inductance M, there is a voltage drop Mdi_2/dt in the primary, from the dotted terminal to the undotted terminal. The two voltages add to produce

$$v_{aa'} = sL_a i_a + sMi_2 \tag{6.38}$$

Because of the symmetry between the lumped elements, the current orientations, and the dot locations, the voltage drop $v_{bb'}$ must be

$$v_{bb'} = sL_b i_2 + sMi_a \tag{6.39}$$

Equations (6.38) and (6.39) describe the behavior of the central portion of the transformer model—the portion between nodes a and b. If the coil resistances are very small (which will be the case if the wires are of relatively large diameter), these two equations describe the transformer behavior fully.

The second terms on the right sides of equations (6.38) and (6.39) each relate the voltage drop at one port to the current at the opposite port. We represent those terms by dependent sources in the circuit model of Figure 6.25(c). The polarity dots are unnecessary in the circuit model because the orientations of the induced voltages are shown explicitly in the dependent sources.

The difference between electrical transformers and other converters is apparent in the circuit model. The proportionality coefficient sM of the dependent sources includes the time-derivative operator s. That is, the mutual coupling is operative only for time-varying signals. For constant source current, the transformer output is zero.

It can be shown that $M^2 \leq L_a L_b$.[10] If either L_a or L_b is very small, then the *signal-transfer* parameter M must be small, and the transformer cannot produce any significant output. Thus, we cannot build an electrical transformer that has negligible self-inductance.

[10] See reference [6.4].

That is why we include the inductors L_a and L_b between the nodes a and b and treat them as part of the *ideal* transformer. The ratio

$$\alpha = \frac{M^2}{L_a L_b} \tag{6.40}$$

is known as the **coupling coefficient** of the electrical transformer. It is a measure of the degree to which the flux generated by one coil couples with the other coil. If the coefficient of coupling between the two inductors of a transformer is nearly 1, we say that the transformer is *tightly coupled*. Network simulation programs permit the user to specify the coupling coefficient for a pair of inductors. We could use the coupling coefficient α as an alternative to M to denote the coupling between coils in Figure 6.25(b).

Because of the magnetic field that surrounds every current, any conductor that carries a time-varying current in the vicinity of another conductor induces voltages in the neighboring conductor. That is, unintended transformer-type coupling occurs between *every* pair of neighboring conductors. This coupling is usually slight, unless the conductors are coiled together to form *inductors* with a shared flux path. On the other hand, a circuit that produces a large di/dt—say, an electric motor drive circuit—will definitely interfere with a neighboring circuit that senses very small voltages—say, a radio or television.

We define the output-to-input **turns ratio** of a transformer by

$$n = \frac{N_b}{N_a} \tag{6.41}$$

where N_a and N_b are the numbers of turns in the primary and secondary coils. The self-inductance of a uniform coil is proportional to the square of the number of turns in the coil (see Figure 4.13(d)). Therefore,

$$n = \sqrt{\frac{L_b}{L_a}} \tag{6.42}$$

We shall show that, for tightly coupled transformers, the output-port variables are related to the input-port variables by the turns ratio n.

To do so, we solve equation (6.38) for i_a and substitute the result into equation (6.39). If we let the coupling be tight ($\alpha \approx 1$), then $M \approx \sqrt{L_a L_b}$, i_2 drops out of the equation, and we are left with

$$v_{bb'} \approx n v_{aa'} \tag{6.43}$$

Alternatively, we can take the ratio of equation (6.38) to equation (6.39) and solve the resulting equation for i_2/i_a as a function of $v_{bb'}/v_{aa'}$. If we let the coupling be tight ($M \approx \sqrt{L_a L_b}$), then $v_{bb'}/v_{aa'}$ drops out of the equation, leaving

$$-i_2 \approx \frac{i_a}{n} \tag{6.44}$$

Equations (6.43) and (6.44) have the form typical of an ideal direct converter with converter constant n. We use these equations (with $i_b \triangleq -i_2$) to represent the ideal electrical transformer in Figure 6.16(c). Thus, an **ideal electrical transformer** is a tightly coupled transformer with negligible coil resistances.

We achieve tight coupling by winding one transformer coil tightly on top of the other. We can achieve high inductances by using many turns of wire. Tight winding of many turns requires that we use wire of small diameter. Then the coils have a lot of resistance.

Winding the coils on an *iron core* can produce extremely tight coupling. An iron core has high magnetic permeability, resists flux leakage, and increases the inductances greatly. The iron core, then, produces high inductances and tight coupling without large winding resistances. The resulting transformer is nearly ideal. For high flux density, however, iron becomes magnetically *saturated* and less permeable. Therefore, the ideal-model equations (6.43) and (6.44) apply to iron-core transformers only if the inductor currents are not too high.

We must be careful in our use of equations (6.43) and (6.44); they apply only for time-varying signals. Example problem 6.6 examines the frequency response of the lumped circuit model of Figure 6.25(c). The problem shows that the band of signal frequencies for which the ideal equations apply is determined by the transformer inductances and resistances.

Example Problem 6.6: Dynamic behavior of an electrical transformer.
Sound engineers often use an electrical transformer to match a sound source with a loudspeaker. Figure 6.26(a) shows such a circuit. The measured parameters of the transformer are $n = 0.1$, $L_a = 125$ mH, $L_b = 1.25$ mH, $R_a = 10\ \Omega$, $R_b = 1\ \Omega$, and $\alpha = 0.9$.
(a) Find the input-to-output transfer function $G(s) = v_{3b'}/v_S$ by using the detailed transformer circuit model of Figure 6.25(c). Also, find G(s) by using the ideal-transformer model of Figure 6.16(c).
(b) Find the frequency response $G(j\omega)$ of each circuit. Also, find the zero-state response $v_{3b'}(t)$ to the unit step $v_S = u_s$. Explain the significance of these two responses. Use the responses to evaluate the adequacy of the two transfer functions found in part (a).
(c) State how you would test the accuracy of the more detailed transformer model.
Solution: **(a)** Figure 6.26(b) uses the circuit model of Figure 6.25(c) to represent the transformer. The loop equations for the two detached subcircuits are

$$v_S = (sL_a + R_S + R_a)i_a + sMi_2 \tag{6.45}$$

$$0 = sMi_a + (sL_b + R_b + R_L)i_2 \tag{6.46}$$

We solve equation (6.46) for i_a and use the result to eliminate i_a from equation (6.45). Then we solve that equation for i_2 in terms of v_S and substitute the result into the load-voltage equation $v_{3b'} = -R_L i_2$ to obtain an equation that relates $v_{3b'}$ to v_S. We rearrange that equation to find the ratio

$$G(s) = \frac{v_{3b'}}{v_S}$$

$$= \frac{R_L M s}{(L_a L_b - M^2)s^2 + [L_a(R_b + R_L) + L_b(R_S + R_a)]s + (R_S + R_a)(R_L + R_b)} \tag{6.47}$$

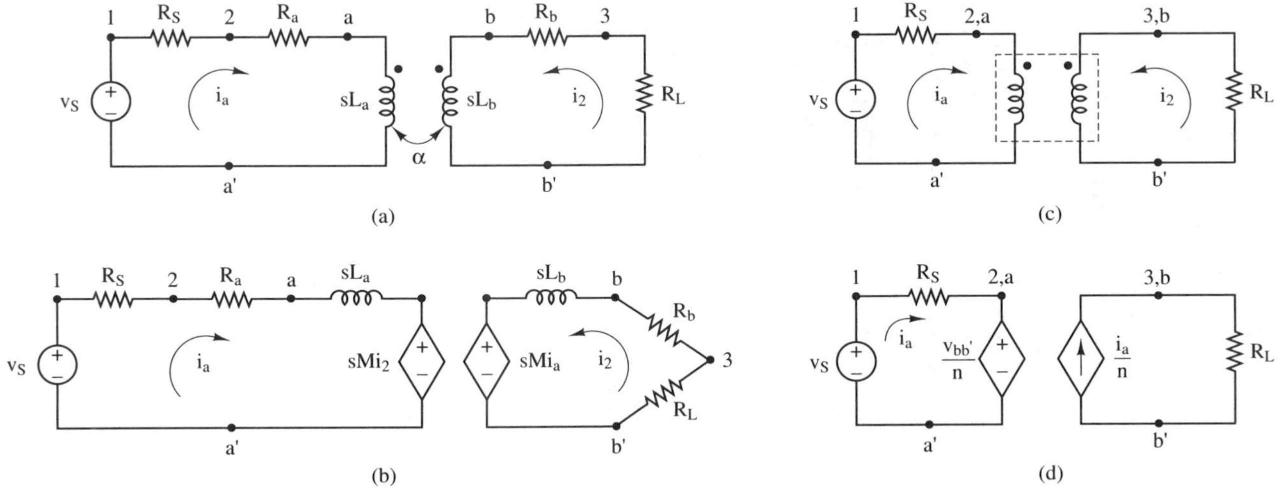

Figure 6.26 A transformer-coupled system. (a) The system with a detailed transformer model. (b) The detailed model in circuit form. (c) The system with an ideal transformer model. (d) The ideal model in circuit form.

For the specific parameter values given, $M = \sqrt{\alpha L_a L_b} = 11.9$ mH, and equation (6.47) yields

$$G(s) = \frac{3{,}050\ s}{s^2 + 7.3 \times 10^4\ s + 1.31 \times 10^8} \tag{6.48}$$

In Figure 6.26(c), we remove the coil resistances and represent the transformer by the ideal-transformer symbol. Note that nodes 3 and b of the circuit are now identical. The equations that describe the ideal transformer are $v_{bb'} = nv_{aa'}$ and $-i_2 = i_a/n$. In Figure 6.26(d), we replace the ideal transformer symbol by a pair of dependent sources that corresponds to this pair of equations. The left subcircuit is described by the loop equation

$$v_S = i_a R_S + \frac{v_{bb'}}{n} \tag{6.49}$$

and the right by the node equation

$$v_{bb'} = \frac{R_L i_a}{n} \tag{6.50}$$

We solve equation (6.50) for i_a and substitute the result into equation (6.49). From the remaining equation, we find the ratio

$$G(s) = \frac{R_L}{nR_S + R_L/n} \tag{6.51}$$

For the specific parameter values given, $G(s) = 0.05$.

The effort required to find the input-output transfer function is much reduced if we use the ideal-transformer model. The resulting transfer function is also much simpler in form—a constant instead of a ratio of polynomials in s. It is not obvious, however, that there is any similarity in the behaviors of the two transfer functions.

(b) We used the computer program CC to compute and plot the frequency response $G(j\omega)$ for the transfer function of equation (6.48). The response is shown in Figure 6.27. The band-pass effect of the electrical transformer is apparent in the frequency response curve. For signal components between 3,000 rad/s and 40,000 rad/s, the system does act nearly ideal, with

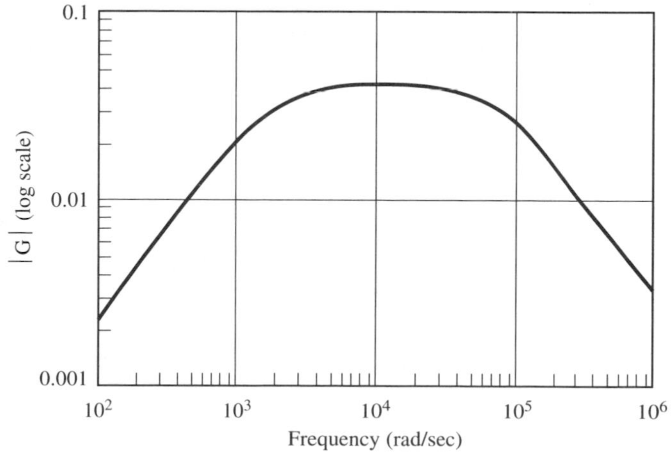

Figure 6.27 Frequency response at port 3, b′ for equation (6.48).

an output-to-input ratio of 0.04 (as compared with the value 0.05 predicted by the ideal transformer). The imperfect coupling and winding resistances cause the 20% reduction in output signal level in the pass band.

Let us define the **pass band** of the system as the band of frequencies for which the response does not drop from its maximum response by more than 3 dB. (A drop of 3 dB corresponds to halving the output power, a noticeable change.) The response drops by 3 dB at the two corner frequencies—the poles of the transfer function. The poles of equation (6.48) lie at $s = 1{,}840$ rad/s and $s = 71{,}200$ rad/s (293 Hz and 11,300 Hz). This transformer circuit, then, has a pass band from 293 Hz to 11.3 kHz.

We used the program CC to compute and plot the zero-state unit-step response of the circuit represented by equation (6.48). That is, we found $v_{3b'}(t)$ for $v_S = u_s$. According to the step-response plot (see Figure 6.28), the transformer output cannot respond instantly to an abrupt input. The *degree of responsiveness* of the transformer circuit is determined by the upper limit of the pass band—the corner frequency $\omega = 71{,}200$ rad/s. The time constant of the rise is $\tau_1 = 1/71{,}200$ rad/s $= 14\ \mu$s.

The output is barely able to reach the 0.04-V level predicted by the pass band of the frequency response before decay becomes appreciable. The transformer cannot sustain a constant output. The *rate of decay* toward zero is determined by the lower limit of the pass band—the corner frequency $\omega = 1{,}840$ rad/s. The time constant of that decay is $\tau_2 = 1/1{,}840$ rad/s $= 0.54$ ms.

It is appropriate to use the simpler ideal-transformer model during the design of the sound system. This model predicts the system behavior adequately for signals in the pass band. We must use the more accurate model, however, to determine the *extent* of the pass band and to predict the effect of the system on signals that have components that lie outside the pass band. If the pass band were much narrower than the one in this example, the circuit would not be able to follow *any* input signal, fast or slow. Then predictions made with the ideal-transformer model would be misleading.

(c) To test the adequacy of the more detailed lumped model for a particular physical transformer, we must compare the measured behavior of the physical transformer with the behavior predicted by the model of the transformer. To begin with, we must measure the parameters

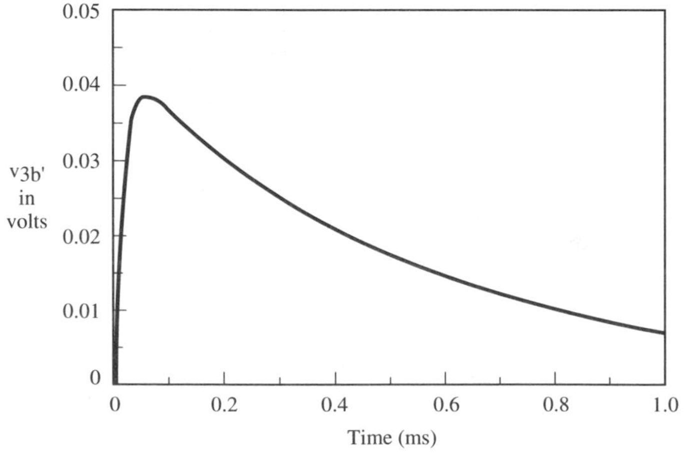

Figure 6.28 Unit-step response at port 3, b′ for equation (6.48).

R_a, L_a, R_b, L_b, and M of the lumped transformer model. (Instead of measuring M, we can measure the turns ratio n and the coupling coefficient α; see equations (6.40) and (6.42).)

We connect the physical transformer in a circuit like that of Figure 6.26. Then we measure the steady-state amplitude response of the transformer for sinusoidal signals of various frequencies to get data points that can be superimposed on the frequency-response curve of Figure 6.27. The differences between the measured and computed frequency-response values indicate how well the model predicts the behavior of the transformer.

An alternative test, using the same circuit, is to measure the response of the transformer to a step input and compare that response with Figure 6.28. Because the response to the step input changes so quickly, it is difficult to record the response values. Therefore, we might choose to measure only key features of the step response—e.g., its peak value and the time constants determining its rise and fall.

In section 5.4, we discovered that the information in the step response is equivalent to the information in the frequency response. Both tests check the *dynamic* operation of the transformer. (The transformer will not operate in a *constant* steady state.) Thus, we need only carry out one test or the other to judge the adequacy of the model.

Exercise 6.10: Carry out the steps to verify equations (6.47) and (6.48).

Exercise 6.11: How could you measure the transformer parameters R_a, L_a, R_b, L_b, and M?

Exercise 6.12: How could you measure the rise and fall time constants for a rapidly changing signal?

The Transformative Effect of Direct Converters

In the translational system of Figures 6.18 and 6.19, we use a lever to transform signal levels. We raise the level of the force at the expense of the level of the velocity or displacement. As a consequence, we also transform the apparent impedances of the source and load, as seen from the opposite side of the lever.

In the hydromechanical system of Figure 6.20, we use a hydraulic cylinder to convert hydraulic actions to mechanical actions. As with the lever, the device transforms signal levels and impedance levels. These transformations are obscured by the fact that the source and load are in different energy domains.

In this section and the next, we examine the transformative effect of linear passive converters (transducers and transformers) on their sources and loads. That is, we examine how the source and load appear as viewed from the opposite side of the converter. The lever is a *direct* converter; the hydraulic cylinder is a *gyrating* converter. We shall find that direct converters and gyrating converters transform the variables differently. We use the context-free notation of potentials and flows (A and T) to expose the fundamental transformative properties of converters.

Figure 6.29(a) shows a line-graph structure for an ideal direct converter. This structure corresponds to the structure of the lever model in Figure 6.19(a). Figure 6.29(b) shows a line-graph structure for an ideal gyrating converter. This structure corresponds to that of the hydraulic cylinder model in Figure 6.20(b). Compare the direct-converter structure of Figure 6.29(a) with the gyrating-converter structure of Figure 6.29(b). We have oriented the flow variables in Figure 6.29 to be consistent with the orientations for the ideal converters of Figures 6.16 and 6.17. Hence, we expect a positive value of $A_{aa'}$ to correspond to positive values of T_a, $A_{bb'}$, and T_b.

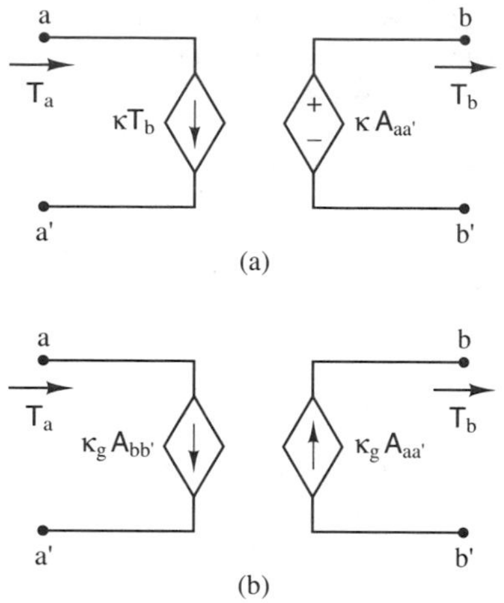

Figure 6.29 Ideal converter structures. (a) Direct converter. (b) Gyrating converter.

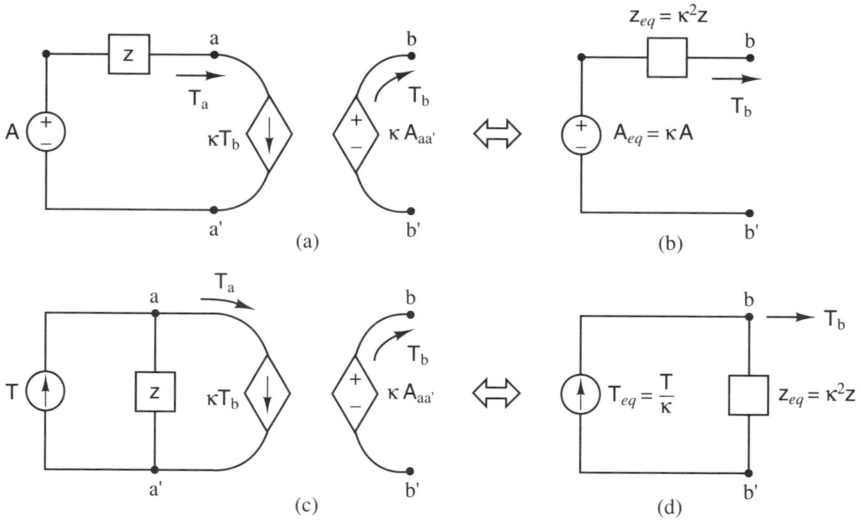

Figure 6.30 Transformative effect of an ideal direct converter. (a) The converter with a Thévenin source. (b) The output Thévenin equivalent. (c) The converter with a Norton source. (d) The output Norton equivalent.

First, we examine the transformative effect of a direct converter. In Figure 6.30(a), the Thévenin equivalent of the source subnet is attached to the left port of the direct converter of Figure 6.29(a). The equation that describes the behavior of the source subnet is

$$A_{aa'} = A - zT_a \tag{6.52}$$

The equations that describe the behavior of the direct converter are

$$A_{bb'} = \kappa A_{aa'} \tag{6.53}$$

$$T_a = \kappa T_b \tag{6.54}$$

where κ is a dimensionless **converter constant.** (It is the dimensionless nature of the converter constant that marks a direct converter, one which transforms potentials to potentials and flow to flows.) We substitute equation (6.52) into equation (6.53) and then substitute equation (6.54) into the result to obtain

$$A_{bb'} = (\kappa A) - (\kappa^2 z)T_b \tag{6.55}$$

Compare equation (6.55) with equation (6.52). The equivalent potential of the source, *as seen through the converter* at port b, b′ is κA. The equivalent impedance as *seen* through the converter is $\kappa^2 z$. Figure 6.30(b) shows the Thévenin equivalent subnet that corresponds to equation (6.55).

If, instead, we represent the source in Norton form, as shown in Figure 6.30(c), the equivalent flow of the source as *seen* at port a is $T = A/z$. The corresponding equivalent flow as *seen* at port b is $\kappa A/\kappa^2 z = T/\kappa$ (see Figure 6.30(d)).

Inserting the direct converter between the source and load produces a change in the *apparent* behavior of the source. As viewed at the load, the converter multiplies source potentials by κ, divides source flows by κ, and multiplies source impedances by κ^2. This knowledge about the effect of the converter guides us to a simpler equivalent network. We can remove the ideal converter from the network model if we *modify every lumped element of the source subnet* according to these transformation rules. Removing the converter from the model does not change the potentials and flows in the *load subnet* if we modify the source subnet in this fashion.

Now let us examine the apparent change in the load as *seen* by the source. If we represent the load by a Thévenin equivalent subnet and carry out an analysis similar to that just carried out for the change in the source as seen by the load, we discover the same form of transformation, but with κ replaced by $1/\kappa$. (The quantity $1/\kappa$ is the converter constant in the reverse direction.) Hence, if the converter increases the apparent impedance of the source, it must decrease the apparent impedance of the load correspondingly.

To obtain a simpler equivalent network, we remove the ideal converter from the network model. Then we divide every potential-difference source in the load subnet by κ, multiply every flow source in the load subnet by κ, and divide every impedance in the load subnet by κ^2. (In general, load subnets can contain sources.) Removing the converter does not change the potentials and flows in the *source subnet* if we modify the load subnet in this fashion.

A source or impedance that is connected in series (parallel) with one side of an ideal *direct* converter can be moved to the other side, in series (parallel) and appropriately scaled, without changing the behavior of the rest of the network. This fact enables us to rearrange the network for our convenience. If a converter is not ideal, the lumped model for the converter includes additional lumped elements attached to the terminals of the ideal model. Those additional elements act as if they are part of the source or load subnets.

Let us apply the foregoing conclusions to the lever of Figures 6.18 and 6.19. The variables are $T_a \equiv F_a$, $A_{aa'} \equiv v_a$, $T_b \equiv F_b$, and $A_{bb'} \equiv v_b$. The converter constant is $\kappa = \ell_b/\ell_a$. We can replace the source subnet and lever by a flow source F_S/κ in parallel with the impedance κ^2/b_S (see Figure 6.31). If the lever is used to magnify the force of the source, then $\kappa < 1$, and the lever appears to reduce the impedance of the source, as

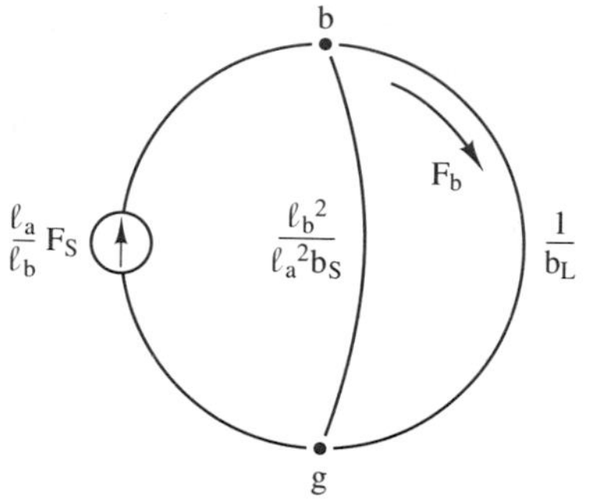

Figure 6.31 A network equivalent to Figure 6.19(a).

seen by the load, by the factor κ^2. The apparent friction constant of the source (the inverse of impedance), as seen by the load, is *increased* to b_S/κ^2. The lever has the opposite effect on the impedance of the load, as seen by the source: The apparent friction constant of the load, as seen by the source, is *decreased* to $\kappa^2 b_L$.

Exercise 6.13: Draw the physical lumped model that corresponds to the line graph of Figure 6.31. Compare the model with Figure 6.18(a).

Suppose that the lever is heavy enough that we must include its mass in the model. Logically, we would attach the mass associated with each end of the lever to the corresponding end of the model. Hence, we attach to Figure 6.19(a) an impedance $1/sm_a$ in parallel with the primary port a, g and an impedance $1/sm_b$ in parallel with the secondary port b, g. Because of the transformative properties of the ideal lever, however, we can move the additional parallel primary-port impedance to the secondary if we transform its value to κ^2/sm_a. Then the two impedances κ^2/sm_a and $1/sm_b$ are in parallel. Accordingly, we can replace them by a single equivalent impedance corresponding to a mass of size

$$m_{eff} = \frac{m_a}{\kappa^2} + m_b \tag{6.56}$$

Exercise 6.14: Attach the lumped masses m_a and m_b to the two ends of the lever of Figure 6.18(a) as just described. Draw the modified lumped model that results when the mass of the lever is represented by equation (6.56).

Exercise 6.15: The mass of the lever can also be represented by a single lumped mass at the primary port. Show that the value of that mass is $m_a + \kappa^2 m_b$.

The Transformative Effect of Gyrating Converters

Now let us examine the transformative effect of a gyrating converter. In Figure 6.32(a), the Thévenin equivalent of the source subnet is attached to the primary side of the converter. The source subnet is described by equation (6.52). The equations that describe the behavior of the gyrating converter are

$$T_b = \kappa_g A_{aa'} \tag{6.57}$$

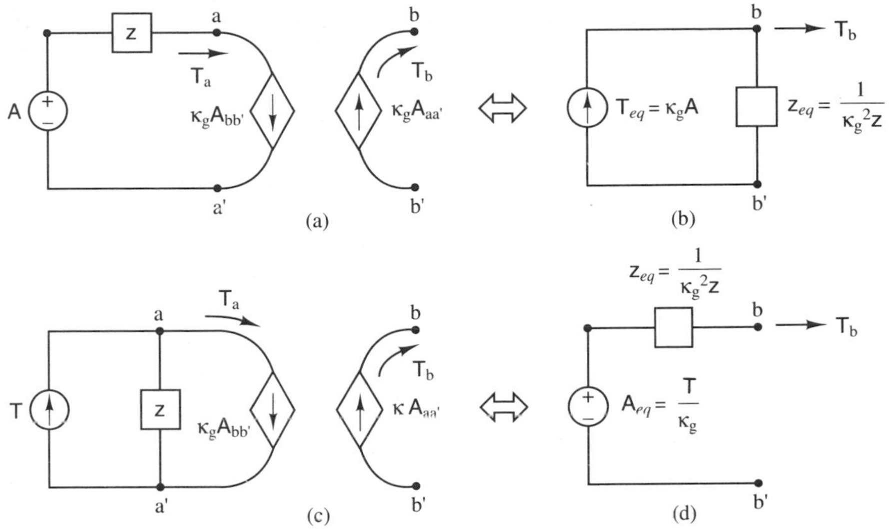

Figure 6.32 Transformative effect of a gyrating converter. (a) The converter with a Thévenin source. (b) The output Norton equivalent. (c) The converter with a Norton source. (d) The output Thévenin equivalent.

$$T_a = \kappa_g A_{bb'} \tag{6.58}$$

where the **converter constant** κ_g has the units of **admittance** (reciprocal impedance). It is the role-reversing nature of the converter constant that marks a gyrating converter, one which transforms potentials to flows and flows to potentials.

We substitute equation (6.52) into equation (6.57) and then substitute equation (6.58) into the result to produce

$$T_b = (\kappa_g A) - (\kappa_g{}^2 z) A_{bb'} \tag{6.59}$$

Equation (6.59) corresponds to a Norton equivalent representation of the source as *seen* through the converter. We show that Norton equivalent in Figure 6.32(b). The transformation is strikingly different from that of Figure 6.30. The gyrating converter makes the potential-difference source A *appear* (to the load) as a flow source of value $\kappa_g A$. It makes the series impedance z *appear* as a parallel admittance of value $\kappa_g{}^2 z$. Thus, it **reverses the roles of all quantities.**

If, instead, we represent the source subnet in Norton form, as shown in Figure 6.32(c), the equivalent flow of the source, as *seen* at port a, is $T = A/z$. The gyrating converter transforms the flow T into an equivalent *potential* of value T/κ_g, as shown in Figure 6.32(d).

The mathematical operations appear the same in Figures 6.30 and 6.32. That is, the effect of the gyrating converter is to multiply all source potentials by κ_g, divide all source flows by κ_g, and multiply all source impedances by κ_g^2. However, the *effect* is quite different for the gyrating converter, owing to the different units for the gyrating converter constant κ_g.

From the *viewpoint* of the source, how does the gyrating converter transform the load? According to equations (6.57) and (6.58), the gyrating converter behaves the same in both directions. Therefore, it multiplies all load potentials by κ_g, divides all load flows by κ_g, and multiplies all load impedances by κ_g^2. Again, this behavior contrasts with that of the direct converter.

A source or impedance that is connected in series (parallel) with one side of an ideal *gyrating* converter can be moved to the other side, in parallel (series) and appropriately scaled, without changing the behavior of the rest of the network. This fact enables us to rearrange the network for our convenience. Note, however, the reversal in network structure during such a move: Series connections become parallel connections and vice versa, impedances become admittances and vice versa, and flows become potentials and vice versa. We demonstrate these network manipulations for the hydraulic cylinder of Figure 6.20. The demonstration shows that a gyrating converter also *reverses the type of an energy-storage element.*

Example Problem 6.7: Transforming a source with a gyrating converter.
Let us modify the hydraulic cylinder of Figure 6.20 to produce Figure 6.33(a). The fluid capacitor C accounts for the compressibility of the fluid in the cylinder. The lumped mass m_p accounts for the mass of the piston.

(a) Use the transformative properties of the hydraulic cylinder to transfer the primary-side (hydraulic subnet) elements, one at a time, to the secondary (mechanical) side of the cylinder.

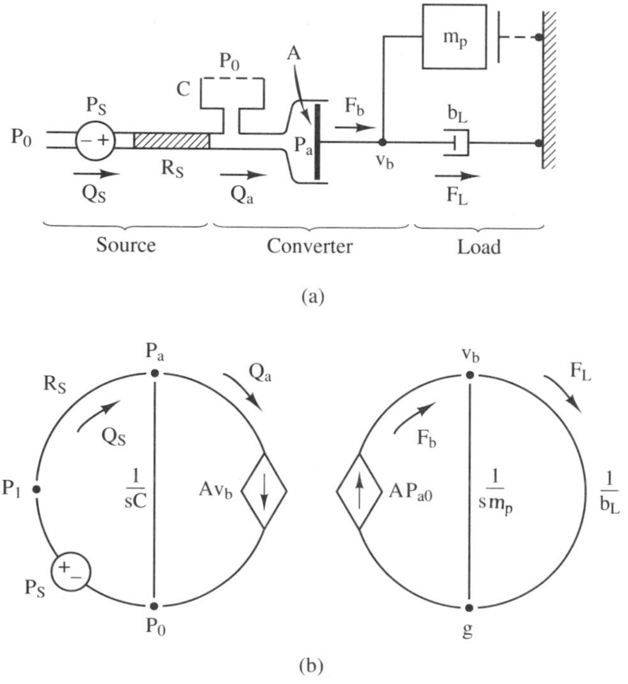

Figure 6.33 Lumped model of a system with a hydraulic cylinder. (a) Lumped model. (b) Line-graph representation.

(b) The sequence of transfers in part (a) replaces all hydraulic elements by mechanical equivalents and eliminates the converter. Draw the equivalent lumped mechanical network, and explain the equivalence intuitively.

Solution: (a) A line-graph representation of the network is shown in Figure 6.33(b). We can move lumped elements from the input port to the output port of the ideal converter if we transform them properly. The variables involved in the transformation are $A_{aa'} \equiv P_{a0}$, $T_a \equiv Q_a$, $A_{bb'} \equiv v_b$, and $T_b \equiv F_b$. The converter constant for the *ideal* cylinder is $\kappa_g = A$.

The fluid capacitor (with impedance 1/sC) is connected in parallel with the input port of the ideal converter. We wish to transfer it to the output port. According to the transformative properties of a gyrating converter, we multiply the impedance of the element that is in parallel with the primary port by κ_g^2 to produce an admittance, and place an element with that admittance in series with the secondary port. The admittance of the transformed fluid capacitor is $A^2/(sC)$. Therefore, we place a lumped element of impedance sC/A^2 in series with the output port of the ideal converter. We show this transformation in Figure 6.34(a). The transformation process moves the time-derivative operator s from the denominator to the numerator. Thus, the parallel capacitor on the hydraulic side transforms to a series spring on the mechanical side.

Moving the fluid capacitor to the secondary side leaves the fluid resistor of impedance R_S in *series* with the primary port of the ideal converter. We multiply the impedance of the element by κ_g^2 to produce the admittance A^2R_S, and we place a lumped element with that admittance in *parallel* with the secondary port of the ideal converter. Hence, the series resistor on the hydraulic side transforms into a parallel damper of impedance $1/(A^2R_S)$. We show this transformation in Figure 6.34(b).

Note that the variables at the terminals of the ideal converter change with each element transfer. The transformed network is *equivalent* to the original network in the sense that the variables *in the unchanged portions* of the attached subnets (F_b, v_b, and Q_S) are unaffected. The variable Q_a does not exist in the new subnet. The velocity v_1 does not exist in the original subnet.

Moving the fluid resistor to the secondary side leaves only the pressure source connected to the primary port of the ideal converter. According to the transformative properties of the gyrating converter, we multiply the potential-difference source by κ_g to produce a flow source. In this instance, that flow source has value AP_S. Hence, the pressure source, as *viewed* through the ideal converter, appears to be a force source of value $F_S \equiv AP_S$. We show this transformation in Figure 6.34(c).

(b) Figure 6.35(a) shows the accumulated effect of the sequence of element transfers in line-graph form. The corresponding mechanical lumped model is given in Figure 6.35(b). According to this lumped network, the pressure source in the original system *appears* as a force source to the mechanical load. The fluid capacitance (owing to the compliance of the fluid in the cylinder) *appears* as springiness (mechanical compliance) to the mechanical load. The fluid energy dissipation in the hydraulic resistor is indistinguishable from mechanical energy dissipation in a damper.

Exercise 6.16: Check the units of each of the three transformed elements in the mechanical lumped model in Figure 6.35(b) to see that they correspond to impedances and a flow, as stated.

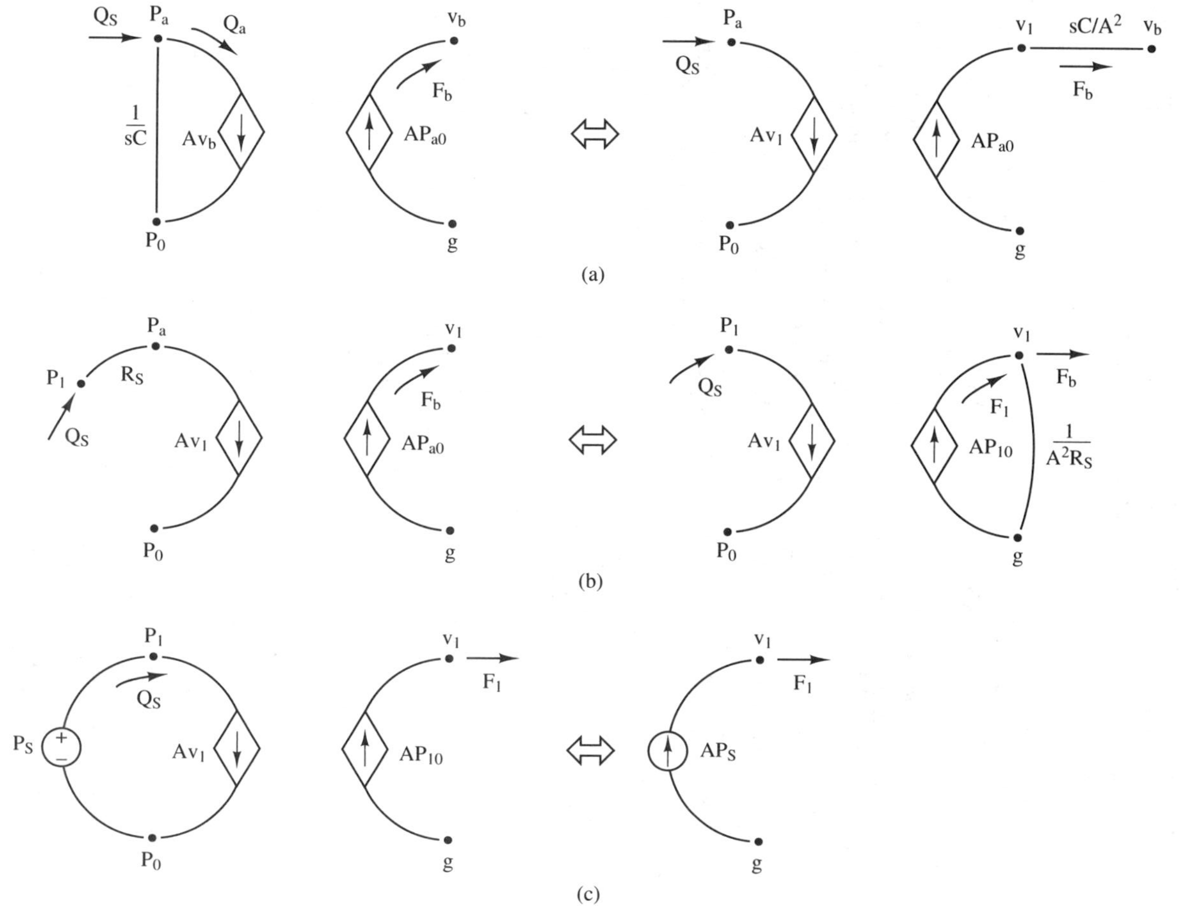

Figure 6.34 Equivalent-network transformations for a gyrating converter. (a) Transfer of a parallel capacitor to the secondary side. (b) Transfer of a series resistor to the secondary side. (c) Transfer of a pressure source to the secondary side.

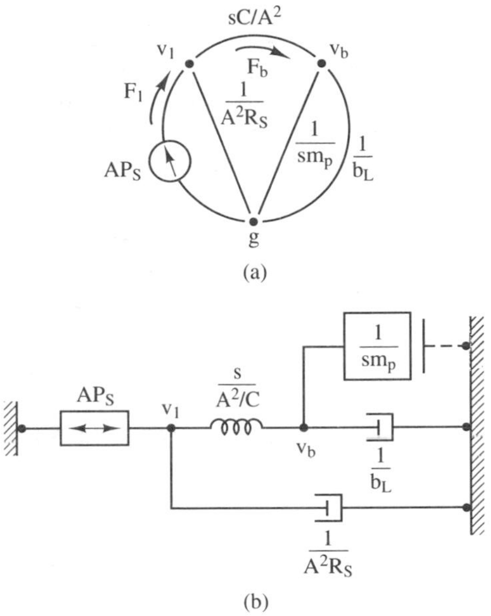

Figure 6.35 Mechanical equivalent of the hydromechanical network of Figure 6.33. (a) Line-graph representation. (b) Lumped network.

6.3 POWER AMPLIFIERS

The independent source signal that energizes a system typically originates in the measurement of some quantity in another system. The measured quantity might be a natural variable, such as wind velocity. Or it might be conceived and executed by a human—say, accelerator pedal motions or words spoken into a microphone. Or it might be information stored by a human, such as magnetic variations on a tape.

The sensor that makes the measurement produces a very weak signal, so weak that the system attached to the sensor will not respond to its command. We can think of the sensor as a **waveform source.** We must use a separate signal strengthener—such as an engine or an electronic amplifier—to force the system to respond as commanded by the waveform source. We call the signal strengthener a **power amplifier.** Together, the waveform source and power amplifier form an ideal independent source.

This section examines power amplifiers—valvelike devices that control the flow of power from an energy reservoir. Unlike other two-port devices, power amplifiers are not passive: They deliver more *signal power* to the output port than they receive at the input (or controlling) port. The extra power comes from the energy reservoir.

In the early parts of this section, we show that mismatching impedances at the sensor and at the load is the key to accurate measurement and to efficient energizing of the load. Then we look at the properties of amplifiers. Finally, we study design details of a few important power-amplifying devices—transistors, hydraulic control valves, and internal combustion engines.

Available Power of a Source

Every physical source is limited in its ability to provide power. We represent that limitation in our source models by means of source impedances. In particular, the source *resistance* characterizes the limited ability of the source to deliver power *on the average*. In discussing average power, we shall determine how to distinguish between a flow source and a potential-difference source and decide what constitutes light loading of a source.

Let us attach a linear dissipative load to a linear source, as shown in Figure 6.36. We denote by node letters i and j the two terminals of the port at which the load is attached to the source. Then A_{ij} is the potential drop across the port. Let T denote the flow at that port.

Figure 6.37 shows the steady-state operating characteristic for the source.[11] The slope of the operating characteristic is $-R_S$. (We use the symbol R and the term *resistance* to describe a linear dissipative element in any energy domain.) The operating characteristic (or load line) for the load, with slope R_L, is also shown in the figure.

The intersection of the source characteristic and the load characteristic is the steady-state operating point (A_{ij}, T). The power delivered by the source to the load is the prod-

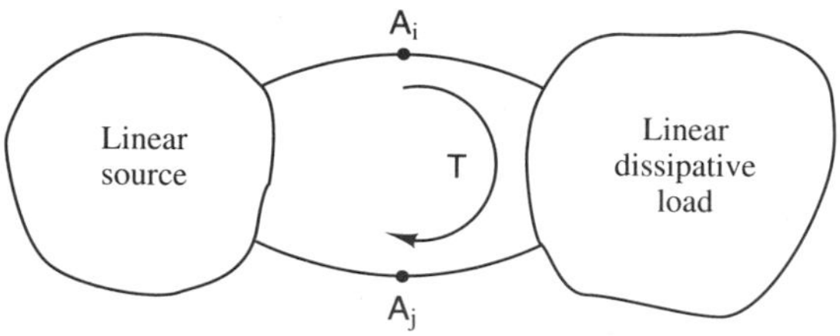

Figure 6.36 A linear source and load.

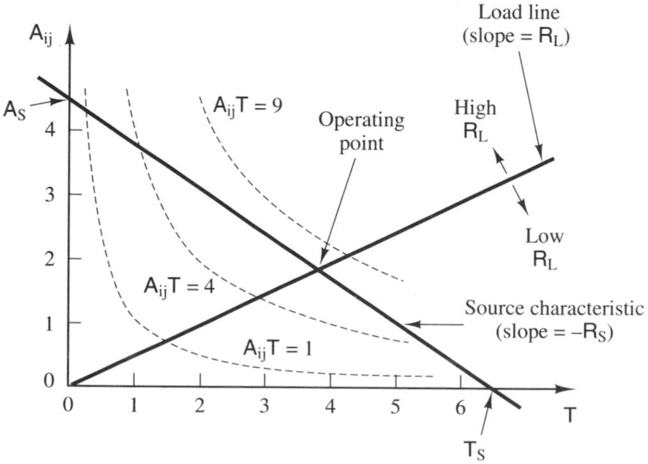

Figure 6.37 Steady-state source and load characteristics.

[11] In a steady state, all network potentials and flows are constant, and each energy-storage element behaves as an open circuit or a short circuit ($s = 0$). Hence, the network behaves as if it contains only dissipative resistances (see sections 3.3 and 4.2).

uct $A_{ij}T$.[12] Superimposed on Figure 6.37 is a set of constant-power curves, i.e., curves of $A_{ij}T$ = constant.

Varying the load (the slope R_L of the load line) changes the power that the load extracts from the source. We showed in example problem 4.9 that, to extract maximum power from a *fixed* linear source, we must match the resistances; that is, we must set $R_L = R_S$. It can be seen from Figure 6.37 that maximum power extraction corresponds to placing the operating point where the source characteristic is tangent to one of the constant-power curves.

We define the **available power of the source** as the maximum power the source can deliver to a dissipative load. We denote the Thévenin equivalent potential of the source by A_S and the corresponding Norton equivalent flow by T_S. Then the available power of the linear source is

$$P_{avail} = \frac{A_S{}^2}{4R_S} = T_S{}^2\frac{R_S}{4} \qquad (6.60)$$

Thus, the available power depends on R_S, the slope of the source characteristic.

If the load resistance R_L is much higher than the source resistance R_S, then the load line is nearly vertical. In that case:

- The power extracted from the source is much less than the available power of the source. We should then view the source as *lightly loaded.*
- The operating point is close to the potential-difference intercept. That is, $A_{ij} \approx A_S$. If we rescale the T-axis to run the load line through the center of the plot, then the source characteristic is nearly horizontal, with a relatively low R_S (compared with the load R_L). It is then appropriate to treat the source as a nearly ideal *potential-difference source.*

On the other hand, if the load resistance R_L is much lower than the source resistance R_S, then the load line is nearly horizontal. In that case:

- The power extracted from the source is much less than the available power of the source. Again, we should view the source as *lightly loaded.*
- The operating point is close to the flow intercept. That is, $T \approx T_S$. If we rescale the A_{ij}-axis to run the load line through the center of the plot, then the source characteristic is nearly vertical, with a relatively high R_S. It is then appropriate to treat the source as a nearly ideal *flow source.*

Exercise 6.17: Represent the source by a Thévenin equivalent subnet. Find the power generated by the Thévenin source, the power consumed in the Thévenin resistance, and the power delivered to the load. Repeat the process for a Norton equivalent subnet. Compare the two results. Note that for a lightly loaded source, one of the equivalent representations is more appropriate. The other requires use of an extremely large source.

[12] In a thermal system, the power delivered by the source is the flow (heat) itself, not the product of potential and flow (see section 6.1).

In general, we call a physical source a *potential-difference source* if its performance characteristic approximates a constant potential difference over the range of flows for which it is used. Then we view its internal resistance as relatively low, and it is best represented in Thévenin form—as an *ideal* potential-difference source in series with its resistance. If a source approximates a constant flow over the usual range of operation, then it has relatively high resistance, we call it a *flow source,* and the Norton representation (*ideal* flow source and parallel resistance) is more appropriate. Neither name is especially appropriate if the source resistance is neither low nor high for the usual range of operation of the source.

Exercise 6.18: Suppose the source and the dissipative load are not linear. Then the constant-power curves of Figure 6.37 are unchanged, but the operating characteristic and the load line are not straight lines. How would you determine the available power of the source? How would you determine whether the source was lightly loaded?

Accurate Measurement of Signals

Sensors produce very weak signals. Moreover, the more power we draw from a sensor, the more we distort the quantity measured. Indeed, it is difficult to get enough power from a sensor to energize a system without greatly distorting the signal that is sensed. Yet measuring quantities accurately is the purpose of a sensor. This section examines design criteria that enable a sensor to produce accurate measurements. We seek the power necessary to energize the system in a later discussion of amplifiers.

To measure the potential difference between two points in a system, we must attach a measuring device between the points, in parallel with the system. (For example, we would connect the terminals of a voltmeter to two points in a circuit; or we would attach a pressure gage to a hole at some point on a pipe and measure the internal hydraulic pressure relative to the ambient air pressure.) Let nodes a and a′ of Figure 6.38(a) denote the two points in question. The measured system acts as a source and the measuring device as a load. We represent the measuring device by an equivalent resistance—its input resistance R_{in}. (For the electrical circuit, R_{in} is the electrical resistance of the meter; for the hydraulic system, it is the steady-state fluid resistance of the pressure gage, typically infinite.) We represent the measured system, as *seen* between the two points by the attached measuring device, by either a Thévenin equivalent or Norton equivalent subnet, as shown in Figures 6.38(b) and (c).[13]

Attaching the measuring device between the nodes introduces an additional flow path to the measured system and changes the values of all its variables. Hence, the measuring device changes the quantity to be measured. We say that the measuring device **loads** the measured system. If the measurement is to be accurate, the measuring device must cause little loading. That is, the added flow T must be very small. (The voltmeter must

[13] To keep the discussion as simple as possible, and to permit the drawing of operating characteristics (see Figure 6.37), we assume that the network is in a steady state. Then all equivalent impedances are dissipative resistances. A similar analysis (without drawing operating characteristics) can be carried out for time-varying behavior with impedances of any type.

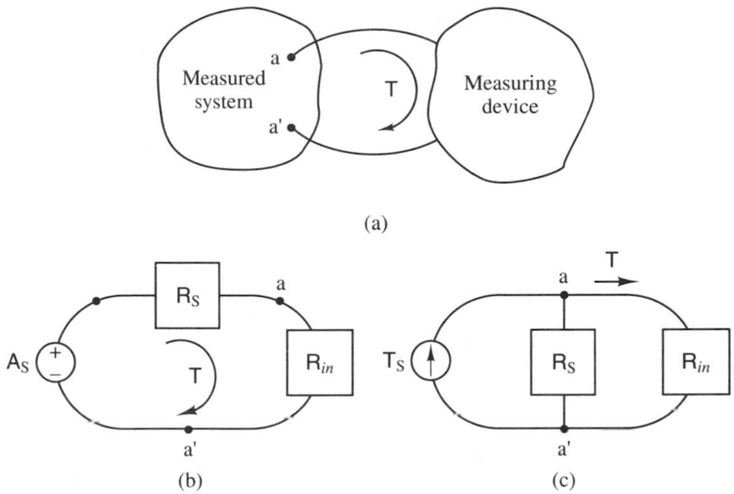

Figure 6.38 Measurement of potential difference and flow. (a) Measurement structure. (b) Thévenin representation of the measured system. (c) Norton representation of the measured system.

have high resistance to current; the pressure gage must have high resistance to fluid flow.) According to Figure 6.38(b), the quantity that we are trying to measure—that we *would* measure if R_{in} were infinite—is A_S.

It is apparent from Figure 6.38(c) that light loading requires that $R_{in} \gg R_S$. Then the power extracted by the measuring device will be much less than the available power of the equivalent source. It follows from the discussion of available power that the Thévenin equivalent is the most appropriate model for the measured system.

Exercise 6.19: Use Figure 6.38(b) to show that 1% accuracy in measurement of the potential difference A_S requires that $R_{in} > 99\ R_S$ (or $P_{load} < 0.04\ P_{avail}$).

If we wish to measure the flow in a branch of a system, we must break that branch and attach a measuring device in series with the branch. (In an electric circuit, for example, we open a branch and insert an ammeter; in a hydraulic circuit, we insert a small fan blade through a hole into the pipe and use the rate of rotation caused by the flow to determine that flow.) Accordingly, let nodes a and a′ of Figure 6.38(a) denote the two ends of the branch breakpoint; that is, a and a′ were originally connected together in the measured system. (In the hydraulic example, a and a′ denote points immediately upstream and downstream from the inserted fan blade.) Again, we represent the measuring device by an equivalent resistance to flow and the measured system by a Thévenin or Norton equivalent subnet, as shown in Figures 6.38(b) and (c).

Inserting the measuring device in the branch changes the resistance to flow in the branch. Therefore, it changes the values of all network variables, including the flow to be measured. Again, the measuring device loads (takes power from) the measured system. To obtain an accurate measurement, we use a measuring device that has a small input resistance R_{in}. (For the electrical circuit, R_{in} is the ammeter resistance: for the hydraulic circuit, it is the flow resistance caused by the fan.) According to Figure 6.38(c), the quantity that we are trying to measure—that we *would* measure if R_{in} were zero—is T_S.

It is apparent from Figure 6.38(b) that light loading requires that $R_{in} \ll R_S$. Then the power extracted by the measuring device will be much less than the available power of the equivalent source. It follows from the discussion of available power that the Norton equivalent is the most appropriate model for the measured system.

Exercise 6.20: Use Figure 6.38(c) to show that 1% accuracy in measurement of the flow T_S requires that $R_{in} < 0.01\ R_S$ (or $P_{load} < 0.04\ P_{avail}$).

In sum, for accurate measurement, the measuring device must not load the measured system significantly. The power that the device extracts from the measured system must be much less than the available power—the power that it *could* extract. To limit the loading, we **mismatch the impedances** (R_{in} and R_S) as much as is practical. To measure a potential difference, we use a sensor with a very large input impedance. To measure a flow, we use a sensor with a very small input impedance.

Sources of Energy

All energy is stored in some form. To produce an independent source signal, we must use some means to draw energy from a storage device. We might also have to convert the energy to another form. We examine the direct use of energy-storage devices as independent sources in the next section. Energy-storage devices also serve as power supplies for the power amplifiers discussed later. This section examines the common sources of stored energy.

Energy is available in the physical world in several forms: as electromagnetic energy radiated from the sun, as heat stored in the earth, as elevated water in streams and lakes (hydraulic potential energy, acquired from the sun), as moving air or water (pneumatic or hydraulic kinetic energy, acquired from the sun), and as nuclear or fossil fuel. Energy is available, then, in chemical, hydraulic, thermal, and electromagnetic forms.

We use transducers to convert energy from one form to another. For example, we use a turbine to convert hydraulic potential energy to rotational energy and then an electric generator to convert the rotational energy to electrical energy for immediate use. Or we use a burner or a reactor to convert fuel to heat and then use the heat to pressurize air or steam (thereby storing pneumatic energy). The pneumatic energy can then be converted to rotational energy by a pneumatic turbine or to translational energy by a pneumatic cylinder (as in a reciprocating engine).

Electrical energy is more easily transported than other forms of energy. A small amount of dc electric power is available from the chemical energy stored in a portable battery. Electric utility companies make electrical energy available to all points in regions of significant population. (To generate this electrical energy, they convert the energy stored in fuel or in water reservoirs.)

Fossil fuels are also transported easily. Natural gas utility companies pipe gas to most locations in urban areas. Gas suppliers also provide bottled gas for use in remote locations. Gasoline for internal combustion engines is easily transported and stored. Hydraulic energy can be transported conveniently in pipes and flexible tubes, but only over short distances. The energy lost to fluid friction is too high to transport it over long distances.

Electric utility companies maintain specified voltages throughout their networks. It is the loads customers attach to the utility outlets that determine the current drawn from the utility. In most countries, the electrical signals generated by the utilities are sinusoidal—*ac* signals. The advantage of sinusoidal signals is that they enable us to use electric transformers to change voltage levels.

Since the electrical loads used by the utility customers vary incessantly, the utility company must adjust the power (torque and speed) of each generator continually to keep the network frequency at 50 or 60 Hz (depending on the country) and to maintain voltage amplitudes near specified values throughout the network. (The utility must also keep the phase of the generated sinusoidal voltage nearly constant in order to maintain the timing of clocks and other devices connected to the network.)

Much of the energy needed for human endeavor is mechanical. Yet mechanical energy is not easily transported. We solve this dilemma by converting easily transported energy to mechanical form. We apply the term **motor** to any transducer that converts energy to mechanical form. Most motors are electric, hydraulic, or pneumatic. (The loudspeaker that transmits acoustic sounds from a radio, television, or telephone is driven by a translational electric motor.) We use the term **engine** to mean a motor that converts *thermal* energy to mechanical energy. The internal combustion engine, for example, converts fuel to heat and pneumatic energy and then uses pneumatic motors—cylinders and pistons—to produce mechanical energy.

The mechanical power that can be provided by a human is typically less than 1 kW. Electric motors with power in the human power range are small enough and light enough to use in portable tools and household appliances. Hydraulic motors in the human power range are about equal in size and weight to their electric counterparts. Since hydraulic power sources are not as portable as electric power sources, however, we seldom see hydraulic tools and appliances in the home.

For mechanical power greater than 1 kW, hydraulic motors provide much greater power (up to 10 times) and torque (up to 1,000 times) for a given size and weight than do electric motors. For that reason, construction equipment and heavy-duty tools are usually powered by hydraulic motors.

Power Supplies

If an energy-storage device has enough storage capacity, it can serve as a constant-valued source of power for a system. Then we refer to it as a **power supply.** We can suddenly connect the power supply directly to the system to produce a powerful (that is, independent) step input. Or we can use the power supply as the source of power for an amplifier. (See the discussions of the transistor amplifier and the hydraulic valve.) In this section, we examine the design of an energy-storage device as a power supply.

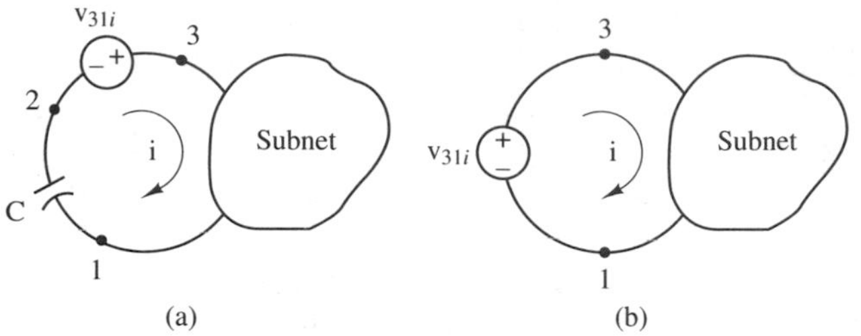

Figure 6.39 An energized storage device as an independent source. (a) A charged capacitor as a source. (b) Equivalent ideal source if C is very large.

In the electrical network of Figure 6.39(a), a capacitor between nodes 3 and 1 is initially charged to voltage v_{31i}. That charged capacitor is represented in the figure by an unenergized capacitor in series with a constant independent voltage source, as discussed in section 4.2. The voltage source causes current i to flow into the attached subnet. Over time, the current i charges the unenergized capacitor of the model to the voltage

$$v_{12}(t) = v_{12}(0) + \frac{1}{C}\int_0^t i\,dt \tag{6.61}$$

where $v_{12}(0) \equiv v_{12i} = 0$. If the capacitance C is very large, the flow of current over time does not build up appreciable voltage, and v_{12} remains essentially zero. Then the charged capacitor can be represented by the simpler independent-source model shown in Figure 6.39(b).

How large must C be in order for the simpler model (with no capacitor) to be accurate? Large enough that the current drawn by the attached system causes little charging of the capacitor over the time interval of interest. Thus, the required size of C depends on i(t) and on t_{max}. Alternatively, for a given size of C and given current i(t), the model is valid for a limited amount of time.

We conclude that a charged capacitor of *relatively* large capacitance acts as a constant independent voltage source. Similarly, a water reservoir of large capacitance (relative to the flow that the pressure induces in the attached hydraulic subnet) acts as an independent pressure source, a rotating object of relatively large inertia acts as a constant rotational-velocity source, and an object of relatively large thermal capacitance acts as a constant-temperature source.

By analogy, we can see that a conducting electrical inductor of relatively large inductance acts as a constant independent current source to the attached subnet. Similarly, a compressed object (spring) of relatively large compliance (1/k) acts as a source of constant force, and a moving fluid of relatively large inertance (say, the oil in a cross-country pipeline) acts as a source of constant hydraulic flow.

We can use any relatively large energized storage device, perhaps with a transducer to convert the energy to a different domain, as a power supply. We can use that power supply as a *constant-valued* independent source for a system. The mechanism that turns the power supply on and off (or connects and disconnects the power supply) is the underlying low-power waveform source.

We define the *regulation* of a power supply to be the fraction by which its value changes during use. Regulation is a measure of the quality of the power supply. A number of sophisticated feedback techniques have been devised to improve regulation above that which can be obtained from the use of a storage element alone. These techniques use a sensor to monitor the potential difference (or flow) of the power supply. When the potential difference (or flow) deviates from the desired value, the error signal initiates the delivery of additional power to reduce the error. This technique is called *feedback regulation* of the power supply. The section on amplifiers examines the manner in which the flow of power can be controlled.

The physical mechanism by means of which power travels from a power supply to an attached system always dissipates some power. We account for that power dissipation by including source resistance in the model for the power supply.

Efficient Energizing of Loads

We found in example problem 4.9 that to extract maximum power from a *fixed source* (in a steady state), we must use a load resistance that matches the source resistance. In this section, we reverse the objective. The subnet that is attached to the load acts as a source to the load. We design that source to provide maximum power to a *fixed load,* as shown in Figure 6.40(a). It is apparent from the constant-power curves of Figure 6.37 that, by increasing the magnitudes of both source-characteristic intercepts (A_S and T_S), we can increase the load power without bound. Hence, we maximize the load power by raising both quantities to their limits.

According to the previous section, the power supply that energizes a system typically has either limited (fixed) potential or limited flow. Limited potential at the source limits the potentials at all points of the system. Similarly, limited flow at the source limits the flows in all branches of the system. Therefore, we fix (at its limit) the value of *either* the Thévenin equivalent source or the Norton equivalent source, as *seen* by the load. Then

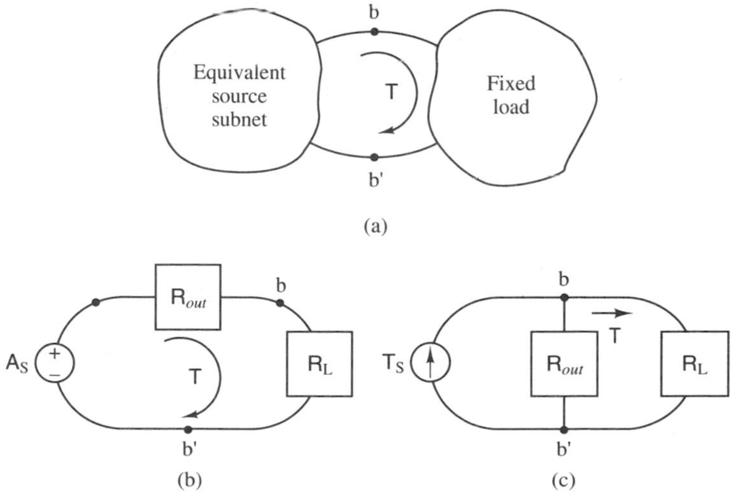

Figure 6.40 Energizing a load. (a) Structure for energizing a load. (b) Thévenin representation of the source. (c) Norton representation of the source.

the load power depends on the output resistance R_{out} of that equivalent source. We show the Thévenin source model in Figure 6.40(b) and the Norton source model in Figure 6.40(c).

Suppose we fix A_S, the Thévenin equivalent potential of the source subnet, and adjust R_{out} to maximize the load power. According to Figure 6.40(b), the load power is maximum if $R_{out} = 0$. (For that value of R_{out}, the Norton equivalent flow is $T_S = \infty$.) The corresponding load power is $P_{load} = A_S^2/R_L$. Of course, it is not possible to make the effective output resistance of the source subnet precisely zero. From a practical standpoint, however, the load power is *nearly* maximum if $R_{out} \ll R_L$.

Similarly, Figure 6.40(c) shows that if we fix T_S, the Norton equivalent flow of the source subnet, and adjust R_{out} to maximize the load power, we should make $R_{out} = \infty$. (For that value of R_{out}, the Thévenin equivalent potential is $A_S = \infty$.) Then $P_{load} = T_S^2 R_L$. Although it is not possible to make the effective output resistance of the source subnet infinite, the load power is nearly maximum if $R_{out} \gg R_L$.

Maximizing the delivery of power to the load (or minimizing the source size needed to deliver a specified power) also maximizes the **efficiency** of delivering power. That is, we simultaneously minimize the power consumed in the supply subsystem and (hence) deliver the highest *fraction* of generated power to the load.

In sum, to maximize the efficiency with which we deliver power to a fixed load, we should **mismatch the source and load resistances** (R_{out} and R_L) to the greatest extent practical—say, by a factor of 10. (This mismatch of resistances means that the source must deliver far less than its available power.) Hence, we make the resistance of a limited-potential source as small as practical. Then the Thévenin representation is the most appropriate model for the source. (The Norton model for the same source would use an artificially large flow source and consume most of its *apparent* power output in the source resistance.) If we make the resistance of a limited-flow source as large as practical, then the Norton representation is the most appropriate model for the source.

Exercise 6.21: Show that, to deliver a *specified* power to a fixed load with a source of minimum equivalent source size (either A_S or T_S), we must, again, mismatch the source and load resistances.

Amplifiers

The independent source signal that energizes a system typically originates in the measurement of a potential or flow in another system. Because the signal produced by the sensor is weak, it cannot energize the load at an adequate power level unless it is strengthened first. We use an amplifier to strengthen the signal.

In this section, we use familiar situations to define what an amplifier is and to help us recognize amplifiers. We examine quantitatively the behavior of one familiar amplifier, a water valve. We find that we can describe the behavior of an amplifier by means of operating characteristics at its input and output ports.

We apply the term **amplifier** to any device in which a weaker signal controls a stronger signal. A tape deck serves as a voltage waveform source for an audio system. The load is a loudspeaker. The power available from the tape deck is much too low to rouse audible sound from the loudspeaker, so we insert an electronic amplifier between the tape deck and the loudspeaker to boost the signal power. The low-power output from the tape deck controls the high-power output from the amplifier.

Suppose a large oceangoing vessel is driven through the water (by its engine and propellers) at a fixed speed. The ship is steered by its rudder. The hydraulic force against the tilted rudder, owing to water resistance, is huge. The helmsman is much too weak to turn the rudder directly. Instead, the helmsman turns a wheel to indicate the desired rudder angle. We insert a hydraulic amplifier between the wheel and the rudder to amplify the power of the helmsman.

We need to propel the vessel through the water at various speeds. The helmsman moves a lever to indicate the desired propelling force. The engine, together with its control mechanism, drive shaft, and propeller, converts the weak lever force of the helmsman to a powerful propelling force. The engine is thus a mechanical amplifier.

In an automobile, the engine, the power steering mechanism, and the power brake mechanism are further examples of amplifiers. The engine amplifies the accelerator pedal force, the power steering system amplifies the human steering force, and the power brake system amplifies the human braking force.

The primary distinguishing feature of an amplifier is its power gain. That is, its output (or controlled) signal is more powerful than its input (or controlling) signal. Amplifier input and output signals need not be in the same energy domain. We can use transducers to change signals to other domains.

In controlling the flow from a pump, the valve in the hydraulic system of Figure 6.41 acts as an amplifier. Little power is needed to operate the valve. In contrast, if the available power of the pump is high, then the hydraulic power controlled by the valve is

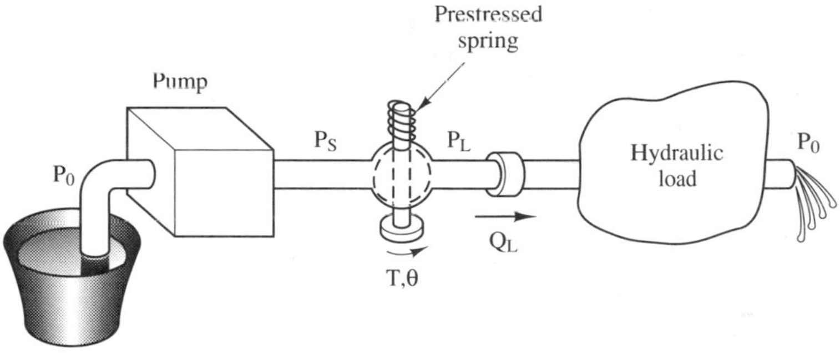

(a)

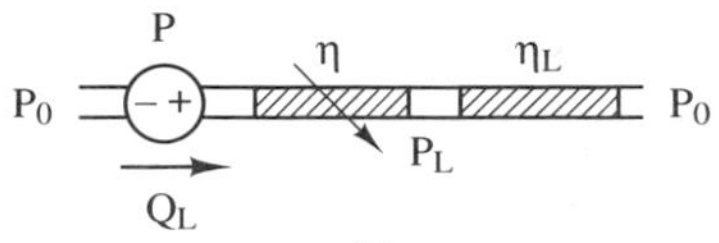

(b)

Figure 6.41 A hydraulic amplifier. (a) Pictorial model. (b) Lumped model.

high. Thus, the low-power valve *rotation* controls the high-power hydraulic *flow*. We shall use this valve to illustrate the properties of amplifiers.

Let us model the pump as an ideal pressure source P in series with a quadratic pump resistance represented by the resistor coefficient η_p. The valve behaves as an adjustable hydraulic resistor with coefficient η_v. Thus, the valve-controlled pump acts as a pressure source with the adjustable resistor coefficient $\eta \triangleq \eta_p + \eta_v$. Figure 6.41(b) shows this lumped model with a resistive load denoted by η_L.

The steady-state operating characteristics of the valve-controlled pump are displayed in Figure 6.42. The curve marked $\theta = 90°$ is the output-port operating characteristic for the valve fully open. (It corresponds, in shape, to the pump curve of Figure 5.23.) Note that the output gage pressure P_{L0} drops quadratically (from the level P) with increasing flow Q_L, as we would expect for turbulent hydraulic flow. That is the reason we use a quadratic resistor coefficient to represent the combined pump and valve. Similarly, the load characteristic shown in the figure has the quadratic shape characteristic of turbulent pipe flow (see section 5.1).

As we rotate the valve stem to close the valve partially, we increase the valve resistance. The curves marked 60° and 30° show the operating characteristics for two levels of valve closure. All the curves have the value P (the pump pressure) if the flow Q_L is zero. At 0°, the valve is fully closed, η is infinite, and $Q_L = 0$, regardless of the load.

For each valve setting, the operating point is the intersection between the load characteristic and the appropriate valve output characteristic. The power output is the product of the values of P_{L0} and Q_L at the operating point. By changing the valve displacement θ, we move the operating point along the load line. Thus, the valve displacement controls the output pressure, output flow, and output power. Changes in output pressure and output flow are roughly proportional to changes in θ, the degree to which the valve is opened.

The valve has two inputs: The hydraulic port, through which the pump delivers water and hydraulic power, and the mechanical port, at which we enter the low-power *signal* that *controls* the flow of hydraulic power. Let us look at the valve from its input *signal* port. Anyone who has used his or her thumb to control the water flow from a hose knows that the valve closure force increases with the degree of closure—a springlike behavior. For the valve in question, the degree of closure is $\hat{\theta} \triangleq 90° - \theta$. The torque T needed to

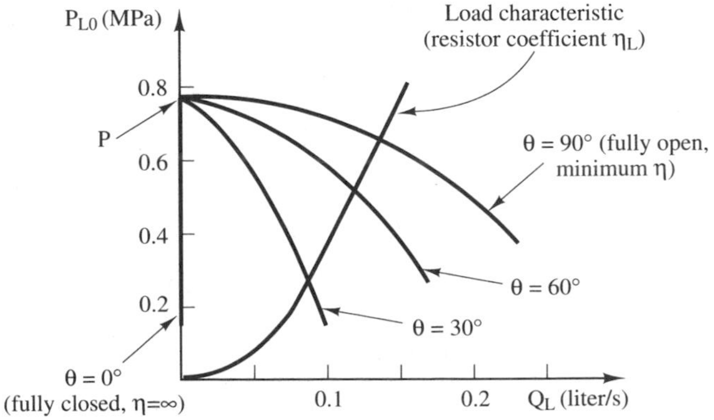

Figure 6.42 Output-port operating characteristics for the valve-controlled pump.

close the valve is proportional to $\hat{\theta}$. Hence, an increase in torque T at the valve input causes a decrease in θ, as well as a decrease in output pressure and flow.

It is less confusing to the person who operates the valve if the output variables *increase* with increasing input torque. Therefore, we attach a prestressed spring to hold the valve firmly closed (see Figure 6.41(a)). The prestressed spring cancels the springlike effect of water pressure on the valve and produces the mechanical behavior shown in Figure 6.43.

The torque required to close the valve increases with the pressure drop P_{SL} across the valve. (It requires no torque to hold the valve closed if $P_{SL} = 0$.) Load-induced variations in Q_L cause variations in P_{SL}. Hence, the slope of the input characteristic varies somewhat during operation of the valve. We make the stiffness of the prestressed spring large enough to mask these variations, and we ignore them in Figure 6.43.

Note that it takes torque, not power, to *hold* the valve at a given setting. The person or device that adjusts the valve must inject *power* only to *change* the setting. Adjustment of the valve requires little power and little energy. (Often a hand-operated valve is designed to have enough static friction to *hold* its setting. Then the user must provide enough torque to overcome the static friction.)

This hydraulic valve illustrates features that are basic to all amplifiers. The valve produces a powerful output hydraulic flow that is proportional to a weak mechanical input torque. The power needed to produce the strong output signal is extracted from an energy supply—the constant-pressure pump. In general, an amplifier is a *three-port device.* The *output signal power* delivered by an amplifier at its **output port** is greater than the *input signal power* at its **control port.** The amplifier has a second input port, its **power supply port,** through which it draws whatever power is needed to satisfy the mandate of the control port. When we speak of the *input port* of an amplifier, we mean its control port. Ultimately, the power output of the amplifier is limited by the *available power* of the supply that is attached to the power supply port.

If we partition a system into two halves, the source half and the load half, each half typically affects the behavior of the other half. To find the behavior of each, we must write the equations that describe both halves and solve them simultaneously. However, as we discovered in the discussion of accurate measurements, if the available power of the source half is much greater than the power drawn by the load half, then we can model the source half as an *ideal* independent source. In other words, the behavior of the (relatively) powerful source half is not much affected by the behavior of the load half. This is the situation that we find if we separate two halves of a system at the *input to an amplifier.*

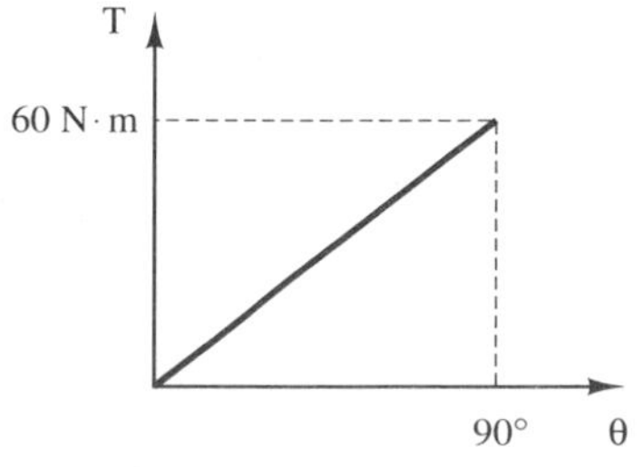

Figure 6.43 Input-port operating characteristic of the spring-loaded valve.

The input (control) port of an amplifier typically consumes much less power than the available power of the subsystem that provides the *control signal.* Therefore, the amplifier does not *load* the controlling subsystem. Hence, we can treat the controlling subsystem as a nearly ideal independent source.

The output port of the amplifier, on the other hand, is quite powerful. It typically can provide much more power than is required by its load. Hence, we can also treat the output port of an amplifier as a nearly ideal independent source. (The *waveform* for that source is provided by the controlling subsystem.)

The steady-state output-port operating characteristics of an amplifier, then, provide all the information necessary to determine the interaction of the amplifier with the system it energizes. As illustrated by Figure 6.42, amplifier operating characteristics are typically presented as a set of curves with an input-port variable as a parameter.

Linear Lumped Amplifier Models

In this section, we derive a linear lumped model for the valve-controlled pump—the amplifier of the previous section. We use this lumped model to guide our thinking about models for all amplifiers. In later sections, we derive linear lumped models for several familiar amplifiers: transistors, hydraulic control valves, and internal combustion engines.

Suppose we energize a hydraulic system with the valve-controlled pump characterized by Figure 6.42. We design the system so that the average operating point (at the valve output) is roughly in the center of the set of operating characteristics. In Figure 6.44, we replace those characteristics by straight-line approximations that are *accurate in the region of intended operation* of the system. The slopes of all the lines are *roughly* the same within the region of operation. Each straight-line approximation can be represented by a linear equation. We express those equations in terms of the Q_L-intercepts because those intercepts are roughly proportional to the corresponding valve-displacement values θ.

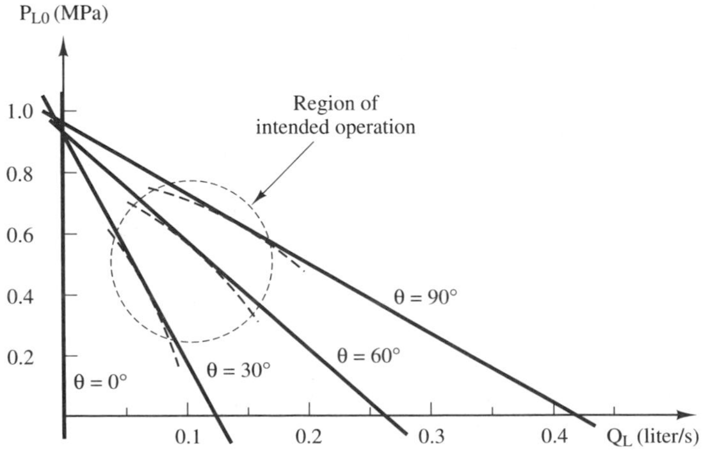

Figure 6.44 Linearized valve output characteristics.

The equation for each straight line is

$$Q_L = Q - \left(\frac{1}{R}\right)P_{L0} \tag{6.62}$$

where $-R$ is the slope of the line and Q is the intercept on the Q_L axis. We treat the lines as if the slopes were *exactly* identical and the Q_L-intercepts were *exactly* proportional to the corresponding θ values. Then, from the horizontal and vertical intercepts for the $\theta = 60°$ line, we find that $R = 3.5 \times 10^9$ N·s/m^5 and $Q = \alpha\theta$, where $\alpha = 4.4 \times 10^{-6}$ m^3/s·degree (θ in degrees). Therefore, the linear equation

$$Q_L = \alpha\theta - \left(\frac{1}{R}\right)P_{L0} \tag{6.63}$$

describes the valve behavior for all operating points (Q_L, P_{L0}) in the region of intended operation. For operation in a different region, we need only change the values of α and R. The linear approximation can be made as accurate as we wish by sufficiently restricting the size of the region of operation.

The linear equation (6.63) corresponds to a linear lumped model in Norton form. We show that model in hydraulic notation in the right half of Figure 6.45(a) and in line-graph notation in the right half of Figure 6.45(b). The only difference between this Norton equivalent model and the Norton equivalent subnets that we derived in section 4.2 is that this flow source is dependent on the input-port variable θ. Therefore, we represent the flow source by a dependent-source symbol.

The lumped model in Figure 6.45 represents only the *steady-state* behavior of the valve output port. To account for the slight compressibility of the water in the valve, we would add a small fluid capacitor to the output port. Instead, we assume that the capacitance is negligible.

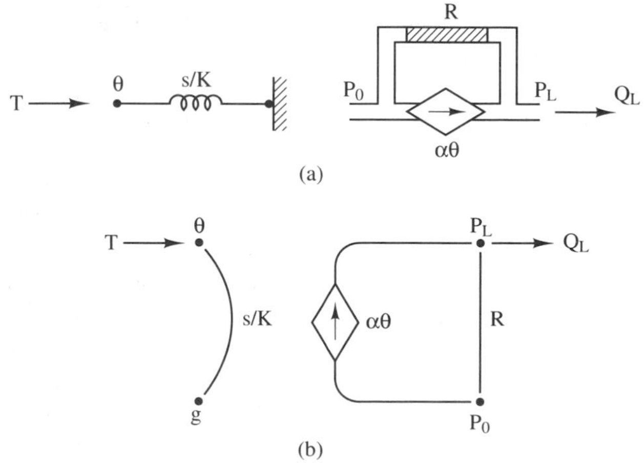

Figure 6.45 Linear lumped model for the valve-controlled pump; $K = 0.67$ N·m/degree, $\alpha = 4.4 \times 10^{-6}$ m^3/s·degree, $R = 3.5 \times 10^9$ N·s/m^5. (a) Lumped model. (b) Line-graph representation.

The valve input characteristic of Figure 6.43 has slope 0.67 N·m/degree. The characteristic corresponds to a lumped spring with equation

$$T = K\theta \tag{6.64}$$

where θ is in degrees and $K = 0.67$ N·m/deg. We show the lumped spring in physical (rotational) notation on the left side of Figure 6.45(a) and in mathematical (line-graph) notation on the left side of Figure 6.45(b). If there were also some friction in the valve mechanism, we would account for it by connecting a damper (linear or nonlinear) in parallel with the spring.

The physical lumped-model notation of Figure 6.45(a) is easy to associate with the physical system. On the other hand, the input and output ports and the impedances *seen* at the ports are easier to recognize in the mathematical line-graph notation of Figure 6.45(b). We shall find that port impedances are important design quantities. For either notation, we use explicit symbols for the potential variables to distinguish energy domains clearly. Similarly, we use impedance notation for all branches to distinguish element types clearly.

The lumped valve model of Figure 6.45 is linear. Thus, the valve-controlled pump can be considered a *linear amplifier* if the region of operation is suitably restricted. Some amplifiers are designed to behave linearly over large regions of operation. We examine a few of these in succeeding sections.

Types of Linear Amplifiers

There are four basic amplifier structures, which may be distinguished by the natures of their input-to-output dependencies. We show linear versions of these structures in line-graph notation in Figure 6.46. Figure 6.46(a) shows a flow amplifier. The *amplification parameter* β characterizes the input-to-output dependency. We shall refer to such an amplifier as a β amp. Similarly, we refer to the other three types of amplifiers shown in Figures 6.46(b), (c), and (d) as μ, r_t, and g_t amps, respectively.

We choose the type of amplifier to meet the system objectives. If we wish to strengthen a measured flow T_a, we use a β amp or an r_t amp. To strengthen a measured potential difference $A_{aa'}$, we use a μ amp or a g_t amp. If we wish to energize a load with a controlled potential-difference waveform $A_{bb'}$, we use a μ amp or an r_t amp. To energize a load with a controlled flow waveform T_b, we use a β amp or a g_t amp. In some situations, we can use a gyrating transducer to convert a potential difference to a flow or a flow to a potential difference. This ability gives us some additional freedom in the type of amplifier we choose.

The amplifier input impedance z_i is the equivalent impedance that is *seen* by the source we attach to the input port. Hence z_i is the *load seen by the source.* For the valve-controlled pump of the previous section, $z_i = s/K$. The amplifier output impedance z_o is the equivalent impedance that is *seen* by the load we attach to the output port. Therefore, z_o is the *source impedance seen by the load.* For the valve-controlled pump, $z_o = R$. The equivalent impedance at the input does not depend on the impedance of the load. Similarly, the equivalent impedance at the output does not depend on the impedance of the source. Thus, the **amplifier isolates the source from the load.** Indeed, one often uses an amplifier precisely to provide such isolation.

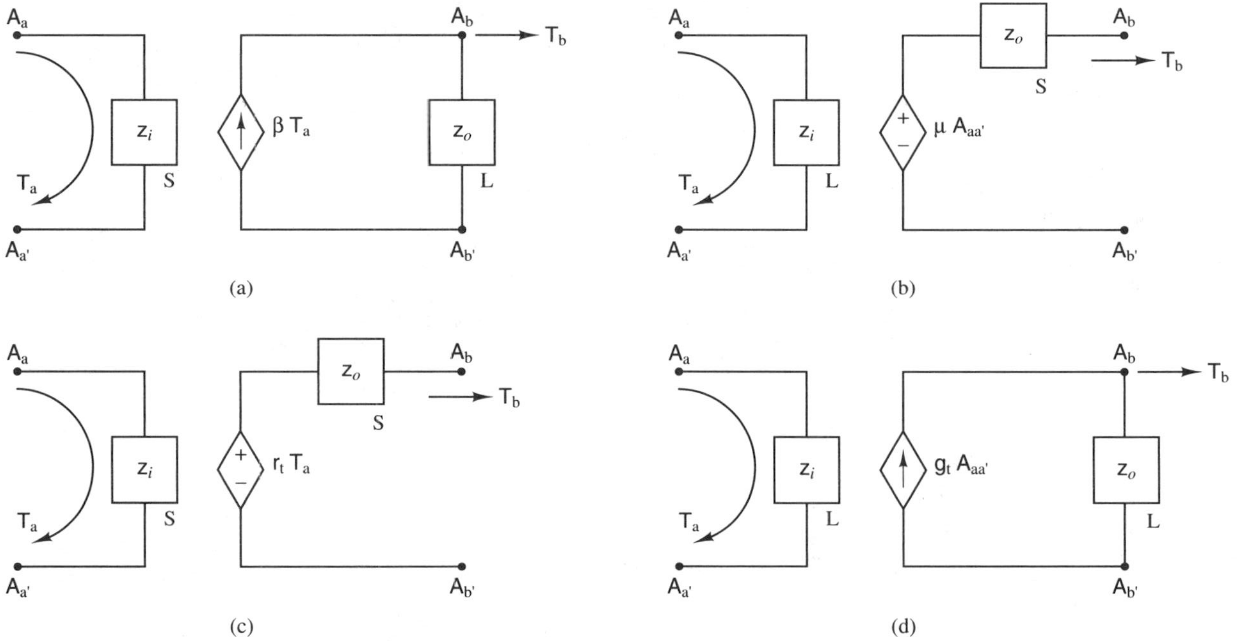

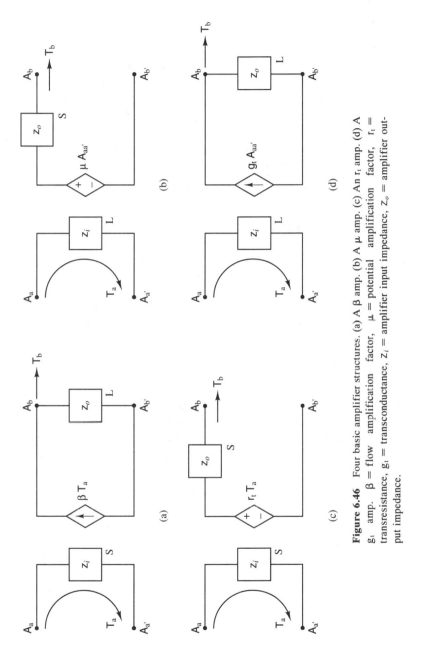

Figure 6.46 Four basic amplifier structures. (a) A β amp. (b) A μ amp. (c) An r_t amp. (d) A g_t amp. β = flow amplification factor, μ = potential amplification factor, r_t = transresistance, g_t = transconductance, z_i = amplifier input impedance, z_o = amplifier output impedance.

Previously, we found that for accurate sensing, the sensor must be lightly loaded. Hence, we make the input impedance Z_i of the amplifier much different than the impedance Z_S of the source. When the control signal to be sensed and amplified is a flow, the amplifier input is in the flow path and light loading means that Z_i is small. Therefore, we want a β amp or an r_t amp to have a small input impedance. When the control signal to be sensed and amplified is a potential difference, the amplifier input provides a parallel flow path, and light loading means that Z_i is large. Thus, we want a μ amp or a g_t amp to have a large input impedance. We note these impedance requirements in Figure 6.46 by the symbol S or L beside each input impedance block.

An amplifier is used to provide power to a load. Earlier, we discovered that for efficient energizing of loads, the output impedance of the subnet that drives the load should be much different than the load impedance. If it is the potential drop across the load that is to be controlled, then we want the output impedance Z_o of the amplifier to be small. Therefore, we want a μ amp or an r_t amp to have a small output impedance. If it is the flow through the load that is to be controlled, then we want the amplifier output impedance Z_o to be large. Hence, we want a β amp or a g_t amp to have a large output impedance. Again, we use the symbols S or L beside each output impedance block in Figure 6.46 to note these requirements.

Exercise 6.22: Ideal amplifiers would have ideal input and output impedances—that is, a zero small impedance and an infinite large impedance. Draw ideal versions of the amplifiers of Figure 6.46 and write the equations that describe them.

We insert an amplifier into a system in order to strengthen a signal. Therefore, we want to know how effective a particular amplifier will be in strengthening signals. Accordingly, we now examine the steady-state amplifying capability of amplifiers. In a steady state, all energy-storage elements behave as open circuits or short circuits, and only dissipative resistances remain. Therefore, we denote the impedances by R rather than Z.

We define the **steady-state power gain** G_P of an amplifier by

$$G_P \triangleq \frac{P_{out}}{P_{in}} = \frac{A_{bb'}T_b}{A_{aa'}T_a} \tag{6.65}$$

where T_a, $A_{aa'}$, T_b, and $A_{bb'}$ are oriented as shown in Figure 6.46. We shall find that the steady-state power gain for a linear amplifier depends on the value of the dependent-source amplification parameter (β, μ, r_t, or g_t) and also on the extent of the mismatch of impedances at the input and output ports.

Figure 6.47 shows a linear β amp connected to a linear source and linear load. (The source and load could be equivalent representations of more complicated source and load subnets.) The flow and potential difference at the amplifier input port are

$$T_a \quad \text{and} \quad A_{aa'} = R_i T_a \tag{6.66}$$

The corresponding output-port flow and potential difference are

$$T_b = \left(\frac{R_o}{R_o + R_L}\right)\beta T_a \quad \text{and} \quad A_{bb'} = \left(\frac{R_o R_L}{R_o + R_L}\right)\beta T_a \tag{6.67}$$

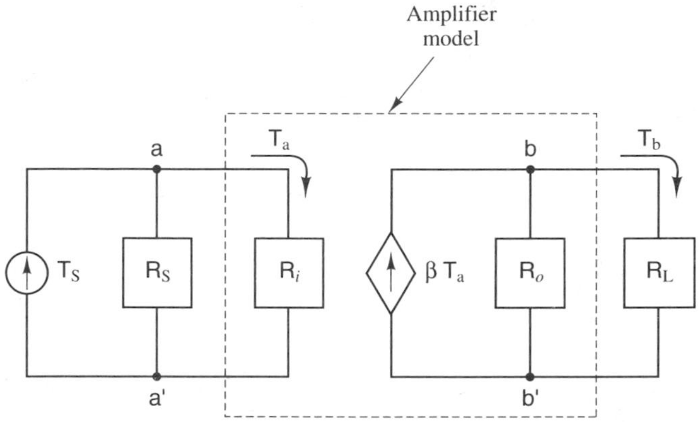

Figure 6.47 Steady-state model for a β amp with source and load.

According to equations (6.65), (6.66), and (6.67), the power gain of the β amp is

$$G_P = \frac{A_{bb'}T_b}{A_{aa'}T_a} = \left(\frac{R_o}{R_o + R_L}\right)^2 \frac{R_L}{R_i}\beta^2 \to \frac{R_L}{R_i}\beta^2 \tag{6.68}$$

where the limiting quantity on the right is the power gain for $R_o \gg R_L$.

Exercise 6.23: Suppose the signal that we *wish* to amplify is T_S. Show that the source-to-load power gain P_L/P_S includes the additional factor $R_S/(R_S + R_i)$. If $R_S \gg R_i$, then this power gain is the same as equation (6.68).

Let us find the power gain for the valve-controlled pump of Figure 6.41. The linear lumped model for this amplifier is shown in Figure 6.45. We let $s = 0$ in the formula $z_i = s/K$ (steady-state operation) to find that $R_i = 0$. The output impedance is $R_o \equiv R = 3.5 \times 10^9$ N·s/m^5. The flow amplification factor satisfies $\beta T = \alpha\theta$. According to the model of the input port, $T = K\theta$. Therefore, $\beta = \alpha/K = (4.4 \times 10^{-6}$ m^3/s·degree)/(0.67 N·m/degree) $= 6.57 \times 10^{-6}$ m^2/N·s.

Now let us find the load resistance. Figure 6.42 shows the nonlinear load characteristic for the attached pipe. Let us approximate that load characteristic by its tangent at the center of the region of operation ($\theta = 60°$ and $P_{L0} = 0.5$ MPa). The equation for the straight-line approximation is $Q_L = Q_0 + P_{L0}/R_L$, where $Q_0 = 0.062$ l/s is the flow intercept and $R_L = 8.3 \times 10^9$ N·s/m^5 (the slope of the line) is the load resistance. Because the model is linear, we can find the effect of the constant source Q_0 separately. Therefore, we ignore it here.

According to equation (6.68), the power gain of the valve-controlled pump is $G_P = \infty$. How can the power gain be infinite when the flow amplification factor is so small? It is owing to the fact that no motion is required at the input port to maintain a flow of power at the output. In equation (6.68), this fact is signaled by the zero value for R_i. On the other hand, suppose the device that holds the valve in position consumes a small amount of power. If we compare the output power to that input power, the ratio will not be infinite.

For many amplifiers, the input and output ports are in the same energy domain. In that case, there are two additional useful quantities associated with a flow amplifier: the *flow gain* and the *potential-difference gain.* Again, we find these from equations (6.66) and (6.67):

$$\text{flow gain} \triangleq \frac{T_b}{T_a} = \left(\frac{R_o}{R_o + R_L}\right)\beta \rightarrow \beta \tag{6.69}$$

$$\text{potential-difference gain} \triangleq \frac{A_{bb'}}{A_{aa'}} = \left(\frac{R_o R_L}{R_o + R_L}\right)\frac{\beta}{R_i} \rightarrow \frac{R_L}{R_i}\beta \tag{6.70}$$

It is apparent that, fundamentally, the β amp is a flow amplifier. It also amplifies the potential difference if the load impedance is not so much smaller than the input impedance that the factor R_L/R_i cancels β. The power gain of equation (6.68) is the product of equations (6.69) and (6.70).

Power gain is often expressed in decibels (dB) as 10 $\log(P_{out}/P_{in})$; flow gain and potential-difference gain are also often expressed in dB, as 20 $\log(A_{bb'}/A_{aa'})$ and 20 $\log(T_b/T_a)$, respectively (see section 5.4).

A similar analysis of power gain can be carried out for each type of amplifier. In each case, the conclusions are similar: The power gain is proportional to the square of the amplification parameter (β, μ, r_t, or g_t). It also depends on the impedance mismatch ratios at the input and output ports.

The steady-state power gain describes the power-amplifying effect of the amplifier only for slowly varying source signals. To determine the effect of an amplifier for rapidly varying signals, we examine the **average sinusoidal power gain** as a function of frequency. For a sinusoidal source signal, each flow and potential difference throughout the network has a specific phase. If the flow and potential difference associated with a particular dissipation element R have different phases, then the power delivered to R, averaged over one cycle of the sinusoid, is less than it would be if they had the same phase.

It can be shown, for sinusoidal steady-state operation, that the average power delivered to the subnet between nodes i and j of a network is

$$P_{ave} = \frac{A_{ijm} T_m}{2} \cos(\phi_A - \phi_T)$$

where A_{ijm} is the amplitude of the potential drop across the port, T_m is the amplitude of the flow through the port in the direction from i to j, and $\phi_A - \phi_T$ is the phase difference between A_{ij} and T. We can find the average power input and output for an amplifier at a specific frequency ω and then take the ratio to find the sinusoidal power gain for that frequency. Typically, the power gain of an amplifier gets smaller as the source frequency increases. A plot of the sinusoidal power gain vs. frequency is called the *frequency response* of the amplifier. Problem 6.35 at the end of the chapter examines the frequency response for a transistor amplifier.

We examine a number of amplifying devices in later sections. We shall see that a *bipolar-junction transistor* amplifies electric current—that is, it is a β amp. A *field-effect transistor,* on the other hand, uses input voltage to control output current. Thus, it is a g_t amp. A *hydraulic control valve* uses electric current to control the flow of hydraulic fluid. Hence, that control valve is a β amp. On the other hand, hydraulic control valves usually drive hydraulic pistons. A piston converts hydraulic flow to mechanical velocity. In terms of that velocity output, the control valve is an r_t amp. Suppose we use a fixed voltage source and a variable resistor to adjust the control current in the hydraulic valve. Then the current is proportional to the mechanical displacement of the resistor adjusting mechanism. Since that displacement is the integral of a velocity, the resistor-controlled valve is an *integrating* g_t amp at the hydraulic output. It is an integrating μ amp at the piston output. Since the accelerator displacement of an automobile engine controls engine torque, the engine is another example of an integrating g_t amp.

We can use transducers to convert the energy at the ports of an amplifying device to the energy domains of the source and load. We can use transformers to adjust the input and output impedances to levels that are appropriate relative to the source and load impedances. The wide variety of transducers and transformers available make it possible to design power amplifiers for a wide variety of applications.

The Diode—an Electrical Flow Controller

In this section, we examine the behavior of a semiconductor diode. A diode permits electric current to flow in only one direction. Devices analogous to the diode exist in the hydraulic domain (a check valve), the mechanical domain (a pawl and ratchet), and the population domain (a restrictive immigration or emigration policy). We use such devices to restrict flow to a preferred direction.

Most electrical amplifiers extract their energy from dc power supplies. The most prevalent source of electrical power, however, is the ac electric utility. We use diodes to obtain dc power from the ac utility.

A semiconductor diode is constructed from an extremely thin wafer of silicon crystal (see Figure 6.48(a)).[14] Portions of the crystal are *doped*—i.e., diffused with minute amounts of selected impurities. In one region of the crystal, we use a kind of impurity each atom of which contributes a free electron. For higher concentrations of these *electron donor* atoms, there are more free electrons and the doped material has higher conductivity. That is, electrons can flow more easily (in the $-$ to $+$ direction) if a voltage source is connected across the material. Material doped with electron donor atoms is known as *n-type* material.

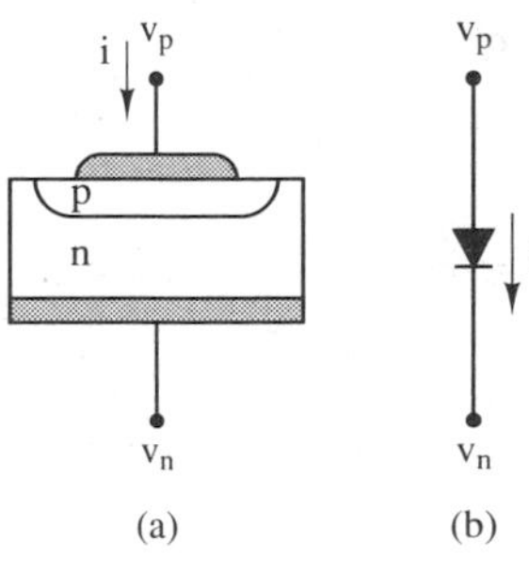

Figure 6.48 A semiconductor diode. (a) Pictorial cross section. (b) Conventional circuit symbol.

[14] Although some diodes are made from other materials, most are made of silicon.

In the neighboring region we use a different kind of impurity, each atom of which removes an electron from the crystal structure and creates a **hole**—a missing electron. There is then a tendency for some other electron near the hole to leave its atom and fill the hole, thereby creating a hole elsewhere. For higher concentrations of these *electron acceptor* atoms, there are more holes and the material has higher conductivity. That is, *holes can flow* more easily (in the + to − direction) if a voltage source is connected across the material. Material doped with electron-acceptor atoms is known as *p-type* material.

Both free electrons and free holes wander randomly from areas where they are highly concentrated to areas where they are not. A boundary between p and n regions is known as a **pn junction.** At the pn junction of a diode, free electrons from the n region tend to wander into the p region, where they immediately combine with holes (see Figure 6.48(a)). The electrons in the n region and the holes in the p region wander toward the junction to replace those electrons and holes that combine. The wandering constitutes a current across the junction. If we apply a voltage across the diode in the forward direction (+ at the p material), we sustain and greatly increase that current. Then we say that the pn junction is *forward biased.* If we apply a voltage across the diode in the reverse direction (+ at the n material), we inhibit the current. In that case, we say that the pn junction is *reverse biased.*

Experiments show that the current through a semiconductor diode varies exponentially with the diode voltage, v_{pn} as shown in Figure 6.49(a). If we apply a voltage larger than 0.6 V across the diode in the forward direction ($v_{pn} > 0$), the resistance becomes small and the current flows freely. If we *reverse-bias* the diode, the resistance becomes large and the current is nearly zero.

The conventional circuit symbol for a semiconductor diode is shown in Figure 6.48(b). The arrow points toward the n-type material, in the direction of easy current flow. As the symbol suggests, the diode conducts almost perfectly in the direction of the arrow, but is nearly nonconducting in the opposing direction. For most purposes, we can treat a diode as if it has the **ideal diode** characteristic of Figure 6.49(b)—as if it conducts perfectly in one direction, but not at all in the other direction.

For a more explicit mathematical representation, we represent the exponential diode characteristic by straight-line approximations in the two regions (see Figure 6.49(c)). The equations for the lines are as follows:

$$\begin{aligned} &\text{Forward direction:} \quad && i \approx \frac{(v_{pn} - v_D)}{R_D} \quad && \text{for } v_{pn} > v_D \\ &\text{Reverse direction:} \quad && i \approx -i_D \quad && \text{for } v_{pn} < 0 \end{aligned} \tag{6.71}$$

The voltage v_D is called the diode **saturation voltage**—the voltage at which the diode is considered fully conductive. The current is nearly zero for $0 < v_{pn} < v_D$; we do not operate a diode in that region. The leakage current i_D is called the diode **saturation current**—the current at full reverse bias. Typical parameter values for a silicon diode are $i_D \approx 60$ nA, $v_D \approx 0.7$ V, and $R_D \approx 500\ \Omega$. The diode current is quite sensitive to temperature; hence, the model parameters i_D and R_D are temperature sensitive.

The circuit of Figure 6.50(a) uses an electrical transformer to reduce the 110-V, 60-Hz signal supplied by the electric utility to a voltage level suitable for electronic equip-

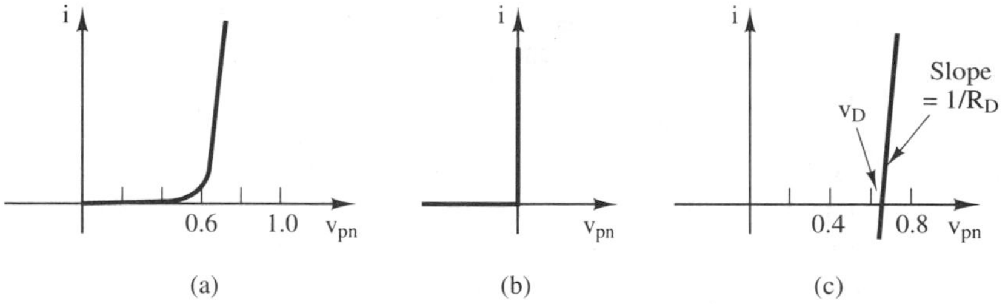

Figure 6.49 Typical silicon diode characteristics; v_{pn} is in volts. (a) Measured characteristic. (b) Ideal characteristic. (c) Linear approximation.

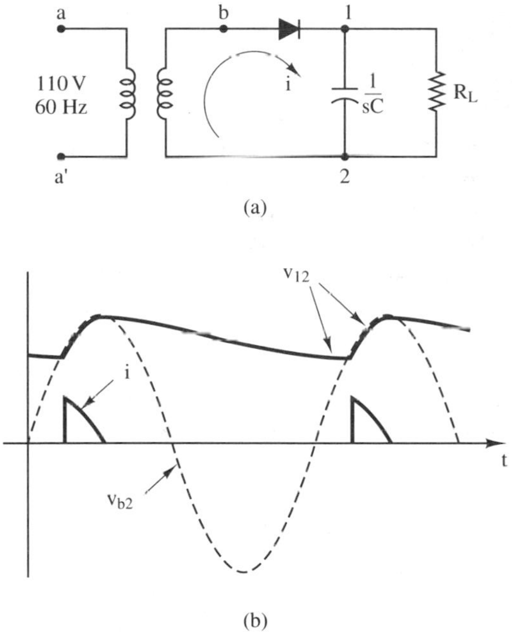

Figure 6.50 An ac-to-dc converter circuit. (a) Lumped model. (b) Steady-state behavior.

ment, say, 10 V. The diode attached to the secondary of the transformer permits current to flow from the transformer in only one direction.

Exercise 6.24: The 110 V value that is quoted by the electric utility is the *effective value* or root-mean-square (rms) value of the voltage, rather than its amplitude. For a resistive load, the sinusoidal signal delivers the same power as a 110-V dc signal would. Show that the amplitude of the sinusoidal voltage is larger than the 110-V effective value by the factor $\sqrt{2}$.

In the steady state, the transformer output voltage v_{b2} exceeds the capacitor voltage v_{12} only when v_{b2} is near its peak value. When $v_{b2} > v_{12}$, the diode becomes forward biased and the current i flows through the diode, primarily into the capacitor. The capacitor charges rapidly enough that v_{12} nearly follows v_{b2} (see Figure 6.50(b)). The magnitude of the diode current is $i \approx C\dot{v}_{12}$.

As the transformer output voltage drops below its peak value, the diode becomes reversed biased and acts like an open circuit. Then the capacitor discharges slowly through the large load resistance R_L. Although the voltage across the load resistor is roughly constant, it does have some *ripple,* as shown in Figure 6.50(b). The capacitor is selected to minimize that ripple; it is known as a *filter capacitor.*

The diode-capacitor combination is known as a **half-wave rectifier.** Selection of an appropriate value for the filter capacitor is examined in problem 6.31 at the end of the chapter.

Exercise 6.25: Show that the time constant for charging the capacitor is R_DC, where R_D is the diode resistance, and that the time constant for discharging the capacitor is R_LC.

The Transistor—an Electrical Amplifier

One of the most useful, inexpensive, and widely available amplification devices is the **bipolar junction transistor.** Figure 6.51(a) shows the exterior of a single transistor. The case of the transistor is about the size of a small pencil eraser. The transistor chip inside the case is about the size of a small pin head. In this section, we examine the behavior of a bipolar junction transistor and develop a linear lumped model that represents that behavior.

A thin wafer of silicon crystal is doped to form a pair of back-to-back pn junctions. The three doped regions are connected to three terminals, as shown in Figure 6.51(b). The regions connected to the terminals labeled E, B, and C are respectively called the **emitter**, the **base**, and the **collector** of the transistor.

The transistor in the figure forms an npn sandwich. It is accordingly known as an *npn transistor.* The n-type emitter and collector regions are heavily doped and relatively conductive. The p-type base region is very thin and lightly doped.[15] Since the base has a

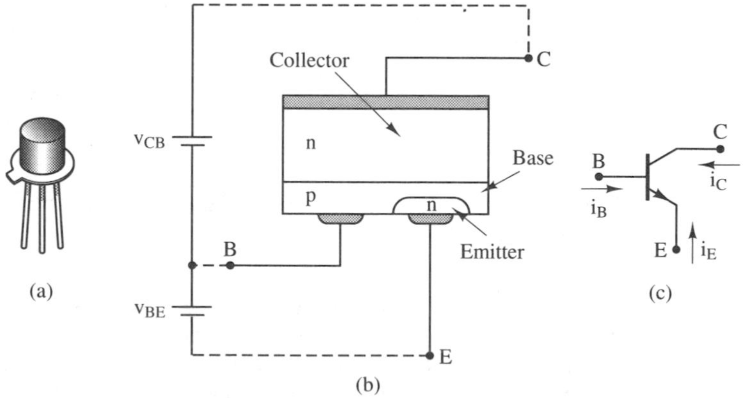

Figure 6.51 An npn bipolar junction transistor. (a) A picture of the transistor case and the protruding leads. (b) A pictorial cross section of the transistor. (c) The conventional circuit symbol for the transistor.

[15] There is also a pnp transistor in which the emitter and collector are heavily doped p regions and the base is a lightly doped n region.

low concentration of holes, it is a poor conductor. There is a tendency for the highly concentrated electrons in both the emitter and the collector to wander into the p-type base. We *reverse-bias* the collector-base junction (make $v_{CB} > 0$) to prevent the flow of collector electrons into the base. We *forward-bias* the emitter-base junction (make $v_{BE} > 0$) to encourage the flow of emitter electrons into the base. (Figure 6.51(b) shows two batteries oriented to provide a proper bias to each junction.)

Since there are few holes in the base (owing to the light doping), little combining of electrons and holes can occur, the base-to-emitter current is small, and most of the electrons that enter the base from the emitter remain there as free electrons. Since the base is very thin, the uncombined free electrons distribute themselves quickly throughout the base. Thus, the base acts as a conductor. Increasing the forward-bias voltage increases the number of free electrons in the base, in turn increasing its conductivity.

Because the emitter and collector are much more conductive than the base, the resistances of the emitter and collector can be ignored and the sum of the two bias voltages ($v_{CB} + v_{BE}$) appears across the base (see Figure 6.51(b)). The collector-to-emitter voltage v_{CE} produces a collector current in proportion to the base resistance. We make the collector current large by using a large reverse-bias voltage v_{CB}. The emitter-base current that controls the base resistance is small. What we have, then, is a small emitter-base current controlling a large collector current. Thus, the transistor is fundamentally a current amplifier. The resistive mechanism for control of the collector current is similar to the resistive mechanism for hydraulic flow control in the valve-controlled pump discussed earlier.

The conventional three-terminal symbol used to represent an npn bipolar junction transistor is shown in Figure 6.51(c). The arrow in the symbol points toward the n-type material at the base-emitter junction. That is, it shows the direction in which the base current flows. For a pnp transistor, the arrow points toward the base, and the voltages and currents reverse sign.

We treat the base-emitter terminal pair as an input port and the collector-emitter terminal pair as an output port. The input port behaves as a diode with a characteristic like that shown in Figure 6.49(a). When the transistor is turned on (that is, $i_B > 0$), we can approximate the input characteristic by a linear equation similar to equation (6.71):

$$v_{BE} \approx v_{BE0} + i_B R_b \tag{6.72}$$

where R_b is the **base-to-emitter resistance** and v_{BE0} is the **junction voltage.** Typical parameter values for a particular version of bipolar junction transistor are $v_{BE0} \approx 0.7$ V and $R_b \approx 1{,}500\ \Omega$. ($v_{BE0} \approx 0.7$ V for any silicon bipolar junction transistor.)

The output port (collector-emitter) operating characteristics of a bipolar junction transistor are curves of i_C vs. v_{CE} for various values of the controlling base current i_B. A typical set of curves for a particular version of transistor is shown in Figure 6.52. (These curves are commonly known as *collector characteristics.*) Over much of the region, the curves are nearly straight lines. Hence, we can represent each curve (in its nearly linear region) by a linear equation of the form $i_C \approx i_{intercept} + v_{CE}/R_c$, where $1/R_c$ is the slope of the line. Figure 6.52 shows the straight line approximation to the curve for $i_B = 0.12$ mA. Note also that all the curves are nearly parallel. Therefore, we treat their linear approximations as if they had the same slope. Finally, for each value of v_{CE}, the collector current i_C

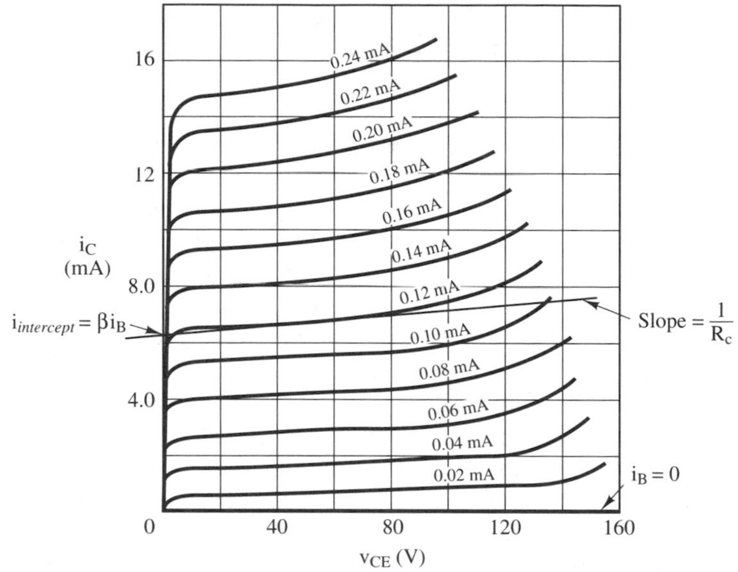

Figure 6.52 Collector characteristics for a particular version of bipolar junction transistor. For this version, $R_b \approx 1500\ \Omega$ and $v_{BE0} \approx 0.7$ V.

is nearly proportional to the base current i_B. Thus, $i_{intercept} \approx \beta i_B$, for some proportionality factor β.

Let us replace the collector characteristics of Figure 6.52 by a set of parallel straight-line approximations. (We made similar straight-line approximations for the output-port characteristics of a hydraulic valve in Figure 6.44.) Then we can represent the set of straight-line approximations by the single linear equation

$$i_C = \beta i_B + \frac{v_{CE}}{R_c} \tag{6.73}$$

where R_c is the **collector-to-emitter resistance** and β is the **current amplification factor.**

Exercise 6.26: Determine from the slope of the approximating straight line in Figure 6.52 that $R_c \approx 110$ kΩ. Determine from the intercept that $\beta \approx 52$. These values change somewhat if we approximate, instead, the curve that corresponds to a different value of i_B.

The linear equations (6.72) and (6.73) describe the behaviors of the input and output ports of the transistor. We represent these equations by lumped linear circuit elements in Figure 6.53. This lumped model is in the form of the β amp model of Figure 6.46(a). The input resistance R_b is small and the output resistance R_c is large, as desired for a β amp.

The linear lumped model represents the transistor behavior only if the emitter-base junction is forward biased (to make $i_B > 0$) and the operating point (i_C, v_{CE}) is in the linear region of the collector characteristics. The circuitry used to establish operation in a particular region of the curves is called the **bias** circuitry. We examine bias circuitry in the next section.

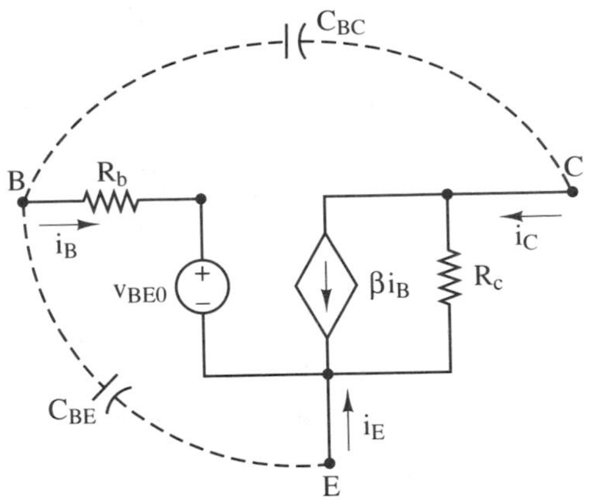

Figure 6.53 Linear lumped model for a bipolar junction transistor.

Exercise 6.27: Sketch, on Figure 6.52, an approximate boundary for the nearly linear region of operation. What is the range of values of i_B within that region? According to equation (6.72), what is the corresponding range of values of input voltage v_{BE}? What is the range of values of output current i_C within the linear region? What is the range of values of output voltage v_{CE} in that region?

The equipotential surfaces near the junctions of a transistor are similar to parallel plates. As a consequence, the junctions have some capacitance. For the transistor of Figure 6.52, the base-emitter capacitance C_{BE} is 80 pF and the base-collector capacitance C_{BC} is 9 pF. These junction capacitances reduce the response of the transistor for high-frequency signals. For high-frequency operation, we would attach these capacitors to the model as suggested by the dashed lines in Figure 6.53. See problem 6.35 at the end of the chapter.

We have examined the behavior of the bipolar junction transistor when it is connected with its emitter terminal common to both the input and output ports. This is known as the **common-emitter** configuration and is the configuration that is usually used. The linear transistor equations (6.72) and (6.73) are expressed in terms of common-emitter variables: i_B and v_{BE} at the input port and i_C and v_{CE} at the output port.

The transistor and its model can be used in other configurations, however. Since i_B controls i_C, we must put the base terminal at the input port and the collector terminal at the output port to get power amplification and input-output isolation. Figure 6.54 compares the common-emitter configuration with the other two configurations in which the transistor acts as an amplifier.

To determine the behavior of the transistor in the common-base or common-collector configuration, we place the linear lumped model of Figure 6.53 in that configuration and rewrite the equations in terms of the new input-port and output-port variables. Problem

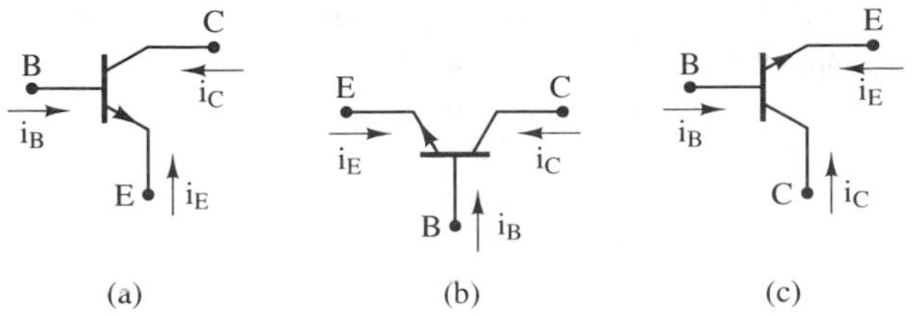

Figure 6.54 Transistor amplifier configurations; for the transistor of Figure 6.52, $R_b \approx 1{,}500\ \Omega$, $R_c \approx 110\ k\Omega$, and $\beta \approx 52$.
(a) Common emitter; $R_i = R_b$, $R_o = R_c$.
(b) Common base; $R_i = R_b/\beta$, $R_o = \infty$.
(c) Common collector; $R_i \sim \beta R_L$, $R_o \approx (R_S + R_b)/\beta$.

6.36 at the end of the chapter shows that the equations for the **common-base** configuration are[16]

$$\begin{aligned} v_{EB} &= \frac{R_c R_b i_E + R_b v_{CB}}{(\beta + 1)R_c + R_b} \approx \left(\frac{R_b}{\beta}\right) i_E \\ i_C &= \frac{v_{CB} - v_{EB}}{(\beta + 1)R_c} - \frac{\beta i_E}{\beta + 1} \approx -i_E \end{aligned} \tag{6.74}$$

Problem 6.37 shows that the equations for the **common-collector** configuration are

$$\begin{aligned} v_{BC} &= R_b i_B + v_{EC} \\ i_E &= -(\beta + 1) i_B + \left(\frac{1}{R_c}\right) v_{EC} \end{aligned} \tag{6.75}$$

Problems 6.36 and 6.37 also find the input and output resistances for the common-base and common-collector configurations. These input and output resistances are compared with the input and output resistances of the common-emitter configuration in Figure 6.54.

The common-base configuration has a much lower input resistance than the common-emitter configuration and has nearly infinite output resistance. Hence, it behaves as a *nearly ideal* β *amp*—a current amplifier. However, its current gain is only 1. The common-base configuration is very effective as an *isolation device.* Because of its low input resistance, it can be used to protect against loading of a current sensor. It can also be used to drive a high impedance load efficiently. Thus, the common-base configuration is sometimes used as the first and/or last stage in a cascaded-amplifier sequence.

The common-collector configuration does not provide total isolation; its input resistance depends on the load resistance and its output resistance depends on the source resistance. Its input resistance is relatively large—larger than the load resistance by the factor β. Its output resistance is very small—smaller than $R_S + R_b$ by the factor β. This configuration does not act like a current amplifier. Rather, it behaves as a *nearly ideal* μ *amp*—a voltage amplifier. The term $R_b i_B$ of equation (6.75) is negligible. Consequently, the output voltage v_{EC} is essentially equal to the input voltage v_{BC}. For that reason, a common-collector amplifier is also known as an *emitter follower* or a *voltage follower.* The common-collector amplifier can be used to drive a low-impedance load efficiently.

Transistor Bias Design

In this section, we show how to design circuitry to *bias* a transistor—that is, to make it operate in a particular region of its operating characteristics. We also demonstrate the use of a transistor as a powerful switch. In the next section, we bias a transistor to operate in its nearly linear region, then use the biased transistor as a linear amplifier of electrical signals.

Figure 6.55 shows typical bias circuitry for a common-emitter amplifier. We select the dc voltage sources (v_{CC} and v_{BB}) and the bias resistors (R_C and R_B) to obtain the desired operating point. Let us design the bias circuitry for the transistor of Figure 6.52. We

[16]The junction voltage v_{BE0} is ignored in deriving equations (6.74) and (6.75).

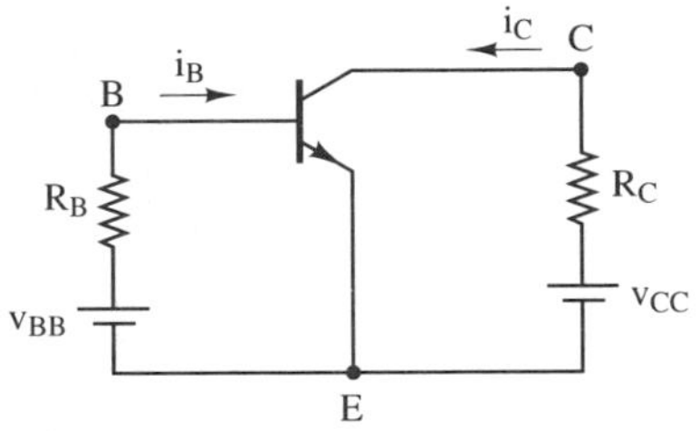

Figure 6.55 Typical common-emitter bias circuitry.

choose to set the operating point of the transistor close to the point $v_{CE} = 40$ V and $i_C = 4$ mA on the collector characteristics. According to Figure 6.52, the base current at this operating point is about $i_B = 0.08$ mA.

We shall design the collector bias subcircuit first. Let us partition the circuit into two halves at the port (C, E). The equation that describes the right half, the collector bias subcircuit, is $v_{CE} = v_{CC} - i_C R_C$, a straight-line relation between i_C and v_{CE}. The left half is described by the collector operating characteristic for $i_B = 0.08$ mA. We can superimpose the collector bias line on the collector operating characteristic in Figure 6.52. However, we have not yet chosen values for v_{CC} and R_C. A number of different straight-line collector bias lines could be drawn through the operating point (40 V, 4 mA). We show one such bias line superimposed on the collector characteristics in Figure 6.56(a). This bias line corresponds to $v_{CC} = 80$ V and $R_C = 10$ kΩ. Note that we cannot make v_{CC} smaller than 40 V, because the bias line must pass through the point (40 V, 4 mA) and R_C must be positive. Once we set v_{CC} and R_C to these values, if we change i_B from its design value of 0.08 mA, we cause the operating point to move up or down that particular collector bias line.

Exercise 6.28: Sketch another collector bias line for the same operating point. What are the corresponding values of v_{CC} and R_C?

Exercise 6.29: We could have used equations instead of graphical lines to find v_{CC} and R_C. Use the linearized transistor output equation (6.73), together with the equation for the collector bias subcircuit, to find values of v_{CC} and R_C for which $i_B = 0.08$ mA, $i_C = 4$ mA, and $v_{CE} = 40$ V.

Now we design the base bias subcircuit. We partition the network into two halves at the port (B, E) of the transistor. In terms of the variables at that port, the equation for the base bias subcircuit is $v_{BE} = v_{BB} - i_B R_B$. Equation (6.72) describes the input-port characteristic of the transistor. We eliminate v_{BE} from the two equations to get

$$v_{BB} - i_B R_B = v_{BE0} + i_B R_b \qquad (6.76)$$

We substitute the typical parameter values and set $i_B = 0.08$ mA to obtain

$$v_{BB} - (0.08 \text{ mA})R_B = 0.7 \text{ V} + (0.08 \text{ mA})(1500\ \Omega)$$

$$= 0.82 \text{ V} \qquad (6.77)$$

It is apparent that we must choose v_{BB} greater than 0.82 V. Let us select $R_B = 10$ kΩ. Then, $v_{BB} = 1.62$ V.

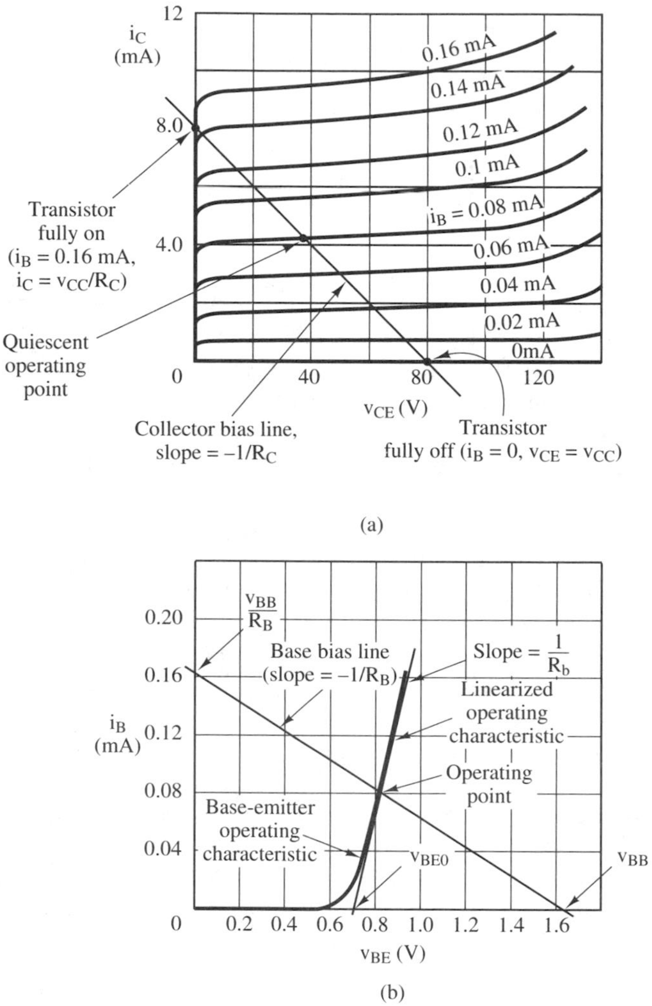

Figure 6.56 Quiescent common-emitter operating conditions for the transistor of Figure 6.52. (a) Output-port operating conditions (b) Input-port operating conditions.

Exercise 6.30: We could, instead, design the base bias subnet by superimposing a base bias line of slope $-1/R_B$ on the measured base-emitter characteristic of the transistor. We adjust the voltage intercept v_{BB} of that line so that the intersection of the line with the measured characteristic occurs at $i_B = 0.08$ mA. Verify that this process is carried out correctly in Figure 6.56(b).

The only purpose of the bias subcircuits is to set the operating point of the network. The bias voltage sources v_{BB} and v_{CC} are not *signal* sources. Rather, we attach a separate signal source to the input (B, E) port. We *might* also attach a separate load to the output

(C, E) port. The intersection of the collector bias line and the collector operating characteristic (for the specified i_B) is known as the *quiescent operating point* of the transistor; i.e., the operating point in the absence of a source signal.

Suppose we attach a current source between nodes B and E of the biased circuit of Figure 6.55. If that current source injects 0.02 mA of current into the base node, then the base current becomes $i_B \approx 0.1$ mA, and the output port of the transistor behaves like the collector characteristic labeled $i_B = 0.1$ mA in Figure 6.56(a). The operating point moves up the collector bias line to the point $i_C = 5.5$ mA, $v_{CE} = 25$ V. That is, a small (0.02-mA) change in base current causes a large (1.5-mA) change in collector current and a large (15-V) change in output-port voltage.

If the attached current source injects 0.07 mA or more at the base, then the total base current (0.15 mA) is large enough to turn the transistor *all the way on.* That is, the operating point moves up the collector bias line to the point where the curves drop sharply. The collector current rises to 8 mA, its maximum value for this collector bias line, and the output-port voltage v_{CE} drops almost to zero.

If the value of the current source is negative, the base current becomes less than 0.08 mA, and the operating point moves *down* the load line. If the source current is -0.08 mA, the total base current is zero, the operating point moves to the $i_B = 0$ curve, and the transistor turns *all the way off.*

Exercise 6.31: Show that the power delivered to the transistor output port by the collector bias subcircuit at the quiescent operating point of Figure 6.56 is 160 mW.

The power delivered *to the transistor* by the collector bias subcircuit causes the transistor to heat up. Because a transistor is so small, it is not unusual for its temperature to rise to 200°C. (The power absorbed per unit volume is about 1,000 times that delivered to the baseplate of a household iron.) However, the power absorbed is nearly zero if the transistor is fully on or fully off. In digital circuits (such as computer circuits), transistors are used as switches: They are turned all the way on or off. Then the power consumption and the temperature rise of the transistor are low.

Although we can set the parameters of the bias subnets accurately, we have only loose control of the transistor parameters R_b, R_c, and β. The transistor manufacturer publishes typical values for the parameters, but the actual values differ from one transistor to the next. Furthermore, the parameters can change by as much as a factor of 2 as the transistor temperature rises during operation. Hence, the bias network will not produce the precise operating values on which we base the design. Yet, the bias network can still be expected to place the operating values in the correct *region* of the collector characteristics. Then the transistor behavior will be *similar to* the behavior of the linear model.

We use an amplifier primarily to strengthen (increase the available power of) a signal waveform. Variations in the transistor parameters change the *magnitude* of the response waveform, but they do not change the *shape* of the response waveform or seriously reduce its available power. The driver of an automobile encounters a similar uncertainty in absolute torque and speed for a given accelerator pedal position, but no uncertainty in the nature of the torque pattern over time. The driver merely monitors the behavior and makes corrections as needed. We find in section 6.5 that feedback of the response signals can be used to reduce the system sensitivity to changes or errors in parameter values.

Example Problem 6.8: Analysis of an electronic switch.
Figure 6.57 shows a weak current source i_S attached to the input port of an npn silicon common-emitter transistor. The source current is proportional to the light that impinges on a light sensor. The load resistance R_L serves as the collector bias resistor R_C.

The load R_L is the resistance of an electromagnetic relay. When light shines on the sensor, the transistor activates the relay and turns off a floodlight. When the sensor receives no light, the transistor releases the relay and the floodlight turns back on. Thus, the transistor serves as an electronic switch.

(a) The collector characteristics of the transistor are shown in Figure 6.58. Superimpose the load line on the collector characteristics. Determine the source current necessary to turn the transistor (and relay) fully on. What is the load current when the transistor is fully on?

(b) Find the input signal power provided to the transistor by the light sensor and the output signal power provided by the transistor to the load when the transistor is fully on. Determine the power gain provided by the electronic switch.

(c) Find the power dissipated in the transistor when it is fully on; half on; fully off.

Solution: **(a)** The collector bias load line is determined by V_{CC} and R_L. It is superimposed on the collector characteristics of the transistor in Figure 6.58. There is no base bias circuit in Figure 6.57. Therefore, $i_B = 0$ when $i_S = 0$. Consequently, when $i_S = 0$, $v_{CE} = 10$ V, the transistor is fully off ($i_C \approx 0$), and the voltage across the load is zero. The transistor is fully on when the operating point is (nearly) at the i_C intercept of the collector bias load line—that is, at $i_C \approx 2$ A and $i_B \approx 37$ mA. The load current is i_C.

(b) The input signal power is $P_{in} = i_S v_{BE}$. When the transistor is turned on, $v_{BE} \approx V_{BE0} \approx 0.7$ V and $i_S \approx 37$ mA. Therefore, $P_{in} \approx 26$ mW. The output signal power is $P_{out} = i_C^2 R_L = (2\text{ A})^2(5\ \Omega) = 20$ W. The power gain is

$$\frac{P_{out}}{P_{in}} = 769\ (28.9\text{ dB}) \tag{6.78}$$

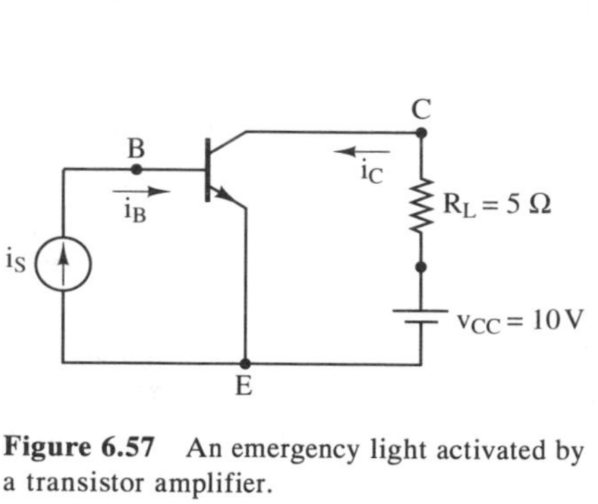

Figure 6.57 An emergency light activated by a transistor amplifier.

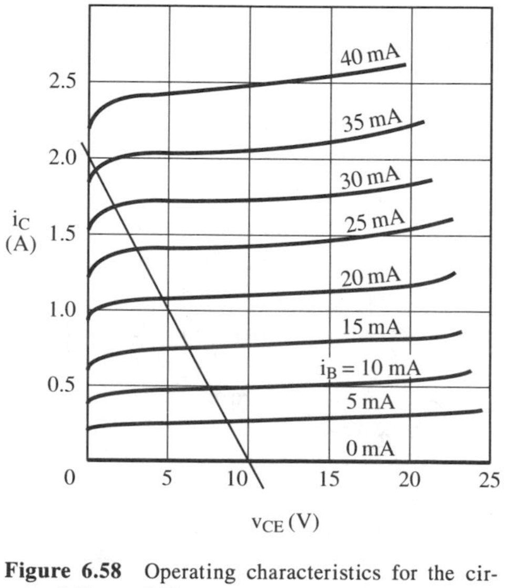

Figure 6.58 Operating characteristics for the circuit of Figure 6.57.

(c) The power dissipated at the base (input port) of the transistor is always negligible relative to that dissipated at the collector (output port). The collector dissipation is $P_C = v_{CE}i_C$. When the transistor is fully on, $i_C \approx 2$ A and $v_{CE} \approx 0.2$ V. Therefore, $P_C \approx 0.4$ W.

When the transistor is half on, $i_C \approx 1$ A. According to Figure 6.58, at this value of i_C, the collector bias load line passes through $v_{CE} = 5$ V. Therefore, the collector dissipation is $P_C = 5$ W.

When the transistor is fully off, the collector bias load line shows that $v_{CE} \approx 10$ V and the current is essentially zero. Therefore, $P_C \approx 0$. Note that the transistor dissipates much less power when it is fully on than it does when it is half on. The dissipated power appears as heat in the tiny transistor chip.

Capacitor-Coupled Amplifier Design

We use amplifiers to enable weak devices to cause powerful actions. In this section we use a transistor amplifier to enable the weak electrical output of a microphone or tape deck to drive a loudspeaker. If we were to attach a tape deck directly to a loudspeaker, its output signals would not cause the loudspeaker to produce audible sounds. A typical power output for a tape deck is 1 μW. The power needed to drive an earphone is about 1 mW. The power needed to drive a room loudspeaker is about 1 W.

Let us bias the transistor of Figure 6.52 as shown in Figures 6.55 and 6.56, with $v_{BB} = 1.62$ V, $v_{CC} = 80$ V, and $R_B = R_C = 10$ kΩ. The biased transistor behaves as a linear amplifier of electrical signals. This amplifier is powerful enough to convert the signal from a tape deck into audible sounds.

A tape deck produces a voltage waveform that is a replica of the magnetization pattern on the tape. If we were to place the tape deck output terminals in series with the bias source v_{BB} of Figure 6.55, then an amplified version of the source signal would appear across the output terminals (C, E). However, because the output resistance of the tape deck would be connected in series with the base bias resistor R_B, that connection would change the base bias line and the operating point. Instead, let us attach the signal source between nodes B and E, in parallel with the bias subcircuit.

Unfortunately, if we were to connect the source directly to node B, we would provide an alternative path for the current from v_{BB} and, again, disturb the operating point established by the base bias subcircuit. To avoid disturbing the operating point, we insert a **coupling capacitor** C_1 between the signal source and node B as shown in Figure 6.59. The constant bias current cannot pass through that capacitor. We make the coupling

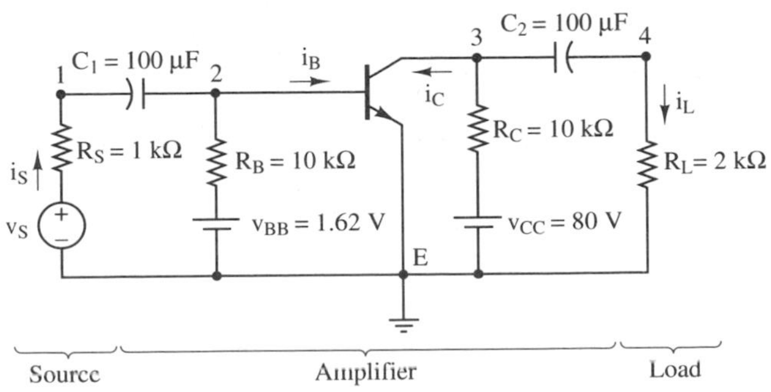

Figure 6.59 A capacitor-coupled common-emitter amplifier.

capacitor large enough (100μF) that its sinusoidal impedance is negligible for audio frequencies; then it will not affect the audio signals. We also use a coupling capacitor C_2 to attach the load (earphone) to the output port, as shown in Figure 6.59. Then the load does not disturb the output-port operating point.

Figure 6.59 can be viewed as a connection of three components: The signal source, the amplifier, and the load. The amplifier includes the transistor, its bias circuitry, and the coupling capacitors. We call it a **capacitor-coupled amplifier.** Note that the coupling capacitors prevent constant source signals from reaching the amplifier or the load. In an application that requires amplification of *dc (constant) signals,* we cannot use capacitor coupling. For that reason, a capacitor-coupled amplifier is sometimes referred to as an *ac (or variable-signal) amplifier.*

Exercise 6.32: Show that for signals of frequency no lower than 80 Hz, the impedance of each coupling capacitor is negligible (compared with the other circuit impedances) if $C_i = 100\ \mu F$.

When the signal source is turned off, the system is said to be *quiescent* (or quiet). That is the reason we call the operating point produced by the biasing process the quiescent operating point. When the source is turned on, it causes the currents i_B and i_C to vary about their quiescent values in proportion to the source signal. When we attach the load to the output port (4, E) of the amplifier, the variations in i_C (at audio frequencies) cause a current to flow through the coupling capacitor into the load in proportion to the source signal. The signal power that flows into the load owing to the source signal is much greater than the power flowing from the source.

For any particular source signal $v_S(t)$, we can apply linear analysis techniques to the linear lumped model shown in Figure 6.59 to determine the response $v_4(t)$ at the load. Ultimately, however, we are not interested in particular signals. We designed the amplifier to produce powerful copies of the source signals at the load. Therefore, we want only to know if the amplifier will make the *copies* of the source signals sufficiently powerful. Therefore, we apply a simple *sinusoidal source signal* v_S, use linear analysis techniques to find the corresponding load signal v_4, and determine the ratio of output signal power to input signal power—the power gain.

To begin the analysis, we replace the transistor in Figure 6.59 by its linear lumped model (shown in Figure 6.53). The resulting network is complicated. In general, the voltage at any point in a network depends on all the sources in the network. There are four sources in this network: v_S, v_{BE0}, v_{BB}, and v_{CC}. Since the lumped model is linear, we can take advantage of the *principle of superposition.* That is, we can find the response v_4 owing to each source separately, and then add those responses. Since the output coupling capacitor acts as an open circuit to each *constant* source, there is no response at node 4 owing to those sources. Thus, the complete response v_4 is the response owing to v_S alone, and we *set to zero the three constant sources* during our analysis.

The coupling capacitors have negligible impedance to the sinusoidal components of audio signals. Accordingly, we short circuit the two coupling capacitors in the model. The junction capacitances C_{BE} and C_{BC} have impedances on the order of 0.5 MΩ at audio fre-

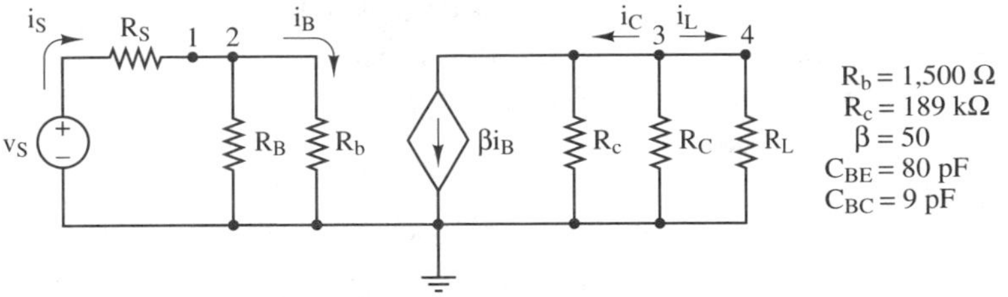

Figure 6.60 Signal model of the circuit of Figure 6.59 for audio frequencies. The transistor parameters are listed for the operating point $i_C = 4$ mA and $v_{CE} = 40$ V.

quencies. Therefore, we open circuit them. We are left with the relatively simple circuit of Figure 6.60. We should view this simplified model of the system as a *signal model.* It describes the relation between source signals and load signals. It is not a complete model. It does not account for the constant components of the voltages and currents within the transistor and the bias circuitry. To obtain a complete model, we must reinsert the coupling capacitors and the three constant sources. (For frequencies above 15 kHz, we would also have to include the junction capacitances C_{BE} and C_{BC}.)

Measured values of the transistor parameters for the specified operating point are shown beside the signal model in Figure 6.60. It is easy to find the signals at the source port and load port of this simple circuit. From those signals we can find the power output of the source and the signal power delivered to the load. The ratio of these two quantities is the power gain. Problem 6.32 at the end of the chapter analyzes this audio-frequency circuit model and shows that the current gain is $i_L/i_S = -35.7$, the voltage gain is $v_4/v_2 = -54.9$, and the power gain is $G_P = 1{,}954$ (33 dB). The increased signal power at the load is derived from the dc source v_{CC}. The power gain is high enough to drive an earphone, but not a loudspeaker.

Although the bipolar junction transistor is fundamentally a current amplifier, this example demonstrates that in the common-emitter orientation it typically provides significant voltage gain as well. The actual gains in voltage and current depend on the load impedance, as shown in equations (6.68)–(6.70). Problem 6.33 at the end of the chapter shows that an electrical transformer can be used to optimize the load impedance. Note that a positive value of the source signal v_S produces negative values for v_4 and i_L. In general, a common-emitter amplifier inverts the input signal.

A capacitor-coupled amplifier is a *band-pass amplifier.* Its gain is reduced at low frequencies because the impedances of the coupling capacitors become significant at those frequencies. At high frequencies, the impedances associated with the junction capacitances become too small to ignore. The base-emitter capacitance C_{BE} reduces i_B by shunting part of the current to ground. The base-collector capacitance C_{BC} reduces i_B by shunting part of the current directly to the collector terminal. Both effects reduce the amount of current multiplied by the current amplification factor β and, therefore, reduce the effective gain of the circuit. The frequency response of this common-emitter amplifier is examined in problem 6.35 at the end of the chapter.

We now summarize the steps we have used to design this transistor amplifier in the form of a design procedure that can be used for many types of amplifiers:

PROCEDURE FOR DESIGNING A LINEAR AMPLIFIER

1. Find a device in which a weak signal controls a strong signal. Associate the weak signal and strong signal with the input port and output port of the amplifier, respectively. The weak signal is the *control variable.*
2. Examine the output-port operating characteristics of the device and select a quiescent operating point in a region where the operating characteristics are nearly linear and nearly equally spaced for equal increments of the control variable.
3. Examine the input-port operating characteristic of the device to determine the quiescent value of the remaining input-port variable. Also determine the range over which the control variable can vary without the device seriously deviating from linear operation.
4. Approximate the output-port operating characteristics by a set of equally spaced straight lines. Approximate the input-port operating characteristic by a single straight line. Write the equations that describe those lines. Construct a linear lumped model that corresponds to those equations.
5. Attach additional devices as necessary to bias the device—to establish quiescent operation at the operating points chosen in steps 2 and 3. (It may be possible to use the source and/or the load as part of the biasing mechanism, as in example problem 6.8.) Construct a lumped model of the biased device.
6. If necessary, determine how to attach the source and load without disturbing the quiescent operation.
7. Determine the power gain of the amplifier for an appropriate source signal: Use the linear model to find the change in load power caused by turning on the source, and then divide that *change in load power* by the source power that causes the change.

The Control Valve—a Hydraulic Amplifier

The hydraulic lift that raises an automobile in a repair shop is a translational hydraulic motor (see Figures 6.17(a) and 6.21). Similar hydraulic motors maneuver bulldozer blades, car crushers, trash compactors, aircraft wing flaps, ship rudders, and rocket engines. Also known as *hydraulic cylinders, hydraulic actuators,* or *hydraulic rams,* these motors are compact and powerful. For other high-power tasks such as rotating a missile launcher on a naval vessel or turning a power windlass for a crane or a catapult, there are rotary hydraulic motors—*rotary actuators.*

Where do we get the constant-pressure hydraulic source that drives the actuator? Just as a constant-voltage source of electric power can be obtained, by rectification, from the ac electric utility, so can a source of constant-pressure hydraulic power be derived from the ac electric utility.

We use an electric motor to drive a positive-displacement pump (see Figure 6.61) at constant speed. (A positive-displacement pump is needed to reach the high pressures typical of hydraulically actuated systems—5 to 35 MPa). The positive-displacement pump

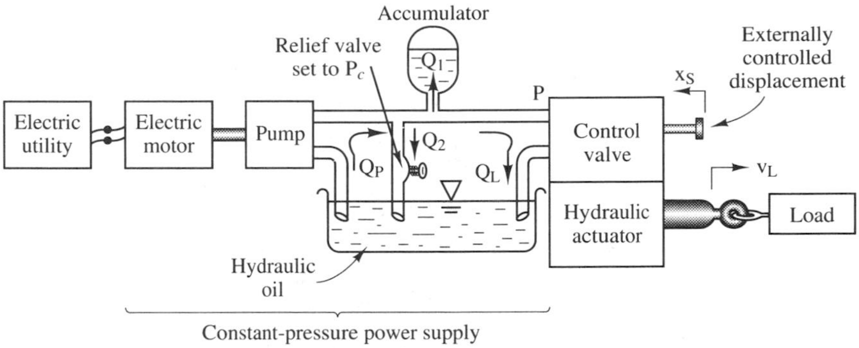

Figure 6.61 A constant-pressure hydraulic power supply system. During start-up, $P < P_c$, $Q_L = 0$, $Q_1 = Q_P$, and $Q_2 = 0$. When ready for use, $P = P_c$, $Q_L = 0$, $Q_2 = Q_P$, and $Q_1 = 0$. During use, $P \approx P_c$, Q_L is controlled, Q_1 is positive or negative, and Q_2 is positive or zero.

provides a constant flow Q_P that we must convert to constant pressure. When the motor is first turned on, the control valve of Figure 6.61 is closed and the pumped hydraulic fluid flows into an air-pressurized tank (called an *accumulator*). The relief valve is spring loaded so it will open when the pressure P at the inlet to the accumulator reaches the desired operating pressure P_c. When the relief valve opens, it provides a path for the continuing pump flow to return to the reservoir. This return flow is labeled Q_2 in the figure.

During operation, if the flow Q_L demanded by the valve-controlled actuator is temporarily greater than the flow Q_P supplied by the pump, then the pressure P drops slightly, the relief valve closes, and some fluid is drawn from the tank. Whenever the flow Q_L is less than the pump flow Q_P, the excess flow refills the accumulator. When the accumulator is full, the pressure P rises slightly and the relief valve opens to return the excess flow to the reservoir. The pressurized tank resists changes in the supply pressure P, just as the electrical capacitor resists changes in the electrical supply voltage in Figure 6.50. Thus, the pressure P at the input to the control valve is held constant at P_c.

Usually we control the motion of the power piston of a hydraulic motor by using a valve to control the flow of hydraulic fluid from a constant-pressure source of hydraulic power. The constant-pressure source is analogous to a dc voltage source. The valve that controls the fluid flow is analogous to a transistor that controls electric current. Just as the transistor amplifies the power of the signal source that controls the base current, so does the valve amplify the power of the signal source that controls the displacement of the valve.

A control valve and actuator are displayed in Figure 6.62. The valve and actuator need not be an integral unit; they need only be connected by a pair of hydraulic lines. The two valve ports labeled P_c are connected to the constant-pressure source. The ambient-pressure return port is labeled P_0. Initially, the sliding spools cover both high-pressure ports and the return port. The horizontal pressure forces on the spools are balanced, and there is no path for fluid to enter or leave the actuator's hydraulic lines. Therefore, the actuator cannot move.

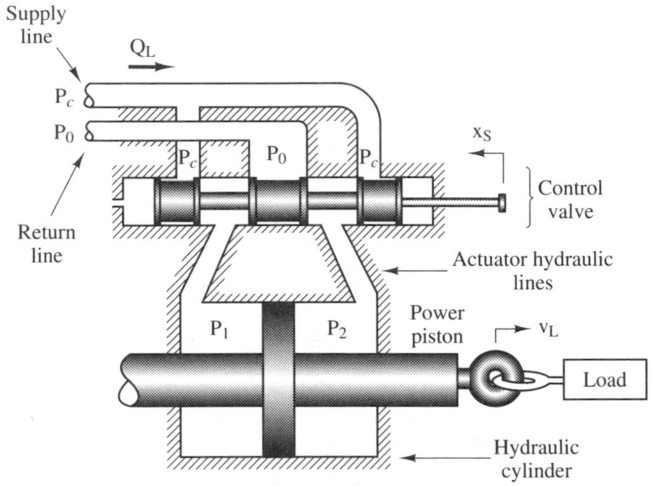

Figure 6.62 A hydraulic control valve and actuator.

Let us displace the sliding spools slightly to the left by an amount x_S. (The displacement can be carried out by an operator or, indirectly, by means of a current-controlled translational electric motor.) The displacement partly uncovers the left pressure port and connects the source pressure P_c to the left side of the actuator. Simultaneously, the displacement partly uncovers the right side of the return port, connects the right side of the actuator to the ambient pressure P_0, and opens a flow path for the hydraulic fluid. The pressure difference $P_c - P_0$ across the actuator piston develops a piston force that is directed toward the right.

Suppose the load permits the actuator to move. Then fluid flows through the valve and actuator (from the source at pressure P_c to the reservoir at pressure P_0) at some rate Q_L. During the motion of the fluid, the pressure difference P_{12} across the piston is less than the supply-to-ambient pressure difference $P_c - P_0$, owing to fluid resistances in the narrow port openings uncovered by the spool. If we increase the spool displacement x_S, we decrease the port resistances, increase the pressure drop across the piston, and increase the flow rate Q_L in proportion to x_S. The spool does not have to move much in order to develop powerful pressure differences and high fluid flow rates.

When the valve is displaced, the pressure difference P_{12} that appears across the piston also appears across the center spool of the valve and produces a spool-restoring force that increases with x_S. If we use a translational electric motor (see Figure 6.17(d)) to displace and hold the spool, then a weak electric current at the input controls a powerful hydraulic flow to the actuator. The ratio of the output flow to the input current is a flow amplification factor β. This factor is one of the measureable parameters of the amplifier.

The electrically controlled valve differs from the transistor in that its input and output flows are in different energy domains, while those of the transistor are in the same energy domain. Also, unlike the dimensionless flow amplification factor of the transistor,

β for the control valve has dimension $m^3/A{\cdot}s$. The resistance R_i at the *control input* of the amplifier is electrical. There is some inductance in the motor winding at that electrical input port, but it is of little importance. The output resistance R_o of the amplifier is hydraulic. Because of compressibility of the fluid in the valve, actuator, and connecting hydraulic lines, there is also some fluid capacitance at the output port. This fluid capacitance limits the speed of response of the control valve.

Modifications to the spool valve produce different valve operating characteristics. Figures 6.63(a)–(c) show measured operating characteristics at the hydraulic output port for three versions of the valve. For each valve, the *load pressure* $P_L \triangleq P_{12}$ is the pressure drop between the output ports of the valve (and across the hydraulic actuator piston). The figures list P_L as a percentage of the *gage pressure* $P_S \triangleq P_c - P_0$ of the hydraulic source. The *load flow* Q_L is the flow through the valve (and through the hydraulic cylinder). The figures list Q_L as a percentage of the *rated flow* Q_R—the maximum flow that the valve is intended to handle.

The performance characteristics for the pressure-flow control valve of Figure 6.63(c) are quite linear. The valve output resistance is the negative inverse slope of the curves: $R_o = P_S/(0.95\ Q_R) = (12\ \text{MPa})/(0.95\ \text{liter/s}) = 1.26 \times 10^{10}\ \text{N}{\cdot}\text{s/m}^5$. The intercepts of the valve characteristics on the flow axis are proportional to the control current i. Specifically, for $P_L = 0$, $Q_{L intercept} = \beta i$. From the uppermost vertical intercept of the valve characteristics, we find that $\beta = Q_{L intercept}/i_{max} = (0.95\ \text{liter/s})/(18\ \text{mA}) = 52.8\ \text{liter/A}{\cdot}\text{s} = 0.0528\ m^3/A{\cdot}s$. The valve input resistance is stated in the figure to be $R_i = 160\ \Omega$.

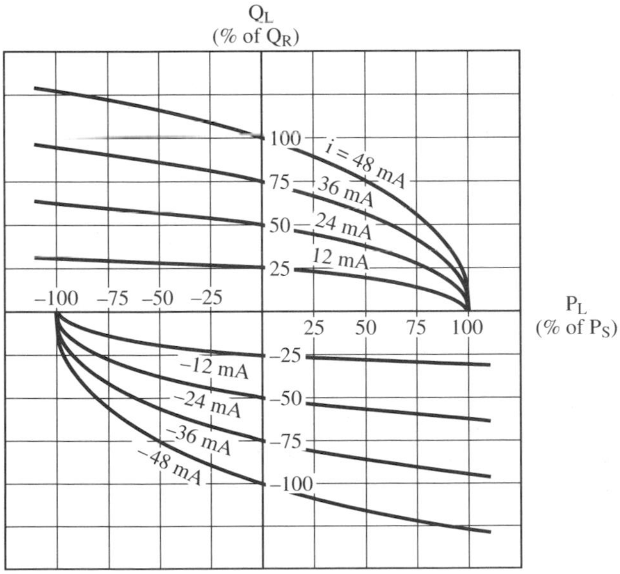

(a)

Figure 6.63 Operating characteristics of three control valves. (a) Flow control valve: $P_S = 8$ MPa, $Q_R = 0.8$ liter/s, $i_{max} = 48$ mA, $R_{coil} = 50\ \Omega$.

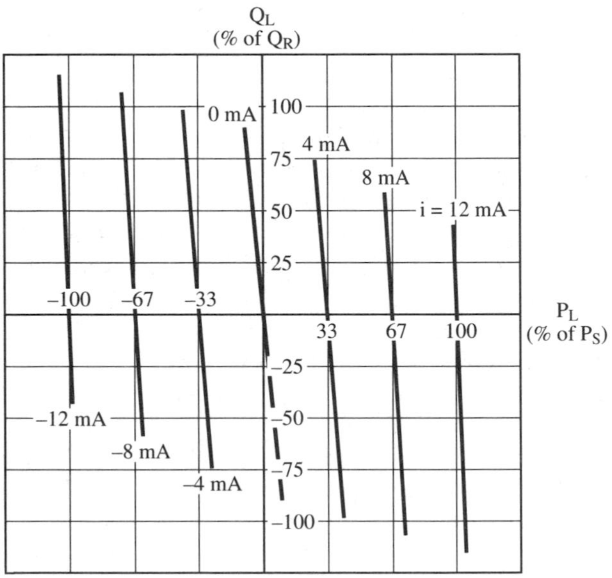

(b)

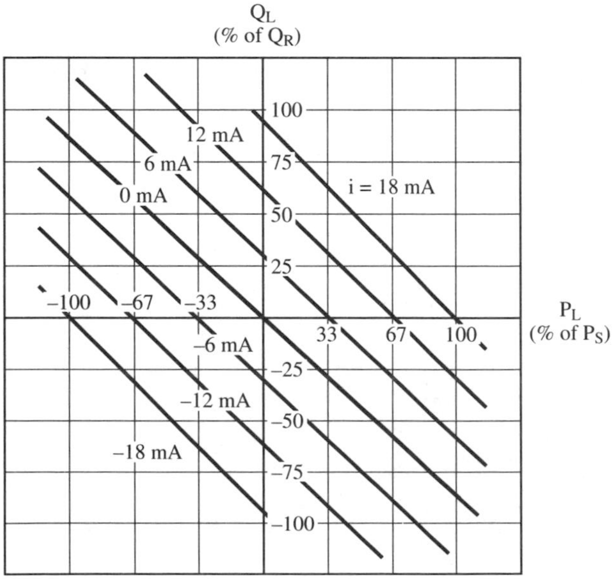

(c)

Figure 6.63 (Continued) Operating characteristics of three control valves. (b) Pressure control valve: $P_S = 16$ MPa, $Q_R = 0.4$ liter/s, $i_{max} = 12$ mA, $R_{coil} = 200\ \Omega$. (c) Pressure-Flow Control Valve: $P_S = 12$ MPa, $Q_R = 1$ liter/s, $i_{max} = 18$ mA, $R_{coil} = 160\ \Omega$.

Exercise 6.33: In steady-state operation, the input port of the electrically operated pressure-flow valve is described by the equation

$$v_{aa'} = R_i i \tag{6.79}$$

Show that the steady-state output-port performance characteristics of the valve can be expressed as

$$Q_L = \beta i - \left(\frac{1}{R_o}\right)P_L \tag{6.80}$$

Equations (6.79) and (6.80) are represented as the two halves of a linear lumped model in Figure 6.64(a). A line-graph representation is shown in Figure 6.64(b). The compliance of the fluid in the valve, cylinder, and connecting hydraulic lines acts as a fluid capacitor connected to the output port. We show that capacitance in dashed lines in the figures. The effect of the fluid compliance is to limit the frequency response of the system. We examine this limiting effect in problem 6.39 at the end of the chapter.

The power consumed at the valve input at maximum current is $P_{in} = i_{max}^2 R_i = 0.052$ W. According to equation (6.60), the *available* power at the output port for maximum current at the input is $P_{oavail} = (\beta i_{max})^2 R_o/4 = (0.0528\ m^3/A{\cdot}s)^2(18\ mA)^2 \times (1.26 \times 10^{10}\ N{\cdot}s/m^5)/4 = 2.85$ kW. The *available* power gain, then, is 54,700 (47 dB).

To find the specific behavior of the control valve, we must know the specific mechanical load. If the load is linear and dissipative, it can be represented by an equivalent linear hydraulic resistor attached to the output port of the valve model in Figure 6.64. If the load is linear but includes storage elements, we can attach corresponding equivalent storage elements to the valve model. If the load is dissipative but not linear, we can superimpose the operating characteristic of the equivalent nonlinear hydraulic resistor on the operating characteristics of the valve to find the steady-state operating point. Then we can linearize the load in the neighborhood of the operating point.

We can derive linear lumped models for the other two types of hydraulic control valves from straight-line approximations to the performance curves. The equations have the same form as equations (6.79) and (6.80), and the lumped models have the same structure as Figure 6.64. Only the parameter values are different. We develop these models in problem 6.40 at the end of the chapter.

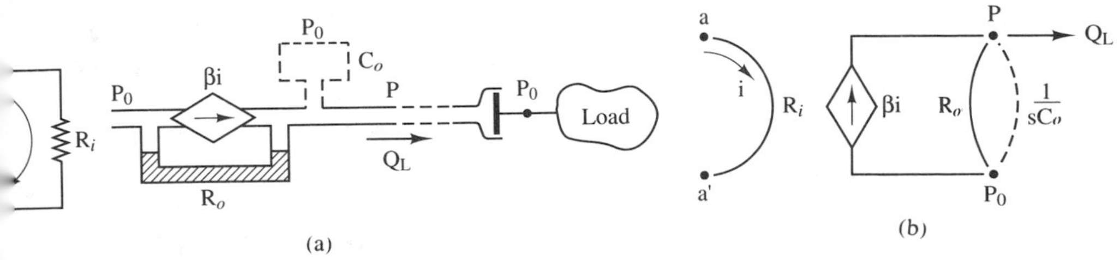

re 6.64 Linear lumped models for the hydraulic control valve of Figure 6.63(c); $P_L = P - P_0$. (a) Lumped l of valve and cylinder. (b) Line-graph representation of valve.

The operating characteristics for the pressure control valve of Figure 6.63(b) are almost vertical. Hence, the output resistance is low. Accordingly, we prefer to treat that valve as if the input current controls the output *pressure.* (That is the reason we call the device a *pressure control* valve.)

Exercise 6.34: Use equation (6.80) to show that the output characteristics of the pressure-flow control valve of Figure 6.63(c) can also be expressed in the form

$$P_L = r_t i - R_o Q_L \tag{6.81}$$

where $r_t = \beta R_o$. This is the equation for the r_t amp of Figure 6.46(c).

It is apparent that the equations for any of the three sets of linear (or linearized) valve characteristics can be expressed in β-amp form or in r_t-amp form. If we supply the control current by means of a voltage source with resistance R_S, we can rewrite the lumped linear equations in μ-amp form. (It would be appropriate to use that form if the source resistance were large compared with the motor coil resistance R_o.) The μ-amp equations can, in turn, be rewritten in g_t-amp form. (It would be appropriate to use that form if the equivalent load resistance R_L, as *seen* at the input to the hydraulic cylinder, were small compared to R_o.) It is the valve input and output impedances, relative to the source and load impedances, that determine which form of the model is most appropriate.

The Internal Combustion Engine—a Mechanical Amplifier

Probably the most familiar high-power source is the automobile engine, a reciprocating internal combustion engine. An *internal combustion engine* converts the chemical energy in fuel to heat and then extracts mechanical energy from the heat. The output port of an engine is rotational. The port variables are the shaft torsion T and rotational velocity Ω. Reciprocating internal combustion engines power not only automobiles, but also electric generators, hydraulic and pneumatic pumps, and various kinds of mechanical equipment. The gas turbine, a nonreciprocating internal combustion engine, is used widely to power aircraft and to drive reserve-power generators for electric utilities.

Unlike an electrical or hydraulic amplifier, an internal combustion engine does not draw its output power from a constant velocity or constant torque supply. Rather, it draws power from a reservoir of fuel. The *power* output of an internal combustion engine is approximately proportional to the input fuel rate. The control-input power is the power necessary to inject the fuel into the engine.

Various internal combustion engines differ in the way they control fuel flow and, hence, power. They also differ in the way they convert fuel energy to heat. In a gas turbine, the fuel burns continuously. In a reciprocating engine, the fuel is burned in chunks to produce pulses of power. A heavy flywheel is used to smooth the pulsatile nature of the reciprocating power source.

The basic principles of operation of internal combustion engines are intuitively understandable, but mathematical analysis is complicated. We shall make no attempt to explain the details of their operation; instead, we shall focus on the measured performance of engines.

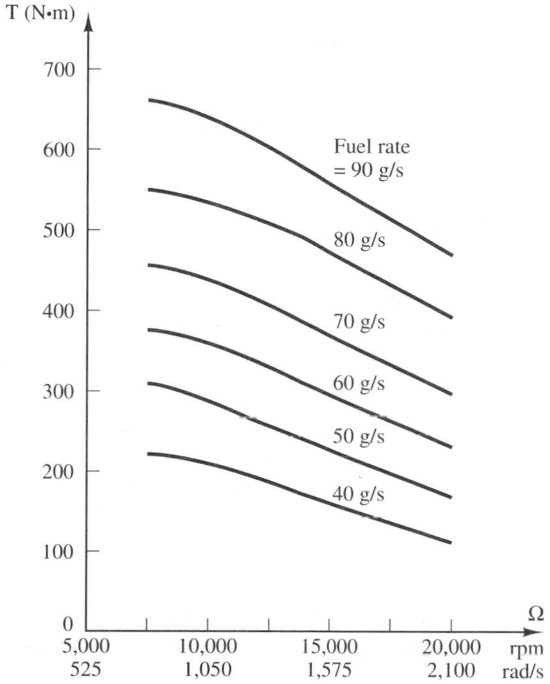

Figure 6.65 Performance characteristics of a gas turbine.

Figure 6.65 shows a set of output-port performance curves for a particular gas turbine. There is one performance curve for each fuel flow rate. The set of curves is comparable to one of the sets of control valve characteristics of Figure 6.63 and to the transistor collector characteristics of Figure 6.52. Note that no data are provided for zero speed because the engine does not operate at that speed. A starting mechanism must accelerate the engine to the operating range before it begins to run.

Fuel is sprayed continuously into the combustion chamber of the turbine. It is injected by an electrically driven positive-displacement pump. The label on each performance curve indicates the fuel mass flow rate. If the fuel were liquid, we could, as well, label each performance curve with the fuel *volume* flow rate Q_{fuel}, with the motor torque needed to drive the pump at that rate, or with the electric current needed to drive the motor for that rate.

The output turbine torque and input pump-motor torque are comparable quantities. We could use them to compute the *torque gain* of the engine. Finding the engine torque gain would be comparable to finding the transistor current gain. However, because engines are not used specifically to provide torque gain, we do not proceed with those computations.

Our immediate objective is to develop a lumped model of the turbine. The performance characteristics are nearly linear. Therefore, we can approximate them by straight lines. Over the region of operation, each characteristic is more like a constant-torque curve than a constant-speed curve. Accordingly, we write the equations as torque equations. The vertical ($\Omega = 0$) intercepts of the straight lines are nearly proportional to the fuel rate. We choose to express these intercepts in terms of the fuel volume flow rate Q_{fuel} because we know how to deal with hydraulic flows. The set of straight-line approximations can then be expressed as

$$T = T_{intercept} - \left(\frac{1}{R_o}\right)\Omega$$
$$= \beta Q_{fuel} - \left(\frac{1}{R_o}\right)\Omega \qquad (6.82)$$

where R_o is the negative inverse slope of the straight-line approximations and β is the proportionality between the fuel volume flow rate Q_{fuel} and the zero-speed torque $T_{intercept}$.[17] (Although the turbine does not operate at zero speed, the straight-line approximations extend to zero speed.)

Exercise 6.35: The density of fuel is 0.9 kg/liter. Use a straight-line approximation to the 70-g/s performance curve to show that $R_o \approx 7.8$ rad/N·m·s and $\beta \approx 7000$ N·m·s/liter.

We represent equation (6.82) by a linear lumped model in the right half of Figure 6.66(a). The gas turbine behaves like a fuel-controlled torsion source with some viscous damping. The figure also shows a lumped model for the fuel delivery mechanism. That model presumes that the positive-displacement pump is set to the desired fuel flow rate (by setting the pump speed), that the spray nozzle through which the fuel is delivered has hydraulic resistance R_i, and that the pump has a leakage resistance R_S. We show a line-graph model of the turbine in Figure 6.66(b).

Let us calculate the steady-state power output and power gain of the turbine. Suppose the operating point is at the center of the operating characteristics—specifically, at the point (1,575 rad/s, 370 N·m) of Figure 6.65(a). In order for the turbine to operate at this point, the load characteristic must pass through the point and fuel must be delivered at 70 g/s (or 78 ml/s). Then the power output of the turbine is $P_{out} = T\Omega = (370 \text{ N·m}) \times (1{,}575 \text{ rad/s}) = 583{,}000$ N·m/s = 583 kW. This output power is extracted from the heat content of the fuel.

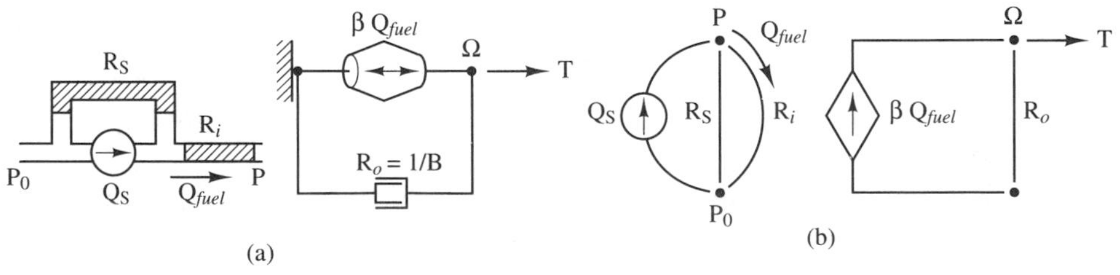

Figure 6.66 A linear lumped model for the gas turbine of Figure 6.65. (a) Lumped model. (b) Line-graph representation.

[17] We must use care in interpreting the output-port resistance R_o of the turbine. Throughout this book, the term *resistance* refers to resistance to flow. In the context of the turbine, $R_o = 1/B_o$, a resistance to the flow that represents the torsion. It is *not* resistance to motion Ω.

To calculate the power gain, we need to know the control-signal power input. We estimate it as follows. Suppose the pump injects the fuel into the nozzle at a gage pressure of 100 kPa. Suppose also that the pump leakage is negligible—that is, $R_S \gg R_i$. Then the input power is $P_{in} = PQ_{fuel} = (100 \text{ kPa})(78 \text{ ml/s}) = 7.8$ W (What is the corresponding value of R_i?) and the power gain is $P_{out}/P_{in} = 75{,}000$ (48.7 dB).

We do, of course, use engines to amplify power. Yet, when we choose and use those engines, we do not *think* in terms of power gain. Rather, we choose an engine that has enough *available* power to energize the loads we are interested in. The power necessary to control the delivery of fuel to the engine need only be small enough that it is well within the capability of a human or a small fuel-delivery device. We adjust the fuel flow (and associated power output) as we need it. (Since we tire of holding down the accelerator pedal of an automobile for a long time, engineers have designed a cruise control mechanism to do this for us. The cruise control is about as powerful as we are.) Our requirements for high enough available engine power and low enough fuel-control power amount to a power gain requirement.

Figure 6.67 shows a single performance curve for each of several reciprocating engines. The volume noted beside each curve is the total volume displaced by one cycle of the pistons in all the cylinders of the engine. This volume is commonly known as the engine *displacement*. It is a measure of the size of the combustion space available in that engine. The *maximum torque* of an engine is roughly proportional to the engine displacement.

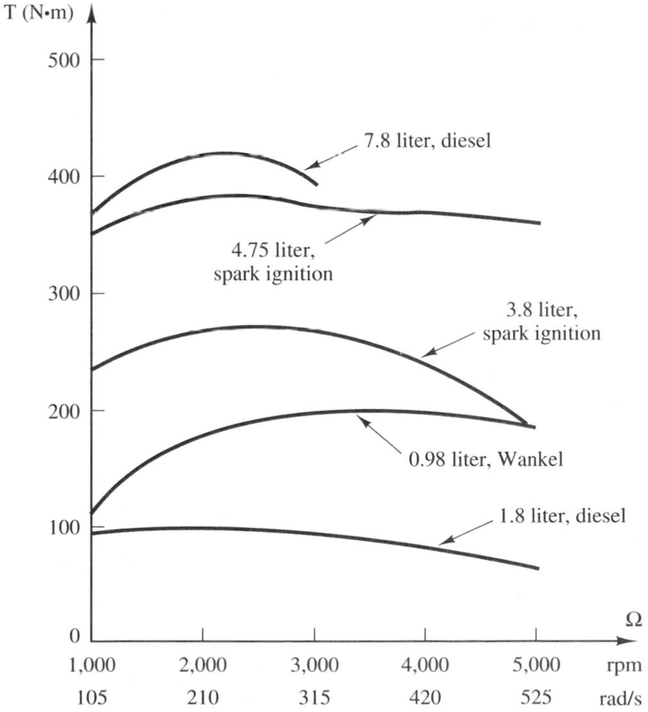

Figure 6.67 Performance characteristics of internal combustion reciprocating engines; all curves for wide-open throttle.

A fuel-air mixture is supplied to each of these engines by means of a carburetor. A throttle valve controls the carburetor opening through which the air is sucked into the engine. Each curve in the figure presupposes a wide-open throttle—that is, the engine turned all the way on. (An automobile engine has a wide-open throttle if the accelerator pedal is pushed to the floor.) These engines cannot operate at low speeds.

If we partially close the throttle valve, the torque-speed characteristic for a particular engine is displaced toward the origin—to a lower level of output power—while keeping the same basic shape. However, part-throttle measurements are not usually made for these engines. The operator of an automobile knows from experience how to adjust the throttle-valve setting to meet his or her needs. Operators of many kinds of equipment open the throttle valve all the way when the equipment is operating (the engine is loaded), but switch to the lowest power setting when the engine is idling (unloaded). The operator can always adjust the throttle slightly if the power output is not quite right.

6.4 DISTRIBUTED MODELS FOR TRANSPORTERS

A transporter is a device or subsystem that is designed to carry power, material, or information from one location to another. Table 6.2 lists a number of examples of transporters. We would like a transporter to be transparent; that is, we would like the input-port variables to be transported instantly, without change, to the output port. But physical transporters are not transparent: Nothing can be transported instantly. Indeed, the fundamental characteristic of transporters is delay, and as the distance of transport increases, the delay increases. The transport process can also change the signal waveform.

TABLE 6.2 TRANSPORTERS IN VARIOUS CONTEXTS

Electrical: Electric power lines carry power from point to point; computer cables carry information from point to point.

Rotational: A speedometer cable carries information from point to point; an automobile drive shaft carries power from point to point; an automobile steering shaft carries information from point to point.

Translational: A valve push rod carries timing information from point to point; an automobile piston rod carries power from point to point.

Hydraulic: An pipeline (or a human artery) carries material (oil, water, blood, concrete) from point to point; a hydraulic control line carries information from point to point; during rapid shutoff, a pipeline carries damaging shock waves from point to point.

Acoustic: An ultrasonic transmitter is used to launch an acoustic traveling wave into a human body. Reflections of the wave from various interfaces return to the transmitter. The round-trip travel times of reflections are used to make a map of the locations of parts of the body. Hence, the body carries acoustic information from point to point. Some ultrasonic scanners are used to break up kidney stones. Then the body carries acoustic power from point to point.

Electromagnetic: Electromagnetic radiation from an antenna or group of antennas carries radio, television, telephone, and radar signals through space. High-power antennas transmit small amounts of electrical power through space.

Thermal: A heating system uses water or air to carry heat from point to point. Heat flow is thermal power. An indoor-outdoor thermometer transports thermal information from point to point.

Transporters exhibit traveling waves, as we observed in the electrical coaxial cable of example problem 4.6. Reflections of a traveling wave from the ends of a transporter interfere with the intended transport process. The pressure pulse in the oil pipeline of example problem 5.5 illustrates this concern. Mismatching of impedances at the ports is the primary cause of wave reflection.

Transporters are distributed devices. Each transporter model developed in previous chapters was a distributed lumped model—a chain of submodels. Distributed lumped models provide for easy simulation and understanding of transporter behavior; however, they do not explain such transporter-related phenomena as signal reflections and signal delay.

In this section, we use a distributed model to understand the behavior of transporters. Then we derive a simple equivalent two-port lumped model for ideal transporters. The fundamental characteristic of ideal transporter behavior is delay.

A Distributed Transporter Model

A transporter has a repetitive dynamic network structure. The term *dynamic* implies that each segment of the distributed model has energy-storage elements. Figure 6.68 shows the forms of the segments of the distributed transporter models derived in example problems of earlier chapters. (We ignore the constant gravity-induced forces in the Slinky model and investigate the dynamic effects alone; since the model is linear, gravity effects can be superposed on the dynamic response.) The last item in the figure is a description of a transporter in general potential-difference and flow notation. We analyze the behaviors of transporters in this general terminology. The reader can maintain an intuitive understanding during the development by interpreting symbols in terms of that transporter context of the figure which seems most familiar.

The symbols z_s and z_p denote the series impedance and parallel impedance, respectively, *per unit length* of the transporter. It will be convenient to refer to z_p in terms of its reciprocal, $y_p \triangleq 1/z_p$, known as the parallel *admittance*. We shall assume that these quantities may be expressed in the forms $z_s = R + sL$ and $y_p = b + sm$, where the symbols R, L, b, and m should be interpreted to mean, respectively, a resistance-type parameter (such as electrical R or translational 1/b), an inductance-type parameter (such as inertance I or compliance 1/k), a conductance-type parameter (such as rotational B or hydraulic $1/R_f$), and a mass-type parameter (such as hydraulic C_f or translational m). The examples of Figure 6.68 are essentially in agreement with these forms, as noted in rows 3 and 4 of Table 6.3. (Since the series resistance in the pipeline is quadratic, the pressure wave in the pipeline does not follow precisely the equations to be derived. However, examination of example problem 5.5 shows that the results are qualitatively correct.)

The relations among the potentials and flows defined in Figure 6.68(e) are

$$\begin{aligned} A_n - A_{n+1} &= T_n z_s \\ T_n - T_{n+1} &= A_{n+1} y_p \end{aligned} \tag{6.83}$$

The quantity $A_n - A_{n+1}$ is the decrease in potential *per segment length* as we proceed along the transporter in the direction of the flow T_n; $T_n - T_{n+1}$ is the decrease in flow *per segment length* as we proceed in the same direction. Let us define that direction as the

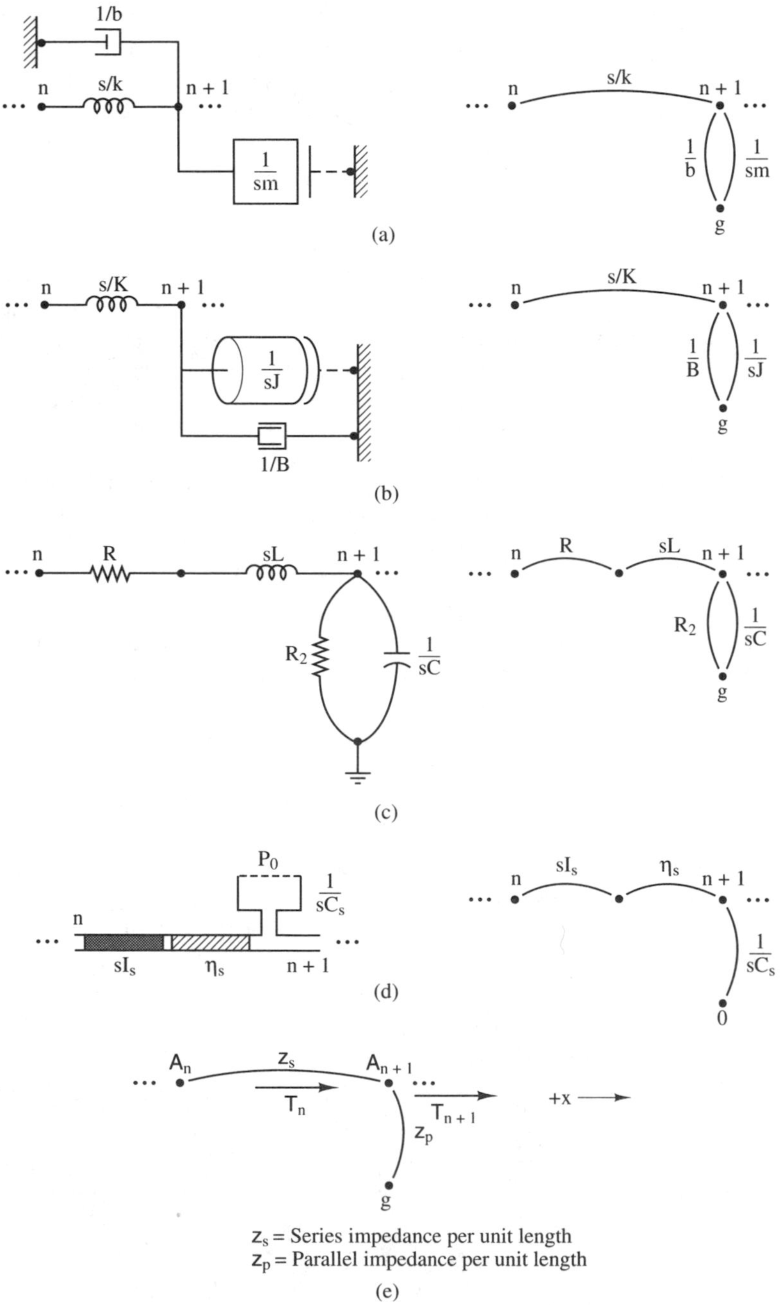

Figure 6.68 Distributed lumped models for transporters. (a) Slinky. (b) Speedometer cable. (c) Coaxial cable. (d) Oil pipeline. (e) General form.

TABLE 6.3 PARAMETERS FOR THE TRANSPORTERS OF FIGURE 6.68

	Slinky	Speedometer Cable	Coaxial Cable	Oil Pipeline
Unit length	1 Slinky length	1 meter	1 meter	1 km
Parameters per unit length	$k = 0.82$ kg/s^2 $b = 0.025$ kg/s $m = 0.25$ kg	$K = 0.11$ kg·m^2/s^2 $B = 2.5 \times 10^{-6}$ kg·m^2/s $J = 9.4 \times 10^{-9}$ kg·m^2	$R = 0.0288\ \Omega$ $L = 0.243\ \mu$H $C = 102$ pF	$\eta = 6.6 \times 10^5$ kg/m^7 $I = 4.4 \times 10^6$ kg/m^4 $C = 1.7 \times 10^{-7}$ m^{4}s^2/kg
z_s	s/k	s/K	$R + sL$	$\eta Q + sI$
y_p	$b + sm$	$B + sJ$	$\approx sC$	sC_f
f_{high}	0.16 Hz	420 Hz	189 kHz	0.1 Hz
β_{high}	$\omega\sqrt{m/k}$	$\omega\sqrt{J/K}$	$\omega\sqrt{LC}$	$\omega\sqrt{IC}$
z_{0high}	$1/\sqrt{km} =$ 2.2 s/kg	$1/\sqrt{KJ} =$ 3.1×10^4 s/kg·m^2	$\sqrt{L/C} = 49\ \Omega$	$\sqrt{I/C} =$ 5.1×10^6 kg/m^{4}s
v_{high}	$\sqrt{k/m} = 1.81$ lengths/s	$\sqrt{K/J} = 3{,}415$ m/s	$1/\sqrt{LC} =$ 2×10^8 m/s	$1/\sqrt{IC} =$ 1.16 km/s

positive direction for distance x along the transporter and think of A and T as functions of time t (owing to the signals imposed at the ends of the transporter) and of distance x (owing to travel of the signals along the transporter). Then rates of *increase* with respect to distance along the transporter represent partial derivatives with respect to x, and, we can rewrite equations (6.83) in the form

$$\frac{\partial A}{\partial x} = -z_s T, \qquad \frac{\partial T}{\partial x} = -y_p A \tag{6.84}$$

where the partial derivatives represent increases in potential and flow, respectively, *per unit length* and where z_s and y_p represent amounts of series impedance and parallel admittance, respectively, *per unit length*. In effect, we have made each segment of the distributed lumped model infinitesimally small; hence, any errors in the model owing to coarseness of the segments has been eliminated.

We differentiate the first equation of (6.84) with respect to x and then substitute the second equation into the first to get the *wave equation*

$$\frac{\partial^2 A}{\partial x^2} = \gamma^2 A \tag{6.85}$$

where $\gamma^2 = z_s y_p = (R + sL)(b + sm)$. The flow T obeys the same equation. (Simply differentiate and substitute the equations in the opposite order.)

We call the parameter γ the **propagation constant** for the transporter. Since z_s and y_p are functions of the time-derivative operator s, γ is also a function of s. We shall find, however, that γ can usually be treated as a constant. That constant is a *system parameter* for the transporter.

Steady-State Sinusoidal Transporter Behavior

In general, the transporter impedance Z_s and admittance Y_p are functions of the time-derivative operator s, and equation (6.85) is a complicated *partial differential equation* in x and t. The transporter behavior is simpler if the source signal is sinusoidal. Since every waveform of interest in a physical system can be treated as a sum of sinusoids of various frequencies (see section 5.3), let us examine the steady-state response of the transporter to one of these sinusoidal components. Because the transporter model is linear, a sinusoidal source signal of frequency ω produces a sinusoidal response of the same frequency. Consequently, we can replace the potential difference A(x, t) and flow T(x, t) by the phasors $\boldsymbol{A}(x)$ and $\boldsymbol{T}(x)$ in equations (6.84) and (6.85). (The variation of the signal with time is now sinusoidal at frequency ω.) We replace the time-derivative operator s by jω in the impedance Z_s, in the admittance Y_p, and in the propagation constant γ. The propagation constant becomes the complex number

$$\gamma = \sqrt{(R + j\omega L)(b + j\omega m)} \triangleq \alpha + j\beta \tag{6.86}$$

where the real part α is called the **attenuation constant** and the imaginary part β the **phase propagation constant.** These two constants should be viewed as system parameters of the transporter.

Calculating γ requires the square root of a complex number. To multiply complex numbers, we multiply their magnitudes and add their angles (see section 5.3). Therefore, to take the square root of a complex number, we take the square root of its magnitude and halve its angle. A complex number has two complex square roots, equal in magnitude and opposite in sign. To get the second root, we reverse the sign on the first root. The roots can be verified by squaring them. For γ, we choose to use the root for which $\alpha > 0$. (The other choice merely reverses the roles of α and β.) Note that γ, α, and β change with the frequency ω.

For high source frequencies ω, we can neglect the real components of Z_s and Y_p. We consider the frequency high if the imaginary component is at least 10 times larger than the real component for both Z_s and Y_p. In that case,

$$\gamma \approx j\omega\sqrt{Lm}, \qquad \beta \approx \omega\sqrt{Lm}, \qquad \alpha \approx 0 \tag{6.87}$$

A transporter is usually operated in its high-frequency range. Table 6.3 shows the lower limit f_{high} of the high-frequency range (in Hz) for the transporters examined in the example problems of previous chapters. The table also expresses the high-frequency value β_{high} of the phase propagation constant, from equation (6.87), in terms of the parameters of each transporter.

Let us assume that the solutions to the linear differential equation (6.85) are exponential functions of the spatial variable x. (This is the same assumption that we made for differential equations in the time variable in section 3.4). That is, we assume solutions are of the form $\boldsymbol{A}(x) = \boldsymbol{D}e^{px}$, where the coefficient $\boldsymbol{D}$ is a phasor. We substitute this form of solution into the differential equation to find that the characteristic roots of the equation are $p = \pm\gamma$. Therefore, the general solution to the equation is

$$\boldsymbol{A}(x) = \boldsymbol{D}_1 e^{-\gamma x} + \boldsymbol{D}_2 e^{\gamma x} \tag{6.88}$$

Differentiating equation (6.88) with respect to x and substituting the result into the first equation of (6.84) yields

$$\boldsymbol{T}(x) = \left(\frac{\gamma}{z_s}\right)(\boldsymbol{D}_1 e^{-\gamma x} - \boldsymbol{D}_2 e^{\gamma x}) \tag{6.89}$$

where both γ and the sinusoidal impedance z_s are complex numbers.

To find the specific response of the transporter owing to a particular sinusoidal source signal of frequency ω, we must determine the complex values of $\boldsymbol{D}_1$ and $\boldsymbol{D}_2$ for that particular source signal. After we find these complex constants, we can obtain the steady-state sinusoidal response functions by the operations $A(x, t) = \mathcal{Re}[\boldsymbol{A}(x)e^{j\omega t}]$ and $T(x, t) = \mathcal{Re}[\boldsymbol{T}(x)e^{j\omega t}]$, as discussed in section 5.3.

The Characteristic Impedance of a Transporter

Intuitively, a transporter acts like a clothesline. If we fasten one end of the clothesline and jerk the free end, a pulse travels the length of the clothesline, reflects from the fastened end, and returns. Suppose that the input end of a transporter is located at $x = 0$ and that the transporter is *infinitely long.* Then a traveling wave launched down the transporter can never return. Let us excite the infinitely long transporter sinusoidally at $x = 0$ and examine the solution given by equations (6.88) and (6.89).

When we defined γ in equation (6.86), we chose $\alpha > 0$. Therefore, the factor $e^{\gamma x} = e^{\alpha x}e^{j\beta x}$ in equations (6.88) and (6.89) must approach infinity as x increases. (It is as if the pulse launched down a long clothesline becomes larger and larger.) No source is powerful enough to make $\boldsymbol{A}$ and $\boldsymbol{T}$ infinite. Therefore, for an infinitely long transporter, $\boldsymbol{D}_2$ must be zero, and the sinusoidal steady-state solution reduces to

$$\boldsymbol{A}(x) = \boldsymbol{D}_1 e^{-\gamma x} \quad \text{and} \quad \boldsymbol{T}(x) = \left(\frac{\gamma}{z_s}\right)\boldsymbol{D}_1 e^{-\gamma x} \tag{6.90}$$

Note that the ratio $\boldsymbol{A}/\boldsymbol{T}$ does not depend on x. That ratio is the *sinusoidal impedance* that we measure at the input to a very long transporter. We call it the **characteristic impedance** z_0. The characteristic impedance is a second *system parameter* of the transporter. It depends only on the per-unit length complex impedances z_s and z_p; specifically,

$$z_0 \triangleq \frac{\boldsymbol{A}}{\boldsymbol{T}} = \frac{z_s}{\gamma} = \sqrt{\frac{z_s}{y_p}} = \sqrt{\frac{R + j\omega L}{b + j\omega m}} \tag{6.91}$$

To calculate z_0 we must, again, take the square root of a complex number. Again, that number changes with the frequency ω.

At $x = 0$, the sinusoidal steady-state solution given by equations (6.90) requires that $\boldsymbol{A}(0) = \boldsymbol{D}_1$ and $\boldsymbol{T}(0) = \boldsymbol{D}_1\gamma/z_s = \boldsymbol{D}_1/z_0 = \boldsymbol{A}(0)/z_0$. Then

$$\boldsymbol{A}(x) = \boldsymbol{A}(0)e^{-\gamma x} \quad \text{and} \quad \boldsymbol{T}(x) = \boldsymbol{T}(0)e^{-\gamma x} \tag{6.92}$$

where $\boldsymbol{A}(0)$ and $\boldsymbol{T}(0)$ are, respectively, the potential-drop phasor and the flow phasor roused at the input port of the transporter by the sinusoidal source. These two values are related by $\boldsymbol{A}(0) = z_0\boldsymbol{T}(0)$. Thus, the infinitely long transporter appears to the source as a single equivalent impedance—the characteristic impedance of the transporter.

For high source frequencies—frequencies above the same high-frequency boundary f_{high} that we discussed in connection with the propagation constant—we find that

$$z_{0high} \approx \sqrt{\frac{L}{m}} \tag{6.93}$$

Thus, the characteristic impedance is real (resistive) at high frequencies. The high-frequency characteristic impedances for each of the transporters of the example problems are expressed in terms of the transporter parameters in Table 6.3. We emphasize, again, that most transporters are operated in their high-frequency ranges.

Propagation Delay in Transporters

Suppose we attach to the infinitely long transporter a sinusoidal source of frequency ω, with potential difference expressed by the phasor $\boldsymbol{A}_S$ and with complex sinusoidal impedance z_S. Figure 6.69 shows a simple model of the system. The node labeled $\boldsymbol{A}(0)$ is at the input port of the transporter. The argument 0 refers to $x = 0$. We label the other node of the port as the reference g. The equivalent impedance of the transporter is its characteristic impedance z_0.

The potential difference across the transporter input port is $\boldsymbol{A}(0) = \boldsymbol{A}_S z_0/(z_S + z_0)$. The flow entering the transporter is $\boldsymbol{T}(0) = \boldsymbol{A}(0)/z_0$. A source impedance is usually real (resistive). For high frequencies, the characteristic impedance of the transporter is also resistive. In that case, the input potential difference $\boldsymbol{A}(0)$ and the input flow $\boldsymbol{T}(0)$ have the same phase, and, according to equation (6.92), it follows that $\boldsymbol{A}(x)$ and $\boldsymbol{T}(x)$ have the same phase at all points along the transporter. That is, for high frequencies, the equivalent impedance of an infinitely long transporter is z_0 as seen from any point within the transporter. (Is this any surprise? If we cut the transporter at any point, it is still infinitely long.)

Equation (6.92) gives only the spatial patterns of the sinusoidal steady-state signals. To get the full signals, we multiply by $e^{j\omega t}$ and take the real part. If we select the reference time ($t = 0$) so that the phase (angle) of $\boldsymbol{A}(0)$ is zero, then $\boldsymbol{A}(0)$ is real and

$$\begin{aligned} A(x,t) &= \boldsymbol{A}(0)e^{-\alpha x}\cos(\omega t - \beta x) \\ T(x,t) &= \left[\frac{\boldsymbol{A}(0)}{|z_0|}\right]e^{-\alpha x}\cos(\omega t - \beta x - \phi) \end{aligned} \tag{6.94}$$

where ϕ is the argument of the complex characteristic impedance z_0. (If z_0 is real, then $\phi = 0$.) At any instant t, these two equations describe sinusoidal patterns in space. At any position x along the transporter, they describe sinusoidal variations in time.

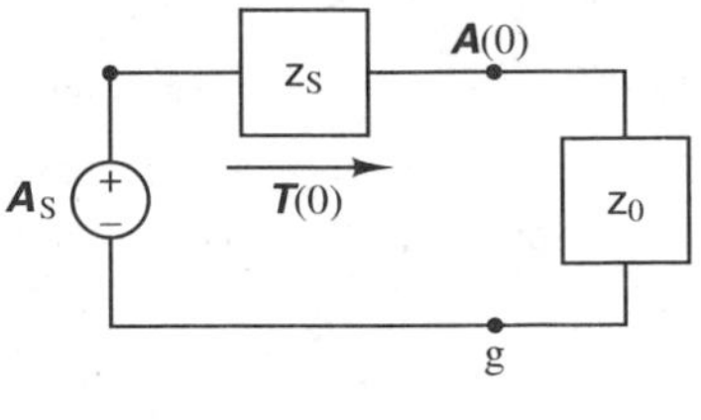

Figure 6.69 A sinusoidal source driving an infinitely long transporter.

Each equation of (6.94) represents a sinusoidal wave traveling in the x-direction. The factor $e^{-\alpha x}$ represents attenuation of the signal with distance along the transporter. That is the reason we call α the attenuation constant of the transporter. It we were to shake sinusoidally the end of a very long clothesline, we would observe sinusoidal waves traveling down the clothesline, attenuating with the travel distance, as indicated by equation (6.94).

The coefficient β determines the rate at which the spatial sinusoid travels in the x-direction. That is the reason it is called the phase propagation constant. For example, at $t = 0$ and $x = 0$, the argument of the cosine factor of $A(x, t)$ is zero and the potential difference is at a peak, with the value $\boldsymbol{A}(0)$. (Recall that we selected the time reference to make $\boldsymbol{A}(0)$ real.) As time progresses, we keep our attention on that peak of the potential-difference wave. That is, we focus on the value of x that keeps the argument of the cosine factor equal to zero—namely, $x = (\omega/\beta)t$. The **phase velocity,** i.e., the velocity of the peak (or the point of constant phase), is

$$v = \frac{\omega}{\beta} \tag{6.95}$$

We define the **wavelength** λ of a traveling wave to be the distance between two adjacent peaks of the wave. Then λ is the change in x that is necessary to increase the argument of the cosine factor by 2π. We let $t = 0$ in that argument to see that $\lambda = 2\pi/\beta$. Since the frequency of the wave is $f = \omega/2\pi$, it follows that $v = \lambda f$.

According to equation (6.87), β is real for high source frequencies, and the high-frequency phase velocity is

$$v_{high} = \frac{1}{\sqrt{Lm}} \tag{6.96}$$

The high-frequency phase velocity is expressed in terms of transporter parameters in Table 6.3, for each of the transporters of the example problems.

Since the sinusoid travels in the x-direction at velocity v, it takes time for a signal that is delivered to one end of a transporter to arrive at the other end. Suppose the transporter length is ℓ. If a sinusoidal signal of frequency ω is delivered to the input ($x = 0$) at $t = 0$, it arrives at the output ($x = \ell$) at

$$t = \frac{\ell}{v} = \frac{\ell\beta}{\omega} \triangleq d \tag{6.97}$$

The travel time d is known as the **propagation delay,** the **transport lag,** or just the **delay.**

The source signal that energizes a transporter is not usually sinusoidal. However, we can view the source signal as a sum of sinusoids of various frequencies. Each frequency component is delayed in traveling along the transporter. Unfortunately, different frequency components of the source waveform travel at different velocities. Hence, we cannot, in general, expect the waveform of the signal that arrives at the output of the transporter to be the same as the source waveform. We say that the frequency components *disperse,* the transporter is *dispersive,* and the signal waveform is *distorted,* or experiences *frequency dispersion.*

However, if all the frequencies components of the source signal are in the high-frequency range of the transporter, as discussed for equation (6.87), then the phase velocity is essentially the same for each component. If it is important that the source signal waveform be preserved, then we must design the transporter (i.e., select the material and geometry) so that its high-frequency boundary f_{high} is lower than all frequency components of the input signals. Then dispersion is not significant.

It can be observed from the phase velocities v_{high} in the examples of Table 6.3 that there is a wide disparity in delay among different types of transporters. Electrical transporter delays are usually measured in microseconds or nanoseconds, whereas mechanical and hydraulic transporter delays are usually measured in milliseconds or seconds. The system designer must select a transporter that can produce an acceptable delay for each application.

In some situations, a precisely controlled delay is needed. In an electronic signal processor, for example, we might wish to delay an electrical signal for 1 millisecond, to compare it (electrically) with a signal arriving from another piece of equipment. We could obtain a 1-ms delay by passing the signal through 1 km of coaxial cable. Alternatively, we could use an electrical-to-translational transducer to convert the signal to mechanical form, use the transducer to vibrate the end of a steel rod, and then use a reverse transducer (mechanical to electrical) to return the signal to electrical form. This alternative method is more complicated; however, we can obtain the 1-ms delay with a 1/3-meter rod instead of a 1-km cable.

Wave Reflections at the Transporter Ends

Equations (6.90) and Figure 6.69 assume an infinite-length transporter. Now shorten the transporter's length to ℓ, and terminate the secondary end of the transporter with a load of impedance z_L, as shown in Figure 6.70. Again, use a sinusoidal source with frequency ω. Then the potential-difference signal $\boldsymbol{A}(x)$ within the transporter is of the form of equation (6.88). We found before that the term $\boldsymbol{D}_1 e^{-\gamma x}$ represents a sinusoidal wave traveling in the x-direction. The term $\boldsymbol{D}_2 e^{\gamma x}$, with the sign of x reversed, represents a sinusoidal wave traveling in the $-x$ direction. The coefficients $\boldsymbol{D}_1$ and $\boldsymbol{D}_2$ are the complex amplitudes of these two waves. (The argument of each complex amplitude represents a shift in phase—a spatial shift of the sinusoid at $t = 0$.)

The outgoing potential-difference wave begins at the input ($x = 0$) at $t = 0$ with complex amplitude $\boldsymbol{D}_1$. It propagates in the x-direction with velocity v and passes the point x at $t = x/v$. It arrives at the end of the transporter ($x = \ell$) at $t = \ell/v = d$.

The incoming potential-difference wave arrives at the input ($x = 0$) at $t = 0$ with complex amplitude $\boldsymbol{D}_2$, moving in the $-x$ direction with velocity v. It previously passed the point x at $t = -x/v$. It departed (earlier) from the end of the transporter at $t = -d$.

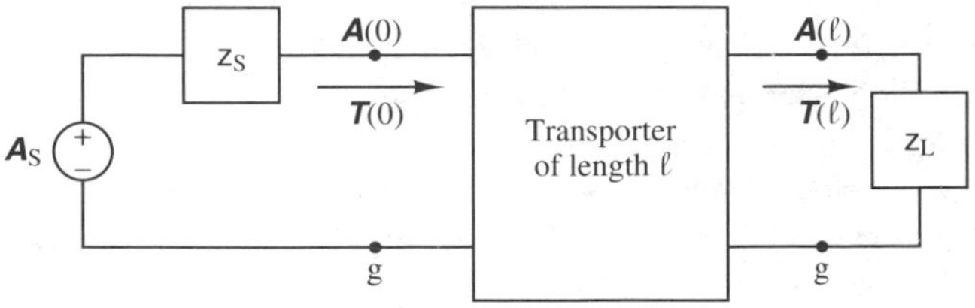

Figure 6.70 A finite-length transporter in a network.

The complex amplitudes $\boldsymbol{D}_1$ and $\boldsymbol{D}_2$ are determined by the boundary conditions at the two ends of the transporter. That is, the potential difference and flow at the input must match the potential difference and flow of the source, and the potential difference and flow at the output must match the potential difference and flow of the load. In effect, we superimpose port characteristics at the two ports to determine $\boldsymbol{D}_1$ and $\boldsymbol{D}_2$. According to equations (6.88) and (6.89), with $z_0 = z_s/\gamma$, the port characteristics at the two ends of the transporter require the following conditions:

$$\begin{aligned} \boldsymbol{A}(0) &= \boldsymbol{D}_1 + \boldsymbol{D}_2 \\ \boldsymbol{T}(0) &= \frac{\boldsymbol{D}_1 - \boldsymbol{D}_2}{z_0} \\ \boldsymbol{A}(\ell) &= \boldsymbol{D}_1 e^{-\gamma\ell} + \boldsymbol{D}_2 e^{\gamma\ell} \\ \boldsymbol{T}(\ell) &= \frac{\boldsymbol{D}_1 e^{-\gamma\ell} - \boldsymbol{D}_2 e^{\gamma\ell}}{z_0} \end{aligned} \tag{6.98}$$

The source and load require that

$$\begin{aligned} \boldsymbol{A}(0) &= \boldsymbol{A}_S - z_S \boldsymbol{T}(0) \\ \boldsymbol{A}(\ell) &= z_L \boldsymbol{T}(\ell) \end{aligned} \tag{6.99}$$

These six equations can be solved to find the coefficients $\boldsymbol{D}_1$ and $\boldsymbol{D}_2$, the input-port variables $\boldsymbol{A}(0)$ and $\boldsymbol{T}(0)$, and the output-port variables $\boldsymbol{A}(\ell)$ and $\boldsymbol{T}(\ell)$, all in terms of the source potential difference $\boldsymbol{A}_S$, the complex source and load impedances z_S and z_L, the transporter system parameters γ and z_0, and the length ℓ of the transporter.

The solution to equations (6.98) and (6.99) is the response to a sinusoidal source signal. To apply the solution to more general source signals, we must use superposition. We break the source signal into frequency components, use the solution to these six equations to find the response for each component, and then add the component responses. We shall not attempt to construct the total solution mathematically. Instead, we shall use the behaviors discovered from the sinusoidal analysis to infer the behavior of the transporter for nonsinusoidal signals.

Exercise 6.36: Solve equations (6.98) and (6.99).

Our purpose in examining the transporter is to understand its end-to-end behavior, not to compute its internal behavior. Just as the potential difference at the input port is the sum of two oppositely directed traveling waves, so is the potential difference at the output (or load) port the sum of two such waves: $\boldsymbol{A}(\ell) = \boldsymbol{D}_1 e^{-\gamma\ell} + \boldsymbol{D}_2 e^{\gamma\ell}$. The only source of energy in the system is the source at the input. Therefore, we view the positive-going component $\boldsymbol{D}_1 e^{-\gamma\ell}$ as arising from the source and the negative-going component $\boldsymbol{D}_2 e^{\gamma\ell}$ as a reflection—at the load—of the positive-going wave. If the purpose of the transporter is to transport signals from the input to the output, reflections are undesirable. Let us examine the seriousness of the reflection at the load.

The complex amplitudes of the incident and reflected potential differences are $\boldsymbol{D}_1 e^{-\gamma\ell}$ and $\boldsymbol{D}_2 e^{\gamma\ell}$ at the load (see the third equation of (6.98)). The load requires that

$z_L = \boldsymbol{A}(\ell)/\boldsymbol{T}(\ell)$. We use the third and fourth equations of (6.98) together with this load equation to find that

$$z_L = \left(\frac{\boldsymbol{D}_1 e^{-\gamma\ell} + \boldsymbol{D}_2 e^{\gamma\ell}}{\boldsymbol{D}_1 e^{-\gamma\ell} - \boldsymbol{D}_2 e^{\gamma\ell}}\right) z_0 \tag{6.100}$$

We define the **potential-difference reflection coefficient** ρ to be the ratio of the reflected and incident potential differences at the load. From equation (6.100), we find that

$$\rho \triangleq \frac{\boldsymbol{D}_2 e^{\gamma\ell}}{\boldsymbol{D}_1 e^{-\gamma\ell}} = \frac{z_L - z_0}{z_L + z_0} \tag{6.101}$$

Note that ρ is a frequency-dependent complex number.

From the fourth equation of (6.98), we conclude that the incident and reflected *flows* are $(\boldsymbol{D}_1/z_0)e^{-\gamma\ell}$ and $(\boldsymbol{D}_2/z_0)e^{\gamma\ell}$, respectively, both variables oriented to the right. (The negative sign in the second term of the equation implies negative flow, that is, flow to the left.) Therefore, the ratio of the reflected to the incident flow at the load is $-\rho$, the *flow reflection coefficient.*

It can be shown that $|\rho|^2$ is the fraction of the positive-going signal power (the power associated with the positive-going potential difference and flow signals) that is reflected from the load. According to equation (6.101), **there is no reflection if** $z_L = z_0$. If the load impedance matches the characteristic impedance of the transporter, then the transporter appears to the source to be infinitely long. The input impedance of the transporter is then z_0.

If the load impedance does not match the characteristic impedance of the transporter, a reflected signal returns to the source. If the source impedance does not match the characteristic impedance, the signal is reflected again and travels once more toward the load. We observe these periodic reflections in the simulations of the speedometer cable in example problem 4.7. Most transporters have low attenuation ($\alpha \approx 0$), so that reflected waves bounce back and forth for a long time. (See the pressure pulse in example problem 5.5, for example.)

Exercise 6.37: Find the reflection coefficient and describe the nature of the reflection at the load if $z_L = 0$ (short circuit) and if $z_L = \infty$ (open circuit).

The preceding analysis shows that reflections of sinusoidal source signals occur at the load if the load impedance does not match the source impedance. According to the previous section, reflected signals are delayed by an interval $d = \ell\beta/\omega$ as they travel from the source to the load and then are delayed an additional interval as they travel back to the source. In general, the power in the reflected sinusoidal wave and the length of the time delay d depend on the source frequency.

A nonsinusoidal source signal is a sum of sinusoids of different frequencies. Since its various frequency components reflect and propagate differently, the signal waveform becomes distorted as it propagates, and the distortion gets worse with time. Thus, no transporter is able to transport arbitrarily shaped signals properly.

On the other hand, if all frequency components of the source signal are in the high-frequency range for the transporter, then the waveform does not distort as it propagates. If the load impedance is essentially dissipative (real) for frequencies in that high-frequency range, then all frequency components of the signal reflect identically. The behaviors that we have discovered for sinusoidal signals apply to nonsinusoidal waveforms that satisfy these two conditions. We use this fact to derive a simple lumped model for transporters in the next section.

If the load and source impedances match the characteristic impedance of the transporter, there can be no reflections at either end. Hence, the only effect of a transporter in a *well-designed* system is to delay signals.

A Lumped Delay Model for an Ideal Transporter

In this section, we devise a two-port model for an **ideal transporter**—a transporter that does not behave dispersively. As usual, whether or not a device acts in an ideal fashion depends on how it is used. The model developed in this section is suitable for source waveforms of arbitrary *shape,* but not of arbitrary rapidity. The ideal-transporter model assumes that:

1. The frequency components of the source signal are within the high-frequency range for the transporter. Then all components are delayed equally. This assumption requires the source signal to be sufficiently rapid.[18]
2. The load impedance is essentially real for the frequency range of the source signal—the high-frequency range for the transporter. The characteristic impedance of the transporter is always real in this range. Then, according to equation (6.101), all frequency components of the signal are reflected in the same proportions, and the reflection coefficient ρ applies to the signal as a whole.
3. The transporter does not attenuate signals.

Although most transporters produce significant delay but little attenuation, we could modify the ideal model to include attenuation. If the other two assumptions are not substantially satisfied, however, then we must revert to a distributed lumped model like the ones in Figure 6.68.

Figure 6.71 shows an ideal transporter with its source and load. In the following paragraphs, we derive a lumped model for the transporter that characterizes its port-to-port behavior. That is, it acts like the physical transporter as seen at its ends, but ignores signals in its interior.

[18] We could also model the transporter for slowly changing source signals. The behavior is obvious from a distributed lumped model and is relatively simple. (See the speedometer model of Figure 3.15 or the coaxial cable model of Figure 4.19, for example.) We just ignore all energy-storage phenomena (by letting $s = 0$ in all impedances).

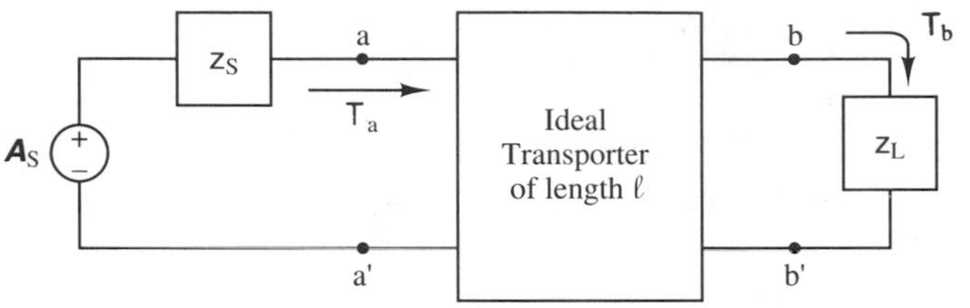

Figure 6.71 An ideal transporter (nondispersive and nonattenuating).

Initially, the source launches an incident wave into the primary port of the transporter. Until a reflected wave returns to the primary port, the transporter behaves no differently than a transporter of infinite length. Therefore, the transporter behaves like its characteristic impedance: The potential-difference signal $A_{aa'}(t) \triangleq A(0,t)$ and the flow signal $T_a(t) \triangleq T(0,t)$ at the input port are related by $A_{aa'}(t) = z_0 T_a(t)$. (The actual waveforms—in the variable t—depend on A_S, z_S, and z_0.) The potential-difference wave $A(x,t)$ and flow wave $T(x,t)$ propagate through the transporter and begin to arrive at the load at $t = d = \ell/v$, where ℓ is the length of the transporter and v is the propagation velocity for the transporter. Thus, the *incident* signals that arrive at the load are $A_{aa'}(t - d)$ and $T_a(t - d)$. These delayed signals are still related by the characteristic impedance z_0.

According to property 11 of the Laplace transformation (see Table 5.6), delaying a waveform by d time units corresponds to multiplying its Laplace transform by e^{-ds}. Therefore, we shall use the symbol e^{-ds} to indicate a **delay operator** with delay interval d, in the same way that we use s to denote a time-derivative operator. Then the *incident* signals that arrive at the load can be denoted $e^{-ds}A_{aa'}$ and $e^{-ds}T_a$. The delay operator is a linear operator. If we were to Laplace-transform the equation, the transformation would convert the delay operator e^{-ds} to the exponential function e^{-ds} of a complex frequency variable s.

We begin development of the model for the ideal transporter by eliminating the coefficients $\boldsymbol{D}_1$ and $\boldsymbol{D}_2$ from the phasor equations (6.98), and then arranging the resulting two equations in the form

$$\begin{aligned} 2\boldsymbol{A}(\ell) &= \boldsymbol{A}(0)(e^{\gamma\ell} + e^{-\gamma\ell}) - z_0\boldsymbol{T}(0)(e^{\gamma\ell} - e^{-\gamma\ell}) \\ 2\boldsymbol{T}(\ell) &= -\frac{\boldsymbol{A}(0)}{z_0}(e^{\gamma\ell} - e^{-\gamma\ell}) + \boldsymbol{T}(0)(e^{\gamma\ell} + e^{-\gamma\ell}) \end{aligned} \tag{6.102}$$

There is such a pair of phasor equations for each frequency component of the transporter signals. According to these equations, each signal phasor at the load port b,b′ has four constituents, each of which is related to a signal phasor at the source port a,a′. For example, $\boldsymbol{A}(0)e^{-\gamma\ell}$ is a version of the signal $\boldsymbol{A}(0)$ that has been delayed by the interval $d = \ell/v$. On the other hand, $\boldsymbol{A}(0)e^{\gamma\ell}$ is a version of the signal $\boldsymbol{A}(0)$ that has been *advanced* in time by the same interval.

In an ideal transporter (see the assumptions above), the delays affect each frequency component identically. Therefore, we can apply the delays inherent in the phasor equations (6.102) to the whole signals. Hence, we make the replacements $\boldsymbol{A}(0) \rightarrow A_{aa'}$, $\boldsymbol{A}(\ell) \rightarrow A_{bb'}$, $\boldsymbol{T}(0) \rightarrow T_a$, and $\boldsymbol{T}(\ell) \rightarrow T_b$. To denote that a signal that has been delayed by the interval d, we premultiply it by e^{-ds}. To denote that a signal that has been advanced by the interval d, we premultiply it by e^{ds}. Accordingly, the signals at the ends of the transporter are related by

$$\begin{aligned} 2A_{bb'} &= (e^{ds} + e^{-ds})A_{aa'} - (e^{ds} - e^{-ds})z_0T_a \\ 2T_b &= -(e^{ds} - e^{-ds})\frac{A_{aa'}}{z_0} + (e^{ds} + e^{-ds})T_a \end{aligned} \tag{6.103}$$

Equations (6.103) relate each output signal to a combination of delayed and advanced versions of the input signals. We now derive from these expressions a relation between the output signals $A_{bb'}$ and T_b and delayed versions of the input signals $A_{aa'}$ and T_a.

Multiply the second equation of (6.103) by $-z_0$ and subtract the result from the first equation. Arrange the resulting expression in the form

$$A_{bb'} = z_0(e^{-ds}T_a - T_b) + e^{-ds}A_{aa'} \tag{6.104}$$

This expression corresponds to the lumped model of the output port shown in Figure 6.72(a).

By performing a Norton-to-Thevenin transformation on the top portion of Figure 6.72(a), then combining the resulting series connection of potential difference sources, we can transform that lumped model into the equivalent series form shown in Figure 6.72(b). (Can you find an equivalent parallel form?)

The transporter is symmetrical in its behavior. Therefore, we can interchange the input and output variables in equations (6.103) and (6.104) if we account for the difference in orientations of the flow variables. That is, we make the substitutions: $A_{bb'} \leftrightarrow A_{aa'}$ and $-T_b \leftrightarrow T_a$. Consequently, the signals at the input port are related to delayed versions of the signals at the output port according to

$$A_{aa'} = z_0(T_a - e^{-ds}T_b) + e^{-ds}A_{bb'} \tag{6.105}$$

This equation corresponds to a lumped model of the input port that is identical in structure to the output-port model in Figure 6.72(a). We show the complete model for the ideal transporter, using conventional signal-delay notation, in Figure (6.73).

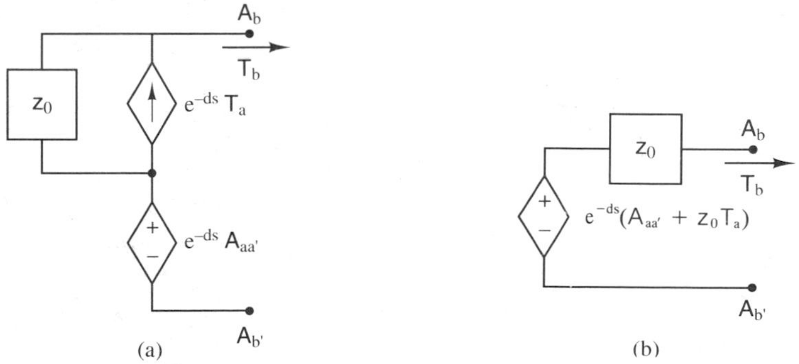

Figure 6.72 Equivalent representations of the transporter output port; $(e^{-ds}T_a)(t) \triangleq T_a(t - d)$ and $(e^{-ds}A_{aa'})(t) \triangleq A_{aa'}(t - d)$. (a) Hybrid representation. (b) Thévenin representation.

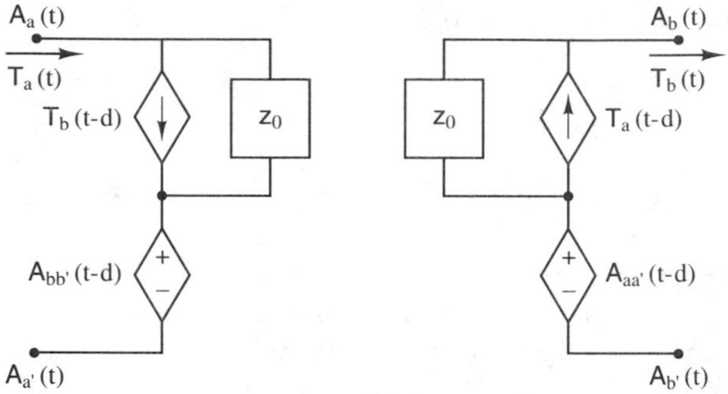

Figure 6.73 A lumped delay model for an ideal transporter.

The lumped delay model of Figure 6.73 can be implemented in network simulation programs. The model does not require a large number of lumped-element segments to simulate the behavior of a transporter accurately. If we are interested in the values of signals at an interior point of the transporter, we merely break the transporter at that point and represent it with two delay models. The transporter parameters used to *compute* γ and z_0 can be expressed in any unit of length: It is the *ratios* of these parameters that determine γ and z_0.

In example problem 4.6, we constructed a four-segment lumped model for a 2-m length of RG-58 coaxial electrical cable. We then used a SPICE-based simulation program to compute the response to a source signal v_S that rose sinusoidally from 0 to 10 V in 20 ns. In Figure 6.74, we replace the four-segment model of Figure 4.19 by the lumped delay model of Figure 6.73. According to Table 6.3, the characteristic impedance is $z_o = 49\ \Omega$ and the propagation velocity is $v_{high} = 2 \times 10^8$ m/s. Thus the delay in the 2-m length of cable is $d = (2\text{ m})/(2 \times 10^8\text{ m/s}) = 10$ ns.

We used a SPICE-based simulation program to compute the response of the delay model of Figure 6.74 to the same rapid source signal used in example problem 4.6. The response is displayed in Figure 6.75. Compare this figure with Figure 4.20, to see that both models give nearly the same response. The delay model, however, does not introduce the slight oscillations that are introduced by the four-segment lumped model. In most situ-

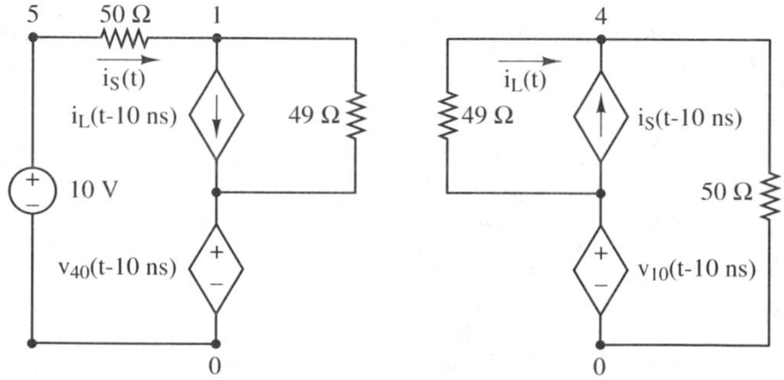

Figure 6.74 Ideal-transporter model for an RG-58 coaxial cable.

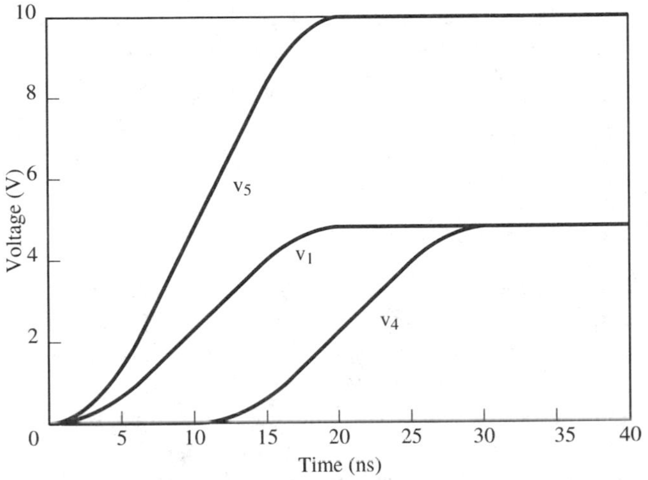

Figure 6.75 Simulated response of the coaxial cable.

ations, the delay model in Figure 6.73 will be simpler than any segmented model that gives comparable accuracy.

6.5 TWO-PORT EQUIVALENT NETWORKS

Let us partition an arbitrary linear lumped network into three pieces, with each partition boundary at a single terminal pair (or port), as shown in Figure 6.76. We think of the three subnets as the source subnet, the load subnet, and a *connector subnet.* We shall sometimes refer to the source and load subnets as *terminating subnets.* We exclude independent sources from the connector subnet by including them in one or the other terminating subnet. Although we shall speak as if the left subnet were the source subnet and the right subnet were the load subnet, we do not actually specify which subnet contains the source and which the load. (In fact, the two terminating subnets could contain competing sources.)

We label the potentials and flows symmetrically at the two ports, as shown in the figure. We use the notations T_1 and T_2 for the flows because T_2 is oriented opposite to the flow T_b used for converters, amplifiers, and transporters throughout sections 6.2 to 6.4. There, we used the T_b orientation to make the constants for most devices positive. We use the T_2 notation here because it is the conventional notation for two-port parameters. The symmetrical T_2 orientation does not presume that the right end of the two-port subnet is the load.

This section examines simple equivalent representations for two-port connector subnets. We show that there are six different representations, in contrast to the two equivalent

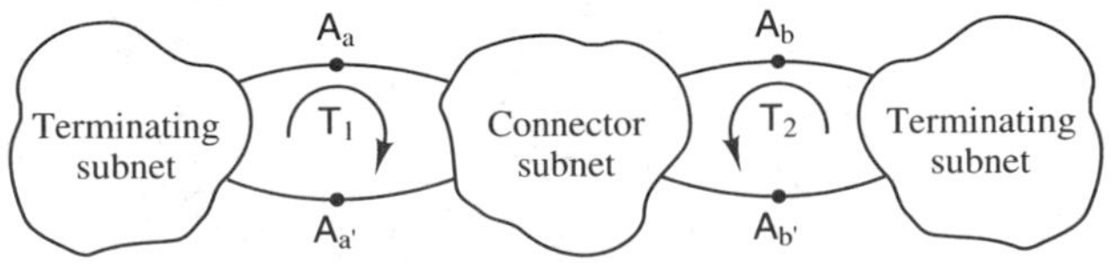

Figure 6.76 A two-port connector subnet.

representations (Thévenin and Norton) for single-port subnets. We also present a matrix notation for the set of (four) parameters that characterizes each representation. Then we find the equivalent representations for each of the ideal two-port devices we have discussed in this chapter—converters, amplifiers, and transporters. Finally, we examine procedures for finding the parameters and the equivalent representations for an arbitrary linear two-port subnet.

Equivalent Representations for Two-Port Subnets

None of the two-port device models of the previous three sections (converters, amplifiers, and transporters) contains *independent* sources. Since we carry out the following discussion primarily to provide a common conceptual framework for these two-port devices, we exclude independent sources from the connector subnet.

The Thévenin and Norton representations of a one-port subnet each include a source and an impedance. The source portion accounts for the presence of independent sources in the subnet. If we were to exclude independent sources from a one-port subnet, the Thévenin and Norton equivalent representations for that subnet would both reduce to a single parameter—the *subnet impedance.*

A linear lumped two-port subnet (without independent sources) can be described by two linear equations that relate the four port variables. We need four parameters to express linear relations among four variables, in contrast to the single parameter (the impedance) needed to relate the port variables of a two-terminal linear subnet. Linear relations among four variables can be expressed in six different ways: thus, we shall present six different equation sets. Each set has conceptual or mathematical advantages relative to the other sets.

In each equation set, we orient the potential drop for each port in the direction of the flow variable at the port. This equation-writing convention is just the downhill-flow convention that we have used for element equations throughout the book. In the notation of Figure 6.76, we write the equations in terms of $\mathsf{A}_{aa'}$ and $\mathsf{A}_{bb'}$. We can choose the *relative* orientations between the two ports arbitrarily. In Figure 6.76, we have chosen to orient the two ports symmetrically, with both flows entering at the top.

In the first equation set, we express the potential-difference variables as functions of the flow variables:

$$\begin{aligned} \mathsf{A}_{aa'} &= z_{11}\mathsf{T}_1 + z_{12}\mathsf{T}_2 \\ \mathsf{A}_{bb'} &= z_{21}\mathsf{T}_1 + z_{22}\mathsf{T}_2 \end{aligned} \tag{6.106}$$

Each of the coefficients z_{ij} is an impedance, a ratio of potential-difference to flow. Hence we refer to these equations as the **impedance equations** for the subnet. The coefficients z_{ij} are known as the *impedance parameters* (or the *z-parameters*) of the subnet. In general, the impedance parameters are functions of the time-derivative operator s. In matrix form, the impedance equations are

$$\begin{bmatrix} \mathsf{A}_{aa'} \\ \mathsf{A}_{bb'} \end{bmatrix} = \begin{bmatrix} z_{11} & z_{12} \\ z_{21} & z_{22} \end{bmatrix} \begin{bmatrix} \mathsf{T}_1 \\ \mathsf{T}_2 \end{bmatrix} \tag{6.107}$$

Inverting the impedance equations produces a second set of equations:

$$\begin{bmatrix} T_1 \\ T_2 \end{bmatrix} = \begin{bmatrix} y_{11} & y_{12} \\ y_{21} & y_{22} \end{bmatrix} \begin{bmatrix} A_{aa'} \\ A_{bb'} \end{bmatrix} \tag{6.108}$$

Equations (6.108) are the **admittance equations** for the subnet. The coefficients y_{ij} are the *admittance parameters* (or the *y-parameters*) of the subnet. The z-parameters and y-parameters are sometimes referred to jointly as *immittance parameters.* We use the symbols **[z]** and **[y]** to denote the 2 × 2 immittance-coefficient matrices.

In the impedance equations (6.107), z_{11} and z_{22} each relate potential difference to flow at a single port—the conventional concept of impedance. In contrast, z_{12} and z_{21} each relate the potential difference at one port to the flow at the other port. We refer to these port-to-port impedances as **transfer impedances.** Similarly, y_{12} and y_{21} are *transfer admittances.* If the two ports are in the same energy domain, all four z or y parameters have the same units. If the two ports are in different energy domains, as they are for a hydraulic cylinder, then all four z or y parameters have different units.

The z-parameter equations correspond to the interconnection of impedances and dependent sources shown in Figure 6.77(a). We call this network the **z-parameter equivalent** of the two-port subnet. The admittance equations (6.108) correspond to a second equivalent network, shown in Figure 6.77(b). (Some authors label the parallel-connected elements with their admittances, y_{11} and y_{22}. **We label all passive elements with their impedances**—in this case, $1/y_{11}$ and $1/y_{22}$.) The z-parameter equivalent is *not* related to the y-parameter equivalent by Thévenin-Norton transformations: The dependent sources in the two subnets involve different variables.

A third arrangement of equations is the following hybrid relation among the potential differences and flows:

$$\begin{aligned} A_{aa'} &= h_{11}T_1 + h_{12}A_{bb'} \\ T_2 &= h_{21}T_1 + h_{22}A_{bb'} \end{aligned} \tag{6.109}$$

The parameter h_{11} is an impedance, h_{22} is an admittance, and h_{12} and h_{21} are ratios of potential differences and flows, respectively. We call this set of parameters the *h-parameters* of the subnet. If the two ports are in the same energy domain, as they are for a lever, then the transfer ratios (h_{12} and h_{21}) are dimensionless. In matrix form, these hybrid relations are

$$\begin{bmatrix} A_{aa'} \\ T_2 \end{bmatrix} = \begin{bmatrix} h_{11} & h_{12} \\ h_{21} & h_{22} \end{bmatrix} \begin{bmatrix} T_1 \\ A_{bb'} \end{bmatrix} \tag{6.110}$$

We invert these equations to obtain another hybrid set, a fourth set of equations:

$$\begin{bmatrix} T_1 \\ A_{bb'} \end{bmatrix} = \begin{bmatrix} g_{11} & g_{12} \\ g_{21} & g_{22} \end{bmatrix} \begin{bmatrix} A_{aa'} \\ T_2 \end{bmatrix} \tag{6.111}$$

The h-parameters and g-parameters are known as **hybrid parameters.** We use the symbols **[h]** and **[g]** to denote the hybrid-parameter matrices. Each hybrid-parameter equation set can be interpreted as a network of lumped elements and dependent sources, as shown

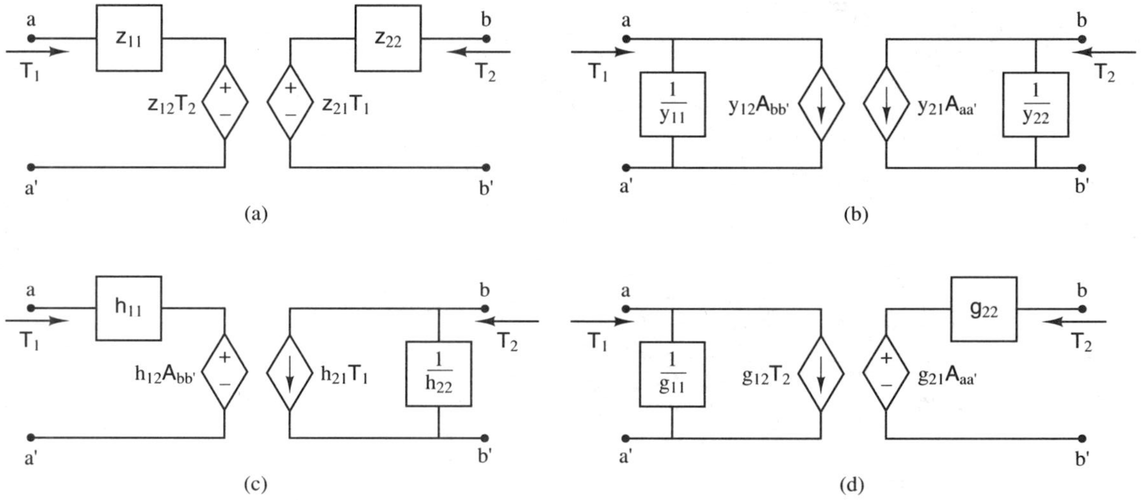

Figure 6.77 Equivalent line-graph representations for linear two-port subnets. (a) z-parameters. (b) y-parameters. (c) h-parameters. (d) g-parameters.

in Figures 6.76(c) and (d). Some authors label the parallel-connected elements with their admittances (h_{22} or g_{11}) rather than their impedances.

Exercise 6.38: Show that Thévenin-Norton transformation of one port of a g-parameter or h-parameter equivalent network produces a structure like the same port of either the z-parameter or y-parameter equivalent network. How many different equivalent structures can be found by such transformations?

In the fifth arrangement of equations, we express the secondary-port variables as functions of the primary-port variables:

$$\begin{aligned} \mathsf{A}_{bb'} &= b_{11}\mathsf{A}_{aa'} + b_{12}(-\mathsf{T}_1) \\ \mathsf{T}_2 &= b_{21}\mathsf{A}_{aa'} + b_{22}(-\mathsf{T}_1) \end{aligned} \tag{6.112}$$

The parameter b_{12} is an impedance, b_{21} is an admittance, and b_{11} and b_{22} are ratios of flows or potential differences. Jointly, we call them the **transmission parameters** or the **b-parameters** of the subnet. If the two ports are in different energy domains, as they are for a hydraulic cylinder, each parameter involves units from both domains. This set of equations is well suited to modeling transporters (see section 6.4).

In matrix form, the b-parameter relations are

$$\begin{bmatrix} \mathsf{A}_{bb'} \\ \mathsf{T}_2 \end{bmatrix} = \begin{bmatrix} b_{11} & b_{12} \\ b_{21} & b_{22} \end{bmatrix} \begin{bmatrix} \mathsf{A}_{aa'} \\ -\mathsf{T}_1 \end{bmatrix} \tag{6.113}$$

Note the negative sign on the *primary* flow T_1 in equations (6.112) and (6.113). These equations are particularly useful for determining the combined effect of a cascade of two-port devices, such as an amplifier followed by a transformer or a motor followed by a gear. The reversed direction of flow implied by the negative sign permits us to assign consistent flows when we connect devices end to end.

We express the primary-port variables in terms of the secondary-port variables to obtain the sixth and final set of equations—the *a-parameter* equations:

$$\begin{bmatrix} \mathsf{A}_{aa'} \\ \mathsf{T}_1 \end{bmatrix} = \begin{bmatrix} a_{11} & a_{12} \\ a_{21} & a_{22} \end{bmatrix} \begin{bmatrix} \mathsf{A}_{bb'} \\ -\mathsf{T}_2 \end{bmatrix} \tag{6.114}$$

In this second set of transmission equations, the negative sign reverses the *secondary* flow.

We use the symbols **[b]** and **[a]** to denote the 2×2 matrices of transmission parameters. The matrices **[a]** and **[b] are not inverses of each other,** because they are defined with reversed directions on different flows. Transmission equations (with b-parameters or a-parameters) do not have *obvious* correspondences to equivalent networks of lumped elements. (Try to associate equation (6.113) or (6.114) with an equivalent network.) However, we shall show a correspondence between the equivalent network for an ideal transporter (shown in Figure 6.73) and the a-parameter matrix for that transporter.

The relationship between two sets of parameters can be found by rearranging the equations for one set so that they are in the form of the equations for the other. These relationships are summarized in Table 6.4. Some two-port devices cannot be represented in

TABLE 6.4 TWO-PORT PARAMETER CONVERSIONS[1]

	z	y	a	b	h	g
z	$\begin{matrix} z_{11} & z_{12} \\ z_{21} & z_{22} \end{matrix}$	$\begin{matrix} \frac{y_{22}}{\lvert y\rvert} & -\frac{y_{12}}{\lvert y\rvert} \\ -\frac{y_{21}}{\lvert y\rvert} & \frac{y_{11}}{\lvert y\rvert} \end{matrix}$	$\begin{matrix} \frac{a_{11}}{a_{21}} & \frac{\lvert a\rvert}{a_{21}} \\ \frac{1}{a_{21}} & \frac{a_{22}}{a_{21}} \end{matrix}$	$\begin{matrix} \frac{b_{22}}{b_{21}} & \frac{1}{b_{21}} \\ \frac{\lvert b\rvert}{b_{21}} & \frac{b_{11}}{b_{21}} \end{matrix}$	$\begin{matrix} \frac{\lvert h\rvert}{h_{22}} & \frac{h_{12}}{h_{22}} \\ -\frac{h_{21}}{h_{22}} & \frac{1}{h_{22}} \end{matrix}$	$\begin{matrix} \frac{1}{g_{11}} & -\frac{g_{12}}{g_{11}} \\ \frac{g_{21}}{g_{11}} & \frac{\lvert g\rvert}{g_{11}} \end{matrix}$
y	$\begin{matrix} \frac{z_{22}}{\lvert z\rvert} & -\frac{z_{12}}{\lvert z\rvert} \\ -\frac{z_{21}}{\lvert z\rvert} & \frac{z_{11}}{\lvert z\rvert} \end{matrix}$	$\begin{matrix} y_{11} & y_{12} \\ y_{21} & y_{22} \end{matrix}$	$\begin{matrix} \frac{a_{22}}{a_{12}} & -\frac{\lvert a\rvert}{a_{12}} \\ -\frac{1}{a_{12}} & \frac{a_{11}}{a_{12}} \end{matrix}$	$\begin{matrix} \frac{b_{11}}{b_{12}} & -\frac{1}{b_{12}} \\ -\frac{\lvert b\rvert}{b_{12}} & \frac{b_{22}}{b_{12}} \end{matrix}$	$\begin{matrix} \frac{1}{h_{11}} & -\frac{h_{12}}{h_{11}} \\ \frac{h_{21}}{h_{11}} & \frac{\lvert h\rvert}{h_{11}} \end{matrix}$	$\begin{matrix} \frac{\lvert g\rvert}{g_{22}} & \frac{g_{12}}{g_{22}} \\ -\frac{g_{21}}{g_{22}} & \frac{1}{g_{22}} \end{matrix}$
a	$\begin{matrix} \frac{z_{11}}{z_{21}} & \frac{\lvert z\rvert}{z_{21}} \\ \frac{1}{z_{21}} & \frac{z_{22}}{z_{21}} \end{matrix}$	$\begin{matrix} -\frac{y_{22}}{y_{21}} & -\frac{1}{y_{21}} \\ -\frac{\lvert y\rvert}{y_{21}} & -\frac{y_{11}}{y_{21}} \end{matrix}$	$\begin{matrix} a_{11} & a_{12} \\ a_{21} & a_{22} \end{matrix}$	$\begin{matrix} \frac{b_{22}}{\lvert b\rvert} & \frac{b_{12}}{\lvert b\rvert} \\ \frac{b_{21}}{\lvert b\rvert} & \frac{b_{11}}{\lvert b\rvert} \end{matrix}$	$\begin{matrix} -\frac{\lvert h\rvert}{h_{21}} & -\frac{h_{11}}{h_{21}} \\ -\frac{h_{22}}{h_{21}} & -\frac{1}{h_{21}} \end{matrix}$	$\begin{matrix} \frac{1}{g_{21}} & \frac{g_{22}}{g_{21}} \\ \frac{g_{11}}{g_{21}} & \frac{\lvert g\rvert}{g_{21}} \end{matrix}$
b	$\begin{matrix} \frac{z_{22}}{z_{12}} & \frac{\lvert z\rvert}{z_{12}} \\ \frac{1}{z_{12}} & \frac{z_{11}}{z_{12}} \end{matrix}$	$\begin{matrix} -\frac{y_{11}}{y_{12}} & -\frac{1}{y_{12}} \\ -\frac{\lvert y\rvert}{y_{12}} & -\frac{y_{22}}{y_{12}} \end{matrix}$	$\begin{matrix} \frac{a_{22}}{\lvert a\rvert} & \frac{a_{12}}{\lvert a\rvert} \\ \frac{a_{21}}{\lvert a\rvert} & \frac{a_{11}}{\lvert a\rvert} \end{matrix}$	$\begin{matrix} b_{11} & b_{12} \\ b_{21} & b_{22} \end{matrix}$	$\begin{matrix} \frac{1}{h_{12}} & \frac{h_{11}}{h_{12}} \\ \frac{h_{22}}{h_{12}} & \frac{\lvert h\rvert}{h_{12}} \end{matrix}$	$\begin{matrix} -\frac{\lvert g\rvert}{g_{12}} & -\frac{g_{22}}{g_{12}} \\ -\frac{g_{11}}{g_{12}} & -\frac{1}{g_{12}} \end{matrix}$
h	$\begin{matrix} \frac{\lvert z\rvert}{z_{22}} & \frac{z_{12}}{z_{22}} \\ -\frac{z_{21}}{z_{22}} & \frac{1}{z_{22}} \end{matrix}$	$\begin{matrix} \frac{1}{y_{11}} & -\frac{y_{12}}{y_{11}} \\ \frac{y_{21}}{y_{11}} & \frac{\lvert y\rvert}{y_{11}} \end{matrix}$	$\begin{matrix} \frac{a_{12}}{a_{22}} & \frac{\lvert a\rvert}{a_{22}} \\ -\frac{1}{a_{22}} & \frac{a_{21}}{a_{22}} \end{matrix}$	$\begin{matrix} \frac{b_{12}}{b_{11}} & \frac{1}{b_{11}} \\ -\frac{\lvert b\rvert}{b_{11}} & \frac{b_{21}}{b_{11}} \end{matrix}$	$\begin{matrix} h_{11} & h_{12} \\ h_{21} & h_{22} \end{matrix}$	$\begin{matrix} \frac{g_{22}}{\lvert g\rvert} & -\frac{g_{12}}{\lvert g\rvert} \\ -\frac{g_{21}}{\lvert g\rvert} & \frac{g_{11}}{\lvert g\rvert} \end{matrix}$
g	$\begin{matrix} \frac{1}{z_{11}} & -\frac{z_{12}}{z_{11}} \\ \frac{z_{21}}{z_{11}} & \frac{\lvert z\rvert}{z_{11}} \end{matrix}$	$\begin{matrix} \frac{\lvert y\rvert}{y_{22}} & \frac{y_{12}}{y_{22}} \\ -\frac{y_{21}}{y_{22}} & \frac{1}{y_{22}} \end{matrix}$	$\begin{matrix} \frac{a_{21}}{a_{11}} & -\frac{\lvert a\rvert}{a_{11}} \\ \frac{1}{a_{11}} & \frac{a_{12}}{a_{11}} \end{matrix}$	$\begin{matrix} \frac{b_{21}}{b_{22}} & -\frac{1}{b_{22}} \\ \frac{\lvert b\rvert}{b_{22}} & \frac{b_{12}}{b_{22}} \end{matrix}$	$\begin{matrix} \frac{h_{22}}{\lvert h\rvert} & -\frac{h_{12}}{\lvert h\rvert} \\ -\frac{h_{21}}{\lvert h\rvert} & \frac{h_{11}}{\lvert h\rvert} \end{matrix}$	$\begin{matrix} g_{11} & g_{12} \\ g_{21} & g_{22} \end{matrix}$

[1]This is Table 4.2.1 of Ruston, H., and Bordogna, J., *Electric Networks: Functions, Filters, Analysis,* (New York: McGraw-Hill, 1966), p. 222. The symbol |·| denotes the determinant of the matrix of coefficients.

all of the ways suggested by equations (6.106)–(6.114). If certain parameters of one set are zero, as they are for an ideal converter, the operations of Table 6.4 that convert that set of parameters to another parameter set may require division by zero (see example problem 6.9).

Exercise 6.39: Derive the impedance parameters z_{ij} as functions of the transmission parameters b_{ij} by rearranging equation (6.112) into the form of equation (6.106). Compare the relationships you find with the corresponding relationships in Table 6.4.

Example Problem 6.9: Parameter matrices for ideal two-port devices.
Find appropriate parameter-matrix representations for:
(a) An ideal direct converter (such as the ideal lever of Figure 6.18);
(b) An ideal gyrating converter (such as the ideal hydraulic cylinder of Figure 6.20);
(c) The four amplifier structures of Figure 6.46; and
(d) The ideal transporter of Figure 6.73.
Solution: **(a)** The converter constant for a direct converter is defined in equation (6.53) as $\kappa \triangleq A_{bb'}/A_{aa'}$. According to equations (6.53) and (6.54) and Figure 6.29(a), the equations that describe the behavior of a direct converter are $T_a = \kappa T_b$ and $A_{bb'} = \kappa A_{aa'}$. The converter flows T_a and T_b of Figure 6.29 are related to the parameter-matrix flows T_1 and T_2 of Figure 6.76 by $T_1 = T_a$ and $T_2 = -T_b$. (For the ideal lever of Figures 6.18 and 6.19, $\kappa = \ell_b/\ell_a$, $T_1 = T_a \equiv F_a$, $T_2 = -T_b \equiv -F_b$, $A_{aa'} \equiv v_{ag}$, and $A_{bb'} \equiv v_{bg}$.) We express the converter equations in the form $A_{aa'} = (1/\kappa)A_{bb'}$ and $(-T_b) = (-1/\kappa)T_a$. These equations have the h-parameter form of equations (6.109) and (6.110). We obtain the g-parameters by inverting the h-parameter matrix. Therefore, we conclude that

$$[\mathbf{h}]_{direct} = \begin{bmatrix} 0 & 1/\kappa \\ -1/\kappa & 0 \end{bmatrix}, \qquad [\mathbf{g}]_{direct} = \begin{bmatrix} 0 & -\kappa \\ \kappa & 0 \end{bmatrix} \tag{6.115}$$

If we compare the h-parameter equivalent network of Figure 6.77(c) with the ideal-lever model of Figure 6.19(b), we see that the two are identical. Similarly, the g-parameter equivalent network of Figure 6.77(d) is identical to the lever model of Figure 6.19(a).

Let us rearrange the two converter equations into the form $A_{bb'} = \kappa A_{aa'}$ and $(-T_b) = (1/\kappa)(-T_a)$. These equations have the b-parameter form of equation (6.112). The alternative arrangement $A_{aa'} = (1/\kappa)A_{bb'}$ and $T_a = \kappa T_b$ has the a-parameter form of equation (6.114). Therefore,

$$[\mathbf{b}]_{direct} = \begin{bmatrix} \kappa & 0 \\ 0 & 1/\kappa \end{bmatrix}, \qquad [\mathbf{a}]_{direct} = \begin{bmatrix} 1/\kappa & 0 \\ 0 & \kappa \end{bmatrix} \tag{6.116}$$

Although **[a]** is the inverse of **[b]** in this instance, this is true only because the two off-diagonal parameters are zero.

An ideal direct converter cannot be represented in terms of immittance parameters. (Try converting one of the foregoing parameter sets to z-or y-form by means of Table 6.4.)
(b) The converter constant for a gyrating converter is defined in equation (6.57) as $\kappa_g = -T_2/A_{aa'}$. According to equations (6.57) and (6.58) and Figure 6.29(b), the equations for a gyrating converter are $T_a = \kappa_g A_{bb'}$ and $T_b = \kappa_g A_{aa'}$. (For the ideal hydraulic cylinder of Figure 6.20, $\kappa_g = A$, $T_1 = T_a \equiv Q_a$, $T_2 = -T_b \equiv -F_b$, $A_{aa'} \equiv P_{a0}$, and $A_{bb'} \equiv v_{bg}$.) We express the converter equations in the form $A_{aa'} = (-1/\kappa_g)(-T_b)$ and $A_{bb'} = (1/\kappa_g)T_a$. These equations have the z-parameter form of equation (6.106). We obtain the y-parameters by inverting the z-parameter matrix. Therefore, we conclude that

$$[\mathbf{z}]_{gyrating} = \begin{bmatrix} 0 & -1/\kappa_g \\ 1/\kappa_g & 0 \end{bmatrix}, \qquad [\mathbf{y}]_{gyrating} = \begin{bmatrix} 0 & \kappa_g \\ -\kappa_g & 0 \end{bmatrix} \tag{6.117}$$

If we compare the z-parameter and y-parameter equivalent networks of Figures 6.77(a) and (b) with the hydraulic cylinder line-graph models in Figures 6.20(c) and (b), we see that they are identical.

Now let us rearrange the two converter equations to the form $A_{aa'} = (1/\kappa_g)T_b$ and $T_a = \kappa_g A_{bb'}$. These equations have the form of the a-parameter equations (6.114). The alternative arrangement $A_{bb'} = (-1/\kappa_g)(-T_a)$ and $-T_b = (-\kappa_g)A_{aa'}$ is in the form of the b-parameter equations (6.113). Therefore,

$$[\mathbf{a}]_{gyrating} = \begin{bmatrix} 0 & 1/\kappa_g \\ \kappa_g & 0 \end{bmatrix}, \qquad [\mathbf{b}]_{gyrating} = \begin{bmatrix} 0 & -1/\kappa_g \\ -\kappa_g & 0 \end{bmatrix} \tag{6.118}$$

Note that $[\mathbf{b}] \neq [\mathbf{a}]^{-1}$.

The ideal hydraulic cylinder cannot be represented in terms of hybrid parameters. (Try to convert equation (6.117) or (6.118) to such parameters.)

(c) Let us compare the four amplifier structures of Figure 6.46 with the four equivalent-network structures of Figure 6.77. We focus on the transfer parameter of the β amp to see that the β-amp structure is in h-parameter form. Comparing the β-amp and h-parameter structures shows that

$$[\mathbf{h}]_\beta = \begin{bmatrix} z_i & 0 \\ -\beta & 1/z_o \end{bmatrix} \rightarrow \begin{bmatrix} 0 & 0 \\ -\beta & 0 \end{bmatrix} \tag{6.119}$$

where the matrix on the right is the limiting value of $[\mathbf{h}]_\beta$ for an *ideal* β *amp* ($z_i = 0$ and $z_o = \infty$). We can use Table 6.4 to convert the h-parameters of equation (6.119) to a-parameters:

$$[\mathbf{a}]_\beta = \begin{bmatrix} \beta z_i/z_o & \beta z_i \\ \beta/z_o & \beta \end{bmatrix} \rightarrow \begin{bmatrix} 0 & 0 \\ 0 & \beta \end{bmatrix} \tag{6.120}$$

Again, the limiting values apply to the ideal β amp.

An attempt to convert $[\mathbf{h}]_\beta$ to other parameter matrices shows that no other parameter matrix exists for the *ideal* β amp (with $z_i = 0$ and $z_o = \infty$). Therefore, we use only the preceding two sets of two-port parameters for β amps. The h-parameters correspond to the β-amp lumped network of Figure 6.46(a). The bipolar junction transistor in the common-emitter configuration is a β amp. The common-emitter transistor model of Figure 6.53 is essentially an h-parameter lumped network.

By making similar comparisons between the other three amplifier types in Figure 6.46 and the other three network structures in Figure 6.77, we find that

$$[\mathbf{g}]_\mu = \begin{bmatrix} 1/z_i & 0 \\ \mu & z_o \end{bmatrix} \rightarrow \begin{bmatrix} 0 & 0 \\ \mu & 0 \end{bmatrix}, \qquad [\mathbf{a}]_\mu = \begin{bmatrix} 1/\mu & z_o/\mu \\ 1/\mu z_i & z_o/\mu z_i \end{bmatrix} \rightarrow \begin{bmatrix} 1/\mu & 0 \\ 0 & 0 \end{bmatrix} \tag{6.121}$$

$$[\mathbf{z}]_{r_t} = \begin{bmatrix} z_i & 0 \\ r_t & z_o \end{bmatrix} \rightarrow \begin{bmatrix} 0 & 0 \\ r_t & 0 \end{bmatrix}, \qquad [\mathbf{a}]_{r_t} = \begin{bmatrix} z_i/r_t & z_i z_o/r_t \\ 1/r_t & z_o/r_t \end{bmatrix} \rightarrow \begin{bmatrix} 0 & 0 \\ 1/r_t & 0 \end{bmatrix} \tag{6.122}$$

$$[\mathbf{y}]_{g_t} = \begin{bmatrix} 1/z_i & 0 \\ -g_t & 1/z_o \end{bmatrix} \rightarrow \begin{bmatrix} 0 & 0 \\ -g_t & 0 \end{bmatrix}, \qquad [\mathbf{a}]_{g_t} = \begin{bmatrix} 1/g_t z_o & 1/g_t \\ 1/g_t z_i z_o & 1/g_t z_i \end{bmatrix} \rightarrow \begin{bmatrix} 0 & 1/g_t \\ 0 & 0 \end{bmatrix} \tag{6.123}$$

(d) We use the lumped model of the ideal transporter in Figure 6.73 to guide the selection of an appropriate two-port parameter set for transporters. Since the transporter is usually intended to carry signals from the input to the output, rather than back and forth, transmission parameters seem most appropriate. The equations that describe the behaviors of the two ports of Figure 6.73 are

$$A_{bb'} = e^{-ds}A_{aa'} + (-T_b + e^{-ds}T_a)Z_0$$

$$A_{aa'} = e^{-ds}A_{bb'} + (T_a - e^{-ds}T_b)Z_0 \tag{6.124}$$

A comparison of Figures 6.73 and 6.76 shows that $T_1 \equiv T_a$ and $T_2 \equiv -T_b$. We make the change in flow notation and rearrange equation (6.124) to the form of equation (6.112) to find the b-parameters:

$$[\mathbf{b}]_{transp} = \begin{bmatrix} \cosh(ds) & z_0\sinh(ds) \\ (1/z_0)\sinh(ds) & \cosh(ds) \end{bmatrix} \tag{6.125}$$

The hyperbolic functions in equation (6.125) have the conventional definitions: $\cosh(ds) = (e^{ds} + e^{-ds})/2$ and $\sinh(ds) = (e^{ds} - e^{-ds})/2$. **These expressions represent operators on signals.** Multiplication of a signal by the cosh(ds) operator, for example, means that we average two versions of the signal, one advanced by d seconds and the other delayed by d seconds.

We can use Table 6.4 to convert the b-parameters of equation (6.125) to other parameter forms. The a-parameter form is

$$[\mathbf{a}]_{transp} = \begin{bmatrix} \cosh(ds) & -z_0\sinh(ds) \\ -(1/z_0)\sinh(ds) & \cosh(ds) \end{bmatrix} \tag{6.126}$$

The a-parameters differ from the b-parameters only in the signs on the off-diagonal terms. This difference in sign owes to the fact that different variables are negated in the defining equations (6.113) and (6.114); the transporter itself acts identically in both directions. Although the other parameter forms exist, they do not have easily determined physical interpretations.

Exercise 6.40: Use equations (6.124) and (6.112) to verify **[b]** of equation (6.125).

Example problem 6.9 shows that each type of two-port device is best characterized by a particular parameter set. Other parameter sets can be inappropriate. However, every such device can be characterized by an a-parameter set.

Each *ideal* device is characterized by one or two *device* parameters. For an ideal converter, the parameter is κ or κ_g. For an ideal amplifier, it is β, μ, r_t, or g_t. An ideal transporter is characterized by d and z_0. Each two-port parameter set is assembled from these device parameters.

A physical transformer, transducer, amplifier, or transporter is typically not ideal. Lumped models for these physical devices include energy-storage elements and energy-dissipating elements. How can we determine the two-port parameters for these physical-device models? In the next section, we show how to combine the parameters for two-port subnets connected end to end. The succeeding section discusses general procedures for finding or measuring two-port parameters.

Cascading of Two-Port Subnets

Figure 6.78 shows two linear two-port subnets connected in cascade—end to end. Suppose each subnet in the cascade is represented by one of the two-port parameter matrices. Each z-, y-, h-, or g-parameter matrix corresponds to a simple two-port equivalent subnet. Thus, we can *think* of Figure 6.78 as two simple models connected in cascade. We now seek to reduce the pair of cascaded two-port models to a single equivalent two-port model.

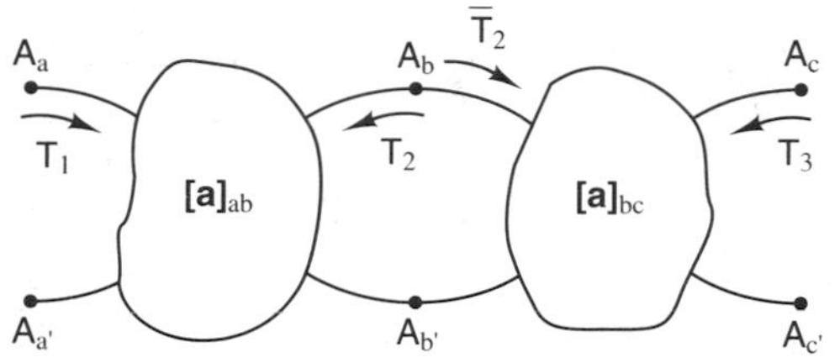

Figure 6.78 A cascade connection of linear two-port subnets.

Transmission parameters are particularly helpful in combining cascaded subnets. Example problem 6.9 showed that all the ideal two-port devices can be represented in the **[a]** form. Therefore, we have labeled the two subnets with their a-parameter matrices. The a-parameter equations (6.114) for the two individual subnets are

$$\begin{bmatrix} A_{aa'} \\ T_1 \end{bmatrix} = [\mathbf{a}]_{ab} \begin{bmatrix} A_{bb'} \\ -T_2 \end{bmatrix}, \qquad \begin{bmatrix} A_{bb'} \\ \overline{T}_2 \end{bmatrix} = [\mathbf{a}]_{bc} \begin{bmatrix} A_{cc'} \\ -T_3 \end{bmatrix} \tag{6.127}$$

where $\overline{T}_2 = -T_2$. We think of $[\mathbf{a}]_{ab}$ as the matrix that produces signals at port b from signals at port a. Substituting the second of these equations into the first produces

$$\begin{bmatrix} A_{aa'} \\ T_1 \end{bmatrix} = [\mathbf{a}]_{ab}[\mathbf{a}]_{bc} \begin{bmatrix} A_{cc'} \\ -T_3 \end{bmatrix} \triangleq [\mathbf{a}]_{ac} \begin{bmatrix} A_{cc'} \\ -T_3 \end{bmatrix} \tag{6.128}$$

Equation (6.128) specifies the procedure for finding the single equivalent model: We merely multiply the a-parameter matrices in the same order in which the ports appear in the cascade connection to produce the a-parameter matrix for the equivalent model:

$$[\mathbf{a}]_{ac} = [\mathbf{a}]_{ab}[\mathbf{a}]_{bc} \tag{6.129}$$

To find a lumped model that corresponds to the a-parameter matrix, we use Table 6.4 to convert to the parameters that correspond to the form we want (see Figure 6.77).

When we construct a realistic lumped model of a two-port device (a converter or amplifier), we attach lumped elements in series or parallel with the ports of an ideal-device model to account for energy storage or energy dissipation. The two-terminal lumped elements that we attach in series or parallel can be viewed as two-port subnets connected in cascade with the ideal converter.

Figure 6.79 shows the two-port structures that correspond to series and parallel impedances, denoted $z_s(s)$ and $z_p(s)$, respectively. We write, by inspection, the equations that relate the variables $A_{aa'}$, $A_{bb'}$, T_1, and T_2. Then we rearrange them into a-parameter

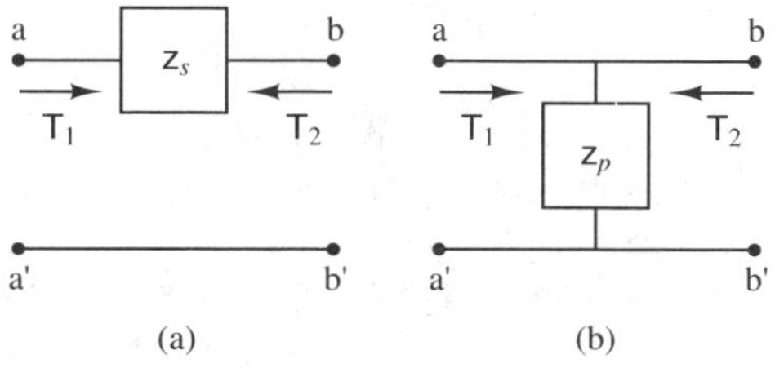

Figure 6.79 Two-port structures for two-terminal elements.
(a) A series-connected element.
(b) A parallel-connected element.

form. The equations for Figure 6.79(a) are $A_{aa'} = A_{bb'} - z_s T_2$ and $T_1 = -T_2$. Therefore, the a-parameter matrix for the series structure is

$$[\mathbf{a}]_{\text{series}} = \begin{bmatrix} 1 & z_s \\ 0 & 1 \end{bmatrix} \tag{6.130}$$

The equations for Figure 6.79(b) are $A_{aa'} = A_{bb'}$ and $(T_1 + T_2)z_p = A_{bb'}$. Therefore, the a-parameter matrix for the parallel structure is

$$[\mathbf{a}]_{\text{parallel}} = \begin{bmatrix} 1 & 0 \\ 1/z_p & 1 \end{bmatrix} \tag{6.131}$$

We demonstrate how to use equation (6.129) by applying it, together with equations (6.130) and (6.131), to the realistic hydraulic cylinder model of Figure 6.80(a). The capacitor C and resistor R represent, respectively, the fluid compressibility and fluid resistance in the cylinder. The mass m_p and damper b_p represent, respectively, the mass of the

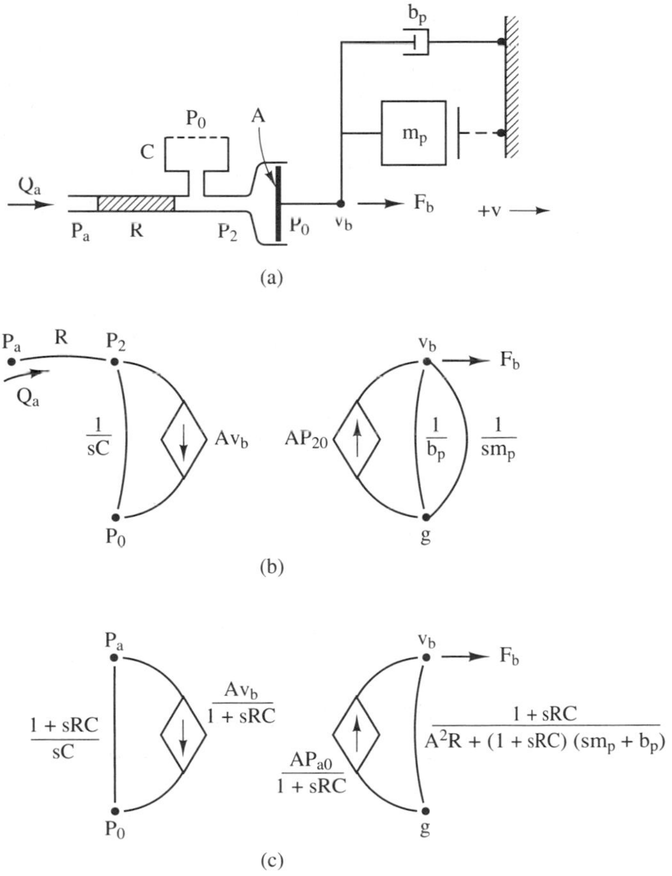

Figure 6.80 A realistic hydraulic cylinder model. (a) Lumped model. (b) Line-graph representation. (c) Line graph for the y-parameter equivalent subnet.

piston and the piston friction at the cylinder wall. The flow Q_a and pressure drop P_{a0} are provided by the hydraulic source. The velocity v_b and compressive force F_b are the reaction of the load.

A line-graph representation of the hydraulic cylinder model is given in Figure 6.80(b). We may view the model as a cascade connection of five two-port devices: A series fluid resistor, a parallel fluid capacitor, an *ideal* hydraulic cylinder, a parallel damper, and a parallel mass. The a-parameter matrices for the series and parallel lumped elements can be obtained from equations (6.130) and (6.131). The a-parameter matrix for an ideal gyrating converter with converter constant κ_g is given in equation (6.118). For the hydraulic cylinder, $\kappa_g = A$. These matrices, presented in the same order as the cascaded elements in Figure 6.80, are

$$\begin{bmatrix} 1 & R \\ 0 & 1 \end{bmatrix} \begin{bmatrix} 1 & 0 \\ sC & 1 \end{bmatrix} \begin{bmatrix} 0 & 1/A \\ A & 0 \end{bmatrix} \begin{bmatrix} 1 & 0 \\ b_p & 1 \end{bmatrix} \begin{bmatrix} 1 & 0 \\ sm_p & 1 \end{bmatrix} \tag{6.132}$$

We multiply the **[a]** matrices of equation (6.132) to produce the hydraulic cylinder parameters

$$[\mathbf{a}]_{cyl} = \frac{1}{A} \begin{bmatrix} A^2R + (1 + sRC)(sm_p + b_p) & 1 + sRC \\ A^2 + sC(sm_p + b_p) & sC \end{bmatrix} \tag{6.133}$$

We use Table 6.4 to convert $[\mathbf{a}]_{cyl}$ to

$$[\mathbf{y}]_{cyl} = \begin{bmatrix} \dfrac{sC}{1 + sRC} & \dfrac{A}{1 + sRC} \\ \dfrac{-A}{1 + sRC} & \dfrac{A^2R + (1 + sRC)(sm_p + b_p)}{1 + sRC} \end{bmatrix} \tag{6.134}$$

The y-parameter matrix in equation (6.134) corresponds to the y-parameter equivalent subnet shown in Figure 6.80(c). The procedure for obtaining the matrix is conceptually simple and computationally straightforward. The matrices in equations (6.133) and (6.134) are mathematically convenient. However, the corresponding equivalent network (see Figure 6.80(c)) does not provide the physical insight associated with the original lumped model.

Finding the Parameters for a Two-Port Subnet

We can use the cascading process of the previous section to find the two-port parameters for any connector subnet. In this section, we present a second method for finding them. The key information needed for this second approach lies in the lumped models in Figure 6.77. The same information is contained in equations (6.106)–(6.111).

Suppose we seek the z-parameters for the two-port connector subnet of Figure 6.76. Figure 6.77(a) is a simple equivalent lumped model of that subnet. The equation that describes the left-port behavior of this simple model is

$$A_{aa'} = z_{11}T_1 + z_{12}T_2 \tag{6.135}$$

If we make $T_2 = 0$ (open-circuit port bb′ of the model), equation (6.135) reduces to a simple relation between $A_{aa'}$ and T_1, namely,

$$z_{11} = \frac{A_{aa'}}{T_1}\bigg|_{T_2=0} \tag{6.136}$$

To find z_{11} for the connector subnet of Figure 6.76, we open-circuit port bb′ of the subnet, attach a potential-difference source A to port aa′ of the subnet, write and solve the subnet equations to find T_1 *as a function of* A, and use equation (6.136) to find $z_{11} = A/T_1$. In general, z_{11} will be a function of the time-derivative operator s. (We could, instead, attach a *flow* source T directed into node a and find $A_{aa'}$ as a function of T.)

If we have access to the *physical two-port subsystem,* we can *measure* z_{11}. To do so, we open-circuit port bb′ physically, apply a potential-difference source A to port aa′, measure T_1, and use equation (6.136) to compute $z_{11} = A/T_1$. To discern the presence of a storage element in the subnet, we use a sinusoidal source, represented by a phasor $\boldsymbol{A}$, measure the phasor $\boldsymbol{T}_1$ vs. the frequency ω, and compute the frequency response $z_{11}(\omega) = \boldsymbol{A}/\boldsymbol{T}_1$. We then use the frequency response to construct $z_{11}(s)$, as discussed in section 5.4.

Similarly, if we set $T_1 = 0$ (open-circuit port aa′), equation (6.135) reduces to a simple relation between $A_{aa'}$ and T_2, namely,

$$z_{12} = \frac{A_{aa'}}{T_2}\bigg|_{T_1=0} \tag{6.137}$$

Therefore, we apply a flow source T directed into node b and find the resulting potential difference $A_{aa'}$ as a function of T. Again, we can base the computation in equation (6.137) on an analysis of the open-circuited subnet or on measurements of the open-circuited physical subsystem. We find z_{21} and z_{22} by repeating the procedures with the two ports reversed.

We can, instead, use corresponding open-circuit and short-circuit procedures to obtain the y-parameters, the h-parameters, or the g-parameters of the two-port subnet. Once we find one set of parameters for the subnet, we can obtain the others by using Table 6.4.

A third approach to finding the equivalent representation of a connector-subnet is to use a sequence of equivalent-network transformations to reduce the model to the form of Figure 6.77(a). The equivalent-network transformations that we use are Thévenin-Norton transformations, series and parallel combinations, and the ideal-converter transformations shown in Figures 6.30 and 6.32. After we reduce the model to the form of the z-parameter equivalent network, we read the z-parameters from the four elements of that network (see Figure 6.77(a)). We can also use equivalent-network transformations to reduce the subnet to y-, h-, or g-parameter form (see Figure 6.77(b), (c), or (d)).

Manufacturers of transformers, transducers, and amplifiers provide data sheets that give typical measured values of lumped-element parameters appropriate to the device. The designer can use these values to construct an accurate lumped model for the device. In some instances, the manufacturer gives the data in the form of two-port parameters. In particular, it is customary for transistor manufacturers to describe transistors by stating their h-parameters.

The measured values in data sheets are not usually specific to any particular copy of the device. Yet, by using these lumped-element values, we can predict the correct *form* of behavior of the device. We can also predict the correct *order of magnitude* of the device variables during operation of the device.

Example Problem 6.10: Finding two-port parameters for a hydraulic cylinder.
Figure 6.80(a) shows a realistic lumped model for a hydraulic cylinder. Find the admittance parameters for the device.

Solution: The line-graph representation of the hydraulic cylinder is shown in Figure 6.80(b). We use the short-circuit approach to find y_{11}. According to equation (6.108), if we set $A_{bb'} = 0$, then $y_{11} = T_1/A_{aa'}$. In this instance, $T_1 \equiv Q_a$ and $A_{aa'} \equiv P_{a0}$. When we short-circuit the secondary port (set $v_b = 0$) and attach a pressure source to the primary port (set $P_{a0} = P$), the line graph becomes Figure 6.81(a). We must solve for Q_a. It is apparent from the left side of the line graph that $P = Q_a(R + 1/(sC)) = Q_a(1 + sRC)/(sC)$. Then

$$y_{11} = \frac{T_1}{A_{aa'}} \equiv \frac{Q_a}{P_{a0}} \equiv \frac{Q_a}{P} = \frac{sC}{1 + sRC} \tag{6.138}$$

Similarly, $y_{21} = T_2/A_{aa'}$ if $A_{bb'} = 0$. Again, we use Figure 6.81(a) and solve for $T_2 \equiv -F_b$. From the loop equation for the left half of the line graph, we find that

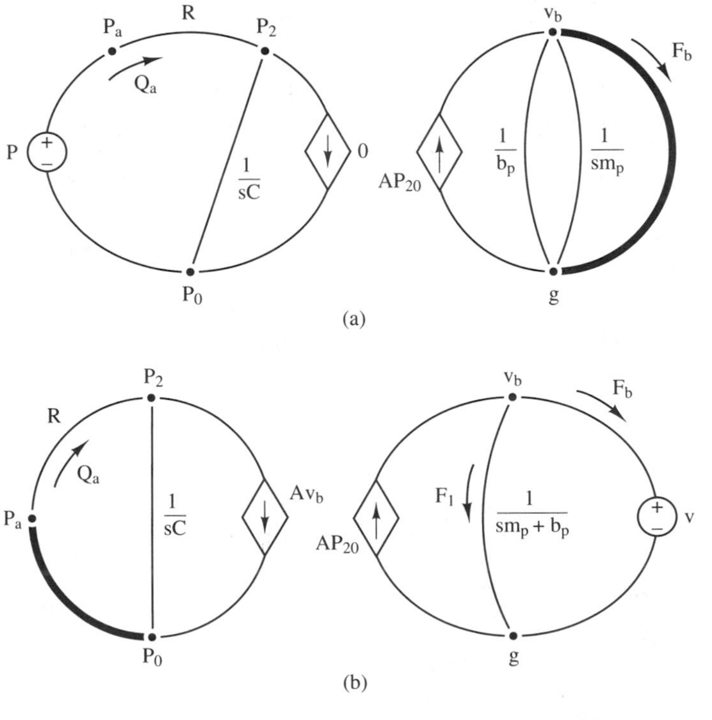

Figure 6.81 Finding the connector-subnet parameters. (a) Source P at side a with short circuit at side b. (b) Source v at side b with short circuit at side a.

$P_{20} = P/(1 + sRC)$. Since $v_b = 0$, there is no flow in b_p or m_p. Therefore, the node equation for the right half of the line graph requires that $F_b = AP_{20} = AP/(1 + sRC)$. Then

$$y_{21} = \frac{T_2}{A_{aa'}} \equiv \frac{-F_b}{P_{a0}} = \frac{-F_b}{P} = \frac{-A}{1 + sRC} \tag{6.139}$$

Now we short-circuit the primary port (set $P_{a0} = 0$) instead and attach a velocity source to the secondary port (set $v_b = v$). The corresponding line graph is Figure 6.81(b). (We have combined the parallel impedances $1/b_p$ and $1/(sm_p)$ to simplify the diagram.) According to equation (6.108),

$$y_{12} = \frac{T_1}{A_{bb'}} = \frac{Q_a}{v} \quad \text{and} \quad y_{22} = \frac{T_2}{A_{bb'}} = \frac{-F_b}{v} \tag{6.140}$$

The parallel impedances R and $1/(sC)$ act as a single impedance of value $R/(1 + sRC)$. The flow Av_b through this equivalent impedance produces the pressure drop $P_{20} = (-Av_b)R/(1 + sRC) = -ARv/(1 + sRC)$. The flow F_1 owing to v is $F_1 = (sm_p + b_p)v$. Therefore, the node equation for the right half of Figure 6.81(b) is $F_b = AP_{20} - F_1 = -A^2Rv/(1 + sRC) - (sm_p + b_p)v$. Then

$$y_{22} = \frac{-F_b}{v} = \frac{A^2R}{1 + sRC} + sm_p + b_p \tag{6.141}$$

The flow through R is $Q_a = -P_{20}/R = Av/(1 + sRC)$. Consequently,

$$y_{12} = \frac{Q_a}{v} = \frac{A}{1 + sRC} \tag{6.142}$$

Compare these y-parameters with the corresponding parameters in equation (6.134). The result is the same. The short-circuit and open-circuit testing procedures are not as straightforward as the **[a]**-matrix multiplications of the cascaded-subsection approach. On the other hand, the manipulations are more insightful. At each stage, we can see the physical relations that the parameters represent. The final equivalent network is shown in Figure 6.80(c).

6.6 EFFECTS OF FEEDBACK

What is a dynamic system? It is a system in which future values of the variables depend not only on future inputs, but also on past inputs—via present values of stored energy. The present values of stored energy constitute the network state (see section 3.4). It is this dependence of the future state on the present state that causes the system equations to be differential rather than algebraic.

In the operational models of section 3.2, the dependence on past inputs is revealed by a feedback structure. The system equation for the first-order network of Figure 3.17 is $\dot{T} = -(K/B)T + K\Omega$. The operational diagram for this network is displayed in Figure 3.21. In the operational diagram, the output signal is multiplied by $(K/B)T$ and *subtracted* from the input signal $K\Omega$. The feedback structure causes signals to circulate in a **feedback loop**. Since there is an integrator in the loop, the present output affects the future output.

The multiplier of the feedback signal is $1/\tau$, the inverse time constant. It is appropriate to think of the *negative feedback* as the *cause* of the exponential decay with the time constant τ. Every signal in the network exhibits the same negative feedback structure with the same feedback-signal multiplier. If we increase the negative feedback (by increasing the multiplier $1/\tau$), we cause all signals in the system to settle more quickly. If we *add* the feedback signal rather than subtract it (positive feedback), the network signals grow exponentially rather than decay exponentially. (Exponential growth is exhibited in the population models of section 6.1.) Systems with multiple feedback loops exhibit multiple time constants and/or sinusoidal (complex exponential) behavior. Because feedback signals have such a profound and fundamental effect on the behavior of systems, we seek ways to enhance or diminish feedback in order to modify that behavior.

Enhanced Feedback

In Figure 6.82(a), an ideal common-collector amplifier energizes an ideal electric motor to drive an inertial load. We show a line graph model for the system in Figure 6.82(b). The ideal-amplifier model is a signal model that corresponds to the common-collector equations (6.75), with $R_b = 0$ and $R_c = \infty$. The ideal-motor model corresponds to equations (6.34)–(6.36).

The back emf inherent in the motor constitutes negative feedback of the motor velocity to the electrical input of the motor, expressed as $v_3 = K_T\Omega_4$. The common-collector amplifier feeds back the amplifier output voltage v_3 to the amplifier input, so that $v_2 = v_3$. The net feedback multiplier from the output Ω_4 to the input v_2 is the product $(1)(K_T) = K_T$.

According to the line graph, the source current is

$$i_S = \frac{v_S - v_3}{R_S} = \frac{v_S - K_T\Omega_4}{R_S} \tag{6.143}$$

The negative term of this equation incorporates the negative feedback owing to both the motor and the amplifier. The velocity of the motor is

$$\Omega_4 = \frac{T_L}{sJ} = \frac{K_T i}{sJ} = \frac{K_T\beta i_S}{sJ} \tag{6.144}$$

We substitute equation (6.143) into equation (6.144) and solve for

$$\Omega_4 = \frac{K_T\beta v_S/(JR_S)}{s + K_T{}^2\beta/(JR_S)} \tag{6.145}$$

The system is first order, with time constant $\tau = JR_S/(K_T^2\beta)$. We can view the numerator constant $K_T\beta/(JR_S)$ as the *forward gain* (amplification) of the system. The magnitude of the characteristic root $1/\tau$ is the product of the forward gain and the feedback multiplier K_T.

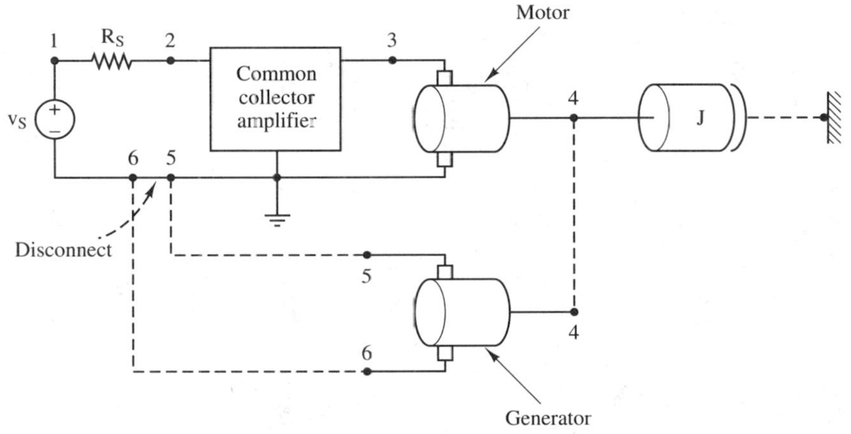

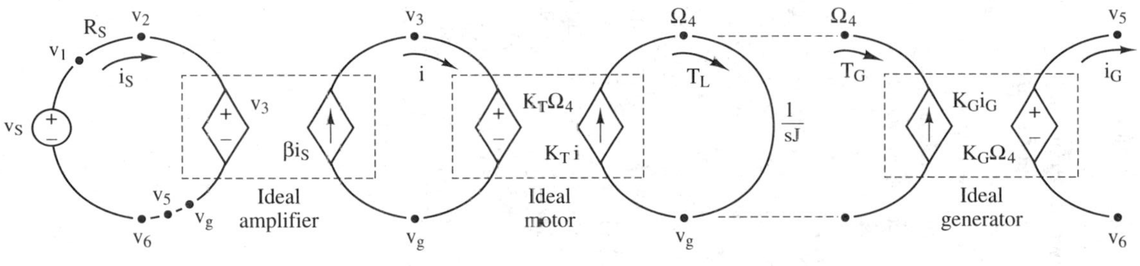

Figure 6.82 An amplifier-driven motor with enhanced feedback. (a) Lumped model. (b) Line-graph representation.

The time constant is a measure of the system's responsiveness. We are limited in our ability to set the value of the time constant by practical choices of (or designs for) the source, amplifier, and load. In this section, we show how additional negative feedback, applied externally, can be used to adjust the time constant.

Let us attach a small electric generator (called a *tachometer*) to the motor shaft, as shown in Figure 6.82(a) and (b). The output of the generator is similar to back emf—a voltage proportional to the velocity of the motor. We assume the generator current i_G is very small; therefore, the generator torque T_G is negligible relative to the load torque T_L, and we can ignore T_G. If we attach the generator output port (terminals 5 and 6) in series with the source at the location shown in the figure, then the voltage generated adds to the back emf and decreases the time constant. Hence, the motor settles more quickly. If we use another transistor to amplify the generator output signal (by a factor A) before we insert it into the source loop, we get a larger decrease in the time constant. If we reverse the orientation of the generator terminals, we provide positive feedback, subtract from the back emf, and increase the time constant. By means of the external feedback, then, we gain complete control over the system's responsiveness.

We represent the motor-driven system of Figure 6.82 by an operational diagram in Figure 6.83(a). The inner feedback loop, obtained from the denominator of equation (6.145), represents the inherent feedback in the motor and common-collector amplifier. The outer feedback loop represents the external (or enhanced) feedback provided by the generator. We can interpret the s-notation to mean either the time derivative or the complex frequency variable. It does not matter whether we represent signals by symbols that denote their values over time or by symbols that denote their Laplace transforms.

The basic negative-feedback structure of operational diagrams is expressed in general transfer-function notation in Figure 6.84. We refer to X_i, X_o, X_f, and X_d as *input, output, feedback,* and *difference* signals, respectively. Since $X_d = X_i - H(s)X_o$ and $X_o = G(s)X_i$, it follows that

$$\frac{X_o}{X_i} = \frac{G(s)}{1 + G(s)H(s)} \tag{6.146}$$

(Equation (6.146) assumes that the feedback is negative; if the feedback were positive, the denominator would become $1 - G(s)H(s)$.)

Equation (6.146) and Figure 6.84 help us to simplify the feedback structures of operational diagrams. For example, suppose we apply equation (6.146) to the inner feedback loop of Figure 6.83(a) to produce the forward transfer function $JR_S/(JR_S s + K_T^2\beta)$ shown in Figure 6.83(b). That forward transfer function contains the inherent time constant $\tau = JR_S/(K_T^2\beta)$ of the motor-driven system. Only the external feedback is left explicit. We can combine the two *forward transfer functions* in Figure 6.83(b) and then apply equation (6.146) a second time to produce the simple forward transfer function shown in Figure 6.83(c). The sequence of diagrams in Figure 6.83 demonstrates that feedback produces the time constant and that the external feedback changes the time constant.

Exercise 6.41: Show that the new time constant after introducing external feedback into Figure 6.83 is the smaller value $\tau = (JR_S)/K_T\beta(K_T + AK_G)$.

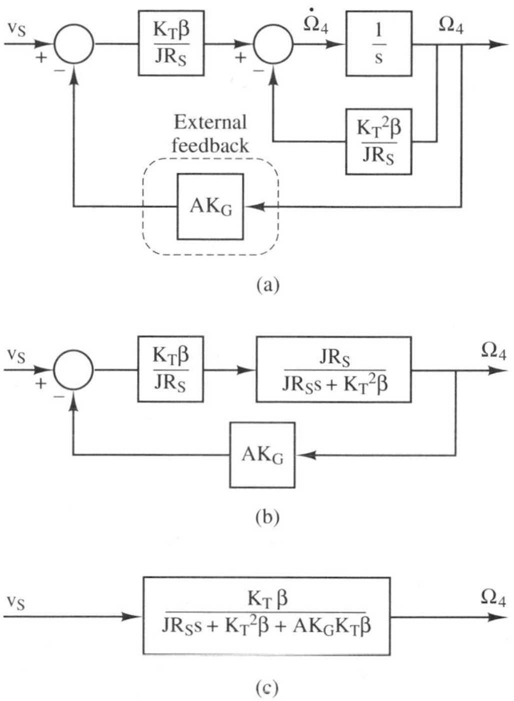

Figure 6.83 Operational diagrams of the system in Figure 6.82. (a) Diagram based on equation (6.145). (b) Inner loop simplified by using equation (6.146). (c) Outer loop simplified by using equation (6.146).

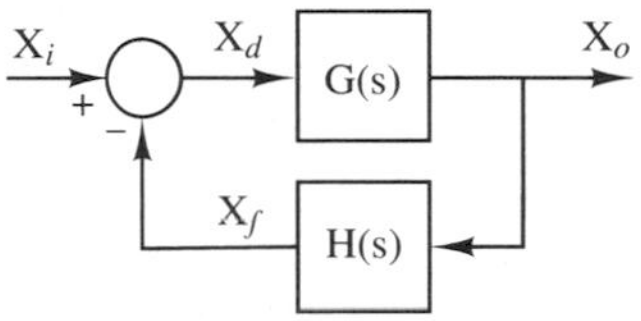

Figure 6.84 A general negative-feedback structure.

Incorporating external feedback into the system changes the behavior of the system. We must distinguish, therefore, between the transfer function before the incorporation of feedback and the transfer function after its incorporation. Let G denote the **forward transfer function,** that is, the input-output transfer function before the incorporation of external feedback. Let H denote the transfer function of the external feedback device. Then GH is the **loop transfer function**—the transfer function from the input to the feedback signal. After the loop is closed (i.e., the feedback signal is subtracted from the input signal), the transfer function is equation (6.146). We call that equation the **closed-loop transfer function.**

Introducing negative feedback with feedback transfer function H(s) around a forward transfer function G(s) has two potential benefits. First, it changes the poles of G(s). Specifically, a system characterized by G(s) is replaced by a system characterized by equation (6.146). This change is beneficial if the poles of equation (6.146) are more desirable than the poles of G(s). Second, negative feedback makes the behavior of the system depend less on the parameters of G(s) by creating a dependence on the parameters of H(s). For example, in the motor-driven system of Figure 6.82, negative feedback reduces the effect on the system behavior of changes in the load J. This effect of feedback is desirable if the parameters of H(s) are more reliable than the parameters of G(s).

In the next section, we show that feedback reduces the sensitivity of the system response to parameter changes in G. We also show that the magnitude of the closed-loop response is much less than that of the system without feedback. This reduction in magnitude must be accounted for during the design of the system.

A potential difficulty associated with negative feedback is that it can make a system less stable or even unstable. For example, the motor in the feedback system of Figure 6.82 is ideal. A physical motor has resistance and inductance, which together add another pole (with a shorter time constant) to the forward transfer function of Figure 6.83(b). For this two-pole system, the external feedback reduces both time constants. It can be shown, however, that if the amplification β is too large, the closed-loop system becomes oscillatory. As the amplification β is increased further, the damping ratio of the pole pair approaches zero. In general, the careless addition of feedback can make a system unstable.

Sensitivity to Parameter Errors

Let the forward transfer function G(s) of Figure 6.84 represent the system in the absence of feedback. That transfer function depends on the system parameters—the lumped-element parameters, amplification factors, etc. (See the motor-driven system in Figure 6.83(b), for example.) Changes or uncertainties in parameter values cause changes or uncertainties in the coefficients in G(s). (For instance, R_S can change with temperature, the load J can deviate from its presumed value, and K_T can diminish for a high motor current.) We refer to such changes or uncertainties as *parameter errors.*

Parameter errors in G(s) cause signal errors in the output signal X_o. We shall show that *feeding back a signal* from X_o *around* G(s) to the input *can reduce the sensitivity* of X_o to parameter errors in G. The feedback transfer function H(s) in Figure 6.84 represents that added feedback. Once we incorporate the feedback represented by H(s), X_o depends on the errors in the parameters of H as well as G. Therefore, to reduce the effect of parameter errors in G, we must use a feedback device that has accurate parameters.

We think of the source signal as a sum of sinusoids of various frequencies, each represented by a phasor (see section 5.4). For each sinusoid, G(s) acts as a complex multiplier $G(j\omega)$ of the corresponding phasor. Parameter errors cause errors in $G(j\omega)$. Typically, each parameter appears as a factor in one of the numerator or denominator coefficients of G. Therefore, a percentage error in a single parameter causes, at worst, the same percentage error in $|G(j\omega)|$. Errors in $G(j\omega)$, in turn, cause signal errors—errors in the system response X_o. First we shall find the sensitivity of X_o to errors in G in the absence of the added feedback, and then we shall find the sensitivities of X_o to errors in both G and H in the presence of the added feedback.

In the original system, without the added feedback, $X_o = GX_i$. The sensitivity of X_o to slight changes in G is the partial derivative $\partial X_o/\partial G$. Let G change by a small amount δG. Then the corresponding change in X_o is

$$\delta X_o = \frac{\partial X_o}{\partial G}\delta G = X_i \delta G \tag{6.147}$$

We divide (6.147) by the input-output relation $X_o = GX_i$ to obtain

$$\frac{\delta X_o}{X_o} = \frac{\delta G}{G} \tag{6.148}$$

That is, the percentage error in the response signal X_o is the same as the percentage error in the original forward transfer function $G(j\omega)$.

After the feedback represented by $H(j\omega)$ is added to the system, the relation between the input X_i and the output X_o is given by equation (6.146). The partial derivatives of X_o with respect to G and H are

$$\frac{\partial X_o}{\partial G} = \frac{X_i}{(1 + GH)^2} \quad \text{and} \quad \frac{\partial X_o}{\partial H} = \frac{-G^2X_i}{(1 + GH)^2} \tag{6.149}$$

Let G and H change by the small amounts δG and δH. Then the corresponding change in X_o is

$$\delta X_o = \frac{X_i\delta G - G^2X_i\delta H}{(1 + GH)^2} \tag{6.150}$$

We divide equation (6.150) by the output X_o, as expressed in equation (6.146), to produce

$$\frac{\delta X_o}{X_o} = \frac{1}{1 + GH}\left(\frac{\delta G}{G} - G\delta H\right) \tag{6.151}$$

If we design the feedback transfer function H so that $|G(j\omega)H(j\omega)| \gg 1$ for the frequencies we are interested in, then the sensitivity of the output signal to errors in G is greatly reduced, $1 + GH \approx GH$, and equation (6.151) becomes

$$\frac{\delta X_o}{X_o} \approx \frac{1}{GH}\frac{\delta G}{G} - \frac{\delta H}{H} \tag{6.152}$$

Thus, the external feedback reduces the effect of a percentage error in G on the percentage error in X_o by the factor $1/(1 + GH)$.

The added feedback reduces the sensitivity of the response to errors in G in inverse proportion to the *loop gain* GH. The feedback transfer function H becomes an additional source of error, however. The sensitivity of the response X_o to errors in H is direct; that is, a percentage error in H causes the same percentage error in X_o. The feedback device (represented by H) must have accurate parameters in order for feedback to reduce the percentage error in X_o.

It is not unreasonable to expect the feedback system to be more accurate than the original system. Let us use the motor-driven system of Figure 6.82 to illustrate this statement. The amplifier and motor must be selected to provide adequate power to the load. The parameters of these two devices typically change with temperature during operation. The feedback device, on the other hand, is a sensor whose only purpose is to observe and report system behavior. Since the sensor need not deliver much power to the system, we can focus on accuracy during its design.

If |GH| $\gg$ 1, then the closed-loop transfer function of equation (6.146) reduces to

$$X_o \approx \frac{X_i}{H} \tag{6.153}$$

and the system behavior is determined by the characteristics of the feedback device alone. Accordingly, to obtain system behavior that is insensitive to variations in the system parameters, we design the system with negative feedback, with a loop gain much greater than 1, and with an accurate feedback device that has a desirable 1/H characteristic. Typically, it is G, which represents the most powerful part of the system, that is used to make |GH| large.

It may seem surprising that by making the most powerful device in the system large in magnitude we make accuracy of that device unimportant. Yet, if a driver has a very powerful car, it is primarily the behavior and skill of the driver (the feedback device) that is apparent; the power of the engine makes it respond well to any command. On the other hand, if the car is underpowered, the skill of the driver cannot compensate for the shortcomings of the engine. In that case, the engine characteristics can be observed in the system behavior.

The price paid for a reduction in the sensitivity of a system is a reduced magnitude of response. According to equation (6.146), the input-output transfer function is reduced in magnitude by the factor 1/(1 + GH). If we use a large loop gain GH to obtain a large reduction in sensitivity, we reduce the output response by the same factor. In practice, then, we must make |H| small enough that |G/(1 + GH)| $\approx$ 1/|H| is sufficiently large.

Operational Amplifiers

Engineers have combined, on a single integrated-circuit chip, a number of transistors connected in series to produce an amplifier that has a very large *voltage* amplification factor ($>10^5$), a very large input resistance (2 MΩ), and a rather small output resistance (75 Ω). It is essentially a high-gain μ amp. The device is called an **operational amplifier**, or *op amp* for short. Operational amplifiers are *direct coupled,* rather than capacitor coupled like the amplifier of Figure 6.59. Therefore, they amplify dc (constant signals). Like most physical systems, an op amp is a low-pass device. Although its voltage gain drops at 20 dB/decade, that gain remains greater than 1 to about 1 MHz. These remarkable features of operational amplifiers make them very versatile system components. Because they are so inexpensive (less than a dollar), it is appropriate to consider their use in all systems. They are particularly appropriate for use in sensing devices and feedback loops, as discussed in the previous section.

A symbolic model for the operational amplifier is shown in Figure 6.85(a). Both positive and negative power supply voltages are required by the internal circuit that biases the transistors. The electrical ground relative to which all signals are measured is provided by that power supply circuit. Since we shall be concerned only with the effects of operational amplifiers on signals, we delete the supply-voltage terminals in the following analyses.

The op amp is a **differential amplifier**, with *ideal* output response

$$v_O = A(v_p - v_n) \tag{6.154}$$

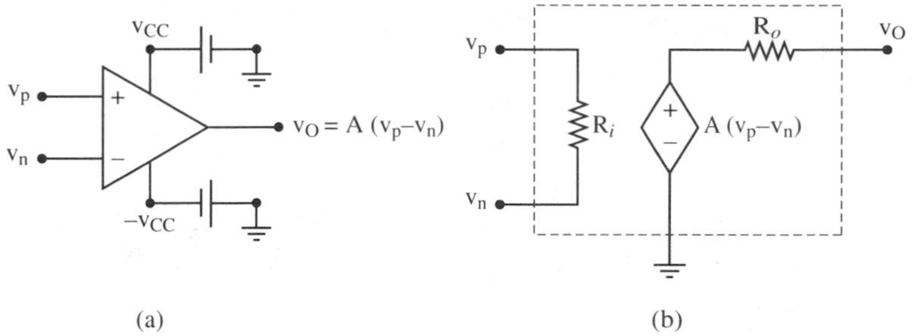

Figure 6.85 Operational amplifier models. (a) Symbolic model. (b) Lumped model.

where A is the voltage amplification factor. (The op amp is described by equation (6.154) only for $-v_{CC} \leq v_O \leq v_{CC}$.) We use the subscripts p and n for the two input signals as a reminder of the positive and negative signs assigned to those inputs in equation (6.154). The two inputs are commonly referred to as the *inverting input* (v_n) and the *noninverting input* (v_p). If one of the inputs is not used, it must be grounded (given a zero signal). A realistic lumped circuit model for the op amp is given in Figure 6.85(b).

Exercise 6.42: Show that the two-port parameter matrices for the operational amplifier just described are

$$[\mathbf{z}] = \begin{bmatrix} R_i & 0 \\ AR_i & R_o \end{bmatrix}, \quad [\mathbf{h}] = \begin{bmatrix} R_i & 0 \\ -AR_i/R_o & 1/R_o \end{bmatrix}, \quad [\mathbf{a}] = \begin{bmatrix} 1/A & R_o/A \\ 1/AR_i & R_o/AR_i \end{bmatrix} \tag{6.155}$$

Op Amp Feedback Circuits

Figure 6.86 shows an operational amplifier feedback circuit that acts as a precise voltage amplifier. The input signal is v_p (relative to ground). Because A is very large, and because v_O cannot exceed v_{CC}, equation (6.154) implies that $v_{pn} \approx 0$; 0.1 mV is typical. Since the input resistance R_i of the op amp is very large, $i_{in} \approx 0$; 5 μA is typical. Since the output resistance R_o of the op amp is very small, it usually can be ignored. Since i_{in} is small, we can usually ignore R_S as well.

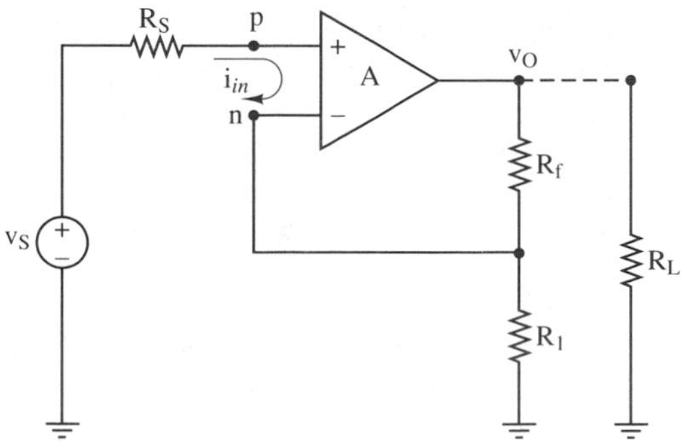

Figure 6.86 A precise voltage amplifier.

Replace the op amp symbol by the lumped model. The voltage divider at the output provides the feedback signal

$$v_n = \frac{R_1}{R_1 + R_f} v_O \tag{6.156}$$

to the input. The feedback is negative because it is applied to the inverting input. (The very small input current i_{in} does not noticeably affect the feedback voltage.) The circuit has the feedback structure of Figure 6.84, with $G = A$ and $H = R_1/(R_1 + R_f)$. According to equation (6.146),

$$\frac{v_O}{v_S} = \frac{A}{1 + AR_1/(R_1 + R_f)} = \frac{R_1 + R_f}{(R_1 + R_f)/A + R_1} \approx \frac{R_1 + R_f}{R_1} \tag{6.157}$$

where the approximation assumes that the loop gain $AR_1/(R_1 + R_f)$ is much greater than 1. Notice that the closed-loop transfer function is essentially the inverse of the feedback transfer function, as indicated in equation (6.153). It does not depend on the large, but uncertain, op amp gain A. If accurate resistors are used for the voltage divider, the voltage gain of the amplifier will be accurate.

The circuit provides for a wide range of voltage amplification factors through an appropriate choice of resistance values. The voltage gain can be made adjustable by using a *potentiometer*—a resistor with a moveable center tap—in place of R_f and R_1. Of course, R_1 and R_f must be much smaller than R_i, the large uncertain input resistance of the op amp, in order that the input current not affect the feedback voltage. Also, R_f must be much larger than R_o, or else we must include the unreliable value of R_o in the final amplification factor. The effects of R_i and R_o are examined in problem 6.49 at the end of the chapter. That problem also shows that this amplifier circuit has nearly infinite input resistance R_I (from point p to ground) and very low output resistance R_O (from point O to ground). The latter feature is important because we want the amplifier to retain the behavior described by equation (6.157) when we attach a load R_L as shown in Figure 6.86. If $R_O \ll R_L$, then R_L does not affect the behavior of the circuit.

An *inverting* feedback circuit is shown in Figure 6.87(a). As in any op amp circuit, $v_{pn} \approx 0$ and $i_{in} \approx 0$. Therefore, the currents in the impedances z_1 and z_f must sum to zero at node p:

$$\frac{v_1}{z_1} + \frac{v_O}{z_f} = 0 \tag{6.158}$$

Then

$$v_O = -\frac{z_f}{z_1} v_1 \tag{6.159}$$

By choosing various impedances for z_f and z_1, we can make this circuit implement various mathematical operations electrically. Note that the output voltage is inverted relative to the input voltage.

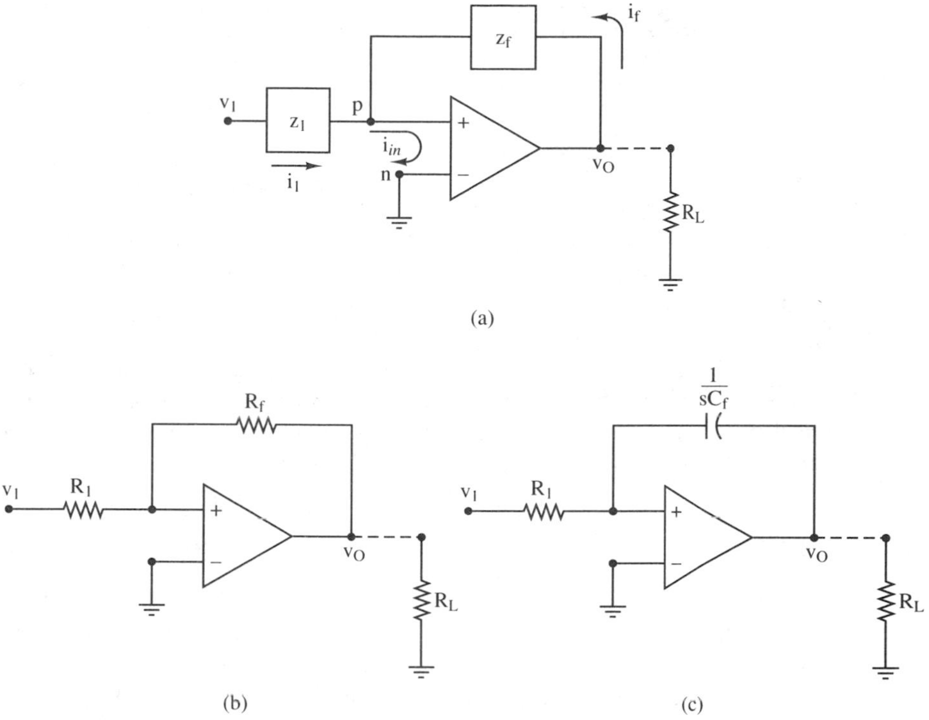

Figure 6.87 Inverting feedback circuits. (a) General inverter circuit. (b) Inverting-scaling circuit. (c) Inverting integrator.

Exercise 6.43: Show that the circuit of Figure 6.87(a) retains the input-output relation (6.159) if the two op amp input terminals are reversed. In other words, the sign cannot be changed by reversing the terminals.

Let both impedances be resistors, as shown in Figure 6.87(b). Then the op amp feedback circuit *inverts* the source signal and multiplies it by the scale factor R_f/R_1. That scale factor can be less than 1 or greater than 1, depending on the sizes chosen for the resistors. The op amp provides the power necessary for voltage amplification. We can set the voltage amplification factor accurately by using precision resistors. As long as the gain A of the op amp is large, its value is unimportant. If $R_f = R_1$, the circuit merely inverts the input signal.

Now let the feedback element be a capacitor, as shown in Figure 6.87(c). Then the op amp circuit *inverts* the source signal, multiplies it by the scale factor $1/(R_1C_f)$, and integrates it with respect to time. Consequently, the circuit is an inverting integrator. The inversion could be removed by cascading the inverting integrator with a simple sign inverter.

The op amp circuit of Figure 6.88 is known as a **buffer**. The output voltage is essentially the same as the input voltage. For this reason, it is also called a *voltage follower*. The advantage of the circuit is that the input resistance R_I is nearly infinite and the output

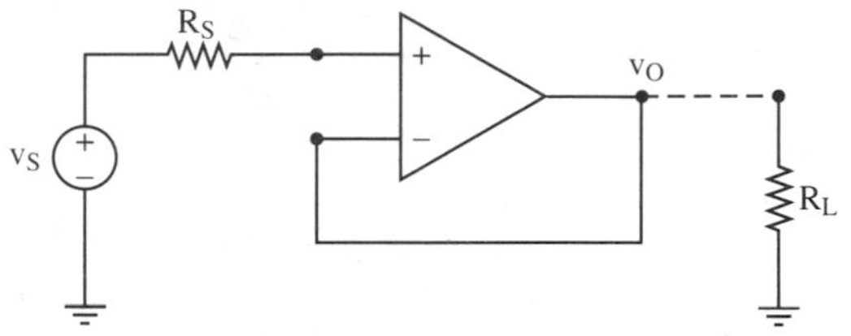

Figure 6.88 A buffer circuit.

resistance R_O is nearly zero. The circuit is used to isolate a source from a load. It allows a low-impedance load R_L to consume whatever power it requires without drawing significant power from the source v_S. This circuit is examined in problem 6.50 at the end of the chapter.

Problems 6.51 and 6.52 examine op amp feedback circuits that convert voltages to currents and currents to voltages. The circuits behave as gyrators and can be thought of as impedance converters. For example, changing a voltage source to a current source is the same as changing a low-impedance source to a high-impedance source.

6.7 SUMMARY

Two-port devices connect and match sources and loads. Each *ideal* transformer or transducer is characterized by a converter constant. An *ideal* amplifier is characterized by an *amplification factor.* An *ideal* transporter has a characteristic impedance. If the load impedance matches the characteristic impedance, the ideal transporter is characterized by a *delay* that is proportional to the length of the transporter. If the load impedance is not matched, signal reflections appear at the load with a period equal to twice the delay.

Physical two-port devices have input and output impedances that are neither zero nor infinite. As a consequence, their behaviors are degraded relative to ideal devices. We account for nonideal behavior by attaching passive lumped storage elements and dissipation elements to the ideal-device models.

We can describe a simple linear two-port device or a complicated linear two-port subnet with three different kinds of parameters: immittance parameters (z or y), hybrid parameters (g or h), and transmission parameters (a or b). Certain of the parameter sets may be especially appropriate or inappropriate for a particular two-port device or subnet. Each set of parameters corresponds to one of the 2×2 matrix equations (6.106)–(6.114). These matrix equations assume that the flow at each port is in the direction of the potential drop at the port.

Each immittance-parameter matrix and hybrid-parameter matrix corresponds directly to a simple equivalent lumped subnet (see Figure 6.77). These equivalent subnets aid in understanding the device or subnet that they represent. The parameters on the diagonal of the matrix are positive and can be interpreted as single-port impedances or admittances. The off-diagonal parameters describe port-to-port relations. The signs of the off-diagonal parameters are affected by the relative orientations of the port variables.

Inserting a two-port device or subnet between a source and load transforms the way the source and load *appear* to each other. Adding external feedback to a system can reduce the sensitivity of the system's signals to variations in parameters and provides the ability to move the poles to desirable locations. Engineers make extensive use of two-port devices and feedback principles to produce desireable features in the systems they design.

The thermal systems and population systems we modeled at the beginning of the chapter have much in common with the systems discussed in earlier chapters. They also have notable differences. Experience with modeling and analysis of these classes of systems greatly expands our ability to deal with new differences as we attempt to modei still other types of systems.

6.8 PROBLEMS

6.1 For one of the linear lumped thermal networks of Figure P6.1:

(a) Define and label the node temperatures and heat flows. Draw a line-graph representation of the network.

(b) Find a set of system equations.

(c) Solve for the values of the variables. (Network simplification techniques and superposition can be used if they are helpful.)

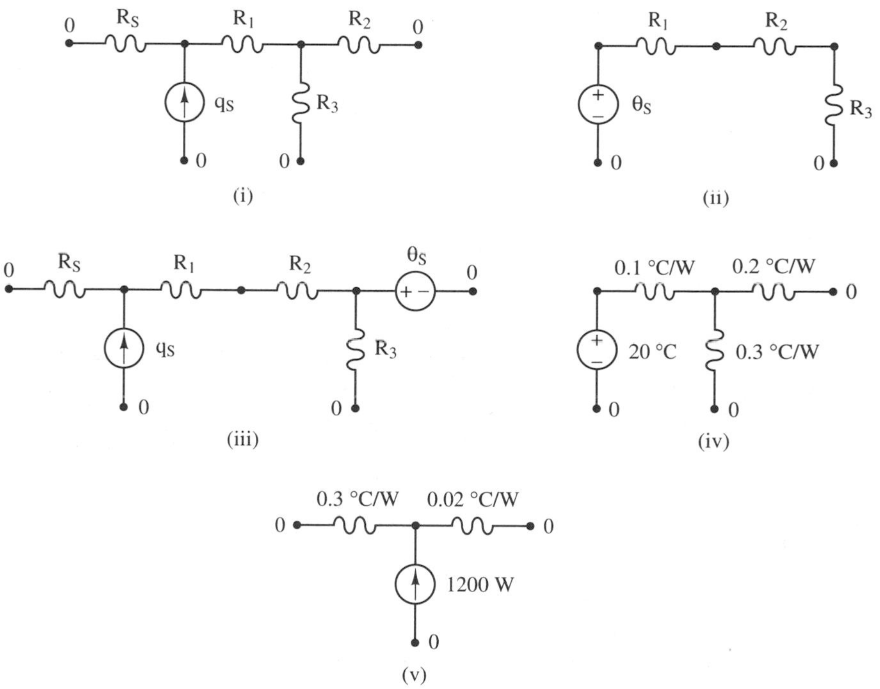

Figure P6.1

6.2 For one of the first-order linear lumped thermal networks of Figure P6.2:

(a) Find a set of system equations.

(b) Find the input-output system equation for the temperature of node 1. Network simplification techniques can be used if they are helpful.

(c) Find the zero-state response $\theta_1(t)$ for an abruptly applied source that has a constant value.

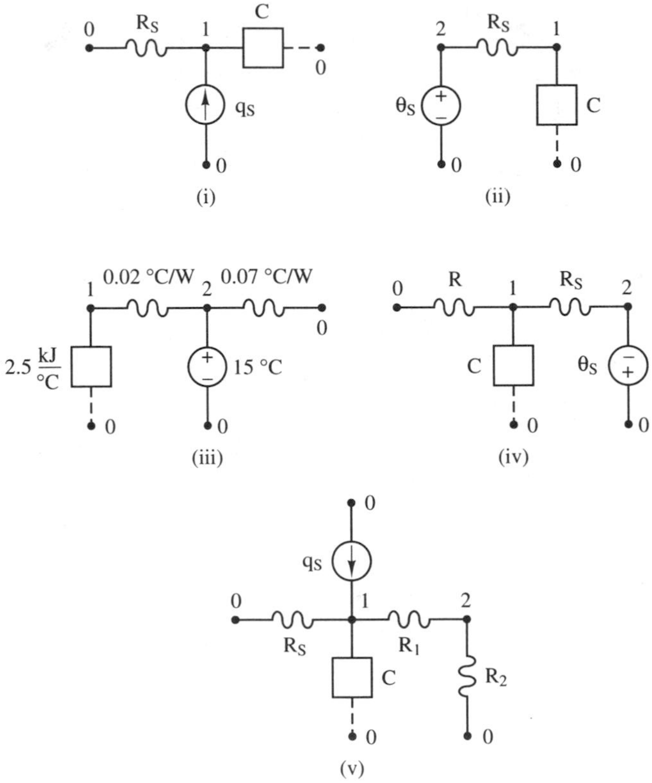

Figure P6.2

6.3 For one of the second-order linear lumped thermal networks of Figure P6.3:

(a) Find a set of system equations.

(b) Find the input-output system equation for the temperature of node 1.

(c) Suppose the network is initially in the zero state and we turn on the source abruptly to a constant value denoted by the subscript *c*. Find the initial value and the final value of v_1. Can you show that the response must be overdamped?

6.4 Derive a *simple* lumped thermal model that describes the main features of one of the following situations or systems. Include a model for the source that energizes the system. Estimate the values of all lumped-element parameters. State your assumptions.

(i) A hot loaf of bread sitting on a table.
(ii) A hot cup of coffee sitting on a table.
(iii) A glass of water and ice sitting on a table.
(iv) An egg frying on the stove.
(v) A pan of water boiling on the stove.
(vi) A potato baking in the oven.
(vii) Walking briskly on a cold day.
(viii) A room being heated by an electric space heater.
(ix) A small room full of people.

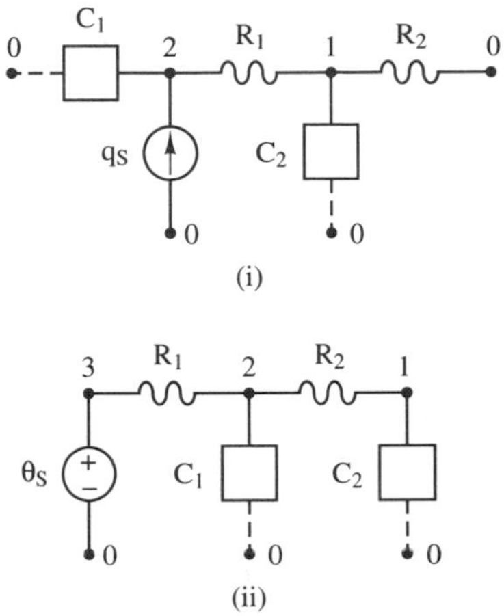

Figure P6.3

6.5 Hot water from a boiler at temperature 80°C is carried at a flow rate of $Q = 0.2$ liter/s via a 3 cm (outside diameter) copper pipe to heat a small building. The pipe travels 30 m through outside air that has temperature 5°C. Figure P6.5 shows a 4-segment steady-state distributed thermal model for the portion of the pipe that passes through the outside air. The resistances (denoted R) account for heat loss from the pipe to the outside air. The dependent sources (labeled rq_i) account for the temperature drop along the pipe owing to that heat loss, as discussed in example problem 6.2.

(a) Calculate R and r under the assumption that the pipe loses heat to the outside air by natural (free) convection. Find the outlet temperature θ_5 and the rate of heat loss to the outside (node 6).

(b) It is proposed to enclose the pipe in a 3 cm thick glass wool jacket to reduce the heat loss. Calculate the new value of R. Find the new rate of heat loss to the outside air. By what fraction is the heat loss reduced?

6.6 In example problem 6.2 we derive a lumped thermal model for a hot-water heating system. In that problem we state that the time constant for transfer of heat from the hot water in the heat exchanger to the fins of the heat exchanger is about 2 s.

(a) Assume the pipe and fins are perfect conductors of heat. Make a lumped model of a 2.67 m segment of the heat exchanger that accounts for the capacitance of the pipe and fins and the resistance to convective heat transfer. Use the data from example problem 6.2 and appropriate constants of materials from appendix A to determine the values of the parameters of your model. Assume that the hot water is inserted into the heat exchanger instantly, that the temperature of the water is held constant (owing to circulation of the hot water), and that no heat transfers from the fins to the room air during the time the pipe and fins are heating.

(b) Use your model to determine the time constant for transfer of heat from the water to the fins.

6.7 In example problem 6.2 we derive a distributed lumped thermal model for a hot water heating system. Figure 6.9 shows that the temperatures in the two rooms respond almost identically

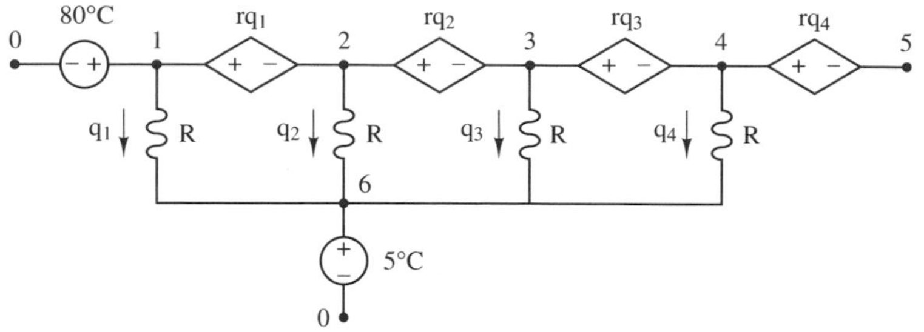

Figure P6.5

to the hot water flow. Therefore, we should be able to make a simpler thermal model for the system.

(a) Make a first-order lumped model for the system. Represent the two rooms by a single node. Represent the heat exchanger by a single node of temperature 75°C. Use the parameter values from the distributed model to determine the parameter values for the first-order model.

(b) Use the model found in part (a) to find, analytically, the room temperature vs. time if that temperature is initially 15.5°C. Compare your result to Figure 6.9.

6.8 In the United States, insulating materials for building construction are rated by an *R number*. The R number is the thermal resistance of the material in Btu/hr/°F per square foot of surface area.

(a) Calculate the R number for glass wool insulation that is 6 inches thick.

(b) Determine the SI equivalent of the U.S. R number. That is, determine the resistance in kw/°C per square meter of surface area for the same 6-inch thickness of glass wool.

6.9 Estimate the steady-state rate of cooling required to maintain the interior of an automobile at 21°C (70°F) on a cloudy day if the outdoor temperature is 32°C (90°F). Assume the automobile contains no people. Treat the interior of the automobile as an empty rectangular box of dimensions 1.8 m × 1.4 m × 1.1 m. Assume that heat transfers from the exterior air to the car body and from the car body to the interior air by *forced* convection and that the thermal resistance of the car body is zero.

6.10 Estimate the cooling load (the required rate of heat removal) owing to one person in an air conditioned space that is maintained at 21°C (70°F). Assume the person has 1 m^2 of skin area that is maintained at a temperature of 32°C (90°F). Assume that the person is covered with clothing that traps a layer of air that averages 3 mm in thickness. Heat transfers from the surface of the clothing to the surrounding air by free convection.

6.11 Estimate the rate of heat loss from a 0.1 m^2 area of skin surface (the approximate area of a bare head) that is maintained at 32°C (90°F) if the temperature of the surrounding air is 0°C under each of the following conditions:

(a) The skin is bare and the air is still.

(b) The skin is covered with a 5 mm layer of clothing and the air is still.

(c) The skin is covered with a 5 mm layer of clothing and the wind is blowing hard.

6.12 Figure P6.12 shows a simple lumped model of a cattle herd. Node 1 represents the size of the adult herd. The herd is half cows (females) and half steers (castrated males). Node 2 represents the size of the calf population. Let αN_1 denote the number of calves produced in a year. We shall assume $\alpha = 0.5$. The manager chooses the number βN_2 of calves to enter the herd the following year. The rest of the calves, $(1 - \beta)N_2$, must be sold or destroyed. The manager must also choose the number γN_1 of cattle to harvest each season. It is through this harvest that the herd produces profit. Thus, β and γ are the decision parameters through which the manager controls herd size and profits. The manager increases the size of the herd when the market is good and decreases it when the market is bad.

(a) Write the system equations for the lumped network. Express them as a pair of difference equations that relates the state (N_1, N_2) at the end of year n + 1 to the state at the end of year n.

(b) Let the initial state be $(N_0, 0)$. Find an expression for the state at the end of year n in terms of the initial state and the decision parameters β and γ. (Hint: This expression can be kept simple if the equations are written in the state-variable form discussed in section 3.4.)

(c) Suppose that we keep β and γ constant from year to year. It can be shown that for a stable herd (constant size from year to year), $\gamma = \alpha\beta = \beta/2$. What happens to the herd if we choose $\gamma > \beta/2$ or $\gamma < \beta/2$? Can you find the steady-state size of the herd in terms of γ?

(d) Suppose we wish to double the size of the herd. Use the line graph and the mathematical equations to *explore* the following issues: How should we set the parameters β and γ to achieve that doubling? What will it cost us to double the size? How long will it take?

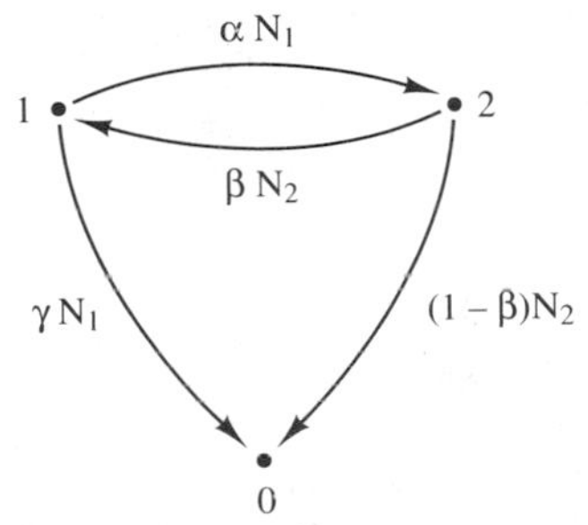

Figure P6.12

6.13 Simplify the three-region rabbit-population model of Figure 6.11 by eliminating region 3. Make the population capacitors and one-way elements implicit by removing them from the line graph. Simplify the notation for the network parameters by defining $\alpha_1 \triangleq \alpha_{12}$, $\alpha_2 \triangleq \alpha_{21}$, $\beta_1 \triangleq d_1 - b_1$, and $\beta_2 \triangleq d_2 - b_2$. Use the simplified line graph and simplified notation to analyze the behavior of the network.

(a) Write a set of system equations for the network and find the input-output system equation for N_1, the population at node 1. Note that there is no input function. The system response is precipitated by the initial state. What is the order of the system?

(b) Find the characteristic equation of the network as a function of α_1, α_2, β_1, and β_2. Assume that the characteristic roots are real. What system behavior would be associated with a positive root? What would be the behavior if all the roots were negative? Find the relations among the network parameters α_1, α_2, β_1, and β_2 in order that the largest root be exactly zero. Interpret the system behavior that would be associated with the network under these conditions.

(c) Suppose the system is in a steady state; that is, the network parameters satisfy the relation that makes the largest root exactly zero. Find the relation between the steady-state values of N_1 and N_2. Note that the steady-state populations are not unique.

6.14 Make a simple population model of the following situation. (Do not put explicit capacitors and one-way devices in the model.) Let node 1 represent the amount of money that a person has in a bank account and node 2 the amount of money invested in the stock market.

(a) Attach branches to represent interest credited to the bank account and dividends credited to the stock account.

(b) Attach branches to represent daily expenses paid from the bank account and losses charged against the stock account.

(c) Attach branches to represent income from employment, decisions to invest further in the stock market, and selling of stocks to meet personal financial needs.

(d) Determine which branch parameters are determined by market forces and which represent decision variables that can be controlled by the owner. Formulate a question concerning personal financial policy in the context of the model.

6.15 Simplify the three-region rabbit-population model of Figure 6.11 by eliminating region 3. Remove the population capacitors and one-way elements to simplify the line graph. Simplify the notation for the network parameters by defining $\alpha_1 \triangleq \alpha_{12}$, $\alpha_2 \triangleq \alpha_{21}$, $\beta_1 \triangleq d_1 - b_1$, and $\beta_2 \triangleq d_2 - b_2$.

(a) Make a model for fox populations in the same two regions. The model should be similar to the rabbit-population model. Denote the corresponding fox populations by N_3 and N_4.

(b) Propose a simple mathematical form for the population-limiting relations between the rabbit-network parameters α_1, α_2, β_1, and β_2 and the fox populations. Define analogous relations between the fox-network parameters and the rabbit populations.

(c) Find the nonlinear equations that determine the steady-state populations. How could you solve them?

(d) Optional: Select specific values for the network parameters and the initial populations and use a SPICE-based network simulator to determine the behavior.

6.16 For one of the following converters:

(a) Derive the equations that describe the behavior of the ideal converter;

(b) Identify a single converter constant and determine its units;

(c) Draw and label a dependent-source line-graph model of the converter.

- **(i)** The hydraulic lever of Figure 6.16(d).
- **(ii)** The hydraulic intensifier of Figure 6.16(e).
- **(iii)** The gear of Figure 6.16(b).
- **(iv)** The wheel of Figure 6.17(b).
- **(v)** The displacement machine of Figure 6.17(e).
- **(vi)** The electric heater of Figure 6.17(f).

6.17 For one of the converters in Figure P6.17:

(a) Define positivity for each port variable;

(b) Derive the equations that describe the behavior of the ideal converter;

(c) Compare the equations found in (b) with the equations for the comparable converter from Figures 6.16 and 6.17.

6.18 For one of the following converters, attach additional elements to the ideal lumped converter model to account for physical limitations of an actual device. State any assumptions that you make.

- **(i)** The hydraulic lever of Figure 6.16(d).
- **(ii)** The hydraulic intensifier of Figure 6.16(e).
- **(iii)** The gear of Figure 6.16(b).
- **(iv)** The wheel of Figure 6.17(b).
- **(v)** The displacement machine of Figure 6.17(e).
- **(vi)** The electric heater of Figure 6.17(f).

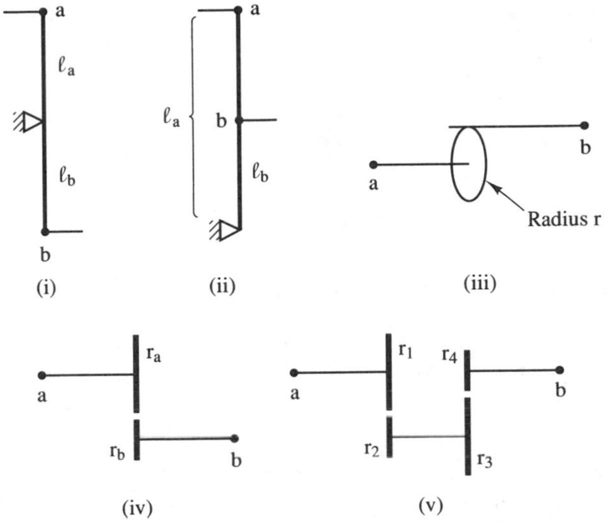

Figure P6.17

6.19 Make a lumped model for one of the following devices. State your assumptions.

(i) A hypodermic needle.
(ii) The drive train of a motorcycle (engine to rear wheel to road).
(iii) A gear pump driven by an electric motor.
(iv) A loudspeaker driven by an electrical source.
(v) A hydraulic brake (foot pedal, hydraulic line, and brake piston).
(vi) A hand-cranked windlass.
(vii) A lever-operated can crusher.

6.20 For one of the lumped networks in Figure P6.20:

(a) Draw a line-graph model;
(b) Use the model to find the input-output system equation for the velocity v_b;
(c) Solve the system equation to find v_b in terms of the source signal and the lumped element parameters.

6.21 For one of the lumped networks of Figure P6.20:

(a) Draw a line-graph model of the network.
(b) Find the Thévenin equivalent representation of the source as *seen* by the load. Represent the simplified equivalent network in terms of translational lumped elements.
(c) Use the simplified equivalent network from part (b) to find the input-output system equation for the velocity v_b.
(d) Find the equivalent impedance of the load as *seen* by the source at the input to the converter. Represent the original source and the simplified equivalent load by lumped elements in the appropriate energy domain.

6.22 For one of the lumped networks of Figure P6.22:

(a) Draw a line-graph model of the network.
(b) Transform the load-side elements into equivalent source-side elements. Redraw the transformed line graph as a lumped model in the appropriate energy domain.
(c) Transform the source-side elements of the line graph into equivalent load-side elements. Redraw the transformed line graph as a lumped model in the appropriate energy domain.

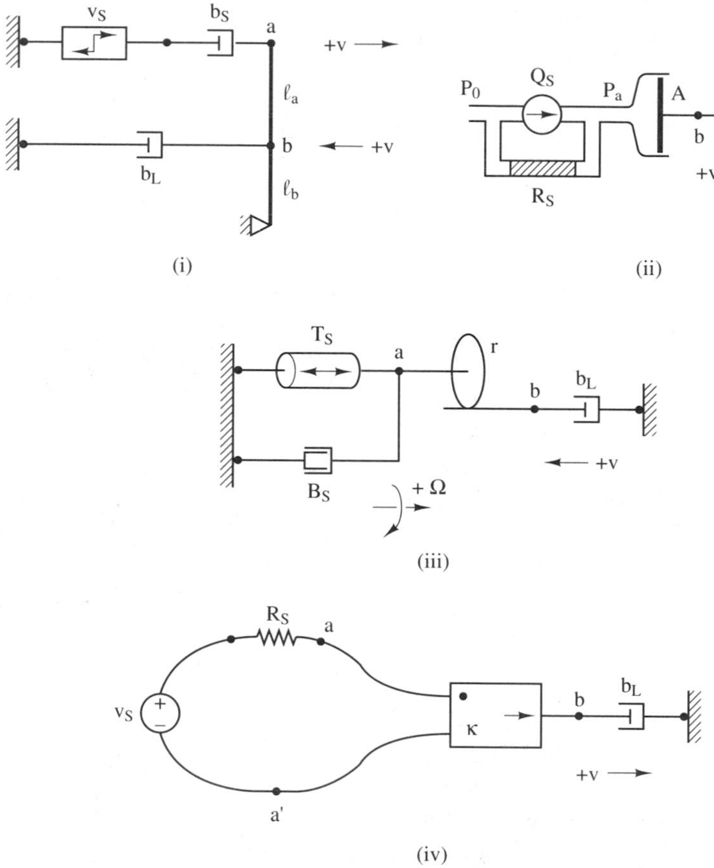

Figure P6.20

6.23 A lumped model and a corresponding line-graph representation for a motor-driven system are given in Figure P6.23; J_m is the inertia of the motor armature and the attached primary-side gear; n is the gear ratio; K_T is the *torque constant* of the motor ($T_m = K_T i$).

(a) Find the load impedance seen by the *ideal* motor at its output shaft. That is, find the Thévenin equivalent of the subnet attached to the ideal motor. According to the equivalent impedance expression, what is the effect of the gear on the inertias in the network?

(b) Suppose the motor and load are initially stationary and the source signal v_S is an abruptly applied constant v_c. Find the initial load acceleration $\dot{\Omega}_5$. Find the gear ratio n that maximizes the initial load acceleration. (This value of n produces *inertial matching.*)

(c) Transform all elements of the network to the load side of the gear. Represent the equivalent network in terms of rotational lumped elements. What is the effect of the gear as displayed in this equivalent model?

6.24 Figure P6.24 shows a lumped translational network with a lever.

(a) Draw a line-graph representation of the network.

(b) Find the equivalent impedance of the leveraged load as *seen* by the source at node 2. Draw the equivalent translational lumped model.

(c) Suppose the source signal v is a step of size v_c. Find the steady-state values of F_1 and x_2 in terms of b, k, ℓ_2, ℓ_3, and v_c.

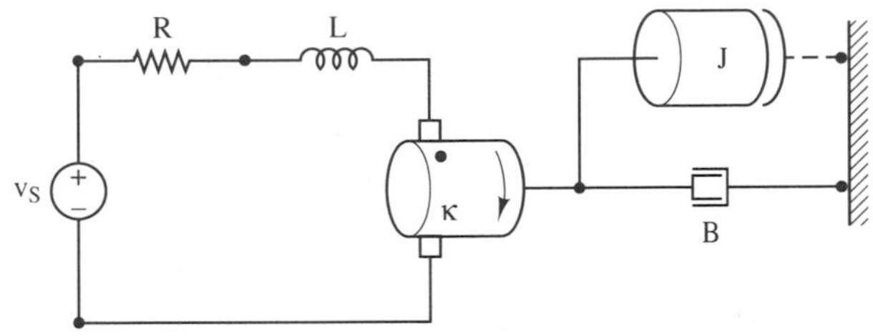

(i)

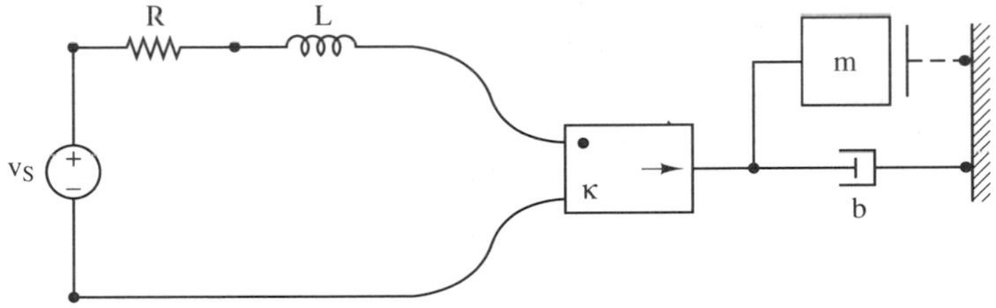

(ii)

Figure P6.22 (Continued on next page)

6.25 A simple lumped model for an ac electrical power generation system is shown in Figure P6.25. All signals in the system are sinusoidal with frequency ω and can be expressed in phasor notation. Find the Thévenin equivalent representation of the power generation system as *seen* by the electrical load.

6.26 Each wheel of a subway car is powered by a separate dc electric motor, as shown in Figure P6.26(a). Each motor accelerates one quarter of the car mass. A lumped linear model of a single-wheel drive motor with its inertial load is shown in Figure P6.26(b). (Although it is the car and motor that move, it appears to the motor as if it is the rail that moves. A mass m equal to one quarter the mass of the car seems attached to the circumference of the wheel. The total inertia J that must be accelerated by the motor is the inertia J_a of the motor armature plus the effective inertia J_L of the apparent mass attached to the wheel.) The following measurements were made on a single-wheel system:

(i) With the car brakes locked, we applied $v_{12} = 500$ V and measured $i = 1000$ A.

(ii) With the voltage source disconnected, we moved the car (by means of the other three wheels) at 45 m/s. This speed corresponded to a motor speed of $\Omega_3 = 104$ rad/s. At this speed we found that the motor generated the voltage $v_{12} = 800$ V.

(iii) We applied $v_{12} = 1200$ V abruptly to all motors of a stationary car. The car eventually reached a steady velocity of 90 m/s ($\Omega_3 = 208$ rad/s). The time required to reach 63.2% of that speed was 1 s.

(a) Write the equations that describe the lumped model of the single-wheel drive system.

(b) Use the equations from part (a) together with the measurements to find the parameters R and κ of the motor and the total effective inertia J of the model.

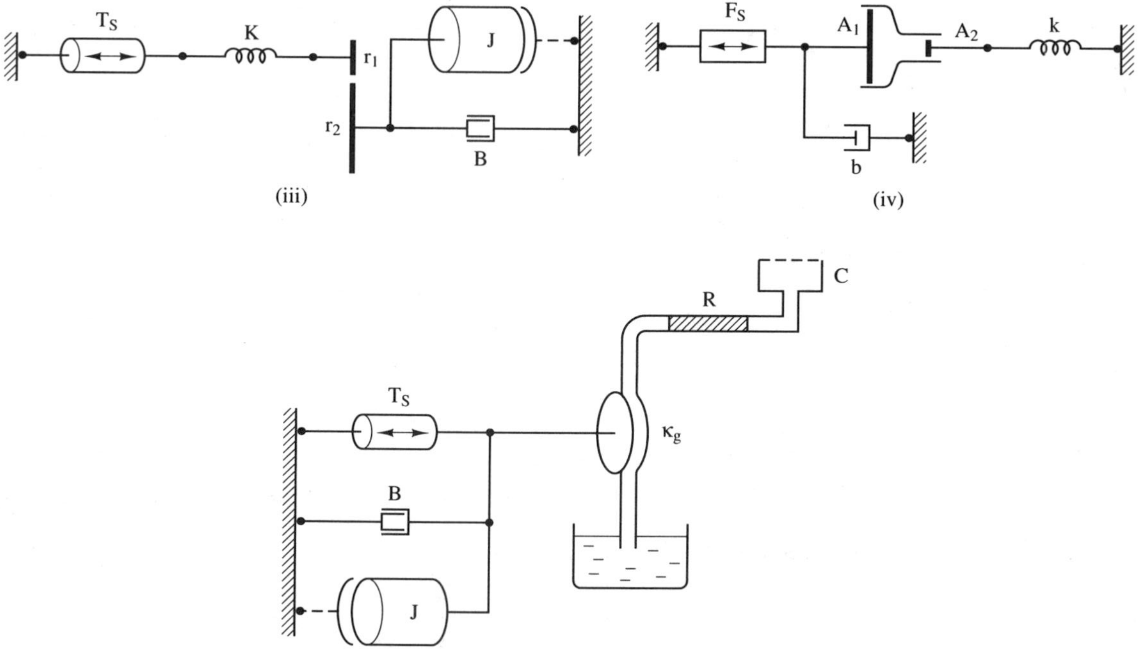

(v)

Figure P6.22 (Continued)

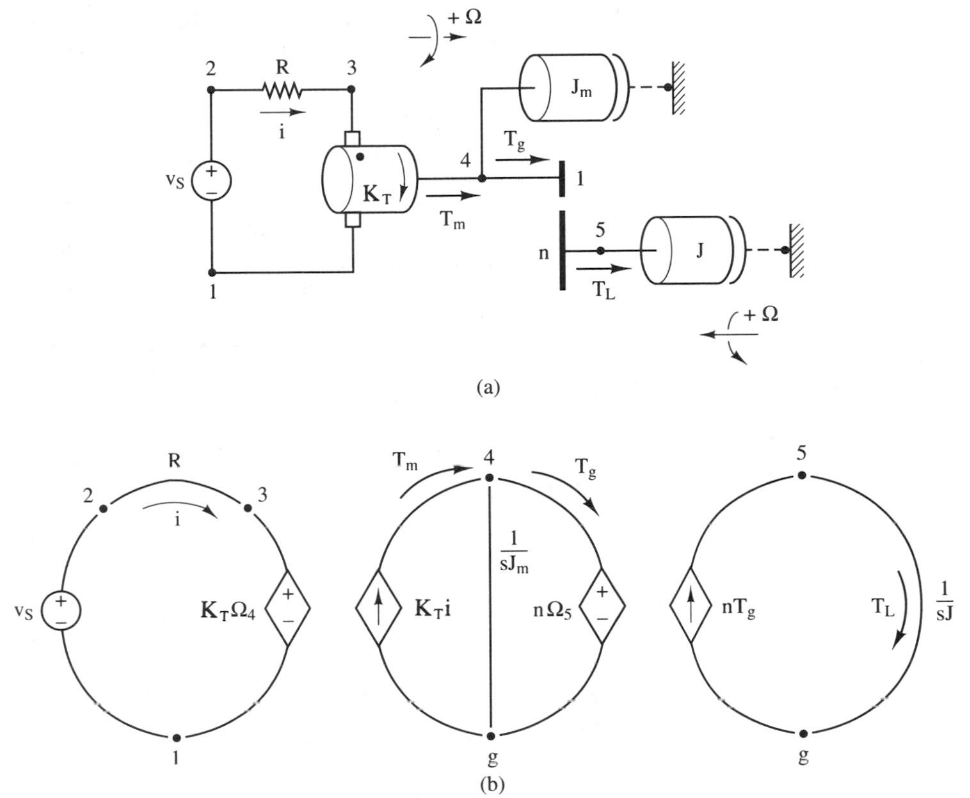

Figure P6.23 (a) Lumped model. (b) Line-graph representation.

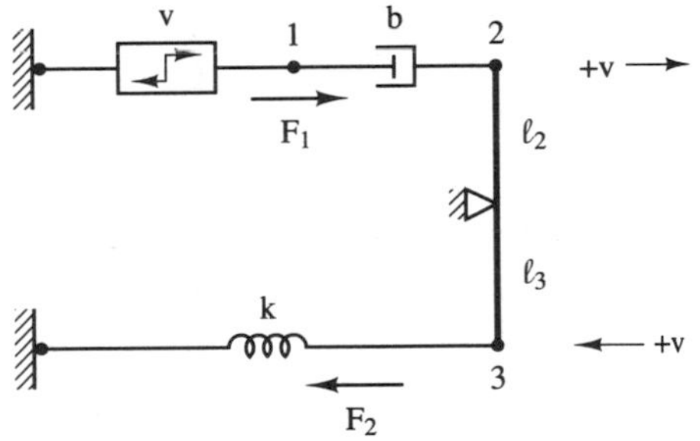

Figure P6.24

6.27 The lumped model in Figure P6.27 represents part of the brake system of an automobile. The source pressure P_S is derived from the foot of the driver. Find the input-output system equation for the flow Q in the hydraulic line.

6.28 Figure P6.28 shows an electric generator that is driven by a rotational mechanical source. The symbol K_G is the *generator constant* ($v_{24} = K_G\Omega_1$).

(a) Find the input impedance z_{in} of the generator—the equivalent mechanical impedance seen by the source at the input to the generator.

(b) Draw the mechanical lumped model that corresponds to the impedance found in (a). That is, replace the generator and electrical load of the figure by the equivalent mechanical subnet.

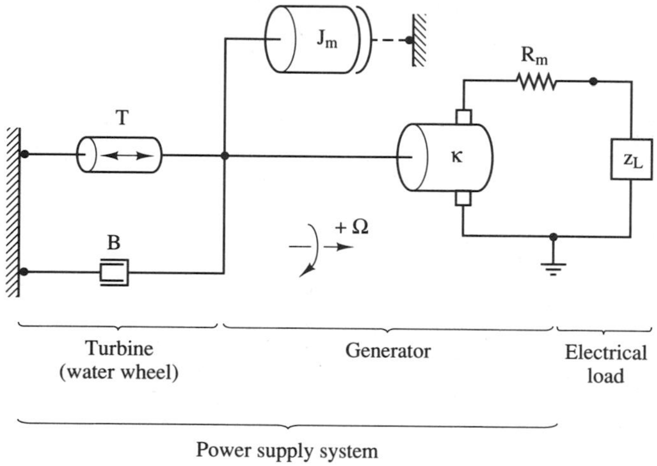

Figure P6.25

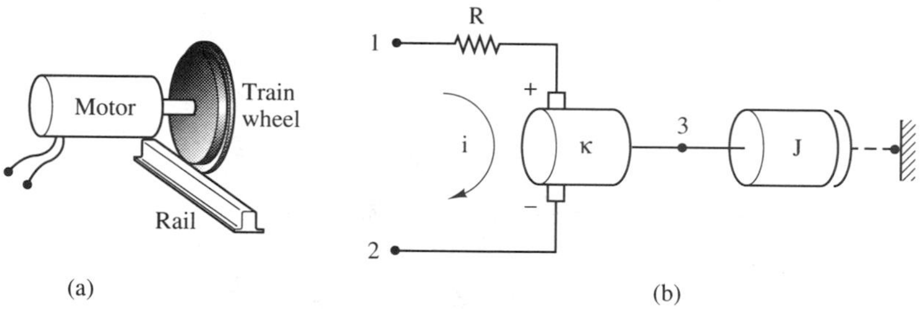

Figure P6.26 (a) Pictorial model. (b) Lumped model.

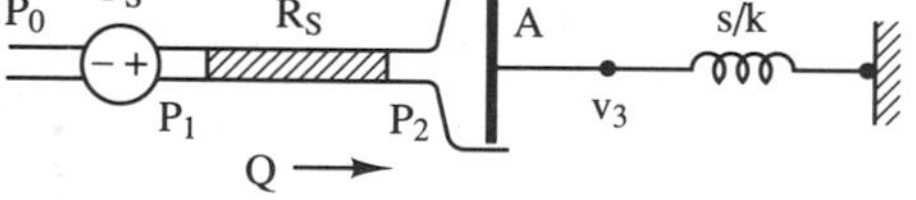

Figure P6.27

6.29 Replace the β amp of Figure 6.47 by an r_t amp. Find the power gain G_P of that amplifier in terms of r_t and the resistances in the network.

6.30 Replace the β amp of Figure 6.47 by a μ amp. Replace the flow source and parallel resistance by a potential-difference source and series resistance. Find the power gain G_P of the amplifier in terms of μ and the resistances in the network.

6.31 Suppose the signal at the output of the electrical transformer of Figure 6.50 is $v_{b2} = 14 \sin(377t)$ volts and the resistive load to the ac-to-dc converter is $R_L = 5\ k\Omega$. The diode in the circuit does not permit the current i to be negative. When $v_{b1} > 0$, the diode acts as a re-

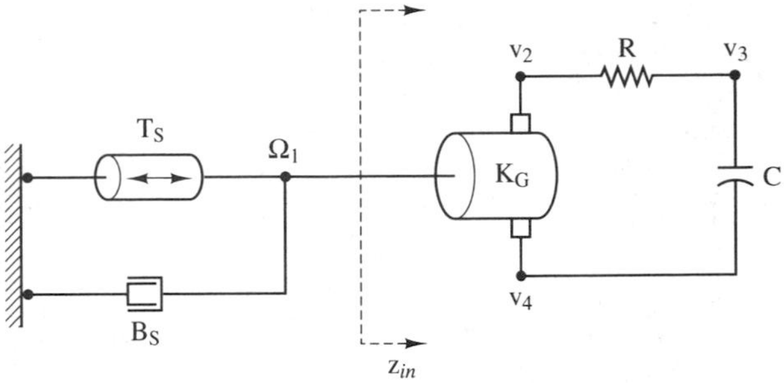

Figure P6.28

sistor $R_D = 500\ \Omega$, the current i is positive, and some of that current charges the capacitor C. When $v_{b1} < 0$, the diode acts as an open circuit and the capacitor delivers part of its charge to the load resistor R_L. Hence, the capacitor smooths the current that flows through R_L, and the ac signals at the input to the transformer are converted to dc signals at the load. We say that the ac signal has been *rectified*.

(a) Calculate the time constant τ_C for the interval in which the diode is conducting and the capacitor is charging.

(b) Calculate the time constant τ_D for the interval in which the diode is open circuited and the capacitor is discharging.

(c) Choose a value for the filter capacitor C which will cause that capacitor to charge quickly during the charging interval and discharge very little during the discharging interval.

(d) Do either task (i) or task (ii) under the assumption that the system has reached steady state, that is, the capacitor has the same charge at the beginning of each discharge cycle.

(i) Use analytical means to find one cycle of the waveform of v_{12}.

(ii) Use a SPICE-based network simulation program to find the steady-state waveform of v_{12}.

6.32 Figure 6.59 shows a common-emitter amplifier circuit. Use a lumped transistor model with the parameters $R_b = 1500\ \Omega$, $\beta = 50$, and $R_c = 189\ k\Omega$. The source signal v_S is sinusoidal. Assume the coupling capacitors (denoted C_1 and C_2) have negligible impedance at the source frequency.

(a) Find the power absorbed by the transistor during *quiescent* operation. That is, find the power delivered by the circuit to the ports (2, g) and (3, g) of the transistor for $v_S = 0$.

(b) Find the input resistance of the amplifier as *seen* by the source. Find the output resistance of the amplifier as *seen* by the load.

(c) Find the load voltage v_4 and load current i_L as functions of the *source signal* voltage v_S. Compare the values of v_4 and i_L to the values that would be obtained if the source were connected directly to the load. (Hint: Ignore the coupling capacitors. Use the superposition principle to ignore the constant voltage sources. Then the circuit reduces to the one in Figure 6.60.)

(d) Find the amplifier current gain i_L/i_S, voltage gain v_4/v_2, and power gain $v_4 i_L/v_2 i_S$.

6.33 Let the source for the amplifier circuit of Figure 6.59 be a tape deck and let the load be a loudspeaker. The impedance of a loudspeaker is typically $8\ \Omega$, a much smaller value than the $2\ k\Omega$ load resistance of that circuit.

(a) Find the power gain v_4/v_S of Figure 6.59 for the transistor parameters shown in the signal model of Figure 6.60, but for $R_S = 5\ k\Omega$ and $R_L = 8\ \Omega$. How does that power gain compare to the 33 dB power gain that we found for Figure 6.60?

(b) The great reduction in power gain found in part (a) is owing primarily to the severe mismatch between the load resistance and the output resistance of the amplifier. One remedy for such a situation is to insert an electrical transformer between the amplifier and load to change the effective load resistance. Let us insert an *ideal* 40-to-1 transformer in front of R_L at port (4, g). Show that the *effective* load resistance *seen* by the amplifier is 12.8 kΩ.

(c) Repeat the power gain calculations of part (a), but with $R_L = 12.8$ kΩ. What has insertion of the transformer done to the power gain?

6.34 The transistor amplifier of Figure P6.34(a) drives a loudspeaker. The *signal* model for the system is shown in Figure P6.34(b). This signal model ignores the constant transistor bias sources, assumes that the capacitor impedance is small enough to ignore for the signal frequencies of interest, and assumes that the bias resistors (R_B and R_C) and the transistor output resistance (R_c) are large enough to neglect. R is the resistance of the speaker coil; m and b are the mass and air resistance of the speaker. κ is the speaker constant.

(a) Find the equivalent impedance of the output subnetwork as *seen* by the transistor at the port (3, g).

(b) Find the transfer function and the differential equation that relate the speaker velocity v_5 to the source voltage v_S.

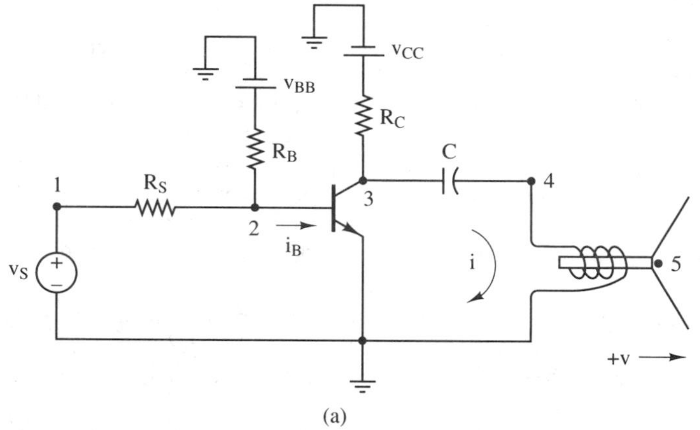

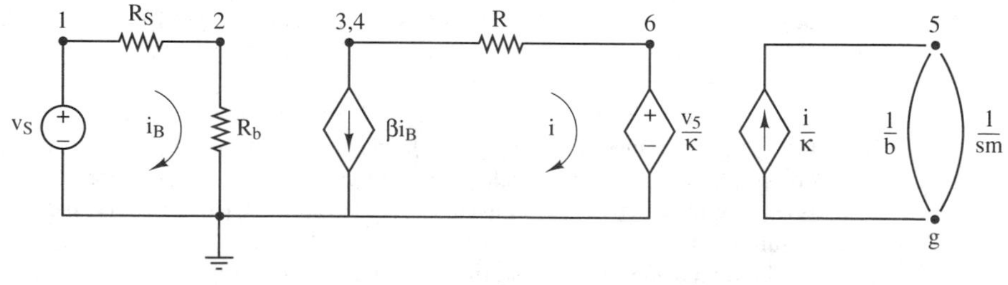

Figure P6.34 (a) An audio amplifier. (b) The signal model.

6.35 Figure 6.60 shows a *signal model* for the common-emitter amplifier circuit of Figure 6.59. The parameter values are given in the figures. The signal model is appropriate for sinusoidal source signals v_S that are in the audio range (50 Hz to 20 kHz). For frequencies that are much higher, the transistor model should also include the capacitor $C_{BC} = 9$ pF between the base and the collector and the capacitor $C_{BE} = 80$ pF between the base and the emitter.

(a) Redraw the signal model with the two transistor capacitances attached.

(b) Find the input-output transfer function $H(s) \triangleq v_4/v_S$, as a function of the circuit parameters.

(c) Insert the parameter values in the transfer function found in part (a). Then sketch the asymptotic log-magnitude frequency response plot for v_4. Determine the upper frequency limit of the amplifier.

(d) What circuit elements determine the lower frequency limit of the amplifier? (Hint: See Figure 6.59.)

(e) Optional: Use a SPICE-based network simulator to find the frequency response of the whole circuit of Figure 6.59.

6.36 Equations (6.72) and (6.73) describe the bipolar-junction transistor in the common-emitter configuration. Let us ignore the junction voltage v_{BE0}; that is, we set it to zero in equation (6.72).

(a) Rearrange those equations to derive the equations (6.74) that describe the transistor in the common-base configuration. (Hint: Use the transistor node equation $I_B + i_E + i_C = 0$ and the transistor loop equation $v_{BC} + v_{CE} + v_{EB} = 0$ to put the common-emitter equations in terms of the common-base port variables v_{EB}, i_E, v_{CB}, and i_C. Use the parameter values noted in Figure 6.54 to determine which terms of the equations can be neglected.)

(b) Draw a lumped linear model that corresponds to the simple approximate version of the common-base equations.

(c) Use the lumped model found in part (b) to find the input impedance R_i and the output impedance R_o, as shown for the common-base transistor in Figure 6.54.

6.37 Equations (6.72) and (6.73) describe the bipolar-junction transistor in the common-emitter configuration. Let us ignore the junction voltage v_{BE0}; that is, we set it to zero in equation (6.72).

(a) Rearrange those equations to derive the equations (6.75) that describe the transistor in the common-collector configuration. (Hint: Use the transistor node equation $i_B + i_E + i_C = 0$ and the transistor loop equation $v_{BC} + v_{CE} + v_{EB} = 0$ to put the common-emitter equations in terms of the common-collector port variables v_{BC}, i_B, v_{EC}, and i_E. Use the parameter values noted in Figure 6.54 to determine which terms can be neglected.)

(b) Draw a lumped linear model that corresponds to the simple approximate version of the common-collector equations.

(c) Attach a source (v_S in series with R_S) and a load R_L to the lumped model found in part (b). Then find the input impedance R_i and the output impedance R_o, as shown for the common-collector transistor in Figure 6.54.

6.38 The amplifier circuit shown in Figure P6.38(a) shows the symbol that is customarily used to represent a *junction field-effect transistor* (JFET). The three terminals of the JFET are called the *source,* the *drain,* and the *gate,* labeled S, D, and G, respectively. The input resistance (gate to source) of the JFET is nearly infinite. The steady-state operating characteristics for the output port of the JFET are displayed in Figure P6.38(b).

(a) Superimpose the operating characteristic of the output-port bias network (determined by v_{DD} and R_D) on the operating characteristics of the JFET output port to determine the *quiescent* operating point ($v_S = 0$).

(b) Construct a lumped linear model for the JFET that is appropriate for operation in the neighborhood of the operating point found in (a). Determine the parameters of that model.

Represent the near-infinite input impedance by an open circuit. (The model for the output port must include a constant current source because i_D is not zero when $v_{GS} = 0$.)

(c) The coupling capacitor C prevents the load resistor from affecting the quiescent operating point. Define the *voltage gain* of the amplifier to be v_3/v_S. Suppose v_S is sinusoidal with frequency high enough that the sinusoidal impedance of C is negligible. Use the linear model to determine the voltage gain of the amplifier. (Hint: Set the constant sources to zero, short circuit the coupling capacitor, and find the response of the linear lumped model to the signal source v_S alone.) What is the current gain, i_L/i_S?

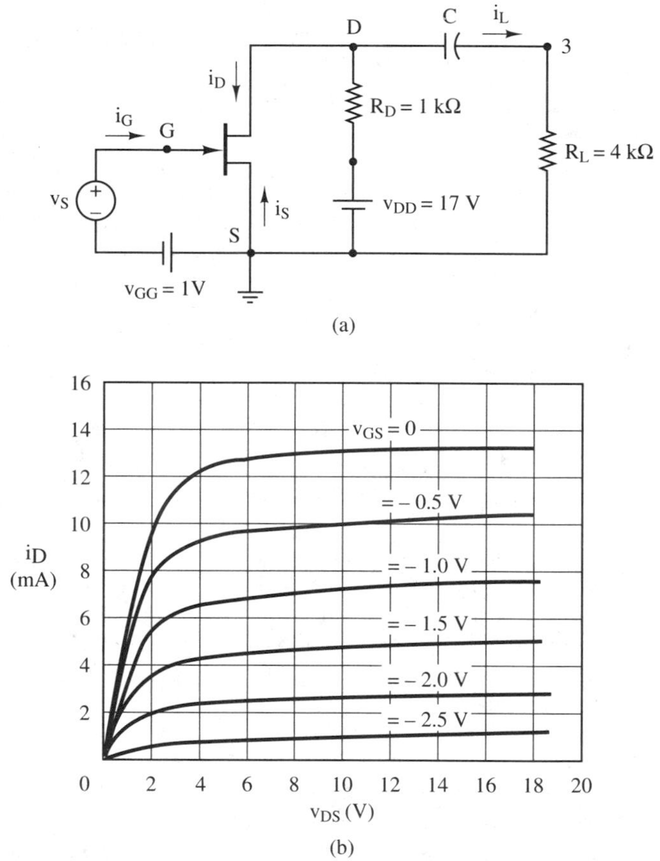

Figure P6.38 (a) A JFET amplifier. (b) JFET operating characteristics.

6.39 Figure 6.64 shows a lumped model for the pressure-flow-control valve for which the operating characteristics are given in Figure 6.63(c). Let us use that control valve to drive two parallel hydraulic cylinders that compact the contents of a trash truck. The operator uses a lever to adjust the valve input current. The diameter of each power piston is 7 cm. The fluid capacitance at the output port of the control valve owing to compressibility of the hydraulic fluid and flexibility of the connecting hoses is $C = 6 \times 10^{-11}$ m^5/N. Represent the resistance of the trash to compaction by a linear damper with friction constant b in parallel with a linear spring of stiffness k.

(a) Make a linear lumped model of the system.
(b) Find the input-output system equation for the displacement of the power piston in one of the hydraulic cylinders, with current as the input variable.
(c) Find the *available* steady-state power at the pistons if the input current for the control valve is set to its maximum value (18 mA).
(d) Find the time constant for acceleration of the power pistons in terms of b and k. What is the no-load time constant ($b = 0$ and $k = 0$).

6.40 Figure 6.63 shows the steady-state output-port operating characteristics for three types of hydraulic control valves. We derived a lumped model for the pressure-flow-control valve in Figure 6.64. For *one* of the following valve types:

(a) Derive a lumped model for the control valve.
(b) Determine the parameters values for the model found in part (a).
(c) Use the model to find the *available* output power for the control valve when the input current is at its specified maximum.
  (i) The flow control valvc of Figurc 6.63(a).
  (ii) The pressure control valve of Figure 6.63(b).

6.41 Figure 6.66 shows a linear lumped model of a gas turbine. That model is derived from the steady-state performance characteristics in Figure 6.65. We use the turbine to drive a dc electric generator. The generator energizes a dc motor that powers the drive shaft and propellors of an ocean vessel.

(a) Make a lumped model of the system. Include the inertia J of the turbine and the attached generator rotor. Also include the inertia J_2 of the drive shaft and propellors. Assume that the generator and motor are ideal (with constants K_G and K_M) and that the propellor friction can be represented by a rotary viscous damper B_P attached to the drive shaft. Treat the fuel flow rate Q_{fuel} as an independent input.
(b) To simplify the lumped model, replace all elements by their equivalents as *seen* at the turbine shaft. Find the system equation for the turbine velocity Ω. Find the system time constant. Find the steady-state turbine velocity as a function of fuel flow rate.
(c) Suppose the propellor friction is, instead, described by a quadratic torque-speed relation of the form $T = \eta\Omega^2$. Transfer that relation to the turbine shaft to find the new system equation for the turbine velocity.

6.42 Figure 6.67 shows the full-throttle performance characteristic for a 3.8 liter spark-ignition engine.

(a) Suppose the engine is operating at full throttle at 4000 rpm. Make a linear approximation to the operating characteristic at that operating point. Write the equation that describes that linear approximation and draw the corresponding linear lumped model of the engine. Determine the values of the parameters of the model.
(b) When the throttle is fully closed, the engine torque is surely zero. Presume that the torque is proportional to the fraction of throttle opening. Convert the source in the lumped model of part (a) to a dependent source that is proportional to the throttle opening. Draw a set of *linearized* operating characteristics for the engine to correspond to the dependent-source model. Over what region of operation (on the torque-speed plot) would you expect the model to be useful?

6.43 The lumped model of Figure P6.43 represents a computer connected to a printer by a coaxial cable. We use the ideal delay structure of Figure 6.72(b) to represent both ports of the cable. Use of a delay model presumes that the computer signals change rapidly enough that all frequency components are in the high-frequency range for the cable.

(a) Use the lumped model to find v_2 as a function of v_S and the model parameters. Recall that z_0 is a real number (a resistance).

(b) Let $R_S = R_L = z_0$ in the solution to part (a) to obtain the result when the source resistance and load resistance match the characteristic impedance of the cable.

(c) Interpret the solution found in part (a). (Hint: Use long division to show that $1/(1 + ae^{-2ds}) = 1 - ae^{-2ds} + a^2e^{-4ds} - \ldots$; multiplying a signal by the factor e^{-ds} corresponds to delaying the signal by d seconds. Therefore, this infinite series produces a sequence of delayed signals.)

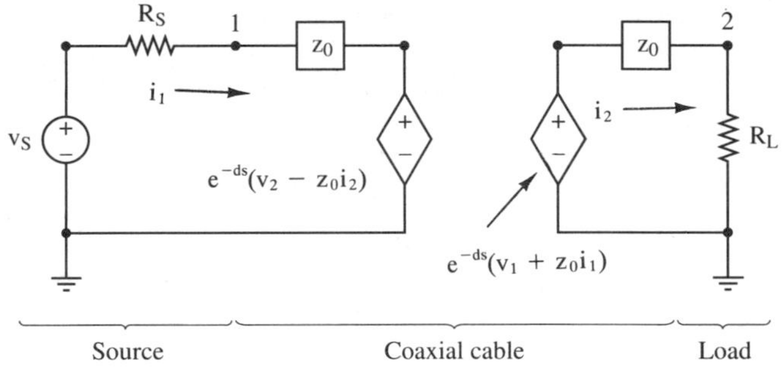

Figure P6.43

6.44 Use a SPICE-based network simulator with an *ideal delay model* to compute the voltage output of an electrical coaxial cable. The parameters of the cable are shown in Table 6.3. Specifically, the characteristic impedance is $z_0 = 49\ \Omega$ and the characteristic delay (for a 2 m cable) is $d = 10$ ns. Let the source signal be the computer-like input signal shown in Figure P6.44. Compute the responses for the following cases:

(a) Impedances matched at both ends: $R_S = R_L = z_0 = 49\ \Omega$.

(b) Mismatched load: $z_0 = 49\ \Omega$, $R_S = 49\ \Omega$, and $R_L = 5\ \Omega$.

(c) Mismatched source: $z_0 = 49\ \Omega$, $R_S = 5\ \Omega$, and $R_L = 49\ \Omega$.

(d) Mismatched at both ends: $z_0 = 49\ \Omega$, $R_S = 5\ \Omega$, and $R_L = 5\ \Omega$.

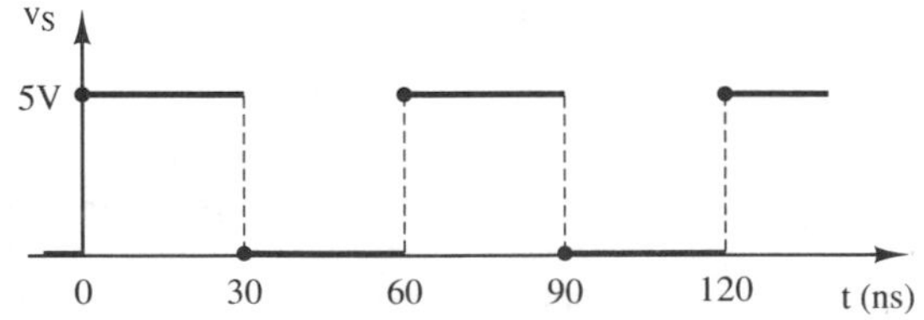

Figure P6.44

6.45 Figure 2.41 shows a distributed model for a toy Slinky. Suppose we insert an ideal velocity source between the Slinky and the support from which it hangs. Then the response of the Slinky is the superposition of the steady-state response owing to the gravity sources and the dynamic response owing to the velocity source. Let us examine the portion of the response owing only to the velocity source. The parameters for the Slinky are shown in Table 6.3. In particular, the characteristic impedance of the Slinky is $z_0 = 2.2$ s/kg and the characteristic delay is $d = 0.552$ s per Slinky length. The source resistance (corresponding to the ideal velocity source) is $R_S \equiv 1/b_S = 0$. The load resistance (corresponding to the free end of the hanging Slinky) is $R_L \equiv 1/b_L = \infty$.

The Slinky model is initially in the zero state. That is, the masses are not moving and the springs are unstressed (except for the static hang which we associate with the separately handled gravity sources). Let the velocity source signal execute the steady-state *displacement* pattern shown in Figure P6.45.

(a) Use the ideal delay model to find *intuitively* the resulting motion at the free end of the Slinky for $0 \le t \le 3$ s.

(b) Use a SPICE-based network simulation program to compute the response for the same interval.

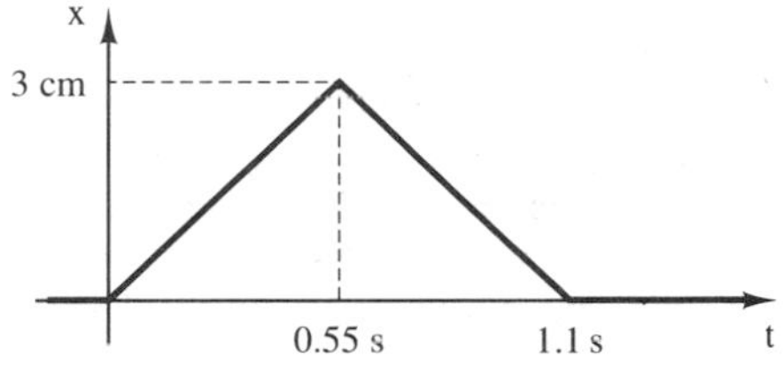

Figure P6.45

6.46 For each of the following configurations of a bipolar junction transistor:

(a) Find the h-parameter representation.

(b) Find the a-parameter representation.

(c) Draw the h-parameter line graph.

(i) The common-emitter configuration; equations (6.72) and (6.73);

(ii) The common-base configuration; equations (6.74);

(iii) The common-collector configuration; equations (6.75).

6.47 Figure 6.59 shows a capacitor-coupled transistor amplifier. When analyzing the effect of the amplifier on source signals, we can set all constant sources to zero and short circuit the coupling capacitors. View the resulting amplifier model as a cascade of three two-port segments: The base-bias segment, the common-emitter transistor segment, and the collector-bias segment.

(a) Find the a-parameter representations of each of the three segments of the amplifier.

(b) Multiply the **[a]** matrices for the three cascaded segments to get the **[a]** matrix for the amplifier as a whole. Write the a-parameter equations for the amplifier in matrix form.

(c) Express the equation for the source subnet of Figure 6.59 in terms of the source-side vector of the a-parameter equation. Use the load equation to express the load-side vector of the a-parameter equation in terms of the load current i_L.

(d) Substitute the vector expressions from part (c) into the matrix expression of part (b) to obtain a relation between the source signal v_S and the load current i_L. Carry out the matrix operations to obtain the input-output system equation for i_L.

6.48 For *one* of the following systems, find a set of two-port parameters in transmission form (**[a]** or **[b]**) and another set in one other appropriate form.

(i) The hydraulic control valve of Figure 6.64.
(ii) The gas turbine of Figure 6.66.
(iii) The electric transformer of Figure 6.25(c).
(iv) The electric motor of Figure 6.23(c), ignoring coulomb friction.

6.49 Suppose we want the noninverting amplifier of Figure 6.86 to provide the precise voltage gain $v_O = 1000v_S$. According to equation (6.157), we need only set $R_f = 999R_1$, regardless of the value of R_1. However, equation (6.157) was derived using the ideal model for the op amp, with input resistance $R_i = \infty$ and output resistance $R_o = 0$.

(a) Use the op amp model of Figure 6.85(b) to rederive the voltage gain v_O/v_S of the amplifier circuit of Figure 6.86.
(b) Find the output resistance R_O of the circuit as *seen* by the load resistor R_L.
(c) Find the input resistance R_I of the circuit as *seen* by the source at node p.
(d) A typical op amp has the parameters $R_i = 2$ MΩ, $R_o = 75$ Ω, and $A = 200{,}000$; R_L, R_S, and R_1 are usually *on the order of* 1 kΩ. Use these typical values to guide elimination of terms from the answers to parts (a)–(c). What is the practical effect of the op amp parameters R_i and R_o on these quantities?

6.50 The circuit of Figure 6.88 acts as a *buffer*. Because it has extremely high input impedance and extremely low output impedance, it isolates the source and the load.

(a) Use the ideal op amp model ($R_i = \infty$ and $R_o = 0$) in the buffer circuit to find the voltage gain v_O/v_S, the input resistance R_I presented to the source, and the output resistance R_O presented to the load.
(b) Use the op amp model of Figure 6.85 in the buffer circuit to find the voltage gain v_O/v_S, the input resistance R_I presented to the source, and the output resistance R_O presented to the load.
(c) A typical op amp has the parameters $R_i = 2$ MΩ, $R_o = 75$ Ω, and $A = 200{,}000$; R_L and R_S are usually *on the order of* 1 kΩ. Use these typical values to guide elimination of terms from the answers to part (b). What is the practical effect of the op amp parameters R_i and R_o on these quantities? What input and output resistances are achievable in a buffer?

6.51 The circuit in Figure P6.51 is intended to convert a source voltage to a proportional load current. Use the ideal op amp model ($R_i = \infty$ and $R_o = 0$) to find the input-output relation, i_L/v_S. Must R_1, R_S, or z_L be restricted in any way for the circuit to work properly?

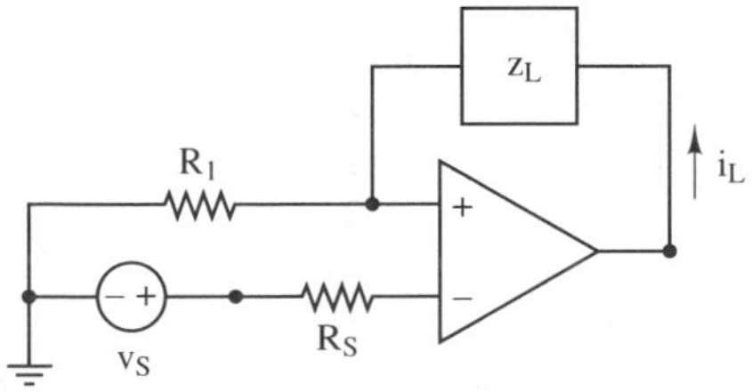

Figure P6.51

6.52 The circuit in Figure P6.52 is intended to convert a source current to a proportional load voltage. Use the ideal op amp model ($R_i = \infty$ and $R_o = 0$) to find the input-output relation, v_O/i_S. Are there any restrictions on the sizes of R_S, R_f, or R_L for the circuit to work properly?

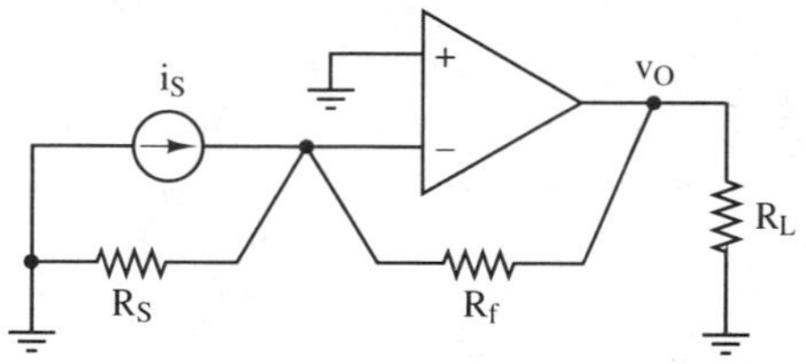

Figure P6.52

6.9 REFERENCES

6.1 Cannon, Robert H., Jr. *Dynamics of Physical Systems.* New York: McGraw-Hill, 1967.

6.2 Forrester, Jay W. *Principles of Systems.* Cambridge, MA: Wright-Allen Press, 1968.

6.3 Forrester, Jay W. *World Dynamics.* Cambridge, MA: Wright-Allen Press, 1971.

6.4 Nilsson, James W. *Electric Circuits.* Reading, MA: Addison-Wesley, 1983.

6.5 Ruston, H., and Bordogna, J. *Electric Networks: Functions, Filters, Analysis.* New York: McGraw-Hill, 1966.

APPENDIX

A

Material Properties and Conversion Factors

Unit Prefixes			Physical Constants
giga	G	10^9	Atmospheric pressure = 0.1 MPa (at sea level)
mega	M	10^6	Gravitational acceleration g = 9.8 m/s^2 (at sea level)
kilo	k	10^3	Speed of light c = 3×10^8 m/s
milli	m	10^{-3}	Permittivity constant ϵ_0 = 8.85 pF/m
micro	μ	10^{-6}	Permeability constant μ_0 = 1.26 μH/m
nano	n	10^{-9}	Stefan-Boltzmann constant σ = 56.7 $nW/m^2 \cdot K^4$
pico	p	10^{-12}	

UNIT CONVERSIONS

Length	Velocity
1 in = 2.54 cm	1 mph = 0.447 m/s
1 ft = 0.305 m	1 kph = 0.278 m/s
1 mile = 1.61 km	1 knot = 0.514 m/s
	1 ft/s = 0.305 m/s

Angle	Angular Velocity
1° = 17.5 mrad	1 rpm = 0.105 rad/s

UNIT CONVERSIONS *(continued)*

Volume	Volume Flow Rate
1 liter = 0.001 m^3	1 gal/min (U.S.) = 0.063 liter/s
1 gallon (U.S.) = 3.79 liter	1 ft^3/min = 0.472 liter/s
1 cubic foot = 28.3 liter	1 m^3/s = 1,000 liter/s
1 barrel (petroleum) = 159 liter	
1 acre·ft = 1.23×10^5 liter	

Energy	Power
1 J = 1 N·m	1 W = 1 J/s
1 ft·lbf = 1.36 J	1 Btu/hour = 0.293 W
1 cal = 4.19 J	1 hp = 746 W
1 Btu = 1.06 kJ	
1 kWh = 36 MJ	

Mass	Force	Pressure
1 lbm = 0.454 kg	1 N = 1 kg·m/s^2	1 Pa = 1 N/m^2
1 slug = 14.6 kg	1 poundal = 0.138 N	1 mm Hg = 133 Pa
	1 lbf = 4.45 N	1 psi = 6,900 Pa
		1 m H_2O = 9,800 Pa
		1 atm = 98 kPa

Temperature
0°C ≡ 32°F ≡ 273 K ≡ 460°R

MATERIAL PROPERTIES

	Young's Modulus (E) (in GPa)	Shear Modulus (G) (in GPa)
Steel	193.	76.
Copper	125.	46.
Fused silica	72.9	31.2
Aluminum	71.	25.
Oak	12.4	
Concrete	31. (in compression)	
Polystyrene	5.3	1.2
Nylon	3.6	1.2
Polyethylene	0.76	0.26
Rat tendon	0.02–0.1	
Rat skin	0.01	
Human aorta	0.001	

MATERIAL PROPERTIES *(continued)*

	Specific Gravity (γ) (density $\rho = \gamma \times 1{,}000$ kg/m^3)	Bulk Modulus (B) (in GPa)
Mercury	13.6	27.6
Lead	11.3	
Steel	7.88	220.
Copper	8.92	129.
Aluminum	2.7	
Cement	2.7–3.0	
Granite	2.7	30.
Glass (common)	2.4–2.8	
Brick	1.4–2.2	
Rubber	1.1	
Nylon	1.1	
Polystyrene	1.06	
Blood	1.06	
Water	1.0	2.2–2.9
Polyethylene	0.9	
Heavy crude oil	0.98	1.9
Light crude oil	0.865	1.44
Hydraulic oil	0.82–0.95	1.6–2.2
Paraffin	0.9	
Dry leather	0.86	
Oak	0.6–0.9	
Cork	0.22–0.26	
Air (STP)	0.0013	

Absolute Viscosity (μ) (in centipoise; 1 centipoise = 0.001 Pa·s)

Heavy machine oil at	15.6°C	661.	Mercury	at 20°C	1.55
	37.8°C	127.		100°C	1.24
Light machine oil at	15.6°C	114.	Water	at 20°C	1.
	37.8°C	34.		100°C	0.28
	100°C	4.9	Air	at 18°C	0.018
Heavy crude oil at	15.6°C	100.		74°C	0.021
	37.8°C	32.	Human blood	at 37°C	~10.
Light crude oil at	15.6°C	8.5	Water slurry		$\sim(1 - x)^{5/2}\mu_{water}$
	37.8°C	4.9	(x = volume fraction of dry solids)		

MATERIAL PROPERTIES *(continued)*

	Specific Heat (c_p) (kJ/kg°C at 25°C)	Thermal Conductivity (k_t) (W/m°C at 20°C)
Water	4.2	0.6
Biological tissue (37°C)	3.6	0.46
Oak	2.4	0.2
Ice (0°C)	1.8	2.2
Air (dry)	1.0	0.026
Aluminum	0.9	204.
Granite	0.84	3.
Brick	0.84	0.4–0.5
Concrete (dry)	0.84	0.13
Glass	0.8	0.75
Glass wool (200 kg/m^3)	0.67	0.04
Light oil (30–100°C)	0.59	0.076
Steel	0.46	54.
Iron	0.45	81.
Copper	0.39	386.
Silver	0.24	427.
Cotton		0.05–0.07
Polystyrene		0.157
Polyurethane foam		0.026

Convection Heat-Transfer Coefficients (h_t) (W/m^2°C)		
Air:	Free convection	5–25
	Forced convection	10–200
Water:	Free convection	20–100
	Forced convection	50–10,000
Boiling water:		3,000–100,000
Condensing water vapor:		5,000–100,000

MATERIAL PROPERTIES *(continued)*

Electrical Resistivity (ρ) (ohm·m)	
Ceramics	10^{16}
Polystyrene	10^{16}
Polyethylene	10^{15}
Nylon	8×10^{12}
Neoprene	8×10^{10}
Glass	10^{6}–10^{9}
Distilled water	10,000
Granite	100–10,000 (moisture dependent)
Dry desert	2,400
Silicon	2,300
Lake or river	100
Shale	1–10 (moisture dependent)
Seawater	0.2
Doped silicon	0.026
Carbon	$3{,}500 \times 10^{-8}$
Steel	10–100×10^{-8}
Nickel	7.8×10^{-8}
Aluminum	2.8×10^{-8}
Copper	1.7×10^{-8}
Silver	1.6×10^{-8}

Relative Permittivity (ϵ_r) ($\epsilon = \epsilon_r \epsilon_0$)	
Air	1
Styrofoam	1.03
Snow	1.2–3.3
Wood	2
Polyethylene	2.3
Wax	2.3–2.7
Polystyrene	2.6
Nylon	3–3.5
Paper	3.5
Silicon dioxide	3.95
Oil	2.6–4.7
Neoprene	6.6
Glass	3.9–8.4
Earth	10–15
Water (fresh or salt)	80

Relative Permeability (μ_r) ($\mu = \mu_r \mu_0$)	
Air	1
Nonferrous materials	1
Nickel	100
Cold rolled steel	180–2,000
Iron	200–5,000
Purified Iron	5,000–180,000
Supermalloy	100,000–800,000

APPENDIX

B

TUTSIM—An Operational-Model Simulator[1]

TUTSIM is a computer program that approximates numerically the operations of operational blocks. The data required for simulation are of four types: (1) *structure data,* which specify the types of blocks and the interconnections among the blocks; (2) *parameter data,* which specify the values of the block parameters; (3) *timing data,* which specify the integration step size δt and the total simulation time T_f; and (4) *plot data,* which specify the variables to be plotted and the scales to use in the plots.

To use TUTSIM, draw an operational block diagram of the system to be simulated. Number the operational blocks in the diagram. (The block numbers assist in specifying the data.) TUTSIM prompts the user during data entry. Run the simulation.

The first section of this appendix lists a few of the operational blocks available in TUTSIM. The next section lists a few basic TUTSIM commands. The final section shows a sample TUTSIM simulation session. For further information, see a TUTSIM user manual.

[1]TUTSIM is a trademark of APPLIED i, Palo Alto, California. For further information about TUTSIM, see a TUTSIM user manual.

TUTSIM OPERATIONAL BLOCKS

We describe operational blocks in TUTSIM in the following form:

Structure Data
Parameter Data

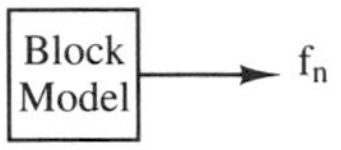

n, i, j = block numbers

f_n = output of block n

$-n$ implies use of the negated signal $-f_n$

Multiple-input blocks can handle more than two inputs.

1. n, CON
   n, c
2. n, PLS
   n, t_1, t_2, p
3. n, SUM, i, −j, . . .
4. n, GAI, i, −j, . . .
5. n, INT, i, −j, . . .
6. n, TIM
7. n, SIN, i, −j, . . .

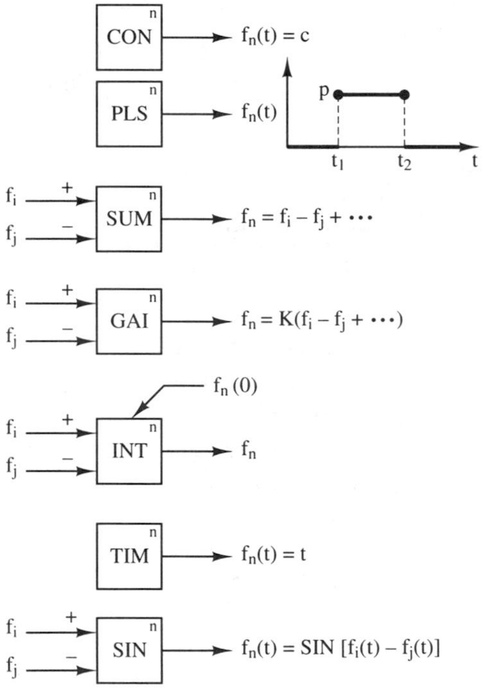

(The argument is in radians)

TUTSIM COMMANDS

SD	Simulate the system and graph the results on the screen.
SN	Simulate the system and list the results numerically on the screen.
L	List the model file on the screen.
LP	Print the model file.
CS	Change the structure of the model.
CP	Change the parameters of the model blocks.
CB	Change the outputs to be plotted and the plotting scales.

CT	Change δt and T_f.
E	Leave the model (restart TUTSIM).
A	Leave TUTSIM (return to DOS).

TUTSIM SAMPLE SESSION

The following TUTSIM session is based on the operational diagram of Figure 3.21. Data entered are shown in boldface. The blocks are labeled as follows, in TUTSIM notation:

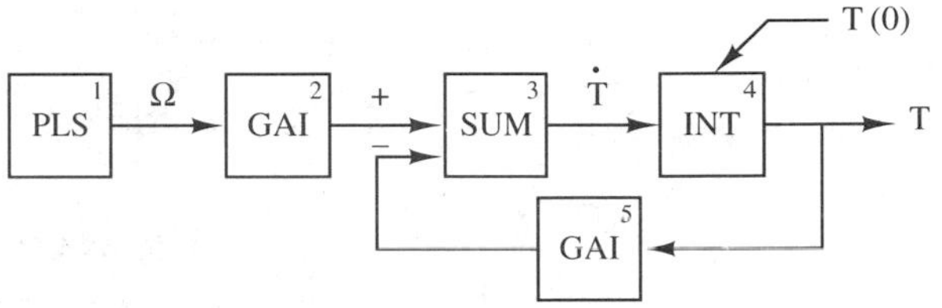

C> **TUTSIM** <cr>
INPUT FROM?:
: **K** <cr> (Keyboard; or use F for File)
MODEL STRUCTURES
FORMAT:BLOCKNBR,TYPE,INPUT1,INPUT2...; COMMENTS
: **1,PLS ; Input velocity pulse** <cr>
: **2,GAI,1 ; Scale factor K** <cr>
: **3,SUM,2,−5** <cr>
: **4,INT,3 ; Integrator** <cr>
: **5,GAI,4 ; Feedback scale factor K/B** <cr>
: <cr> (Ends input of model structure)
MODEL PARAMETERS
FORMAT: BLOCKNBR,PARAMETER1,PARAMETER2
: **1,0,1,1** <cr> (1 rad/s)
: **2,100** <cr> (K = 100 N/rad)
: **4,0** <cr> (T(0) = 0)
: **5,2** <cr> (K/B = 2 rad/s)
: <cr> (Ends input of parameters)
PLOTS BLOCKS
FORMAT: BLOCKNBR,PLOT-MIN,PLOT-MAX (Block 0 denotes time axis)

Horz: **0,0,2** <cr> (0 < t < 2 s)
Y1: **1,0,2** <cr> (0 < input < 2 rad/s)
Y2: **4,0,100** <cr> (0 < T < 100 newtons)
Y3: <cr> <cr> (No outputs 3 and 4)
TIMING DATA
FORMAT: DELTA, FINAL TIME (in appropriate time units)
: **0.05,2** <cr> (in seconds)
COMMAND:
GD <cr> (Draws grid)
SD <cr> (Simulates system and displays output on screen)

The printed graphical output of this session is shown in Figure 3.26.

APPENDIX

C

SPICE—A Network Simulator

The program SPICE was designed to simulate numerically the behaviors of electric networks. It is equally suitable for simulating behaviors of other types of linear networks—mechanical, hydraulic, hydromechanical, etc. It also can handle some nonlinear elements.

SPICE can compute the network response for step inputs or for arbitrarily shaped input waveforms. It can also find the frequency response of the network—the sinusoidal steady-state behavior. It is not necessary for the user to find the system equations for the network. One need only specify the network elements and the connection pattern. The program formulates and solves the equations automatically.

The SPICE algorithm begins with an initial guess at a set of node potentials. It uses these node potentials together with the element equations to compute the branch flows. Then it adjusts the node potentials iteratively until the flows balance to within a specified threshhold at each node. Thereafter, it changes the node potentials to maintain flow balance over a sequence of time steps. SPICE continually increases or decreases the internal time increment to maintain accuracy without performing unnecessary calculations.

The SPICE input data and output data use electrical terminology. To help the user apply SPICE to nonelectrical networks, this appendix relates SPICE terminology to the general network terminology of the text. There are a number of SPICE-based programs designed to run on the IBM PC. PSPICE is one of them.

We describe the SPICE data as required by PSPICE.[1] Most of the following material applies to the original SPICE program and to other SPICE-based programs. To prepare a network for PSPICE simulation:

1. Draw a line-graph representation of the network. This line graph is analogous to an electric circuit.
2. Label the nodes, using letters or numerals. The number 0 is predefined to mean the reference (ground) node.
3. Choose the positive direction for flow in each passive branch.

The PSPICE data for the network must be arranged as a sequence of statements in an ASCII text file. The first statement in the file is a title line of arbitrary content that is used only as a descriptor of the problem. The last statement is

```
.END
```

The remaining statements are either data statements or commands. The ordering of these statements is arbitrary. The data statements specify the network elements. The node labels in the data statements show how the elements are interconnected. The commands specify the analyses to be performed. Lines in the file that begin with * are treated as comments.

THE FORMAT FOR ELEMENT-DATA STATEMENTS

Each network element is specified by:

(a) A name that identifies the element. Begin the name with the symbol for the analogous electrical element—R, L, C, I, or V. PSPICE uses only the first symbol in the name to identify the element *type*. We can use additional symbols (letters or numbers) to distinguish the elements.

(b) An element value, in SI units.

(c) A pair of node labels (N+ and N−) that indicates the orientation of the flow variable for the element. The positive direction for flow is from N+ to N−.
- In a passive element (R, L, and C), then, the potential drops from N+ to N−.
- For a flow source (I), the direction N+ to N− is the direction of the *specified* flow. Hence, the potential usually rises from N+ to N− in a flow source.
- For a potential-difference source (V), the specified potential difference is the potential drop from N+ to N−. Hence, the flow through a potential-difference source is usually from N− to N+, the direction of potential rise. As a consequence, the flow through a potential-difference source is usually negative.

(d) Additional data if appropriate.

[1] PSPICE is a trademark of MicroSim Corp., Irvine, California. For further information about PSPICE, see a PSPICE user manual.

The PSPICE formats for element data are described in the following paragraphs. PSPICE assumes that each element is an ideal electrical element. To specify data for a nonelectrical element, we use the symbol for the analogous electrical element. The analogies in Table 4.3 can help with preparing a PSPICE data file and interpreting PSPICE output.

The element *value* can be followed immediately by one of the following suffixes (in upper or lower case) to denote multiplication by a scale factor: T = 1E12; G = 1E9; MEG = 1E6; k = 1E3; m = 1E−3; u = 1E−6; n = 1E−9; p = 1E−12; f = 1E−15. Any character not in this specific list is ignored. Additional attached characters are also ignored. (These other characters can be used to denote units.)

Resistor-Type Elements

General form:	RXXXXXXX	N+	N−	value
Examples:	RC1	1	2	100Kohms
	Rdamper	21	101	12:meter/newtonsec

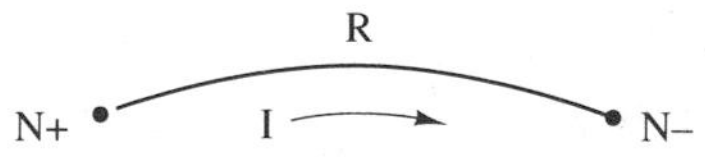

Capacitor-Type or Inductor-Type Elements

General form:	CXXXXXXX	N+	N−	value	[IC=initialval]
	LXXXXXXX	N+	N−	value	[IC=initialval]
Examples:	CAB1	13	0	1UF	
	C2	3	1	10UF	IC=1.5volt
	L3	7	3	3MH	IC=3Mamp
	Cmass	3	0	2.3:kg	
	Lspring	5	1	50:meter/newton	

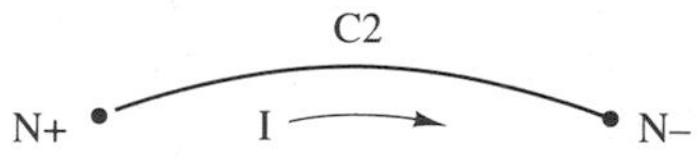

Brackets [ ] denote the optional use of a nonzero initial value (t = 0) of the potential difference across a capacitor-type element or of the flow through an inductor-type element during computation of the response. If no initial value is specified, 0 is assumed. These initial values have no effect on the network behavior, unless they are activated by the key word UIC in the .TRAN command introduced later.

Independent Voltage-Type and Current-Type Sources

General form:

VXXXXXXX	N+	N−	[DC, value]	[waveform]	[AC, mag, [phase]]
IXXXXXXX	N+	N−	[DC, value]	[waveform]	[AC, mag, [phase]]

Examples:

VS	7	0	DC	6volt	
Vvelocity	13	0	DC	6:meter/sec	
I1	5	21	DC	0.003:amp	
Vin	2	1	EXP(−1V 1V 2Ms 1Ms 1Ms 2Ms)		
Itorsion	4	0	PWL(1Ms 0:N 2Ms 3:N 3Ms 4:N 5Ms 2:N)		
Vin	2	1	AC	6volt	45:degrees
VS	2	1	AC	6:meter/sec	0:degrees

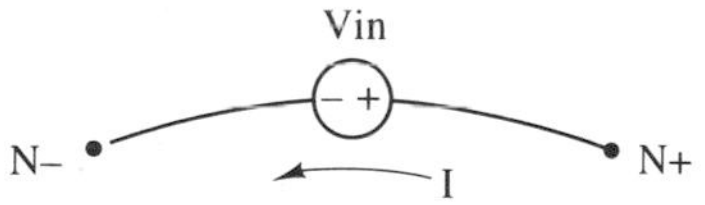

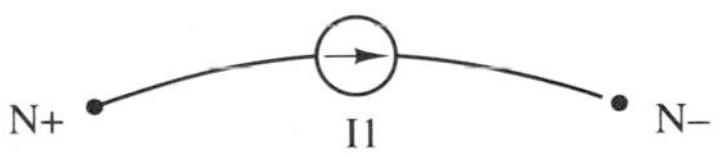

Note that we specify the *signal* data in the source data statement; we specify how to *use* those data in a separate *compute* command. The [DC] and [waveform] data are used to compute the values of network variables versus time. The [AC] magnitude and phase data are used to calculate the sinusoidal steady-state response of network variables in phasor notation. (The phase is expressed in *degrees;* the frequency is specified in a separate compute command.) Several forms of source waveform are available. We describe three.

DC (Constant Value). If UIC is specified in the .TRAN command, the effect is to compute a step response. If UIC is not specified in the .TRAN command, the constant steady-state values of all the network variables are computed.

A Piecewise-Linear Signal.

General Form: PWL(t_1 i_1 t_2 i_2 t_3 i_3 . . . t_n i_n)

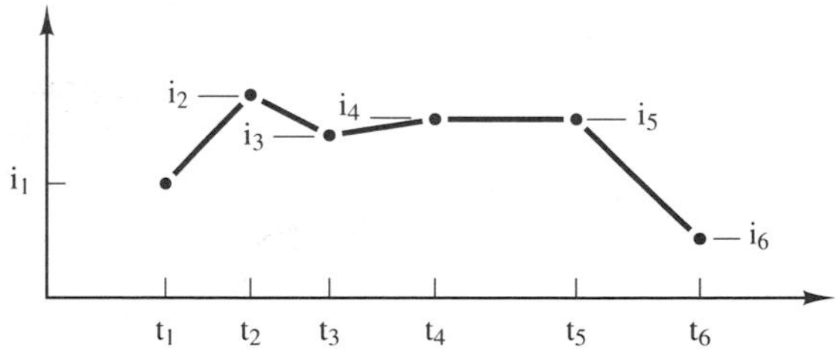

A Pulse with Exponential Rise and Fall.

General Form: EXP(i_{off} i_{peak} t_{dr} t_{cr} t_{df} t_{cf})

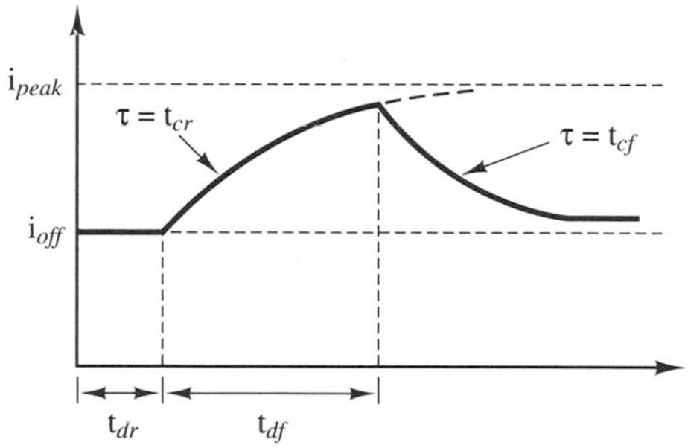

Dependent Sources

A dependent source is a source that has a value proportional to the value of another network variable. (Nonlinear dependencies can also be used.)

Potential-Controlled Voltage-Type Sources.

General Form:	EXXXXXXX	N+	N−	a	b	vgain
Examples:	E2	0	2	1	3	20volt/volt
	E3pressure	2	0	3	0	1.2:pascal/pascal

Note that $V_{N+} - V_{N-} = \text{vgain} * (V_a - V_b)$.

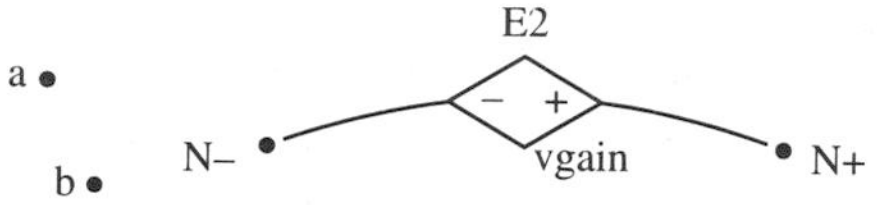

Flow-Controlled Current-Type Sources.

General Form:	FXXXXXXX	N+	N−	Vsense	flowgain
Examples:	Fout	3	2	Vsense	15:amps/amp
	Vsense	2	1	0volts	
	Fforce	3	1	V2sense	2.7:newton/newton
	V2sense	2	0	0:meter/s	

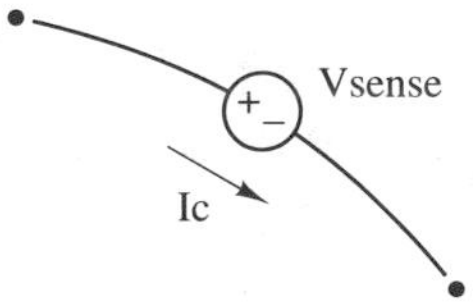

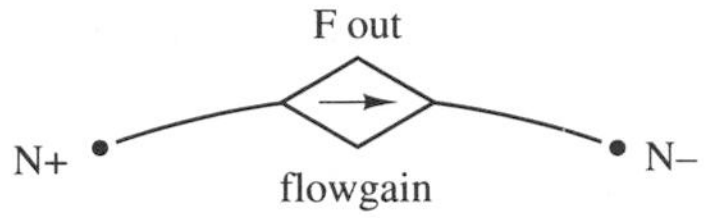

Note the need to insert a (zero-potential) source (Vsense) to serve as a flow meter in the branch that carries the controlling flow Ic. Then the flow from N+ to N− in the dependent source is the product of *flowgain* and the flow Ic from + to − in the flow meter.

Potential-Controlled Current-Type Sources

General Form:	GXXXXXXX	N+	N−	a	b	transgain
Examples:	G2	3	2	1	0	25:amps/volt
	Gmotor	4	1	3	2	7.5:newtonmeter/amp
	Gpump	2	1	4	3	23.5liter/sec/rad/sec

Note that the flow from N+ to N− in the dependent source is transgain * $(V_a - V_b)$.

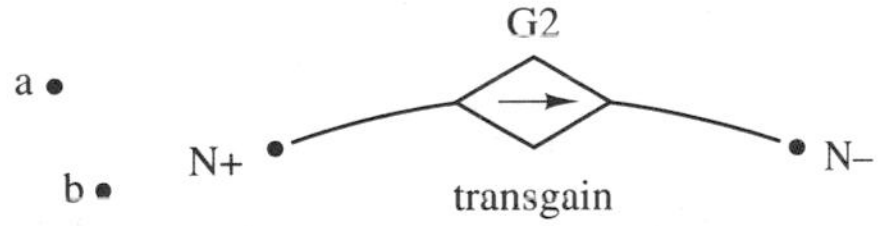

Flow-Controlled Voltage-Type Sources

General Form:	HXXXXXXX	N+	N−	Vsense	transgain
Examples:	Hcont	2	3	Vsense	15volts/amp
	Vsense	4	1	0volts	
	Hhydromech	3	0	V2sense	0.05:meter/sec/liter/sec
	V2sense	2	4	0:pascal	

Note the need to insert a (zero-potential) source (Vsense) to serve as a flow meter in the branch that carries the controlling flow Ic. Then $(V_{N+} - V_{N-})$ is the product of *transgain* and the flow Ic from + to − in the meter.

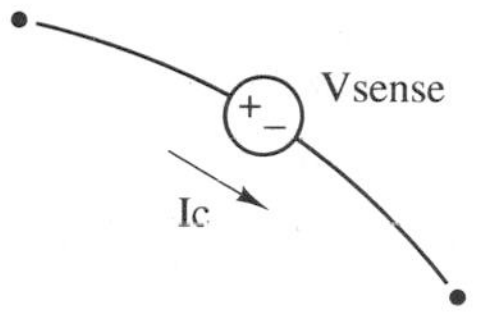

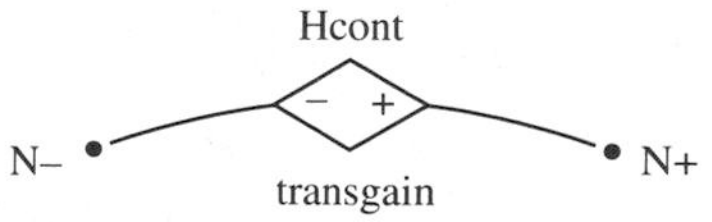

COMMANDS TO COMPUTE NETWORK RESPONSE

Response Versus Time

General Form:	.TRAN	stepval	finaltimeval	[UIC]
Example:	.TRAN	1ms	50ms	UIC

This command computes the *transient response*—i.e., the response versus time—*of all network variables* to the source waveform specified in the source-element description(s). In the example, the response is calculated for 50 milliseconds.[2] The UIC notation in the command *activates* the initial conditions stated in the energy-storage element descriptions (or zero-default initial conditions). The .TRAN command does not produce any output; output is produced by an output command.

If the UIC is not included in the .TRAN statement, PSPICE calculates steady-state (*bias point* or *operating point*) values of all variables, under the condition that the *sources* are held steady at their initial values, and then uses the resulting states of the energy-storage devices as their initial conditions.

The potentials at certain nodes can be set to particular values and then released at the start of the transient computation by means of the command

```
.IC   V(N2)=N2val   V(N5)=N5val
```

This command produces a different behavior than does use of the UIC notation in the .TRAN command. UIC specifies the use of the initial values of flows through or potential differences across the energy-storage elements, whereas .IC specifies the initial absolute potentials at certain nodes.

DC Sweep

```
General Form:   .DC   variable          list
Example:        .DC   Vin      0mV      1mV     5mV
```

For this example, PSPICE calculates the steady-state (DC) solutions corresponding to three values of the potential-difference source Vin. Again, no output is produced unless there is a separate output command. This type of computation can be used to find the effect of a nonlinear element on the network variables.

Sinusoidal Frequency Response

```
General Form:   .AC   [LIN] [OCT] [DEC]    numfreqs    startfreq    endfreq
Example:        .AC   DEC     10     1kHz     100kHz
```

For this example, PSPICE calculates the sinusoidal frequency response (amplitude and phase) of the network at 10 frequencies per decade, *spaced logarithmically,* between 1 kHz and 100 kHz. Exactly one of the frequency-spacing commands LIN (linear), OCT (by octaves), and DEC (by decades) must be used. During the AC analysis, the only independent sources that have nonzero amplitude are those with AC specifications. The magnitudes and phases of the sources are constant at the values specified in the AC specifications of the source statements.

[2] PSPICE does not use the information "stepval = 1 ms" as an integration step size in the manner of the operational-diagram simulator TUTSIM. In PSPICE, it is used only as the data spacing for output.

OUTPUT COMMANDS

```
.PRINT   [DC]   [AC]   [TRAN]   [var1, var2]
.PLOT    [DC]   [AC]   [TRAN]   [var1, var2] [low, high]
.PROBE
```

These three commands generate program output. The .PRINT command produces numerical lists of response data. The .PLOT command uses text symbols to make approximate plots of response waveforms. The .PROBE command writes all computed data to a file for later graphics postprocessing. PROBE can plot the waveforms of all network variables and mathematical modifications (such as derivatives and integrals) of those waveforms.

Examples:

```
.PLOT  TRAN  V(2)  V(5,3)  (-1,1)  I(R2)
.PRINT  AC  VDB(2)  VP(2)
.PLOT  DC  V(2)  V(3,5)  I(R2)  I(Vin)
```

The first example plots the transient values of the potential at node 2 and the potential drop from node 5 to node 3 on a scale from −1 to 1. It also plots the flow through element R2 on a scale determined by PSPICE. The time increments used in the plot are determined by stepval in the .TRAN command.

In the second example, the suffixes DB and P on the variable V in the .PRINT statement designate the magnitude response in dB and the phase response *in degrees*, respectively. If V(2) were specified without a suffix, only the magnitude response would be plotted.

The third example plots the potential at node 2, the potential drop between nodes 3 and 5, and the flows through element R2 and through the source Vin during a DC sweep.

SAMPLE PSPICE DATA FILE

```
Example of a Mass-Damper Transient Response
*Use N to denote newtons.
R 1 0 1:m/Nsec
C 1 0 1:kg
*Apply a 1-newton step input, via a DC flow source,
*with initial conditions specified in the .TRAN command.
*Also, apply a sinusoidal input, via an AC flow source, of
*magnitude 1 newton and zero phase, in order to generate
*a frequency response with the .AC command.
I 0 1 DC 1:N AC 1:N 0degrees
*Generate a transient response for 5 s at 0.5-s intervals.
.TRAN 0.5s 5s UIC
*Plot the transient output velocity and the forces in b and m.
.PLOT TRAN V(1) I(R) I(C)
*Also, print these transient variables.
.PRINT TRAN V(1) I(R) I(C)
*Compute the frequency response at 5 frequencies per
*decade, over the range 0.01 Hz to 100 Hz.
.AC DEC 5 0.01Hz 100Hz
*Plot the frequency response: amplitude (in dB) and phase.
.PLOT AC VDB(1) VP(1)
*Save all computed results for use by the PROBE postprocessor
.PROBE
.END
```

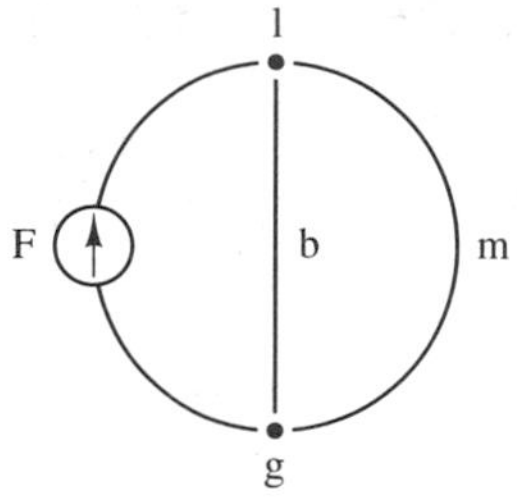

APPENDIX

D

CC—A Program for Analyzing Transfer Functions[1]

CC is a computer program for analyzing and designing linear time-invariant systems that are described by input-output transfer functions. The program provides for easy entering and manipulating of transfer functions. It also displays transfer functions symbolically for easy checking. The program can translate back and forth between the polynomial form and various factored forms of the transfer function. The root locus routine incorporated in CC can compute the paths of the poles of a transfer function as a function of onc of the parameters of the characteristic equation.

CC generates the step response and the impulse response of the system represented by the transfer function. To obtain the response to another input signal, premultiply the system transfer function with a signal-transforming transfer function that will act on the impulse or step.

CC generates the time response by means of partial fraction expansion and Laplace transform inversion. Hence, the time response is not approximate. It can be generated either in closed form or as graphical output on a screen or on paper.

CC computes the frequency response of the system represented by the transfer function and presents it in various forms of linear, logarithmic, and polar display. Interactive features permit the user to tailor both the placement of computed frequency points and the structure of the display for clear presentation prior to generating hard-copy output.

[1] Program CC is a trademark of Systems Technology, Inc., Hawthorne, California. For further information about CC, see a Program CC user manual.

DATA ENTRY AND MANIPULATION

A transfer function can be entered as a ratio of polynomials in BASIC notation—for example, G1 = 1/(s + 2). It can also be generated by combining previously entered transfer functions or transfer function factors—for example, G3 = 1 + 2 * G1/G2 or G3 = G1/(s + 1). The following commands are used to manipulate the transfer function:

CANCEL, G1: Cancels common numerator and denominator factors of the transfer function G1.

DISPLAY, G1: Displays the transfer function G1 symbolically on the screen.

PZF, G1: Converts the transfer function G1 to pole-zero form.

SINGLE, G1: Converts the transfer function G1 to a ratio of two polynomials.

UNITARY, G1: Converts the transfer function G1 to a ratio of two polynomials for which the coefficients on the highest powers of s are normalized to 1.

PFE, G1: Performs a partial fraction expansion of the transfer function G1.

It is also possible to RESCALE, as in G2(s) = G1(a * s), to SUBSTITUTE, as in G3(s) = G2(G1(s)), and to CHANGE one or more coefficients in G(s).

SYSTEM ANALYSIS COMMANDS

ILT, G1: Performs an inverse Laplace transform on the transfer function G1. A closed-form expression for the time response to a unit impulse input is produced.

TIME, G1: Plots the time response of the system to a unit-impulse or unit-step input. Select the *open-loop response* to indicate that the transfer function G1 directly multiplies the transform of the impulse or step to produce the output-signal transform.

BODE, G1: Plots the frequency response of the transfer function G1. Various forms of linear and logarithmic axes are available. BODE requires prior specification, by means of the FREQUENCY command, of the frequencies at which the response should be computed. Additional frequencies can be added interactively to improve the appearance of the plot.

LIST, G1: Lists the points specified by the current frequency file.

POINT, G1: Computes the frequency response of the transfer function G1 for a single frequency.

ROOTLOCUS, G1: Determines the paths of the roots of the characteristic equation for the transfer function G1 as a function of one of the parameters of the system.

ZEROS: Finds the zeros of a polynomial.

SAMPLE SESSION

Data entered are shown in boldface:

C> **cc** <enter>
CC> **g1=(s+100)/((s+1)*(s+10))** <enter>
CC> **display,g1** <enter>

$$G1(s) = \frac{s + 100}{(s + 1)(s + 10)}$$

CC> **ilt,g1** <enter>

Causal inverse Laplace transform

G1(t) = −10*exp(−10t) + 11*exp(−1t) for t>0
        0 for t<0

CC> **frequency** <enter>
Enter transfer function Gi > **g1** <enter>
Enter low freq, high freq, # points > **0.1,1e5,100** <enter>
Enter 0=log10 scale, 1=linear scale > 0 <enter>
CC> **bode** <enter>

The program then guides the user through a number of choices to design the plot and scale the axes. The plot is as follows:

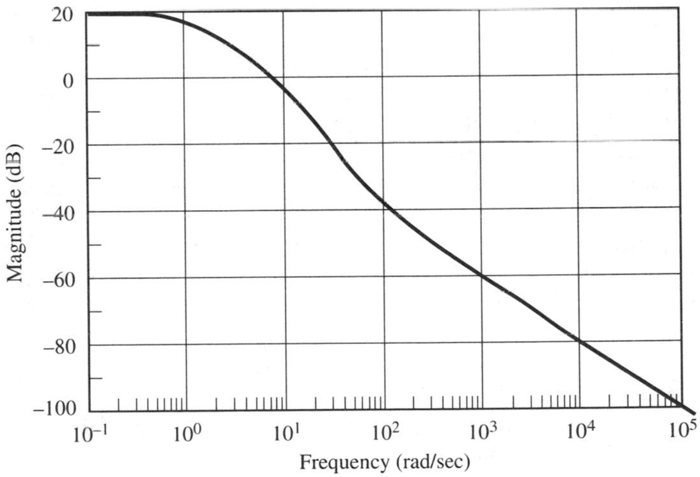

Answers to Selected Problems

1.1(i) **(a)** Signals: The motions of the two ends and the tension in the cable. **(b)** The cable tension is a flow. The end motions are potentials. **(c)** The cable is dynamic. The stretched cable cannot retract instantly, owing to the mass of cable.
1.4 **(b)** $v(t) = 64e^{-1.386t}$ for $t \geq 2$.
2.1(i) **(a)** Orient $+v$ to the right. Label the node v_1. Define F_1 in the source and F_2 in the spring, both oriented to the right. **(b)** $F_2 = (k/s)v_{1g} = kx_{1g}$; $v_{1g} = v$ (or $x_{1g} = x$, where x is the displacement produced by the velocity source); $F_1 = F_2$. **(c)** $P = F_2 v_{1g} = kxv$.
2.2(iv) **(a)** Orient $+v$ to the right. Label the node v_1. Define F_1 in the damper, F_2 in the mass, and F_3 in the source, all oriented to the right. **(b)** $v_{1g} = -v$, $F_1 + F_2 = F_3$, $F_1 = bv_{g1}$, and $F_2 = smv_{g1}$.
2.3 A typical spring length is 1 m. A typical spring width is 8 cm. A typical spring sag owing to car weight is 3 cm.
2.4 **(a)** $k = 44.4$ g/cm $= 43.6$ N/m; **(c)** 5% of full scale uncalibrated.
2.5 **(a)** $b = 37.7$ N·s/m.
2.7 **(i)** Use a velocity source and a damper. The damper is not viscous. **(ii)** Use a velocity source and a viscous damper. **(iii)** Use a force source, a mass, and a damper. **(iv)** Use a force source, two masses, two dampers, and a spring. **(viii)** Use two force sources, a spring, two masses, and two dampers. The dampers are not viscous. **(x)** Use a force source, a mass, and a damper. Treat the river as a velocity source relative to the fixed earth. **(xii)** Use two force sources, a mass, and a damper.
2.15(i) **(a)** Orient $+v$ to the right. Define the forces F_1 in b_1, F_2 in b_2, F_3 in b_3, and F_4 in the source, all oriented to the right. **(b)** $F_1 = b_1 v_{g1}$, $F_2 = b_2 v_{g2}$, $F_3 = b_3 v_{2g}$, $v_{12} = -v_c$, $F_4 = F_1$, and $F_4 + F_2 = F_3$. **(c)** $F_1 = \dfrac{b_1(b_2 + b_3)v_c}{(b_1 + b_2 + b_3)}$, $v_1 = -\dfrac{(b_2 + b_3)v_c}{(b_1 + b_2 + b_3)}$.

2.18 **(a)** $k_e = \dfrac{(k_1k_2 + k_1k_3 + k_2k_3)}{(k_2 + k_3)}$. **(b)** Use the sequence of computations: $x_1 = \dfrac{F}{k_e}$, $F_1 = k_1x_1$, $F_2 = \dfrac{k_2k_3x_1}{(k_2 + k_3)}$, $F_3 = F_2$, and $x_2 = \dfrac{F_2}{k_3}$.

2.21 **(a)** Orient +v to the right. In the damper parallel to the velocity source, the scalar force oriented to the right is $F_1 = (2\ \text{N·s/cm})(-0.1\ \text{m/s}) = -20$ N; that scalar force is tensile. Combine the parallel dampers to form a single damper with $b_{eq} = 3$ N·s/cm. The scalar force in that damper is 1 N, oriented to the right. Therefore, $v_{12} = (1\ \text{N})/(3\ \text{N·s/cm}) = 0.333$ cm/s. Then, $v_2 = -0.1\ \text{m/s} + 0.333\ \text{cm/s} = -9.67$ cm/s. The compressive force (oriented to the right) in the upper damper of the parallel pair is $F_2 = (2\ \text{N·s/cm})(0.333\ \text{cm/s}) = 0.667$ N.

2.22 **(b)** $v_1 = -\dfrac{(b_2 + b_3)v}{(b_1 + b_2 + b_3 + b_4)}$.

2.23 **(b)** The compressive force (oriented to the right) in the source is $\dfrac{(b_1 + b_4)(b_2 + b_3)v}{(b_1 + b_2 + b_3 + b_4)}$.

2.24 **(b)** Orient +v to the right. Then, $x_1 = -F/(k_1 + k_4)$ and $x_2 = F/(k_2 + k_3)$.

2.25 **(b)** The counterclockwise loop flow in the mesh formed by k_2 and k_3 is $k_3F/(k_2 + k_3)$.

2.27 **(c)** $[(b_1 + b_2)s + k]F_1 = b_1b_2s(v_a + v_b)$ or $(b_1 + b_2)\dot{F}_1 + kF_1 = b_1b_2(\dot{v}_a + \dot{v}_b)$.

2.30 **(b)** $v_1 = \dfrac{(s^2m + k)F}{(bms^2 + kms + kb)}$, or $bm\ddot{v}_1 + km\dot{v}_1 + kbv_1 = m\ddot{F} + kF$.

3.1(i) **(a)** Orient linear $+\Omega$ to the right. Label the node Ω_1. Define a single flow T (identical to the source signal) oriented to the right. **(b)** T(t) is specified; $\Omega_1 = T/B$. **(c)** $P = T^2/B$ watts. **(d)** $E = T_c^2/B$ joules.

3.2(ii) **(a)** Orient linear $+\Omega$ to the right. Define the flow T in the source, oriented to the right. Define T_1 in K and T_2 in J, oriented to the right.
(b) T(t) is specified; $T = T_1 + T_2$; $T_1 = K\theta_1$; $T_2 = J\dot{\Omega}_1$. **(c)** $J\ddot{\theta}_1 + K\theta_1 = T$.

3.3 For Figure P3.2(ii): **(a)** Same as 3.2(ii)(a). **(b)** Same as 3.2(ii)(c). **(c)** $\phi = 0$, $\theta_{1c} = T_c/(K - J\omega^2)$. **(d)** If the frequency ω of the applied torque is near $\sqrt{K/J}$, then the amplitude θ_{1c} of the rotation approaches ∞.

3.6 A typical spring sag owing to car weight is 3 cm.

3.7 **(b)** K = 75,500 N·m/rad, J = 2.42 kg·m². **(c)** $\ddot{\theta}_1 + 31{,}200\theta_1 = 0.413T$.

3.8 **(b)** If the flywheel is represented by node 3, then $\dot{\Omega}_3 = T/J$.

3.11 **(a)** $K_{eff} = K_1(K_2 + K_3)/(K_1 + K_2 + K_3)$. **(b)** $\theta_1 = T/K_{eff}$; $\theta_2 = T/(K_2 + K_3)$.

3.18(ii) **(b)** $v_3(t) = v_c(1 - e^{-(k/b)t})$; $F(t) = (v_c/b)(1 - e^{-(k/b)t})$.

3.18(iv) **(b)** $v_1(t) = (F_c/b)(1 - e^{-(b/m)t})$; in the mass, $F_1(t) = F_ce^{-(b/m)t}$; in the damper, $F_2(t) = F_c(1 - e^{-(b/m)t})$.

3.21(iii) **(a)** $\ddot{\Omega}_2 + (B/J)\dot{\Omega}_2 + (K/J)\Omega_2 = (B/J)\dot{\Omega} + (K/J)\Omega$. **(c)** $\Omega_2(t) = \Omega_c[1 - 1.15e^{-0.5t} \times \cos(0.866t + 0.524)]$ rad/s.

3.21(v) **(a)** $\ddot{v}_2 + (b/m)\dot{v}_2 + (k/m)v_2 = (b/m)\dot{v}$. **(c)** $v_2(t) = 1.15v_ce^{-0.5t}\sin(0.866t)$ m/s.

3.25 **(a)** v_3, x_{23}, and $\int F$ are prototype variables for the network.

3.29 for Figure P3.21(v): **(a)** F_1 is a prototype variable; $\ddot{F}_1 + (b/m)\dot{F}_1 + (k/m)F_1 = (kb/m)v$. **(b)** First row of **[A]** = (0 1); second row of **[A]** = (−k/m −b/m); **[B]** = $(0\ \ kb/m)^T$; **[C]** = (0 1/k).

3.30 for Figure P3.21(v): **(b)** b = 50. **(c)** $(s + 4.38)v_2 = 1.1sv$; $v_2(t) = 1.1e^{-t/0.228}$ m/s. **(d)** Eliminate the mass. Then, $(s + 4)v_2 = sv$; $v_2(t) = e^{-t/0.25}$ m/s.

4.1(iv) **(a)** Label the top node 1 and the bottom node 2; orient i_1 downward in the 1 kΩ resistor and i_2 downward in the 4 kΩ resistor. **(b)** $i_1 + i_2 = 5$ mA, $v_{12} = i_1(1\ \text{k}\Omega)$, and $v_{12} = i_2(4\ \text{k}\Omega)$. **(c)** $v_{12} = 4$ V, $i_1 = 4$ mA, and $i_2 = 1$ mA.

4.2(iii) **(a)** Label nodes 1, 2, and 3, clockwise from the top left. **(b)** $sCv_{12} = i$, $v_{23} = Ri$, and $v_{13} = v_{12} + v_{23}$. **(c)** $v_{23} = iR$.
4.2(vi) **(a)** Label the top node 1 and the bottom node 2; orient the currents i_1, i_2, and i_3 downward in the resistor, inductor, and capacitor, respectively. **(b)** $i_1 + i_2 + i_3 = i$; $v_{12} = Ri_1 = (sL)i_2 = (1/sC)i_3$. **(c)** $(RCLs^2 + Ls + R)v_{12} = RLsi$.
4.4(a) $R_1 = R_2 = 2000\ \Omega$.
4.5(c) $v_{rms} = v_{max}/\sqrt{2}$.
4.6(i) $v_1 = 70.6$ V and $v_2 = 34.1$ V.
4.12(ii) $v_1(t) = i_cR_S[1 - \exp(-t/R_SC)]$.
4.12(v) $\tau = 0.529$ ms, $v_{1i} = 0.167v_c$, and $v_{1f} = 0.118v_c$.
4.13(ii) **(a)** $s = -1300 \pm j557$. **(b)** $v_{1i} = Ri_c$ volts, $\dot{v}_{1i} = -(R^2/L)i_c$ volts/s, and $v_{1f} = 0$ volts; $R \triangleq 100\ \Omega$ and $R_L \triangleq 30\ \Omega$. **(d)** $v_1(t) = i_c[100\mathcal{B}(t) - 141\mathcal{C}(t)]$; use $\zeta = 0.92$ to select the correct $\mathcal{B}$ and $\mathcal{C}$ from Figures 3.53 and 3.54; use $\omega_n = 1414$ rad/s to set the time scales.
4.13(iv) **(a)** $s = -4125 \pm j6745$ rad/s. **(b)** $v_{1i} = 0$ volts, $\dot{v}_{1i} = 5 \times 10^5 i_c$ volts/s, and $v_{1f} = 16i_c$ volts. **(c)** $v_1(t) = 16i_c - 66.3i_c\cos(6745t - 1.33)$ volts.
4.14(i) $[s^2LC + s(L/R_S + R_LC) + (1 + R_L/R_S)]v_1 = [sL/R_S + R_L/R_S]v_S$.
4.21 **(c)** $i_2(t) = -[v_S/(R_S + R_L)]e^{-(R_L+R_2)t/L}$.
4.26(ii) **(a)** $v_{eq}(s) = 20i/(1 + 2 \times 10^{-9}s)$ volts; $z_{eq}(s) = (30 + 2 \times 10^{-8}s)/(1 + 2 \times 10^{-9}s)\ \Omega$. **(b)** $i_{eq}(s) = 20/(30 + 2 \times 10^{-8}s)$ amps. **(c)** $(s + 9 \times 10^8)i_L = 4 \times 10^8 i$.
4.27(iv) **(a)** $v_{eq}(s) = 0.2F$ m/s; $z_{eq}(s) = (s + 0.167)/(5s)$ m/N·s. **(b)** $F_{eq} = sF/(s + 0.167)$ N. **(c)** $(s^2 + 6s + 1)F_L = 6sF$.
4.34 **(a)** $v_3(t) = 15 - 3e^{-t/(25\mu s)}$ volts.
4.40 **(a)** $v_2(t) = v_c[1 - e^{-t/RC}]u_s(t) - v_c[1 - e^{-(t-\tau)/RC}]u_s(t - \tau)$ for $0 \le t < 2\tau$. **(b)** $v_2(t) = v_c[1 - e^{-t/RC}]$ for $0 \le t < \tau$; $v_2(t) = 0.632v_ce^{-t/RC}$ for $\tau \le t < 2\tau$. **(c)** $v_{2c} = 0.269v_c$.
4.41 **(b)** $v_2(t) = 0$ for $t < 0$; $v_2(t) = 5[1 - e^{-t/(0.1\text{ ms})}]$ for $0 \le t < 0.1$ ms; $v_2(t) = 8.6e^{-t/(0.1\text{ ms})}$ for $0.1\text{ ms} \le t$.
5.1(ii) **(a)** Label the ends of the pipe with the symbol P_0 (reference pressure); label the pipe connection between the pressure source and the fluid resistor with the symbol P_1; orient the flow Q to the right. **(b)** $P_{10} = P_S$, $P_{10} = \eta_LQ^2$. **(c)** $Q = \sqrt{P_S/\eta_L}$.
5.2(ii) **(a)** Label the left and right ends of the pressure source with the symbols P_1 and P_2; $P_S = Q/(sC) + \eta Q^2$ or $2C\eta Q\dot{Q} + Q = C\dot{P}_S$. **(b)** $Q_i = \sqrt{P_S/\eta}$. **(c)** $Q_f = 0$.
5.5 **(a)** Represent the water line and valve by quadratic resistors η and η_v; treat the water main as a constant pressure source P_S. **(b)** $\eta = 1.06 \times 10^{12}$ kg/m^7. **(c)** $\eta_v \approx 1.97 \times 10^{13}$ kg/m^7. **(d)** $Q \approx 0.158$ liter/s.
5.6(i) $P_{10} = P_S/[1 + \eta_S(1/\sqrt{\eta_1} + 1/\sqrt{\eta_2})^2]$.
5.7 **(b)** $\eta_{eff} = \eta_1 + \eta_2 + \eta_S = 1.51 \times 10^{12}$ kg/m^7; $Q_{shower} = 0.576$ liter/s.
(c) $\eta_{eff} = 7.73 \times 10^{11}$ kg/m^7; $Q_{inlet} = 0.804$ liter/s; $Q_{shower} = 0.309$ liter/s.
(d) $\eta_{eff} = 7.59 \times 10^{11}$ kg/m^7; $Q_{inlet} = 0.812$ liter/s; $Q_{shower} = 0.145$ liter/s.
5.8 **(b)** $C_1 = 1.12 \times 10^{-3}$ m^5/N; $\eta_{pipe1} = 2.46 \times 10^9$ kg/m^7; $\eta_{contraction1} + \eta_{elbow1} + \eta_{tee1} = 4.03 \times 10^8$ kg/m^7; $\eta_{valve} = 1.92 \times 10^7$ kg/m^7; $\rho gh_1 = 29.4$ kPa. **(c)** $Q_{valve} = 15$ liter/s.
5.11(i) $4.36\angle 36.6°$ or $4.36\cos(\omega t + 0.639)$.
5.11(v) $4.17\angle 135°$ or $4.17\cos[(\pi/2)t + (3\pi/4)]$.
5.12(ii) $\boldsymbol{Q}_S = 1\angle -90°$ liter/s; $\boldsymbol{P}_{30} = -5\angle 0°$ kPa; $\boldsymbol{P}_{20} = 5\angle 0°$ kPa; $\boldsymbol{P}_{10} = 11.2\angle -63.4°$ kPa.
5.13 **(a)** $\boldsymbol{V}_S = -150 + j150\text{ V} = 212\angle 45°$ V; **(c)** $H(j5000) = 0.527\angle -18.4°$.
5.14(iii) **(a)** $\boldsymbol{T}_2 = 10\angle 0°$ N·m; $\boldsymbol{\Omega}_2 = 5\angle -90°$ rad/s; $\boldsymbol{T}_3 = 10\angle -90°$ N·m; $\boldsymbol{T}_1 = 14.14\angle -45°$ N·m; $\boldsymbol{\Omega}_1 = 14.1\angle 135°$ rad/s; **(c)** $\Omega(t) = 18\cos(t - 0.983)$ rad/s.
5.18 **(a)** $P = 7.2$ kW. **(b)** We use effective values; multiply by $\sqrt{2}$ to obtain conventional phasors: $\boldsymbol{I}_R = 60\angle 0°$ A; $\boldsymbol{I}_L = 60\angle -90°$ A; $\boldsymbol{I} = 84.9\angle -45°$ A; $P = 1.8$ kW.

5.20(iii) (b) The corner points lie at $\omega = 2$ and $\omega = 20$; At $\omega = 1$ rad/s, the value of the low-frequency asymptote is $-20 \log[(2)(20)] = -32$ dB; the slope of the low-frequency asymptote is 0 dB/decade; the slope of the high-frequency asymptote is -40 dB/decade.
5.21(iii) (b) ω_n = corner frequency = 10 rad/s; $\hat{M}_r = 2.5$ and $\zeta = 0.2$. **(d)** $\omega_r \approx 10$ rad/s, $\Delta\omega = 4$ rad/s, $Q = 2.5$, and $\zeta = 0.2$; we cannot determine ω_n accurately from the approximate value of ω_r.
5.26 (a) $H(s) = 1/(ICs^2 + RCs + 1)$. **(b)** $\omega_{tide} = 1.45 \times 10^{-4}$ rad/s and $|H(j\omega_{tide})| \approx 1$; $\omega_{wave} = 0.524$ rad/s and $|H(j\omega_{wave})| = 0.132$. **(c)** Examine the frequency response plot; adjust A to move the corner frequency to the left by an appropriate amount.
5.31(iv) $F(s) = a(1 - e^{-t_1 s} - st_1 e^{-t_1 s})/(t_1 s^2)$.
5.33(ii) $f(t) = e^{-4t}(\cos 6t - \frac{1}{3} \sin 6t) = 1.054e^{-4t}\cos(6t + 0.32)$.
5.35(i) (c) $y(t) = \frac{1}{2} + \frac{1}{2}e^{-t}\cos t + \frac{1}{2}e^{-t}\sin t$ for $t \geq 0$.
5.35(ii) (c) $y(t) = \frac{1}{3} + 2e^{-2t} - (4/3)e^{-3t}$ for $t \geq 0$.
5.37(iii) (a) $H(s) \triangleq \boldsymbol{\Omega_1/T} = s/(Bs + K)$. **(b)** $h(t) = (1/B)[\delta(t) - (K/B)e^{-(K/B)t}]$; the zero-state unit-step response is $\Omega_1(t) = (1/B)e^{-(K/B)t}$ for $t \geq 0$. **(c)** the zero-input response is $\Omega_1(t) = \left(\frac{\sqrt{2K}}{B}\right)e^{-(K/B)t}$ for $t \geq 0$.
6.1(iv) (a) Label the top of the temperature source as node 1 and the other node as node 2; orient q_1 upward in the source; orient q_2 downward through the 0.3°C/W resistor; orient q_3 to the right in the remaining resistor. **(c)** $\theta_2 = 10.9$°C, $q_1 = 90.9$ W, $q_2 = 36.4$ W, and $q_3 = 54.6$ W.
6.2(iv) (b) $(sCR_{eq} + 1)\theta_1 = -R\theta_S/(R + R_S)$, where $R_{eq} = RR_S/(R + R_S)$.
6.5 (a) Heat loss = 5.14 kW for $h_t = 1000$ W/m²°C.
6.6 (b) $\tau = 1.87$ s for $h_t = 1000$ W/m²°C.
6.7 (a) $R = 0.0347$°C/W; $C = 94.2$ kJ/°C; $R_O = 0.0264$°C/W; **(b)** $\tau = 23.5$ min.
6.12 (a) $\mathbf{N}(n + 1) = (\mathbf{I} + \mathbf{D})\mathbf{N}(n)$, where $\mathbf{N}(n) \triangleq (N_1(n), N_2(n))^T$, the state vector at the end of year n, **I** is the 2 × 2 identity matrix, and **D** is the matrix which has the first row $(-\gamma \quad \beta)$ and the second row $(\alpha \quad -1)$.
6.13 (a) $[s^2 + (\beta_1 + \alpha_1 + \beta_2 + \alpha_2)s + (\beta_1 + \alpha_1)(\beta_2 + \alpha_2) - \alpha_1\alpha_2]N_1 = 0$. **(b)** $\beta_1\beta_2 + \beta_1\alpha_2 + \alpha_1\beta_2 = 0$.
6.15 (b) Let $\beta_1 = b_1N_3$, $\beta_2 = b_2N_4$, $\beta_3 = b_3/N_1$, and $\beta_4 = b_4/N_2$.
6.16(iii) (a) Orient the $+\Omega$ directions to agree with the polarity arrows; orient T_a into node a and T_b out of node b (into the load); $T_a/r_a = T_b/r_b$ and $r_a\Omega_a = r_b\Omega_b$. **(b)** Define $\kappa \triangleq r_a/r_b$ (dimensionless).
6.17(iii) (c) The equations are identical to those of Figure 6.17(b); however, the physical interpretation (compressive or tensile) of the scalar force variable F_b is reversed from that of Figure 6.17(b).
6.18(iv) Insert a torsion spring in series between the rotational source and the primary wheel shaft (node a); also attach a rotary damper and an inertia between node a (the wheel shaft) and ground; alternatively, attach translational counterparts of these rotational elements at the secondary end.
6.20(ii) (b) $(A^2R_S + b_L)v_b = AR_SQ_S$, an algebraic equation.
6.21(ii) (b) A series connection of a velocity source with value Q_S/A and a damper with friction coefficient A^2R_S. **(d)** A hydraulic resistor of value $R_{eqL} = b_L/A^2$.
6.22(i) (b) Replace the motor and mechanical load as seen at the motor input terminals by a parallel connection of a resistor with resistance $1/(\kappa^2B)$ and a capacitor with capacitance κ^2J. **(c)** Replace the motor and electrical source as seen at the motor shaft by a series connection of a rotational velocity source with value κv_S, a rotary damper with friction coefficient $1/(\kappa^2R)$, and a torsion spring with stiffness $1/(\kappa^2L)$.
6.23 (b) $n = \sqrt{J/J_m}$.
6.26 (b) $R = 0.5\ \Omega$, $\kappa = 0.13$ rad/V·s $\equiv 0.13$ A/N·m, and $J = 118$ N·m·s².

6.31 **(a)** $\tau_C = CR_DR_L/(R_D + R_L)$. **(b)** $\tau_D = R_LC$. **(c)** $C = 5.53\ \mu F$ provides a suitable compromise.
6.32 **(b)** $R_i = 1{,}300\ \Omega$. **(c)** $v_4 = -31v_S$. **(d)** $G_P = 1954$ (32.9 dB).
6.38 **(b)** The lumped model for the output port (D, S) of the JFET is a parallel connection of a constant-current source with value $i_{intQ} = 6.4$ mA, a resistor with value $R_d = 11.5\ k\Omega$, and a dependent current source with value $g_t v_{GS}$, where $g_t = 5.6$ mA/V. **(c)** $v_3/v_S = 4.2$.
6.43 **(a)** $v_2 = 2z_0v_S/[(R_S + z_0)(1 + z_0/R_L)e^{ds} + (z_0 - R_S)(1 - z_0/R_L)e^{-ds}]$.
6.46(ii) **(a)** $[h] = \begin{bmatrix} R_b/\beta & 0 \\ -1 & 0 \end{bmatrix}$ **(b)** $[a] = \begin{bmatrix} 0 & R_b/\beta \\ 0 & 1 \end{bmatrix}$
6.50 **(a)** $v_O/v_S = A/(A + 1)$, $R_I = \infty\ \Omega$, and $R_O = 0\ \Omega$; **(b)** $v_O/v_S \approx 1$, $R_I \approx AR_i$, and $R_O \approx R_o/A$.

Index